IMPORTANT SYMBOLS

Symbol	*Description*	*Page*
$\bar{z}$	conjugate of complex number z	**A3**
$\arg z$	argument of complex number z	**A7**
$m \times n$	size of a matrix	**37**
$[a_{ij}]$	matrix	**38**
A^T	transpose of A	**45**
$\mathbf{u} \cdot \mathbf{v}$	dot product	**48, 170**
I, I_n	identity matrix	**50**
$\operatorname{tr}(A)$	trace of matrix A	**338**
A^{-1}	inverse of matrix A	**64**
$\det A$	determinant of matrix A	**117**
$M_{ij}(A)$	(i,j)—minor of matrix A	**115**
$C_{ij}(A)$	(i,j)—cofactor of matrix A	**116**
$\operatorname{adj}(A)$	adjoint of matrix A	**133**
$c_A(x)$	characteristic polynomial of A	**331**
$\lVert\mathbf{v}\rVert$	length or norm of **v**	**149, 282**
$\overrightarrow{P_1P_2}$	vector from P_1 to P_2	**153**
$\mathbf{v}_1 \times \mathbf{v}_2$	cross product	**181**
$\mathbb{R}$	real numbers	**200**
$\mathbb{R}^n$	space of n–tuples	**200**
$\mathbf{M}_{mn}$	space of $m \times n$ matrices	**204**
$\deg[p(x)]$	degree of $p(x)$	**204**
$\mathbf{P}$	space of polynomials	**205**
$\mathbf{P}_n$	space of polynomials of degree at most n	**205**
$[a,b]$	interval from a to b	**205**
$\mathbf{F}[a,b]$	space of functions on $[a,b]$	**205**
$\operatorname{span}\{\mathbf{v}_1,\ldots,\mathbf{v}_n\}$	span of a set of vectors	**219**
$\subseteq$	set containment	**221**
$\dim V$	dimension of V	**230**

Elementary Linear Algebra

with Applications

The Prindle, Weber & Schmidt Series in Mathematics

Althoen and Bumcrot, *Introduction to Discrete Mathematics*
Brown and Sherbert, *Introductory Linear Algebra with Applications*
Buchthal and Cameron, *Modern Abstract Algebra*
Burden and Faires, *Numerical Analysis*, Fourth Edition
Cullen, *Linear Algebra and Differential Equations*
Cullen, *Mathematics for the Biosciences*
Eves, *In Mathematical Circles*
Eves, *Mathematical Circles Adieu*
Eves, *Mathematical Circles Revisited*
Eves, *Mathematical Circles Squared*
Eves, *Return to Mathematical Circles*
Fletcher and Patty, *Foundations of Higher Mathematics*
Gantner and Gantner, *Trigonometry*
Geltner and Peterson, *Geometry for College Students*
Gilbert and Gilbert, *Elements of Modern Algebra*, Second Edition
Gobran, *Beginning Algebra*, Fourth Edition
Gobran, *College Algebra*
Gobran, *Intermediate Algebra*, Fourth Edition
Gordon, *Calculus and the Computer*
Hall, *Algebra for College Students*
Hall and Bennett, *College Algebra with Applications*, Second Edition
Hartfiel and Hobbs, *Elementary Linear Algebra*
Kaufmann, *Algebra for College Students*, Third Edition
Kaufmann, *Algebra with Trigonometry for College Students*, Second Edition
Kaufmann, *College Algebra*, Second Edition
Kaufmann, *College Algebra and Trigonometry*, Second Edition
Kaufmann, *Elementary Algebra for College Students*, Third Edition
Kaufmann, *Intermediate Algebra for College Students*, Third Edition
Kaufmann, *Precalculus*
Kaufmann, *Trigonometry*
Laufer, *Discrete Mathematics and Applied Modern Algebra*
Nicholson, *Elementary Linear Algebra with Applications*, Second Edition
Powers, *Elementary Differential Equations*
Powers, *Elementary Differential Equations with Boundary-Value Problems*
Powers, *Elementary Differential Equations with Linear Algebra*
Proga, *Arithmetic and Algebra*, Second Edition
Proga, *Basic Mathematics*, Second Edition
Rice and Strange, *Plane Trigonometry*, Fifth Edition
Schelin and Bange, *Mathematical Analysis for Business and Economics*, Second Edition
Strnad, *Introductory Algebra*

Swokowski, *Algebra and Trigonometry with Analytic Geometry,* Seventh Edition
Swokowski, *Calculus with Analytic Geometry,* Second Alternate Edition
Swokowski, *Calculus with Analytic Geometry,* Fourth Edition
Swokowski, *Fundamentals of Algebra and Trigonometry,* Seventh Edition
Swokowski, *Fundamentals of College Algebra,* Seventh Edition
Swokowski, *Fundamentals of Trigonometry,* Seventh Edition
Swokowski, *Precalculus: Functions and Graphs,* Sixth Edition
Tan, *Applied Calculus,* Second Edition
Tan, *Applied Finite Mathematics,* Third Edition
Tan, *Calculus for the Managerial, Life, and Social Sciences,* Second Edition
Tan, *College Mathematics,* Second Edition
Trim, *Applied Partial Differential Equations*
Venit and Bishop, *Elementary Linear Algebra,* Third Edition
Venit and Bishop, *Elementary Linear Algebra,* Alternate Second Edition
Willard, *Calculus and Its Applications,* Second Edition
Wood and Capell, *Arithmetic*
Wood, Capell, and Hall, *Developmental Mathematics,* Fourth Edition
Wood and Capell, *Intermediate Algebra*
Zill, *A First Course in Differential Equations with Applications,* Fourth Edition
Zill, *Calculus with Analytic Geometry,* Second Edition
Zill, *Differential Equations with Boundary-Value Problems,* Second Edition

The Prindle, Weber & Schmidt Series in Advanced Mathematics

Brabenec, *Introduction to Real Analysis*
Eves, *Foundations and Fundamental Concepts of Mathematics,* Third Edition
Keisler, *Elementary Calculus: An Infinitesimal Approach,* Second Edition
Kirkwood, *An Introduction to Real Analysis*

Elementary Linear Algebra

with Applications

Second Edition

W. Keith Nicholson
University of Calgary

PWS Publishing Company
Boston

PWS
Publishing Company

20 Park Plaza
Boston, Massachusetts 02116

PWS Publishing Company is a division of Wadsworth, Inc.

Library of Congress Cataloging-in-Publication Data

Nicholson, W. Keith.
Elementary linear algebra, with applications / W. Keith Nicholson.—2nd ed.
p. cm.
ISBN 0-534-92189-2
1. Algebra, Linear. I. Title.
QA184.N53 1990
512′.5—dc20 89-48121
CIP

Printed in the United States of America

7 8 9 — 94 93

Sponsoring editor: Steve Quigley
Production editor: Pamela Rockwell
Manufacturing coordinator: Peter Leatherwood
Interior design: Pamela Rockwell
Cover design: Pineiro Design Associates
Typesetting: Weimer Typesetting
Cover printing: Henry F. Sawyer
Printing and binding: R. R. Donnelley & Sons

PREFACE

This textbook is a basic introduction to the ideas and techniques of linear algebra for first- or second-year students who have a working knowledge of high school algebra. Its aim is to achieve a balance among the computational skills, theory, and applications of linear algebra, while keeping the level suitable for beginning students. The contents are arranged to permit enough flexibility to allow the presentation of a traditional introduction to the subject, or to allow a more applied course. Calculus is not a prerequisite; places where it is mentioned are clearly marked and may be omitted.

Linear algebra has wide application to the mathematical and natural sciences, to engineering, to computer science, and (increasingly) to management and the social sciences. As a rule, students of linear algebra learn the subject by studying examples and solving problems. More than 340 solved examples are included here, many of a computational nature, and are keyed to a wide variety of exercises. In addition, there are a number of applications. These are optional, but they are included at the end of the relevant chapters (rather than at the end of the book) to encourage students to browse.

The examples also play a role in motivating theorems, although most proofs are included at a level appropriate to the student. This means that the book can be used to give a course emphasizing computation and examples (and omitting many proofs) or to give a more rigorous treatment. Some longer proofs are omitted altogether or are deferred to the end of the chapter.

Features

- Presentation of techniques in examples, with an emphasis on concrete computations.
- Over 340 solved examples, keyed to the exercises.

- A wide variety of exercises, beginning with routine, computational problems, and proceeding to the more theoretical exercises.
- Applications (optional) at the end of each chapter, where linear algebra yields new insight rather than merely playing a descriptive role.
- Emphasis on the algorithmic nature of several of the techniques.
- Flexibility in the ordering of chapters (see chart on page x). In particular Chapter 4 can be omitted, and diagonalization can be done before linear transformations.
- Section on complex matrices (optional).
- Sections on LP-factorization and LU-factorization (optional).
- Appendix on linear programming (requires only Chapter 1).
- Appendices on complex numbers and mathematical induction.
- Answers to the even-numbered computational exercises and to selected others.
- Instructor's manual containing answers or solutions to all exercises.
- Computer package available. (A Pascal program MAX: MAtriX Algebra Calculator) is available on request.

Chapter Summaries

Chapter 1: A standard treatment of Gaussian elimination is given. Manipulation of the solutions of a homogeneous system is introduced.

Chapter 2: The operations of matrix algebra (including transposition) are introduced, and matrix inverses are defined and studied through the use of elementary matrices. The relationship of matrix algebra to linear equations is emphasized, and block multiplication is introduced to simplify matrix computations. An optional section on LU-factorization is included for more applied courses.

Chapter 3: Determinants are defined inductively. The Laplace expansion is stated first (motivated by examples and the 2×2 case), so the students begin by computing determinants using familiar row and column operations (the proof is given later). The usual rules are deduced from the Laplace expansion, and the adjoint formula is given.

Chapter 4: Vector operations are defined (motivated by examples) and used to solve (primarily geometric) problems. Then coordinates are introduced in $\mathbb{R}^2$, $\mathbb{R}^3$, and straight lines and planes are described via the dot and cross products.

This chapter can be omitted with no loss of continuity.

Chapter 5: The basic theory of finite dimensional vector spaces is given. The prototype example throughout is $\mathbb{R}^n$. Many examples are

sets, linear independence, and dimension. Examples involving matrices and polynomials are also given, as are examples of spaces of functions (examples requiring calculus are clearly marked). The pace is slow because this is the first acquaintance many students have had with an abstract system.

Chapter 6: General inner products are introduced (the prototype example being $\mathbb{R}^n$), and distance, norm, and the Schwarz inequality are discussed. The Gram–Schmidt algorithm is given, projections are introduced, and the approximation theorem is proved.

Chapter 7: Eigenvalues and the characteristic polynomial are introduced. Then similarity is defined, the diagonalization algorithm is given, and the principal axes theorem is proved. Numerical examples are provided at every stage. In order to allow for a more applied course, linear transformations are not required. An optional section on diagonalization of complex matrices is included.

Chapter 8: Linear transformations are introduced, motivated by many examples from geometry, matrix theory, and calculus (clearly marked). The kernel and image are defined, and the dimension theorem is proved. Then isomorphisms are discussed, and the vector space of linear transformations is related to the space of matrices.

Chapter 9: Linear operators are discussed and the connection with similar matrices is made clear. The idea of a similarity invariant is brought in. Then invariant subspaces and direct sums are introduced, leading to the theorem that every symmetric operator is diagonalizable.

Chapter Dependencies

The chart on the next page suggests how the material introduced in each chapter draws on concepts covered in certain earlier chapters. A solid arrow means that ready assimilation of ideas and techniques presented in the later chapter depends on familiarity with the earlier chapter. A broken arrow indicates that some reference to the earlier chapter is made but the chapter need not be covered.

Suggested Course Outlines

1. *Two-Semester Course.* Much of the book can be covered in two 35-lecture semesters, with time left for some applications. The following outline is based on class experience and includes three applications. The pace in the first semester is more leisurely.

Chapter 1	Sections 1.1–1.3	4 lectures
Applications	Sections 1.4–1.5	1 lecture
Chapter 2	Sections 2.1–2.3	9 lectures

Application	Section 2.5.1	3 lectures
Chapter 3	Sections 3.1–3.2	6 lectures
Chapter 4	Sections 4.1–4.2	10 lectures
Application	Section 4.3	2 lectures
Chapter 5	Sections 5.1–5.5	10 lectures
Chapter 6	Sections 6.1–6.2	5 lectures
Chapter 7	Sections 7.1–7.2	5 lectures
Application	Section 7.4.1	3 lectures
Chapter 8	Sections 8.1–8.2	6 lectures
Chapter 9	Sections 9.1–9.3	6 lectures

2. *One-Semester Applied Course.* This 35-lecture outline goes directly to diagonalization and its applications. The sections marked with an asterisk are intended as alternatives.

Chapter 1	Sections 1.1–1.3	3 lectures
Chapter 2	Sections 2.1–2.3	6 lectures
LU-factorization*	Section 2.4	2 lectures*
Chapter 3	Sections 3.1–3.2	4 lectures

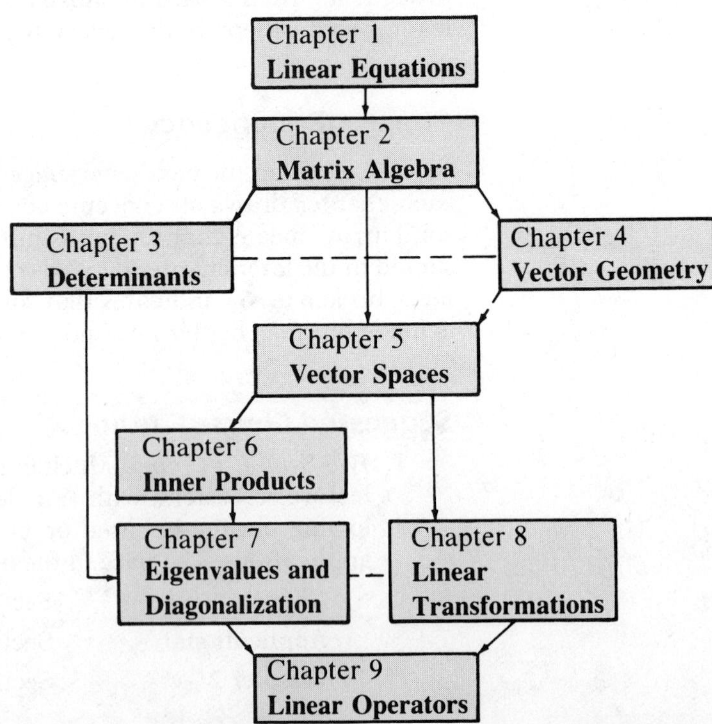

Chapter 5	Sections 5.1–5.4	8 lectures
Chapter 6	Sections 6.1–6.2	5 lectures
Chapter 7	Sections 7.1–7.2	5 lectures
Complex Matrices*	Section 7.3	2 lectures*
Application	Section 7.4.1 or 7.4.2	2 lectures

New Features in the Second Edition

- Definitions are highlighted for easier reference.
- New material on Euclidean n-space has been added to Chapter 5 to help effect the transition to abstract vector spaces.
- The sections on rank and on the Gram–Schmidt algorithm have been rewritten to improve the exposition.
- The sections on isomorphisms and on composition have been combined, and less emphasis is placed on the abstract definition of inverse transformations.
- New exercises and examples have been inserted.
- In response to the increasing interest in numerical methods spawned by the use of computers, a new section on LU-factorization has been added.
- A new section on complex matrices has been added.
- The appendix on complex numbers has been expanded to include more material on quadratics. New exercises have been added.
- Applications to graphs, economic models, and Euclidean geometry have been converted to examples and exercises. The theoretical application to differential equations has been deleted from Chapter 8, and the discussion of isometries has been deleted from Chapter 9.

Acknowledgments

I would like to record my appreciation to the following people for their useful comments and suggestions:

G. D. Allen, Texas A & M University
R. F. V. Anderson, University of British Columbia
S. Bulman-Fleming, Wilfred Laurier University
D. Burbulla, University of Toronto
I. Gombos, Dawson College
E. W. Johnson, University of Iowa
B. J. Kirby, Queens University
S. O. Kochman, York University

M. Chacron,
Carleton University

G. A. Chambers,
University of Alberta

V. F. Connolly,
Worcester Polytechnic Institute

P. Cook II,
Furman University

H. Cunsolo,
University of Guelph

H. P. Decell, Jr.,
University of Houston

G. J. Etgen,
University of Houston

J. B. Florence,
University of Western Ontario

P. Lancaster,
University of Calgary

A. H. Low,
University of New South Wales

J. Petro,
Western Michigan University

J. Repka,
University of Toronto

L. G. Roberts,
University of British Columbia

P. N. Stewart,
Dalhousie University

B. F. Wyman,
Ohio State University

I would like to thank several colleagues at the University of Calgary for their suggestions, especially P. A. Binding, N. H. Choksy, P. F. Elhers, D. Gunderson, A. Schaer, J. Schaer, and M. G. Stone.

I would also like to thank Harry Campbell for his assistance on the first edition and Dave Geggis, Thomas Stone, and Steve Quigley for their work on the second edition. Thanks are also due to the production staff at PWS-KENT, to Joanne Longworth and Gisele Vezina who typed the first edition, to Jason Brown who helped with the exercises, to Nasli Choksy for proofreading, and to Jason and Mark Nicholson for typing and proofreading. Finally, I want to thank my wife, Kathleen, for her unfailing support.

W. Keith Nicholson

CONTENTS

1 Systems of Linear Equations

SECTION 1.1 Introduction

One of the great historical motivations for the development of mathematics has been to find a way to analyze and solve practical problems. This need prompted the use of numbers and of geometry, two of the most basic mathematical systems. Certain types of problems led naturally to systems of linear equations that, when solved, gave useful practical information. Linear algebra arose from attempts to find systematic methods for solving these systems, so it is natural to begin this book by studying linear equations.

If a, b, and c are real numbers, the graph of an equation of the form

$$ax + by = c$$

is a straight line (provided that a and b are not both zero). Accordingly, such an equation is called a linear equation in the variables x and y. When only two or three variables are present, they are usually denoted by x, y, and z. However, it is often convenient to write the variables as $x_1, x_2, \ldots, x_n$, particularly when more than three variables are involved.

DEFINITION

An equation of the form

$$a_1x_1 + a_2x_2 + \cdots + a_nx_n = b$$

is called a **linear equation** in the n variables $x_1, x_2, \ldots, x_n$. Here $a_1, a_2, \ldots, a_n$ denote real numbers (called the **coefficients** of $x_1, x_2, \ldots, x_n$, respectively) and b is also a number (called the **constant term** of the equation). A finite collection of linear equations in the variables $x_1, x_2, \ldots, x_n$ is called a **system of linear equations** in these variables.

Hence,

$$2x_1 - 3x_2 + 5x_3 = 7 \quad \text{and}$$

$$x_1 + x_2 + x_3 + x_4 = 0$$

are both linear equations. Note that each variable in a linear equation occurs to the first power only, so the following are *not* linear equations.

$$x_1^2 + 3x_2 - 2x_3 = 5$$

$$x_1 + x_1x_2 + 2x_3 = 1$$

$$\sqrt{x_1} + x_2 - x_3 = 0$$

DEFINITION

An **ordered *n*-tuple** $\begin{bmatrix} s_1 \\ s_2 \\ \vdots \\ s_n \end{bmatrix}$ is an ordered sequence of n numbers s_1, $s_2, \ldots, s_n$ (called the **entries** of the n-tuple). Two such n-tuples are defined to be **equal** only when corresponding entries are equal.

$$\begin{bmatrix} s_1 \\ s_2 \\ \vdots \\ s_n \end{bmatrix} = \begin{bmatrix} t_1 \\ t_2 \\ \vdots \\ t_n \end{bmatrix} \quad \text{means} \quad s_1 = t_1, s_2 = t_2, \ldots, s_n = t_n$$

For example, $\begin{bmatrix} 2 \\ 1 \\ 3 \\ 0 \end{bmatrix} \neq \begin{bmatrix} 2 \\ 3 \\ 1 \\ 0 \end{bmatrix}$ whereas $\begin{bmatrix} 1 \\ 2 \\ x \\ 0 \end{bmatrix} = \begin{bmatrix} 1 \\ 2 \\ 5 \\ 0 \end{bmatrix}$ if and only if $x = 5$. Ordered 2- and 3-tuples are called **ordered pairs** and **ordered triples,** respectively. Incidentally, ordered n-tuples can also be written as rows.

DEFINITION

Given a linear equation $a_1x_1 + a_2x_2 + \cdots + a_nx_n = b$, an ordered n-tuple $\begin{bmatrix} s_1 \\ s_2 \\ \vdots \\ s_n \end{bmatrix}$ is called a **solution** to the equation if $a_1s_1 + a_2s_2 + \cdots + a_ns_n = b$—that is, if the equation is satisfied when the substitutions $x_1 = s_1, x_2 = s_2, \ldots, x_n = s_n$ are made. An ordered n-tuple is called a **solution to a system** of equations if it is a solution to every equation in the system.

A system may have no solution at all, or it may have an infinite family of solutions. For example, the system $x + y = 2$, $x + y = 3$ has no solution, whereas Example 1 exhibits a system with infinitely many solutions. The aim in general is to **solve** the system of linear equations—that is, to find *all* solutions to the system. This chapter is devoted primarily to developing a systematic method for doing this.

EXAMPLE 1

Show that $\begin{bmatrix} t - s - 1 \\ t + s + 1 \\ s \\ t \end{bmatrix}$ is a solution to the system

$$\begin{aligned} x_1 - 2x_2 + 3x_3 + x_4 &= -3 \\ 2x_1 - x_2 + 3x_3 - x_4 &= -3 \end{aligned}$$ for any values of s and t.

Solution Simply substitute $x_1 = t - s - 1, x_2 = t + s + 1, x_3 = s$, and $x_4 = t$ in each equation.

$$x_1 - 2x_2 + 3x_3 + x_4 = (t - s - 1) - 2(t + s + 1) + 3s + t = -3$$

$$2x_1 - x_2 + 3x_3 - x_4 = 2(t - s - 1) - (t + s + 1) + 3s - t = -3$$

Because both equations are satisfied, it is a solution for all s and t.

The solutions given in Example 1 can be written as follows:

$$\begin{aligned} x_1 &= t - s - 1 \\ x_2 &= t + s + 1 \\ x_3 &= s \\ x_4 &= t \end{aligned} \qquad (s \text{ and } t \text{ arbitrary})$$

This means that, for any choice of s and t, the values of x_1, x_2, x_3, and x_4 given by these formulas will satisfy the equations. The quantities s and t are called **parameters,** and this set of solutions, described in this way, is said to be given in **parametric form.** It turns out that solutions to systems of linear equations quite often appear in this form and that such descriptions arise naturally. The following examples show how this comes about in the simplest systems where only one equation is present.

EXAMPLE 2 Describe all solutions to $3x - y = 4$ in parametric form.

Solution The equation can be written in the form

$$y = 3x - 4$$

Thus, if t denotes *any* number at all, we can quite arbitrarily set $x = t$ and then obtain $y = 3t - 4$. This is clearly a solution to our equation for any value of t. On the other hand, *every* solution to $3x - y = 4$ arises in this way (t is just the value of x). Hence the set of *all* solutions can be described parametrically as

$$\begin{bmatrix} x \\ y \end{bmatrix} = \begin{bmatrix} t \\ 3t - 4 \end{bmatrix} \qquad t \text{ arbitrary}$$

Note that there are *infinitely* many distinct solutions, one for each choice of the parameter t.

It is important to realize that the solutions to $3x - y = 4$ can be given in parametric form in several ways. We found the foregoing solution by observing that $y = 3x - 4$ and then choosing $x = t$, t a parameter. However, we could have found x in terms of y:

$$x = \frac{1}{3}(y + 4)$$

and then chosen $y = s$ (s a parameter). Hence the solutions are

$$\begin{bmatrix} x \\ y \end{bmatrix} = \begin{bmatrix} \frac{1}{3}(s + 4) \\ s \end{bmatrix} \qquad s \text{ arbitrary}$$

This is also a correct parametric representation of the solutions to $3x - y = 4$. In fact, the parameters are related by $s = 3t - 4$ (or $t = \frac{1}{3}(s + 4)$).

EXAMPLE 3 Describe all solutions to $3x - y + 2z = 6$ in parametric form.

Solution Solving the equation for y in terms of x and z, we get $y = 3x + 2z - 6$. If s and t are arbitrary, then, setting $x = s$, $z = t$, we get solutions

$$\begin{bmatrix} s \\ 3s + 2t - 6 \\ t \end{bmatrix} \qquad s \text{ and } t \text{ arbitrary}$$

Of course we could have solved for x.

$$x = \frac{1}{3}(y - 2z + 6)$$

Then, if we take $y = p, z = q$, the solutions are represented as follows:

$$\begin{bmatrix} \frac{1}{3}(p - 2q + 6) \\ p \\ q \end{bmatrix} \qquad p \text{ and } q \text{ arbitrary}$$

The same family of solutions can "look" quite different!

When only two variables are involved, the solutions to systems of linear equations can be described geometrically because the graph of a linear equation $ax + by = c$ is a straight line. Moreover, a point $P(s, t)$ with coordinates s and t lies on the line if and only if $as + bt = c$—that is, when $\begin{bmatrix} s \\ t \end{bmatrix}$ is a solution to the equation. Hence solutions $\begin{bmatrix} s \\ t \end{bmatrix}$ to a *system* of linear equations correspond to the points $P(s, t)$ that lie on *all* the lines in question. In particular, if the system consists of just one equation (as in Example 2), there must be infinitely many solutions because there are infinitely many points on a line. If the system has two equations, there are three possibilities for the corresponding straight lines.

1. They intersect in a single point. Then the system has a *unique solution* corresponding to that point.
2. They are parallel (and distinct) and so do not intersect. Then the system has *no solution.*
3. They are identical. Then the system has *infinitely many solutions*—one for each point on the (common) line.

These three situations are illustrated in Figure 1.1. In each case the graphs of two specific lines are plotted and the corresponding equations

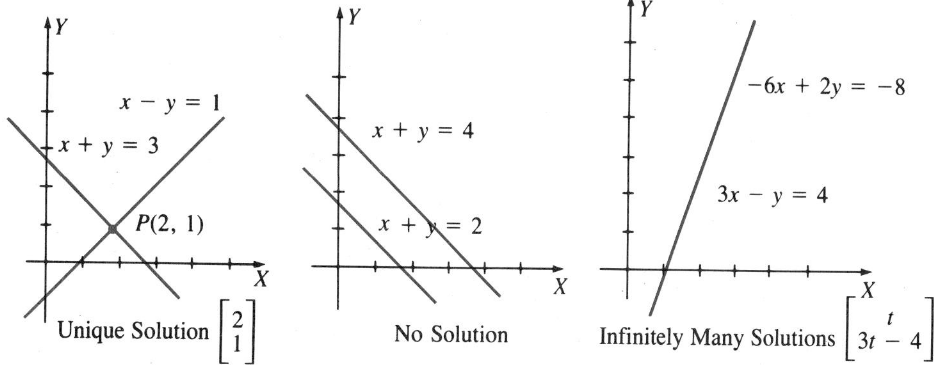

FIGURE 1.1

indicated. In the last case, the equations are $3x - y = 4$ (treated in Example 2) and $-6x + 2y = -8$, which have identical graphs.

A similar situation occurs when three variables are present. The graph of an equation $ax + by + cz = d$ can be shown to be a plane (provided not all of a, b, and c are zero), and consists of all points $P(r, s, t)$ in space such that $\begin{bmatrix} r \\ s \\ t \end{bmatrix}$ is a solution to the equation. Hence, as we found for lines, this plane provides a "picture" of the set of solutions of the equation. In particular, the solutions to a system of three linear equations in three variables correspond to the points common to all three planes. The same possibilities arise as before: no solution, a unique solution, or infinitely many solutions (see Figure 1.2). In fact, it is not difficult to show that one of these possibilities results no matter how the planes are oriented (Exercise 8).

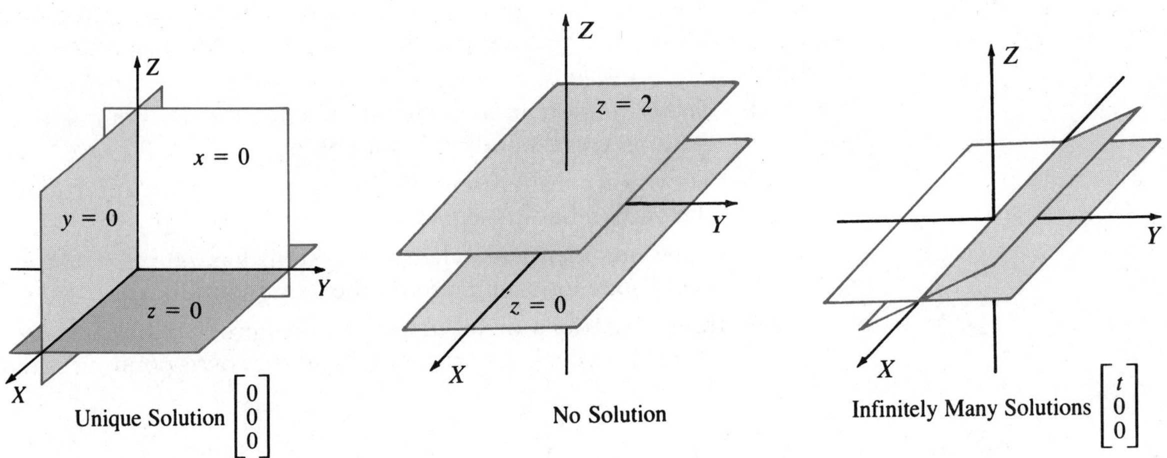

FIGURE 1.2

These illustrations indicate that, at least in the case of two or three variables, an examination of the graphs of the equations can give useful insight into the nature of the solutions. However, this graphical method has its limitations; when more than three variables are involved, no physical image of the graphs (called hyperplanes) is possible. It is necessary to turn to more "algebraic" methods of solution.

EXERCISES 1.1

1. In each case verify that the following are solutions for all values of s and t.

   (a) $\begin{bmatrix} 19t - 35 \\ 25 - 13t \\ t \end{bmatrix}$ is a solution of $\begin{aligned} 2x + 3y + z &= 5 \\ 5x + 7y - 4z &= 0 \end{aligned}$

   (b) $\begin{bmatrix} 2s + 12t + 13 \\ s \\ -s - 3t - 3 \\ t \end{bmatrix}$ is a solution of $\begin{aligned} 2x_1 + 5x_2 + 9x_3 + 3x_4 &= -1 \\ x_1 + 2x_2 + 4x_3 \quad &= 1 \end{aligned}$

2. Find all solutions to the following in parametric form in two ways.

   (a) $3x + y = 2$ **(b)** $2x + 3y = 1$ **(c)** $3x - y + 2z = 5$ **(d)** $x - 2y + 5z = 1$

3. Regarding $2x = 5$ as the equation $2x + 0y = 5$ in two variables, find all solutions in parametric form.

4. Regarding $4x - 2y = 3$ as the equation $4x - 2y + 0z = 3$ in three variables, find all solutions in parametric form.

5. Find all solutions to the general system $ax = b$ of one equation in one variable **(a)** when $a = 0$ and **(b)** when $a \neq 0$.

6. Show that a system consisting of exactly one linear equation can have no solution, one solution, or infinitely many solutions. Give examples.

7. By examining the possible positions of lines in the plane, show that three equations in two variables can have zero, one, or infinitely many solutions.

8. By examining the possible positions of planes in space, show that three equations in three variables can have zero, one, or infinitely many solutions.

9. Can two equations in three variables have a unique solution? Give reasons for your answer.

10. Solve the system: $\begin{aligned} 3x + 2y &= 5 \\ 7x + 5y &= 1 \end{aligned}$ by changing variables: $\begin{aligned} x &= 5x' - 2y' \\ y &= -7x' + 3y' \end{aligned}$

    and solving the resulting equations for x' and y'.

SECTION 1.2 Gaussian Elimination

1.2.1 Equivalent Systems of Equations

In this section a general method (called Gaussian elimination) for solving systems of linear equations will be introduced. We begin by solving the system

$$\begin{aligned} x + 2y &= -2 \\ 2x + 3y &= 7 \end{aligned}$$

in rather complete detail and analyzing the techniques with an eye to generalizing them. If twice the first equation is subtracted from the second, the result is the system

$$\begin{aligned} x + 2y &= -2 \\ -y &= 11 \end{aligned}$$

This new system has the same solutions as the original system, but this one is easier to solve. Clearly, $y = -11$ and we can substitute this back into $x + 2y = -2$ (the other equation would do just as well) to get $x = 20$. Hence $\begin{bmatrix} 20 \\ -11 \end{bmatrix}$ is the solution.

Before proceeding, it is useful to introduce an easier way of describing systems of linear equations. The system we just solved is written again below, along with the array of numbers that occur in the equations.

$$\begin{aligned} x + 2y &= -2 \\ 2x + 3y &= 7 \end{aligned} \qquad \begin{bmatrix} 1 & 2 & -2 \\ 2 & 3 & 7 \end{bmatrix}$$

The rectangular array of numbers is called the **augmented matrix** of the system of equations. Each row of the matrix consists of the coefficients of the variables (in order) from the corresponding equation, together with the constant term. Clearly the augmented matrix is just a different way of describing the system of equations. (In general, a rectangular array of numbers is called a **matrix**. Such matrices will be discussed in their own right in Chapter 2, where the reason for the adjective *augmented* will be explained.)

Returning to our example, it is very important to observe that the manipulations by which we solved the system of equations should be thought of as manipulations of the *system* rather than changes to the *individual equations* of the system. Furthermore, the changes to the system that are allowed are such that the new system has the *same set of solutions* as the original (see Theorem 1). Such a system is said to be **equivalent** to the original system. The point of the method is this: Because the aim is to find all solutions to the original system, and because each system that is equivalent to the original has the same set of solutions, it is enough to solve some equivalent system. The goal, then, is to find a system of equations that is both equivalent to the original system and easy to solve. We proceed by writing down a sequence of systems of equations, starting with the original system, and such that each system is equivalent to the previous one (hence all are equivalent to the original system). The aim is to end up at some stage with a system that is easy to solve.

Let us solve the foregoing system once again by writing down the sequence of equivalent systems. At each stage, the corresponding augmented matrix is displayed. The original system is

$$\begin{aligned} x + 2y &= -2 \\ 2x + 3y &= 7 \end{aligned} \qquad \begin{bmatrix} 1 & 2 & -2 \\ 2 & 3 & 7 \end{bmatrix}.$$

The first manipulation we carried out was to subtract twice the first equation from the second. The resulting system is

$$\begin{aligned} x + 2y &= -2 \\ -y &= 11 \end{aligned} \qquad \begin{bmatrix} 1 & 2 & -2 \\ 0 & -1 & 11 \end{bmatrix}$$

and this is equivalent to the original. At this stage we obtained $y = -11$. The present point of view insists that we modify the *system.* Hence we leave the first equation alone and multiply the second equation by -1. The result is the equivalent system

$$\begin{aligned} x + 2y &= -2 \\ y &= -11 \end{aligned} \qquad \begin{bmatrix} 1 & 2 & -2 \\ 0 & 1 & -11 \end{bmatrix}$$

Next we subtract twice the second equation from the first to get another equivalent system.

$$\begin{aligned} x &= 20 \\ y &= -11 \end{aligned} \qquad \begin{bmatrix} 1 & 0 & 20 \\ 0 & 1 & -11 \end{bmatrix}$$

Now *this* system is easy to solve! And because it is equivalent to the original system, it provides the solution to that system.

Observe that, at each stage, a certain operation is performed on the system (and thus on the augmented matrix) to produce an equivalent system—that is, a system with the same solutions. The operations that satisfy this requirement are the following **elementary operations**:

I. Interchange two equations.
II. Multiply one equation by a nonzero number.
III. Add a multiple of one equation to a different equation.

THEOREM 1 Suppose an elementary operation is performed on a system of linear equations. Then the resulting system has the same set of solutions as the original. That is, the two systems are equivalent.

Proof We prove it only for operations of type III. Let

$$c_1x_1 + c_2x_2 + \cdots + c_nx_n = d \tag{*}$$

and

$$a_1x_1 + a_2x_2 + \cdots + a_nx_n = b \tag{**}$$

denote two different equations in the system in question, and suppose k times equation (*) is added to equation (**). Then the new system is identical to the original except that (**) is replaced by

$$(a_1 + kc_1)x_1 + \cdots + (a_n + kc_n)x_n = b + kd \tag{***}$$

If $\begin{bmatrix} s_1 \\ s_2 \\ \vdots \\ s_n \end{bmatrix}$ is a solution to the new system, then it satisfies (***) and (*). [This is where we use the fact that (*) and (**) are *different* equations in the original system.] Hence $(a_1 + kc_1)s_1 + \cdots + (a_n + kc_n)s_n = b + kd$ and $c_1s_1 + \cdots + c_ns_n = d$. If we subtract k times the second of these from the first, we see that $\begin{bmatrix} s_1 \\ s_2 \\ \vdots \\ s_n \end{bmatrix}$ satisfies (**) and so is a solution to the original system. Similarly, every solution to the original system satisfies the new one. ■

Manipulations performed on a system of equations (multiplying by a number, subtracting one from another, and so on) produce corresponding manipulations on the *rows* of the augmented matrix. Thus multiplying a row of a matrix by a number k means multiplying *every entry* of the row by k. Adding one row to another row means adding *each entry* of that row to the corresponding entry of the other row. Subtracting two rows is done similarly.

When dealing with a specific example, we usually manipulate the rows of the augmented matrix rather than the equations. For this reason, and because we shall be manipulating the matrices themselves in Chapter 2 with little or no reference to equations, let us restate these elementary operations for matrices.

DEFINITION The following are called **elementary row operations** on a matrix.

I. Interchange two rows.
II. Multiply one row by a nonzero number.
III. Add a multiple of one row to a different row.

Clearly, performing a sequence of elementary row operations on the augmented matrix of a system of linear equations corresponds to performing the same sequence of elementary operations on the equations. In the case of three equations in three variables, the goal is to produce a matrix of the form

$$\begin{bmatrix} 1 & 0 & 0 & a \\ 0 & 1 & 0 & b \\ 0 & 0 & 1 & c \end{bmatrix}$$

Then the corresponding equations are

$$\begin{aligned} x \qquad\qquad &= a \\ y \qquad &= b \\ z &= c \end{aligned}$$

and, because this system is equivalent to the original system, the solution is apparent.

EXAMPLE 1 Find all solutions to the following system of equations.

$$\begin{aligned} 3x + 4y + z &= 1 \\ 2x + 3y \qquad &= 0 \\ 4x + 3y - z &= -2 \end{aligned}$$

Solution The sequence of augmented matrices appears below. At the right of each matrix, it is indicated how that matrix was derived from the preceding one. (Note that all bracketed instructions refer to the rows of the *preceding* matrix.) The augmented matrix of the original system is

$$\begin{bmatrix} 3 & 4 & 1 & 1 \\ 2 & 3 & 0 & 0 \\ 4 & 3 & -1 & -2 \end{bmatrix}$$

The first step is to eliminate x from all but one equation and then to place that equation first (if it is not already first). We begin by subtracting the second equation from the first. The result is

$$\begin{bmatrix} 1 & 1 & 1 & 1 \\ 2 & 3 & 0 & 0 \\ 4 & 3 & -1 & -2 \end{bmatrix} \qquad [(\text{row } 1) - (\text{row } 2)]$$

This creates the 1 in the upper left corner, which is used next to "clean up" the first column. For convenience, two row operations are done in one step.

$$\begin{bmatrix} 1 & 1 & 1 & 1 \\ 0 & 1 & -2 & -2 \\ 0 & -1 & -5 & -6 \end{bmatrix} \quad \begin{matrix} \\ [(\text{row } 2) - 2(\text{row } 1)] \\ [(\text{row } 3) - 4(\text{row } 1)] \end{matrix}$$

Now x occurs only in the first of the corresponding equations. The next step is to obtain y only in the second equation. This is easy here: We use the 1 in the second position of the second row to "clean up" the second column.

$$\begin{bmatrix} 1 & 0 & 3 & 3 \\ 0 & 1 & -2 & -2 \\ 0 & 0 & -7 & -8 \end{bmatrix} \quad \begin{matrix} [(\text{row } 1) - (\text{row } 2)] \\ \\ [(\text{row } 3) + (\text{row } 2)] \end{matrix}$$

Note that these manipulations *did not affect* the first column (the second row has a zero there), so our previous effort there has not been undermined. Finally we "clean up" the last column.

$$\begin{bmatrix} 1 & 0 & 3 & 3 \\ 0 & 1 & -2 & -2 \\ 0 & 0 & 1 & \frac{8}{7} \end{bmatrix} \quad \begin{matrix} \\ \\ [-1/7\,(\text{row } 3)] \end{matrix}$$

$$\begin{bmatrix} 1 & 0 & 0 & -\frac{3}{7} \\ 0 & 1 & 0 & \frac{2}{7} \\ 0 & 0 & 1 & \frac{8}{7} \end{bmatrix} \quad \begin{matrix} [(\text{row } 1) - 3(\text{row } 3)] \\ [(\text{row } 2) + 2(\text{row } 3)] \\ \\ \end{matrix}$$

The corresponding equations are $x = -\frac{3}{7}$, $y = \frac{2}{7}$, and $z = \frac{8}{7}$, which give the solution.

Unfortunately, it is not always possible to finish up with a matrix (system) that is quite as nice as the one in Example 1. The next example illustrates a situation where the best we can do is get an augmented matrix of the form

$$\begin{bmatrix} 1 & 0 & p & a \\ 0 & 1 & q & b \\ 0 & 0 & 0 & 0 \end{bmatrix}$$

EXAMPLE 2 Solve the following system of equations.

$$\begin{aligned} x + 2y - z &= 2 \\ 2x + 5y + 2z &= -1 \\ 7x + 17y + 5z &= -1 \end{aligned}$$

Solution The sequence of augmented matrices follows. Operations are indicated. As before, they refer to the rows of the preceding matrix.

$$\begin{bmatrix} 1 & 2 & -1 & 2 \\ 2 & 5 & 2 & -1 \\ 7 & 17 & 5 & -1 \end{bmatrix} \quad \text{[original]}$$

$$\begin{bmatrix} 1 & 2 & -1 & 2 \\ 0 & 1 & 4 & -5 \\ 0 & 3 & 12 & -15 \end{bmatrix} \quad \begin{matrix} \\ [(\text{row } 2) - 2(\text{row } 1)] \\ [(\text{row } 3) - 7(\text{row } 1)] \end{matrix}$$

$$\begin{bmatrix} 1 & 2 & -1 & 2 \\ 0 & 1 & 4 & -5 \\ 0 & 0 & 0 & 0 \end{bmatrix} \quad \begin{matrix} \\ \\ [(\text{row } 3) - 3(\text{row } 2)] \end{matrix}$$

Something remarkable has happened. The third equation has disappeared! The reason is that it is *redundant* in the sense that it provides no new information about the solutions to the system. In fact, the third equation is just the sum of the first equation plus three times the second, so any solution to the first two equations is *automatically* a solution to the third. (We will discuss redundancy after we complete the solution to the system.) It is convenient to do one more operation in order to eliminate y from the first equation.

$$\begin{bmatrix} 1 & 0 & -9 & 12 \\ 0 & 1 & 4 & -5 \\ 0 & 0 & 0 & 0 \end{bmatrix} \quad \begin{matrix} [(\text{row } 1) - 2(\text{row } 2)] \\ \\ \\ \end{matrix}$$

The corresponding system of equations is

$$\begin{aligned} x \qquad - 9z &= 12 \\ y + 4z &= -5 \\ 0 &= 0 \end{aligned}$$

Now the solution can be readily given. If z is any number at all, say $z = t$ where t is an arbitrary parameter, we obtain $x = 9t + 12$ and $y = -4t - 5$. Hence $\begin{bmatrix} 9t + 12 \\ -4t - 5 \\ t \end{bmatrix}$ represents all solutions where t is a parameter.

This is an important example. It shows how solutions in parametric form are generated, and it is quite typical of what happens when more variables are involved. It is instructive to see how one can *discover* the fact that the third equation is equal to the first plus three times the second. We do this by "keeping track" of the row operations performed in the solution to Example 2. Let R_1, R_2, R_3, denote the three rows of the original augmented matrix.

$$R_1 = [1 \quad 2 \quad -1 \quad 2]$$

$$R_2 = [2 \quad 5 \quad 2 \quad -1]$$

$$R_3 = [7 \quad 17 \quad 5 \quad -1]$$

Then the first manipulation above can be described as follows:

$$\begin{bmatrix} 1 & 2 & -1 & 2 \\ 2 & 5 & 2 & -1 \\ 7 & 17 & 5 & -1 \end{bmatrix} \begin{matrix} R_1 \\ R_2 \\ R_3 \end{matrix}$$

$$\begin{bmatrix} 1 & 2 & -1 & 2 \\ 0 & 1 & 4 & -5 \\ 0 & 3 & 12 & -15 \end{bmatrix} \begin{matrix} R_1 \\ R_2 - 2R_1 \\ R_3 - 7R_1 \end{matrix}$$

Now it is evident that, in this last matrix, the last row is three times the second. Hence

$$R_3 - 7R_1 = 3(R_2 - 2R_1)$$

This can be manipulated algebraically to give

$$R_3 = 3R_2 + R_1$$

In other words, the original third row, R_3, equals the first row plus three times the second. This is the relationship mentioned (for equations and without motivation) in the solution to Example 2. Note that we have been manipulating the symbols R_1, R_2, and R_3 as though they were numbers or variables. This is a typical *vector* calculation. Vector calculations will be treated in detail later in this book.

The next example illustrates a different type of situation that we may encounter when solving systems of equations.

EXAMPLE 3 Solve the following system of equations.

$$\begin{aligned} x \qquad + 10z &= 5 \\ 3x + y - 4z &= -1 \\ 4x + y + 6z &= 1 \end{aligned}$$

Solution We manipulate the augmented matrix, "keeping track" of the manipulations for reference. We use R_1, R_2, R_3 to denote the three rows, as before. Then row operations give

$$\begin{bmatrix} 1 & 0 & 10 & 5 \\ 3 & 1 & -4 & -1 \\ 4 & 1 & 6 & 1 \end{bmatrix} \begin{matrix} R_1 \\ R_2 \\ R_3 \end{matrix}$$

$$\begin{bmatrix} 1 & 0 & 10 & 5 \\ 0 & 1 & -34 & -16 \\ 0 & 1 & -34 & -19 \end{bmatrix} \quad \begin{matrix} R_1 \\ R_2 - 3R_1 \\ R_3 - 4R_1 \end{matrix}$$

$$\begin{bmatrix} 1 & 0 & 10 & 5 \\ 0 & 1 & -34 & -16 \\ 0 & 0 & 0 & -3 \end{bmatrix} \quad \begin{matrix} R_1 \\ R_2 - 3R_1 \\ (R_3 - 4R_1) - (R_2 - 3R_1) \end{matrix}$$

But this means that the following equations

$$\begin{aligned} x \qquad + 10z &= 5 \\ y - 34z &= -16 \\ 0 &= -3 \end{aligned}$$

are equivalent to the original system. In other words, the two have the *same solutions*. But this last system clearly has *no* solution (to satisfy the last equation, x, y, and z would have to be found with the property that $0x + 0y + 0z = -3$; obviously no such x, y, and z exist). Hence the original system has *no* solution.

Because we have "kept track" of the operations performed, it is possible to give a clear explanation of *why* there is no solution. The offending equation $0 = -3$ corresponds to the row $(0, 0, 0, -3)$ in the last matrix. In terms of the rows R_1, R_2, and R_3 of the original matrix, the last row is

$$(0, 0, 0, -3) = (R_3 - 4R_1) - (R_2 - 3R_1) = R_3 - (R_1 + R_2)$$

The fact that this is "almost zero" suggests that we compare R_3 with $R_1 + R_2$ or, what is the same thing, that we compare the third equation with the first plus the second.

Third equation:	$4x + y + 6z = 1$
First equation plus the second:	$4x + y + 6z = 4$

Now it is obvious that the system has no solution, because any solution of the first two equations would have to be a solution of $4x + y + 6z = 4$ and so could not be a solution of the third equation. Systems of linear equations that have no solution are called **inconsistent systems.** Not surprisingly, systems that have at least one solution are said to be **consistent.**

EXAMPLE 4 Find a condition on the numbers a, b, and c such that the following system of equations is consistent. When that condition is satisfied, find all solutions (in terms of a, b, and c).

$$\begin{aligned} x + 3y + z &= a \\ -x - 2y + z &= b \\ 3x + 7y - z &= c \end{aligned}$$

Solution The procedure is just as before, except that now the augmented matrix has entries a, b, and c as well as known numbers.

$$\begin{bmatrix} 1 & 3 & 1 & a \\ -1 & -2 & 1 & b \\ 3 & 7 & -1 & c \end{bmatrix}$$

$$\begin{bmatrix} 1 & 3 & 1 & a \\ 0 & 1 & 2 & b + a \\ 0 & -2 & -4 & c - 3a \end{bmatrix}$$

$$\begin{bmatrix} 1 & 0 & -5 & -2a - 3b \\ 0 & 1 & 2 & b + a \\ 0 & 0 & 0 & c - (a - 2b) \end{bmatrix}$$

Now the whole thing depends on the quantity $c - (a - 2b)$. The last row corresponds to an equation $0 = c - (a - 2b)$. Thus, if $c - (a - 2b)$ is *not* zero, there is no solution (just as in Example 3). Hence the condition for consistency is $c = a - 2b$. Then the last matrix becomes

$$\begin{bmatrix} 1 & 0 & -5 & -2a - 3b \\ 0 & 1 & 2 & a + b \\ 0 & 0 & 0 & 0 \end{bmatrix}$$

Taking $z = t$, t a parameter, gives the solutions $\begin{bmatrix} 5t - (2a + 3b) \\ -2t + (a + b) \\ t \end{bmatrix}$. Hence these are the solutions if $c = a - 2b$. If $c \neq a - 2b$, the system is inconsistent.

EXERCISES 1.2.1

1. Write down the augmented matrix for each of the following systems of linear equations.

(a) $\begin{aligned} x - 3y &= 5 \\ 2x + y &= 1 \end{aligned}$ **(b)** $\begin{aligned} x + 2y &= 0 \\ y &= 1 \end{aligned}$ **(c)** $\begin{aligned} x - y + z &= 2 \\ x - z &= 1 \\ y + 2x &= 0 \end{aligned}$ **(d)** $\begin{aligned} x + y &= 1 \\ y + z &= 0 \\ z - x &= 2 \end{aligned}$

2. Write down a system of linear equations that has each of the following augmented matrices.

(a) $\begin{bmatrix} 1 & -1 & 6 & 0 \\ 0 & 1 & 0 & 3 \\ 2 & -1 & 0 & 1 \end{bmatrix}$ **(b)** $\begin{bmatrix} 2 & -1 & 0 & -1 \\ -3 & 2 & 1 & 0 \\ 0 & 1 & 1 & 3 \end{bmatrix}$

3. Find all solutions (if any) to each of the following systems of linear equations.

(a) $\begin{aligned} x - 2y &= 1 \\ 4y - x &= -2 \end{aligned}$ **(b)** $\begin{aligned} 3x - y &= 0 \\ 2x - 3y &= 1 \end{aligned}$ **(c)** $\begin{aligned} 2x + y &= 5 \\ 3x + 2y &= 6 \end{aligned}$

(d) $\begin{aligned} 3x - y &= 2 \\ 2y - 6x &= -4 \end{aligned}$ **(e)** $\begin{aligned} 3x - y &= 4 \\ 2y - 6x &= 1 \end{aligned}$ **(f)** $\begin{aligned} 2x - 3y &= 5 \\ 3y - 2x &= 2 \end{aligned}$

4. Find all solutions (if any) to each of the following systems of linear equations.

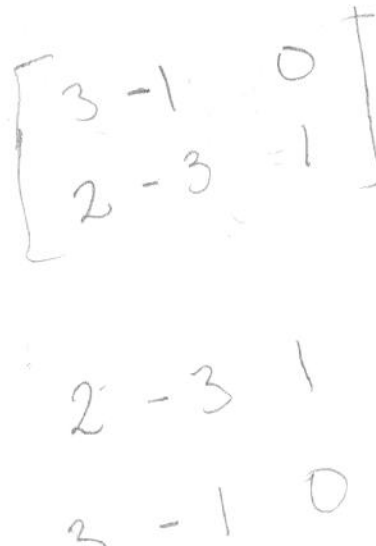

(a) $\begin{aligned} x + y + 2z &= 1 \\ 3x - y + z &= -1 \\ -x + 3y + 4z &= 1 \end{aligned}$ **(b)** $\begin{aligned} -2x + 3y + 3z &= -9 \\ 3x - 4y + z &= 5 \\ -5x + 7y + 2z &= -14 \end{aligned}$

(c) $\begin{aligned} x + y - z &= 3 \\ -x + 4y + 5z &= -2 \\ x + 6y + 3z &= 4 \end{aligned}$ **(d)** $\begin{aligned} x + 2y - z &= 2 \\ 2x + 5y - 3z &= 1 \\ x + 4y - 3z &= 3 \end{aligned}$

(e) $\begin{aligned} 5x + y &= 2 \\ 3x - y + 2z &= 1 \\ x + y - z &= 5 \end{aligned}$ **(f)** $\begin{aligned} 3x - 2y + z &= -2 \\ x - y + 3z &= 5 \\ -x + y + z &= -1 \end{aligned}$

(g) $\begin{aligned} x + y + z &= 4 \\ x + z &= 5 \\ 2x + 5y + 2z &= 5 \end{aligned}$ **(h)** $\begin{aligned} x + 2y - 4z &= 10 \\ 2x - y + 2z &= 5 \\ x + y - 2z &= 7 \end{aligned}$

5. Find all solutions to each of the following systems and express the last equation of each system as a sum of multiples of the first two.

(a) $\begin{aligned} x_1 + x_2 + x_3 &= 1 \\ 2x_1 - x_2 + 3x_3 &= 3 \\ 3x_1 - 3x_2 + 5x_3 &= 5 \end{aligned}$ **(b)** $\begin{aligned} 2x_1 + x_2 - 3x_3 &= -3 \\ 3x_1 + x_2 - 5x_3 &= 5 \\ -2x_1 + x_2 + 5x_3 &= -35 \end{aligned}$

6. In each of the following, find conditions on a and b such that the system has no solution, one solution, and infinitely many solutions.

(a) $\begin{aligned} x - 2y &= 1 \\ ax + by &= 5 \end{aligned}$ **(b)** $\begin{aligned} x + by &= -1 \\ ax + 2y &= 5 \end{aligned}$ **(c)** $\begin{aligned} x - by &= -1 \\ x + ay &= 3 \end{aligned}$ **(d)** $\begin{aligned} ax + y &= 1 \\ 2x + y &= b \end{aligned}$

(e) Invent two such problems and solve them.

7. In each of the following, find (if possible) conditions on a, b, and c such that the system has no solution, one solution, and infinitely many solutions.

(a) $\begin{aligned} 3x + y - z &= a \\ x - y + 2z &= b \\ 5x + 3y - 4z &= c \end{aligned}$ **(b)** $\begin{aligned} 2x + y - z &= a \\ 2y + 3z &= b \\ x - z &= c \end{aligned}$

(c) $\begin{aligned} -x + 3y + 2z &= -8 \\ x + z &= 2 \\ 3x + 3y + az &= b \end{aligned}$ **(d)** $\begin{aligned} x + ay &= 0 \\ y + bz &= 0 \\ z + cx &= 0 \end{aligned}$

(e) $\begin{aligned} 3x - y + 2z &= 3 \\ x + y - z &= 2 \\ 2x - 2y + 3z &= b \end{aligned}$ **(f)** $\begin{aligned} x + ay - z &= 1 \\ -x + (a-2)y + z &= -1 \\ 2x + 2y + (a-2)z &= 1 \end{aligned}$

(g) $\begin{aligned} x + 2y - 4z &= 4 \\ 3x - y + 13z &= 2 \\ 4x + y + a^2z &= a + 3 \end{aligned}$

(h) $\begin{aligned} x + y + 3z &= a \\ ax + y + 5z &= 4 \\ x + ay + 4z &= a \end{aligned}$

8. Find the circle $x^2 + y^2 + ax + by + c = 0$ passing through the following points.
   (a) $(-2, 1)$, $(5, 0)$, and $(4, 1)$
   (b) $(1, 1)$, $(5, -3)$, and $(-3, -3)$

9. Find the quadratic $y = ax^2 + bx + c$ passing through the following points:
   (a) $(1, 10)$, $(-1, 12)$, and $(2, 24)$
   (b) $(2, -2)$, $(-2, -2)$, and $(3, 3)$

10. A man is ordered by his doctor to take 5 units of vitamin A, 13 units of vitamin B, and 23 units of vitamin C each day. Three brands of vitamin pills are available, and the numbers of units of each vitamin per pill are shown in the accompanying table.

	Vitamin		
Brand	*A*	*B*	*C*
1	1	2	4
2	1	1	3
3	0	1	1

   (a) Find all combinations of pills that provide exactly the required amount of vitamins (no partial pills allowed).
   (b) If brands 1, 2, and 3 cost 3¢, 2¢, and 5¢ per pill, respectively, find the least expensive treatment.

11. A restaurant owner plans to use x tables seating 4, y tables seating 6, and z tables seating 8, for a total of 20 tables. When fully occupied, the tables seat 108 customers. If only half of the x tables, half of the y tables, and one-fourth of the z tables are used, each fully occupied, then 46 customers will be seated. Find x, y, and z.

12. Three Nissans, two Fords, and four Chevrolets can be rented for \$106 per day. At the same rates two Nissans, four Fords, and three Chevrolets cost \$107 per day; whereas four Nissans, three Fords, and two Chevrolets cost \$102 per day. Find the rental rates for all three kinds of cars.

13. A school has three clubs, and each student is required to belong to exactly one club. One year the students switched club membership as follows:

    Club A. $\frac{4}{10}$ remain in A, $\frac{1}{10}$ switch to B, $\frac{5}{10}$ switch to C.
    Club B. $\frac{7}{10}$ remain in B, $\frac{2}{10}$ switch to A, $\frac{1}{10}$ switch to C.
    Club C. $\frac{6}{10}$ remain in C, $\frac{2}{10}$ switch to A, $\frac{2}{10}$ switch to B.

    If the fraction of the student population in each club is unchanged, find each of these fractions.

14. An amusement park charges \$7 for adults, \$2 for youths, and \$.50 for children. If 150 people enter and pay a total of \$100, find the numbers of adults, youths, and children. [*Hint:* These numbers are nonnegative *integers*.]

15. Solve the following system of equations for x and y.

$$\begin{aligned} x^2 + xy - y^2 &= 1 \\ 2x^2 - xy + 3y^2 &= 13 \\ x^2 + 3xy + 2y^2 &= 0 \end{aligned}$$

[*Hint:* These equations are linear in the new variables $x_1 = x^2$, $x_2 = xy$, and $x_3 = y^2$.]

1.2.2 Row-Echelon Matrices

In Section 1.2.1, a method was introduced for solving systems of linear equations. In its full generality this process is called **Gaussian elimination**. The name honors Karl Friedrich Gauss (1777–1855), one of the greatest mathematicians of all time. The usual application of the method is to modify the augmented matrix (rather than the set of equations itself) with the aim of ending up with a "nice" matrix for which the corresponding equations are easily solved. In the case of three equations in three variables, the best we can hope for is to end up, as in Example 1, with a matrix of the form

$$\begin{bmatrix} 1 & 0 & 0 & * \\ 0 & 1 & 0 & * \\ 0 & 0 & 1 & * \end{bmatrix}$$

(Here the asterisks represent numbers, not necessarily zero.) This is "nice" because the corresponding equations can be solved immediately. However, in the case of Example 2, things were not quite so pleasant. The best that could be achieved was a matrix of the form

$$\begin{bmatrix} 1 & 0 & * & * \\ 0 & 1 & * & * \\ 0 & 0 & 0 & 0 \end{bmatrix}$$

DEFINITION A matrix is said to be in **row-echelon form** (and will be called a **row-echelon matrix**) if it satisfies the following conditions:

1. All **zero rows** (consisting entirely of zeros) are at the bottom.
2. The first nonzero entry from the left in each nonzero row is a 1, called the **leading 1** for that row.
3. Each leading 1 is to the right of all leading 1's in the rows above it.

If, in addition, the matrix satisfies the following condition, it is said to be in **reduced row-echelon form** (and will be called a **reduced row-echelon matrix**).

4. Each leading 1 is the only nonzero entry in its column.

The row-echelon matrices have a sort of "staircase" form, as indicated by the following example.

$$\begin{bmatrix} 0 & 1 & * & * & * & * & * \\ 0 & 0 & 0 & 1 & * & * & * \\ 0 & 0 & 0 & 0 & 1 & * & * \\ 0 & 0 & 0 & 0 & 0 & 0 & 1 \\ 0 & 0 & 0 & 0 & 0 & 0 & 0 \end{bmatrix}$$

The leading 1's proceed "down and to the right" through the matrix. Entries above and to the right of the leading 1's are arbitrary, but all entries below and to the left of them are zero. Hence a matrix in row-echelon form is in reduced form if, in addition, the entries directly above each leading 1 are zero. In particular, the following matrices are in row-echelon form (for any choice of numbers in *-positions).

$$\begin{bmatrix} 1 & * & * \\ 0 & 0 & 1 \end{bmatrix} \quad \begin{bmatrix} 0 & 1 & * & * \\ 0 & 0 & 1 & * \\ 0 & 0 & 0 & 0 \end{bmatrix} \quad \begin{bmatrix} 1 & * & * & * \\ 0 & 1 & * & * \\ 0 & 0 & 0 & 1 \end{bmatrix} \quad \begin{bmatrix} 1 & * & * \\ 0 & 1 & * \\ 0 & 0 & 1 \end{bmatrix}$$

The following, on the other hand, are in reduced row-echelon form.

$$\begin{bmatrix} 1 & * & 0 \\ 0 & 0 & 1 \end{bmatrix} \quad \begin{bmatrix} 0 & 1 & 0 & * \\ 0 & 0 & 1 & * \\ 0 & 0 & 0 & 0 \end{bmatrix} \quad \begin{bmatrix} 1 & 0 & * & 0 \\ 0 & 1 & * & 0 \\ 0 & 0 & 0 & 1 \end{bmatrix} \quad \begin{bmatrix} 1 & 0 & 0 \\ 0 & 1 & 0 \\ 0 & 0 & 1 \end{bmatrix}$$

Clearly the choice of the positions for the leading 1's determines the (reduced) row-echelon form (apart from the numbers in *-positions).

Now suppose an arbitrary matrix is presented to you and you start to apply elementary row operations to it. Here is a procedure by which it can be brought to row-echelon form.

GAUSSIAN ALGORITHM

1. If the matrix consists entirely of zeros, there is nothing to do—it is *already* in row-echelon form.
2. Otherwise, find the first column from the left containing a nonzero entry (call it a), and move the row containing that entry to the top position.

3. Now multiply that row by $\frac{1}{a}$ to create a leading 1.
4. By subtracting multiples of that row from rows below it, make each entry below the leading 1 zero.

This completes the first row, and all further row operations are carried out on the other rows. The fact that the column below the leading 1 consists of zeros means that these manipulations do not affect that column. Now proceed to the remaining rows.

5. Repeat steps 1–4 on the remaining rows.

Of course, if these rows are all zero, you are finished—the matrix is in row-echelon form. If not, the process creates the second leading 1 in row 2, which is *below and to the right* of the first leading 1. This completes the second row. Clearly this procedure can now be repeated with the remaining rows, and it ultimately leads to a matrix in row-echelon form.

Observe that the Gaussian algorithm is recursive: When the first leading 1 has been obtained, the procedure is repeated on the remaining rows of the matrix. This makes the algorithm easy to use on a computer.

If it is desired to carry the matrix to reduced row-echelon form, step 4 can be modified to make each entry above the leading 1 zero. However, it is numerically more efficient to go to row-echelon form first and *then* create the zeros above the leading 1's, starting with the first leading 1 from the right and working from right to left. This is an important consideration when one is solving a large system.

In any case, only elementary row operations are used, so this proves

THEOREM 2 Every matrix can be brought to (reduced) row-echelon form by the use of elementary row operations.

Several examples of this process were given (for augmented matrices of systems of equations) in the previous section. Here is another example.

EXAMPLE 5 Bring

$$\begin{bmatrix} 0 & 1 & 3 & -1 & 2 & 1 \\ 0 & -2 & -6 & 2 & 0 & 2 \\ 0 & 0 & 0 & 1 & -1 & 7 \\ 0 & 2 & 6 & -2 & 2 & 0 \\ 0 & 3 & 9 & -2 & 2 & 7 \end{bmatrix}$$

to reduced row-echelon form, using elementary row operations.

Solution The leading 1 in the first row is already in place, so we use it to create zeros below it in the first column.

$$\begin{bmatrix} 0 & 1 & 3 & -1 & 2 & 1 \\ 0 & 0 & 0 & 0 & 4 & 4 \\ 0 & 0 & 0 & 1 & -1 & 7 \\ 0 & 0 & 0 & 0 & -2 & -2 \\ 0 & 0 & 0 & 1 & -4 & 4 \end{bmatrix} \quad \begin{matrix} \\ [(\text{row } 2) + 2(\text{row } 1)] \\ \\ [(\text{row } 4) - 2(\text{row } 1)] \\ [(\text{row } 5) - 3(\text{row } 1)] \end{matrix}$$

This completes the work on the first two columns. The next leading 1 will *not* appear in the third column because only zeros appear below row 1. This means we move to the fourth column where two 1's present themselves. Choosing the 1 in the third row (quite arbitrarily), we put that row in the second position (creating the second leading 1) and use it to introduce zeros in its column.

$$\begin{bmatrix} 0 & 1 & 3 & 0 & 1 & 8 \\ 0 & 0 & 0 & 1 & -1 & 7 \\ 0 & 0 & 0 & 0 & 4 & 4 \\ 0 & 0 & 0 & 0 & -2 & -2 \\ 0 & 0 & 0 & 0 & -3 & -3 \end{bmatrix} \quad \begin{matrix} [(\text{row } 1) + (\text{row } 3)] \\ [\text{interchange rows 2 and 3}] \\ \\ \\ [(\text{row } 5) - (\text{row } 3)] \end{matrix}$$

We create the next leading 1 by multiplying the third row by $\frac{1}{4}$. If it is then used to create zeros in the fifth column, the result is

$$\begin{bmatrix} 0 & 1 & 3 & 0 & 0 & 7 \\ 0 & 0 & 0 & 1 & 0 & 8 \\ 0 & 0 & 0 & 0 & 1 & 1 \\ 0 & 0 & 0 & 0 & 0 & 0 \\ 0 & 0 & 0 & 0 & 0 & 0 \end{bmatrix}$$

This is now in reduced row-echelon form.

As was illustrated in Section 1.2.1, we solve a system of equations by first reducing the augmented matrix to reduced row-echelon form. Variables corresponding to columns containing a leading 1 are called **leading variables.** The nonleading variables (if any) end up as parameters in the final solution, and the leading variables are given (by the equations) in terms of these parameters.

EXAMPLE 6 Solve the following system of equations.

$$\begin{aligned} x_1 - 3x_2 \phantom{{}+ 2x_3} - x_4 \phantom{{}+ x_5} &= -1 \\ -x_1 + 3x_2 + x_3 + 3x_4 \phantom{{}+ x_5} &= 3 \\ 2x_1 - 6x_2 + x_3 \phantom{{}+ 5x_4} - x_5 &= -1 \\ -x_1 + 3x_2 + 2x_3 + 5x_4 + x_5 &= 6 \end{aligned}$$

Solution The reduction of the augmented matrix to reduced row-echelon form is as follows (the details are omitted).

$$\begin{bmatrix} 1 & -3 & 0 & -1 & 0 & -1 \\ -1 & 3 & 1 & 3 & 0 & 3 \\ 2 & -6 & 1 & 0 & -1 & -1 \\ -1 & 3 & 2 & 5 & 1 & 6 \end{bmatrix} \to \begin{bmatrix} 1 & -3 & 0 & -1 & 0 & -1 \\ 0 & 0 & 1 & 2 & 0 & 2 \\ 0 & 0 & 0 & 0 & 1 & 1 \\ 0 & 0 & 0 & 0 & 0 & 0 \end{bmatrix}$$

Hence the leading variables are x_1, x_3, and x_5. If the nonleading variables are assigned as parameters $x_2 = s$, $x_4 = t$, the corresponding equations yield

$$x_1 = 3s + t - 1$$

$$x_3 = -2t + 2$$

$$x_5 = 1$$

Hence $\begin{bmatrix} 3s + t - 1 \\ s \\ -2t + 2 \\ t \\ 1 \end{bmatrix}$ gives all solutions.

Suppose A is any matrix. Theorem 2 shows that a sequence of elementary row operations can be found that carries A to a matrix (call it R) in row-echelon form. Suppose R has r leading 1's (that is, r nonzero rows). Of course it is possible that some other sequence of elementary row operations will carry A to a *different* row-echelon matrix R'. The remarkable thing is that R' will *also* have r leading 1's (this will be proved in Chapter 5).[†] Hence this number r depends only on A and not on the way in which A is carried to row-echelon form.

DEFINITION If a matrix A is carried to a row-echelon matrix R by elementary row operations, the number of leading 1's in R is called the **rank** of A and is denoted rank A.

EXAMPLE 7

Compute the rank of $A = \begin{bmatrix} 1 & 1 & -1 & 4 \\ 2 & 1 & 3 & 0 \\ 0 & 1 & -5 & 8 \end{bmatrix}$.

[†] If R' is in *reduced* row-echelon form, the whole matrix R' is uniquely determined by A.

Solution The reduction of A to row-echelon form is

$$A = \begin{bmatrix} 1 & 1 & -1 & 4 \\ 2 & 1 & 3 & 0 \\ 0 & 1 & -5 & 8 \end{bmatrix} \rightarrow \begin{bmatrix} 1 & 1 & -1 & 4 \\ 0 & -1 & 5 & -8 \\ 0 & 1 & -5 & 8 \end{bmatrix} \rightarrow \begin{bmatrix} 1 & 1 & -1 & 4 \\ 0 & 1 & -5 & 8 \\ 0 & 0 & 0 & 0 \end{bmatrix}$$

Because this row-echelon matrix has two leading 1's, rank $A = 2$.

The notion of rank of a matrix has a nice application to equations.

THEOREM 3 Suppose a system of m equations in n variables has a solution. If the rank of the augmented matrix is r, the set of solutions involves exactly $n - r$ parameters.

Proof The fact that the rank of the augmented matrix is r means there are exactly r leading variables, and hence exactly $n - r$ nonleading variables. These nonleading variables are all assigned as parameters (as in Example 6), so the set of solutions involves exactly $n - r$ parameters. ■

In particular, this shows that, for any system of linear equations, exactly three possibilities exist:

1. *No solution.*
2. *A unique solution.* This occurs when *every* variable is a leading variable.
3. *Infinitely many solutions.* This occurs when there is at least one nonleading variable, so a parameter is involved.

EXERCISES 1.2.2

1. Which of the following matrices are in reduced row-echelon form? Which are in row-echelon form?

(a) $\begin{bmatrix} 1 & -1 & 2 \\ 0 & 0 & 0 \\ 0 & 0 & 1 \end{bmatrix}$ **(b)** $\begin{bmatrix} 2 & 1 & -1 & 3 \\ 0 & 0 & 0 & 0 \end{bmatrix}$ **(c)** $\begin{bmatrix} 1 & -2 & 3 & 5 \\ 0 & 0 & 0 & 1 \end{bmatrix}$

(d) $\begin{bmatrix} 1 & 0 & 0 & 3 & 1 \\ 0 & 0 & 0 & 1 & 1 \\ 0 & 0 & 0 & 0 & 1 \end{bmatrix}$ **(e)** $\begin{bmatrix} 1 & 1 \\ 0 & 1 \end{bmatrix}$ **(f)** $\begin{bmatrix} 0 & 0 & 1 \\ 0 & 0 & 1 \\ 0 & 0 & 1 \end{bmatrix}$

2. Carry each of the following matrices to reduced row-echelon form.

(a) $\begin{bmatrix} 0 & -1 & 2 & 1 & 2 & 1 & -1 \\ 0 & 1 & -2 & 2 & 7 & 2 & 4 \\ 0 & -2 & 4 & 3 & 7 & 1 & 0 \\ 0 & 3 & -6 & 1 & 6 & 4 & 1 \end{bmatrix}$ **(b)** $\begin{bmatrix} 0 & -1 & 3 & 1 & 3 & 2 & 1 \\ 0 & -2 & 6 & 1 & -5 & 0 & -1 \\ 0 & 3 & -9 & 2 & 4 & 1 & -1 \\ 0 & 1 & -3 & -1 & 3 & 0 & 1 \end{bmatrix}$

3. The augmented matrix of a system of linear equations has been carried to the following by row operations. In each case solve the system.

(a) $\begin{bmatrix} 1 & 2 & 0 & 3 & 1 & 0 & -1 \\ 0 & 0 & 1 & -1 & 1 & 0 & 2 \\ 0 & 0 & 0 & 0 & 0 & 1 & 3 \\ 0 & 0 & 0 & 0 & 0 & 0 & 0 \end{bmatrix}$ (b) $\begin{bmatrix} 1 & -2 & 0 & 2 & 0 & 1 & 1 \\ 0 & 0 & 1 & 5 & 0 & -3 & -1 \\ 0 & 0 & 0 & 0 & 1 & 6 & 1 \\ 0 & 0 & 0 & 0 & 0 & 0 & 0 \end{bmatrix}$

(c) $\begin{bmatrix} 1 & 2 & 1 & 3 & 1 & 1 \\ 0 & 1 & -1 & 0 & 1 & 1 \\ 0 & 0 & 0 & 1 & -1 & 0 \\ 0 & 0 & 0 & 0 & 0 & 0 \end{bmatrix}$ (d) $\begin{bmatrix} 1 & -1 & 2 & 4 & 6 & 2 \\ 0 & 1 & 2 & 1 & -1 & -1 \\ 0 & 0 & 0 & 1 & 0 & 1 \\ 0 & 0 & 0 & 0 & 0 & 0 \end{bmatrix}$

4. Find all solutions to the following systems.

(a) $\begin{aligned} 3x_1 + 8x_2 - 3x_3 - 14x_4 &= 1 \\ 2x_1 + 3x_2 - x_3 - 2x_4 &= 2 \\ x_1 - 2x_2 + x_3 + 10x_4 &= 3 \\ x_1 + 5x_2 - 2x_3 - 12x_4 &= -1 \end{aligned}$ (b) $\begin{aligned} x_1 - x_2 + x_3 - x_4 &= 0 \\ -x_1 + x_2 + x_3 + x_4 &= 0 \\ x_1 + x_2 - x_3 + x_4 &= 0 \\ x_1 + x_2 + x_3 + x_4 &= 0 \end{aligned}$

(c) $\begin{aligned} x_1 - x_2 + x_3 - 2x_4 &= 3 \\ -x_1 + x_2 + x_3 + x_4 &= 2 \\ -x_1 + 2x_2 + 3x_3 - x_4 &= 9 \\ x_1 - x_2 + 2x_3 + x_4 &= 2 \end{aligned}$ (d) $\begin{aligned} x_1 + x_2 + 2x_3 - x_4 &= 4 \\ 3x_2 - x_3 + 4x_4 &= 2 \\ x_1 + 2x_2 - 3x_3 + 5x_4 &= 0 \\ x_1 + x_2 - 5x_3 + 6x_4 &= -3 \end{aligned}$

(e) $\begin{aligned} x_1 + x_2 + x_3 - x_4 &= 3 \\ 3x_1 + 5x_2 - 2x_3 + x_4 &= 1 \\ -3x_1 - 7x_2 + 7x_3 - 5x_4 &= 7 \\ x_1 + 3x_2 - 4x_3 + 3x_4 &= -5 \end{aligned}$ (f) $\begin{aligned} x_1 + 4x_2 - x_3 + x_4 &= 2 \\ 3x_1 + 2x_2 + x_3 + 2x_4 &= 5 \\ x_1 - 6x_2 + 3x_3 &= 1 \\ x_1 + 14x_2 - 5x_3 + 2x_4 &= 3 \end{aligned}$

5. In each case, show that the reduced row-echelon form is as given.

(a) $\begin{bmatrix} p & 0 & a \\ b & 0 & 0 \\ q & c & r \end{bmatrix}$ with $abc \neq 0$ $\begin{bmatrix} 1 & 0 & 0 \\ 0 & 1 & 0 \\ 0 & 0 & 1 \end{bmatrix}$

(b) $\begin{bmatrix} 1 & a & b+c \\ 1 & b & c+a \\ 1 & c & a+b \end{bmatrix}$ where $c \neq a$ or $b \neq a$ $\begin{bmatrix} 1 & 0 & * \\ 0 & 1 & * \\ 0 & 0 & 0 \end{bmatrix}$

6. Show that $\left\{\begin{aligned} ax + by + cz &= 0 \\ a_1x + b_1y + c_1z &= 0 \end{aligned}\right\}$ always has a solution other than $\begin{bmatrix} 0 \\ 0 \\ 0 \end{bmatrix}$.

7. Find the rank of each of the matrices in Exercise 1.

8. Find the rank of each of the following matrices.

(a) $\begin{bmatrix} 1 & 1 & 2 \\ 3 & -1 & 1 \\ -1 & 3 & 4 \end{bmatrix}$ (b) $\begin{bmatrix} -2 & 3 & 3 \\ 3 & -4 & 1 \\ -5 & 7 & 2 \end{bmatrix}$

(c) $\begin{bmatrix} 1 & 1 & -1 & 3 \\ -1 & 4 & 5 & -2 \\ 1 & 6 & 3 & 4 \end{bmatrix}$ (d) $\begin{bmatrix} 3 & -2 & 1 & -2 \\ 1 & -1 & 3 & 5 \\ -1 & 1 & 1 & -1 \end{bmatrix}$

9. Show that any two rows of a matrix can be interchanged by elementary row transformations of the other two types.

10. Find a sequence of row operations carrying $\begin{bmatrix} b_1 + c_1 & b_2 + c_2 & b_3 + c_3 \\ c_1 + a_1 & c_2 + a_2 & c_3 + a_3 \\ a_1 + b_1 & a_2 + b_2 & a_3 + b_3 \end{bmatrix}$

to $\begin{bmatrix} a_1 & a_2 & a_3 \\ b_1 & b_2 & b_3 \\ c_1 & c_2 & c_3 \end{bmatrix}$.

11. Given points (p_1, q_1), (p_2, q_2), and (p_3, q_3) in the plane with p_1, p_2, and p_3 distinct, show that they lie on some curve with equation $y = a + bx + cx^2$. [*Hint:* Solve for a, b, and c.]
12. The scores of three players in a tournament have been lost. The only information available is the total of the scores for players 1 and 2, the total for players 2 and 3, and the total for players 3 and 1.
    (a) Show that the individual scores can be rediscovered.
    (b) Is this true with four players (knowing the totals for players 1 and 2, 2 and 3, 3 and 4, and 4 and 1)?
13. **(a)** Show that a matrix with two rows and two columns that is in reduced row-echelon form must have one of the following forms:

$$\begin{bmatrix} 1 & 0 \\ 0 & 1 \end{bmatrix} \quad \begin{bmatrix} 0 & 1 \\ 0 & 0 \end{bmatrix} \quad \begin{bmatrix} 0 & 0 \\ 0 & 0 \end{bmatrix} \quad \begin{bmatrix} 1 & * \\ 0 & 0 \end{bmatrix}$$

    [*Hint:* The leading 1 in the first row must be in column 1 or 2 or not exist.]
    (b) List the seven reduced row-echelon forms for matrices with two rows and three columns.
    (c) List the four reduced row-echelon forms for matrices with three rows and two columns.

SECTION 1.3 Homogeneous Equations

When we are solving systems of linear equations, there is always the annoying possibility that there is no solution. One situation wherein a solution *always* exists is when the constant terms are all zero; it is clear that such equations are satisfied when every variable is set equal to 0. In general, a system of equations in the variables $x_1, x_2, \ldots, x_n$ is called **homogeneous** if all the constant terms are zero—that is, if each equation of the system has the form

$$a_1x_1 + a_2x_2 + \cdots + a_nx_n = 0$$

Clearly $\begin{bmatrix} 0 \\ 0 \\ \vdots \\ 0 \end{bmatrix}$ is a solution to such a system; it is called the **trivial** solution. Sometimes (see Example 2) it is of interest to know that a homogeneous system has a nontrivial solution—that is, a solution in which at least one variable has a nonzero value.

EXAMPLE 1 Show that the following homogeneous system has nontrivial solutions.

$$\begin{aligned} x_1 - x_2 + 2x_3 + x_4 &= 0 \\ 2x_1 + 2x_2 \qquad - x_4 &= 0 \\ 3x_1 + x_2 + 2x_3 + x_4 &= 0 \end{aligned}$$

Solution The reduction of the augmented matrix to reduced row-echelon form is outlined below.

$$\begin{bmatrix} 1 & -1 & 2 & 1 & 0 \\ 2 & 2 & 0 & -1 & 0 \\ 3 & 1 & 2 & 1 & 0 \end{bmatrix} \rightarrow \begin{bmatrix} 1 & -1 & 2 & 1 & 0 \\ 0 & 4 & -4 & -3 & 0 \\ 0 & 4 & -4 & -2 & 0 \end{bmatrix} \rightarrow \begin{bmatrix} 1 & 0 & 1 & 0 & 0 \\ 0 & 1 & -1 & 0 & 0 \\ 0 & 0 & 0 & 1 & 0 \end{bmatrix}$$

The leading variables are x_1, x_2, and x_4, so x_3 is assigned as a parameter—say $x_3 = t$. Then the general solution is $\begin{bmatrix} -t \\ t \\ t \\ 0 \end{bmatrix}$, so $\begin{bmatrix} -1 \\ 1 \\ 1 \\ 0 \end{bmatrix}$ is a nontrivial solution.

The existence of a nontrivial solution (infinitely many, in fact) was ensured by the presence of a parameter in the solution. This, in turn, was guaranteed by the fact that some variable was *not* leading. But this *had* to occur here because there were four variables but only three equations (and hence at *most* three leading variables). This discussion generalizes to a proof of the following useful theorem.

THEOREM 1 If a homogeneous system of linear equations has more variables than equations, then it has a nontrivial solution (in fact, infinitely many).

Note that the converse of Theorem 1 is not true (Exercise 2(b)).

The next example provides an illustration of how Theorem 1 is used.

EXAMPLE 2 We call the graph of an equation $ax^2 + bxy + cy^2 + dx + ey + f = 0$ a **conic** if the numbers a, b, c, d, e, and f are not all zero. Show that there is at least one conic through any five points in the plane.

Solution Let the coordinates of the five points be (p_1, q_1), (p_2, q_2), (p_3, q_3), (p_4, q_4), and (p_5, q_5). The conic $ax^2 + bxy + cy^2 + dx + ey + f$ passes through (p_i, q_i) if

$$ap_i^2 + bp_iq_i + cq_i^2 + dp_i + eq_i + f = 0$$

This gives five equations, linear in the six variables a, b, c, d, e, and f. Hence there is a nontrivial solution by Theorem 1.

The set of solutions to a homogeneous system of equations can be described in a compact way. In order to do so, it is necessary to define three new operations on n-tuples. Given two n-tuples $\boldsymbol{s} = \begin{bmatrix} s_1 \\ s_2 \\ \vdots \\ s_n \end{bmatrix}$ and $\boldsymbol{t} = \begin{bmatrix} t_1 \\ t_2 \\ \vdots \\ t_n \end{bmatrix}$, their **sum $\boldsymbol{s} + \boldsymbol{t}$**, their **difference $\boldsymbol{s} - \boldsymbol{t}$**, and the **scalar product** of $\boldsymbol{t}$ times a number k are defined as follows:

$$\boldsymbol{s} + \boldsymbol{t} = \begin{bmatrix} s_1 + t_1 \\ s_2 + t_2 \\ \vdots \\ s_n + t_n \end{bmatrix} \qquad \boldsymbol{s} - \boldsymbol{t} = \begin{bmatrix} s_1 - t_1 \\ s_2 - t_2 \\ \vdots \\ s_n - t_n \end{bmatrix} \qquad k\boldsymbol{t} = \begin{bmatrix} kt_1 \\ kt_2 \\ \vdots \\ kt_n \end{bmatrix}.$$

In other words, to add (or subtract) two n-tuples, add (or subtract) corresponding entries; to multiply an n-tuple by a number, multiply each entry by that number.

EXAMPLE 3 If $\boldsymbol{s} = \begin{bmatrix} 3 \\ 2 \\ 0 \\ -1 \end{bmatrix}$ and $\boldsymbol{t} = \begin{bmatrix} 1 \\ -1 \\ 2 \\ 5 \end{bmatrix}$, compute $\boldsymbol{s} + \boldsymbol{t}$, $-3\boldsymbol{t}$, and $2\boldsymbol{s} - \boldsymbol{t}$.

Solution $$\boldsymbol{s} + \boldsymbol{t} = \begin{bmatrix} 3 \\ 2 \\ 0 \\ -1 \end{bmatrix} + \begin{bmatrix} 1 \\ -1 \\ 2 \\ 5 \end{bmatrix} = \begin{bmatrix} 3 + 1 \\ 2 - 1 \\ 0 + 2 \\ -1 + 5 \end{bmatrix} = \begin{bmatrix} 4 \\ 1 \\ 2 \\ 4 \end{bmatrix}$$

$$-3\boldsymbol{t} = -3\begin{bmatrix} 1 \\ -1 \\ 2 \\ 5 \end{bmatrix} = \begin{bmatrix} -3 \\ 3 \\ -6 \\ -15 \end{bmatrix}$$

$$2\boldsymbol{s} - \boldsymbol{t} = 2\begin{bmatrix} 3 \\ 2 \\ 0 \\ -1 \end{bmatrix} - \begin{bmatrix} 1 \\ -1 \\ 2 \\ 5 \end{bmatrix} = \begin{bmatrix} 6 - 1 \\ 4 + 1 \\ 0 - 2 \\ -2 - 5 \end{bmatrix} = \begin{bmatrix} 5 \\ 5 \\ -2 \\ -7 \end{bmatrix}$$

THEOREM 2

Suppose $\mathbf{s} = \begin{bmatrix} s_1 \\ s_2 \\ \vdots \\ s_n \end{bmatrix}$ and $\mathbf{t} = \begin{bmatrix} t_1 \\ t_2 \\ \vdots \\ t_n \end{bmatrix}$ are both solutions to a system of homogeneous equations in n variables $x_1, x_2, \ldots, x_n$. Then

1. $\mathbf{s} + \mathbf{t}$ is also a solution.
2. $k\mathbf{s}$ is also a solution for any number k.

Proof Each equation in the system has the form

$$a_1x_1 + a_2x_2 + \cdots + a_nx_n = 0 \qquad (*)$$

and the fact that $\mathbf{s}$ and $\mathbf{t}$ are solutions means that

$$\begin{aligned} a_1s_1 + a_2s_2 + \cdots + a_ns_n &= 0 \\ a_1t_1 + a_2t_2 + \cdots + a_nt_n &= 0 \end{aligned}$$

Adding these equations gives

$$a_1(s_1 + t_1) + a_2(s_2 + t_2) + \cdots + a_n(s_n + t_n) = 0$$

which asserts that $\mathbf{s} + \mathbf{t}$ is a solution to (*) and hence to the system. Similarly, $k\mathbf{s}$ is a solution to the system. ■

A sum of scalar multiples of n-tuples is called a **linear combination** of those n-tuples. For example, if $\mathbf{s} = \begin{bmatrix} 1 \\ -1 \end{bmatrix}$ and $\mathbf{t} = \begin{bmatrix} 2 \\ 0 \end{bmatrix}$, then

$$3\mathbf{s} - 2\mathbf{t} = 3\begin{bmatrix} 1 \\ -1 \end{bmatrix} - 2\begin{bmatrix} 2 \\ 0 \end{bmatrix} = \begin{bmatrix} 3 \\ -3 \end{bmatrix} - \begin{bmatrix} 4 \\ 0 \end{bmatrix} = \begin{bmatrix} -1 \\ -3 \end{bmatrix}$$

is a linear combination of $\mathbf{s}$ and $\mathbf{t}$. Note that every linear combination of n-tuples is again an n-tuple.

Theorem 2 implies that linear combinations of solutions of a system of linear homogeneous equations are again solutions to that system. Conversely, the next example illustrates how to express any solution as a linear combination of certain basic solutions.

EXAMPLE 4 Describe the solutions to the following homogeneous system, using sums and scalar multiples.

$$\begin{aligned} x_1 - 2x_2 + x_3 + x_4 &= 0 \\ -x_1 + 2x_2 + x_4 &= 0 \\ 2x_1 - 4x_2 + x_3 &= 0 \end{aligned}$$

Solution The augmented matrix is reduced as follows:

$$\begin{bmatrix} 1 & -2 & 1 & 1 & 0 \\ -1 & 2 & 0 & 1 & 0 \\ 2 & -4 & 1 & 0 & 0 \end{bmatrix} \to \begin{bmatrix} 1 & -2 & 1 & 1 & 0 \\ 0 & 0 & 1 & 2 & 0 \\ 0 & 0 & -1 & -2 & 0 \end{bmatrix} \to \begin{bmatrix} 1 & -2 & 0 & -1 & 0 \\ 0 & 0 & 1 & 2 & 0 \\ 0 & 0 & 0 & 0 & 0 \end{bmatrix}$$

so the solution is $\begin{bmatrix} 2s+t \\ s \\ -2t \\ t \end{bmatrix}$, where s and t are parameters. We separate this into parts involving s and t only.

$$\begin{bmatrix} 2s+t \\ s \\ -2t \\ t \end{bmatrix} = \begin{bmatrix} 2s \\ s \\ 0 \\ 0 \end{bmatrix} + \begin{bmatrix} t \\ 0 \\ -2t \\ t \end{bmatrix} = s\begin{bmatrix} 2 \\ 1 \\ 0 \\ 0 \end{bmatrix} + t\begin{bmatrix} 1 \\ 0 \\ -2 \\ 1 \end{bmatrix}$$

Hence all solutions can be "built up" as linear combinations of the two basic solutions $\begin{bmatrix} 2 \\ 1 \\ 0 \\ 0 \end{bmatrix}$ and $\begin{bmatrix} 1 \\ 0 \\ -2 \\ 1 \end{bmatrix}$. Note the role played by the parameters; they are the coefficients in these linear combinations of solutions.

EXERCISES 1.3

1. In each of the following, find all values of a for which the system has nontrivial solutions, and determine all solutions in each case.

(a) $\begin{aligned} x - 2y + z &= 0 \\ x + ay - 3z &= 0 \\ -x + 6y - 5z &= 0 \end{aligned}$

(b) $\begin{aligned} x + 2y + z &= 0 \\ x + 3y + 6z &= 0 \\ 2x + 3y + az &= 0 \end{aligned}$

(c) $\begin{aligned} x + y - z &= 0 \\ ay - z &= 0 \\ x + y + az &= 0 \end{aligned}$

(d) $\begin{aligned} ax + y + z &= 0 \\ x + y - z &= 0 \\ x + y + az &= 0 \end{aligned}$

2. (a) Does Theorem 1 imply that the system $\begin{aligned} -x + 3y &= 0 \\ 2x - 6y &= 0 \end{aligned}$ has nontrivial solutions?

(b) Show that the converse to Theorem 1 is not true. That is, show that the existence of nontrivial solutions does *not* imply that there are more variables than equations.

3. The graph of an equation $ax + by + cz = 0$ is a plane through the origin (provided that not all of a, b, and c are zero). Use Theorem 1 to show that two planes through the origin have a point in common other than the origin $(0, 0, 0)$.

4. (a) Show that there is a line through any pair of points in the plane. [*Hint:* Every line has equation $ax + by + c = 0$, where a, b, and c are not all zero.]
   (b) Generalize and show that there is a plane $ax + by + cz + d = 0$ through any three points in space.
5. In each of the following, write the solution as a linear combination of specific solutions.

(a)
$$\begin{aligned} x + y + z &= 0 \\ -2x - 3y &= 0 \\ x + 2y - z &= 0 \end{aligned}$$

(b)
$$\begin{aligned} x + 2y - 3z &= 0 \\ 2x + y + 3z &= 0 \\ x + 5y - 12z &= 0 \end{aligned}$$

(c)
$$\begin{aligned} x_1 + 2x_2 + x_3 - x_4 + 3x_5 &= 0 \\ x_1 + 2x_2 + 2x_3 + x_4 + 2x_5 &= 0 \\ 2x_1 + 4x_2 + 2x_3 - x_4 + 7x_5 &= 0 \end{aligned}$$

(d)
$$\begin{aligned} x_1 + x_2 - 2x_3 + 3x_4 + 2x_5 &= 0 \\ 2x_1 - x_2 + 3x_3 + 4x_4 + x_5 &= 0 \\ -x_1 - 2x_2 + 3x_3 + x_4 &= 0 \\ 3x_1 + x_3 + 7x_4 + 3x_5 &= 0 \end{aligned}$$

(e)
$$\begin{aligned} x_1 + x_2 + 2x_4 &= 0 \\ x_1 - x_2 + x_3 + 2x_4 &= 0 \\ 2x_1 + x_2 - x_3 + 4x_4 &= 0 \\ -x_1 + 2x_2 - 2x_4 &= 0 \end{aligned}$$

(f)
$$\begin{aligned} x_1 - x_2 + 2x_3 - x_4 + x_5 &= 0 \\ 4x_1 - 3x_2 - x_3 + x_4 - 2x_5 &= 0 \\ 3x_1 - 2x_2 - 3x_3 + 2x_4 - 3x_5 &= 0 \\ 6x_1 - 5x_2 + 3x_3 - x_4 &= 0 \end{aligned}$$

6. Give an example of a homogeneous system with exactly one solution. Can such a system have exactly two solutions? Why?
7. Consider a homogeneous system of linear equations in n variables, and suppose that the augmented matrix has rank r. If $n > r$, show that the system has nontrivial solutions.

SECTION 1.4 Applications of Systems of Equations (Optional)†

1.4.1 Network Flow Problems

Many types of problems concern a network of conductors along which some sort of flow is observed. Examples include an irrigation network and a network of streets or freeways. There are generally points in the system at which a net flow either enters or leaves the system. The basic principle behind the analysis of such systems is that flow is conserved—that is, the total flow into the system must equal the total flow out. But more important, it means that, at each of the junctions in the network, the total flow *into that junction* must equal the flow out. This requirement gives a linear equation relating the flows in conductors emanating from that junction.

EXAMPLE 1 A network of one-way streets is shown in the diagram on page 32. The rate of flow of cars into intersection A is 500 cars per hour, and 400 and 100 cars per hour emerge from B and C, respectively. Find the possible flows along each street.

†The two applications in this section are independent and may be taken in any order.

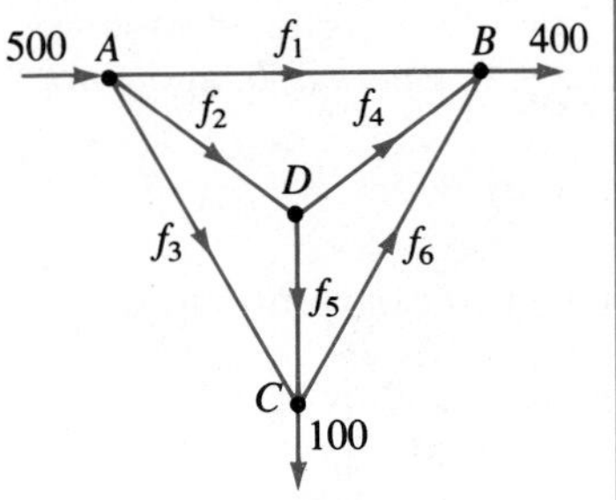

Solution Suppose the flows along the streets are f_1, f_2, f_3, f_4, f_5, and f_6 cars per hour in the directions shown. Then, equating the flow in with the flow out at each intersection, we get

Intersection A	$500 = f_1 + f_2 + f_3$
Intersection B	$f_1 + f_4 + f_6 = 400$
Intersection C	$f_3 + f_5 = f_6 + 100$
Intersection D	$f_2 = f_4 + f_5$

These give four equations in the six variables $f_1, f_2, \ldots, f_6$.

$$\begin{aligned} f_1 + f_2 + f_3 \qquad\qquad\qquad &= 500 \\ f_1 \qquad\quad + f_4 \quad + f_6 &= 400 \\ f_3 \quad + f_5 - f_6 &= 100 \\ f_2 \quad - f_4 - f_5 \qquad &= 0 \end{aligned}$$

The reduction of the augmented matrix is

$$\begin{bmatrix} 1 & 1 & 1 & 0 & 0 & 0 & 500 \\ 1 & 0 & 0 & 1 & 0 & 1 & 400 \\ 0 & 0 & 1 & 0 & 1 & -1 & 100 \\ 0 & 1 & 0 & -1 & -1 & 0 & 0 \end{bmatrix} \to \begin{bmatrix} 1 & 0 & 0 & 1 & 0 & 1 & 400 \\ 0 & 1 & 0 & -1 & -1 & 0 & 0 \\ 0 & 0 & 1 & 0 & 1 & -1 & 100 \\ 0 & 0 & 0 & 0 & 0 & 0 & 0 \end{bmatrix}$$

Hence, when we use f_4, f_5, and f_6 as parameters, the general solution is

$$f_1 = 400 - f_4 - f_6$$

$$f_2 = f_4 + f_5$$

$$f_3 = 100 - f_5 + f_6$$

This gives all solutions to the system of equations and hence all the possible flows.

Of course, not all these solutions may be acceptable in the real situation. For example, the flows $f_1, f_2, \ldots, f_6$ are all *positive* in the present context (if one came out negative, it would mean traffic flowed in the opposite direction). This imposes constraints on the flows: $f_1 \geq 0$ and $f_3 \geq 0$ become

$$f_4 + f_6 \leq 400$$

$$f_5 - f_6 \leq 100$$

Further constraints might be imposed by imposing maximum values on the flow in each street.

EXERCISES 1.4.1

1. Find the possible flows in each of the following networks of pipes.

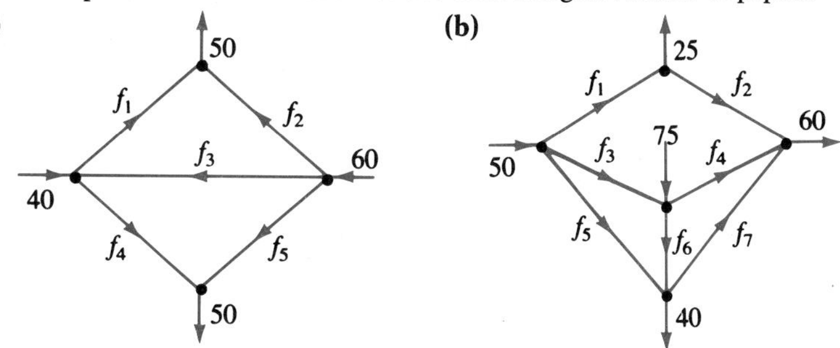

2. A proposed network of irrigation canals is diagrammed below. At peak demand, the flows at interchanges A, B, C, and D are as shown.
   (a) Find the possible flows.
   (b) If canal BC is closed, what range of flow on AD must be maintained so that no canal carries a flow of more than 30?

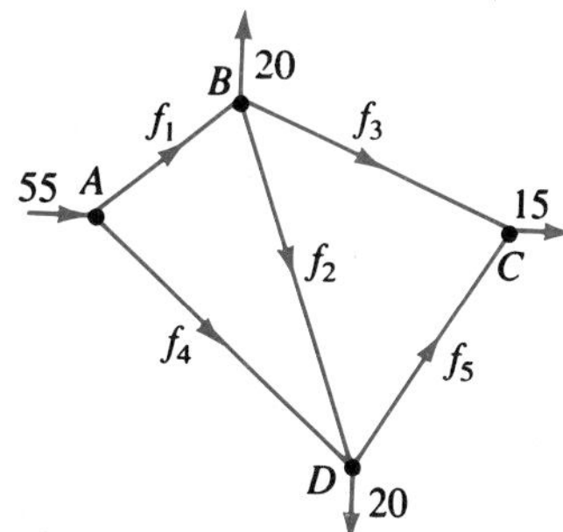

3. A traffic circle has five one-way streets, and vehicles enter and leave as shown below.
   (a) Compute the possible flows.
   (b) What road in the circle has the heaviest flow?

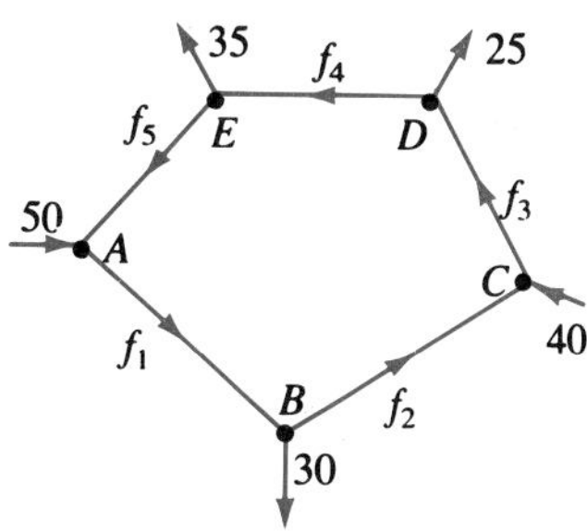

1.4.2 Electrical Networks

In an electrical network it is often necessary to find the current in amperes (A) flowing in various parts of the network. These networks usually contain resistors that retard the current. The resistors are indicated by a symbol -ʌʌʌ-, and the resistance is measured in ohms (Ω). Also, the current is increased at various points by voltage sources (for example, a battery). The voltage of these sources is measured in volts (V), and they are represented by the symbol ⊣|⊢. We assume these voltage sources have no resistance. The flow of current is governed by the following principles.

OHM'S LAW

The current I and the voltage drop V across a resistance R are related by the equation $V = RI$.

KIRCHHOFF'S LAWS

1. (Junction Rule) The current flow into a junction equals the current flow out of that junction.
2. (Circuit Rule) The algebraic sum of the voltage drops (due to resistances) around any closed circuit of the network must equal the sum of the voltage increases around the circuit.

When applying rule 2, select a direction (clockwise or counterclockwise) around the closed circuit and then consider all voltages and currents positive when in this direction and negative when in the opposite direction. This is why the term *algebraic sum* is used in rule 2. Here is an example.

EXAMPLE 2 Find the various currents in the circuit below.

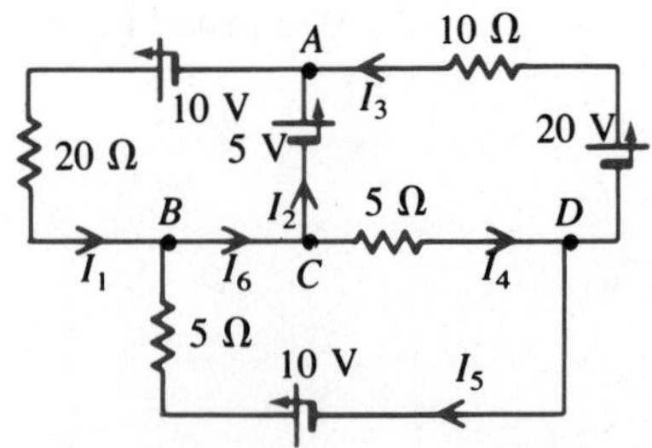

Solution First apply the junction rule at junctions A, B, C, and D to obtain

$$\begin{aligned} &\text{Junction A} & I_1 &= I_2 + I_3 \\ &\text{Junction B} & I_6 &= I_1 + I_5 \\ &\text{Junction C} & I_2 + I_4 &= I_6 \\ &\text{Junction D} & I_3 + I_5 &= I_4 \end{aligned}$$

Note that these equations are not independent (in fact, the third is an easy consequence of the other three).

Next apply the circuit rule to each of the three closed circuits. By Ohm's Law, the voltage loss across a resistance R (in the direction of the current I) is RI. Going counterclockwise around three closed circuits yields

$$\begin{aligned} &\text{Upper left} & 10 + 5 &= 20I_1 \\ &\text{Upper right} & -5 + 20 &= 10I_3 + 5I_4 \\ &\text{Lower} & -10 &= -5I_5 - 5I_4 \end{aligned}$$

Hence, disregarding the redundant equation obtained at junction C, we have six equations in the six unknowns $I_1, \ldots, I_6$. The solution is

$$I_1 = \frac{15}{20} \qquad I_4 = \frac{28}{20}$$

$$I_2 = \frac{-1}{20} \qquad I_5 = \frac{12}{20}$$

$$I_3 = \frac{16}{20} \qquad I_6 = \frac{27}{20}$$

The fact that I_2 is negative means, of course, that this current is in the opposite direction, with a magnitude of $\frac{1}{20}$ amperes.

EXERCISES 1.4.2

Find the currents in each of the following circuits.

1.

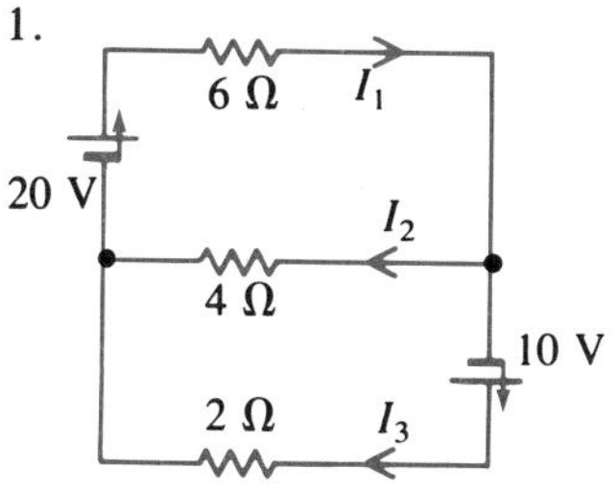

2.

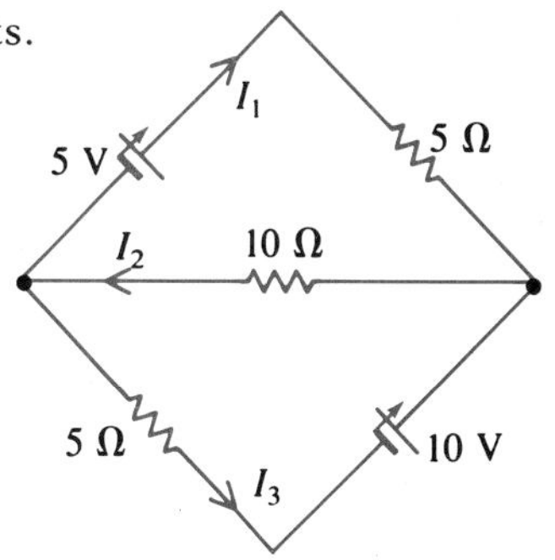

3.

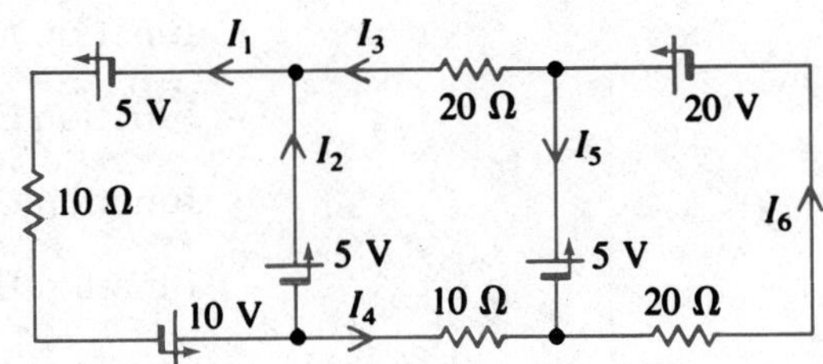

4. All resistances are 10Ω.

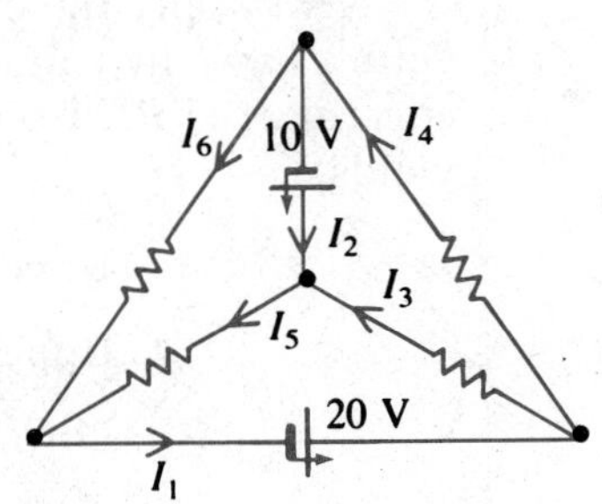

2 Matrix Algebra

SECTION 2.1 Matrix Addition, Scalar Multiplication, and Transposition

In the study of systems of linear equations in Chapter 1, we found it convenient to manipulate the "augmented matrix" of the system. Our aim was to reduce it to row-echelon form (using elementary row operations) and hence to write down all solutions to the system. In the present chapter, we will consider matrices for their own sake, although some of the motivation comes from linear equations. This subject is quite old. It was first studied systematically in 1851 by Arthur Cayley (1821–1895). The term *matrix* was first used in 1850 by James Sylvester (1814–1897).

DEFINITION

A rectangular array of numbers is called a **matrix** (the plural is **matrices**), and the numbers are called the **entries** of the matrix.

Matrices are usually denoted by upper-case letters: A, B, C, and so on. Hence

$$A = \begin{bmatrix} 1 & -1 \\ 0 & 2 \end{bmatrix} \qquad B = \begin{bmatrix} 1 & 2 & -1 \\ 0 & 5 & 6 \end{bmatrix} \qquad C = \begin{bmatrix} 1 \\ 3 \\ 2 \end{bmatrix}$$

are matrices. Clearly matrices come in various shapes depending on the number of **rows** and **columns**. In general, a matrix with m rows and n columns is referred to as an $\boldsymbol{m \times n}$ **matrix** (which is read "an m by n matrix"). Thus matrices A, B, and C are 2×2, 2×3, and 3×1 matrices, respectively. An $n \times n$ matrix is called, not surprisingly, a **square matrix**. Two matrices are said to be of the same **size** if they both have the same number of rows and the same number of columns. Thus no two of matrices A, B, and C are the same size. A matrix of size $1 \times n$

is called a **row matrix**, whereas one of size $n \times 1$ is called a **column matrix**. (These are both referred to as ordered n-tuples in Chapter 1.)

Each entry of a matrix is identified by the row and column in which it lies. The rows are numbered from the top down, and the columns are numbered from left to right. Then the **(i,j)-entry** of a matrix is the number lying simultaneously in row i and column j. For example, the (1,2)-entry of $\begin{bmatrix} 1 & -1 \\ 0 & 1 \end{bmatrix}$ is -1, and the (2,3)-entry of $\begin{bmatrix} 1 & 2 & -1 \\ 0 & 5 & 6 \end{bmatrix}$ is 6. A special notation has been devised for the entries of a matrix. If A is an $m \times n$ matrix, and if the (i,j)-entry of A is denoted as a_{ij}, then A is displayed as follows:

$$A = \begin{bmatrix} a_{11} & a_{12} & a_{13} & \cdots & a_{1n} \\ a_{21} & a_{22} & a_{23} & \cdots & a_{2n} \\ \vdots & \vdots & \vdots & & \vdots \\ a_{m1} & a_{m2} & a_{m3} & \cdots & a_{mn} \end{bmatrix}$$

This is usually denoted simply as $A = [a_{ij}]$. If matrix A is a square (so $m = n$), the **main diagonal** of the matrix A consists of the elements

$$a_{11}, a_{22}, a_{33}, \ldots, a_{nn}$$

When A is written as above, these elements are in a line extending from the upper left corner of A to the lower right corner.

It is worth pointing out a convention regarding rows and columns: *Rows are mentioned before columns.* For example, if a matrix has size $m \times n$, it has m rows and n columns; if we speak of the (i,j)-entry of a matrix, it lies in row i and column j; if this entry is denoted a_{ij}, the first subscript i refers to the row and the second subscript j to the column in which a_{ij} lies.

The $n \times 1$ matrices are just the ordered n-tuples discussed in Section 1.1, and two of these ordered n-tuples are equal if and only if corresponding entries are equal. This rule is extended to all matrices as follows. Two matrices A and B are called **equal** (written $A = B$) if and only if

1. They have the same size.
2. Corresponding entries are equal.

If the entries of A and B are written in the form $A = [a_{ij}]$, $B = [b_{ij}]$ described above, then the second condition takes the following form:

$$[a_{ij}] = [b_{ij}] \text{ means } a_{ij} = b_{ij} \text{ for all } i \text{ and } j$$

EXAMPLE 1 Given $A = \begin{bmatrix} a & b \\ c & d \end{bmatrix}$, $B = \begin{bmatrix} 1 & 2 & -1 \\ 3 & 0 & 1 \end{bmatrix}$, and $C = \begin{bmatrix} 1 & 0 \\ -1 & 2 \end{bmatrix}$, discuss the possibility that $A = B$, $B = C$, $A = C$.

Solution $A = B$ is impossible, because A and B are of different sizes: A is 2×2 whereas B is 2×3. Similarly, $B = C$ is impossible. $A = C$ is possible, however, provided that corresponding entries are equal: $\begin{bmatrix} a & b \\ c & d \end{bmatrix} = \begin{bmatrix} 1 & 0 \\ -1 & 2 \end{bmatrix}$ means $a = 1, b = 0, c = -1$, and $d = 2$.

The addition of ordered n-tuples by adding corresponding entries, which we discussed in Section 1.3, has a natural extension to matrices.

DEFINITION

If A and B are matrices of the same size, their **sum** $A + B$ is defined to be the matrix formed by adding corresponding entries. If $A = [a_{ij}]$ and $B = [b_{ij}]$, this takes the form

$$A + B = [a_{ij} + b_{ij}]$$

Note that addition is *not* defined for matrices of different sizes.

EXAMPLE 2

If $A = \begin{bmatrix} 2 & 1 & 3 \\ -1 & 2 & 0 \end{bmatrix}$ and $B = \begin{bmatrix} 1 & 1 & -1 \\ 2 & 0 & 6 \end{bmatrix}$, compute $A + B$.

Solution $A + B = \begin{bmatrix} 2+1 & 1+1 & 3-1 \\ -1+2 & 2+0 & 0+6 \end{bmatrix} = \begin{bmatrix} 3 & 2 & 2 \\ 1 & 2 & 6 \end{bmatrix}$

EXAMPLE 3

Find a, b, and c if $\begin{bmatrix} a & -a \\ c & b \end{bmatrix} + \begin{bmatrix} -b & c \\ a & -a \end{bmatrix} = \begin{bmatrix} 4 & -3 \\ 3 & -4 \end{bmatrix}$.

Solution If the matrices on the left side are added, the equation becomes

$$\begin{bmatrix} a-b & -a+c \\ c+a & b-a \end{bmatrix} = \begin{bmatrix} 4 & -3 \\ 3 & -4 \end{bmatrix}$$

Because corresponding entries must be equal, this gives four equations:

$$a - b = 4 \qquad -a + c = -3$$
$$c + a = 3 \qquad b - a = -4$$

Solving these yields $a = 3, b = -1, c = 0$.

Two basic facts about matrix addition are expressed in the following equations. If A, B, and C are any matrices of *the same size,* then

$$A + B = B + A \qquad \text{(Commutative Law)}$$

$$A + (B + C) = (A + B) + C \qquad \text{(Associative Law)}$$

In fact, if $A = [a_{ij}]$ and $B = [b_{ij}]$, then the (i,j)-entries of $A + B$ and $B + A$ are $a_{ij} + b_{ij}$ and $b_{ij} + a_{ij}$, respectively, and so are equal. Hence

$$A + B = [a_{ij} + b_{ij}] = [b_{ij} + a_{ij}] = B + A$$

The other rule is verified similarly.

The number 0 plays an important role in ordinary arithmetic because it is "neutral" with respect to addition. That is to say, $0 + x = x$ holds for every number x. Given m and n, the $m \times n$ matrix each of whose entries is zero is called the **zero matrix** and is denoted as 0 (or 0_{mn} if it is important to emphasize the size). Clearly,

$$0 + X = X$$

holds for all $m \times n$ matrices X.

If A is an $m \times n$ matrix, the **negative** of A (written $-A$) is defined to be the $m \times n$ matrix obtained by multiplying each entry of A by -1. If $A = [a_{ij}]$, this becomes

$$-A = [-a_{ij}]$$

Clearly,

$$A + (-A) = 0$$

holds for all matrices A where, of course, 0 is the zero matrix of the same size as A.

A closely related notion is that of subtracting matrices. If A and B are two $m \times n$ matrices, their **difference** $A - B$ is defined to be the $m \times n$ matrix formed by subtracting corresponding entries. In symbols, if $A = [a_{ij}]$ and $B = [b_{ij}]$, then

$$A - B = [a_{ij} - b_{ij}]$$

Note that A and B must be the same size and that

$$A - B = A + (-B)$$

EXAMPLE 4 $A = \begin{bmatrix} 3 & -1 & 0 \\ 1 & 2 & -4 \end{bmatrix}$, $B = \begin{bmatrix} 1 & -1 & 1 \\ -2 & 0 & 6 \end{bmatrix}$, and $C = \begin{bmatrix} 1 & 0 & -2 \\ 3 & 1 & 1 \end{bmatrix}$. Compute $-A$, $A - B$, and $A + B - C$.

Solution

$$-A = \begin{bmatrix} -3 & 1 & 0 \\ -1 & -2 & 4 \end{bmatrix}$$

$$A - B = \begin{bmatrix} 3-1 & -1-(-1) & 0-1 \\ 1-(-2) & 2-0 & -4-6 \end{bmatrix} = \begin{bmatrix} 2 & 0 & -1 \\ 3 & 2 & -10 \end{bmatrix}$$

$$A + B - C = \begin{bmatrix} 3+1-1 & -1-1-0 & 0+1+2 \\ 1-2-3 & 2+0-1 & -4+6-1 \end{bmatrix} = \begin{bmatrix} 3 & -2 & 3 \\ -4 & 1 & 1 \end{bmatrix}$$

EXAMPLE 5 Solve $\begin{bmatrix} 3 & 2 \\ -1 & 1 \end{bmatrix} + X = \begin{bmatrix} 1 & 0 \\ -1 & 2 \end{bmatrix}$, where X is a matrix.

Solution 1 X must be a 2×2 matrix. If $X = \begin{bmatrix} x & y \\ z & w \end{bmatrix}$, the equation reads

$$\begin{bmatrix} 1 & 0 \\ -1 & 2 \end{bmatrix} = \begin{bmatrix} 3 & 2 \\ -1 & 1 \end{bmatrix} + \begin{bmatrix} x & y \\ z & w \end{bmatrix} = \begin{bmatrix} 3+x & 2+y \\ -1+z & 1+w \end{bmatrix}$$

The rule of matrix equality gives $1 = 3 + x$, $0 = 2 + y$, $-1 = -1 + z$, and $2 = 1 + w$. Thus $X = \begin{bmatrix} -2 & -2 \\ 0 & 1 \end{bmatrix}$.

Solution 2 We solve a numerical equation $a + x = b$ by subtracting the number a from both sides to obtain $x = b - a$. This also works for matrices. To solve $\begin{bmatrix} 3 & 2 \\ -1 & 1 \end{bmatrix} + X = \begin{bmatrix} 1 & 0 \\ -1 & 2 \end{bmatrix}$, simply subtract the matrix $\begin{bmatrix} 3 & 2 \\ -1 & 1 \end{bmatrix}$ from both sides to get

$$X = \begin{bmatrix} 1 & 0 \\ -1 & 2 \end{bmatrix} - \begin{bmatrix} 3 & 2 \\ -1 & 1 \end{bmatrix} = \begin{bmatrix} 1-3 & 0-2 \\ -1+1 & 2-1 \end{bmatrix} = \begin{bmatrix} -2 & -2 \\ 0 & 1 \end{bmatrix}$$

This is the same solution as before.

The two solutions here are really different ways of doing the same thing. However, the first obtains four numerical equations, one for each entry, and solves them to get the four entries of X. The second solves the single matrix equation directly via matrix subtraction, and manipulation of entries comes in only at the end. The matrices themselves are manipulated. This ability to work with matrices as entities lies at the heart of matrix algebra.

It is important to note that the size of X in Example 5 was inferred *from the context:* X had to be a 2×2 matrix, because otherwise the equation itself would not make sense. This type of situation occurs frequently; the sizes of matrices involved in some calculation are often determined by the context. For example, if

$$A + C = \begin{bmatrix} 1 & 3 & -1 \\ 2 & 0 & 1 \end{bmatrix}$$

then A and C must be the same size (in order that $A + C$ make sense) and that size must be 2×3 (in order that the sum be 2×3). For simplicity, we shall often omit reference to such facts when they are clear from the context.

When a matrix A is added to itself three times, it is natural to call the result $3A$. Hence $3A = A + A + A$, so each entry of $3A$ is obtained by multiplying the corresponding entry of A by 3. This suggests the following definition:

DEFINITION

If A is any matrix and k is any number, the **scalar product** kA is defined to be the matrix obtained from A by multiplying each entry of A by k. If $A = [a_{ij}]$, this is

$$kA = [ka_{ij}]$$

The term *scalar* arises here because the set of numbers from which the entries are drawn is usually referred to as the set of scalars. We have been using the real numbers as scalars, but we could equally well have been using complex numbers.

EXAMPLE 6

If $A = \begin{bmatrix} 3 & -1 & 4 \\ 2 & 0 & 6 \end{bmatrix}$ and $B = \begin{bmatrix} 1 & 2 & -1 \\ 0 & 3 & 2 \end{bmatrix}$, compute $5A$, $\frac{1}{2}B$, and $3A - 2B$.

Solution $5A = \begin{bmatrix} 15 & -5 & 20 \\ 10 & 0 & 30 \end{bmatrix}$

$$\frac{1}{2}B = \begin{bmatrix} \frac{1}{2} & 1 & \frac{-1}{2} \\ 0 & \frac{3}{2} & 1 \end{bmatrix}$$

$$3A - 2B = \begin{bmatrix} 9 & -3 & 12 \\ 6 & 0 & 18 \end{bmatrix} - \begin{bmatrix} 2 & 4 & -2 \\ 0 & 6 & 4 \end{bmatrix} = \begin{bmatrix} 7 & -7 & 14 \\ 6 & -6 & 14 \end{bmatrix}$$

If A is any matrix, it is clear that kA is the same size as A for all scalars k. It is also evident that

$$0A = 0 \qquad \text{and} \qquad k0 = 0$$

The converse of these properties is also true.

EXAMPLE 7 If $kA = 0$, show that either $k = 0$ or $A = 0$.

Solution Write $A = [a_{ij}]$ so that $kA = 0$ means $ka_{ij} = 0$ for all i and j. If $k = 0$ there is nothing to do. If $k \neq 0$, then $a_{ij} = 0$ for all i and j; that is, $A = 0$.

For future reference, the basic properties of matrix addition and scalar multiplication are enumerated in Theorem 1.

THEOREM 1 Let A, B, and C denote arbitrary $m \times n$ matrices where m and n are fixed. Let k and p denote arbitrary real numbers. Then

1. $A + B = B + A$
2. $A + (B + C) = (A + B) + C$
3. There is an $m \times n$ matrix 0 such that $0 + A = A$ for each A.
4. For each A there is an $m \times n$ matrix $-A$ such that $A + (-A) = 0$.
5. $k(A + B) = kA + kB$
6. $(k + p)A = kA + pA$
7. $(kp)A = k(pA)$
8. $1A = A$

Proof Properties 1–4 were given above. To check property 5, let $A = [a_{ij}]$ and $B = [b_{ij}]$ denote matrices of the same size. Then $A + B = [a_{ij} + b_{ij}]$, as before, so the (i,j)-entry of $k(A + B)$ is $k(a_{ij} + b_{ij}) = ka_{ij} + kb_{ij}$. But this is just the (i,j)-entry of $kA + kB$, and it follows that $k(A + B) = kA + kB$. The other properties can be similarly verified; the details are left to the reader. ■

These properties enable one to do calculations with matrices in much the same way that numerical calculations are carried out. To begin with, property 2 implies that the sum $(A + B) + C = A + (B + C)$ is the same no matter how it is formed and so will be written as $A + B + C$. Similarly, the sum $A + B + C + D$ is independent of how it is formed (for example, it equals both $(A + B) + (C + D)$ and $A + [B + (C + D)]$). Furthermore, property 1 ensures that, for example, $B + D + A + C = A + B + C + D$. In other words, the *order* in which the matrices are added does not matter. A similar remark applies to sums of five (or more) matrices.

Properties 5 and 6 in Theorem 1 extend to sums of more than two matrices. For example, $k(A + B + C) = kA + kB + kC$ and $(k + p + m)A = kA + pA + mA$. Similar observations hold for more than three summands. These facts, together with properties 7 and 8, enable us to simplify expressions by collecting like terms, expanding, and taking common factors in exactly the same way that algebraic expressions involving variables are manipulated. The following examples illustrate these techniques.

EXAMPLE 8 Simplify $2(A + 3C) - 3(2C - B) - 3[2(2A + B - 4C) - 4(A - 2C)]$ where A, B, and C are all matrices of the same size.

Solution The reduction proceeds as though A, B, and C were variables.

$$\begin{aligned} &2(A + 3C) - 3(2C - B) - 3[2(2A + B - 4C) - 4(A - 2C)] \\ &\quad = 2A + 6C - 6C + 3B - 3[4A + 2B - 8C - 4A + 8C] \\ &\quad = 2A + 3B - 3[2B] \\ &\quad = 2A - 3B \end{aligned}$$

EXAMPLE 9 Find 2×2 matrices X and Y such that

$$X + 2Y = \begin{bmatrix} 1 & -1 \\ 0 & 1 \end{bmatrix}$$

$$2X + 3Y = \begin{bmatrix} 2 & 1 \\ -1 & 0 \end{bmatrix}$$

Solution If we write $A = \begin{bmatrix} 1 & -1 \\ 0 & 1 \end{bmatrix}$ and $B = \begin{bmatrix} 2 & 1 \\ -1 & 0 \end{bmatrix}$, the equations become $X + 2Y = A$ and $2X + 3Y = B$. The manipulations used to solve these equations when X, Y, A, and B are numbers all apply in the present context. Hence, when the second equation is subtracted from twice the first, the result is

$$Y = 2A - B = \begin{bmatrix} 2 & -2 \\ 0 & 2 \end{bmatrix} - \begin{bmatrix} 2 & 1 \\ -1 & 0 \end{bmatrix} = \begin{bmatrix} 0 & -3 \\ 1 & 2 \end{bmatrix}$$

Substituting $Y = 2A - B$ in the first equation $(X + 2Y = A)$ gives

$$\begin{aligned} X &= A - 2Y = A - 2(2A - B) = 2B - 3A \\ &= \begin{bmatrix} 4 & 2 \\ -2 & 0 \end{bmatrix} - \begin{bmatrix} 3 & -3 \\ 0 & 3 \end{bmatrix} = \begin{bmatrix} 1 & 5 \\ -2 & -3 \end{bmatrix} \end{aligned}$$

This is the required solution.

Many results about a matrix A (for example, Theorem 2 in Section 1.2) involve the *rows* of A, and the corresponding result for columns is derived in an analogous way, essentially by replacing the word *row* by the word *column* throughout. The following definition is made with such applications in mind.

DEFINITION If A is an $m \times n$ matrix, the **transpose** of A, written A^T, is the $n \times m$ matrix whose rows are just the columns of A in the same order.

More precisely, the first row of A^T is the first column of A, the second row of A^T is the second column of A, and so on.

EXAMPLE 10 Write down the transpose of each of the following matrices.

$$A = \begin{bmatrix} 1 \\ 3 \\ 2 \end{bmatrix} \quad B = [5 \ 2 \ 6] \quad C = \begin{bmatrix} 1 & 2 \\ 3 & 4 \\ 5 & 6 \end{bmatrix} \quad D = \begin{bmatrix} 3 & 1 & -1 \\ 1 & 3 & 2 \\ -1 & 2 & 1 \end{bmatrix}$$

Solution $A^T = [1 \ 3 \ 2]$ $\quad B^T = \begin{bmatrix} 5 \\ 2 \\ 6 \end{bmatrix}$ $\quad C^T = \begin{bmatrix} 1 & 3 & 5 \\ 2 & 4 & 6 \end{bmatrix}$ $\quad D^T = D$

If $A = [a_{ij}]$ is a matrix, write $A^T = [b_{ij}]$. Then b_{ij} is the jth element of the ith row of A^T and so is the jth element of the ith *column* of A. This means $b_{ij} = a_{ji}$, so the definition of A^T can be stated as follows:

$$\text{If } A = [a_{ij}], \text{ then } A^T = [a_{ji}].$$

This is useful in verifying the following properties of transposition.

THEOREM 2 Let A and B denote matrices of the same size, and let k denote a scalar.

1. If A is an $m \times n$ matrix, then A^T is an $n \times m$ matrix.
2. $(A^T)^T = A$
3. $(kA)^T = kA^T$
4. $(A + B)^T = A^T + B^T$

Proof We prove only property 3. If $A = [a_{ij}]$, then $kA = [ka_{ij}]$, so

$$(kA)^T = [ka_{ji}] = k[a_{ji}] = kA^T$$

which proves property 3. ■

The matrix D in Example 10 has the property that $D = D^T$. In general, a matrix A is called **symmetric** if $A = A^T$. Such matrices are necessarily square (if A is $m \times n$, then A^T is $n \times m$, so $A = A^T$ forces $n = m$), and the name comes from the fact that they exhibit symmetry about the main diagonal. That is, elements in positions that are directly across the main diagonal from each other are equal. Thus, $\begin{bmatrix} a & b & c \\ b' & d & e \\ c' & e' & f \end{bmatrix}$ is symmetric if $b = b'$, $c = c'$, and $e = e'$.

EXAMPLE 11 If A and B are symmetric $n \times n$ matrices, show that $A + B$ is symmetric.

Solution We have $A^T = A$ and $B^T = B$, so, by Theorem 2, $(A + B)^T = A^T + B^T = A + B$.

EXERCISES 2.1

1. Find a, b, c, and d if:

(a) $\begin{bmatrix} a & b \\ c & d \end{bmatrix} = \begin{bmatrix} c - 3d & -d \\ 2a + d & a + b \end{bmatrix}$

(b) $\begin{bmatrix} a - b & b - c \\ c - d & d - a \end{bmatrix} = 2\begin{bmatrix} 1 & 1 \\ -3 & 1 \end{bmatrix}$

(c) $3\begin{bmatrix} a \\ b \end{bmatrix} + 2\begin{bmatrix} a \\ b \end{bmatrix} = \begin{bmatrix} 1 \\ 2 \end{bmatrix}$

(d) $\begin{bmatrix} a & b \\ c & d \end{bmatrix} = \begin{bmatrix} b & c \\ d & a \end{bmatrix}$

2. Compute the following.

(a) $\begin{bmatrix} 3 & 2 & 1 \\ 5 & 1 & 0 \end{bmatrix} - 5\begin{bmatrix} 3 & 0 & -2 \\ 1 & -1 & 2 \end{bmatrix}$

(b) $3\begin{bmatrix} 3 \\ -1 \end{bmatrix} - 5\begin{bmatrix} 6 \\ 2 \end{bmatrix} + 7\begin{bmatrix} 1 \\ -1 \end{bmatrix}$

(c) $\begin{bmatrix} -2 & 1 \\ 3 & 2 \end{bmatrix} - 4\begin{bmatrix} 1 & -2 \\ 0 & -1 \end{bmatrix} + 3\begin{bmatrix} 2 & -3 \\ -1 & -2 \end{bmatrix}$

(d) $[3 \ \ -1 \ \ 2] - 2[9 \ \ 3 \ \ 4] + [3 \ \ 11 \ \ -6]$

(e) $\begin{bmatrix} 1 & -5 & 4 & 0 \\ 2 & 1 & 0 & 6 \end{bmatrix}^T$

(f) $\begin{bmatrix} 0 & -1 & 2 \\ 1 & 0 & -4 \\ -2 & 4 & 0 \end{bmatrix}^T$

(g) $\begin{bmatrix} 3 & -1 \\ 2 & 1 \end{bmatrix} - 2\begin{bmatrix} 1 & -2 \\ 1 & 1 \end{bmatrix}^T$

(h) $3\begin{bmatrix} 2 & 1 \\ -1 & 0 \end{bmatrix}^T - 2\begin{bmatrix} 1 & -1 \\ 2 & 3 \end{bmatrix}$

3. Let $A = \begin{bmatrix} 2 & 1 \\ 0 & -1 \end{bmatrix}$, $B = \begin{bmatrix} 3 & -1 & 2 \\ 0 & 1 & 4 \end{bmatrix}$, $C = \begin{bmatrix} 3 & -1 \\ 2 & 0 \end{bmatrix}$, $D = \begin{bmatrix} 1 & 3 \\ -1 & 0 \\ 1 & 4 \end{bmatrix}$, and $E = \begin{bmatrix} 1 & 0 & 1 \\ 0 & 1 & 0 \end{bmatrix}$. Compute the following (where possible).

(a) $3A - 2B$ **(b)** $5C$ **(c)** $3E^T$
(d) $B + D$ **(e)** $4A^T - 3C$ **(f)** $(A + C)^T$
(g) $2B - 3E$ **(h)** $A - D$ **(i)** $(B - 2E)^T$

4. Find A if:

(a) $5A - \begin{bmatrix} 1 & 0 \\ 2 & 3 \end{bmatrix} = 3A - \begin{bmatrix} 5 & 2 \\ 6 & 1 \end{bmatrix}$ **(b)** $3A + \begin{bmatrix} 2 \\ 1 \end{bmatrix} = 5A - 2\begin{bmatrix} 3 \\ 0 \end{bmatrix}$

5. Find A in terms of B if:

(a) $A + B = 3A + 2B$ **(b)** $2A - B = 5(A + 2B)$

6. If X, Y, A, and B are matrices of the same size, solve the following equations to obtain X and Y in terms of A and B.

(a) $5X + 3Y = A$
$2X + Y = B$

(b) $4X + 3Y = A$
$5X + 4Y = B$

7. Find all matrices X and Y such that:

(a) $3X - 2Y = [3, -1]$ **(b)** $2X - 5Y = [1, 2]$

8. Simplify the following expressions where A, B, and C represent matrices.

(a) $2[9(A - B) + 7(2B - A)] - 2[3(2B + A) - 2(A + 3B) - 5(A + B)]$

(b) $5[3(A - B + 2C) - 2(3C - B) - A] + 2[3(3A - B + C) + 2(B - 2A) - 2C]$

9. If A is any 2×2 matrix, show that:

(a) $A = a\begin{bmatrix} 1 & 0 \\ 0 & 0 \end{bmatrix} + b\begin{bmatrix} 0 & 1 \\ 0 & 0 \end{bmatrix} + c\begin{bmatrix} 0 & 0 \\ 1 & 0 \end{bmatrix} + d\begin{bmatrix} 0 & 0 \\ 0 & 1 \end{bmatrix}$ for some numbers $a, b, c,$ and d

(b) $A = p\begin{bmatrix} 1 & 0 \\ 0 & 1 \end{bmatrix} + q\begin{bmatrix} 1 & 1 \\ 0 & 0 \end{bmatrix} + r\begin{bmatrix} 1 & 0 \\ 1 & 0 \end{bmatrix} + s\begin{bmatrix} 0 & 1 \\ 1 & 0 \end{bmatrix}$ for some numbers $p, q, r,$ and s

10. Let $A = [1 \ 1 \ -1]$, $B = [0 \ 1 \ 2]$, and $C = [3 \ 0 \ 1]$. If $rA + sB + tC = 0$ for some scalars r, s and t, show that necessarily $r = s = t = 0$.

11. **(a)** If $Q + A = A$ holds for some $m \times n$ matrix A, show that $Q = 0_{mn}$.

(b) If A is an $m \times n$ matrix and $A + A' = 0_{mn}$, show that $A' = -A$.

12. If A denotes an $m \times n$ matrix, show that $A = -A$ if and only if $A = 0$.

13. A square matrix is called a **diagonal** matrix if all the entries off the main diagonal are zero. If A and B are diagonal matrices, show that the following matrices are also diagonal.

(a) $A + B$ **(b)** $A - B$ **(c)** kA for any number k

14. Let A and B be symmetric (of the same size). Show that each of the following is symmetric.

(a) $A - B$ **(b)** kA for any scalar k

15. Show that $A + A^T$ is symmetric for *any* square matrix A.

16. A square matrix W is called **skew-symmetric** if $W^T = -W$. Let A be any square matrix.

(a) Show that $A - A^T$ is skew-symmetric.

(b) Find a symmetric matrix S and a skew-symmetric matrix W such that $A = S + W$.

(c) Show that S and W in part (b) are uniquely determined by A.

17. If W is skew-symmetric (Exercise 16), show that the entries on the main diagonal are zero.

18. Prove the following parts of Theorem 1.

(a) $(k + p)A = kA + pA$ **(b)** $(kp)A = k(pA)$

19. Let $A, A_1, A_2, \ldots, A_n$ denote matrices of the same size. Use induction on n to verify the following extensions of properties 5 and 6 of Theorem 1.

(a) $k(A_1 + A_2 + \cdots + A_n) = kA_1 + kA_2 + \cdots + kA_n$ for any number k

(b) $(k_1 + k_2 + \cdots + k_n)A = k_1A + k_2A + \cdots + k_nA$ for any numbers $k_1, k_2, \ldots, k_n$

SECTION 2.2 Matrix Multiplication

Matrix multiplication is a little more complicated than matrix addition or scalar multiplication, but it is well worth the extra effort. It provides a new way to look at systems of linear equations and has a wide variety of other applications as well, some of which are discussed in Section 2.5.

Before giving the general definition, we introduce the following useful notion. If $\mathbf{u} = [u_1, u_2, \ldots, u_n]$ and $\mathbf{v} = [v_1, v_2, \ldots, v_n]$ are ordered n-tuples,[†] the number

$$\mathbf{u} \cdot \mathbf{v} = u_1v_1 + u_2v_2 + \cdots + u_nv_n$$

is called the **dot product** of $\mathbf{u}$ and $\mathbf{v}$. Thus $\mathbf{u} \cdot \mathbf{v}$ is calculated by multiplying corresponding entries of $\mathbf{u}$ and $\mathbf{v}$ and adding the results.

EXAMPLE 1 Compute $\mathbf{u} \cdot \mathbf{v}$ if (a) $\mathbf{u} = [3\ \ -1\ \ 7]$, $\mathbf{v} = [1\ \ 2\ \ 0]$ and (b) $\mathbf{u} = [3\ \ -1\ \ 5\ \ -2\ \ 4]$, $\mathbf{v} = [2\ \ 9\ \ 1\ \ 7\ \ 3]$.

Solution (a) $\mathbf{u} \cdot \mathbf{v} = 3 \cdot 1 + (-1)2 + 7 \cdot 0 = 3 - 2 + 0 = 1$

(b) $\mathbf{u} \cdot \mathbf{v} = 3 \cdot 2 + (-1)9 + 5 \cdot 1 + (-2)7 + 4 \cdot 3 = 0$

DEFINITION If A is an $m \times n$ matrix and B is an $n \times k$ matrix, the **product** AB of A and B is defined to be the $m \times k$ matrix whose (i,j)-entry is the dot product of the ith row of A and the jth column of B. Computing the (i,j)-entry of AB involves going *across* the ith row of A and *down* the jth column of B and forming the dot product.

EXAMPLE 2 Given $A = \begin{bmatrix} 3 & -1 & 2 \\ 0 & 1 & 4 \end{bmatrix}$ and $B = \begin{bmatrix} 2 & 1 \\ 0 & 2 \\ -1 & 0 \end{bmatrix}$, compute the products AB and BA.

[†]We shall frequently denote n-tuples (written as rows or columns) in boldface type.

Solution We write out the calculation of the dot products explicitly.

$$AB = \begin{bmatrix} 3 & -1 & 2 \\ 0 & 1 & 4 \end{bmatrix} \begin{bmatrix} 2 & 1 \\ 0 & 2 \\ -1 & 0 \end{bmatrix}$$

$$= \begin{bmatrix} 3\cdot 2 + (-1)\cdot 0 + 2(-1) & 3\cdot 1 + (-1)\cdot 2 + 2\cdot 0 \\ 0\cdot 2 + 1\cdot 0 + 4(-1) & 0\cdot 1 + 1\cdot 2 + 4\cdot 0 \end{bmatrix}$$

$$= \begin{bmatrix} 4 & 1 \\ -4 & 2 \end{bmatrix}$$

$$BA = \begin{bmatrix} 2 & 1 \\ 0 & 2 \\ -1 & 0 \end{bmatrix} \begin{bmatrix} 3 & -1 & 2 \\ 0 & 1 & 4 \end{bmatrix}$$

$$= \begin{bmatrix} 2\cdot 3 + 1\cdot 0 & 2(-1) + 1\cdot 1 & 2\cdot 2 + 1\cdot 4 \\ 0\cdot 3 + 2\cdot 0 & 0(-1) + 2\cdot 1 & 0\cdot 2 + 2\cdot 4 \\ (-1)3 + 0\cdot 0 & (-1)(-1) + 0\cdot 1 & (-1)2 + 0\cdot 4 \end{bmatrix}$$

$$= \begin{bmatrix} 6 & -1 & 8 \\ 0 & 2 & 8 \\ -3 & 1 & -2 \end{bmatrix}$$

Clearly the rows of A and the columns of B must be the same length in order that their dot products may be formed. The following mnemonic is a useful way to remember when the product of A and B may be formed and what the size of the product matrix is.

$$\underset{A}{m \times n} \quad \underset{B}{n' \times p}$$

The product is defined only when $n = n'$; in this case, the product matrix AB is of size $m \times p$.

EXAMPLE 3 If $A = [1, 3, 2]$ and $B = \begin{bmatrix} 5 \\ 6 \\ 4 \end{bmatrix}$, compute A^2, AB, BA, and B^2 when they are defined.

Solution Here A is a 1×3 matrix and B is a 3×1 matrix, so A^2 and B^2 are not defined. However, the mnemonic reads

$$\underset{1 \times 3}{A} \quad \underset{3 \times 1}{B} \qquad \text{and} \qquad \underset{3 \times 1}{B} \quad \underset{1 \times 3}{A}$$

so both AB and BA may be formed and these are 1×1 and 3×3 matrices, respectively.

$$AB = [1\ 3\ 2]\begin{bmatrix}5\\6\\4\end{bmatrix} = [1 \cdot 5 + 3 \cdot 6 + 2 \cdot 4] = [31]$$

$$BA = \begin{bmatrix}5\\6\\4\end{bmatrix}[1\ 3\ 2] = \begin{bmatrix}5 \cdot 1 & 5 \cdot 3 & 5 \cdot 2\\6 \cdot 1 & 6 \cdot 3 & 6 \cdot 2\\4 \cdot 1 & 4 \cdot 3 & 4 \cdot 2\end{bmatrix} = \begin{bmatrix}5 & 15 & 10\\6 & 18 & 12\\4 & 12 & 8\end{bmatrix}$$

Unlike numerical multiplication, matrix products AB and BA *need not be equal.* In fact they need not even be the same size, as Example 3 shows. It turns out to be relatively rare that $AB = BA$ (although it is by no means impossible), and A and B are said to **commute** when this happens.

EXAMPLE 4 Let $A = \begin{bmatrix}6 & 9\\-4 & -6\end{bmatrix}$ and $B = \begin{bmatrix}1 & 2\\-1 & 0\end{bmatrix}$. Compute A^2, AB, and BA.

Solution $A^2 = \begin{bmatrix}6 & 9\\-4 & -6\end{bmatrix}\begin{bmatrix}6 & 9\\-4 & -6\end{bmatrix} = \begin{bmatrix}0 & 0\\0 & 0\end{bmatrix}$, so $A^2 = 0$ can occur even if $A \neq 0$. Next $AB = \begin{bmatrix}6 & 9\\-4 & -6\end{bmatrix}\begin{bmatrix}1 & 2\\-1 & 0\end{bmatrix} = \begin{bmatrix}-3 & 12\\2 & -8\end{bmatrix}$, whereas $BA = \begin{bmatrix}1 & 2\\-1 & 0\end{bmatrix}\begin{bmatrix}6 & 9\\-4 & -6\end{bmatrix} = \begin{bmatrix}-2 & -3\\-6 & -9\end{bmatrix}$. Hence $AB \neq BA$, even though AB and BA are the same size.

The number 1 plays a neutral role in numerical multiplication in the sense that $1 \cdot a = a$ and $a \cdot 1 = a$ for all numbers a. An analogous role for matrix multiplication is played by square matrices of the following types:

$$\begin{bmatrix}1 & 0\\0 & 1\end{bmatrix} \quad \begin{bmatrix}1 & 0 & 0\\0 & 1 & 0\\0 & 0 & 1\end{bmatrix} \quad \begin{bmatrix}1 & 0 & 0 & 0\\0 & 1 & 0 & 0\\0 & 0 & 1 & 0\\0 & 0 & 0 & 1\end{bmatrix} \quad \text{and so on.}$$

In general, an **identity matrix** I is a *square matrix with 1's on the main diagonal and zeros elsewhere.* If it is important to stress the size of an $n \times n$ identity matrix, we shall denote it by I_n; however, these matrices are usually written simply as I. Identity matrices play a neutral role with respect to matrix multiplication in the sense that

$$AI = A \qquad \text{and} \qquad IB = B$$

whenever the products are defined.

Before proceeding, let us state the definition of matrix multiplication more formally. If A is $m \times n$ and B is $n \times p$, we write

$$A = [a_{ij}] \qquad \text{and} \qquad B = [b_{ij}]$$

Then the ith row of A and the jth column of B are, respectively,

$$[a_{i1}\ a_{i2} \cdots a_{in}] \qquad \text{and} \qquad \begin{bmatrix} b_{1j} \\ b_{2j} \\ \vdots \\ b_{nj} \end{bmatrix}$$

so the (i,j)-entry of the product matrix AB is their dot product

$$a_{i1}b_{1j} + a_{i2}b_{2j} + \cdots + a_{in}b_{nj} = \sum_{k=1}^{n} a_{ik}b_{kj}$$

where summation notation has been introduced for convenience. This is useful in verifying facts about matrix multiplication.

THEOREM 1 Assume that k is an arbitrary scalar and that A, B, and C are matrices of sizes such that the indicated operations can be performed.

1. $IA = A, \qquad BI = B$
2. $A(BC) = (AB)C$
3. $A(B + C) = AB + AC, \qquad A(B - C) = AB - AC$
4. $(B + C)A = BA + CA, \qquad (B - C)A = BA - CA$
5. $k(AB) = (kA)B = A(kB)$
6. $(AB)^T = B^TA^T$

Proof We prove properties 3 and 6, leaving the rest as exercises.

Property 3. Write $A = [a_{ij}]$, $B = [b_{ij}]$, and $C = [c_{ij}]$, and assume that A is $m \times n$ and that B and C are $n \times p$. Then $B + C = [b_{ij} + c_{ij}]$, so the (i,j)-entry of $A(B + C)$ is

$$\sum_{k=1}^{n} a_{ik}(b_{kj} + c_{kj}) = \sum_{k=1}^{n} (a_{ik}b_{kj} + a_{ik}c_{kj}) = \sum_{k=1}^{n} a_{ik}b_{kj} + \sum_{k=1}^{n} a_{ik}c_{kj}$$

This is the (i,j)-entry of $AB + AC$ because the sums on the right are the (i,j)-entries of AB and AC, respectively. Hence $A(B + C) = AB + AC$.

Property 6. Write $A^T = [a'_{ij}]$ and $B^T = [b'_{ij}]$, where $a'_{ij} = a_{ji}$ and $b'_{ij} = b_{ji}$. If B^T and A^T are $p \times n$ and $n \times m$, respectively, the (i,j)-entry of B^TA^T is

$$\sum_{k=1}^{n} b'_{ik}a'_{kj} = \sum_{k=1}^{n} b_{ki}a_{jk} = \sum_{k=1}^{n} a_{jk}b_{ki}$$

This is the (j,i)-entry of AB—that is, the (i,j)-entry of $(AB)^T$. Hence $B^TA^T = (AB)^T$. ■

Property 2 asserts that the relationship $A(BC) = (AB)C$ holds for all matrices (if the products are defined) and so the product is the same no matter which way it is formed. For this reason it is simply written ABC. This extends: The product $ABCD$ of four matrices can be formed several ways—for example, $(AB)(CD)$, $[A(BC)]D$, and $A[B(CD)]$—but property 2 implies that they are all equal and so are written simply as $ABCD$. A similar remark applies to products of more than four matrices. The situation for matrix addition is analogous, and the two basic properties are

$$A(BC) = (AB)C \qquad \text{and} \qquad A + (B + C) = (A + B) + C$$

We describe these properties by saying that matrix multiplication and addition are **associative** operations. They allow us to write products and sums unambiguously with no parentheses.

However, a note of caution about matrix multiplication is in order. The fact that AB and BA need *not* be equal means that the *order* of the factors is important in a product of matrices. For example, $ABCD$ and $ADCB$ may *not* be equal. We emphasize this important difference between matrix multiplication and numerical multiplication as follows:

Warning: If the order of the factors in a product of matrices is changed, the product matrix may change (or may not exist).

Ignoring this is a source of many errors by students of linear algebra! The analogous warning for matrix sums is not necessary, because $A + B = B + A$ for any two matrices (of the same size). This is the **commutative** property of matrix addition; the warning about matrix products results from the fact that matrix multiplication is *not* commutative.

Properties 3 and 4 in Theorem 1 are called the **distributive** properties, and they extend to more than two terms. For example,

$$A(B - C + D - E) = AB - AC + AD - AE$$

$$(A + C - D)B = AB + CB - DB$$

Note again that the warning is in effect: For example, $A(B - C)$ need *not* equal $AB - CA$. Together with property 5 of Theorem 1, the distributive properties make possible a lot of simplification of matrix expressions.

EXAMPLE 5 Simplify each of the following expressions.

(a) $A(3B - C) + (A - 2B)C + 2B(C + 2A)$
(b) $A(BC - CD) + A(C - B)D - AB(C - D)$

Solution

(a)
$$\begin{aligned} A(3B - C) + (A - 2B)C + 2B(C + 2A) &= 3AB - AC + AC - 2BC + 2BC + 4BA \\ &= 3AB + 4BA \end{aligned}$$

(b)
$$\begin{aligned} A(BC - CD) + A(C - B)D - AB(C - D) &= ABC - ACD + (AC - AB)D - ABC + ABD \\ &= ABC - ACD + ACD - ABD - ABC + ABD \\ &= 0 \end{aligned}$$

The following examples show how we can use the properties in Theorem 1 to deduce facts about matrix multiplication.

EXAMPLE 6 Suppose that A, B, and C are $n \times n$ matrices and that both A and B commute with C; that is, $AC = CA$ and $BC = CB$. Show that AB commutes with C.

Solution Showing that AB commutes with C means verifying that $(AB)C = C(AB)$. The computation uses property 2 of Theorem 1 and the given facts that $AC = CA$ and $BC = CB$.

$$(AB)C = A(BC) = A(CB) = (AC)B = (CA)B = C(AB)$$

EXAMPLE 7 Show that $AB = BA$ if and only if $(A - B)(A + B) = A^2 - B^2$.

Solution Theorem 1 shows that the following always holds:

$$(A - B)(A + B) = A(A + B) - B(A + B) = A^2 + AB - BA - B^2 \quad (*)$$

Hence if $AB = BA$, then $(A - B)(A + B) = A^2 - B^2$ follows. Conversely, if this last equation holds, then (*) becomes

$$A^2 - B^2 = A^2 + AB - BA - B^2$$

This gives $0 = AB - BA$, and $AB = BA$ follows.

One of the most important motivations for matrix multiplication results from the fact that any system of linear equations can be written as a single matrix equation. The next example illustrates how this is done.

EXAMPLE 8 Write the following system of linear equations in matrix form.

$$\begin{aligned} 3x_1 - 2x_2 + x_3 &= b_1 \\ 2x_1 + x_2 - x_3 &= b_2 \end{aligned}$$

Solution The two linear equations can be written as a single matrix equation as follows:

$$\begin{bmatrix} 3x_1 - 2x_2 + x_3 \\ 2x_1 + x_2 - x_3 \end{bmatrix} = \begin{bmatrix} b_1 \\ b_2 \end{bmatrix} \qquad (*)$$

Indeed, this matrix equality means that the top entries must be equal (this is the first linear equation) and that the lower entries must also be equal (the second linear equation). However, the matrix on the left can be factored as a product of matrices. Write

$$A = \begin{bmatrix} 3 & -2 & 1 \\ 2 & 1 & -1 \end{bmatrix}, \mathbf{x} = \begin{bmatrix} x_1 \\ x_2 \\ x_3 \end{bmatrix}, \text{ and } \mathbf{b} = \begin{bmatrix} b_1 \\ b_2 \end{bmatrix}, \text{ and compute}$$

$$A\mathbf{x} = \begin{bmatrix} 3 & -2 & 1 \\ 2 & 1 & -1 \end{bmatrix} \begin{bmatrix} x_1 \\ x_2 \\ x_3 \end{bmatrix} = \begin{bmatrix} 3x_1 - 2x_2 + x_3 \\ 2x_1 + x_2 - x_3 \end{bmatrix}$$

This is just the left side of (*), so (*) becomes

$$A\mathbf{x} = \mathbf{b}$$

This matrix equality is precisely equivalent to the two equations of the system.

Hence a system of two linear equations in two variables becomes a *single* matrix equation. The same is true for *any* such system. If the variables involved are $x_1, x_2, \ldots, x_n$, then any system of m linear equations in these variables takes the form

$$\begin{aligned} a_{11}x_1 + a_{12}x_2 + \cdots + a_{1n}x_n &= b_1 \\ a_{21}x_1 + a_{22}x_2 + \cdots + a_{2n}x_n &= b_2 \\ \vdots \qquad \vdots \qquad\quad \vdots \qquad & \quad \vdots \\ a_{m1}x_1 + a_{m2}x_2 + \cdots + a_{mn}x_n &= b_m \end{aligned}$$

If $A = [a_{ij}]$, $\mathbf{x} = \begin{bmatrix} x_1 \\ x_2 \\ \vdots \\ x_n \end{bmatrix}$, and $\mathbf{b} = \begin{bmatrix} b_1 \\ b_2 \\ \vdots \\ b_m \end{bmatrix}$, these equations are equivalent to the single matrix equation

$$A\mathbf{x} = \mathbf{b}$$

The matrix A is called the **coefficient matrix** of the system, and $\mathbf{b}$ is the **constant matrix**. We say that $A\mathbf{x} = \mathbf{b}$ is the **matrix form** of the system of equations. Note that the augmented matrix introduced in Section 1.2 is just the present coefficient matrix together with an extra column, the column of constants.

Suppose a system of equations is given in matrix form, $A\mathbf{x} = \mathbf{b}$. The new system

$$A\mathbf{x} = \mathbf{0}$$

obtained by replacing all the constant terms by zero is called the **associated homogeneous system**. Now it can happen that the nonhomogeneous system $A\mathbf{x} = \mathbf{b}$ can have no solution (see Example 3 in Section 1.2), but this does not happen for homogeneous systems. However, if the nonhomogeneous system *does* have a solution, there is a close relationship between its solutions and those of the associated homogeneous system: Every solution to $A\mathbf{x} = \mathbf{b}$ can be obtained from any particular one by adding a solution to $A\mathbf{x} = \mathbf{0}$. The proof is a nice application of matrix algebra.

THEOREM 2

Suppose $\mathbf{s}_0$ is a particular solution to the system $A\mathbf{x} = \mathbf{b}$ of linear equations.

1. If $\mathbf{t}$ is any solution to the associated homogeneous system $A\mathbf{x} = \mathbf{0}$, then $\mathbf{s}_0 + \mathbf{t}$ is a solution to $A\mathbf{x} = \mathbf{b}$.
2. Every solution to $A\mathbf{x} = \mathbf{b}$ arises in this way for some solution $\mathbf{t}$ of $A\mathbf{x} = \mathbf{0}$.

Proof The fact that $\mathbf{s}_0$ is a solution to $A\mathbf{x} = \mathbf{b}$ means that $A\mathbf{s}_0 = \mathbf{b}$. Similarly, $A\mathbf{t} = \mathbf{0}$, so

$$A(\mathbf{s}_0 + \mathbf{t}) = A\mathbf{s}_0 + A\mathbf{t} = \mathbf{b} + \mathbf{0} = \mathbf{b}$$

Hence $\mathbf{s}_0 + \mathbf{t}$ is indeed a solution to $A\mathbf{x} = \mathbf{b}$, proving property 1. Now suppose $\mathbf{s}$ is *any* solution to $A\mathbf{x} = \mathbf{b}$, so that $A\mathbf{s} = \mathbf{b}$. Write $\mathbf{t} = \mathbf{s} - \mathbf{s}_0$ and compute

$$A\mathbf{t} = A(\mathbf{s} - \mathbf{s}_0) = A\mathbf{s} - A\mathbf{s}_0 = \mathbf{b} - \mathbf{b} = \mathbf{0}$$

Thus $\mathbf{t}$ is a solution to the associated homogeneous system and $\mathbf{s} = \mathbf{s}_0 + \mathbf{t}$, as asserted in property 2. ■

EXAMPLE 9 Express every solution of the following system as the sum of a specific solution and a solution of the associated homogeneous system.

$$\begin{aligned} x - y - z &= 2 \\ 2x - y - 3z &= 6 \\ x - 2z &= 4 \end{aligned}$$

Solution The reduction of the augmented matrix for the system is

$$\begin{bmatrix} 1 & -1 & -1 & 2 \\ 2 & -1 & -3 & 6 \\ 1 & 0 & -2 & 4 \end{bmatrix} \to \begin{bmatrix} 1 & -1 & -1 & 2 \\ 0 & 1 & -1 & 2 \\ 0 & 1 & -1 & 2 \end{bmatrix} \to \begin{bmatrix} 1 & 0 & -2 & 4 \\ 0 & 1 & -1 & 2 \\ 0 & 0 & 0 & 0 \end{bmatrix}$$

Hence, taking $z = t$, we find that the general solution is

$$\mathbf{s} = \begin{bmatrix} 4 + 2t \\ 2 + t \\ t \end{bmatrix} = \begin{bmatrix} 4 \\ 2 \\ 0 \end{bmatrix} + t\begin{bmatrix} 2 \\ 1 \\ 1 \end{bmatrix}$$

Thus $\mathbf{s}_0 = \begin{bmatrix} 4 \\ 2 \\ 0 \end{bmatrix}$ is a specific solution, and it is easily verified that $t\begin{bmatrix} 2 \\ 1 \\ 1 \end{bmatrix}$ gives *all* solutions to the associated homogeneous system.

$$\begin{aligned} x - y - z &= 0 \\ 2x - y - 3z &= 0 \\ x - 2z &= 0 \end{aligned}$$

It is often convenient to view a matrix as a row of columns or as a column of rows. These are special cases of a more general way of looking at matrices that, among its other uses, can greatly simplify matrix multiplications. Consider the matrices

$$A = \left[\begin{array}{cc|ccc} 1 & 0 & 0 & 0 & 0 \\ 0 & 1 & 0 & 0 & 0 \\ \hline 2 & -1 & 4 & 2 & 1 \\ 3 & 1 & -1 & 7 & 5 \end{array}\right] \quad \text{and} \quad B = \left[\begin{array}{cc} 4 & -2 \\ 5 & 6 \\ \hline 7 & 3 \\ -1 & 0 \\ 1 & 6 \end{array}\right]$$

which have been **partitioned into blocks** as indicated. This is a natural way to think of A in view of the blocks I_2 and 0_{23} that occur. These matrices are written as follows

$$A = \begin{bmatrix} I_2 & 0_{23} \\ P & Q \end{bmatrix} \qquad B = \begin{bmatrix} X \\ Y \end{bmatrix}$$

where the blocks are

$$P = \begin{bmatrix} 2 & -1 \\ 3 & 1 \end{bmatrix} \quad Q = \begin{bmatrix} 4 & 2 & 1 \\ -1 & 7 & 5 \end{bmatrix} \quad X = \begin{bmatrix} 4 & -2 \\ 5 & 6 \end{bmatrix} \quad Y = \begin{bmatrix} 7 & 3 \\ -1 & 0 \\ 1 & 6 \end{bmatrix}$$

This notation is particularly useful when we are multiplying the matrices A and B, because the product AB can be computed in block form as follows:

$$AB = \left[\begin{array}{c|c} I & 0 \\ \hline P & Q \end{array}\right] \left[\begin{array}{c} X \\ \hline Y \end{array}\right] = \left[\begin{array}{c} IX + 0Y \\ \hline PX + QY \end{array}\right] = \left[\begin{array}{c} X \\ \hline PX + QY \end{array}\right] = \left[\begin{array}{cc} 4 & -2 \\ 5 & 6 \\ \hline 30 & 8 \\ 8 & 27 \end{array}\right]$$

This is easily checked to be the product AB, computed in the conventional manner. In other words, *we can compute the product by ordinary matrix multiplication, using blocks as entries.* The only requirement is that the blocks be **compatible**. That is, *the sizes of the blocks must be such that all (matrix) products of blocks that occur make sense.* This means that the number of columns in each block of A must equal the number of rows in the corresponding block of B. The general result is as expressed in Theorem 3.

THEOREM 3
Block Multiplication

Let matrices A and B be partitioned into blocks as follows:

$$A = \begin{bmatrix} A_{11} & A_{12} & \cdots & A_{1s} \\ A_{21} & A_{22} & \cdots & A_{2s} \\ \vdots & \vdots & & \vdots \\ A_{r1} & A_{r2} & \cdots & A_{rs} \end{bmatrix} \qquad B = \begin{bmatrix} B_{11} & B_{12} & \cdots & B_{1t} \\ B_{21} & B_{22} & \cdots & B_{2t} \\ \vdots & \vdots & & \vdots \\ B_{s1} & B_{s2} & \cdots & B_{st} \end{bmatrix}$$

where the columns of A and the rows of B are partitioned in the same way (so that $A_{ik}B_{kj}$ can be formed for all i, j, and k). Then the product AB can be computed in block form as

$$AB = \begin{bmatrix} C_{11} & C_{12} & \cdots & C_{1t} \\ C_{21} & C_{22} & \cdots & C_{2t} \\ \vdots & \vdots & & \vdots \\ C_{r1} & C_{r2} & \cdots & C_{rt} \end{bmatrix} \text{ where } C_{ij} = A_{i1}B_{1j} + A_{i2}B_{2j} + \cdots + A_{is}B_{sj}$$

In other words, we compute AB by matrix multiplication, using the blocks as entries.

The proof is not difficult and is left as Exercise 34.

EXAMPLE 10 Let A and B be the matrices given before Theorem 3. Compute AB by partitioning into blocks in a different way.

Solution Write

$$A = \left[\begin{array}{ccc|cc} 1 & 0 & 0 & 0 & 0 \\ 0 & 1 & 0 & 0 & 0 \\ \hline 2 & -1 & 4 & 2 & 1 \\ 3 & 1 & -1 & 7 & 5 \end{array}\right] = \begin{bmatrix} J & 0 \\ M & N \end{bmatrix} \qquad B = \left[\begin{array}{cc} 4 & -2 \\ 5 & 6 \\ 7 & 3 \\ \hline -1 & 0 \\ 1 & 6 \end{array}\right] = \begin{bmatrix} E \\ F \end{bmatrix}$$

Then

$$AB = \begin{bmatrix} J & 0 \\ M & N \end{bmatrix} \begin{bmatrix} E \\ F \end{bmatrix} = \begin{bmatrix} JE + 0F \\ ME + NF \end{bmatrix}$$

$$= \left[\begin{array}{c} \begin{bmatrix} 4 & -2 \\ 5 & 6 \end{bmatrix} \\ \hline \begin{bmatrix} 31 & 2 \\ 10 & -3 \end{bmatrix} + \begin{bmatrix} -1 & 6 \\ -2 & 30 \end{bmatrix} \end{array}\right] = \left[\begin{array}{cc} 4 & -2 \\ 5 & 6 \\ \hline 30 & 8 \\ 8 & 27 \end{array}\right]$$

This is the same result as before.

Suppose $A = [A_{ij}]$ is a square matrix that has been partitioned into blocks. If A^2 is to be computed, the diagonal blocks must all be square matrices (possibly of different sizes). Then the partitioning can be used to compute higher powers of A. Here is an example.

EXAMPLE 11 Let $A = \begin{bmatrix} 1 & -1 & 0 \\ -1 & 1 & 0 \\ 1 & -1 & -2 \end{bmatrix}$. Compute A^8 using block multiplication.

Solution Write $A = \begin{bmatrix} X & 0 \\ Y & Z \end{bmatrix}$, where $X = \begin{bmatrix} 1 & -1 \\ -1 & 1 \end{bmatrix}$, $Y = [1, -1]$, and $Z = [-2]$. Then

$$A^2 = \begin{bmatrix} X & 0 \\ Y & Z \end{bmatrix} \begin{bmatrix} X & 0 \\ Y & Z \end{bmatrix} = \begin{bmatrix} X^2 & 0 \\ YX + ZY & Z^2 \end{bmatrix}$$

Now $X^2 = \begin{bmatrix} 2 & -2 \\ -2 & 2 \end{bmatrix} = 2X$, and $YX + ZY = [2, -2] + [-2, 2] = [0, 0]$. Hence

$$A^2 = \begin{bmatrix} X^2 & 0 \\ 0 & Z^2 \end{bmatrix}$$

$$A^4 = (A^2)^2 = \begin{bmatrix} X^2 & 0 \\ 0 & Z^2 \end{bmatrix}\begin{bmatrix} X^2 & 0 \\ 0 & Z^2 \end{bmatrix} = \begin{bmatrix} X^4 & 0 \\ 0 & Z^4 \end{bmatrix}$$

$$A^8 = (A^4)^2 = \begin{bmatrix} X^4 & 0 \\ 0 & Z^4 \end{bmatrix}\begin{bmatrix} X^4 & 0 \\ 0 & Z^4 \end{bmatrix} = \begin{bmatrix} X^8 & 0 \\ 0 & Z^8 \end{bmatrix}$$

Because $X^2 = 2X$, we get $X^4 = (2X)^2 = 4X^2 = 8X$, and $X^8 = (8X)^2 = 64X^2 = 128X$. Clearly $Z^8 = [-2]^8 = [256]$. Hence

$$A^8 = \begin{bmatrix} 128X & 0 \\ 0 & 256 \end{bmatrix} = 128\begin{bmatrix} X & 0 \\ 0 & 2 \end{bmatrix} = 128\begin{bmatrix} 1 & -1 & 0 \\ -1 & 1 & 0 \\ 0 & 0 & 2 \end{bmatrix}$$

as desired.

Theorem 3 is useful in computing products of matrices in a computer with limited memory capacity. The matrices are partitioned into blocks in such a way that each product of blocks can be handled. Then the blocks are stored in auxiliary memory (on tape, for example) and the products are computed one by one.

EXERCISES 2.2

1. Compute the following matrix products.

(a) $\begin{bmatrix} 1 & 3 \\ 0 & -2 \end{bmatrix}\begin{bmatrix} 2 & -1 \\ 0 & 1 \end{bmatrix}$

(b) $\begin{bmatrix} 1 & -1 & 2 \\ 2 & 0 & 4 \end{bmatrix}\begin{bmatrix} 2 & 3 & 1 \\ 1 & 9 & 7 \\ -1 & 0 & 2 \end{bmatrix}$

(c) $\begin{bmatrix} 5 & 0 & -7 \\ 1 & 5 & 9 \end{bmatrix}\begin{bmatrix} 3 \\ 1 \\ -1 \end{bmatrix}$

(d) $\begin{bmatrix} 1 & 3 & -3 \end{bmatrix}\begin{bmatrix} 3 & 0 \\ -2 & 1 \\ 0 & 6 \end{bmatrix}$

(e) $\begin{bmatrix} 1 & 0 & 0 \\ 0 & 1 & 0 \\ 0 & 0 & 1 \end{bmatrix}\begin{bmatrix} 3 & -2 \\ 5 & -7 \\ 9 & 7 \end{bmatrix}$

(f) $\begin{bmatrix} 1 & -1 & 3 \end{bmatrix}\begin{bmatrix} 2 \\ 1 \\ -8 \end{bmatrix}$

(g) $\begin{bmatrix} 2 \\ 1 \\ -7 \end{bmatrix}\begin{bmatrix} 1 & -1 & 3 \end{bmatrix}$

(h) $\begin{bmatrix} 3 & 1 \\ 5 & 2 \end{bmatrix}\begin{bmatrix} 2 & -1 \\ -5 & 3 \end{bmatrix}$

(i) $\begin{bmatrix} 2 & 3 & 1 \\ 5 & 7 & 4 \end{bmatrix}\begin{bmatrix} a & 0 & 0 \\ 0 & b & 0 \\ 0 & 0 & c \end{bmatrix}$

2. In each of the following cases, find all possible products A^2, AB, AC, and so on.

(a) $A = \begin{bmatrix} 1 & 2 & 3 \\ -1 & 0 & 0 \end{bmatrix}$ $\quad B = \begin{bmatrix} 1 & -2 \\ \frac{1}{2} & 3 \end{bmatrix}$ $\quad C = \begin{bmatrix} -1 & 0 \\ 2 & 5 \\ 0 & 3 \end{bmatrix}$

(b) $A = \begin{bmatrix} 1 & 2 & 4 \\ 0 & 1 & -1 \end{bmatrix}$ $\quad B = \begin{bmatrix} -1 & 6 \\ 1 & 0 \end{bmatrix}$ $\quad C = \begin{bmatrix} 2 & 0 \\ -1 & 1 \\ 1 & 2 \end{bmatrix}$

3. Find a, b, a_1, and b_1 if:

(a) $\begin{bmatrix} a & b \\ a_1 & b_1 \end{bmatrix}\begin{bmatrix} 3 & -5 \\ -1 & 2 \end{bmatrix} = \begin{bmatrix} 1 & -1 \\ 2 & 0 \end{bmatrix}$ **(b)** $\begin{bmatrix} 2 & 1 \\ -1 & 2 \end{bmatrix}\begin{bmatrix} a & b \\ a_1 & b_1 \end{bmatrix} = \begin{bmatrix} 7 & 2 \\ -1 & 4 \end{bmatrix}$

4. Verify that $A^2 - A - 6I = 0$ if:

(a) $A = \begin{bmatrix} 3 & -1 \\ 0 & -2 \end{bmatrix}$ **(b)** $A = \begin{bmatrix} 2 & 2 \\ 2 & -1 \end{bmatrix}$

5. Given $A = \begin{bmatrix} 1 & -1 \\ 0 & 1 \end{bmatrix}$, $B = \begin{bmatrix} 1 & 0 & -2 \\ 3 & 1 & 0 \end{bmatrix}$, $C = \begin{bmatrix} 1 & 0 \\ 2 & 1 \\ 5 & 8 \end{bmatrix}$, and $D = \begin{bmatrix} 3 & -1 & 2 \\ 1 & 0 & 5 \end{bmatrix}$, verify the following facts from Theorem 1.

(a) $A(B - D) = AB - AD$ **(b)** $A(BC) = (AB)C$ **(c)** $(CD)^T = D^TC^T$

6. Write each of the following systems of linear equations in matrix form.

(a)
$$\begin{aligned} 3x_1 + 2x_2 - x_3 + x_4 &= 1 \\ x_1 - x_2 + 3x_4 &= 0 \\ 2x_1 - x_2 - x_3 &= 5 \end{aligned}$$

(b)
$$\begin{aligned} -x_1 + 2x_2 - x_3 + x_4 &= 6 \\ 2x_1 + x_2 - x_3 + 2x_4 &= 1 \\ 3x_1 - 2x_2 + x_4 &= 0 \end{aligned}$$

7. In each case, express every solution of the system as a sum of a specific solution plus a solution of the associated homogeneous system.

(a)
$$\begin{aligned} x + y + z &= 2 \\ 2x + y &= 3 \\ x - y - 3z &= 0 \end{aligned}$$

(b)
$$\begin{aligned} x - y - 4z &= -4 \\ x + 2y + 5z &= 2 \\ x + y + 2z &= 0 \end{aligned}$$

(c)
$$\begin{aligned} x_1 + x_2 - x_3 - 5x_5 &= 2 \\ x_2 + x_3 - 4x_5 &= -1 \\ x_2 + x_3 + x_4 - x_5 &= -1 \\ 2x_1 - 4x_3 + x_4 + x_5 &= 6 \end{aligned}$$

(d)
$$\begin{aligned} 2x_1 + x_2 - x_3 - x_4 &= -1 \\ 3x_1 + x_2 + x_3 - 2x_4 &= -2 \\ -x_1 - x_2 + 2x_3 + x_4 &= 2 \\ -2x_1 - x_2 + 2x_4 &= 3 \end{aligned}$$

8. Show that, if a (possibly nonhomogeneous) system of equations is consistent and has more variables than equations, then it must have infinitely many solutions. [*Hint:* Use Theorem 2 in this section and Theorem 1 in Section 1.3.]

9. Prove Theorem 2 in Section 1.3, using matrix algebra.

10. Assume that a system $A\mathbf{x} = \mathbf{b}$ of linear equations has at least two distinct solutions $\mathbf{s}$ and $\mathbf{t}$.

(a) Show that $\mathbf{s}_k = \mathbf{s} + k(\mathbf{s} - \mathbf{t})$ is a solution for every k.

(b) Show that $\mathbf{s}_k = \mathbf{s}_m$ implies $k = m$. [*Hint:* See Example 7 in Section 2.1.]

(c) Deduce that $A\mathbf{x} = \mathbf{b}$ has infinitely many solutions.

11. (a) If A^2 can be formed, what can be said about the size of A?

(b) If AB and BA can both be formed, describe the sizes of A and B.

(c) If ABC can be formed, A is 3×3, and C is 5×5, what size is B?

12. (a) Find two 2×2 matrices A such that $A^2 = 0$.

(b) Find three 2×2 matrices A such that **(i)** $A^2 = I$; **(ii)** $A^2 = A$.

(c) Find 2×2 matrices A and B such that $AB = 0$ but $BA \neq 0$.

13. (a) Compute AB, using the indicated block partitioning.

$$A = \left[\begin{array}{cc|cc} 2 & -1 & 3 & 1 \\ 1 & 0 & 1 & 2 \\ \hline 0 & 0 & 1 & 0 \\ 0 & 0 & 0 & 1 \end{array}\right] \qquad B = \left[\begin{array}{cc|c} 1 & 2 & 0 \\ -1 & 0 & 0 \\ \hline 0 & 5 & 1 \\ 1 & -1 & 0 \end{array}\right]$$

(b) Partition A and B in part **(a)** differently and compute AB again.

(c) Find A^2 using the partitioning in part **(a)** and then again using a different partitioning.

14. In each case compute all powers of A, using the block decomposition indicated.

(a) $A = \left[\begin{array}{c|cc} 1 & 0 & 0 \\ \hline 1 & 1 & -1 \\ 1 & -1 & 1 \end{array}\right]$ **(b)** $A = \left[\begin{array}{cc|cc} 1 & -1 & 2 & -1 \\ 0 & 1 & 0 & 0 \\ \hline 0 & 0 & -1 & 1 \\ 0 & 0 & 0 & 1 \end{array}\right]$.

15. Let $R_1, R_2, \ldots, R_m$ denote the rows (in order) of an $m \times n$ matrix A, and let $C_1, C_2, \ldots, C_n$ denote the columns (in order).

(a) If P is $n \times k$, show that $AP = \begin{bmatrix} R_1P \\ R_2P \\ \vdots \\ R_mP \end{bmatrix}$. That is, show that R_iP is the ith row of AP.

(b) If Q is $l \times m$, show that $QA = [QC_1 \quad QC_2 \quad \ldots \quad QC_n]$. That is, show that QC_j is the jth column of QA.

16. (a) If A has a row of zeros, show that the same is true of AB for any B.

(b) If B has a column of zeros, show that the same is true of AB for any A.

17. Let A denote an $m \times n$ matrix.

(a) If $AX = 0$ for every $n \times 1$ matrix X, show that $A = 0$.

(b) If $YA = 0$ for every $1 \times m$ matrix Y, show that $A = 0$.

18. Simplify the following expressions where A, B, and C represent matrices.

(a) $A(B + C - D) + B(C - A + D) - (A + B)C + (A - B)D$

(b) $AB(BC - CB) + (CA - AB)BC + CA(A - B)C$

19. If A and B commute with C, show that the same is true of:

(a) $A + B$ **(b)** kA, k any scalar

20. (a) If A is any matrix, show that AA^T and A^TA are symmetric.

(b) If A and B are symmetric, show that AB is symmetric if and only if $AB = BA$.

(c) If A is a 2×2 matrix, show that $A^TA = AA^T$ if and only if A is symmetric or $A = \begin{bmatrix} a & b \\ -b & a \end{bmatrix}$ for some a and b.

21. Show that there exist no 2×2 matrices A and B such that $AB - BA = I$. [*Hint:* Examine the (1,1)- and (2,2)-entries.]

22. Let B be an $n \times n$ matrix. Suppose $AB = 0$ for some nonzero $m \times n$ matrix A. Show that *no* $n \times n$ matrix C exists such that $BC = I$.

23. (a) Let A be a 3×3 matrix with all entries on and below the main diagonal zero. Show that $A^3 = 0$.

(b) Generalize to the $n \times n$ case and prove your answer.

24. **(a)** If A and B are 2×2 matrices whose rows sum to 1, show that the rows of AB also sum to 1.

(b) Repeat part **(a)** for the case where A and B are $n \times n$.

25. The **trace** of a square matrix A, denoted tr A, is the sum of the elements on the main diagonal. Show that, if A and B are $n \times n$ matrices:

(a) $\operatorname{tr}(A + B) = \operatorname{tr} A + \operatorname{tr} B$

(b) $\operatorname{tr}(kA) = k \operatorname{tr}(A)$ for any number k

(c) $\operatorname{tr}(A^T) = \operatorname{tr}(A)$

(d) $\operatorname{tr}(AB) = \operatorname{tr}(BA)$

(e) $\operatorname{tr}(AA^T)$ is the sum of the squares of all entries of A.

26. A square matrix P is called an **idempotent** if $P^2 = P$. Show that:

(a) 0 and I are idempotents.

(b) $\begin{bmatrix} 1 & 1 \\ 0 & 0 \end{bmatrix}$, $\begin{bmatrix} 1 & 0 \\ 1 & 0 \end{bmatrix}$, and $\frac{1}{\sqrt{2}}\begin{bmatrix} 1 & 1 \\ 1 & 1 \end{bmatrix}$ are idempotents.

(c) If P is an idempotent, so is $I - P$, and $P(I - P) = 0$.

(d) If P is an idempotent, so is P^T.

(e) If P is an idempotent, so is $Q = P + AP - PAP$ for any square matrix A (of the same size as P).

(f) If A is $n \times m$ and B is $m \times n$, and if $AB = I_n$, then BA is an idempotent.

27. Let A and B be $n \times n$ **diagonal matrices** (all entries off the main diagonal are zero).

(a) Show that AB is diagonal and $AB = BA$.

(b) Formulate a rule for calculating XA if X is $m \times n$.

(c) Formulate a rule for calculating AY if Y is $n \times k$.

28. If A and B are $n \times n$ matrices, show that:

(a) $AB = BA$ if and only if $(A + B)^2 = A^2 + 2AB + B^2$.

(b) $AB = BA$ if and only if $(A + B)(A - B) = (A - B)(A + B)$.

(c) $AB = BA$ if and only if $A^TB^T = B^TA^T$.

29. Let A be a 2×2 matrix.

(a) If A commutes with $\begin{bmatrix} 0 & 1 \\ 0 & 0 \end{bmatrix}$, show that $A = \begin{bmatrix} a & b \\ 0 & a \end{bmatrix}$ for some a and b.

(b) If A commutes with $\begin{bmatrix} 0 & 0 \\ 1 & 0 \end{bmatrix}$, show that $A = \begin{bmatrix} a & 0 \\ c & a \end{bmatrix}$ for some a and c.

(c) Show that A commutes with *every* 2×2 matrix if and only if $A = \begin{bmatrix} a & 0 \\ 0 & a \end{bmatrix}$ for some a.

30. Let I_{pq} denote the $n \times n$ matrix with (p,q)-entry equal to 1 and all other entries 0. Show that:

(a) $I_n = I_{11} + I_{22} + \cdots + I_{nn}$

(b) $I_{pq} I_{rs} = \begin{cases} I_{ps} & \text{if } q = r \\ 0 & \text{if } q \neq r \end{cases}$

(c) If $A = [a_{ij}]$, then $A = \sum_{i=1}^{n} \sum_{j=1}^{n} a_{ij} I_{ij}$.

(d) If $A = [a_{ij}]$, then $I_{pq} A I_{rs} = a_{qr} I_{ps}$ for all p, q, r, and s.

31. A matrix of the form aI_n, where a is a number, is called an $n \times n$ **scalar matrix**.

(a) Show that each $n \times n$ scalar matrix commutes with every $n \times n$ matrix.

(b) Show that A is a scalar matrix if it commutes with every $n \times n$ matrix. [*Hint:* See part **(d)** of Exercise 30.]

32. Prove the following parts of Theorem 1.

(a) Part 1 (b) Part 2 (c) Part 4 (d) Part 7

33. Let $M = \begin{bmatrix} A & B \\ C & D \end{bmatrix}$, where A, B, C, and D are all $n \times n$ and each commutes with all the others. If $M^2 = 0$, show that $(A + D)^3 = 0$. [*Hint:* First show that $A^2 = -BC = D^2$ and that $B(A + D) = 0 = C(A + D)$.]

34. Prove Theorem 3.

SECTION 2.3 Matrix Inverses

2.3.1 Definition and Basic Properties

Three basic operations on matrices were discussed in Sections 2.1 and 2.2: scalar multiplication, matrix addition, and matrix multiplication. The latter two are analogs for matrices of the same operations for numbers. Moreover, an analog of numerical subtraction exists for matrices, so it is natural to ask whether the process of division of numbers has a useful analog for matrices.

To begin, consider how a numerical equation

$$ax = b$$

is solved when a and b are known numbers. If $a = 0$, there is no solution (unless $b = 0$). But if $a \neq 0$, we can multiply both sides by the inverse a^{-1} to obtain the solution $x = a^{-1}b$. This multiplication by a^{-1} is commonly called dividing by a, and it is here that we can make the generalization to matrices. The property of a^{-1} that makes this work is that $a^{-1}a = 1$. Moreover, we saw in Section 2.2 that the role that 1 plays in arithmetic is played in matrix algebra by the identity matrix I. This suggests the following definition.

DEFINITION

If A is a square matrix, a matrix B is called an **inverse** of A if and only if

$$AB = I = BA$$

A matrix A that has an inverse is called an **invertible matrix.**

EXAMPLE 1

Show that $B = \begin{bmatrix} -1 & 1 \\ 1 & 0 \end{bmatrix}$ is an inverse of $A = \begin{bmatrix} 0 & 1 \\ 1 & 1 \end{bmatrix}$.

Solution Compute AB and BA.

$$AB = \begin{bmatrix} 0 & 1 \\ 1 & 1 \end{bmatrix}\begin{bmatrix} -1 & 1 \\ 1 & 0 \end{bmatrix} = \begin{bmatrix} 1 & 0 \\ 0 & 1 \end{bmatrix} \quad BA = \begin{bmatrix} -1 & 1 \\ 1 & 0 \end{bmatrix}\begin{bmatrix} 0 & 1 \\ 1 & 1 \end{bmatrix} = \begin{bmatrix} 1 & 0 \\ 0 & 1 \end{bmatrix}$$

Hence $AB = I = BA$, so B is indeed an inverse of A.

EXAMPLE 2

Show that $A = \begin{bmatrix} 0 & 0 \\ 1 & 3 \end{bmatrix}$ has no inverse.

Solution Let $B = \begin{bmatrix} a & b \\ c & d \end{bmatrix}$ denote an arbitrary 2×2 matrix. Then

$$AB = \begin{bmatrix} 0 & 0 \\ 1 & 3 \end{bmatrix}\begin{bmatrix} a & b \\ c & d \end{bmatrix} = \begin{bmatrix} 0 & 0 \\ a + 3c & b + 3d \end{bmatrix}$$

so AB has a row of zeros. Hence AB cannot equal I for any B.

Example 2 shows that *it is possible for a nonzero matrix to have no inverse.* But if a matrix *does* have an inverse, it has only one.

THEOREM 1 If a square matrix has an inverse, that inverse is unique.

Proof Suppose B and C are two (possibly different) inverses of A. Then $AB = I = BA$ and $AC = I = CA$, so

$$B = IB = (CA)B = C(AB) = CI = C$$

Hence $B = C$, and A has only one inverse. ■

The uniqueness in Theorem 1 entitles us to use a special notation: The inverse of A is denoted as A^{-1} when it exists. Hence A^{-1} (when it exists) is a square matrix of the same size as A and is characterized by the equations

$$AA^{-1} = I = A^{-1}A$$

In other words, if somehow a matrix B can be found such that $AB = I = BA$, then A is invertible and B is the inverse of A; in symbols, $B = A^{-1}$. This gives us a way of verifying that the inverse of a matrix exists—and sometimes provides a formula for it. The next four examples offer illustrations.

EXAMPLE 3 If $A = \begin{bmatrix} 0 & -1 \\ 1 & -1 \end{bmatrix}$, show that $A^3 = I$ and so find A^{-1}.

Solution We have $A^2 = \begin{bmatrix} 0 & -1 \\ 1 & -1 \end{bmatrix}\begin{bmatrix} 0 & -1 \\ 1 & -1 \end{bmatrix} = \begin{bmatrix} -1 & 1 \\ -1 & 0 \end{bmatrix}$, and so

$$A^3 = A^2A = \begin{bmatrix} -1 & 1 \\ -1 & 0 \end{bmatrix}\begin{bmatrix} 0 & -1 \\ 1 & -1 \end{bmatrix} = \begin{bmatrix} 1 & 0 \\ 0 & 1 \end{bmatrix} = I$$

Hence $A^3 = I$, as asserted. This can be written as $A^2A = I = AA^2$, so it shows that A^2 is the inverse of A. That is, $A^{-1} = A^2 = \begin{bmatrix} -1 & 1 \\ -1 & 0 \end{bmatrix}$.

EXAMPLE 4 Show that $A = \begin{bmatrix} 2 & 4 \\ 3 & 1 \end{bmatrix}$ satisfies $A^2 - 3A - 10I = 0$. Hence find A^{-1}.

Solution The equation is verified directly as follows:

$$A^2 - 3A - 10I = \begin{bmatrix} 16 & 12 \\ 9 & 13 \end{bmatrix} - \begin{bmatrix} 6 & 12 \\ 9 & 3 \end{bmatrix} - \begin{bmatrix} 10 & 0 \\ 0 & 10 \end{bmatrix} = \begin{bmatrix} 0 & 0 \\ 0 & 0 \end{bmatrix}$$

Hence $A^2 - 3A = 10I$, or $A(A - 3I) = 10I$. Multiply both sides by $\frac{1}{10}$ to get $A[\frac{1}{10}(A - 3I)] = I$. Similarly, $[\frac{1}{10}(A - 3I)]A = I$, so

$$A^{-1} = \frac{1}{10}[A - 3I] = \frac{1}{10}\left[\begin{pmatrix} 2 & 4 \\ 3 & 1 \end{pmatrix} - \begin{pmatrix} 3 & 0 \\ 0 & 3 \end{pmatrix}\right] = \frac{1}{10}\begin{bmatrix} -1 & 4 \\ 3 & -2 \end{bmatrix}$$

EXAMPLE 5 If A is an invertible matrix, show that the transpose A^T is also invertible. Show further that the inverse of A^T is just the transpose of A^{-1}; in symbols, $(A^T)^{-1} = (A^{-1})^T$.

Solution A^{-1} exists (by assumption). Its transpose $(A^{-1})^T$ is the candidate proposed for the inverse of A^T. We test it as follows:

$$A^T(A^{-1})^T = (A^{-1}A)^T = I^T = I$$
$$(A^{-1})^TA^T = (AA^{-1})^T = I^T = I$$

Hence $(A^{-1})^T$ is indeed the inverse of A^T; that is, $(A^T)^{-1} = (A^{-1})^T$.

EXAMPLE 6 If A and B are invertible $n \times n$ matrices, show that their product AB is also invertible and $(AB)^{-1} = B^{-1}A^{-1}$.

Solution We are given a candidate for the inverse of AB, namely $B^{-1}A^{-1}$. We test it as follows:

$$(B^{-1}A^{-1})(AB) = B^{-1}(A^{-1}A)B = B^{-1}IB = B^{-1}B = I$$
$$(AB)(B^{-1}A^{-1}) = A(BB^{-1})A^{-1} = AIA^{-1} = AA^{-1} = I$$

Hence $B^{-1}A^{-1}$ is indeed the inverse of AB; in symbols, $(AB)^{-1} = B^{-1}A^{-1}$.

The following example gives a useful formula for the inverse of a 2×2 matrix.

EXAMPLE 7 If $A = \begin{bmatrix} a & b \\ c & d \end{bmatrix}$ and $ad - bc \neq 0$, show that $A^{-1} = \frac{1}{ad - bc}\begin{bmatrix} d & -b \\ -c & a \end{bmatrix}$

Solution We verify that $AA^{-1} = I$ and leave $A^{-1}A = I$ to the reader.

$$\begin{bmatrix} a & b \\ c & d \end{bmatrix}\left(\frac{1}{ad - bc}\begin{bmatrix} d & -b \\ -c & a \end{bmatrix}\right) = \frac{1}{ad - bc}\begin{bmatrix} a & b \\ c & d \end{bmatrix}\begin{bmatrix} d & -b \\ -c & a \end{bmatrix}$$
$$= \frac{1}{ad - bc}\begin{bmatrix} ad - bc & 0 \\ 0 & ad - bc \end{bmatrix} = I$$

We now collect several basic properties of matrix inverses for reference.

THEOREM 2 All the following matrices are square matrices of the same size.

1. I is invertible and $I^{-1} = I$.
2. If A is invertible, so is A^{-1}, and $(A^{-1})^{-1} = A$.
3. If A and B are invertible, so is AB, and $(AB)^{-1} = B^{-1}A^{-1}$.
4. If $A_1, A_2, \ldots, A_k$ are all invertible, so is their product $A_1A_2 \ldots A_k$, and $(A_1A_2 \cdots A_k)^{-1} = A_k^{-1} \cdots A_2^{-1} A_1^{-1}$.
5. If A is invertible, so is A^k for $k \geq 1$, and $(A^k)^{-1} = (A^{-1})^k$.
6. If A is invertible and $a \neq 0$ is a number, then aA is invertible and $(aA)^{-1} = \frac{1}{a}A^{-1}$.
7. If A is invertible, so is its transpose A^T, and $(A^T)^{-1} = (A^{-1})^T$.

Proof

1. This is an immediate consequence of the formula $I^2 = I$.
2. A and A^{-1} are related by the equations $AA^{-1} = I = A^{-1}A$. But these also show that A is the inverse of A^{-1}; in symbols, $(A^{-1})^{-1} = A$.
3. This is Example 6.
4. Use induction on k. If $k = 1$, there is nothing to prove because the conclusion reads $(A_1)^{-1} = A_1^{-1}$. If $k = 2$, the result is just

property 3. If $k > 2$, assume inductively that $(A_1A_2 \cdots A_{k-1})^{-1} = A_{k-1}{}^{-1} \cdots A_2{}^{-1}A_1{}^{-1}$. We apply this fact together with property 3 as follows:

$$\begin{aligned}[A_1A_2 \cdots A_{k-1}A_k]^{-1} &= [(A_1A_2 \cdots A_{k-1})A_k]^{-1} \\ &= A_k{}^{-1}(A_1A_2 \cdots A_{k-1})^{-1} \\ &= A_k{}^{-1}(A_{k-1}{}^{-1} \cdots A_2{}^{-1}A_1{}^{-1})\end{aligned}$$

Here property 3 is applied to get the second equality. This is the conclusion for k matrices, so the proof by induction is complete.

5. This is a special case of property 4 with $A_1 = A_2 = \cdots = A_k = A$.
6. This is left as Exercise 28.
7. This is Example 5. ■

The reversal of the order of the inverses in properties 3 and 4 of Theorem 2 is a consequence of the fact that matrix multiplication is not commutative. Another manifestation of this comes when matrix equations are dealt with. If a matrix equation $B = C$ is given, it can be *left-multiplied* by a matrix A to yield $AB = AC$. Similarly, *right-multiplication* gives $BA = CA$. However, one cannot mix the two: If $B = C$, it need *not* be the case that $AB = CA$. The next examples illustrate how such manipulations are used.

EXAMPLE 8
Cancellation Laws

Let A be an invertible matrix. Show that:

(a) If $AB = AC$, then $B = C$.

(b) If $BA = CA$, then $B = C$.

Solution Given the equation $AB = AC$, *left*-multiply both sides by A^{-1} to obtain $A^{-1}AB = A^{-1}AC$. This gives $IB = IC$—that is, $B = C$. This proves part (a), and the proof of part (b) is similar.

One application of cancellation is as follows: If A is invertible, then the only matrix X such that $AX = 0$ is $X = 0$. This follows directly from Example 8, because $AX = 0$ can be written $AX = A0$. (Alternatively, left-multiply $AX = 0$ by A^{-1} to get $X = A^{-1}(AX) = A^{-1}0 = 0$.) Of course, $YA = 0$ implies $Y = 0$ in the same way, and these facts give a useful method of showing that a matrix is *not* invertible.

EXAMPLE 9

Show that $A = \begin{bmatrix} 6 & 8 \\ 3 & 4 \end{bmatrix}$ has no inverse.

Solution Observe that $AX = 0$ if $X = \begin{bmatrix} 4 \\ -3 \end{bmatrix}$. Hence, if A had an inverse, it would mean $X = 0$ (as above), which is not the case. So A has no inverse.

Matrix inverses can be used to solve certain systems of linear equations. Recall (Example 8 in Section 2.2) that a *system* of linear equations can be written as a *single* matrix equation:

$$A\mathbf{x} = \mathbf{b}$$

where A and $\mathbf{b}$ are known matrices and $\mathbf{x}$ is to be determined. The analogous numerical equation is $ax = b$, which is solved (if $a \neq 0$) by multiplying both sides by a^{-1} to get $x = a^{-1}b$. The matrix equation can be solved in an analogous fashion, provided that A has an inverse. In this case, we left-multiply each side of the equation $A\mathbf{x} = \mathbf{b}$ by A^{-1} to get

$$A^{-1}A\mathbf{x} = A^{-1}\mathbf{b}$$
$$I\mathbf{x} = A^{-1}\mathbf{b}$$
$$\mathbf{x} = A^{-1}\mathbf{b}$$

A and $\mathbf{b}$ are known, so this equation gives $\mathbf{x}$ and hence yields the solution to the system of equations (the reader should verify that $\mathbf{x} = A^{-1}\mathbf{b}$ really does satisfy $A\mathbf{x} = \mathbf{b}$). Furthermore, the argument shows that, if $\mathbf{x}$ is a solution, then necessarily $\mathbf{x} = A^{-1}\mathbf{b}$, so the solution is unique (if it exists). Of course the technique works only when the coefficient matrix A has an inverse. This proves Theorem 3.

THEOREM 3 Suppose a system of n equations in n variables is written in matrix form as

$$A\mathbf{x} = \mathbf{b}$$

If the $n \times n$ coefficient matrix A is invertible, the system has the unique solution

$$\mathbf{x} = A^{-1}\mathbf{b}$$

EXAMPLE 10 If $A = \begin{bmatrix} 1 & -2 & 2 \\ 2 & 1 & 1 \\ 1 & 0 & 1 \end{bmatrix}$, show that $A^{-1} = \begin{bmatrix} 1 & 2 & -4 \\ -1 & -1 & 3 \\ -1 & -2 & 5 \end{bmatrix}$ and use it to solve the following system of linear equations.

$$\begin{aligned} x_1 - 2x_2 + 2x_3 &= 3 \\ 2x_1 + x_2 + x_3 &= 0 \\ x_1 \qquad + x_3 &= -2 \end{aligned}$$

Solution Verification that the inverse of A is as given is left to the reader. The matrix form of the system of equations is $A\mathbf{x} = \mathbf{b}$, where A is as above and

$$\mathbf{x} = \begin{bmatrix} x_1 \\ x_2 \\ x_3 \end{bmatrix} \qquad \mathbf{b} = \begin{bmatrix} 3 \\ 0 \\ -2 \end{bmatrix}$$

Theorem 3 gives the solution

$$\mathbf{x} = A^{-1}\mathbf{b} = \begin{bmatrix} 1 & 2 & -4 \\ -1 & -1 & 3 \\ -1 & -2 & 5 \end{bmatrix} \begin{bmatrix} 3 \\ 0 \\ -2 \end{bmatrix} = \begin{bmatrix} 11 \\ -9 \\ -13 \end{bmatrix}$$

It is clear that, given a particular $n \times n$ matrix A, it is desirable to have an efficient technique that will determine whether A has an inverse and, if so, will find that inverse. For simplicity, we shall derive the technique for 2×2 matrices; the $n \times n$ case is entirely analogous.

Given the invertible 2×2 matrix A, we determine A^{-1} from the equation $AA^{-1} = I$. Write

$$A^{-1} = \begin{bmatrix} x_1 & x_2 \\ y_1 & y_2 \end{bmatrix}$$

where x_1, y_1, x_2, and y_2 are to be determined. Equating columns in the equation $AA^{-1} = I$ gives

$$A\begin{bmatrix} x_1 \\ y_1 \end{bmatrix} = \begin{bmatrix} 1 \\ 0 \end{bmatrix} \quad \text{and} \quad A\begin{bmatrix} x_2 \\ y_2 \end{bmatrix} = \begin{bmatrix} 0 \\ 1 \end{bmatrix}$$

These are systems of linear equations, each with A as coefficient matrix. The fact that A has an inverse means there is a sequence of elementary row operations carrying A to the 2×2 identity matrix I. This sequence carries the augmented matrices to reduced row-echelon form:

$$\left[\begin{array}{c|c} A & \begin{matrix} 1 \\ 0 \end{matrix} \end{array}\right] \to \left[\begin{array}{c|c} I & \begin{matrix} x_1 \\ y_1 \end{matrix} \end{array}\right] \qquad \left[\begin{array}{c|c} A & \begin{matrix} 0 \\ 1 \end{matrix} \end{array}\right] \to \left[\begin{array}{c|c} I & \begin{matrix} x_2 \\ y_2 \end{matrix} \end{array}\right]$$

and so solves the systems. Hence we can do *both* calculations simultaneously.

$$\left[\begin{array}{c|cc} A & 1 & 0 \\ & 0 & 1 \end{array}\right] \to \left[\begin{array}{c|cc} I & x_1 & x_2 \\ & y_1 & y_2 \end{array}\right]$$

This can be written more compactly as follows:

$$[A \quad I] \to [I \quad A^{-1}]$$

In other words, the sequence of row operations that carries A to I also carries I to A^{-1}. This procedure is easily programmed for a computer, so it is the desired algorithm.

MATRIX INVERSION ALGORITHM

If A is a (square) invertible matrix, there exists a sequence of elementary row operations that carry A to the identity matrix I of the same size, written $A \to I$. This same sequence of row operations carries I to A^{-1}; that is, $I \to A^{-1}$. The algorithm can be summarized as follows:

$$[A \quad I] \to [I \quad A^{-1}]$$

where the row operations on A and I are carried out simultaneously.

EXAMPLE 11 Use the inversion algorithm to find the inverse of the matrix

$$A = \begin{bmatrix} 2 & 7 & 1 \\ 1 & 4 & -1 \\ 1 & 3 & 0 \end{bmatrix}$$

Solution Apply elementary row operations to the "double" matrix

$$[A \quad I] = \begin{bmatrix} 2 & 7 & 1 & 1 & 0 & 0 \\ 1 & 4 & -1 & 0 & 1 & 0 \\ 1 & 3 & 0 & 0 & 0 & 1 \end{bmatrix}$$

so as to carry A to I. The steps are as follows. (Operations refer to the preceding matrix.)

$$\begin{bmatrix} 1 & 4 & -1 & 0 & 1 & 0 \\ 2 & 7 & 1 & 1 & 0 & 0 \\ 1 & 3 & 0 & 0 & 0 & 1 \end{bmatrix} \quad \text{[interchange rows 1 and 2]}$$

$$\begin{bmatrix} 1 & 4 & -1 & 0 & 1 & 0 \\ 0 & -1 & 3 & 1 & -2 & 0 \\ 0 & -1 & 1 & 0 & -1 & 1 \end{bmatrix} \quad \begin{matrix} \text{[row 2 − 2(row 1)]} \\ \text{[row 3 − row 1]} \end{matrix}$$

$$\begin{bmatrix} 1 & 0 & 11 & 4 & -7 & 0 \\ 0 & 1 & -3 & -1 & 2 & 0 \\ 0 & 0 & -2 & -1 & 1 & 1 \end{bmatrix} \quad \begin{matrix} \text{[row 1 + 4(row 2)]} \\ \text{[−(row 2)]} \\ \text{[row 3 − row 2]} \end{matrix}$$

$$\begin{bmatrix} 1 & 0 & 0 & \frac{-3}{2} & \frac{-3}{2} & \frac{11}{2} \\ 0 & 1 & 0 & \frac{1}{2} & \frac{1}{2} & \frac{-3}{2} \\ 0 & 0 & 1 & \frac{1}{2} & \frac{-1}{2} & \frac{-1}{2} \end{bmatrix} \quad \begin{matrix} \left[\text{row } 1 + \frac{11}{2}(\text{row } 3)\right] \\ \left[\text{row } 2 - \frac{3}{2}(\text{row } 3)\right] \\ \left[\frac{-1}{2}(\text{row } 3)\right] \end{matrix}$$

Hence $A^{-1} = \frac{1}{2}\begin{bmatrix} -3 & -3 & 11 \\ 1 & 1 & -3 \\ 1 & -1 & -1 \end{bmatrix}$, as is readily verified.

Given any $n \times n$ matrix A, Theorem 2 in Section 1.2 shows that A can be carried by elementary row operations to a matrix R in reduced row-echelon form. If $R = I$, the algorithm produces A^{-1}. If $R \neq I$, then R has a row of zeros (it is square), and no system of linear equations $A\mathbf{x} = \mathbf{b}$ can have a unique solution. But then A is not invertible by Theorem 3. Hence the algorithm is effective in the sense conveyed in Theorem 4.

THEOREM 4 If A is an $n \times n$ matrix, either A can be reduced to I by elementary row operations or it cannot. In the first case, the algorithm produces A^{-1}; in the second case, A^{-1} does not exist.

EXERCISES 2.3.1

1. In each case, show that the matrices are inverses of each other.

(a) $\begin{bmatrix} 3 & 5 \\ 1 & 2 \end{bmatrix}, \begin{bmatrix} 2 & -5 \\ -1 & 3 \end{bmatrix}$ **(b)** $\begin{bmatrix} 3 & 0 \\ 1 & -4 \end{bmatrix}, \frac{1}{12}\begin{bmatrix} 4 & 0 \\ 1 & -3 \end{bmatrix}$

(c) $\begin{bmatrix} 1 & 2 & 0 \\ 0 & 2 & 3 \\ 1 & 3 & 1 \end{bmatrix}, \begin{bmatrix} 7 & 2 & -6 \\ -3 & -1 & 3 \\ 2 & 1 & -2 \end{bmatrix}$ **(d)** $\begin{bmatrix} 3 & 0 \\ 0 & 5 \end{bmatrix}, \begin{bmatrix} \frac{1}{3} & 0 \\ 0 & \frac{1}{5} \end{bmatrix}$

2. Find the inverse of each of the following matrices.

(a) $\begin{bmatrix} 3 & -1 \\ -3 & 2 \end{bmatrix}$ **(b)** $\begin{bmatrix} 4 & 1 \\ 3 & 2 \end{bmatrix}$ **(c)** $\begin{bmatrix} 1 & -1 & 0 \\ 3 & 0 & 2 \\ -1 & 0 & -1 \end{bmatrix}$

(d) $\begin{bmatrix} 1 & -1 & 2 \\ -5 & 7 & -11 \\ -2 & 3 & -5 \end{bmatrix}$ **(e)** $\begin{bmatrix} 3 & 5 & 0 \\ 1 & 2 & 1 \\ 3 & 7 & 1 \end{bmatrix}$ **(f)** $\begin{bmatrix} 3 & 1 & -1 \\ 2 & 1 & 0 \\ 1 & 5 & -1 \end{bmatrix}$

(g) $\begin{bmatrix} 2 & 3 & 4 \\ 4 & 3 & 1 \\ 1 & 2 & 4 \end{bmatrix}$ (h) $\begin{bmatrix} 3 & 1 & -1 \\ 5 & 2 & 0 \\ 1 & 1 & -1 \end{bmatrix}$ (i) $\begin{bmatrix} 2 & 1 & 3 \\ 3 & -1 & 1 \\ 4 & 2 & 1 \end{bmatrix}$

(j) $\begin{bmatrix} -1 & 4 & 5 & 2 \\ 0 & 0 & 0 & -1 \\ 1 & -2 & -2 & 0 \\ 0 & -1 & -1 & 0 \end{bmatrix}$ (k) $\begin{bmatrix} 1 & -1 & 5 & 2 \\ 1 & 0 & 7 & 5 \\ 0 & 1 & 3 & 6 \\ 1 & -1 & 5 & 1 \end{bmatrix}$ (l) $\begin{bmatrix} 1 & 2 & 0 & 0 & 0 \\ 0 & 1 & 3 & 0 & 0 \\ 0 & 0 & 1 & 5 & 0 \\ 0 & 0 & 0 & 1 & 7 \\ 0 & 0 & 0 & 0 & 1 \end{bmatrix}$

3. In each case, solve the equations by finding the inverse of the coefficient matrix.

(a) $\begin{aligned} 3x - y &= 5 \\ 2x + 3y &= 1 \end{aligned}$ **(b)** $\begin{aligned} 2x - 3y &= 0 \\ x - 4y &= 1 \end{aligned}$

(c) $\begin{aligned} x + y + 2z &= 5 \\ x + y + z &= 0 \\ x + 2y + 4z &= -2 \end{aligned}$ **(d)** $\begin{aligned} x + 4y + 2z &= 1 \\ 2x + 3y + 3z &= -1 \\ 4x + y + 4z &= 0 \end{aligned}$

(e) $\begin{aligned} x + y \qquad - w &= 1 \\ -x + y - z \qquad &= -1 \\ y + z + w &= 0 \\ x \qquad - z + w &= 1 \end{aligned}$ **(f)** $\begin{aligned} x + y + z + w &= 1 \\ x + y \qquad &= 0 \\ y \qquad + w &= -1 \\ x \qquad + w &= 2 \end{aligned}$

4. Given $A^{-1} = \begin{bmatrix} 1 & -1 & 3 \\ 2 & 0 & 5 \\ -1 & 1 & 0 \end{bmatrix}$:

(a) Solve the system of equations $A\mathbf{x} = \begin{bmatrix} 1 \\ -1 \\ 3 \end{bmatrix}$.

(b) Find a matrix B such that $AB = \begin{bmatrix} 1 & -1 & 2 \\ 0 & 1 & 1 \\ 1 & 0 & 0 \end{bmatrix}$.

(c) Find a matrix C such that $CA = \begin{bmatrix} 1 & 2 & -1 \\ 3 & 1 & 1 \end{bmatrix}$.

5. Find A when:

(a) $(3A)^{-1} = \begin{bmatrix} 1 & -1 \\ 0 & 1 \end{bmatrix}$ **(b)** $(2A)^T = \begin{bmatrix} 1 & -1 \\ 2 & 3 \end{bmatrix}^{-1}$

(c) $(I + 2A)^{-1} = \begin{bmatrix} 2 & 0 \\ 1 & 1 \end{bmatrix}$ **(d)** $(I - 2A^T)^{-1} = \begin{bmatrix} 2 & 1 \\ 1 & 1 \end{bmatrix}$

(e) $\left(A\begin{bmatrix} 1 & -1 \\ 0 & 1 \end{bmatrix}\right)^{-1} = \begin{bmatrix} 2 & 3 \\ 1 & 2 \end{bmatrix}$ **(f)** $\left(\begin{bmatrix} 1 & 0 \\ 2 & 1 \end{bmatrix}A\right)^{-1} = \begin{bmatrix} 1 & 0 \\ 2 & 2 \end{bmatrix}$

6. Find A when:

(a) $A^{-1} = \begin{bmatrix} 1 & -1 & 3 \\ 2 & 1 & 1 \\ 0 & 2 & -2 \end{bmatrix}$ **(b)** $A^{-1} = \begin{bmatrix} 0 & 1 & -1 \\ 1 & 2 & 1 \\ 1 & 0 & 1 \end{bmatrix}$

7. Given $\begin{bmatrix} x_1 \\ x_2 \\ x_3 \end{bmatrix} = \begin{bmatrix} 3 & -1 & 2 \\ 1 & 0 & 4 \\ 2 & 1 & 0 \end{bmatrix} \begin{bmatrix} y_1 \\ y_2 \\ y_3 \end{bmatrix}$ and $\begin{bmatrix} z_1 \\ z_2 \\ z_3 \end{bmatrix} = \begin{bmatrix} 1 & -1 & 1 \\ 2 & -3 & 0 \\ -1 & 1 & -2 \end{bmatrix} \begin{bmatrix} y_1 \\ y_2 \\ y_3 \end{bmatrix}$, express the variables x_1, x_2, and x_3 in terms of z_1, z_2, and z_3.

8. **(a)** In the system $\begin{matrix} 3x + 4y = 7 \\ 4x + 5y = 1 \end{matrix}$, substitute the new variables x' and y' given by $\begin{matrix} x = -5x' + 4y' \\ y = 4x' - 3y' \end{matrix}$. Then find x and y.

(b) Explain part **(a)** by writing the equations as $A\begin{bmatrix} x \\ y \end{bmatrix} = \begin{bmatrix} 7 \\ 1 \end{bmatrix}$ and $\begin{bmatrix} x \\ y \end{bmatrix} = B\begin{bmatrix} x' \\ y' \end{bmatrix}$. What is the relationship between A and B? Generalize.

9. Find 2×2 invertible matrices A and B such that $A + B$ is not invertible.

10. If A, B, and C are square matrices and $AB = I = CA$, show that $B = C = A^{-1}$.

11. Suppose $CA = I_m$, where C is $m \times n$ and A is $n \times m$. Consider the system $A\mathbf{x} = \mathbf{b}$ of n equations in m variables.

(a) Show that this system has a unique solution $C\mathbf{b}$.

(b) If $C = \begin{bmatrix} 0 & -5 & 1 \\ 3 & 0 & -1 \end{bmatrix}$ and $A = \begin{bmatrix} 2 & -3 \\ 1 & -2 \\ 6 & -10 \end{bmatrix}$, find $\mathbf{x}$ when:

(i) $\mathbf{b} = \begin{bmatrix} 1 \\ 0 \\ 3 \end{bmatrix}$ **(ii)** $\mathbf{b} = \begin{bmatrix} 7 \\ 4 \\ 22 \end{bmatrix}$

12. Verify that $A = \begin{bmatrix} 1 & -1 \\ 0 & 2 \end{bmatrix}$ satisfies $A^2 - 3A + 2I = 0$, and use this fact to show that $A^{-1} = \frac{1}{2}(3I - A)$.

13. Let $Q = \begin{bmatrix} a & -b & -c & -d \\ b & a & -d & c \\ c & d & a & -b \\ d & -c & b & a \end{bmatrix}$. Compute QQ^T and so find Q^{-1}.

14. Let $U = \begin{bmatrix} 0 & 1 \\ 1 & 0 \end{bmatrix}$. Show that each of U, $-U$, and $-I_2$ is its own inverse and that the product of any two of these is the third.

15. Consider $A = \begin{bmatrix} 1 & 1 \\ -1 & 0 \end{bmatrix}$, $B = \begin{bmatrix} 0 & -1 \\ 1 & 0 \end{bmatrix}$, and $C = \begin{bmatrix} 0 & 1 & 0 \\ 0 & 0 & 1 \\ 5 & 0 & 0 \end{bmatrix}$. Find the inverses by computing **(a)** A^6, **(b)** B^4, and **(c)** C^3.

16. In each case find A^{-1} in terms of c.

(a) $\begin{bmatrix} c & 1 \\ -1 & c \end{bmatrix}$ **(b)** $\begin{bmatrix} 2 & -c \\ c & 3 \end{bmatrix}$ **(c)** $\begin{bmatrix} 1 & c & 0 \\ c & -1 & c \\ 2 & c & 1 \end{bmatrix}$ **(d)** $\begin{bmatrix} 1 & 0 & 1 \\ c & 1 & c \\ 3 & c & 2 \end{bmatrix}$

17. If $c \neq 0$, find the inverse of $\begin{bmatrix} 1 & -1 & 1 \\ 2 & -1 & 2 \\ 0 & 2 & c \end{bmatrix}$ in terms of c.

18. Find the inverse of $\begin{bmatrix} \sin\theta & \cos\theta \\ -\cos\theta & \sin\theta \end{bmatrix}$ for any real number θ.

19. Show that A has no inverse when **(a)** A has a row of zeros; **(b)** A has a column of zeros; **(c)** Each row of A sums to 0; **(d)** each column of A sums to 0.

20. Let A denote a square matrix.

(a) Let $YA = 0$ for some matrix $Y \neq 0$. Show that A has no inverse.

(b) Use part **(a)** to show that **(i)** $\begin{bmatrix} 1 & -1 & 1 \\ 0 & 1 & 1 \\ 1 & 0 & 2 \end{bmatrix}$ and **(ii)** $\begin{bmatrix} 2 & 1 & -1 \\ 1 & 1 & 0 \\ 1 & 0 & -1 \end{bmatrix}$ have no inverse. [*Hint:* For part **(ii)** compare row 3 with the difference between row 1 and row 2.]

21. If A is invertible, show that **(a)** $A^2 \neq 0$; **(b)** $A^k \neq 0$ for all $k = 1, 2, \ldots$.

22. Suppose $AB = 0$, where A and B are square matrices. Show that:

(a) If one of A and B has an inverse, the other is zero.

(b) It is impossible for both A and B to have inverses.

(c) $(BA)^2 = 0$

23. **(a)** Show that $\begin{bmatrix} a & 0 \\ 0 & b \end{bmatrix}$ is invertible if and only if $a \neq 0$ and $b \neq 0$. Describe the inverse.

(b) Show that a diagonal matrix is invertible if and only if all the main diagonal entries are nonzero. Describe the inverse.

(c) If A and B are square matrices, show that **(i)** the block matrix $\begin{bmatrix} A & 0 \\ 0 & B \end{bmatrix}$ is invertible if and only if A and B are both invertible, and

(ii) $\begin{bmatrix} A & 0 \\ 0 & B \end{bmatrix}^{-1} = \begin{bmatrix} A^{-1} & 0 \\ 0 & B^{-1} \end{bmatrix}$.

(d) Use part **(c)** to find the inverses of:

(i) $\begin{bmatrix} 1 & 0 & 0 \\ 0 & 2 & -1 \\ 0 & 1 & -1 \end{bmatrix}$ **(ii)** $\begin{bmatrix} 3 & 1 & 0 \\ 5 & 2 & 0 \\ 0 & 0 & -1 \end{bmatrix}$

(iii) $\begin{bmatrix} 2 & 1 & 0 & 0 \\ 1 & 1 & 0 & 0 \\ 0 & 0 & 1 & -1 \\ 0 & 0 & 1 & -2 \end{bmatrix}$ **(iv)** $\begin{bmatrix} 3 & 4 & 0 & 0 \\ 2 & 3 & 0 & 0 \\ 0 & 0 & 1 & 3 \\ 0 & 0 & 0 & -1 \end{bmatrix}$

(e) Extend part **(c)** to **block diagonal matrices**—that is, matrices with square blocks down the main diagonal and zero blocks elsewhere.

24. **(a)** Show that $\begin{bmatrix} a & x \\ 0 & b \end{bmatrix}$ is invertible if and only if $a \neq 0$ and $b \neq 0$.

(b) If A and B are square and invertible, show that **(i)** the block matrix $\begin{bmatrix} A & X \\ 0 & B \end{bmatrix}$ is invertible for any X and

(ii) $\begin{bmatrix} A & X \\ 0 & B \end{bmatrix}^{-1} = \begin{bmatrix} A^{-1} & -A^{-1}XB^{-1} \\ 0 & B^{-1} \end{bmatrix}$.

(c) Use part (b) to invert: (i) $\begin{bmatrix} 1 & 2 & 1 & 1 \\ 0 & -1 & -1 & 0 \\ 0 & 0 & 2 & 1 \\ 0 & 0 & 1 & 1 \end{bmatrix}$ and (ii) $\begin{bmatrix} 3 & 1 & 3 & 0 \\ 2 & 1 & -1 & 1 \\ 0 & 0 & 5 & 2 \\ 0 & 0 & 3 & 1 \end{bmatrix}$.

25. If A and B are invertible symmetric matrices such that $AB = BA$, show that A^{-1}, AB, AB^{-1}, and $A^{-1}B^{-1}$ are also invertible and symmetric.

26. **(a)** Let $A = \begin{bmatrix} 1 & 1 \\ 0 & 1 \end{bmatrix}$, $B = \begin{bmatrix} 0 & 0 \\ 1 & 2 \end{bmatrix}$, and $C = \begin{bmatrix} 1 & 1 \\ 1 & 1 \end{bmatrix}$. Verify that $AB = CA$ and that A is invertible but $B \neq C$. (Compare with Example 8.)

(b) Find 2×2 matrices P, Q, and R such that $PQ = PR$, P is not invertible, and $Q \neq R$. (Compare with Example 8.)

27. Let A be an $n \times n$ matrix and let I be the $n \times n$ identity matrix.

(a) If $A^2 = 0$, verify that $(I - A)(I + A) = I$, so $(I - A)^{-1} = I + A$.

(b) If $A^3 = 0$, verify that $(I - A)(I + A + A^2) = I$, so $(I - A)^{-1} = I + A + A^2$.

(c) Using part **(b)**, find the inverse of $\begin{bmatrix} 1 & 2 & -1 \\ 0 & 1 & 3 \\ 0 & 0 & 1 \end{bmatrix}$.

(d) Generalize to the case $A^n = 0$ and find the formula for $(I - A)^{-1}$ in this case.

28. Prove property 6 of Theorem 2: If A is invertible and $a \neq 0$, then aA is invertible and $(aA)^{-1} = \frac{1}{a}A^{-1}$.

29. Let A, B, and C denote $n \times n$ matrices. Show that:

(a) If A and AB are both invertible, B is invertible.

(b) If AB and BA are both invertible, A and B are both invertible. [*Hint:* See Exercise 10.]

(c) If A, C, and ABC are all invertible, B is invertible.

30. Let A and B denote invertible $n \times n$ matrices.

(a) If $A^{-1} = B^{-1}$, does it mean that $A = B$? Explain.

(b) Show that $A = B$ if and only if $A^{-1}B = I$.

31. Let A, B, and C be $n \times n$ matrices, with A and B invertible. Show that:

(a) If A commutes with C, then A^{-1} commutes with C.

(b) If A commutes with B, then A^{-1} commutes with B^{-1}.

32. Let A and B be square matrices of the same size.

(a) Show that $(AB)^2 = A^2B^2$ if $AB = BA$.

(b) If A and B are invertible and $(AB)^2 = A^2B^2$, show that $AB = BA$.

(c) If $A = \begin{bmatrix} 1 & 0 \\ 0 & 0 \end{bmatrix}$ and $B = \begin{bmatrix} 1 & 1 \\ 0 & 0 \end{bmatrix}$, show that $(AB)^2 = A^2B^2$ but $AB \neq BA$.

33. An $n \times n$ matrix P is called an *idempotent* if $P^2 = P$. Show that:

(a) I is the only invertible idempotent.

(b) P is an idempotent if and only if $I - 2P$ is self-inverse.

(c) U is self-inverse if and only if $U = I - 2P$ for some idempotent P.

(d) $I - aP$ is invertible for any $a \neq 1$, and $(I - aP)^{-1} = I + \left(\frac{a}{1-a}\right)P$.

34. If $A^2 = kA$, where $k \neq 0$, show that A is invertible if and only if $A = kI$.

35. Let A and B denote $n \times n$ invertible matrices.

(a) Show that $A^{-1} + B^{-1} = A^{-1}(A + B)B^{-1}$.

(b) If $A + B$ is also invertible, show that $A^{-1} + B^{-1}$ is invertible and find a formula for $(A^{-1} + B^{-1})^{-1}$.

36. Let A and B be $n \times n$ matrices, and let I be the $n \times n$ identity matrix.

(a) Verify that $A(I + BA) = (I + AB)A$ and that $(I + BA)B = B(I + AB)$.

(b) If $I + AB$ is invertible, verify that $I + BA$ is also invertible and that $(I + BA)^{-1} = I - B(I + AB)^{-1}A$.

2.3.2 Elementary Matrices

It is now evident that elementary row operations play a fundamental role in linear algebra. They provide a general method for solving systems of linear equations, and this leads to the matrix inversion algorithm. It turns out that these elementary row operations can be effected by left-multiplication by certain invertible matrices (called elementary matrices). Section 2.3.2 is devoted to a discussion of this useful fact and some of its consequences.

Recall that the elementary row operations are of three types:

Type I: Interchange two rows.

Type II: Multiply a row by a nonzero number.

Type III: Add a multiple of a row to a different row.

DEFINITION

An $n \times n$ matrix is called an **elementary matrix** if it is obtained from the $n \times n$ identity matrix by an elementary row operation.

The elementary matrix so constructed is said to be of type I, II, or III when the corresponding row operation is of type I, II, or III.

EXAMPLE 12

Verify that $E_1 = \begin{bmatrix} 0 & 1 & 0 \\ 1 & 0 & 0 \\ 0 & 0 & 1 \end{bmatrix}$, $E_2 = \begin{bmatrix} 1 & 0 & 0 \\ 0 & 1 & 0 \\ 0 & 0 & 9 \end{bmatrix}$, and $E_3 = \begin{bmatrix} 1 & 0 & 5 \\ 0 & 1 & 0 \\ 0 & 0 & 1 \end{bmatrix}$ are elementary matrices.

Solution E_1 is obtained from the 3×3 identity I_3 by interchanging the first two rows, so it is an elementary matrix of type I. Similarly, E_2 comes from multiplying the third row of I_3 by 9 and so is an elementary matrix of type II. Finally, E_3 is an elementary matrix of type III; it is obtained by adding 5 times the third row of I_3 to the first row.

Now consider the following three 2×2 elementary matrices E_1, E_2, and E_3 obtained by doing the indicated elementary row operations to I_2.

$$E_1 = \begin{bmatrix} 0 & 1 \\ 1 & 0 \end{bmatrix} \quad \text{Interchange rows 1 and 2 of } I_2$$

$$E_2 = \begin{bmatrix} 1 & 0 \\ 0 & k \end{bmatrix} \quad \text{Multiply row 2 of } I_2 \text{ by } k \neq 0.$$

$$E_3 = \begin{bmatrix} 1 & k \\ 0 & 1 \end{bmatrix} \quad \text{Add } k \text{ times row 2 of } I_2 \text{ to row 1.}$$

If $A = \begin{bmatrix} a & b & c \\ p & q & r \end{bmatrix}$ is an arbitrary 2×3 matrix, compute E_1A, E_2A, and E_3A.

$$E_1A = \begin{bmatrix} 0 & 1 \\ 1 & 0 \end{bmatrix} \begin{bmatrix} a & b & c \\ p & q & r \end{bmatrix} = \begin{bmatrix} p & q & r \\ a & b & c \end{bmatrix}$$

$$E_2A = \begin{bmatrix} 1 & 0 \\ 0 & k \end{bmatrix} \begin{bmatrix} a & b & c \\ p & q & r \end{bmatrix} = \begin{bmatrix} a & b & c \\ kp & kq & kr \end{bmatrix}$$

$$E_3A = \begin{bmatrix} 1 & k \\ 0 & 1 \end{bmatrix} \begin{bmatrix} a & b & c \\ p & q & r \end{bmatrix} = \begin{bmatrix} a + kp & b + kq & c + kr \\ p & q & r \end{bmatrix}$$

Observe that E_1A is the matrix resulting from interchanging rows 1 and 2 of A and that this row operation is the one that was used to produce E_1 from I_2. Similarly, E_2A is obtained from A by the same row operation that produced E_2 from I_2 (multiplying row 2 by k). Finally, the same is true of E_3A: It is obtained from A by the same operation that produced E_3 from I_2 (adding k times row 2 to row 1). This phenomenon holds for arbitrary $m \times n$ matrices A.

THEOREM 5 Let A denote any $m \times n$ matrix, and let E be the $m \times m$ elementary matrix obtained by performing some elementary row operation on the $m \times m$ identity matrix I. If the same elementary row operation is performed on A, the resulting matrix is EA.

Proof We prove it only if E is of type III. The other verifications are similar and are left as Exercise 16. Assume that E is obtained by adding k times row p of I to row q. We must show that EA is obtained from A in the same way. If $R_1, R_2, \ldots, R_m$ and $K_1, K_2, \ldots, K_m$ denote the rows of E and I, respectively, then

$$R_q = K_q + kK_p$$

$$R_i = K_i \qquad \text{if } i \neq q$$

Now recall that row i of EA equals $R_i A$ and that row i of $IA = A$ is $K_i A$. Hence

$$\text{row } i \text{ of } EA = R_i A = K_i A = \text{row } i \text{ of A} \qquad \text{if } i \neq q$$

whereas

$$\text{row } q \text{ of } EA = R_q A = (K_q + kK_p)A = K_q A + kK_p A$$

This is row q of A plus k times row p of A, which is what we wanted. ■

EXAMPLE 13

Given $A = \begin{bmatrix} 4 & 1 & 2 & 1 \\ 3 & 0 & 1 & 6 \\ 5 & 7 & 9 & 8 \end{bmatrix}$, find an elementary matrix E such that EA is the result of subtracting 7 times row 1 from row 3.

Solution The elementary matrix is $E = \begin{bmatrix} 1 & 0 & 0 \\ 0 & 1 & 0 \\ -7 & 0 & 1 \end{bmatrix}$, obtained by doing the given row operation to I_3. The product

$$EA = \begin{bmatrix} 1 & 0 & 0 \\ 0 & 1 & 0 \\ -7 & 0 & 1 \end{bmatrix} \begin{bmatrix} 4 & 1 & 2 & 1 \\ 3 & 0 & 1 & 6 \\ 5 & 7 & 9 & 8 \end{bmatrix} = \begin{bmatrix} 4 & 1 & 2 & 1 \\ 3 & 0 & 1 & 6 \\ -23 & 0 & -5 & 1 \end{bmatrix}$$

is indeed the result of applying the operation to A.

The first application of Theorem 5 is to show that every elementary matrix is invertible. In fact, more is true: The inverse of an elementary matrix is again elementary. The reason is that each elementary row operation is "reversible" by another such operation of the same type.

THEOREM 6 Every elementary matrix E is invertible, and the inverse is an elementary matrix of the same type. More precisely:

Type I — If E is obtained by interchanging two rows of I, then $E^{-1} = E$.

Type II — If E is obtained by multiplying row p of I by $k \neq 0$, then E^{-1} is obtained by multiplying row p of I by $\frac{1}{k}$.

Type III — If E is obtained by adding k times row p of I to row q, then E^{-1} is obtained by subtracting k times row p of I from row q.

Proof The verifications are straightforward matrix calculations and are left to the reader. However, a proof using Theorem 5 is instructive. Let $E_{pq}(k)$ denote the elementary matrix obtained by adding k times row p of I to row q. The effect of adding k times row p of a matrix A to row q, and then adding $-k$ times row p to row q, is to leave A unchanged. Using Theorem 5 (twice), this means that $E_{pq}(-k)E_{pq}(k)A = A$. In particular, if $A = I$, this gives $E_{pq}(-k)E_{pq}(k) = I$. The verification that $E_{pq}(k)E_{pq}(-k) = I$ is similar, so $E_{pq}(k)^{-1} = E_{pq}(-k)$. This is the result for type III; types I and II are dealt with in the same way. ■

EXAMPLE 14 Write down the inverses of the elementary matrices E_1, E_2, and E_3 in Example 12.

Solution The matrices are $E_1 = \begin{bmatrix} 0 & 1 & 0 \\ 1 & 0 & 0 \\ 0 & 0 & 1 \end{bmatrix}$, $E_2 = \begin{bmatrix} 1 & 0 & 0 \\ 0 & 1 & 0 \\ 0 & 0 & 9 \end{bmatrix}$, and $E_3 = \begin{bmatrix} 1 & 0 & 5 \\ 0 & 1 & 0 \\ 0 & 0 & 1 \end{bmatrix}$, so they are of types I, II, and III, respectively. Hence Theorem 6 gives

$$E_1^{-1} = E_1 = \begin{bmatrix} 0 & 1 & 0 \\ 1 & 0 & 0 \\ 0 & 0 & 1 \end{bmatrix} \qquad E_2^{-1} = \begin{bmatrix} 1 & 0 & 0 \\ 0 & 1 & 0 \\ 0 & 0 & \frac{1}{9} \end{bmatrix} \qquad E_3^{-1} = \begin{bmatrix} 1 & 0 & -5 \\ 0 & 1 & 0 \\ 0 & 0 & 1 \end{bmatrix}$$

Now suppose A is any $m \times n$ matrix, and assume that a series of elementary row operations are performed on A. Let $E_1, E_2, \ldots, E_{k-1}, E_k$ denote the $m \times m$ elementary matrices associated with these row operations. Then Theorem 5 asserts that A is carried to E_1A under the first operation; in symbols, $A \to E_1A$. Now the second row operation is applied to E_1A (not to A) and the result is $E_2(E_1A)$, again by Theorem 5. Hence the reduction can be described as follows:

$$A \to E_1A \to E_2E_1A \to E_3E_2E_1A \to \cdots \to E_kE_{k-1} \cdots E_2E_1A$$

In other words, the net effect of the *sequence* of elementary row operations is to left-multiply by the *product* $U = E_kE_{k-1} \ldots E_2E_1$ of the corresponding elementary matrices (note the order). The result is

$$A \to UA \qquad \text{where} \qquad U = E_kE_{k-1} \cdots E_2E_1$$

Moreover, the matrix U can be easily constructed. Apply the same sequence of elementary operations to the $n \times n$ identity matrix I in place of A. The result is

$$I \to UI = U$$

In other words, the sequence of elementary row operations that carries $A \to UA$ also carries $I \to U$. Hence it carries the "double" matrix

$$[A \quad I] \to [UA \quad U]$$

just as in the matrix inversion algorithm. This simple observation is surprisingly useful, and we record it as Theorem 7.

THEOREM 7

Let A be an $m \times n$ matrix, and assume that A can be carried to a matrix B by elementary row operations.

1. $B = UA$ where U is an invertible $m \times m$ matrix.
2. $U = E_k E_{k-1} \dots E_2 E_1$ where $E_1, E_2, \dots, E_{k-1}, E_k$ are the elementary matrices corresponding (in order) to the elementary row operations that carry $A \to B$.
3. U can be constructed without finding the E_i by

$$[A \quad I] \to [UA \quad U]$$

In other words, the operations that carry $A \to UA$ also carry $I \to U$.

Proof All that remains is to verify that U is invertible. But this is immediate by Theorems 6 and 2. ■

EXAMPLE 15

Find the reduced row-echelon form R of $A = \begin{bmatrix} 2 & 3 & 1 \\ 1 & 2 & 1 \end{bmatrix}$ and express it as $R = UA$, where U is invertible.

Solution Use the usual row-reduction $A \to R$, but carry out $I \to U$ simultaneously in the format $[A \quad I] \to [\text{R} \quad \text{U}]$.

$$\left[\begin{array}{ccc|cc} 2 & 3 & 1 & 1 & 0 \\ 1 & 2 & 1 & 0 & 1 \end{array}\right] \to \left[\begin{array}{ccc|cc} 1 & 2 & 1 & 0 & 1 \\ 2 & 3 & 1 & 1 & 0 \end{array}\right]$$

$$\to \left[\begin{array}{ccc|cc} 1 & 2 & 1 & 0 & 1 \\ 0 & -1 & -1 & 1 & -2 \end{array}\right]$$

$$\to \left[\begin{array}{ccc|cc} 1 & 0 & -1 & 2 & -3 \\ 0 & 1 & 1 & -1 & 2 \end{array}\right]$$

Hence $R = \begin{bmatrix} 1 & 0 & -1 \\ 0 & 1 & 1 \end{bmatrix}$ and $U = \begin{bmatrix} 2 & -3 \\ -1 & 2 \end{bmatrix}$.

The next example shows how the factorization of U into elementary matrices (as in property 2 of Theorem 7) can be obtained.

EXAMPLE 16

Bring $A = \begin{bmatrix} 1 & 2 & -1 \\ 1 & 1 & 5 \end{bmatrix}$ to a reduced row-echelon matrix R by elementary row operations, and find elementary matrices E_1, E_2, and E_3 such that $R = E_3E_2E_1A$.

Solution The reduction is as follows:

$$A = \begin{bmatrix} 1 & 2 & -1 \\ 1 & 1 & 5 \end{bmatrix} \to \begin{bmatrix} 1 & 2 & -1 \\ 0 & -1 & 6 \end{bmatrix} \to \begin{bmatrix} 1 & 2 & -1 \\ 0 & 1 & -6 \end{bmatrix} \to \begin{bmatrix} 1 & 0 & 11 \\ 0 & 1 & -6 \end{bmatrix} = R$$

Each row operation can be effected by left-multiplication by an elementary matrix obtained by performing that row operation on the 2×2 identity matrix. The three reductions and the corresponding elementary matrices are

$$\begin{bmatrix} 1 & 2 & -1 \\ 1 & 1 & 5 \end{bmatrix} = A$$

$$\downarrow$$

$$\begin{bmatrix} 1 & 2 & -1 \\ 0 & -1 & 6 \end{bmatrix} = \begin{bmatrix} 1 & 0 \\ -1 & 1 \end{bmatrix}\begin{bmatrix} 1 & 2 & -1 \\ 1 & 1 & 5 \end{bmatrix} = E_1A$$

$$\downarrow$$

$$\text{where } E_1 = \begin{bmatrix} 1 & 0 \\ -1 & 1 \end{bmatrix}$$

$$\begin{bmatrix} 1 & 2 & -1 \\ 0 & 1 & -6 \end{bmatrix} = \begin{bmatrix} 1 & 0 \\ 0 & -1 \end{bmatrix}\begin{bmatrix} 1 & 2 & -1 \\ 0 & -1 & 6 \end{bmatrix} = E_2(E_1A)$$

$$\downarrow$$

$$\text{where } E_2 = \begin{bmatrix} 1 & 0 \\ 0 & -1 \end{bmatrix}$$

$$\begin{bmatrix} 1 & 0 & 11 \\ 0 & 1 & -6 \end{bmatrix} = \begin{bmatrix} 1 & -2 \\ 0 & 1 \end{bmatrix}\begin{bmatrix} 1 & 2 & -1 \\ 0 & 1 & -6 \end{bmatrix} = E_3(E_2E_1A)$$

$$\text{where } E_3 = \begin{bmatrix} 1 & -2 \\ 0 & 1 \end{bmatrix}$$

This gives $R = E_3E_2E_1A$, as required.

These techniques are very useful when applied to square matrices. In particular, Theorem 7 provides a way to deduce several conditions for invertibility. Recall that the rank of a matrix A is the number of nonzero rows in any row-echelon matrix to which it can be carried (Section 1.2.2).[†]

THEOREM 8 The following conditions are equivalent for an $n \times n$ matrix A:

1. A is invertible.
2. If $YA = 0$ where Y is $1 \times n$, then necessarily $Y = 0$.
3. A has rank n.
4. A can be carried to the $n \times n$ identity matrix by elementary row operations.
5. A is a product of elementary matrices.

Proof We show that, if any of the statements is true, then the next one is necessarily true, and also that the truth of property 5 implies the truth of property 1. Hence, if any statement is true, they all are true.

(1) implies (2). Assume A is invertible so that A^{-1} exists. If $YA = 0$, right-multiplication by A^{-1} gives $Y = YAA^{-1} = 0A^{-1} = 0$.

(2) implies (3). Assume property 2 holds. Now A can be carried to a matrix R in reduced row-echelon form that, by Theorem 7, can be written as $R = UA$ for some invertible matrix U. We must show that R has n nonzero rows. If not, the last row of R consists of zeros (R is $n \times n$), so $YR = 0$ where $Y = [0, 0, \ldots, 0, 1]$. But then $YUA = 0$, so $YU = 0$ by property 2. Because U is invertible, this implies $Y = 0$, a contradiction. Hence R has n nonzero rows, and property 3 follows.

(3) implies (4). A can be carried to a matrix R in reduced row-echelon form, and R has n nonzero rows by property 3. Hence $R = I$, so property 4 follows.

(4) implies (5). Given property 4, Theorem 7 implies that $I = UA$, where U is an invertible matrix that can be factored as a product $U = E_kE_{k-1} \ldots E_2E_1$ of elementary matrices. Hence

$$A = U^{-1} = (E_kE_{k-1} \cdots E_2E_1)^{-1} = E_1{}^{-1}E_2{}^{-1} \cdots E_{k-1}{}^{-1}E_k{}^{-1}$$

and property 5 follows from Theorem 6.

(5) implies (1). Clear by Theorem 2, because elementary matrices are invertible. ■

We have seen that the matrix products AB and BA need *not* be equal, even if A and B are both square matrices. However, if $AB = I$, then necessarily $BA = I$ as well. The proof requires Theorem 8.

[†]The proof that the number of rows is the same in each row-echelon matrix to which A can be carried will be given in Section 5.4. This proof makes no use of Theorem 8.

THEOREM 9 Let A and B be $n \times n$ matrices. If $AB = I$, then also $BA = I$, so A and B are invertible, $A = B^{-1}$, and $B = A^{-1}$.

Proof Assume $AB = I$. We use property 2 of Theorem 8. If $YA = 0$, where Y is $1 \times n$, then right-multiplication by B gives

$$0 = (YA)B = Y(AB) = YI = Y$$

Hence $YA = 0$ implies $Y = 0$, so A is invertible by Theorem 8. But then left-multiply $AB = I$ by A^{-1} to get

$$B = IB = A^{-1}AB = A^{-1}I = A^{-1}$$

This shows that B is invertible (A^{-1} is invertible) and that $B^{-1} = (A^{-1})^{-1} = A$. ■

EXAMPLE 17 Express $A = \begin{bmatrix} -2 & 3 \\ 1 & 0 \end{bmatrix}$ as a product of elementary matrices.

Solution We reduce A to I and write down the elementary matrix at each stage.

$$\begin{bmatrix} -2 & 3 \\ 1 & 0 \end{bmatrix} = A$$

$$\downarrow$$

$$\begin{bmatrix} 1 & 0 \\ -2 & 3 \end{bmatrix} = E_1A \qquad \text{where } E_1 = \begin{bmatrix} 0 & 1 \\ 1 & 0 \end{bmatrix}$$

$$\downarrow$$

$$\begin{bmatrix} 1 & 0 \\ 0 & 3 \end{bmatrix} = E_2(E_1A) \qquad \text{where } E_2 = \begin{bmatrix} 1 & 0 \\ 2 & 1 \end{bmatrix}$$

$$\downarrow$$

$$\begin{bmatrix} 1 & 0 \\ 0 & 1 \end{bmatrix} = E_3(E_2E_1A) \qquad \text{where } E_3 = \begin{bmatrix} 1 & 0 \\ 0 & 1/3 \end{bmatrix}$$

Hence $E_3E_2E_1A = I$ and so, by Theorem 9, $A = (E_3E_2E_1)^{-1}$. This means that $A = E_1^{-1}E_2^{-1}E_3^{-1} = \begin{bmatrix} 0 & 1 \\ 1 & 0 \end{bmatrix} \begin{bmatrix} 1 & 0 \\ -2 & 1 \end{bmatrix} \begin{bmatrix} 1 & 0 \\ 0 & 3 \end{bmatrix}$ by Theorem 6. This is the desired factorization.

Finally, we give two conditions for invertibility of a square matrix A in terms of systems of equations with A as coefficient matrix. The first condition is that a homogeneous system with A as coefficient matrix has only the trivial solution. The second condition asserts that every such nonhomogeneous system *has* a solution (actually unique by Theorem 3). The proof is left as Exercise 19.

THEOREM 10 Let A be an $n \times n$ matrix. The following conditions are equivalent:

1. A is invertible.
2. The homogeneous system $A\mathbf{x} = \mathbf{0}$ has only the trivial solution $\mathbf{x} = \mathbf{0}$.
3. The system $A\mathbf{x} = \mathbf{b}$ has a solution for every $n \times 1$ matrix $\mathbf{b}$.

EXERCISES 2.3.2

1. For each of the following elementary matrices, describe the corresponding elementary row operation and write down the inverse.

(a) $E = \begin{bmatrix} 1 & 0 & 3 \\ 0 & 1 & 0 \\ 0 & 0 & 1 \end{bmatrix}$ **(b)** $E = \begin{bmatrix} 0 & 0 & 1 \\ 0 & 1 & 0 \\ 1 & 0 & 0 \end{bmatrix}$ **(c)** $E = \begin{bmatrix} 1 & 0 & 0 \\ 0 & 1/2 & 0 \\ 0 & 0 & 1 \end{bmatrix}$

(d) $E = \begin{bmatrix} 1 & 0 & 0 \\ -2 & 1 & 0 \\ 0 & 0 & 1 \end{bmatrix}$ **(e)** $E = \begin{bmatrix} 0 & 1 & 0 \\ 1 & 0 & 0 \\ 0 & 0 & 1 \end{bmatrix}$ **(f)** $E = \begin{bmatrix} 1 & 0 & 0 \\ 0 & 1 & 0 \\ 0 & 0 & 5 \end{bmatrix}$

2. In each case find an elementary matrix E such that $B = EA$.

(a) $A = \begin{bmatrix} 2 & 1 \\ 3 & -1 \end{bmatrix}, B = \begin{bmatrix} 2 & 1 \\ 1 & -2 \end{bmatrix}$ **(b)** $A = \begin{bmatrix} -1 & 2 \\ 0 & 1 \end{bmatrix}, B = \begin{bmatrix} 1 & -2 \\ 0 & 1 \end{bmatrix}$

(c) $A = \begin{bmatrix} 1 & 1 \\ -1 & 2 \end{bmatrix}, B = \begin{bmatrix} -1 & 2 \\ 1 & 1 \end{bmatrix}$ **(d)** $A = \begin{bmatrix} 4 & 1 \\ 3 & 2 \end{bmatrix}, B = \begin{bmatrix} 1 & -1 \\ 3 & 2 \end{bmatrix}$

(e) $A = \begin{bmatrix} -1 & 1 \\ 1 & -1 \end{bmatrix}, B = \begin{bmatrix} -1 & 1 \\ -1 & 1 \end{bmatrix}$ **(f)** $A = \begin{bmatrix} 2 & 1 \\ -1 & 3 \end{bmatrix}, B = \begin{bmatrix} -1 & 3 \\ 2 & 1 \end{bmatrix}$

3. Let $A = \begin{bmatrix} 1 & 2 \\ -1 & 1 \end{bmatrix}$ and $C = \begin{bmatrix} -1 & 1 \\ 2 & 1 \end{bmatrix}$.

(a) Find elementary matrices E_1 and E_2 such that $C = E_2E_1A$.

(b) Show that there is *no* elementary matrix E such that $C = EA$.

4. If E is elementary, show that A and EA differ in at most two rows.

5. **(a)** Is I an elementary matrix? Explain.

(b) Is 0 an elementary matrix? Explain.

6. In each case find an invertible matrix U such that $UA = R$ is in reduced row-echelon form, and express U as a product of elementary matrices.

(a) $A = \begin{bmatrix} 1 & -1 & 2 \\ -2 & 1 & 0 \end{bmatrix}$ **(b)** $A = \begin{bmatrix} 1 & 2 & 1 \\ 5 & 12 & -1 \end{bmatrix}$

(c) $A = \begin{bmatrix} 1 & 2 & -1 & 0 \\ 3 & 1 & 1 & 2 \\ 1 & -3 & 3 & 2 \end{bmatrix}$ **(d)** $A = \begin{bmatrix} 2 & 1 & -1 & 0 \\ 3 & -1 & 2 & 1 \\ 1 & -2 & 3 & 1 \end{bmatrix}$

7. In each case find an invertible matrix U such that $UA = B$, and express U as a product of elementary matrices.

(a) $A = \begin{bmatrix} 2 & 1 & 3 \\ -1 & 1 & 2 \end{bmatrix}, B = \begin{bmatrix} 1 & -1 & -2 \\ 3 & 0 & 1 \end{bmatrix}$

(b) $A = \begin{bmatrix} 2 & -1 & 0 \\ 1 & 1 & 1 \end{bmatrix}$, $B = \begin{bmatrix} 3 & 0 & 1 \\ 2 & -1 & 0 \end{bmatrix}$

8. In each case factor A as a product of elementary matrices.

(a) $A = \begin{bmatrix} 1 & 1 \\ 2 & 1 \end{bmatrix}$ (b) $A = \begin{bmatrix} 2 & 3 \\ 1 & 2 \end{bmatrix}$

(c) $A = \begin{bmatrix} 1 & 0 & 2 \\ 0 & 1 & 1 \\ 2 & 1 & 6 \end{bmatrix}$ (d) $A = \begin{bmatrix} 1 & 0 & -3 \\ 0 & 1 & 4 \\ -2 & 2 & 15 \end{bmatrix}$

9. Let E be an elementary matrix. Show that E^T is also elementary of the same type.

10. Show that every matrix A can be factored as $A = UR$ where U is invertible and R is in reduced row-echelon form.

11. Let A and B be $m \times n$ and $n \times m$ matrices, respectively.

(a) If $m > n$, show that AB is not invertible. [*Hint:* Use Theorem 1 of Section 1.3 to find $Y \neq 0$ with $YA = 0$.]

(b) Show that the only invertible matrices are square; that is, $AB = I_m$ and $BA = I_n$ imply $m = n$. (But see Exercises 12 and 13.)

12. Let $A = \frac{1}{7}\begin{bmatrix} 2 & 6 & -3 \\ 3 & 2 & 6 \end{bmatrix}$. Show that $AA^T = I_2$ but $A^TA \neq I_3$.

13. If $A = \begin{bmatrix} 1 & 3 & 2 \\ 1 & 2 & 2 \end{bmatrix}$ and $B = \begin{bmatrix} 0 & 3 \\ 1 & -1 \\ -1 & 0 \end{bmatrix}$, verify that $AB = I_2$ but $BA \neq I_3$.

14. Show that the following are equivalent for $n \times n$ matrices A and B:

(a) A and B are both invertible.

(b) AB is invertible.

15. Consider $A = \begin{bmatrix} 1 & 3 & -1 \\ 2 & 1 & 5 \\ 1 & -7 & 13 \end{bmatrix}$, $B = \begin{bmatrix} 1 & 1 & 2 \\ 3 & 0 & -3 \\ -2 & 5 & 17 \end{bmatrix}$.

(a) Show that A is not invertible by finding a nonzero 1×3 matrix Y such that $YA = 0$. [*Hint*: Row 3 of A equals 2(row 2) − 3(row 1).]

(b) Show that B is not invertible by showing column 3 = 3(column 2) − column 1.

16. Prove Theorem 5 for elementary matrices of: **(a)** type I; **(b)** type II.

17. Define an *elementary column operation* on a matrix to be one of the following: (I) Interchange two columns. (II) Multiply a column by a nonzero scalar. (III) Add a multiple of a column to another column. Show that:

(a) If an elementary column operation is done to an $m \times n$ matrix A, the result is AF, where F is an $n \times n$ elementary matrix.

(b) Given any $m \times n$ matrix A, there exist $m \times m$ elementary matrices $E_1, \ldots, E_k$ and $n \times n$ elementary matrices $F_1, \ldots, F_p$ such that, in block form,

$$E_k \cdots E_1 A F_1 \cdots F_p = \begin{bmatrix} I_r & 0 \\ 0 & 0 \end{bmatrix}$$

18. Suppose B is obtained from A by: **(a)** interchanging rows i and j; **(b)** multiplying row i by $k \neq 0$; **(c)** adding k times row i to row $j (i \neq j)$. In each case describe how to obtain B^{-1} from A^{-1}. [*Hint:* See part **(a)** of the preceding exercise.]

19. Prove Theorem 10. [*Hints:* To show that 2 implies 1, use property 2 of Theorem 8 on A^T. To show that 3 implies 1, solve $AB = I$ by writing $B = [X_1\ X_2\ \ldots\ X_n]$ in terms of its columns and showing that each column X_j can be found.]

20. Two matrices A and B of the same size are called **row-equivalent** (written $A \overset{r}{\sim} B$) if there is a sequence of elementary row operations carrying A to B.

(a) Show that $A \overset{r}{\sim} B$ if and only if $A = UB$ for some invertible matrix U.

(b) Show that: **(i)** $A \overset{r}{\sim} A$ for all matrices A.
(ii) If $A \overset{r}{\sim} B$, then $B \overset{r}{\sim} A$.
(iii) If $A \overset{r}{\sim} B$ and $B \overset{r}{\sim} C$, then $A \overset{r}{\sim} C$.

(c) Show that, if A and B are both row-equivalent to some third matrix, then $A \overset{r}{\sim} B$.

(d) Show that $\begin{bmatrix} 1 & -1 & 3 & 2 \\ 0 & 1 & 4 & 1 \\ 1 & 0 & 8 & 6 \end{bmatrix}$ and $\begin{bmatrix} 1 & -1 & 4 & 5 \\ -2 & 1 & -11 & -8 \\ -1 & 2 & 2 & 2 \end{bmatrix}$ are row-equivalent. [*Hint:* Consider part **(c)** and Theorem 2 of Section 1.2.2.]

(e) Find all matrices that are row-equivalent to:

(i) $\begin{bmatrix} 0 & 0 & 0 \\ 0 & 0 & 0 \end{bmatrix}$ **(ii)** $\begin{bmatrix} 0 & 0 & 0 \\ 0 & 0 & 1 \end{bmatrix}$ **(iii)** $\begin{bmatrix} 1 & 0 & 0 \\ 0 & 1 & 0 \end{bmatrix}$

(f) Show that an $n \times n$ matrix is invertible if and only if it is row-equivalent to I_n.

(g) Which of the following pairs of matrices are row-equivalent?

(i) $\begin{bmatrix} 1 & -1 \\ -1 & 1 \end{bmatrix}, \begin{bmatrix} 1 & 2 \\ 2 & 4 \end{bmatrix}$ **(ii)** $\begin{bmatrix} 1 & 2 \\ -1 & 3 \end{bmatrix}, \begin{bmatrix} 3 & 1 \\ 2 & 1 \end{bmatrix}$

(iii) $\begin{bmatrix} 1 & 3 & -1 & 4 \\ 2 & 1 & 0 & 2 \\ 1 & 1 & -1 & 3 \end{bmatrix}, \begin{bmatrix} 0 & -5 & 2 & -6 \\ 0 & -2 & 0 & -1 \\ 1 & 0 & 1 & -1 \end{bmatrix}$

(iv) $\begin{bmatrix} 1 & -1 & 1 \\ 2 & 1 & 0 \\ 1 & 2 & -1 \end{bmatrix}, \begin{bmatrix} 1 & 2 & -1 \\ 2 & -1 & 1 \\ -1 & 3 & -2 \end{bmatrix}$

SECTION 2.4 LU-Factorization (Optional)†

In this section the Gaussian algorithm will be used to show that any matrix A can be written as a product of matrices of a particularly nice type. This is useful in numerical calculations.

†This section is not used later, so it may be omitted with no loss of continuity.

The matrices that arise here are defined as follows: An $m \times n$ matrix $A = [a_{ij}]$ is called **upper triangular** if $a_{ij} = 0$ whenever $i > j$. If, as for square matrices, the elements $a_{11}, a_{22}, \ldots$ are called the **main diagonal** of A, then A is upper triangular if and only if each entry of A below and to the left of the main diagonal is zero. Hence the matrices

$$\begin{bmatrix} 1 & -1 & 0 & 3 \\ 0 & 2 & 1 & 1 \\ 0 & 0 & -3 & 0 \end{bmatrix} \quad \begin{bmatrix} 0 & 2 & 1 & 0 & 5 \\ 0 & 0 & 0 & 3 & 1 \\ 0 & 0 & 1 & 0 & 1 \end{bmatrix} \quad \begin{bmatrix} 1 & 1 & 1 \\ 0 & -1 & 1 \\ 0 & 0 & 0 \\ 0 & 0 & 0 \end{bmatrix}$$

are all upper triangular. Clearly, each row-echelon matrix is upper triangular.

By analogy, a matrix is called **lower triangular** if its transpose is upper triangular—that is, each entry above and to the right of the main diagonal is zero. A matrix is called **triangular** if it is either upper or lower triangular.

One reason for the importance of triangular matrices is the ease with which systems of linear equations can be solved when the coefficient matrix is triangular.

EXAMPLE 1 Solve the system

$$\begin{bmatrix} 1 & 2 & -3 & -1 & 5 \\ 0 & 0 & 5 & 1 & 1 \\ 0 & 0 & 0 & 0 & 2 \end{bmatrix} \begin{bmatrix} x_1 \\ x_2 \\ x_3 \\ x_4 \\ x_5 \end{bmatrix} = \begin{bmatrix} 3 \\ 8 \\ 6 \end{bmatrix}$$

equivalently,

$$\begin{aligned} x_1 + 2x_2 - 3x_3 - x_4 + 5x_5 &= 3 \\ 5x_3 + x_4 + x_5 &= 8 \\ 2x_5 &= 6 \end{aligned}$$

Solution As is the case for a row-echelon matrix, let $x_2 = s$ and $x_4 = t$. Then solve for x_5, x_3, and x_1 successively as follows:

$$x_5 = 6/2 = 3$$

Substitution in the second equation gives

$$x_3 = 1 - \tfrac{1}{5}t$$

Finally substitution of both x_5 and x_3 in the first equation gives

$$x_1 = -9 - 2s + \tfrac{2}{5}t$$

The method used in Example 1 is called **back substitution** for obvious reasons. It works because the matrix is upper triangular, and it provides an efficient method for finding the solutions (when they exist). In particular, it can be used in Gaussian elimination because row-echelon matrices are upper triangular. Similarly, if the matrix of a system of equations is lower triangular, the system can be solved (if a solution exists) by **forward substitution.** Here each equation is used to solve for one variable by substituting values already found for earlier variables.

Suppose now that an arbitrary matrix A is given, and consider the system

$$A\mathbf{x} = \mathbf{b}$$

of linear equations with A as coefficient matrix. If A can be factored as $A = LU$, where L is lower triangular and U is upper triangular, the system can be solved in two stages as follows:

1. Solve $L\mathbf{y} = \mathbf{b}$ for $\mathbf{y}$ by forward substitution.
2. Solve $U\mathbf{x} = \mathbf{y}$ for $\mathbf{x}$ by back substitution.

Then $\mathbf{x}$ is a solution to $A\mathbf{x} = \mathbf{b}$, because $A\mathbf{x} = LU\mathbf{x} = L\mathbf{y} = \mathbf{b}$. Moreover, every solution arises in this way (take $\mathbf{y} = U\mathbf{x}$). This focuses attention on obtaining such factorizations $A = LU$ of matrices. The Gaussian algorithm provides a method of obtaining these factorizations in some cases. The method exploits the following facts about triangular matrices.

THEOREM 1

The product of two lower triangular matrices (or two upper triangular matrices) is again lower triangular (upper triangular).

Proof Let A and B be lower triangular (the other case is similar). We use induction on the number n of columns of A. If $n = 1$, it is clear (every column is lower triangular). In general, write A and B in block form as $A = \begin{bmatrix} a & 0 \\ X & A_1 \end{bmatrix}$ and $B = \begin{bmatrix} b & 0 \\ Y & B_1 \end{bmatrix}$. Then A_1 and B_1 are lower triangular, and block multiplication gives $AB = \begin{bmatrix} ab & 0 \\ bX + A_1Y & A_1B_1 \end{bmatrix}$. Because A_1 has fewer columns than A, A_1B_1 is lower triangular by induction, and hence AB is lower triangular. ■

THEOREM 2

Let A be an $n \times n$ lower triangular (or upper triangular) matrix. Then A is invertible if and only if no main diagonal entry is zero. In this case, A^{-1} is also lower (upper) triangular.

Proof We use induction on n, the case $n = 1$ being clear. If $n > 1$ and A is invertible, write $A = \begin{bmatrix} a & 0 \\ X & A_1 \end{bmatrix}$ and $A^{-1} = \begin{bmatrix} b & Z \\ Y & B \end{bmatrix}$ in block form. Because A_1 is lower triangular, it suffices (by induction) to show that $a \neq 0$, $Z = 0$, and A_1 is invertible. But this follows from $\begin{bmatrix} 1 & 0 \\ 0 & I \end{bmatrix} = AA^{-1} = \begin{bmatrix} ab & aZ \\ bX + A_1Y & XZ + A_1B \end{bmatrix}$. Conversely, assume that each diagonal entry of A is nonzero. If $A = \begin{bmatrix} a & 0 \\ X & A_1 \end{bmatrix}$ as before, this means that $a \neq 0$ and (by induction) A_1 is invertible with A_1^{-1} lower triangular. The reader can verify that

$$A^{-1} = \begin{bmatrix} a^{-1} & 0 \\ -a^{-1}A_1^{-1}X & A_1^{-1} \end{bmatrix}$$

so A is invertible and A^{-1} is lower triangular. The upper triangular case is analogous. ■

Now let A be any $m \times n$ matrix. The Gaussian algorithm produces a sequence of row operations that carry A to a row-echelon matrix U. However, because we are not insisting on *reduced* row-echelon form, no multiple of a row need ever be added to a row *above* it. This procedure will be called the **Gaussian lower algorithm**. The point is that, apart from row interchanges,[†] the only row operations needed are such that the corresponding elementary matrix is *lower triangular.* This observation gives the following theorem.

THEOREM 3

Suppose that, via the Gaussian lower algorithm, a matrix A can be carried to a row-echelon matrix U with no row interchanges being used. Then

$$A = LU$$

[†]Any row interchange can be accomplished by row operations of other types (Exercise 5), but these must involve adding a multiple of some row to a row *above* it.

where L is lower triangular and invertible and U is row-echelon (and upper triangular).

Proof The hypotheses imply that there exist lower triangular, elementary matrices $E_1, E_2, \ldots, E_k$ such that $U = (E_k \cdots E_2E_1)A$. Hence $A = LU$, where $L = E_1^{-1}E_2^{-1} \cdots E_k^{-1}$ is lower triangular and invertible by Theorems 1 and 2. ■

A factorization such as that in Theorem 3 is called an **LU-factorization** of the matrix A. Such a factorization may not exist (Exercise 4), because at least one row interchange is required in the Gaussian lower algorithm. A procedure for dealing with this situation will be outlined below. However, if an LU-factorization does exist, the row-echelon matrix U in Theorem 3 is obtained by Gaussian elimination. Moreover, the Gaussian lower algorithm also yields a simple procedure for writing down the matrix L. The following example illustrates the technique.

EXAMPLE 2

Find an LU-factorization of the matrix $A = \begin{bmatrix} 0 & 2 & -6 & -2 & 4 \\ 0 & -1 & 3 & 3 & 2 \\ 0 & -1 & 3 & 7 & 10 \end{bmatrix}$.

Solution We are assuming that we can carry A to a row-echelon matrix U as above, using no row interchanges. The steps in the Gaussian lower algorithm are given below and, at each stage, the corresponding elementary matrix is computed. The reason for the circled entries will be apparent shortly.

$$\begin{bmatrix} 0 & \textcircled{2} & -6 & -2 & 4 \\ 0 & \textcircled{-1} & 3 & 3 & 2 \\ 0 & \textcircled{-1} & 3 & 7 & 10 \end{bmatrix} = A$$

$$\begin{bmatrix} 0 & 1 & -3 & -1 & 2 \\ 0 & -1 & 3 & 3 & 2 \\ 0 & -1 & 3 & 7 & 10 \end{bmatrix} = E_1A \qquad E_1 = \begin{bmatrix} 1/2 & 0 & 0 \\ 0 & 1 & 0 \\ 0 & 0 & 1 \end{bmatrix}$$

$$\begin{bmatrix} 0 & 1 & -3 & -1 & 2 \\ 0 & 0 & 0 & 2 & 4 \\ 0 & -1 & 3 & 7 & 10 \end{bmatrix} = E_2E_1A \qquad E_2 = \begin{bmatrix} 1 & 0 & 0 \\ 1 & 1 & 0 \\ 0 & 0 & 1 \end{bmatrix}$$

$$\begin{bmatrix} 0 & 1 & -3 & -1 & 2 \\ 0 & 0 & 0 & \textcircled{2} & 4 \\ 0 & 0 & 0 & \textcircled{6} & 12 \end{bmatrix} = E_3E_2E_1A \qquad E_3 = \begin{bmatrix} 1 & 0 & 0 \\ 0 & 1 & 0 \\ 1 & 0 & 1 \end{bmatrix}$$

$$\begin{bmatrix} 0 & 1 & -3 & -1 & 2 \\ 0 & 0 & 0 & 1 & 2 \\ 0 & 0 & 0 & 6 & 12 \end{bmatrix} = E_4E_3E_2E_1A \qquad E_4 = \begin{bmatrix} 1 & 0 & 0 \\ 0 & 1/2 & 0 \\ 0 & 0 & 1 \end{bmatrix}$$

$$U = \begin{bmatrix} 0 & 1 & -3 & -1 & 2 \\ 0 & 0 & 0 & 1 & 2 \\ 0 & 0 & 0 & 0 & 0 \end{bmatrix} = E_5E_4E_3E_2E_1A \qquad E_5 = \begin{bmatrix} 1 & 0 & 0 \\ 0 & 1 & 0 \\ 0 & -6 & 1 \end{bmatrix}$$

Thus (as in the proof of Theorem 3), the LU-factorization of A is $A = LU$, where

$$L = E_1^{-1}E_2^{-1}E_3^{-1}E_4^{-1}E_5^{-1} = \begin{bmatrix} 2 & 0 & 0 \\ -1 & 2 & 0 \\ -1 & 6 & 1 \end{bmatrix}$$

Now observe that the first two columns of L can be obtained from the columns circled during the execution of the algorithm.

The procedure in Example 2 works in general. Moreover, we can construct the matrix L as we go along, one column at a time, starting from the left. This is suitable for use in a computer program, because the circled columns can be stored in memory as they are created. To describe the process in general, the following notation is useful. Given positive integers $m \geq r$, let $\mathbf{c}_1, \mathbf{c}_2, \ldots, \mathbf{c}_r$ be columns of decreasing lengths $m, m-1, \ldots, m-r+1$. Then let

$$L_m[\mathbf{c}_1, \ldots, \mathbf{c}_r]$$

denote the $m \times m$ lower triangular matrix obtained from the identity matrix by replacing the bottom j entries of column j by $\mathbf{c}_j$ for each $j = 1, 2, \ldots, r$. Thus the matrix L in Example 2 has this form:

$$L = L_3[\mathbf{c}_1, \mathbf{c}_2] = \begin{bmatrix} 2 & 0 & 0 \\ -1 & 2 & 0 \\ -1 & 6 & 1 \end{bmatrix} \text{ where } \mathbf{c}_1 = \begin{bmatrix} 2 \\ -1 \\ -1 \end{bmatrix} \text{ and } \mathbf{c}_2 = \begin{bmatrix} 2 \\ 6 \end{bmatrix}.$$

Here is another example.

EXAMPLE 3

If $\mathbf{c}_1 = \begin{bmatrix} 3 \\ -1 \\ 0 \\ 2 \end{bmatrix}$ and $\mathbf{c}_2 = \begin{bmatrix} 5 \\ -1 \\ 7 \end{bmatrix}$, then $L_4[\mathbf{c}_1, \mathbf{c}_2] = \begin{bmatrix} 3 & 0 & 0 & 0 \\ -1 & 5 & 0 & 0 \\ 0 & -1 & 1 & 0 \\ 2 & 7 & 0 & 1 \end{bmatrix}$.

Note that if $r < m$, the last $m - r$ columns of $L_m[\mathbf{c}_1, \ldots, \mathbf{c}_r]$ are the corresponding columns of the identity matrix I_m.

Now the general version of the procedure in Example 2 can be stated. Given a nonzero matrix A, call the first nonzero column of A (from the left) the **leading column** of A.

LU-ALGORITHM

Let A be an $m \times n$ matrix that, via the Gaussian lower algorithm, can be carried to a row-echelon matrix U with no row interchanges being used. An LU-factorization $A = LU$ can be obtained as follows:

1. If $A = 0$, take $L = I_m$ and $U = 0$.
2. If $A \neq 0$, let $\mathbf{c}_1$ be the leading column of A, and do row operations (with no row interchanges) to bring A to the following block form:

$$\left[\begin{array}{c|c|c} 0 & 1 & X_1 \\ \hline 0 & 0 & A_1 \end{array}\right]$$

3. If $A_1 \neq 0$ let $\mathbf{c}_2$ be the leading column of A_1, and apply step 2 to bring A_1 to block form:

$$\left[\begin{array}{c|c|c} 0 & 1 & X_2 \\ \hline 0 & 0 & A_2 \end{array}\right]$$

4. Continue in this way until all the rows below the last leading 1 created consist of zeros. Take U to be the (row-echelon) matrix just created, and take

$$L = L_m[\mathbf{c}_1, \mathbf{c}_2, \ldots, \mathbf{c}_r]$$

where $\mathbf{c}_1, \mathbf{c}_2, \mathbf{c}_3, \ldots$ are the leading columns of the matrices A, $A_1, A_2, \ldots$.

Proof Proceed by induction on n. If $n = 1$, it is left to the reader. If $n > 1$, let $\mathbf{c}_1$ denote the leading column of A, and let $\mathbf{e}_1$ denote the first column of the $m \times m$ identity matrix. There exist elementary matrices $E_1, \ldots, E_k$ such that, in block form,

$$(E_k \cdots E_2E_1)A = \left[\begin{array}{c|c|c} 0 & \mathbf{e}_1 & \begin{array}{c} X_1 \\ \hline A_1 \end{array} \end{array}\right] \quad \text{where } (E_k \cdots E_2E_1)\mathbf{c}_1 = \mathbf{e}_1.$$

Moreover, each E_j may be taken to be lower triangular (by assumption). Write

$$L_0 = (E_k \cdots E_2E_1)^{-1} = E_1^{-1}E_2^{-1} \cdots E_k^{-1}$$

Then L_0 is lower triangular, and $L_0\mathbf{e}_1 = \mathbf{c}_1$. Moreover, because $E_1, \ldots, E_k$ correspond in order to the row operations that carry $\mathbf{c}_1$ to $\mathbf{e}_1$, then $E_k^{-1}, \ldots, E_1^{-1}$ correspond in order to the operations carrying $\mathbf{e}_1$ to $\mathbf{c}_1$. Hence

$$L_0 = \left[\begin{array}{c|c} \mathbf{c}_1 & \begin{array}{c} 0 \\ \hline I_{m-1} \end{array} \end{array}\right]$$

in block form. Now, by induction, let $A_1 = L_1U_1$ be an LU-factorization of A_1, where $L_1 = L_{m-1}[\mathbf{c}_2, \ldots, \mathbf{c}_r]$ and U_1 is row-echelon. Then block multiplication gives

$$L_0^{-1}A = \left[\begin{array}{c|c|c} 0 & \mathbf{e}_1 & \begin{array}{c} X_1 \\ \hline L_1U_1 \end{array} \end{array}\right] = \left[\begin{array}{c|c} 1 & 0 \\ \hline 0 & L_1 \end{array}\right]\left[\begin{array}{c|c|c} 0 & 1 & X_1 \\ \hline 0 & 0 & U_1 \end{array}\right]$$

Hence, $A = LU$, where $U = \left[\begin{array}{c|c|c} 0 & 1 & X_1 \\ \hline 0 & 0 & U_1 \end{array}\right]$ is row-echelon and

$$L = \left[\begin{array}{c|c} \mathbf{c}_1 & \begin{array}{c} 0 \\ \hline I_{m-1} \end{array} \end{array}\right]\left[\begin{array}{c|c} 1 & 0 \\ \hline 0 & L_1 \end{array}\right] = \left[\begin{array}{c|c} \mathbf{c}_1 & \begin{array}{c} 0 \\ \hline L_1 \end{array} \end{array}\right] = L_m[\mathbf{c}_1, \mathbf{c}_2, \ldots, \mathbf{c}_r].$$

This completes the proof. ■

Of course the integer r in the LU-algorithm is the number of leading 1's in the row-echelon matrix U, so it is the rank of A.

EXAMPLE 4

Find an LU-factorization for $A = \begin{bmatrix} 5 & -5 & 10 & 0 & 5 \\ -3 & 3 & 2 & 2 & 1 \\ -2 & 2 & 0 & -1 & 0 \\ 1 & -1 & 10 & 2 & 5 \end{bmatrix}$.

Solution The reduction to row-echelon form is

$$\begin{bmatrix} \boxed{5} & -5 & 10 & 0 & 5 \\ \boxed{-3} & 3 & 2 & 2 & 1 \\ \boxed{-2} & 2 & 0 & -1 & 0 \\ \boxed{1} & -1 & 10 & 2 & 5 \end{bmatrix} \rightarrow \begin{bmatrix} 1 & -1 & 2 & 0 & 1 \\ 0 & 0 & \boxed{8} & 2 & 4 \\ 0 & 0 & \boxed{4} & -1 & 2 \\ 0 & 0 & \boxed{8} & 2 & 4 \end{bmatrix}$$

$$\rightarrow \begin{bmatrix} 1 & -1 & 2 & 0 & 1 \\ 0 & 0 & 1 & \frac{1}{4} & \frac{1}{2} \\ 0 & 0 & 0 & \boxed{-2} & 0 \\ 0 & 0 & 0 & \boxed{0} & 0 \end{bmatrix}$$

$$\rightarrow \begin{bmatrix} 1 & -1 & 2 & 0 & 1 \\ 0 & 0 & 1 & \frac{1}{4} & \frac{1}{2} \\ 0 & 0 & 0 & 1 & 0 \\ 0 & 0 & 0 & 0 & 0 \end{bmatrix}.$$

If U denotes this row-echelon matrix, then $A = LU$, where

$$L = \begin{bmatrix} 5 & 0 & 0 & 0 \\ -3 & 8 & 0 & 0 \\ -2 & 4 & -2 & 0 \\ 1 & 8 & 0 & 1 \end{bmatrix}.$$

The next example deals with a case where no row of zeros is present in U (in fact, A is invertible).

EXAMPLE 5

Find an LU-factorization for $A = \begin{bmatrix} 2 & 4 & 2 \\ 1 & 1 & 2 \\ -1 & 0 & 2 \end{bmatrix}$.

Solution The reduction to row-echelon form is

$$\begin{bmatrix} 2 & 4 & 2 \\ 1 & 1 & 2 \\ -1 & 0 & 2 \end{bmatrix} \rightarrow \begin{bmatrix} 1 & 2 & 1 \\ 0 & -1 & 1 \\ 0 & 2 & 3 \end{bmatrix} \rightarrow \begin{bmatrix} 1 & 2 & 1 \\ 0 & 1 & -1 \\ 0 & 0 & 5 \end{bmatrix} \rightarrow \begin{bmatrix} 1 & 2 & 1 \\ 0 & 1 & -1 \\ 0 & 0 & 1 \end{bmatrix} = U$$

so $L = \begin{bmatrix} 2 & 0 & 0 \\ 1 & -1 & 0 \\ -1 & 2 & 5 \end{bmatrix}$.

The factorization in Theorem 3 is not unique. For example,

$$\begin{bmatrix} 1 & 0 \\ 3 & 2 \end{bmatrix} \begin{bmatrix} 1 & -2 & 3 \\ 0 & 0 & 0 \end{bmatrix} = \begin{bmatrix} 1 & 0 \\ 3 & 1 \end{bmatrix} \begin{bmatrix} 1 & -2 & 3 \\ 0 & 0 & 0 \end{bmatrix}$$

However, the fact that the row-echelon matrix here has a row of zeros is no coincidence. Recall that the rank of a matrix A is the number of nonzero rows in any row-echelon matrix U to which A can be carried by row operations. Thus if A is $m \times n$, the matrix U has no row of zeros if and only if A has rank m.

THEOREM 4

Let A be an $m \times n$ matrix that has an LU-factorization

$$A = LU$$

If A has rank m (that is, U has no row of zeros), then L and U are uniquely determined by A.

Proof Suppose $A = MV$ is another LU-factorization of A, so M is lower triangular and invertible and V is row-echelon. Hence $LU = MV$, and we must show that $L = M$ and $U = V$. We write $N = M^{-1}L$. Then N is lower triangular and invertible (Theorems 1 and 2) and $NU = V$, so it suffices to prove that $N = I$. If N is $m \times m$, we use induction on m. The case $m = 1$ is left to the reader. If $m > 1$, observe first that column 1 of V is N times column 1 of U. Thus if either column is zero, so is the other (N is invertible). Hence we may assume (by deleting zero columns) that the (1,1)-entry is 1 in both U and V. Now we write $N = \begin{bmatrix} a & 0 \\ X & N_1 \end{bmatrix}$, $U = \begin{bmatrix} 1 & Y \\ 0 & U_1 \end{bmatrix}$, and $V = \begin{bmatrix} 1 & Z \\ 0 & V_1 \end{bmatrix}$ in block form. Then $NU = V$ becomes $\begin{bmatrix} a & aY \\ X & XY + N_1U_1 \end{bmatrix} = \begin{bmatrix} 1 & Z \\ 0 & V_1 \end{bmatrix}$. Hence $a = 1$, $Y = Z$, $X = 0$, and $N_1U_1 = V_1$. But $N_1U_1 = V_1$ implies $N_1 = I$ by induction, whence $N = I$. ■

If A is an $m \times m$ invertible matrix, then A has rank m by Theorem 8 of Section 2.3. Hence we get the following important special case of Theorem 4.

COROLLARY If an invertible matrix A has an LU-factorization $A = LU$, then L and U are uniquely determined by A.

Of course, in this case U is an upper triangular matrix with 1's along the main diagonal.

There are matrices (for example, $\begin{bmatrix} 0 & 1 \\ 1 & 0 \end{bmatrix}$) that have no LU-factorization and so require at least one row interchange when being carried to row-echelon form via the Gaussian lower algorithm. However, it turns out that if all the row interchanges encountered in the algorithm are carried out first, the resulting matrix requires no interchanges and so has an LU-factorization. Here is the precise result.

THEOREM 5 Suppose an $m \times n$ matrix A is carried to a row-echelon matrix U via the Gaussian lower algorithm. Let $P_1, P_2, \ldots, P_s$ be the elementary matrices corresponding (in order) to the row interchanges used, and write $P = P_s \cdots P_2P_1$. (If no interchanges are used, take $P = I_m$.)

Then:

1. PA is the matrix obtained from A by doing these interchanges (in order) to A.
2. PA has an LU-factorization.

The proof is given at the end of this section.

A matrix P that is the product of elementary matrices corresponding to row interchanges is called a **permutation matrix**. Hence such a matrix can be obtained from the identity matrix by arranging the rows in a different order, so it has exactly one 1 in each row and each column and has zeros elsewhere. We regard the identity matrix as a permutation matrix. The elementary permutation matrices are those obtained from I by a single row interchange, and every permutation matrix is a product of elementary ones.

EXAMPLE 6

If $A = \begin{bmatrix} 0 & 0 & -1 & 2 \\ -1 & -1 & 1 & 2 \\ 2 & 1 & -3 & 6 \\ 0 & 1 & -1 & 4 \end{bmatrix}$, find a permutation matrix P such that PA has an LU-factorization, and then find the factorization.

Solution Apply the Gaussian lower algorithm.

$$A \to \begin{bmatrix} -1 & -1 & 1 & 2 \\ 0 & 0 & -1 & 2 \\ 2 & 1 & -3 & 6 \\ 0 & 1 & -1 & 4 \end{bmatrix} \to \begin{bmatrix} 1 & 1 & -1 & -2 \\ 0 & 0 & -1 & 2 \\ 0 & -1 & -1 & 10 \\ 0 & 1 & -1 & 4 \end{bmatrix} \to \begin{bmatrix} 1 & 1 & -1 & -2 \\ 0 & -1 & -1 & 10 \\ 0 & 0 & -1 & 2 \\ 0 & 1 & -1 & 4 \end{bmatrix}$$

$$\to \begin{bmatrix} 1 & 1 & -1 & -2 \\ 0 & 1 & 1 & -10 \\ 0 & 0 & -1 & 2 \\ 0 & 0 & -2 & 14 \end{bmatrix} \to \begin{bmatrix} 1 & 1 & -1 & -2 \\ 0 & 1 & 1 & -10 \\ 0 & 0 & 1 & -2 \\ 0 & 0 & 0 & 1 \end{bmatrix}.$$

Two row interchanges were needed, first rows 1 and 2 and then rows 2 and 3. Hence

$$P = \begin{bmatrix} 1 & 0 & 0 & 0 \\ 0 & 0 & 1 & 0 \\ 0 & 1 & 0 & 0 \\ 0 & 0 & 0 & 1 \end{bmatrix} \begin{bmatrix} 0 & 1 & 0 & 0 \\ 1 & 0 & 0 & 0 \\ 0 & 0 & 1 & 0 \\ 0 & 0 & 0 & 1 \end{bmatrix} = \begin{bmatrix} 0 & 1 & 0 & 0 \\ 0 & 0 & 1 & 0 \\ 1 & 0 & 0 & 0 \\ 0 & 0 & 0 & 1 \end{bmatrix}$$

If we do these interchanges (in order) to A, the result is PA.

$$PA = \begin{bmatrix} -1 & -1 & 1 & 2 \\ 2 & 1 & -3 & 6 \\ 0 & 0 & -1 & 2 \\ 0 & 1 & -1 & 4 \end{bmatrix} \to \begin{bmatrix} 1 & 1 & -1 & -2 \\ 0 & -1 & -1 & 10 \\ 0 & 0 & -1 & 2 \\ 0 & 1 & -1 & 4 \end{bmatrix} \to \begin{bmatrix} 1 & 1 & -1 & -2 \\ 0 & 1 & 1 & -10 \\ 0 & 0 & -1 & 2 \\ 0 & 0 & -2 & 14 \end{bmatrix}$$

$$\rightarrow \begin{bmatrix} 1 & 1 & -1 & -2 \\ 0 & 1 & 1 & -10 \\ 0 & 0 & 1 & -2 \\ 0 & 0 & 0 & \textcircled{10} \end{bmatrix} \rightarrow \begin{bmatrix} 1 & 1 & -1 & -2 \\ 0 & 1 & 1 & -10 \\ 0 & 0 & 1 & -2 \\ 0 & 0 & 0 & 1 \end{bmatrix} = U.$$

Hence $PA = LU$, where $L = \begin{bmatrix} -1 & 0 & 0 & 0 \\ 2 & -1 & 0 & 0 \\ 0 & 0 & -1 & 0 \\ 0 & 1 & -2 & 10 \end{bmatrix}$

and $U = \begin{bmatrix} 1 & 1 & -1 & -2 \\ 0 & 1 & 1 & -10 \\ 0 & 0 & 1 & -2 \\ 0 & 0 & 0 & 1 \end{bmatrix}$.

Theorem 5 provides an important general factorization theorem for matrices. If A is any $m \times n$ matrix, it asserts that there exists a permutation matrix P and an LU-factorization $PA = LU$. Moreover, it shows that $P = P_s \cdots P_2P_1$, where the P_i are elementary permutation matrices. Now observe that $P_i^{-1} = P_i$ for each i. Thus $P^{-1} = P_1P_2 \cdots P_s$, so the matrix A can be factored as

$$A = P^{-1}LU$$

where P^{-1} is a permutation matrix, L is lower triangular and invertible, and U is a row-echelon matrix. This is called a **PLU-factorization** of A.

Proof of Theorem 5 Let A be a nonzero $m \times n$ matrix, and let $\mathbf{e}_j$ denote column j of I_m. There is a permutation matrix P_1 (where either P_1 is elementary or $P_1 = I_m$) such that the first nonzero column $\mathbf{c}_1$ of P_1A has a nonzero entry on top. Then, as in the LU-algorithm,

$$L_m[\mathbf{c}_1]^{-1} \cdot P_1 \cdot A = \left[\begin{array}{c|c|c} 0 & 1 & X_1 \\ \hline 0 & 0 & A_1 \end{array}\right]$$

in block form. Then let P_2 be a permutation matrix (either elementary or I_m) such that

$$P_2 \cdot L_m[\mathbf{c}_1]^{-1} \cdot P_1 \cdot A = \left[\begin{array}{c|c|c} 0 & 1 & X_1 \\ \hline 0 & 0 & A_1' \end{array}\right]$$

and the first nonzero column $\mathbf{c}_2$ of A_1' has a nonzero entry on top. Then

$$L_m[\mathbf{e}_1, \mathbf{c}_2]^{-1} \cdot P_2 \cdot L_m[\mathbf{c}_1]^{-1} \cdot P_1 \cdot A = \left[\begin{array}{c|c|ccc} 0 & 1 & & X_1 & \\ \hline \multirow{2}{*}{0} & \multirow{2}{*}{0} & 0 & 1 & X_2 \\ \cline{3-5} & & 0 & 0 & A_2 \end{array}\right]$$

in block form. Continue to obtain elementary permutation matrices P_1, $P_2, \ldots, P_r$ and columns $\mathbf{c}_1, \mathbf{c}_2, \ldots, \mathbf{c}_r$ of lengths $m, m-1, \ldots$, such that

$$(L_r P_r L_{r-1} P_{r-1} \cdots L_2 P_2 L_1 P_1)A = U$$

where U is a row-echelon matrix and $L_j = L_m[\mathbf{e}_1, \ldots, \mathbf{e}_{j-1}, \mathbf{c}_j]^{-1}$ for each j where the notation means the first $j-1$ columns are those of I_m. It is not hard to verify that each L_j has the form $L_j = L_m[\mathbf{e}_1, \ldots, \mathbf{e}_{j-1}, \mathbf{c}_j]$ where $\mathbf{c}_j$ is a column of length $m-j+1$. We now claim that each permutation matrix P_k can be "moved past" each matrix L_j to the right of it, in the sense that

$$P_k L_j = L_j' P_k$$

where $L_j' = L_m[\mathbf{e}_1, \ldots, \mathbf{e}_{j-1}, \mathbf{c}_j'']$ for some column $\mathbf{c}_j''$ of length $m-j+1$. Given that this is true, we obtain a factorization of the form

$$(L_r L_{r-1}' \cdots L_2' L_1')(P_r P_{r-1} \cdots P_2 P_1)A = U$$

If we write $P = P_r P_{r-1} \cdots P_2 P_1$, this shows that PA has an LU-factorization, because $L_r L_{r-1}' \cdots L_2' L_1'$ is lower triangular and invertible. All that remains is to prove the following rather technical result. ■

LEMMA Let P_k result from interchanging row k of I_m with a row below it. If $j < k$, let $\mathbf{c}_j$ be a column of length $m-j+1$. Then there is another column $\mathbf{c}_j'$ of length $m-j+1$ such that

$$P_k \cdot L_m[\mathbf{e}_1, \ldots, \mathbf{e}_{j-1}, \mathbf{c}_j] = L_m[\mathbf{e}_1, \ldots, \mathbf{e}_{j-1}, \mathbf{c}_j'] \cdot P_k$$

The proof is left as Exercise 9.

EXERCISES 2.4

1. Find an LU-factorization of the following matrices.

(a) $\begin{bmatrix} 2 & 6 & -2 & 0 & 2 \\ 3 & 9 & -3 & 3 & 1 \\ -1 & -3 & 1 & -3 & 1 \end{bmatrix}$ (b) $\begin{bmatrix} 2 & 4 & 2 \\ 1 & -1 & 3 \\ -1 & 7 & -7 \end{bmatrix}$

(c) $\begin{bmatrix} 2 & 6 & -2 & 0 & 2 \\ 1 & 5 & -1 & 2 & 5 \\ 3 & 7 & -3 & -2 & 5 \\ -1 & -1 & 1 & 2 & 3 \end{bmatrix}$ (d) $\begin{bmatrix} -1 & -3 & 1 & 0 & -1 \\ 1 & 4 & 1 & 1 & 1 \\ 1 & 2 & -3 & -1 & 1 \\ 0 & -2 & -4 & -2 & 0 \end{bmatrix}$

(e) $\begin{bmatrix} 2 & 2 & 4 & 6 & 0 & 2 \\ 1 & -1 & 2 & 1 & 3 & 1 \\ -2 & 2 & -4 & -1 & 1 & 6 \\ 0 & 2 & 0 & 3 & 4 & 8 \\ -2 & 4 & -4 & 1 & -2 & 6 \end{bmatrix}$ (f) $\begin{bmatrix} 2 & 2 & -2 & 4 & 2 \\ 1 & -1 & 0 & 2 & 1 \\ 3 & 1 & -2 & 6 & 3 \\ 1 & 3 & -2 & 2 & 1 \end{bmatrix}$

2. Find a permutation matrix P and an LU-factorization of PA if A is:

(a) $\begin{bmatrix} 0 & 0 & 2 \\ 0 & -1 & 4 \\ 3 & 5 & 1 \end{bmatrix}$ (b) $\begin{bmatrix} 0 & -1 & 2 \\ 0 & 0 & 4 \\ -1 & 2 & 1 \end{bmatrix}$

(c) $\begin{bmatrix} 0 & -1 & 2 & 1 & 3 \\ -1 & 1 & 3 & 1 & 4 \\ 1 & -1 & -3 & 6 & 2 \\ 2 & -2 & -4 & 1 & 0 \end{bmatrix}$ (d) $\begin{bmatrix} -1 & -2 & 3 & 0 \\ 2 & 4 & -6 & 5 \\ 1 & 1 & -1 & 3 \\ 2 & 5 & -10 & 1 \end{bmatrix}$

3. In each case use the given LU-decomposition of A to solve the system $A\mathbf{x} = \mathbf{b}$ by finding $\mathbf{y}$ such that $L\mathbf{y} = \mathbf{b}$, and then $\mathbf{x}$ such that $U\mathbf{x} = \mathbf{y}$:

(a) $A = \begin{bmatrix} 2 & 0 & 0 \\ 0 & -1 & 0 \\ 1 & 1 & 3 \end{bmatrix} \begin{bmatrix} 1 & 0 & 0 & 1 \\ 0 & 0 & 1 & 2 \\ 0 & 0 & 0 & 1 \end{bmatrix}; \mathbf{b} = \begin{bmatrix} 1 \\ -1 \\ 2 \end{bmatrix}$

(b) $A = \begin{bmatrix} 2 & 0 & 0 \\ 1 & 3 & 0 \\ -1 & 2 & 1 \end{bmatrix} \begin{bmatrix} 1 & 1 & 0 & -1 \\ 0 & 1 & 0 & 1 \\ 0 & 0 & 0 & 0 \end{bmatrix}; \mathbf{b} = \begin{bmatrix} -2 \\ -1 \\ 1 \end{bmatrix}$

(c) $A = \begin{bmatrix} -2 & 0 & 0 & 0 \\ 1 & -1 & 0 & 0 \\ -1 & 0 & 2 & 0 \\ 0 & 1 & 0 & 2 \end{bmatrix} \begin{bmatrix} 1 & -1 & 2 & -1 \\ 0 & 1 & 1 & -4 \\ 0 & 0 & 1 & -\frac{1}{2} \\ 0 & 0 & 0 & 1 \end{bmatrix}; \mathbf{b} = \begin{bmatrix} 1 \\ -1 \\ 2 \\ 0 \end{bmatrix}$

(d) $A = \begin{bmatrix} 2 & 0 & 0 & 0 \\ 1 & -1 & 0 & 0 \\ -1 & 1 & 2 & 0 \\ 3 & 0 & 1 & -1 \end{bmatrix} \begin{bmatrix} 1 & -1 & 0 & 1 \\ 0 & 1 & -2 & -1 \\ 0 & 0 & 1 & 1 \\ 0 & 0 & 0 & 0 \end{bmatrix}; \mathbf{b} = \begin{bmatrix} 4 \\ -6 \\ 4 \\ 5 \end{bmatrix}$

4. Show that $\begin{bmatrix} 0 & 1 \\ 1 & 0 \end{bmatrix}$ has no LU-factorization.

5. Show that we can accomplish any row interchange by using only row operations of other types.

6. Let E and F be the elementary matrices obtained from the identity matrix by adding multiples of row k to rows p and q. If $k \neq p$ and $k \neq q$, show that $EF = FE$. [*Hint:* See Exercise 30(b) of Section 2.2.]

7. **(a)** Let L and L_1 be invertible lower triangular matrices, and let U and U_1 be invertible upper triangular matrices. Show that $LU = L_1U_1$ if and only if there exists an invertible diagonal matrix D such that $L_1 = LD$ and $U_1 = D^{-1}U$. [*Hint:* Scrutinize $L^{-1}L_1 = U\,U_1^{-1}$.]

   (b) Use part **(a)** to prove Theorem 4 in the case that A is invertible.

8. Let $\mathbf{c}_1, \mathbf{c}_2, \ldots, \mathbf{c}_r$ be columns of lengths $m, m-1, \ldots, m-r+1$. If $\mathbf{e}_j$ denotes column j of I_m, show that $L_m[\mathbf{c}_1, \mathbf{c}_2, \ldots, \mathbf{c}_r] = L_m[\mathbf{c}_1]\, L_m[\mathbf{e}_1, \mathbf{c}_2]\, L_m[\mathbf{e}_1, \mathbf{e}_2, \mathbf{c}_3] \cdots L_m[\mathbf{e}_1, \mathbf{e}_2, \ldots, \mathbf{e}_{r-1}, \mathbf{c}_r]$. The notation is as in the proof of Theorem 5. [*Hint:* Use induction on m and block multiplication.]

9. Prove the lemma following the proof of Theorem 5. [*Hint:* $P_k^{-1} = P_k$. Write $P_k = \begin{bmatrix} I_k & 0 \\ 0 & P_0 \end{bmatrix}$ in block form where P_0 is an $(m-k) \times (m-k)$ permutation matrix.]

SECTION 2.5 Applications of Matrix Algebra (Optional)[†]

2.5.1 Input–Output Economic Models

In 1974 Wassily Leontief was awarded the Nobel prize in economics for his work on mathematical models. Roughly speaking, an economic system in this model consists of several industries, each of which produces a product and each of which uses some of the production of the other industries. The following example is typical.

EXAMPLE 1 A primitive society has three basic needs: food, shelter, and clothing. There are thus three industries in the society—the farming, housing, and garment industries—that produce these commodities. Each of these industries consumes a certain proportion of the total output of each commodity according to the following table.

		Output		
		Farming	*Housing*	*Garment*
Consumption	*Farming*	.4	.2	.3
	Housing	.2	.6	.4
	Garment	.4	.2	.3

Find the annual prices that each industry must charge in order that its income will equal its expenditures.

Solution Let p_1, p_2, and p_3 be the prices charged per year by the farming, housing, and garment industries, respectively, for their total output. To see how these prices are determined, consider the farming industry. It receives p_1 for its production in any year. But it *consumes* products from all these industries in the following amounts (from row 1 of the table): 40% of the food, 20% of the housing, and 30% of the clothing. Hence the expenditures of the farming industry are $0.4p_1 + 0.2p_2 + 0.3p_3$, so

$$0.4p_1 + 0.2p_2 + 0.3p_3 = p_1$$

A similar analysis of the other two industries leads to the following system of equations.

$$0.4p_1 + 0.2p_2 + 0.3p_3 = p_1$$

$$0.2p_1 + 0.6p_2 + 0.4p_3 = p_2$$

[†]The two applications in this section are independent and may be taken in any order.

$$0.4p_1 + 0.2p_2 + 0.3p_3 = p_3$$

This has the matrix form $EP = P$, where

$$E = \begin{bmatrix} 0.4 & 0.2 & 0.3 \\ 0.2 & 0.6 & 0.4 \\ 0.4 & 0.2 & 0.3 \end{bmatrix} \quad \text{and} \quad P = \begin{bmatrix} p_1 \\ p_2 \\ p_3 \end{bmatrix}$$

The equations can be written as the homogeneous system

$$(I - E)P = 0$$

where I is the 3×3 identity matrix, and the solutions are

$$P = \begin{bmatrix} 2t \\ 3t \\ 2t \end{bmatrix}$$

where t is a parameter. Thus the pricing must be such that the total output of the farming industry has the same value as the total output of the garment industry, whereas the total value of the housing industry must be $\frac{3}{2}$ as much.

In general, suppose an economy has n industries, each of which uses some (possibly none) of the production of the other industries (possibly its own production). We assume that the economy is **closed** (that is, no product is exported or imported) and that all product is used. Given two industries i and j, let e_{ij} denote the proportion of the total annual output of industry j that is consumed by industry i. Then $E = [e_{ij}]$ is called the **input–output** matrix for the economy. Clearly,

1. $0 \le e_{ij} \le 1$ for all i and j

Moreover, all the output from industry j is used by *some* industry (the model is closed), so

2. $e_{1j} + e_{2j} + \cdots + e_{nj} = 1$ for each j

Condition 2 asserts that each column of E sums to 1. Matrices satisfying conditions 1 and 2 are called **stochastic matrices**.

As in Example 1, let p_i denote the price of the total annual production of industry i. Then p_i is the annual revenue of industry i. On the other hand, industry i spends $e_{i1}p_1 + e_{i2}p_2 + \cdots + e_{in}p_n$ annually for the product it uses ($e_{ij}p_j$ is the cost for product from industry j). The economic system is said to be in **equilibrium** if the annual expenditure equals the annual revenue for each industry—that is, if

$$e_{i1}p_1 + e_{i2}p_2 + \cdots + e_{in}p_n = p_i \qquad \text{for each } i = 1, 2, \cdots, n$$

If we write $P = \begin{bmatrix} p_1 \\ p_2 \\ \vdots \\ p_n \end{bmatrix}$, these equations can be written as the matrix equation

$$EP = P$$

This is called the **equilibrium condition**, and the solutions P are called **equilibrium price structures**. The equilibrium condition can be written as

$$(I - E)P = 0$$

which is a system of homogeneous equations for P. Moreover, there is always a nontrivial solution P. Indeed, the column sums of $I - E$ are all 0 (because E is stochastic), so the row-echelon form of $I - E$ has a row of zeros. In fact, more is true.

THEOREM 1 Let E be any $n \times n$ stochastic matrix (each entry is nonnegative and each of the column sums is 1). Then there is a nonzero $1 \times n$ matrix P with nonnegative entries such that $EP = P$. If all the entries of E are positive, the matrix P can be chosen with all entries positive.

Theorem 1 guarantees the existence of an equilibrium price structure for any closed input–output system of the type discussed here. The proof is beyond the scope of this book. The interested reader is referred to P. Lancaster's *Theory of Matrices* (New York: Academic Press, 1969) or to E. Seneta's *Non-negative Matrices* (New York: Wiley, 1973).

EXAMPLE 2 Find the equilibrium price structures for four industries if the input–output matrix is

$$E = \begin{bmatrix} 0.6 & 0.2 & 0.1 & 0.1 \\ 0.3 & 0.4 & 0.2 & 0 \\ 0.1 & 0.3 & 0.5 & 0.2 \\ 0 & 0.1 & 0.2 & 0.7 \end{bmatrix}$$

Find the prices if the total value of business is \$1000.

Solution If $P = \begin{bmatrix} p_1 \\ p_2 \\ p_3 \\ p_4 \end{bmatrix}$ is the equilibrium price structure, then the equilibrium condition is $EP = P$. When we write this as $(I - E)P = 0$, the methods of Chapter 1 yield the following family of solutions:

$$P = \begin{bmatrix} 44t \\ 39t \\ 51t \\ 47t \end{bmatrix}$$

where t is a parameter. If we insist that $p_1 + p_2 + p_3 + p_4 = \$1000$, then $t = 5.525$ (to four figures). Hence

$$P = \begin{bmatrix} 243.1 \\ 215.5 \\ 281.8 \\ 259.7 \end{bmatrix}$$

to four figures.

EXERCISES 2.5.1

1. Find the possible equilibrium price structures when the input–output matrices are:
   (a) $\begin{bmatrix} .1 & .2 & .3 \\ .6 & .2 & .3 \\ .3 & .6 & .4 \end{bmatrix}$
   (b) $\begin{bmatrix} .5 & 0 & .5 \\ .1 & .9 & .2 \\ .4 & .1 & .3 \end{bmatrix}$
   (c) $\begin{bmatrix} .3 & .1 & .1 & .2 \\ .2 & .3 & .1 & 0 \\ .3 & .3 & .2 & .3 \\ .2 & .3 & .6 & .5 \end{bmatrix}$
   (d) $\begin{bmatrix} .5 & 0 & .1 & .1 \\ .2 & .7 & 0 & .1 \\ .1 & .2 & .8 & .2 \\ .2 & .1 & .1 & .6 \end{bmatrix}$
2. Three industries A, B, and C are such that all the output of A is used by B, all the output of B is used by C, and all the output of C is used by A. Find the possible equilibrium price structures.
3. Find the possible equilibrium price structures for three industries where the input–output matrix is $\begin{bmatrix} 1 & 0 & 0 \\ 0 & 0 & 1 \\ 0 & 1 & 0 \end{bmatrix}$. Discuss why there are two parameters here.
4. Prove Theorem 1 for a 2×2 stochastic matrix E by first writing it in the form $E = \begin{bmatrix} a & b \\ 1 - a & 1 - b \end{bmatrix}$, where $0 \le a \le 1$ and $0 \le b \le 1$.
5. If E is an $n \times n$ stochastic matrix and C is an $n \times 1$ matrix, show that the sum of the entries of C equals the sum of the entries of the $n \times 1$ matrix EC.
6. Let $W = [1, 1, 1, \ldots, 1]$. Let E and F denote $n \times n$ matrices with nonnegative entries.
   (a) Show that E is a stochastic matrix if and only if $WE = W$.
   (b) Use part **(a)** to deduce that, if E and F are both stochastic matrices, then EF is also stochastic.

2.5.2 Markov Chains

Many natural phenomena progress through various stages and can be in a variety of states at each stage. For example, the weather in a given city progresses day by day and, on any given day, may be "sunny" or "rainy." Here the states are "sun" or "rain" and the weather progresses from one state to another in daily stages. Another example might be a football team: the stages of its evolution are the games it plays, and the possible states are "win," "draw," and "loss."

The general set-up is as follows: A "system" evolves through a series of "stages," and at any stage it can be in any one of a finite number of "states." At any given stage the state to which it will go at the next stage depends on the past and present history of the system—that is, on the sequence of states it has occupied to date. A **Markov chain** is such an evolving system wherein the state to which it will go next depends *only* on its *present* state and does not depend on the earlier history of the system.

Even in the case of a Markov chain, the state the system will occupy at any stage is determined only in terms of probabilities. In other words, chance plays a role. For example, if a football team wins a particular game, we do not know whether it will win, draw, or lose the next game. On the other hand, we may know that the team tends to persist in "winning streaks"—for example, if it wins one game it may win the next game $\frac{1}{2}$ of the time, lose $\frac{4}{10}$ of the time and draw $\frac{1}{10}$ of the time. These fractions are called the **probabilities** of these various possibilities. Similarly, if the team loses, it may lose the next game with probability $\frac{1}{2}$ (that is, half the time), win with probability $\frac{1}{4}$, and draw with probability $\frac{1}{4}$. The probability of the various outcomes after a drawn game will also be known. We shall treat probabilities informally here: *The probability that a given event will occur is the long-run proportion of the time that the event does indeed occur.* Hence all probabilities are numbers between 0 and 1. A probability of 0 means the event is impossible and never occurs; events with probability 1 are certain to occur.

If a Markov chain is in a particular state, the probabilities that it goes to the various states at the next stage of its evolution are called the **transition probabilities** for the chain, and they are assumed to be known quantities. To motivate the general considerations below, consider the following simple example. Here the "system" is a woman, the "stages" are her successive outings, and the "states" are the two escorts with whom she goes out.

EXAMPLE 3 A woman has two male escorts A and B. She never goes out with A twice in a row. However, if she goes out with B, she is three times as likely to go out with B next time as with A. Initially she is equally likely to go out with either escort.

(a) What is the probability that she goes out with A on the third outing after the initial one?

(b) What proportion of her time does she spend with A?

		Present Escort	
		A	B
Next Escort	A	0	.25
	B	1	.75

Solution The table of transition probabilities is displayed above. The first column indicates the fact that, if she goes with A, she never goes with him on the next outing and is certain to go with B. The second column shows that, if she goes with B, she goes with him again on the next outing $\frac{3}{4}$ of the time and switches to A only $\frac{1}{4}$ of the time.

The escort she has on a given outing is not determined. The most that we can expect is to know the probability that she will be with A or B on that outing. Let $S_m = \begin{bmatrix} s_1^{(m)} \\ s_2^{(m)} \end{bmatrix}$ denote the *state vector* for outing m. Here $s_1^{(m)}$ denotes the probability that she is with A on outing m, and $s_2^{(m)}$ is the probability that she is with B. It is convenient to let S_0 correspond to the initial outing. Because she is equally likely to be with A or with B on that initial outing, $s_1^{(0)} = .5$ and $s_2^{(0)} = .5$, so $S_0 = \begin{bmatrix} .5 \\ .5 \end{bmatrix}$.

Now let

$$P = \begin{bmatrix} 0 & .25 \\ 1 & .75 \end{bmatrix}$$

denote the "transition matrix." We claim that the relationship

$$S_{m+1} = PS_m$$

holds for all m. This will be derived later; for now, we will use it as follows to successively compute $S_1, S_2, S_3, \ldots$.

$$S_1 = PS_0 = \begin{bmatrix} 0 & .25 \\ 1 & .75 \end{bmatrix} \begin{bmatrix} .5 \\ .5 \end{bmatrix} = \begin{bmatrix} .125 \\ .875 \end{bmatrix}$$

$$S_2 = PS_1 = \begin{bmatrix} 0 & .25 \\ 1 & .75 \end{bmatrix} \begin{bmatrix} .125 \\ .875 \end{bmatrix} = \begin{bmatrix} .21875 \\ .78125 \end{bmatrix}$$

$$S_3 = PS_2 = \begin{bmatrix} 0 & .25 \\ 1 & .75 \end{bmatrix} \begin{bmatrix} .21875 \\ .78125 \end{bmatrix} = \begin{bmatrix} .1953125 \\ .8046875 \end{bmatrix}$$

Hence the probability that the third outing is with A is approximately .195, whereas the probability that it is with B is .805.

If we carry these calculations on, the next state vectors are (to five figures)

$$S_4 = \begin{bmatrix} .20117 \\ .79883 \end{bmatrix} \qquad S_5 = \begin{bmatrix} .19971 \\ .80029 \end{bmatrix}$$

$$S_6 = \begin{bmatrix} .20007 \\ .79993 \end{bmatrix} \qquad S_7 = \begin{bmatrix} .19998 \\ .80002 \end{bmatrix}$$

Moreover, the higher values of S_m get closer and closer to $\begin{bmatrix} .2 \\ .8 \end{bmatrix}$. Hence the probability that she is with A is 0.2 for all outings after the seventh, so, in the long run, she spends 20% of her time with A and 80% with B.

Example 3 incorporates most of the essential features of all Markov chains. The general model is as follows: The "system" evolves through various "stages" and, at each stage, can be in exactly one of n distinct "states." It progresses through a sequence of states as time goes on.

DEFINITION If a Markov chain is in state j at a particular stage of its development, the probability p_{ij} that it goes to state i at the next stage is called the **transition probability.** The $n \times n$ matrix $P = [p_{ij}]$ is called the **transition matrix** for the Markov chain.

The intuitive meaning can be described as follows: In the long term, among all the times the system is in state j, the the fraction of the time it will then go to state i is p_{ij}. Thus $p_{ij} = 0$ means that it is *impossible* to go from state j to state i in one transition, whereas $p_{ij} = 1$ means that, if the system is in state j, it is *certain* to go to state i at the next stage. If $p_{ij} = \frac{1}{2}$, then, if the system is in state j, it is equally likely to go to state i or not at the next stage. Clearly, $0 \le p_{ij} \le 1$ holds for each i and j.

Consider the jth column of the transition matrix P.

$$\begin{bmatrix} p_{1j} \\ p_{2j} \\ \vdots \\ p_{nj} \end{bmatrix}$$

If the system is in state j at some stage of its evolution, the transition probabilities $p_{1j}, p_{2j}, \ldots, p_{nj}$ represent the long-term fraction of the time that the system will move to state 1, state 2, ... state n at the next stage. We assume that it has to go to *some* state at each transition, so the sum of these probabilities equals 1:

$$p_{1j} + p_{2j} + \cdots + p_{nj} = 1 \qquad \text{for each } j$$

Thus the columns of P all sum to 1 and the entries of P lie between 0 and 1. A matrix with these properties is called a **stochastic matrix** (stochastic matrices also arose in Section 2.5.1 in connection with input–output matrices).

As in Example 3, we introduce the following notation: Let $s_i^{(m)}$ denote the probability that the system is in state i after m transitions. In other words, $s_i^{(m)}$ is the long-term fraction of the time that the system will be in state i after m transitions. The $n \times 1$ matrices

$$S_m = \begin{bmatrix} s_1^{(m)} \\ s_2^{(m)} \\ \vdots \\ s_n^{(m)} \end{bmatrix} \qquad m = 0, 1, 2, \ldots$$

are called the **state vectors** for the Markov chain. Note that the sum of the entries of S_m must equal 1, because the system must be in *some* state after m transitions. The matrix S_0 is called the **initial state vector** for the Markov chain and is given as part of the data of the particular chain. For example, if the chain has only two states, then an initial vector $S_0 = \begin{bmatrix} 1 \\ 0 \end{bmatrix}$ means that it started in state 1. If it started in state 2, the initial vector would be $\begin{bmatrix} 0 \\ 1 \end{bmatrix}$. If $S_0 = \begin{bmatrix} .5 \\ .5 \end{bmatrix}$, it is equally likely that the system started in state 1 or in state 2.

We make one important assumption about the transition matrix $P = [p_{ij}]$: It does *not* depend on which stage the process is in. For example, the probability p_{21} of going to state 2 from state 1 is the same whether the system is in its initial stage of evolution or has already progressed through several stages. And this is assumed to be the case for each p_{ij}. This assumption means that the transition probabilities are *independent of time* — that is, they do not change as time goes on. It is this assumption that distinguishes Markov chains in the literature of this subject.

THEOREM 2 Let P be the transition matrix for an n-state Markov chain. If S_m is the state vector at stage m, then

$$S_{m+1} = PS_m$$

for each $m = 0, 1, 2, \ldots$

Heuristic Proof Suppose that the Markov chain has been run N times, each time starting with the same initial state vector. Recall that p_{ij} is the proportion of the time the system goes from state j at some stage to state i at the next stage, whereas $s_i^{(m)}$ is the proportion of the time it is in state i at stage m. Hence $s_i^{(m+1)}N$ is (approximately) the number of times the system is in state i at stage $m + 1$. We are going to calculate this number another way. The system got to state i at stage $m + 1$ through *some* other state (say state j) at stage m. The number of times it was *in* state j at that stage is (approximately) $s_j^{(m)}N$, so the number of times it got to state i via state j is $p_{ij}(s_j^{(m)}N)$. Summing over j gives the number of times the system is in state i (at stage $m + 1$). That is,

$$s_i^{(m+1)}N = p_{i1}s_1^{(m)}N + p_{i2}s_2^{(m)}N + \cdots + p_{in}s_n^{(m)}N$$

Canceling N gives $s_i^{(m+1)} = p_{i1}s_1^{(m)} + p_{i2}s_2^{(m)} + \cdots + p_{in}s_n^{(m)}$ for each i, and this can be expressed as the matrix equation $S_{m+1} = PS_m$. ■

If the initial probability vector S_0 and the transition matrix P are given, Theorem 2 gives $S_1, S_2, S_3, \ldots$, one after the other, as follows:

$$S_1 = PS_0$$

$$S_2 = PS_1$$

$$S_3 = PS_2$$

Hence the state vector S_m is completely determined for each $m = 0, 1, 2, \ldots$ by P and S_0.

EXAMPLE 4 A wolf always hunts in one of three regions R_1, R_2, and R_3. His hunting habits are as follows:

1. If he hunts in one region one day, he is as likely as not to hunt there again the next day.
2. If he hunts in R_1, he never hunts in R_2 the next day.
3. If he hunts in R_2 or R_3, he is equally likely to hunt in each of the other regions the next day.

If he hunts in R_1 on Monday, find the probability that he hunts there on Thursday.

Solution The "stages" of this process are the successive days; the "states" are the three regions. The transition matrix P is determined as follows (see the table): The first habit asserts that $p_{11} = p_{22} = p_{33} = \frac{1}{2}$. Now column 1 displays what happens when he starts in R_1: He never goes to state 2, so $p_{21} = 0$ and, because the column must sum to 1, $p_{31} = \frac{1}{2}$. Column 2 describes what happens if he starts in R_2: $p_{22} = \frac{1}{2}$ and p_{12} and p_{32} are equal (by habit 3), so $p_{12} = p_{32} = \frac{1}{4}$ because the column sum must equal 1. Column 3 is filled in a similar way.

	R_1	R_2	R_3
R_1	$\frac{1}{2}$	$\frac{1}{4}$	$\frac{1}{4}$
R_2	0	$\frac{1}{2}$	$\frac{1}{4}$
R_3	$\frac{1}{2}$	$\frac{1}{4}$	$\frac{1}{2}$

Now let Monday be the initial stage. Then $S_0 = \begin{bmatrix} 1 \\ 0 \\ 0 \end{bmatrix}$ because he hunts in R_1 on that day. Then S_1, S_2, and S_3 describe Tuesday, Wednesday, and Thursday, respectively, and we compute them using Theorem 2.

$$S_1 = PS_0 = \begin{bmatrix} \frac{1}{2} \\ 0 \\ \frac{1}{2} \end{bmatrix} \qquad S_2 = PS_1 = \begin{bmatrix} \frac{3}{8} \\ \frac{1}{8} \\ \frac{4}{8} \end{bmatrix} \qquad S_3 = PS_2 = \begin{bmatrix} \frac{11}{32} \\ \frac{6}{32} \\ \frac{15}{32} \end{bmatrix}$$

Hence the probability that he hunts in Region R_1 on Thursday is 11/32.

Another phenomenon that was observed in Example 3 can be expressed in general terms. The state vectors $S_0, S_1, S_2, \ldots$ were calculated in that example and were found to "approach" $S = \begin{bmatrix} .2 \\ .8 \end{bmatrix}$. That means that the first component of S_m becomes and remains very close to .2 as m becomes large, whereas the second component approaches .8 as m increases. When this is the case, we say that S_m **converges** to S. For large m, then, there is very little error in taking $S_m = S$, so the "long-term" probability that the system is in state 1 is .2, whereas the probability that it is in state 2 is .8. In Example 3, enough state vectors were computed for the limiting vector S to be apparent. However, there is a better way to do this that works in most cases.

Suppose P is the transition matrix of a Markov chain, and assume that the state vectors S_m converge to a limiting vector S. Then S_m is "very close" to S for sufficiently large m, so S_{m+1} is also "very close" to S. Thus the equation $S_{m+1} = PS_m$ from Theorem 2 is closely approximated by

$$S = PS$$

so it is not surprising that S should be a solution to this matrix equation. Moreover, it is easily solved because it can be written as a system of linear equations

$$(I - P)S = 0$$

with the entries of S as variables.

In Example 3 where $P = \begin{bmatrix} 0 & .25 \\ 1 & .75 \end{bmatrix}$, the general solution is $S = \begin{bmatrix} t \\ 4t \end{bmatrix}$, where t is a parameter. But if we insist that the entries of S sum to 1 (as must be true of all state vectors), we find $t = .2$ and

$S = \begin{bmatrix} .2 \\ .8 \end{bmatrix}$ as before.

All this is predicated on the existence of a limiting vector for the sequence of state vectors of the Markov chain, and such a vector may not always exist. However, it does in one commonly occurring situation. A stochastic matrix P is called **regular** if some power P^m of P has every entry positive. The matrix $P = \begin{bmatrix} 0 & .25 \\ 1 & .75 \end{bmatrix}$ of Example 3 is regular (in this case, each entry of P^2 is positive), and the general theorem is as follows:

THEOREM 3 Let P be the transition matrix of a Markov chain and assume that P is regular. Then there is a unique column matrix S satisfying the following conditions.

1. $PS = S$.
2. The entries of S are positive and sum to 1.

Moreover, condition 1 can be written as $(I - P)S = 0$ and so gives a homogeneous system of linear equations for S. Finally, the sequence of state vectors $S_0, S_1, S_2, \ldots$ converges to S in the sense that if m is large enough, each entry of S_m is closely approximated by the corresponding entry of S.

This theorem will not be proved here. The interested reader can find an elementary proof in J. Kemeny, H. Mirkil, J. Snell, and G. Thompson, *Finite Mathematical Structures* (Englewood Cliffs, N.J.: Prentice-Hall, 1958).

If P is a regular transition matrix of a Markov chain, the column S satisfying conditions 1 and 2 of Theorem 3 is called the **steady-state vector** for the Markov chain. The entries of S are the long-term probabilities that the chain will be in each of the various states.

EXAMPLE 5 A man eats one of three soups—beef, chicken, and vegetable—each day. He never eats the same soup two days in a row. If he eats beef soup on a certain day, he is equally likely to eat each of the others the next day; if he does not eat beef soup, he is twice as likely to eat it the next day as the alternative.

(a) If he has beef soup one day, what is the probability that he has it again two days later?

(b) What are the long-run probabilities that he eats each of the three soups?

	B	C	V
B	0	$\frac{2}{3}$	$\frac{2}{3}$
C	$\frac{1}{2}$	0	$\frac{1}{3}$
V	$\frac{1}{2}$	$\frac{1}{3}$	0

Solution The states here are B, C, and V, the three soups. The transition matrix P is given in the table. (Recall that for each state, the corresponding column lists the probabilities for the next state.) If he has beef soup initially, then the initial state vector is

$$S_0 = \begin{bmatrix} 1 \\ 0 \\ 0 \end{bmatrix}$$

Then two days later the state vector is S_2. If P is the transition matrix, then

$$S_1 = PS_0 = \frac{1}{2}\begin{bmatrix} 0 \\ 1 \\ 1 \end{bmatrix}, \qquad S_2 = PS_1 = \frac{1}{6}\begin{bmatrix} 4 \\ 1 \\ 1 \end{bmatrix}$$

so he eats beef soup two days later with probability $\frac{2}{3}$. This answers (a) and also shows that he eats chicken and vegetable soup each with probability $\frac{1}{6}$.

To find the long-run probabilities, we must find the steady-state vector S. Theorem 3 applies because P is regular (P^2 has positive entries), so S satisfies $PS = S$. That is, $(I - P)S = 0$ where

$$I - P = \frac{1}{6}\begin{bmatrix} 6 & -4 & -4 \\ -3 & 6 & -2 \\ -3 & -2 & 6 \end{bmatrix}$$

So $S = \begin{bmatrix} 4t \\ 3t \\ 3t \end{bmatrix}$, where t is a parameter, and we use $S = \begin{bmatrix} .4 \\ .3 \\ .3 \end{bmatrix}$ because the entries of S must sum to 1. Hence, in the long run, he eats beef soup 40% of the time and each of chicken soup and vegetable soup 30% of the time.

EXERCISES 2.5.2

1. Which of the following stochastic matrices is regular?

   (a) $\begin{bmatrix} 0 & 0 & \frac{1}{2} \\ 1 & 0 & \frac{1}{2} \\ 0 & 1 & 0 \end{bmatrix}$ (b) $\begin{bmatrix} \frac{1}{2} & 0 & \frac{1}{3} \\ \frac{1}{4} & 1 & \frac{1}{3} \\ \frac{1}{4} & 0 & \frac{1}{3} \end{bmatrix}$

2. In each case find the steady-state vector and, assuming that it starts in state 1, find the probability that it is in state 2 after 3 transitions.

   (a) $\begin{bmatrix} .5 & .3 \\ .5 & .7 \end{bmatrix}$ (b) $\begin{bmatrix} \frac{1}{2} & 1 \\ \frac{1}{2} & 0 \end{bmatrix}$ (c) $\begin{bmatrix} 0 & \frac{1}{2} & \frac{1}{4} \\ 1 & 0 & \frac{1}{4} \\ 0 & \frac{1}{2} & \frac{1}{2} \end{bmatrix}$

(d) $\begin{bmatrix} .4 & .1 & .5 \\ .2 & .6 & .2 \\ .4 & .3 & .3 \end{bmatrix}$ (e) $\begin{bmatrix} .8 & 0 & .2 \\ .1 & .6 & .1 \\ .1 & .4 & .7 \end{bmatrix}$ (f) $\begin{bmatrix} .1 & .3 & .3 \\ .3 & .1 & .6 \\ .6 & .6 & .1 \end{bmatrix}$

3. A fox hunts in three territories A, B, and C. He never hunts in the same territory on two successive days. If he hunts in A, then he hunts in C the next day. If he hunts in B or C, he is twice as likely to hunt in A the next day as in the other territory.
   (a) What proportion of his time does he spend in A, in B, and in C?
   (b) If he hunts in A on Monday (C on Monday), what is the probability that he will hunt in B on Thursday?
4. Assume that there are three classes—upper, middle, and lower—and that social mobility behaves as follows:

   70% of the children of upper-class parents remain upper-class, whereas 10% become middle-class and 20% become lower-class.

   Of the children of middle-class parents, 80% remain middle-class, whereas the others are evenly split between the upper class and the lower class.

   For the children of lower-class parents, 60% remain lower-class, whereas 30% become middle-class and 10% upper-class.
   (a) Find the probability that the grandchild of lower-class parents becomes upper-class.
   (b) Find the long-term breakdown of society into classes.
5. The Prime Minister says she will call an election. This gossip is passed from person to person with a probability $p \neq 0$ that the information is passed incorrectly at any stage. Assume that, when a person hears the gossip, he or she passes it to one other person. Find the long-term probability that a person will hear that there is going to be an election.
6. John makes it to work on time one Monday out of four. On other work days his behavior is as follows: If he is late one day, he is twice as likely to come to work on time the next day as to be late. If he is on time one day, he is as likely to be late as not the next day. Find the probability of his being late and that of his being on time Wednesdays.
7. Suppose you have 1¢ and match coins with a friend. At each match you either win or lose 1¢ with equal probability. If you go broke or ever get 4¢, you quit. Assume your friend never quits. If the states are 0, 1, 2, 3, and 4 representing your wealth, show that the corresponding transition matrix P is not regular. Find the probability that you will go broke after 3 matches.

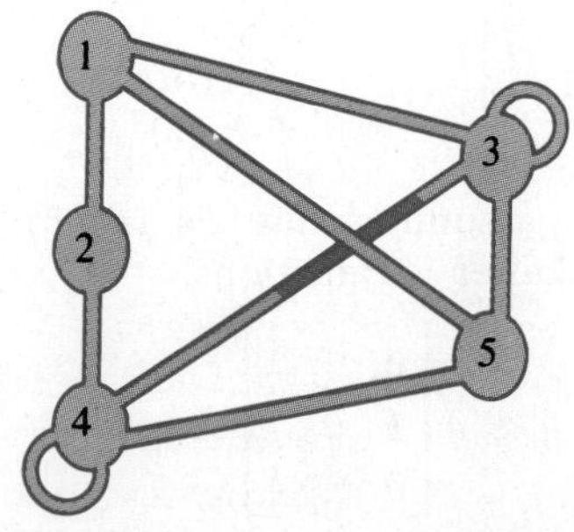

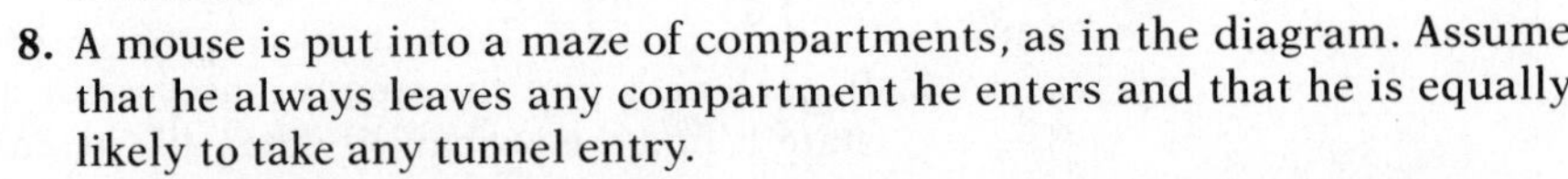

8. A mouse is put into a maze of compartments, as in the diagram. Assume that he always leaves any compartment he enters and that he is equally likely to take any tunnel entry.
   (a) If he starts in compartment 1, find the probability that he is in compartment 4 after 3 moves.
   (b) Find the compartment in which he spends most of his time if he is left for a long time.

9. If a stochastic matrix has a 1 on its main diagonal, show that it cannot be regular.

10. If S_m is the stage-m state vector for a Markov chain, show that $S_{m+k} = P^k S_m$ holds for all $m \geq 1$ and $k \geq 1$ (where P is the transition matrix).

11. A stochastic matrix is **doubly stochastic** if all the row sums also equal 1. Find the steady-state vector for a doubly stochastic matrix.

12. Consider the 2×2 stochastic matrix $P = \begin{bmatrix} 1-p & q \\ p & 1-q \end{bmatrix}$, where $0 < p < 1$ and $0 < q < 1$.

(a) Show that $\dfrac{1}{p+q}\begin{bmatrix} q \\ p \end{bmatrix}$ is the steady-state vector for P.

(b) Show that P^m converges to the matrix $\dfrac{1}{p+q}\begin{bmatrix} q & q \\ p & p \end{bmatrix}$ by first verifying inductively that

$$P^m = \frac{1}{p+q}\begin{bmatrix} q & q \\ p & p \end{bmatrix} + \frac{(1-p-q)^m}{p+q}\begin{bmatrix} p & -q \\ -p & q \end{bmatrix}$$

for $m = 1, 2, \ldots$. (It can be shown that the sequence of powers P, P^2, P^3, ... of any regular transition matrix converges to the matrix each of whose columns equals the steady-state vector for P.)

3 Determinants

SECTION 3.1 The Laplace Expansion

It is a well-known fact that division by a number a is allowed only if $a \neq 0$. In other words a^{-1} exists if and only if $a \neq 0$. Let A denote a square matrix. The **determinant** of A is a number det A, associated with A, with the property that A is invertible if and only if $\det A \neq 0$. In this section det A will be defined, and some methods for computing it will be given.

There is no difficulty if A is 1×1.

DEFINITION Given a 1×1 matrix $[a]$, define $\det[a] = a$.

Then $[a]$ has an inverse if and only if $\det[a] \neq 0$. In fact the inverse is $[1/a]$.

DEFINITION If $A = \begin{bmatrix} a & b \\ c & d \end{bmatrix}$, define $\det A = ad - bc$.

For example, $\det\begin{bmatrix} 3 & 5 \\ 2 & 4 \end{bmatrix} = 12 - 10 = 2$ and $\det\begin{bmatrix} -1 & -1 \\ 0 & 3 \end{bmatrix} = -3 - 0 = -3$.

The fact that a 2×2 matrix A has an inverse if and only if $\det A \neq 0$ can be seen easily. Write $A = \begin{bmatrix} a & b \\ c & d \end{bmatrix}$ and consider the matrix $B = \begin{bmatrix} d & -b \\ -c & a \end{bmatrix}$, called the **adjoint** of A. If AB and BA are computed explicitly, the result is

$$AB = (\det A)I = BA$$

If $\det A \neq 0$, we may multiply through by $\dfrac{1}{\det A}$ to obtain the formula

$$A^{-1} = \frac{1}{\det A}B.$$

On the other hand, if A^{-1} exists we claim that $\det A \neq 0$. For if $\det A = 0$, then $AB = (\det A)I = 0$. This implies that $B = A^{-1}AB = A^{-1}0 = 0$, and hence that $a = b = c = d = 0$. But then $A = 0$, contrary to the assumption that A is invertible.

In the case where A is 1×1 or 2×2, then, we do indeed obtain the result that A is invertible if and only if $\det A \neq 0$. In the next section this result will be proved for any square matrix A (Theorem 2).

EXAMPLE 1 In each case, calculate the determinant of A and find the inverse (if it exists) from the formula.

(a) $A = \begin{bmatrix} 5 & 4 \\ 2 & 3 \end{bmatrix}$ (b) $A = \begin{bmatrix} 2 & 3 \\ 6 & 9 \end{bmatrix}$

Solution (a) $\det A = \det\begin{bmatrix} 5 & 4 \\ 2 & 3 \end{bmatrix} = 5 \cdot 3 - 4 \cdot 2 = 7$, so A^{-1} exists. The adjoint of A is $B = \begin{bmatrix} 3 & -4 \\ -2 & 5 \end{bmatrix}$, so $A^{-1} = \dfrac{1}{\det A}B = \dfrac{1}{7}\begin{bmatrix} 3 & -4 \\ -2 & 5 \end{bmatrix}$.

(b) In this case, $\det A = \det\begin{bmatrix} 2 & 3 \\ 6 & 9 \end{bmatrix} = 2 \cdot 9 - 3 \cdot 6 = 0$. Hence A has no inverse.

Here is a procedure for defining the determinant of any $n \times n$ matrix. Once we know how to define determinants of 2×2 matrices, we then give a rule by which the determinant of any 3×3 matrix can be defined (in terms of certain determinants of 2×2 matrices). Next we do 4×4 matrices in terms of 3×3 matrices, and so on. At each stage we give a rule for defining the determinant of a square matrix in terms of determinants of square matrices one size smaller.

Before stating this rule, we must introduce minors and cofactors of a square matrix.

DEFINITION Suppose A is an $n \times n$ matrix and it has been specified how to compute determinants of $(n - 1) \times (n - 1)$ matrices. Then the ***(i,j)*-minor**

of A, denoted $M_{ij}(A)$, is defined to be the determinant of the $(n-1) \times (n-1)$ matrix formed from A by deleting row i and column j. Next the **(i,j)-cofactor** of A, denoted $C_{ij}(A)$, is defined by

$$C_{ij}(A) = (-1)^{i+j}M_{ij}(A)$$

Clearly $C_{ij}(A)$ equals either $M_{ij}(A)$ or $-M_{ij}(A)$, depending on the choice of i and j. The number $(-1)^{i+j}$ is called the **sign** of the (i,j)-position.

The following diagram is a useful mnemonic for calculating the sign of a position.

$$\begin{bmatrix} +1 & -1 & +1 & -1 & +1 & \cdots \\ -1 & +1 & -1 & +1 & -1 & \cdots \\ +1 & -1 & +1 & -1 & +1 & \cdots \\ -1 & +1 & -1 & +1 & -1 & \cdots \\ \vdots & \vdots & \vdots & \vdots & \vdots & \end{bmatrix}$$

Note that the signs alternate along each row and column and that the sign of position (1,1) is $+1$. We have already decided how to compute the determinant of a 2×2 matrix, so we can find the minors and cofactors for any 3×3 matrix.

EXAMPLE 2 Find the minors and cofactors of positions (1,2), (3,1), and (2,3) in the following matrix.

$$A = \begin{bmatrix} 5 & -1 & 6 \\ 5 & 2 & 7 \\ 8 & 9 & 4 \end{bmatrix}$$

Solution The (1,2)-minor is the determinant of the matrix $\begin{bmatrix} 5 & 7 \\ 8 & 4 \end{bmatrix}$ that remains when row 1 and column 2 are deleted. The sign of position (1,2) is $(-1)^{1+2} = -1$ (this is also the (1,2)-entry of the mnemonic), so the (1,2)-minor and the (1,2)-cofactor are

$$M_{12}(A) = \det\begin{bmatrix} 5 & 7 \\ 8 & 4 \end{bmatrix} = 5 \cdot 4 - 7 \cdot 8 = -36$$

$$C_{12}(A) = (-1)^{1+2}M_{12}(A) = (-1)(-36) = 36$$

Turning to position (3,1), we find that the minor is the determinant of the matrix $\begin{bmatrix} -1 & 6 \\ 2 & 7 \end{bmatrix}$ that remains when row 3 and column 1 are deleted from A. Hence

$$M_{31}(A) = \det\begin{bmatrix} -1 & 6 \\ 2 & 7 \end{bmatrix} = (-1) \cdot 7 - 6 \cdot 2 = -19$$

$$C_{31}(A) = (-1)^{3+1}M_{31}(A) = (+1)(-19) = -19$$

Finally, the (2,3)-minor and the (2,3)-cofactor are

$$M_{23}(A) = \det\begin{bmatrix} 5 & -1 \\ 8 & 9 \end{bmatrix} = 5 \cdot 9 - (-1) \cdot 8 = 53$$

$$C_{23}(A) = (-1)^{2+3}M_{23}(A) = (-1) \cdot 53 = -53$$

Clearly other minors and cofactors can be found—there are nine in all, one for each position in the matrix.

With the notion of minor and cofactor in hand, we can formulate the rule for finding the determinant of an $n \times n$ matrix A. Recall that the idea is to do this inductively—that is, to find a way to compute $\det A$ in terms of determinants of certain $(n - 1) \times (n - 1)$ matrices.

DEFINITION

Given $n \geq 2$, assume that $\det M$ has been defined for any $(n - 1) \times (n - 1)$ matrix M. If A is $n \times n$, define

$$\det A = a_{11}C_{11}(A) + a_{21}C_{21}(A) + \cdots + a_{n1}C_{n1}(A)$$

In other words, the definition says that $\det A$ can be found by multiplying each entry a_{i1} in the first column by the corresponding cofactor $C_{i1}(A)$ and adding the results. This is called the **Laplace expansion** of $\det A$ along the first column. The astonishing thing is that $\det A$ can be computed by taking the Laplace expansion along *any* column: Simply multiply the entries of the column by the corresponding cofactors and add the results. Even more remarkably, $\det A$ can be found from the Laplace expansion along any row.

THEOREM 1
Laplace Expansion

The determinant of an $n \times n$ matrix A can be computed by using the Laplace expansion along any row or column of A. More precisely, if $A = [a_{ij}]$ so that a_{ij} is the (i,j)-entry of A, then the expansion along row i is

$$\det A = a_{i1}C_{i1}(A) + a_{i2}C_{i2}(A) + a_{i3}C_{i3}(A) + \cdots + a_{in}C_{in}(A)$$

The expansion along column j is given by

$$\det A = a_{1j}C_{1j}(A) + a_{2j}C_{2j}(A) + a_{3j}C_{3j}(A) + \cdots + a_{nj}C_{nj}(A)$$

The proof will be given in Section 3.3.

EXAMPLE 3 Compute the determinant of $A = \begin{bmatrix} 3 & 4 & 5 \\ 1 & 7 & 2 \\ 9 & 8 & -6 \end{bmatrix}$.

Solution The Laplace expansion along the first row is as follows:

$$\begin{aligned} \det A &= 3C_{11}(A) + 4C_{12}(A) + 5C_{13}(A) \\ &= 3M_{11}(A) - 4M_{12}(A) + 5M_{13}(A) \\ &= 3\det\begin{bmatrix} 7 & 2 \\ 8 & -6 \end{bmatrix} - 4\det\begin{bmatrix} 1 & 2 \\ 9 & -6 \end{bmatrix} + 5\det\begin{bmatrix} 1 & 7 \\ 9 & 8 \end{bmatrix} \\ &= 3(-58) - 4(-24) + 5(-55) \\ &= -353 \end{aligned}$$

Now we compute det A by expanding along the first column.

$$\begin{aligned} \det A &= 3C_{11}(A) + 1C_{21}(A) + 9C_{31}(A) \\ &= 3M_{11}(A) - M_{21}(A) + 9M_{31}(A) \\ &= 3\det\begin{bmatrix} 7 & 2 \\ 8 & -6 \end{bmatrix} - \det\begin{bmatrix} 4 & 5 \\ 8 & -6 \end{bmatrix} + 9\det\begin{bmatrix} 4 & 5 \\ 7 & 2 \end{bmatrix} \\ &= 3(-58) - (-64) + 9(-27) \\ &= -353 \end{aligned}$$

The reader is invited to verify that det A can be computed by expanding along the second or third row or along the second or third column.

The fact that the Laplace expansion along *any row or column* of a matrix A always gives the same result (the determinant of A) is remarkable to say the least. The choice of a particular row or column can simplify the calculation.

EXAMPLE 4 Compute det A where $A = \begin{bmatrix} 3 & 0 & 0 & 0 \\ 5 & 1 & 2 & 0 \\ 2 & 6 & 0 & -1 \\ -6 & 3 & 1 & 0 \end{bmatrix}$.

Solution The first choice we must make is which row or column to use in the Laplace expansion. The expansion involves multiplying entries by cofactors, so the work is reduced when the row or column has as many zeros in it as possible. Row 1 is clearly the best in the present situation, and the expansion is

$$\det A = 3C_{11}(A) + 0C_{12}(A) + 0C_{13}(A) + 0C_{14}(A)$$

$$= 3M_{11}(A) + 0 + 0 + 0$$

$$= 3 \det \begin{bmatrix} 1 & 2 & 0 \\ 6 & 0 & -1 \\ 3 & 1 & 0 \end{bmatrix}$$

This is the first stage of the calculation, and we have succeeded in expressing the determinant of (the 4×4 matrix) A in terms of the determinant of a 3×3 matrix. The next stage involves this 3×3 matrix. Again, we may use any row or column for the Laplace expansion. The third column is preferred (with two zeros), so

$$\begin{aligned} \det A &= 3\left(0 \det \begin{bmatrix} 6 & 0 \\ 3 & 1 \end{bmatrix} - (-1) \det \begin{bmatrix} 1 & 2 \\ 3 & 1 \end{bmatrix} + 0 \det \begin{bmatrix} 1 & 2 \\ 6 & 0 \end{bmatrix}\right) \\ &= 3[0 + 1(-5) + 0] \\ &= -15 \end{aligned}$$

This completes the calculation.

Computing the determinant of matrix A can be tedious, even by the Laplace expansion. For example, if A is a 4×4 matrix, the Laplace expansion along any row or column involves calculating four minors, each of which is itself the determinant of a 3×3 matrix. And if A is 5×5, the expansion involves five determinants of 4×4 matrices! There is a clear need for some techniques to cut down the work.

The motivation for the method (see Example 4) is the observation that calculating a determinant is simplified a great deal when a row or column consists mostly of zeros. (In fact, when a row or column consists *entirely* of zeros, the determinant is zero—simply expand along that row or column.) Recall next that one method of *creating* zeros in a matrix is to apply elementary row operations to it. Hence a natural question to ask is what effect such a row operation has on the determinant of the matrix. It turns out that the effect is easy to determine and that elementary *column* operations can be used in the same way. These observations lead to a technique for evaluating determinants that considerably reduces the labor involved. The necessary information is given in Theorem 2.

THEOREM 2 Let A denote an $n \times n$ matrix.

1. If A has a row or column of zeros, $\det A = 0$.
2. If two distinct rows (or columns) of A are interchanged, the determinant of the resulting matrix is $-\det A$.
3. If a row (or column) of A is multiplied by a constant u, the determinant of the resulting matrix is $u(\det A)$.

4. If two distinct rows (or columns) of A are identical, $\det A = 0$.
5. If a multiple of one row of A is added to a different row (or if a multiple of a column is added to a different column), the determinant of the resulting matrix is $\det A$.

Proof We prove properties 2, 4, and 5 and leave the rest as exercises.

Property 2. If A is $n \times n$, this follows by induction on n. If $n = 2$, the verification is left to the reader. If $n > 2$ and two rows are interchanged, let B denote the resulting matrix. Expand $\det A$ and $\det B$ along a row *other than* the two that were interchanged. The entries in this row are the same for both A and B, but the cofactors in B are the negatives of those in A (by induction) because the corresponding $(n-1) \times (n-1)$ matrices have two rows interchanged. Hence $\det B = -\det A$, as required. A similar argument works if two columns are interchanged.

Property 4. If two rows of A are equal, let B be the matrix obtained by interchanging them. Then $B = A$, so $\det B = \det A$. But $\det B = -\det A$ by property 2, so $\det A = \det B = 0$. Again, the same argument works for columns.

Property 5. Let B be obtained from $A = [a_{ij}]$ by adding u times row p to row q. Then row q of B is $(a_{q1} + ua_{p1}, a_{q2} + ua_{p2}, \ldots, a_{qn} + ua_{pn})$. The cofactors of these elements in B are the same as in A (they do not involve row q); in symbols, $C_{qj}(B) = C_{qj}(A)$ for each j. Hence, expanding B along row q gives

$$\begin{aligned}\det B &= \sum_{j=1}^{n} (a_{qj} + ua_{pj})C_{qj}(B) \\ &= \sum_{j=1}^{n} a_{qj}C_{qj}(A) + u \sum_{j=1}^{n} a_{pj}C_{qj}(A) \\ &= \det A + u \det C\end{aligned}$$

where C is the matrix obtained from A by replacing row q by row p (and both expansions are along row q). Because rows p and q of C are equal, $\det C = 0$ by property 4. Hence $\det B = \det A$, as required. As before, a similar proof holds for columns. ■

To illustrate Theorem 2, consider the following matrices.

$$\det\begin{bmatrix} 3 & -1 & 2 \\ 2 & 5 & 1 \\ 0 & 0 & 0 \end{bmatrix} = 0$$ (because the last row consists of zeros)

$$\det\begin{bmatrix} 3 & -1 & 5 \\ 2 & 8 & 7 \\ 1 & 2 & -1 \end{bmatrix} = -\det\begin{bmatrix} 5 & -1 & 3 \\ 7 & 8 & 2 \\ -1 & 2 & 1 \end{bmatrix}$$ (because two columns are interchanged)

$$\det\begin{bmatrix} 8 & 1 & 2 \\ 3 & 0 & 9 \\ 1 & 2 & -1 \end{bmatrix} = 3\det\begin{bmatrix} 8 & 1 & 2 \\ 1 & 0 & 3 \\ 1 & 2 & -1 \end{bmatrix}$$ (because the second row of the matrix on the left is 3 times the second row of the matrix on the right)

$$\det\begin{bmatrix} 2 & 1 & 2 \\ 4 & 0 & 4 \\ 1 & 3 & 1 \end{bmatrix} = 0$$ (because two columns are identical)

$$\det\begin{bmatrix} 2 & 5 & 2 \\ -1 & 2 & 9 \\ 3 & 1 & 1 \end{bmatrix} = \det\begin{bmatrix} 0 & 9 & 20 \\ -1 & 2 & 9 \\ 3 & 1 & 1 \end{bmatrix}$$ (because twice the second row of the matrix on the left was added to the first row)

The following example illustrates how Theorem 2 is used to evaluate determinants.

EXAMPLE 5

Evaluate det A when $A = \begin{bmatrix} 1 & -1 & 3 \\ 1 & 0 & -1 \\ 2 & 1 & 6 \end{bmatrix}$.

Solution The matrix does have zero entries, so expansion along (say) the second row would involve somewhat less work. However, a column operation can be used to get a zero in position (2,3)—namely, add column 1 to column 3. Because this does not change the value of the determinant, we obtain

$$\det A = \det\begin{bmatrix} 1 & -1 & 3 \\ 1 & 0 & -1 \\ 2 & 1 & 6 \end{bmatrix} = \det\begin{bmatrix} 1 & -1 & 4 \\ 1 & 0 & 0 \\ 2 & 1 & 8 \end{bmatrix} = -\det\begin{bmatrix} -1 & 4 \\ 1 & 8 \end{bmatrix} = 12$$

where we expanded the second 3×3 matrix along row 2. Alternatively, we could add the third row of A to the first and then expand along column 2.

$$\det A = \det\begin{bmatrix} 1 & -1 & 3 \\ 1 & 0 & -1 \\ 2 & 1 & 6 \end{bmatrix} = \det\begin{bmatrix} 3 & 0 & 9 \\ 1 & 0 & -1 \\ 2 & 1 & 6 \end{bmatrix} = -\det\begin{bmatrix} 3 & 9 \\ 1 & -1 \end{bmatrix} = 12$$

Of course the result is the same as before.

EXAMPLE 6

If $\det\begin{bmatrix} a & b & c \\ p & q & r \\ x & y & z \end{bmatrix} = 6$, evaluate det A

where $A = \begin{bmatrix} a+x & b+y & c+z \\ 3x & 3y & 3z \\ -p & -q & -r \end{bmatrix}$

Solution First take common factors out of rows 2 and 3.

$$\det A = 3(-1)\det\begin{bmatrix} a+x & b+y & c+z \\ x & y & z \\ p & q & r \end{bmatrix}$$

Now subtract the second row from the first and interchange the last two rows.

$$\det A = -3\det\begin{bmatrix} a & b & c \\ x & y & z \\ p & q & r \end{bmatrix} = 3\det\begin{bmatrix} a & b & c \\ p & q & r \\ x & y & z \end{bmatrix} = 3 \cdot 6 = 18$$

If A is an $n \times n$ matrix, forming uA means multiplying *every* row of A by u. Applying property 3 of Theorem 2 n times gives the following useful result.

THEOREM 3 If A is an $n \times n$ matrix, then $\det(uA) = u^n \det A$ for any number u.

The determinant of a matrix is a sum of products of its entries. In particular, if these entries are polynomials in x, then the determinant itself is a polynomial in x. It is often of interest to determine which values of x make the determinant zero, so it is very useful if the determinant is given in factored form. Theorem 2 can help.

EXAMPLE 7

Find the values of x for which $\det A = 0$, where $A = \begin{bmatrix} 1 & x & x \\ x & 1 & x \\ x & x & 1 \end{bmatrix}$.

Solution To evaluate $\det A$, subtract x times row 1 from rows 2 and 3.

$$\det A = \det\begin{bmatrix} 1 & x & x \\ x & 1 & x \\ x & x & 1 \end{bmatrix} = \det\begin{bmatrix} 1 & x & x \\ 0 & 1-x^2 & x-x^2 \\ 0 & x-x^2 & 1-x^2 \end{bmatrix}$$

$$= \det\begin{bmatrix} 1-x^2 & x-x^2 \\ x-x^2 & 1-x^2 \end{bmatrix}$$

At this stage we could simply evaluate the determinant (the result is $2x^3 - 3x^2 + 1$). Then we would have to factor this polynomial in order to find the values of x that make it zero. However, this factorization can be obtained directly by first factoring each entry in the determinant and taking a common factor of $(1 - x)$ from each row.

$$\det A = \det\begin{bmatrix} 1 - x^2 & x - x^2 \\ x - x^2 & 1 - x^2 \end{bmatrix} = \det\begin{bmatrix} (1 - x)(1 + x) & x(1 - x) \\ x(1 - x) & (1 - x)(1 + x) \end{bmatrix}$$

$$= (1 - x)^2 \det\begin{bmatrix} 1 + x & x \\ x & 1 + x \end{bmatrix} = (1 - x)^2(2x + 1)$$

Hence $\det A = 0$ means $(1 - x)^2(2x + 1) = 0$, so $x = 1$ or $x = -\frac{1}{2}$.

EXAMPLE 8 If a_1, a_2 and a_3 are given, show that

$$\det\begin{bmatrix} 1 & 1 & 1 \\ a_1 & a_2 & a_3 \\ a_2{}^2 & a_2{}^2 & a_3{}^2 \end{bmatrix} = (a_3 - a_2)(a_3 - a_1)(a_2 - a_1)$$

Solution Begin by subtracting the second column from the third, and then subtract the first from the second.

$$\det\begin{bmatrix} 1 & 1 & 1 \\ a_1 & a_2 & a_3 \\ a_2{}^2 & a_2{}^2 & a_3{}^2 \end{bmatrix} = \det\begin{bmatrix} 1 & 0 & 0 \\ a_1 & a_2 - a_1 & a_3 - a_2 \\ a_1{}^2 & a_2{}^2 - a_1{}^2 & a_3{}^2 - a_2{}^2 \end{bmatrix}$$

$$= \det\begin{bmatrix} a_2 - a_1 & a_3 - a_2 \\ a_2{}^2 - a_1{}^2 & a_3{}^2 - a_2{}^2 \end{bmatrix}$$

Now $(a_2 - a_1)$ and $(a_3 - a_2)$ are common factors in the first and second columns, so

$$\det\begin{bmatrix} 1 & 1 & 1 \\ a_1 & a_2 & a_3 \\ a_2{}^2 & a_2{}^2 & a_3{}^2 \end{bmatrix} = (a_3 - a_2)(a_2 - a_1)\det\begin{bmatrix} 1 & 1 \\ a_2 + a_1 & a_3 + a_2 \end{bmatrix}$$

$$= (a_3 - a_2)(a_2 - a_1)(a_3 - a_1)$$

The matrix in Example 8 is called a **Vandermonde matrix**, and the formula for its determinant can be generalized to the $n \times n$ case (see Theorem 2 in Section 3.4).

A square matrix is called an **upper triangular matrix** if all entries below the main diagonal are zero. Similarly, a **lower triangular matrix** is one for which all entries above the main diagonal are zero. A **triangular matrix** is one that is either upper or lower triangular. Theorem 4 gives an easy rule for calculating the determinant of any triangular matrix. The proof is Exercise 21.

THEOREM 4 If A is a triangular matrix, then det A is the product of the entries on the main diagonal.

This theorem is useful in computer calculations, because it is a routine matter to carry a matrix to triangular form using row operations. Here is an example.

EXAMPLE 9

Evaluate det A where $A = \begin{bmatrix} 3 & 2 & 5 \\ -1 & 4 & 6 \\ 2 & -9 & -2 \end{bmatrix}$.

Solution We carry out row operations to put the matrix in upper triangular form. First interchange rows 1 and 2, and then "clean up" the first column.

$$\det A = -\det\begin{bmatrix} -1 & 4 & 6 \\ 3 & 2 & 5 \\ 2 & -9 & -2 \end{bmatrix} = -\det\begin{bmatrix} -1 & 4 & 6 \\ 0 & 14 & 23 \\ 0 & -1 & 10 \end{bmatrix}$$

Now interchange rows 2 and 3, and then complete the reduction to upper triangular form.

$$\det A = \det\begin{bmatrix} -1 & 4 & 6 \\ 0 & -1 & 10 \\ 0 & 14 & 23 \end{bmatrix} = \det\begin{bmatrix} -1 & 4 & 6 \\ 0 & -1 & 10 \\ 0 & 0 & 163 \end{bmatrix}$$

Hence det $A = (-1)(-1)(163) = 163$ by Theorem 4.

A matrix A is called a **block upper triangular matrix** if it can be partitioned into blocks

$$A = \begin{bmatrix} A_{11} & A_{12} & A_{13} & \cdots & A_{1n} \\ 0 & A_{22} & A_{23} & \cdots & A_{2n} \\ 0 & 0 & A_{33} & \cdots & A_{3n} \\ \vdots & \vdots & \vdots & & \vdots \\ 0 & 0 & 0 & \cdots & A_{nn} \end{bmatrix}$$

where each "diagonal block" $A_{11}, A_{22}, \ldots, A_{nn}$ is a square matrix and every block below the main diagonal is zero. If every block above the main diagonal is zero, the matrix is called a **block lower triangular matrix.** If the matrix is one or the other, it is called simply a **block triangular matrix.** The matrices

$$A = \left[\begin{array}{cc|ccc} 3 & 1 & 0 & 0 & 0 \\ 2 & 1 & 0 & 0 & 0 \\ \hline 0 & 1 & 3 & -1 & 2 \\ 2 & -1 & 5 & 0 & 0 \\ 3 & 1 & 1 & 0 & 1 \end{array}\right] \quad \text{and} \quad B = \left[\begin{array}{cc|c|cc} 3 & 1 & 0 & 0 & 0 \\ 2 & 1 & 0 & 0 & 0 \\ \hline 3 & 0 & 5 & 0 & 0 \\ \hline 1 & 5 & 4 & 1 & 1 \\ -1 & 0 & 2 & 1 & 1 \end{array}\right]$$

are both block lower triangular. In particular, a triangular matrix is nothing but a block triangular matrix whose diagonal blocks are all 1×1. Theorem 4 asserts that the determinant of a triangular matrix is the product of the main diagonal entries. This generalizes as follows.

THEOREM 5 Let A decompose in block upper triangular form as

$$A = \begin{bmatrix} A_{11} & A_{12} & A_{13} & \cdots & A_{1n} \\ 0 & A_{22} & A_{23} & \cdots & A_{2n} \\ 0 & 0 & A_{33} & \cdots & A_{3n} \\ \vdots & \vdots & \vdots & & \vdots \\ 0 & 0 & 0 & \cdots & A_{nn} \end{bmatrix}$$

where the diagonal blocks $A_{11}, A_{22}, \ldots, A_{nn}$ are square (but not necessarily of the same size). Then

$$\det A = (\det A_{11})(\det A_{22}) \ldots (\det A_{nn})$$

The analogous result holds for block lower triangular matrices.

Proof Proceed by induction on n, where A is $n \times n$. If $n = 1$ or $n = 2$, it is easily verified. In general, compute $\det A$ using the Laplace expansion along the first column.

$$\det A = a_{11}C_{11}(A) + a_{21}C_{21}(A) + \cdots + a_{k1}C_{k1}(A) \qquad (*)$$

where $a_{11}, a_{21}, \ldots, a_{k1}$ denote the entries in the first column of A_{11}. We have

$$C_{i1}(A) = (-1)^{i+1}M_{i1}(A)$$

where the minor $M_{i1}(A)$ is the determinant of the matrix obtained by deleting row i and column 1 of A. This matrix is also block upper triangular (one size smaller than A), and the upper left block is obtained by deleting row i and column 1 of A_{11}. So, by induction,

$$M_{i1}(A) = M_{i1}(A_{11})(\det A_{22}) \cdots (\det A_{nn})$$

Hence

$$C_{i1}(A) = (-1)^{i+1}M_{i1}(A_{11})(\det A_{22}) \cdots (\det A_{nn})$$
$$= C_{i1}(A_{11})(\det A_{22}) \cdots (\det A_{nn})$$

and so the product $(\det A_{22}) \cdots (\det A_{nn})$ is a common factor in every term in the sum (*). Hence the sum becomes

$$\det A = [a_{11}C_{11}(A_{11}) + a_{21}C_{21}(A_{11}) + \cdots + a_{k1}C_{k1}(A_{11})](\det A_{22}) \cdots (\det A_{nn})$$
$$= (\det A_{11})(\det A_{22}) \ldots (\det A_{nn})$$

because the expression in square brackets is the Laplace expansion of $\det(A_{11})$ along column 1. This completes the induction. ■

EXAMPLE 10 Compute the determinants of the matrices preceding Theorem 5.

Solution $\det A = \det\begin{bmatrix} 3 & 1 \\ 2 & 1 \end{bmatrix} \det\begin{bmatrix} 3 & -1 & 2 \\ 5 & 0 & 0 \\ 1 & 0 & 1 \end{bmatrix} = 1 \cdot 5 = 5$

$$\det B = \det\begin{bmatrix} 3 & 1 \\ 2 & 1 \end{bmatrix} \det[5] \det\begin{bmatrix} 1 & 1 \\ 1 & 1 \end{bmatrix} = 1 \cdot 5 \cdot 0 = 0$$

EXERCISES 3.1

1. Compute the determinants of the following matrices.

(a) $\begin{bmatrix} 2 & -1 \\ 3 & 2 \end{bmatrix}$

(b) $\begin{bmatrix} 6 & 9 \\ 8 & 12 \end{bmatrix}$

(c) $\begin{bmatrix} a^2 & ab \\ ab & b^2 \end{bmatrix}$

(d) $\begin{bmatrix} a+1 & a \\ a & a-1 \end{bmatrix}$

(e) $\begin{bmatrix} \cos\theta & -\sin\theta \\ \sin\theta & \cos\theta \end{bmatrix}$

(f) $\begin{bmatrix} 2 & 0 & -3 \\ 1 & 2 & 5 \\ 0 & 3 & 0 \end{bmatrix}$

(g) $\begin{bmatrix} 1 & 2 & 3 \\ 4 & 5 & 6 \\ 7 & 8 & 9 \end{bmatrix}$

(h) $\begin{bmatrix} 0 & a & 0 \\ b & c & d \\ 0 & e & 0 \end{bmatrix}$

(i) $\begin{bmatrix} 1 & b & c \\ b & c & 1 \\ c & 1 & b \end{bmatrix}$

(j) $\begin{bmatrix} 0 & a & b \\ a & 0 & c \\ b & c & 0 \end{bmatrix}$

(k) $\begin{bmatrix} 1 & 0 & -1 & 0 \\ 0 & 3 & 0 & 2 \\ 1 & 0 & 2 & 1 \\ 0 & 5 & 0 & 7 \end{bmatrix}$

(l) $\begin{bmatrix} 1 & 0 & 3 & 1 \\ 2 & 2 & 6 & 0 \\ -1 & 0 & -3 & 1 \\ 4 & 1 & 12 & 0 \end{bmatrix}$

(m) $\begin{bmatrix} 3 & 1 & -5 & 2 \\ 1 & 0 & 5 & 2 \\ 1 & 3 & 0 & 1 \\ 1 & 1 & 2 & -1 \end{bmatrix}$

(n) $\begin{bmatrix} 4 & -1 & 3 & -1 \\ 3 & 1 & 0 & 2 \\ 0 & 1 & 2 & 2 \\ 1 & 2 & -1 & 1 \end{bmatrix}$

(o) $\begin{bmatrix} 1 & -1 & 5 & 2 \\ 3 & 1 & 2 & 4 \\ -1 & -3 & 8 & 0 \\ 1 & 1 & 2 & -1 \end{bmatrix}$ (p) $\begin{bmatrix} 0 & 0 & 0 & a \\ 0 & 0 & b & p \\ 0 & c & q & k \\ d & s & t & u \end{bmatrix}$

2. Show that $\det A = 0$ if A has a row or column consisting of zeros.
3. Show that the sign of the position in the last row and the last column is always $+1$.
4. Show that $\det I = 1$ for any identity matrix I.
5. Evaluate each determinant by reducing it to upper triangular form.

(a) $\begin{bmatrix} 1 & -1 & 2 \\ 3 & 1 & 1 \\ 2 & -1 & 3 \end{bmatrix}$ (b) $\begin{bmatrix} -1 & 3 & 1 \\ 2 & 5 & 3 \\ 1 & -2 & 1 \end{bmatrix}$

(c) $\begin{bmatrix} -1 & -1 & 1 & 0 \\ 2 & 1 & 1 & 3 \\ 0 & 1 & 1 & 2 \\ 1 & 3 & -1 & 2 \end{bmatrix}$ (d) $\begin{bmatrix} 2 & 3 & 1 & 1 \\ 0 & 2 & -1 & 3 \\ 0 & 5 & 1 & 1 \\ 1 & 1 & 2 & 5 \end{bmatrix}$

6. Evaluate by inspection: (a) $\det\begin{bmatrix} a & b & c \\ a+1 & b+1 & c+1 \\ a-1 & b-1 & c-1 \end{bmatrix}$

(b) $\det\begin{bmatrix} a & b & c \\ a+b & 2b & c+b \\ 2 & 2 & 2 \end{bmatrix}$

7. If $\det\begin{bmatrix} a & b & c \\ p & q & r \\ x & y & z \end{bmatrix} = -1$, compute:

(a) $\det\begin{bmatrix} -x & -y & -z \\ 3p+a & 3q+b & 3r+c \\ 2p & 2q & 2r \end{bmatrix}$ (b) $\det\begin{bmatrix} -2a & -2b & -2c \\ 2p+x & 2q+y & 2r+z \\ 3x & 3y & 3z \end{bmatrix}$

8. Show that: (a) $\det\begin{bmatrix} p+x & q+y & r+z \\ a+x & b+y & c+z \\ a+p & b+q & c+r \end{bmatrix} = 2\det\begin{bmatrix} a & b & c \\ p & q & r \\ x & y & z \end{bmatrix}$

(b) $\det\begin{bmatrix} 2a+p & 2b+q & 2c+r \\ 2p+x & 2q+y & 2r+z \\ 2x+a & 2y+b & 2z+c \end{bmatrix} = 9\det\begin{bmatrix} a & b & c \\ p & q & r \\ x & y & z \end{bmatrix}$

9. Compute the determinants of each matrix, using Theorem 5.

(a) $\begin{bmatrix} 1 & -1 & 2 & 0 & -2 \\ 0 & 1 & 0 & 4 & 1 \\ 1 & 1 & 3 & 0 & 0 \\ 0 & 0 & 0 & 3 & -1 \\ 0 & 0 & 0 & 1 & 1 \end{bmatrix}$ (b) $\begin{bmatrix} 1 & 2 & 0 & 3 & 0 \\ -1 & 3 & 1 & 4 & 0 \\ 0 & 0 & 2 & 1 & 1 \\ 0 & 0 & -1 & 0 & 2 \\ 0 & 0 & 3 & 0 & 1 \end{bmatrix}$

(c) $\begin{bmatrix} 1 & 2 & 0 & 0 & 1 & -2 \\ 3 & -1 & 0 & 0 & 4 & 3 \\ 0 & 0 & 2 & 4 & 0 & 0 \\ 0 & 0 & 6 & 1 & 0 & 0 \\ 0 & 0 & 0 & 0 & 1 & -1 \\ 0 & 0 & 0 & 0 & 2 & 0 \end{bmatrix}$ (d) $\begin{bmatrix} 1 & 2 & 1 & 0 & 2 & 0 \\ 3 & 0 & 2 & 0 & 1 & 0 \\ -1 & 0 & 2 & 0 & 6 & 0 \\ 0 & 0 & 0 & 2 & 4 & 1 \\ 0 & 0 & 0 & 1 & 0 & 2 \\ 0 & 0 & 0 & 1 & 1 & 1 \end{bmatrix}$

10. Evaluate by first adding all other rows to the first row.

(a) $\det\begin{bmatrix} x-1 & 2 & 3 \\ 2 & -3 & x-2 \\ -2 & x & -2 \end{bmatrix}$ **(b)** $\det\begin{bmatrix} x-1 & -3 & 1 \\ 2 & -1 & x-1 \\ -3 & x+2 & -2 \end{bmatrix}$

11. **(a)** Find b if $\det\begin{bmatrix} 3 & -1 & x \\ 2 & 6 & y \\ -5 & 4 & z \end{bmatrix} = ax + by + cz$.

(b) Find c if $\det\begin{bmatrix} 2 & x & -1 \\ 1 & y & 3 \\ -3 & z & 4 \end{bmatrix} = ax + by + cz$.

12. Find the real numbers x and y such that $\det A = 0$ if:

(a) $A = \begin{bmatrix} 0 & x & y \\ y & 0 & x \\ x & y & 0 \end{bmatrix}$ **(b)** $A = \begin{bmatrix} 1 & x & x \\ -x & -2 & x \\ -x & -x & -3 \end{bmatrix}$

(c) $A = \begin{bmatrix} 1 & x & x^2 & x^3 \\ x & x^2 & x^3 & 1 \\ x^2 & x^3 & 1 & x \\ x^3 & 1 & x & x^2 \end{bmatrix}$ **(d)** $A = \begin{bmatrix} x & y & 0 & 0 \\ 0 & x & y & 0 \\ 0 & 0 & x & y \\ y & 0 & 0 & x \end{bmatrix}$

13. Show that $\det\begin{bmatrix} 0 & 1 & 1 & 1 \\ 1 & 0 & x & x \\ 1 & x & 0 & x \\ 1 & x & x & 0 \end{bmatrix} = -3x^2$.

14. Show that $\det\begin{bmatrix} 1 & x & x^2 & x^3 \\ a & 1 & x & x^2 \\ p & b & 1 & x \\ q & r & c & 1 \end{bmatrix} = (1-ax)(1-bx)(1-cx)$.

15. Show that $\det\begin{bmatrix} x & -1 & 0 & 0 \\ 0 & x & -1 & 0 \\ 0 & 0 & x & -1 \\ a & b & c & x+d \end{bmatrix} = a + bx + cx^2 + dx^3 + x^4$.

(This matrix is called the **companion matrix** of the polynomial $a + bx + cx^2 + dx^3 + x^4$.)

16. Show that $\det\begin{bmatrix} a+x & b+x & c+x \\ b+x & c+x & a+x \\ c+x & a+x & b+x \end{bmatrix} = (a+b+c+3x)[(ab+ac+bc) - (a^2+b^2+c^2)]$.

17. If $C_1, C_2, \ldots, C_n$ denote the columns of a matrix A, write $A = [C_1, C_2, \ldots, C_n]$ in block form.

(a) Show that $\det[C_1 + C'_1, C_2, \ldots, C_n] = \det[C_1, C_2, \ldots, C_n] + \det[C'_1, C_2, \ldots, C_n]$.

(b) Prove the analogous result when column i has the form $C_i + C'_i$.

(c) Prove the result for rows instead of columns.

18. By expanding along the first column, show that

$$\det\begin{bmatrix} 1 & 1 & 0 & 0 & \dots & 0 & 0 \\ 0 & 1 & 1 & 0 & \dots & 0 & 0 \\ 0 & 0 & 1 & 1 & \dots & 0 & 0 \\ \vdots & \vdots & \vdots & \vdots & & \vdots & \vdots \\ 0 & 0 & 0 & 0 & \dots & 1 & 1 \\ 1 & 0 & 0 & 0 & \dots & 0 & 1 \end{bmatrix} = 1 + (-1)^{n+1} \text{ if the matrix is } n \times n, n \geq 2$$

19. Form matrix B from a matrix A by writing the columns of A in reverse order. Express $\det B$ in terms of $\det A$.

20. Prove property 3 of Theorem 2 by expanding along the row (or column) in question.

21. Prove Theorem 4 by expanding along row or column 1 and using induction on n.

SECTION 3.2 Determinants and Matrix Inverses

In this section, several theorems about determinants will be derived. One consequence of these theorems will be that a square matrix A is invertible if and only if $\det A \neq 0$. Moreover, determinants can be used to give a formula for A^{-1} that, in turn, yields a formula (called Cramer's rule) for the solution of any system of linear equations with an invertible coefficient matrix.

In Section 2.3 it was shown that the invertible square matrices are just the ones that are products of elementary matrices. Consequently, it is not surprising that, before determinants and invertibility can be related, the determinant of an elementary matrix must be computed. To this end, recall that each elementary matrix E is obtained by performing an elementary row operation on the identity matrix I. Because $\det I = 1$, Theorem 2 in Section 3.1 provides the following information:

$$\det E = \begin{cases} -1 \text{ if } E \text{ results from interchanging two rows of } I \text{ (type I)} \\ u \text{ if } E \text{ results from multiplying a row of } I \text{ by } u \neq 0 \text{ (type II)} \\ 1 \text{ if } E \text{ results from adding a multiple of one row of } I \text{ to} \\ \quad\quad \text{another row (type III)} \end{cases}$$

Next recall that when the elementary row operation that produced E from I is performed on any $n \times n$ matrix A, the resulting matrix is EA (Theorem 5 in Section 2.3). But then, in the three cases above, $\det(EA)$ equals $-\det A$, $u \det A$, and $\det A$, respectively (again by Theorem 2). Combining this with the formulas for $\det E$, we have

$$\det(EA) = \det E \det A$$

This can be extended as follows: If E_1 and E_2 are both elementary, then

$$\det(E_2E_1A) = \det E_2 \det(E_1A) = \det E_2 \det E_1 \det A$$

This process continues in the obvious way to produce Lemma 1.

LEMMA 1

If $E_1, E_2, \ldots, E_k$ are $n \times n$ elementary matrices, then

$$\det(E_kE_{k-1} \cdots E_2E_1A) = \det E_k \det E_{k-1} \cdots \det E_2 \det E_1 \det A$$

holds for any $n \times n$ matrix A.

This formula is the key to proving three important theorems about determinants. The following preliminary result is needed.

LEMMA 2

If A is a noninvertible square matrix, then $\det A = 0$.

Proof Because A is *not* invertible, it *cannot* be carried to the identity matrix by row operations (Theorem 8 in Section 2.3). Therefore, if R is the reduced row-echelon form of A, R must have a row of zeros. But then $\det R = 0$ (Theorem 2 in Section 3.1). On the other hand, R can be obtained from A by row operations, so (by Theorem 7 in Section 2.3) there exist elementary matrices $E_1, E_2, \ldots, E_k$ such that $R = E_kE_{k-1} \cdots E_2E_1A$. But then Lemma 1 gives

$$0 = \det R = \det E_k \det E_{k-1} \cdots \det E_2 \det E_1 \det A$$

so $\det A = 0$ (each $\det E_i \neq 0$ by the remark above). This establishes Lemma 2. ■

The three theorems now follow easily.

THEOREM 1
Product Theorem

If A and B are $n \times n$ matrices, then $\det(AB) = \det A \det B$.

Proof We split the argument into two cases.

Case 1. A is not invertible. In this case, $\det A = 0$ by Lemma 2. Now observe that AB is not invertible either (if $(AB)^{-1}$ did exist, then $A[B(AB)^{-1}] = I$, so A would be invertible by Theorem 9 in Section 2.3). But then $\det(AB) = 0$ by Lemma 2 (applied to AB in place of A), so $\det A \det B = 0 \det B = 0 = \det(AB)$, as required.

Case 2. A is invertible. Then A is a product of elementary matrices by Theorem 8 in Section 2.3; say, $A = E_kE_{k-1} \cdots E_2E_1$. Then Lemma 1 with $A = I$ gives

$$\det A = \det(E_kE_{k-1} \cdots E_2E_1) = \det E_k \det E_{k-1} \cdots \det E_2 \det E_1.$$

Now use Lemma 1 once more to obtain

$$\begin{aligned}\det(AB) &= \det(E_kE_{k-1}\cdots E_2E_1B)\\ &= \det E_k \det E_{k-1}\cdots \det E_2 \det E_1 \det B\\ &= \det A \det B\end{aligned}$$

Hence the result holds in this case too. ■

It should be pointed out that Theorem 1 extends to $\det(ABC) = \det A \det B \det C$ and, by induction, to

$$\det(A_1A_2\cdots A_{k-1}A_k) = \det A_1 \det A_2 \cdots \det A_{k-1} \det A_k$$

for any square matrices $A_1, \ldots, A_k$ of the same size. In particular, if each $A_i = A$, we obtain

$$\det(A^k) = (\det A)^k \qquad \text{for any } k \geq 1$$

THEOREM 2

> An $n \times n$ matrix A is invertible if and only if $\det A \neq 0$. When this is the case, $\det(A^{-1}) = \dfrac{1}{\det A}$.

Proof If $\det A \neq 0$, then A is invertible by Lemma 2. On the other hand, if A is invertible, then $AA^{-1} = I$ so, using Theorem 1,

$$1 = \det I = \det(AA^{-1}) = \det A \det A^{-1}$$

Hence $\det A \neq 0$ and $\det A^{-1} = \dfrac{1}{\det A}$. This completes the proof. ■

THEOREM 3

> If A is any square matrix, $\det A = \det A^T$.

Proof Consider first the case of an elementary matrix E. If E is of type I or II it is symmetric—that is, $E = E^T$—so certainly $\det E = \det E^T$. If E is of type III, then E^T is also of type III and, by the above formula, $\det E = 1 = \det E^T$. Hence $\det E = \det E^T$ for every elementary matrix E. Now, if A is any square matrix, consider two cases as in the proof of Theorem 1.

Case 1. A is invertible. Then $A = E_kE_{k-1}\cdots E_2E_1$, where the E_i are elementary matrices (Theorem 8 in Section 2.3), so $A^T = E_1^TE_2^T\cdots E_{k-1}^TE_k^T$ by Theorem 1 in Section 2.2. Hence

$$\det A = \det E_k \det E_{k-1}\cdots \det E_2 \det E_1$$

$$= \det E_k^T \det E_{k-1}^T \cdots \det E_1^T = \det A^T$$

Case 2. A is not invertible. Then A^T is not invertible either (by Theorem 2 in Section 2.3 because $A = (A^T)^T$), so $\det A = 0 = \det A^T$ by Lemma 2. ■

The following examples give some indication of how these theorems are used.

EXAMPLE 1 For which values of c does $A = \begin{bmatrix} 1 & 0 & -c \\ -1 & 3 & 1 \\ 0 & 2c & -4 \end{bmatrix}$ have an inverse?

Solution Compute $\det A$ by first adding c times column 1 to column 3 and then expanding along row 1.

$$\det A = \det \begin{bmatrix} 1 & 0 & -c \\ -1 & 3 & 1 \\ 0 & 2c & -4 \end{bmatrix} = \det \begin{bmatrix} 1 & 0 & 0 \\ -1 & 3 & 1-c \\ 0 & 2c & -4 \end{bmatrix} = 2(c+2)(c-3).$$

Hence $\det A = 0$ if $c = -2$ or $c = 3$, so A has an inverse if $c \neq -2$ and $c \neq 3$.

EXAMPLE 2 Re-prove Theorem 9 in Section 2.3: If $AB = I$, then A and B are invertible, $A = B^{-1}$, $B = A^{-1}$, and $BA = I$.

Solution Take determinants to obtain $1 = \det I = \det(AB) = \det A \det B$. This implies $\det A \neq 0$ and $\det B \neq 0$, so A^{-1} and B^{-1} exist. Then left-multiplying $AB = I$ by A^{-1} yields $B = A^{-1}$, and $A = B^{-1}$ is derived similarly. Finally, $BA = BB^{-1} = I$.

EXAMPLE 3 If $\det A = 2$ and $\det B = 5$, calculate $\det(A^3B^{-1}A^TB^2)$.

Solution We use several of the facts derived above.

$$\begin{aligned} \det(A^3B^{-1}A^TB^2) &= \det(A^3) \det B^{-1} \det A^T \det (B^2) \\ &= (\det A)^3 \frac{1}{\det B} \det A \, (\det B)^2 \\ &= 2^3 \cdot \frac{1}{5} \cdot 2 \cdot 5^2 \\ &= 80 \end{aligned}$$

EXAMPLE 4 A square matrix is called **orthogonal** if $A^{-1} = A^T$. What are the possible values of $\det A$ if A is orthogonal?

Solution If A is orthogonal we have $I = AA^T$. Take determinants to obtain $1 = \det I = \det(AA^T) = \det A \det A^T = (\det A)^2$. Hence $\det A = \pm 1$.

Define the adjoint of a 2×2 matrix $A = \begin{bmatrix} a & b \\ c & d \end{bmatrix}$ to be $\operatorname{adj}(A) = \begin{bmatrix} d & -b \\ -c & a \end{bmatrix}$. At the beginning of Section 3.1 we verified that $A(\operatorname{adj} A) = (\det A)I = (\operatorname{adj} A)A$ and hence that if $\det A \neq 0$, $A^{-1} = \frac{1}{\det A} \operatorname{adj} A$. It is now possible to define the adjoint of an arbitrary square matrix and to show that this formula for the inverse is valid (when the inverse exists).

Recall that the (i,j)-cofactor $C_{ij}(A)$ of a square matrix A is a number defined for each position (i,j) in the matrix.

DEFINITION

If A is a square matrix, the **cofactor matrix of A** is defined to be the matrix $[C_{ij}(A)]$ whose (i,j)-entry is the (i,j)-cofactor of A. The **adjoint** of A, denoted $\operatorname{adj}(A)$, is the transpose of this cofactor matrix; in symbols,

$$\operatorname{adj}(A) = [C_{ij}(A)]^T$$

It is easily verified that this agrees with the earlier definition for a 2×2 matrix A (recall that the determinant of a 1×1 matrix is given by $\det[a] = a$). Here is a numerical example.

EXAMPLE 5

Compute the adjoint of $A = \begin{bmatrix} 1 & 3 & -2 \\ 0 & 1 & 5 \\ -2 & -6 & 7 \end{bmatrix}$ and calculate $A(\operatorname{adj} A)$ and $(\operatorname{adj} A)A$.

Solution We first find the cofactor matrix.

$$\begin{bmatrix} C_{11}(A) & C_{12}(A) & C_{13}(A) \\ C_{21}(A) & C_{22}(A) & C_{23}(A) \\ C_{31}(A) & C_{32}(A) & C_{33}(A) \end{bmatrix} = \begin{bmatrix} \det\begin{bmatrix} 1 & 5 \\ -6 & 7 \end{bmatrix} & -\det\begin{bmatrix} 0 & 5 \\ -2 & 7 \end{bmatrix} & \det\begin{bmatrix} 0 & 1 \\ -2 & -6 \end{bmatrix} \\ -\det\begin{bmatrix} 3 & -2 \\ -6 & 7 \end{bmatrix} & \det\begin{bmatrix} 1 & -2 \\ -2 & 7 \end{bmatrix} & -\det\begin{bmatrix} 1 & 3 \\ -2 & -6 \end{bmatrix} \\ \det\begin{bmatrix} 3 & -2 \\ 1 & 5 \end{bmatrix} & -\det\begin{bmatrix} 1 & -2 \\ 0 & 5 \end{bmatrix} & \det\begin{bmatrix} 1 & 3 \\ 0 & 1 \end{bmatrix} \end{bmatrix}$$

$$= \begin{bmatrix} 37 & -10 & 2 \\ -9 & 3 & 0 \\ 17 & -5 & 1 \end{bmatrix}$$

Then the adjoint of A is the transpose of this cofactor matrix.

$$\text{adj } A = \begin{bmatrix} 37 & -10 & 2 \\ -9 & 3 & 0 \\ 17 & -5 & 1 \end{bmatrix}^T = \begin{bmatrix} 37 & -9 & 17 \\ -10 & 3 & -5 \\ 2 & 0 & 1 \end{bmatrix}$$

The computation of $A(\text{adj } A)$ gives

$$A(\text{adj } A) = \begin{bmatrix} 1 & 3 & -2 \\ 0 & 1 & 5 \\ -2 & -6 & 7 \end{bmatrix} \begin{bmatrix} 37 & -9 & 17 \\ -10 & 3 & -5 \\ 2 & 0 & 1 \end{bmatrix} = \begin{bmatrix} 3 & 0 & 0 \\ 0 & 3 & 0 \\ 0 & 0 & 3 \end{bmatrix} = 3I$$

and the reader may verify that also $(\text{adj } A)A = 3I$. Moreover, $\det A = 3$, as is indicated by analogy with the situation above for 2×2 matrices.

The relationship $A(\text{adj } A) = (\det A)I$ holds for any square matrix A. To see why this is so, consider the general 3×3 case. Writing $C_{ij}(A) = C_{ij}$ for short, we have

$$\text{adj } A = \begin{bmatrix} C_{11} & C_{12} & C_{13} \\ C_{21} & C_{22} & C_{23} \\ C_{31} & C_{32} & C_{33} \end{bmatrix}^T = \begin{bmatrix} C_{11} & C_{21} & C_{31} \\ C_{12} & C_{22} & C_{32} \\ C_{13} & C_{23} & C_{33} \end{bmatrix}$$

If $A = [a_{ij}]$ in the usual notation, we are to verify that $A(\text{adj } A) = (\det A)I$. That is,

$$A(\text{adj } A) = \begin{bmatrix} a_{11} & a_{12} & a_{13} \\ a_{21} & a_{22} & a_{23} \\ a_{31} & a_{32} & a_{33} \end{bmatrix} \begin{bmatrix} C_{11} & C_{21} & C_{31} \\ C_{12} & C_{22} & C_{32} \\ C_{13} & C_{23} & C_{33} \end{bmatrix} \overset{?}{=} \begin{bmatrix} \det A & 0 & 0 \\ 0 & \det A & 0 \\ 0 & 0 & \det A \end{bmatrix}$$

Consider the (1,1)-entry in the product. It is given by $a_{11}C_{11} + a_{12}C_{12} + a_{13}C_{13}$, and this is just the Laplace expansion of $\det A$ along the first row of A. Similarly, the (2,2)-entry and the (3,3)-entry are the Laplace expansions of $\det A$ along rows 2 and 3, respectively.

So it remains to be seen why the off-diagonal elements in the matrix product $A(\text{adj } A)$ are all zero. Consider the (1,2)-entry of the product. It is given by $a_{11}C_{21} + a_{12}C_{22} + a_{13}C_{23}$. This *looks* like the Laplace expansion of the determinant of *some* matrix. To see which, observe that C_{21}, C_{22}, and C_{23} are all computed by *deleting* row 2 of A and one of the columns, so they remain the same if row 2 of A is changed. In particular, if row 2 of A is replaced by row 1, we obtain

$$a_{11}C_{21} + a_{12}C_{22} + a_{13}C_{23} = \det \begin{bmatrix} a_{11} & a_{12} & a_{13} \\ a_{11} & a_{12} & a_{13} \\ a_{31} & a_{32} & a_{33} \end{bmatrix} = 0$$

where the expansion is along row 2, and where the determinant is zero because two rows are identical. A similar argument shows that the other off-diagonal entries are zero.

This argument works in general and yields the first part of Theorem 4. The second assertion follows from the first by multiplying through by the scalar $\frac{1}{\det A}$.

THEOREM 4
Adjoint Formula

If A is any square matrix, then

$$A(\text{adj } A) = (\det A)I = (\text{adj } A)A$$

In particular, if $\det A \neq 0$, the inverse of A is given by

$$A^{-1} = \frac{1}{\det A} \text{adj } A$$

It is important to note that this theorem is *not* an efficient way to find the inverse of matrix A. For example, if A were 10×10, the calculation of adj A would require computing $10^2 = 100$ determinants of 9×9 matrices! On the other hand, the matrix inversion algorithm would find A^{-1} with about the same effort as finding det A. Clearly Theorem 4 is not a *practical* result; its main virtue is that it gives a formula for A^{-1} that is useful for *theoretical* purposes.

EXAMPLE 6

Use Theorem 4 to find the inverse of $A = \begin{bmatrix} 1 & 1 & a \\ -a & 1 & -a \\ a & -1 & 1 \end{bmatrix}$ for the values of a for which it exists.

Solution The adjoint is computed as follows:

$$\text{adj } A = \begin{bmatrix} C_{11} & C_{12} & C_{13} \\ C_{21} & C_{22} & C_{23} \\ C_{31} & C_{32} & C_{33} \end{bmatrix}^T = \begin{bmatrix} 1-a & a-a^2 & 0 \\ -1-a & 1-a^2 & 1+a \\ -2a & a-a^2 & 1+a \end{bmatrix}^T$$

$$= \begin{bmatrix} 1-a & -1-a & -2a \\ a-a^2 & 1-a^2 & a-a^2 \\ 0 & 1+a & 1+a \end{bmatrix}$$

The reader can verify that $A(\text{adj } A) = (1-a^2)I = (\text{adj } A)A$, and this shows that $\det A = 1 - a^2$ (as may be separately verified). Hence A^{-1} exists if $1 - a^2 \neq 0$ (that is, $a \neq \pm 1$), and in this case

$$A^{-1} = \frac{1}{1-a^2}\begin{bmatrix} 1-a & -1-a & -2a \\ a-a^2 & 1-a^2 & a-a^2 \\ 0 & 1+a & 1+a \end{bmatrix}$$

Theorem 4 has a nice application to linear equations. Suppose

$$A\mathbf{x} = \mathbf{b}$$

is a system of n equations in n variables $x_1, x_2, \ldots, x_n$. Here A is the $n \times n$ coefficient matrix, and $\mathbf{x}$ and $\mathbf{b}$ are the columns

$$\mathbf{x} = \begin{bmatrix} x_1 \\ x_2 \\ \vdots \\ x_n \end{bmatrix} \qquad \mathbf{b} = \begin{bmatrix} b_1 \\ b_2 \\ \vdots \\ b_n \end{bmatrix}$$

of variables and scalars, respectively. If $\det A \neq 0$, we left-multiply by A^{-1} to obtain the solution $\mathbf{x} = A^{-1}\mathbf{b}$. When we use the adjoint formula, this becomes

$$\begin{aligned} \begin{bmatrix} x_1 \\ x_2 \\ \vdots \\ x_n \end{bmatrix} &= \frac{1}{\det A} \operatorname{adj} A\, \mathbf{b} \\ &= \frac{1}{\det A} \begin{bmatrix} C_{11}(A) & C_{21}(A) & \ldots & C_{n1}(A) \\ C_{12}(A) & C_{22}(A) & \ldots & C_{n2}(A) \\ \vdots & \vdots & & \vdots \\ C_{1n}(A) & C_{2n}(A) & \ldots & C_{nn}(A) \end{bmatrix} \begin{bmatrix} b_1 \\ b_2 \\ \vdots \\ b_n \end{bmatrix} \end{aligned}$$

Hence the variables $x_1, x_2, \ldots, x_n$ are given by

$$\begin{aligned} x_1 &= \frac{1}{\det A}[b_1C_{11}(A) + b_2C_{21}(A) + \cdots + b_nC_{n1}(A)] \\ x_2 &= \frac{1}{\det A}[b_1C_{12}(A) + b_2C_{22}(A) + \cdots + b_nC_{n2}(A)] \\ &\vdots \qquad\qquad\qquad \vdots \\ x_n &= \frac{1}{\det A}[b_1C_{1n}(A) + b_2C_{2n}(A) + \cdots + b_nC_{nn}(A)] \end{aligned}$$

Now the quantity $b_1C_{11}(A) + b_2C_{21}(A) + \cdots + b_nC_{n1}(A)$ occurring in the formula for x_1 looks like the Laplace expansion of the determinant of a matrix. The cofactors involved are $C_{11}(A), C_{21}(A), \ldots, C_{n1}(A)$, corresponding to the first column of A. If A_1 is obtained from A by replacing the first column of A by $\mathbf{b}$, then $C_{i1}(A_1) = C_{i1}(A)$ for each i, so expanding $\det(A_1)$ by the first column gives

$$\begin{aligned} \det A_1 &= b_1C_{11}(A_1) + b_2C_{21}(A_1) + \cdots + b_nC_{n1}(A_1) \\ &= b_1C_{11}(A) + b_2C_{21}(A) + \cdots + b_nC_{n1}(A) \\ &= \det A\ x_1 \end{aligned}$$

Hence $x_1 = \dfrac{\det A_1}{\det A}$, and similar results hold for the other variables:

THEOREM 5
Cramer's Rule

If A is an invertible $n \times n$ matrix, the solution to the system

$$A\mathbf{x} = \mathbf{b}$$

of n equations in the variables $x_1, x_2, \ldots, x_n$ is given by

$$x_1 = \frac{\det A_1}{\det A}, x_2 = \frac{\det A_2}{\det A}, \ldots, x_n = \frac{\det A_n}{\det A}$$

where, for each k, A_k is the matrix obtained from A by replacing column k by $\mathbf{b}$.

EXAMPLE 7 Find x_1, given the following system of equations.

$$\begin{aligned} 5x_1 + x_2 - x_3 &= 4 \\ 9x_1 + x_2 - x_3 &= 1 \\ x_1 - x_2 + 5x_3 &= 2 \end{aligned}$$

Solution Compute the determinant of the coefficient matrix A and the matrix A_1 obtained from it by replacing the first column by the column of constants.

$$\det A = \det \begin{bmatrix} 5 & 1 & -1 \\ 9 & 1 & -1 \\ 1 & -1 & 5 \end{bmatrix} = -16$$

$$\det A_1 = \det \begin{bmatrix} 4 & 1 & -1 \\ 1 & 1 & -1 \\ 2 & -1 & 5 \end{bmatrix} = 12$$

Hence $x_1 = (\det A_1)/\det A = -3/4$.

Note that Cramer's rule enabled us to calculate x_1 here without computing x_2 or x_3. Although this might seem an advantage, the truth of the matter is that for large systems of equations, the number of computations needed to find *all* the variables by the Gaussian algorithm is comparable to the number required to find *one* of the determinants involved in Cramer's rule. Furthermore, the algorithm works when the matrix of the system is not invertible, and even when the coefficient matrix is not square. Like the adjoint formula, then, Cramer's rule is *not* a practical numerical technique; its virtue is theoretical.

EXERCISES 3.2

1. Find the adjoint of each of the following matrices.

(a) $\begin{bmatrix} 3 & 1 & 2 \\ -1 & 1 & 3 \\ 1 & 3 & 8 \end{bmatrix}$ **(b)** $\begin{bmatrix} 1 & -1 & 2 \\ 3 & 1 & 0 \\ 0 & -1 & 1 \end{bmatrix}$

(c) $\begin{bmatrix} 1 & 0 & 1 \\ 1 & 1 & 0 \\ 0 & 1 & 1 \end{bmatrix}$ **(d)** $\frac{1}{3}\begin{bmatrix} -1 & 2 & 2 \\ 2 & -1 & 2 \\ 2 & 2 & -1 \end{bmatrix}$

2. Use determinants to find which real values of c make each of the following matrices invertible.

(a) $\begin{bmatrix} 1 & 0 & 2 \\ 3 & -4 & c \\ 2 & 5 & 8 \end{bmatrix}$ **(b)** $\begin{bmatrix} 0 & c & -c \\ -1 & 2 & -1 \\ c & -c & c \end{bmatrix}$ **(c)** $\begin{bmatrix} c & 1 & 0 \\ 0 & 2 & c \\ -1 & c & 1 \end{bmatrix}$

(d) $\begin{bmatrix} 4 & c & 3 \\ c & 2 & c \\ 5 & c & 4 \end{bmatrix}$ **(e)** $\begin{bmatrix} 1 & 2 & 4 \\ -1 & -1 & c \\ 2 & c & 1 \end{bmatrix}$ **(f)** $\begin{bmatrix} 1 & c & -1 \\ c & 1 & 1 \\ 0 & 1 & c \end{bmatrix}$

3. Let A, B, and C denote $n \times n$ matrices and assume that $\det A = -1$, $\det B = 2$, and $\det C = 3$. Evaluate:

(a) $\det(A^2BC^TB^{-1})$ **(b)** $\det(B^2C^{-1}AB^{-1}C^T)$

4. Let A and B be invertible $n \times n$ matrices. Evaluate:

(a) $\det(B^{-1}AB)$ **(b)** $\det(A^{-1}B^{-1}AB)$

5. Let $A = \begin{bmatrix} a & b & c \\ p & q & r \\ u & v & w \end{bmatrix}$ and assume that $\det A = 3$. Compute:

(a) $\det(3B^{-1})$ where $B = \begin{bmatrix} 4u & 2a & -p \\ 4v & 2b & -q \\ 4w & 2c & -r \end{bmatrix}$

(b) $\det(2C^{-1})$ where $C = \begin{bmatrix} 2p & -a+u & 3u \\ 2q & -b+v & 3v \\ 2r & -c+w & 3w \end{bmatrix}$

6. Solve each of the following by Cramer's rule:

(a) $\begin{aligned} 2x + y &= 1 \\ 3x + 7y &= -2 \end{aligned}$ **(b)** $\begin{aligned} 3x + 4y &= 9 \\ 2x - y &= -1 \end{aligned}$

(c) $\begin{aligned} 5x + y - z &= -7 \\ 2x - y - 2z &= 6 \\ 3x + 2z &= -7 \end{aligned}$ **(d)** $\begin{aligned} 4x - y + 3z &= 1 \\ 6x + 2y - z &= 0 \\ 3x + 3y + 2z &= -1 \end{aligned}$

7. Use Theorem 4 to find the (2,3)-entry of A^{-1} if:

(a) $A = \begin{bmatrix} 3 & 2 & 1 \\ 1 & 1 & 2 \\ -1 & 2 & 1 \end{bmatrix}$ **(b)** $A = \begin{bmatrix} 1 & 2 & -1 \\ 3 & 1 & 1 \\ 0 & 4 & 7 \end{bmatrix}$

8. Explain what can be said about $\det A$ if:

(a) $A^2 = A$ **(b)** $A^2 = I$ **(c)** $A^3 = A$

(d) $PA = P$, P invertible **(e)** $A^2 = uA$, A is $n \times n$

(f) $A = -A^T$, A is $n \times n$ **(g)** $A^2 + I = 0$, A is $n \times n$

9. Let A be $n \times n$. Show that $uA = (uI)A$, and use this with Theorem 1 to deduce the result in Theorem 3 of Section 3.1: $\det(uA) = u^n \det A$.

10. If A and B are $n \times n$ matrices, $AB = -BA$, and n is odd, show that either A or B has no inverse.

11. Show that $\det AB = \det BA$ holds for any two $n \times n$ matrices A and B.

12. If $A^k = 0$ for some $k \geq 1$, show that A is not invertible using Theorem 2.

13. If ABC is invertible (A, B, and C all square), show that B is invertible using Theorem 2.

14. Let $A = \begin{bmatrix} a & b \\ -b & a \end{bmatrix}$ and $B = \begin{bmatrix} c & d \\ -d & c \end{bmatrix}$. Compute $\det AB = \det A \det B$ to prove that $(a^2 + b^2)(c^2 + d^2) = (ac - bd)^2 + (ad + bc)^2$.

15. Show that $\det(A + B^T) = \det(A^T + B)$ for any $n \times n$ matrices A and B.

16. Let A and B be invertible $n \times n$ matrices. Show that $\det A = \det B$ if and only if $A = UB$, where U is a matrix with $\det U = 1$.

17. For each of the matrices in Exercise 2, find the inverse for those values of c for which it exists.

18. If $A = \begin{bmatrix} 1 & a & b \\ -a & 1 & c \\ -b & -c & 1 \end{bmatrix}$, show that $\det A = 1 + a^2 + b^2 + c^2$. Hence find A^{-1} for any a, b, and c.

19. **(a)** Show that $A = \begin{bmatrix} a & p & q \\ 0 & b & r \\ 0 & 0 & c \end{bmatrix}$ has an inverse if and only if $abc \neq 0$, and find A^{-1} in that case.

(b) Show that if an upper triangular matrix is invertible, the inverse is again upper triangular.

20. Let A be a matrix each of whose entries are integers. Show that each of the following conditions implies the other.

(1) A is invertible and A^{-1} also has integer entries.

(2) $\det A = 1$ or -1

21. **(a)** If A is 3×3 and $\det A = 2$, find $\det(A^{-1} + 4 \operatorname{adj} A)$.

(b) If $A^{-1} = \begin{bmatrix} 3 & 0 & 1 \\ 0 & 2 & 3 \\ 3 & 1 & -1 \end{bmatrix}$, find $\operatorname{adj} A$.

22. Show that $\operatorname{adj}(uA) = u^{n-1} \operatorname{adj} A$ for all $n \times n$ matrices A.

23. Let A and B denote $n \times n$ matrices. Show that:

(a) $\det A = 0$ if and only if $\det(\operatorname{adj} A) = 0$ [*Hint:* Use property 2 of Theorem 8 in Section 2.3.]

(b) $\det(\operatorname{adj} A) = (\det A)^{n-1}$ [*Hint:* If $\det A = 0$, use part **(a)**.]

24. Let A and B denote invertible $n \times n$ matrices. Show that:

(a) $\operatorname{adj}(\operatorname{adj} A) = (\det A)^{n-2} A$ (here $n \geq 2$) [*Hint:* The preceding exercise.]

(b) $\operatorname{adj}(A^{-1}) = (\operatorname{adj} A)^{-1}$

(c) $\text{adj}(A^T) = (\text{adj } A)^T$

(d) $\text{adj}(AB) = (\text{adj } B)(\text{adj } A)$
[*Hint:* Show that $AB \text{ adj}(AB) = AB \text{ adj } B \text{ adj } A$.]

SECTION 3.3 Proof of the Laplace Expansion (Optional)

Recall that our definition of the term *determinant* is an inductive one: The determinant of any 1×1 matrix is defined first; then it is used to define the determinants of 2×2 matrices. Then that is used for the 3×3 case, and so on. The case of a 1×1 matrix $[a]$ poses no problem. We simply define

$$\det[a] = a$$

as was done in Section 3.1. Given an $n \times n$ matrix A, define A_{ij} to be the $(n - 1) \times (n - 1)$ matrix obtained from A by deleting row i and column j. Now assume that the determinant of any $(n - 1) \times (n - 1)$ matrix has been defined. Then the **determinant** of A is defined to be

$$\begin{aligned} \det A &= a_{11} \det A_{11} - a_{21} \det A_{21} + \cdots \pm a_{n1} \det A_{n1} \\ &= \sum_{i=1}^{n} (-1)^{i+1} a_{i1} \det A_{i1} \end{aligned}$$

Observe that, in the terminology of Section 3.1, this is just the Laplace expansion of $\det A$ along the first column, that $\det A_{ij}$ is just the (i,j)-minor of A (denoted as $M_{ij}(A)$ above), and that $(-1)^{i+j} \det A_{ij}$ is the (i,j)-cofactor (denoted as $C_{ij}(A)$ above). To illustrate the definition, consider the 2×2 matrix $A = \begin{bmatrix} a_{11} & a_{12} \\ a_{21} & a_{22} \end{bmatrix}$. Then the definition gives

$$\det \begin{bmatrix} a_{11} & a_{12} \\ a_{21} & a_{22} \end{bmatrix} = a_{11} \det[a_{22}] - a_{21} \det[a_{12}] = a_{11}a_{22} - a_{21}a_{12}$$

and this is the same as the definition given in Section 3.1.

Of course, the task now is to use this definition to *prove* that the Laplace expansion along *any* row or column yields $\det A$ (this is Theorem 1 in Section 3.1). The proof proceeds by first establishing the properties of determinants stated in Theorem 2 of Section 3.1, but for *rows* only (Lemma 2 below). This being done, the full proof of Theorem 1 in Section 3.1 is not difficult. The proof of Lemma 2 requires the following preliminary result.

LEMMA 1 Let A, B, and C be $n \times n$ matrices that are identical except that the pth row of A is the sum of the pth rows of B and C. Then $\det A = \det B + \det C$.

Proof We proceed by induction on n, the cases $n = 1$ and $n = 2$ being easily checked. Consider a_{i1} and A_{i1}:

1. If $i \neq p$,

$$a_{i1} = b_{i1} = c_{i1} \qquad \text{and} \qquad \det A_{i1} = \det B_{i1} + \det C_{i1}$$

by induction because A_{i1}, B_{i1}, and C_{i1} are identical except that one row of A_{i1} is the sum of the corresponding rows of B_{i1} and C_{i1}.

2. If $i = p$,

$$a_{p1} = b_{p1} + c_{p1} \qquad \text{and} \qquad A_{p1} = B_{p1} = C_{p1}$$

Now write out the defining sum for $\det A$, splitting off the pth term for special attention.

$$\begin{aligned} \det A &= \sum_{i \neq p} a_{i1}(-1)^{i+1} \det A_{i1} + a_{p1}(-1)^{p+1} \det A_{p1} \\ &= \sum_{i \neq p} a_{i1}(-1)^{i+1} [\det B_{i1} + \det C_{i1}] + (b_{p1} + c_{p1})(-1)^{p+1} \det A_{p1} \end{aligned}$$

But the terms here involving B_{i1} and b_{p1} add up to $\det B$, because $a_{i1} = b_{i1}$ if $i \neq p$ and $A_{p1} = B_{p1}$. Similarly, the terms involving C_{i1} and c_{p1} add up to $\det C$. Hence $\det A = \det B + \det C$, as required. ■

LEMMA 2 Let $A = [a_{ij}]$ denote an $n \times n$ matrix.

1. If $B = [b_{ij}]$ is formed from A by multiplying a row of A by a number u, then $\det B = u \det A$.
2. If A contains a row of zeros, then $\det A = 0$.
3. If $B = [b_{ij}]$ is formed by interchanging two rows of A, then $\det B = -\det A$.
4. If A contains two identical rows, then $\det A = 0$.
5. If $B = [b_{ij}]$ is formed by adding a multiple of one row of A to a different row, then $\det B = \det A$.

Proof For reference below, the defining sums for $\det A$ and $\det B$ are as follows:

$$\det A = \sum_{i=1}^{n} a_{i1}(-1)^{i+1} \det A_{i1} \quad (*)$$

$$\det B = \sum_{i=1}^{n} b_{i1}(-1)^{i+1} \det B_{i1}. \quad (**)$$

Property 1. The proof is by induction on n, the cases $n = 1$ and $n = 2$ being easily verified. Suppose that row p of A is multiplied by u. Consider the ith term in the sum (**) for det B.

a. If $i \neq p$, then $b_{i1} = a_{i1}$ and $\det B_{i1} = u \det A_{i1}$ by induction, because B_{i1} comes from A_{i1} by multiplying a row by u.

b. If $i = p$, then $b_{p1} = ua_{p1}$ and $B_{p1} = A_{p1}$.

In either case, each term in (**) is u times the corresponding term in (*), so it is clear that $\det B = u \det A$.

Property 2. This is clear by property 1, because the row of zeros has a common factor $u = 0$.

Property 3. Observe first that it suffices to prove property 3 for interchanges of adjacent rows. (Rows p and q ($q > p$) can be interchanged by carrying out $2(q - p) - 1$ adjacent changes, which results in an *odd* number of sign changes in the determinant.) So suppose that rows p and $p + 1$ of A are interchanged to obtain B. Again consider the ith term in (**).

a. If $i \neq p$ and $i \neq p + 1$, then $b_{i1} = a_{i1}$ and $\det B_{i1} = -\det A_{i1}$ by induction, because B_{i1} results from interchanging adjacent rows in A_{i1}. Hence the ith term in (**) is the negative of the ith term in (*), and so $\det B = -\det A$.

b. If $i = p$ or $i = p + 1$, then $b_{p1} = a_{p+1\,1}$ and $B_{p1} = A_{p+1\,1}$, whereas $b_{p+1\,1} = a_{p1}$ and $B_{p+1\,1} = A_{p1}$. Hence terms p and $p + 1$ in (**) are

$$b_{p1}(-1)^{p+1} \det B_{p1} = -a_{p+1\,1}(-1)^{(p+1)+1}\det(A_{p+1\,1})$$

$$b_{p+1\,1}(-1)^{(p+1)+1}\det(B_{p+1\,1}) = -a_{p1}(-1)^{p+1}\det A_{p1}$$

This means that terms p and $p + 1$ in (**) are the same as these terms in (*), except that the order is reversed and the signs are changed. Thus the sum (**) is the negative of the sum (*); that is, $\det B = -\det A$.

Property 4. If rows p and q in A are identical, let B be obtained from A by interchanging these rows. Then $B = A$, so $\det A = \det B$. But $\det B = -\det A$ by property 3, so $\det A = -\det A$. This implies that $\det A = 0$.

Property 5. Suppose B results from adding u times row q of A to row p. Then Lemma 1 applies to B to show that $\det B = \det A + \det C$, where C is obtained from A by replacing row p by u times row q. It now follows from properties 1 and 4 that $\det C = 0$, so $\det B = \det A$, as asserted. ■

These facts are enough to enable us to prove Theorem 1 in Section 3.1. For convenience, it is restated here in the notation of the foregoing lemmas. The only difference between the notations is that the (i,j)-cofactor of an $n \times n$ matrix A was denoted earlier by

$$C_{ij}(A) = (-1)^{i+j}\det A_{ij}$$

THEOREM 1 If $A = [a_{ij}]$ is an $n \times n$ matrix, then:

1. $\det A = \sum_{i=1}^{n} a_{ij}(-1)^{i+j}\det A_{ij}$ (Laplace expansion along column j)
2. $\det A = \sum_{j=1}^{n} a_{ij}(-1)^{i+j}\det A_{ij}$ (Laplace expansion along row i).

Here A_{ij} denotes the matrix obtained from A by deleting row i and column j.

Proof Lemma 2 establishes the truth of Theorem 2 in Section 3.1 for *rows*. With this information, the arguments in Section 3.2 proceed exactly as written to establish that $\det A = \det A^T$ holds for any $n \times n$ matrix A. Now suppose B is obtained from A by interchanging two columns. Then B^T is obtained from A^T by interchanging two rows, so, by property 3 of Lemma 2,

$$\det B = \det B^T = -\det A^T = -\det A$$

Hence property 3 of Lemma 2 holds for *columns* too.

This enables us to prove the Laplace expansion for columns. Given an $n \times n$ matrix $A = [a_{ij}]$, let $B = [b_{ij}]$ be obtained by moving column j to the left side, using $j - 1$ interchanges of adjacent columns. Then $\det B = (-1)^{j-1} \det A$ and, because $B_{i1} = A_{ij}$ and $b_{i1} = a_{ij}$ for all i, we obtain

$$\det A = (-1)^{j-1}\det B = (-1)^{j-1} \sum_{i=1}^{n} b_{i1}(-1)^{1+i}\det B_{i1}$$

$$= \sum_{i=1}^{n} a_{ij}(-1)^{i+j}\det A_{ij}$$

This is the Laplace expansion of det A along column j.

Finally, to prove the row expansion, write $B = A^T$. Then $B_{ij} = A_{ji}{}^T$ and $b_{ij} = a_{ji}$ for all i and j. Expanding det B along column j gives

$$\det A = \det A^T = \det B = \sum_{i=1}^{n} b_{ij}(-1)^{i+j}\det B_{ij}$$

$$= \sum_{i=1}^{n} a_{ji}(-1)^{j+i}\det(A_{ji}{}^T) = \sum_{i=1}^{n} a_{ji}(-1)^{j+i}\det A_{ji}$$

This is the required expansion of det A along row j. ■

EXERCISES 3.3

1. Prove Lemma 1 for columns.
2. Verify that interchanging rows p and q ($q > p$) can be accomplished using $2(q - p) - 1$ adjacent interchanges.
3. If u is a number and A is an $n \times n$ matrix, prove that $\det(uA) = u^n \det A$ by induction on n, using only the definition of det A.

SECTION 3.4 An Application to Polynomial Interpolation (Optional)

There often arise situations wherein two variables x and y are related but the actual functional form $y = f(x)$ of the relationship is unknown. Suppose that, for certain values $x_1, x_2, \ldots, x_n$ of x, the corresponding values $y_1, y_2, \ldots, y_n$ are known (say from experimental measurements). One way to estimate the value of y corresponding to some other value of x is to find a polynomial $p(x)$ that "fits" the data, that is, $p(x_i) = y_i$ holds for each $i = 1, 2, \ldots, n$. Then the estimate for y is $p(x)$. Such a polynomial always exists if the x_i are distinct.

THEOREM 1 Let n data pairs $(x_1,y_1), (x_2,y_2), \ldots, (x_n,y_n)$ be given, and assume that the x_i are distinct. Then there exists a unique polynomial

$$p(x) = r_0 + r_1x + r_2x^2 + \cdots + r_{n-1}x^{n-1}$$

such that $p(x_i) = y_i$ for each $i = 1, 2, \ldots, n$.

Proof The conditions that $p(x_i) = y_i$ are

$$\begin{array}{ccccccccccc}
r_0 & + & r_1x_1 & + & r_2x_1{}^2 & + & \cdots & + & r_{n-1}x_1{}^{n-1} & = & y_1 \\
r_0 & + & r_1x_2 & + & r_2x_2{}^2 & + & \cdots & + & r_{n-1}x_2{}^{n-1} & = & y_2 \\
\vdots & & \vdots & & \vdots & & & & \vdots & & \vdots \\
r_0 & + & r_1x_n & + & r_2x_n{}^2 & + & \cdots & + & r_{n-1}x_n{}^{n-1} & = & y_n
\end{array}$$

In matrix form, this is

$$\begin{bmatrix} 1 & x_1 & x_1^2 & \dots & x_1^{n-1} \\ 1 & x_2 & x_2^2 & \dots & x_2^{n-1} \\ \vdots & \vdots & \vdots & & \vdots \\ 1 & x_n & x_n^2 & \dots & x_n^{n-1} \end{bmatrix} \begin{bmatrix} r_0 \\ r_1 \\ \vdots \\ r_{n-1} \end{bmatrix} = \begin{bmatrix} y_1 \\ y_2 \\ \vdots \\ y_n \end{bmatrix}$$

It can be shown (see Theorem 2 below) that the determinant of the coefficient matrix equals the product of all terms $(x_i - x_j)$ with $i > j$ and so is nonzero (because the x_i are distinct). Hence the equations have a unique solution $r_0, r_1, \dots, r_{n-1}$ and the theorem follows. ■

The polynomial in Theorem 1 is called the **interpolating polynomial** for the data.

EXAMPLE 1 Find a polynomial $p(x)$ of degree 2 such that $p(0) = 1$, $p(1) = 3$, and $p(2) = 2$.

Solution Write $p(x) = r_0 + r_1x + r_2x^2$. The conditions are

$$\begin{aligned} p(0) &= r_0 &&= 1 \\ p(1) &= r_0 + r_1 + r_2 &&= 3 \\ p(2) &= r_0 + 2r_1 + 4r_2 &&= 2 \end{aligned}$$

The solution is $r_0 = 1$, $r_1 = \frac{7}{2}$, and $r_2 = -\frac{3}{2}$, so $p(x) = \frac{1}{2}(2 + 7x - 3x^2)$.

The next example shows how Theorem 1 is used in interpolation.

EXAMPLE 2 Given the data values

$$(0,1.21), (1,3.53), (2,5.01), (3,3.79)$$

use polynomial interpolation to estimate the value of y corresponding to $x = 1.5$.

Solution We find the polynomial $p(x)$ of degree 3 that fits these data. If $p(x) = r_0 + r_1x + r_2x^2 + r_3x^3$, the conditions are

$$\begin{aligned} r_0 &&&= 1.21 \\ r_0 + r_1 + r_2 + r_3 &&&= 3.53 \\ r_0 + 2r_1 + 4r_2 + 8r_3 &&&= 5.01 \\ r_0 + 3r_1 + 9r_2 + 27r_3 &&&= 3.79 \end{aligned}$$

The solution is $r_0 = 1.21$, $r_1 = 2.12$, $r_2 = 0.51$, and $r_3 = -0.31$, so the interpolating polynomial is $p(x) = 1.21 + 2.12x + 0.51x^2 - 0.31x^3$. Hence the estimated value of y corresponding to $x = 1.5$ is $p(1.5) = 4.49$.

As a final example, we construct a polynomial that will approximate a known function. This type of approximation is often useful in practical situations, because polynomials are easy to compute.

EXAMPLE 3 Find a cubic polynomial $p(x)$ that approximates the function $\sin x$ on the interval $0 \le x \le \pi/2$ (x in radians). Use the following values of $\sin x$.

$$\sin 0 = 0, \quad \sin(\pi/6) = 0.5000, \quad \sin(\pi/3) = .8660, \quad \sin(\pi/2) = 1$$

Then use $p(x)$ to approximate $\sin(0.5)$.

Solution If $p(x) = r_0 + r_1x + r_2x^2 + r_3x^3$, use $x_0 = 0$, $x_1 = \pi/6$, $x_2 = \pi/3$, and $x_3 = \pi/2$, and $y_i = \sin(x_i)$ as given.

$$\begin{aligned}
p(0) &= r_0 &&= 0\\
p\left(\frac{\pi}{6}\right) &= r_0 + r_1\left(\frac{\pi}{6}\right) + r_2\left(\frac{\pi}{6}\right)^2 + r_3\left(\frac{\pi}{6}\right)^3 &&= 0.5000\\
p\left(\frac{\pi}{3}\right) &= r_0 + r_1\left(\frac{\pi}{3}\right) + r_2\left(\frac{\pi}{3}\right)^2 + r_3\left(\frac{\pi}{3}\right)^3 &&= 0.8660\\
p\left(\frac{\pi}{2}\right) &= r_0 + r_1\left(\frac{\pi}{2}\right) + r_2\left(\frac{\pi}{2}\right)^2 + r_3\left(\frac{\pi}{2}\right)^3 &&= 1.0000
\end{aligned}$$

Clearly $r_0 = 0$, so multiplying the remaining equations by $6/\pi$, $3/\pi$, and $2/\pi$, respectively, gives

$$\begin{aligned}
r_1 + \frac{r_2\pi}{6} + \frac{r_3\pi^2}{36} &= 0.9549\\
r_1 + \frac{r_2\pi}{3} + \frac{r_3\pi^2}{9} &= 0.8270\\
r_1 + \frac{r_2\pi}{2} + \frac{r_3\pi^2}{4} &= 0.6366
\end{aligned}$$

If these are regarded as equations in r_1, $r_2\pi$, and $r_3\pi^2$, the coefficient matrix has an inverse.

$$\begin{bmatrix} 1 & \frac{1}{6} & \frac{1}{36} \\ 1 & \frac{1}{3} & \frac{1}{9} \\ 1 & \frac{1}{2} & \frac{1}{4} \end{bmatrix}^{-1} = \begin{bmatrix} 3 & -3 & 1 \\ -15 & 24 & -9 \\ 18 & -36 & 18 \end{bmatrix}$$

This leads to $r_1 = 1.0203$, $r_2\pi = -0.2049$, and $r_3\pi^2 = -1.1250$. Finally, then, $r_2 = -0.0652$ and $r_3 = -0.1140$, so

$$p(x) = 1.0203x - 0.0652x^2 - 0.1140x^3$$

This gives $p(0.5) = 0.4796$ as the approximation to $\sin(0.5)$. The true value is $\sin(0.5) = 0.4794$, to four decimal places. This is quite good, and even better approximations are achieved with polynomials of higher degree.

We conclude this section by evaluating the determinant of the matrix that arose in the proof of Theorem 1. If $a_1, a_2, \ldots, a_n$ are numbers, the $n \times n$ matrix

$$\begin{bmatrix} 1 & 1 & 1 & \cdots & 1 & 1 \\ a_1 & a_2 & a_3 & \cdots & a_{n-1} & a_n \\ a_1^2 & a_2^2 & a_3^2 & \cdots & a_{n-1}^2 & a_n^2 \\ \vdots & \vdots & \vdots & & \vdots & \vdots \\ a_1^{n-1} & a_2^{n-1} & a_3^{n-1} & \cdots & a_{n-1}^{n-1} & a_n^{n-1} \end{bmatrix}$$

(or its transpose) is called a **Vandermonde matrix,** and its determinant is called a **Vandermonde determinant.** There is a very nice formula for this determinant. If $n = 2$, the determinant is $(a_2 - a_1)$; if $n = 3$, it is $(a_3 - a_2)(a_3 - a_1)(a_2 - a_1)$ by Example 8, Section 3.1. The general result is as follows:

THEOREM 2 Vandermonde Determinant

Let $a_1, a_2, \ldots, a_n$ be real numbers, $n \geq 2$. Then the corresponding Vandermonde determinant is given by

$$\det\begin{bmatrix} 1 & 1 & 1 & \cdots & 1 \\ a_1 & a_2 & a_3 & \cdots & a_n \\ a_1^2 & a_2^2 & a_3^2 & \cdots & a_n^2 \\ \vdots & \vdots & \vdots & & \vdots \\ a_1^{n-1} & a_2^{n-1} & a_3^{n-1} & \cdots & a_n^{n-1} \end{bmatrix} = \prod_{1 \leq j < i \leq n} (a_i - a_j)$$

where $\prod_{1 \leq j < i \leq n} (a_i - a_j)$ means the product of all factors $(a_i - a_j)$ where $1 \leq j < i \leq n$.

Proof We may assume that the a_i are distinct, because otherwise both sides would be zero. Proceed by induction on $n \geq 2$, and assume inductively that the theorem is true for $n - 1$. The trick is to replace a_n by a variable x and consider the determinant.

$$p(x) = \det\begin{bmatrix} 1 & 1 & \cdots & 1 & 1 \\ a_1 & a_2 & \cdots & a_{n-1} & x \\ a_1^2 & a_2^2 & \cdots & a_{n-1}^2 & x^2 \\ \vdots & \vdots & & \vdots & \vdots \\ a_1^{n-1} & a_2^{n-1} & \cdots & a_{n-1}^{n-1} & x^{n-1} \end{bmatrix}$$

If this is expanded along the last column, it is clear that $p(x)$ is a polynomial in x of degree at most $n - 1$. Moreover, $p(a_1) = 0$ (because $x = a_1$ produces identical columns), so $x - a_1$ is a factor of $p(x)$ by the factor theorem — say, $p(x) = (x - a_1)p_1(x)$. Then the fact that $p(a_2) = 0$ and $a_2 \neq a_1$ means that $p_1(a_2) = 0$. So, again by the factor theorem, $p_1(x) = (x - a_2)p_2(x)$. This gives $p(x) = (x - a_1)(x - a_2)p_2(x)$. The process continues (the a_i are distinct) to give

$$p(x) = (x - a_1)(x - a_2) \dots (x - a_{n-1})d \qquad (*)$$

where d is a constant. In fact, d is the coefficient of x^{n-1} in $p(x)$ and so, by the Laplace expansion, d is the (n,n)-cofactor of the matrix:

$$d = (-1)^{n+n} \det \begin{bmatrix} 1 & 1 & \dots & 1 \\ a_1 & a_2 & \dots & a_{n-1} \\ a_1^2 & a_2^2 & \dots & a_{n-1}^2 \\ \vdots & \vdots & & \vdots \\ a_1^{n-2} & a_2^{n-2} & \dots & a_{n-1}^{n-2} \end{bmatrix}$$

Because $(-1)^{n+n} = 1$, the induction hypothesis shows that d is the product of all terms $(a_i - a_j)$ where $1 \leq j < i \leq n - 1$, and the result now follows from (*) by substituting $x = a_n$ in $p(x)$. ■

EXERCISES 3.4

1. Find a polynomial $p(x)$ of degree 2 such that:
   (a) $p(0) = 2, p(1) = 3, p(3) = 8$
   (b) $p(0) = 5, p(1) = 3, p(2) = 5$
2. Find a polynomial $p(x)$ of degree 3 such that:
   (a) $p(0) = p(1) = 1, p(-1) = 4, p(2) = -5$
   (b) $p(0) = p(1) = 1, p(-1) = 2, p(-2) = -3$
3. Given the following data pairs, find the interpolating polynomial of degree 3 and estimate the value of y corresponding to $x = 1.5$.
   (a) (0,1), (1,2), (2,5), (3,10)
   (b) (0,1), (1,1.49), (2, −0.42), (3, −11.33)
   (c) (0,2), (1,2.03), (2, −0.40), (−1,0.89)
4. Use the polynomial $p(x)$ in Example 3 to approximate **(a)** sin(.3) and **(b)** sin(.7).
5. Find a quadratic polynomial $p(x)$ approximating e^{3x} on the range $0 \leq x \leq \frac{1}{2}$. (Use $x_0 = 0$, $x_1 = \frac{1}{4}$, and $x_2 = \frac{1}{2}$ so $y_0 = 1$, $y_1 = 2.117$, and $y_2 = 4.482$.) Use $p(x)$ to estimate $e^{9/8}$.

4 Vector Geometry

SECTION 4.1 Geometric Vectors and Lines

4.1.1 Geometric Vectors

Many quantities in nature are completely specified by one number (called the *magnitude* of the quantity) and are usually referred to as scalar quantities. Some examples are temperature, time, length, and net assets. However, certain quantities require both a magnitude and a direction to specify them. Consider displacement: To say that a boat sailed 10 kilometers does not specify where it went. It is necessary to give the direction too; perhaps it sailed 10 kilometers northwest. Quantities that require both a magnitude and a direction to describe them are called **vector** quantities. Examples include displacement, velocity, and force. Vector quantities will be denoted by boldface type: $\mathbf{u}$, $\mathbf{v}$, $\mathbf{w}$, and so on. The **magnitude** of a vector $\mathbf{v}$ will be denoted by $\|\mathbf{v}\|$ and is a non-negative number sometimes referred to as the **length** of $\mathbf{v}$. This terminology comes from a very useful and suggestive geometrical representation of vectors as arrows (see Figure 4.1). The magnitude of $\mathbf{v}$ is represented by the length of the arrow, and the direction the arrow points indicates the direction of $\mathbf{v}$. The head and tail of the arrow are called the **terminal point** and the **initial point** of the vector.

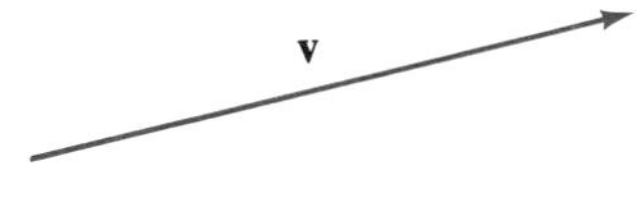

FIGURE 4.1

Two vectors $\mathbf{v}$ and $\mathbf{w}$ are **equal** (written $\mathbf{v} = \mathbf{w}$) if they have the same length and the same direction, regardless of the position of the terminal point. There is only one vector with zero length. It is called the **zero vector** and is denoted by $\mathbf{0}$. In other words,

$$\mathbf{v} = \mathbf{0} \qquad \text{if and only if } \|\mathbf{v}\| = 0$$

No direction is assigned to the vector $\mathbf{0}$. Two nonzero vectors are called **parallel** if they have the same or opposite directions (Figure 4.2a). Given a vector $\mathbf{v}$, the vector with the same magnitude as $\mathbf{v}$ but the opposite direction is called the **negative** of $\mathbf{v}$ and is denoted $-\mathbf{v}$ (Figure 4.2(b)). Because $\mathbf{0}$ is the only vector with length 0, it follows that $-\mathbf{0} = \mathbf{0}$. Clearly $\mathbf{v}$ and $-\mathbf{v}$ are parallel if $\mathbf{v} \neq \mathbf{0}$.

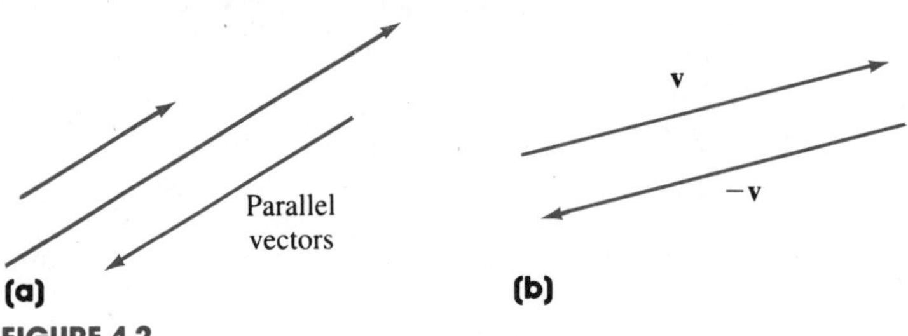

FIGURE 4.2

Probably the most useful aspect of vectors is that they can be added and multiplied by a number in such a way that these operations reflect physical facts when the vectors represent velocity, force, or some other physical quantity. Consider the following example.

EXAMPLE 1 Find the displacement resulting from a 1-km walk northeast and a 1-km walk east.

Solution Let $\mathbf{w}_1$ and $\mathbf{w}_2$ denote the northeast and east displacements, respectively. If $\mathbf{w}_1$ is done first, the resulting displacement $\mathbf{v}$ is given by the first diagram. The second diagram shows the resulting displacement when $\mathbf{w}_2$ is done first. It is evident that the resulting displacement is the same in both cases, so it is designated $\mathbf{v}$ in both. It is also quite clear that $\|\mathbf{v}\|$ and the direction of $\mathbf{v}$ (given by θ) can be calculated (Exercise 1), so $\mathbf{v}$ is determined completely by $\mathbf{w}_1$ and $\mathbf{w}_2$.

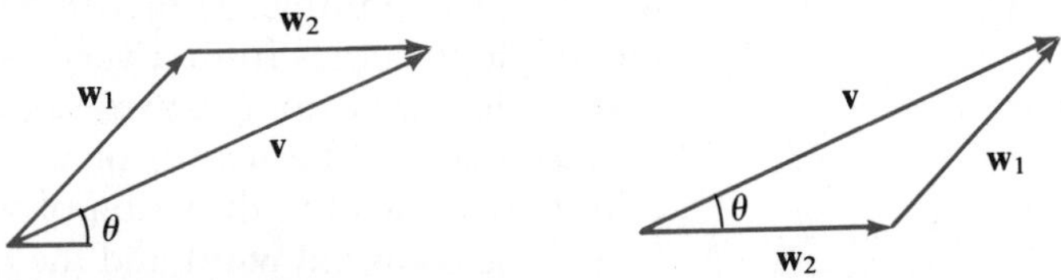

The situation in Example 1 is typical of many vector quantities other than displacements, and this leads to a general notion of vector addition.

DEFINITION If $\mathbf{u}$ and $\mathbf{w}$ are two vectors, their **sum** $\mathbf{u} + \mathbf{w}$ is defined as follows: Position $\mathbf{u}$ and $\mathbf{w}$ so that they emanate from a common point P. They determine a parallelogram, and the diagonal drawn from P represents $\mathbf{u} + \mathbf{w}$. This is called the **parallelogram rule.**

The situation is shown in Figure 4.3(a). It is clear that $\mathbf{u} + \mathbf{w} = \mathbf{w} + \mathbf{u}$, because the two vectors enter into the definition in a symmetric fashion. However, it is often convenient to regard $\mathbf{u} + \mathbf{w}$ as first $\mathbf{u}$ and then $\mathbf{w}$ so that the equation $\mathbf{u} + \mathbf{w} = \mathbf{w} + \mathbf{u}$ expresses a very useful property of vector addition. Figures 4.3(b) and 4.3(c) illustrate the situation. The fact that a vector is completely determined by its magnitude and direction means that the point at which the vector acts is unimportant. In depicting $\mathbf{u} + \mathbf{w}$ in Figure 4.3(b), we placed the initial point of $\mathbf{w}$ at the terminal point of $\mathbf{u}$. In Figure 4.3(c), however, depicting $\mathbf{w} + \mathbf{u}$, we placed the initial point of $\mathbf{u}$ at the terminal point of $\mathbf{w}$. The resulting vector has the same magnitude and direction in both cases, so it is represented as $\mathbf{u} + \mathbf{w} = \mathbf{w} + \mathbf{u}$.

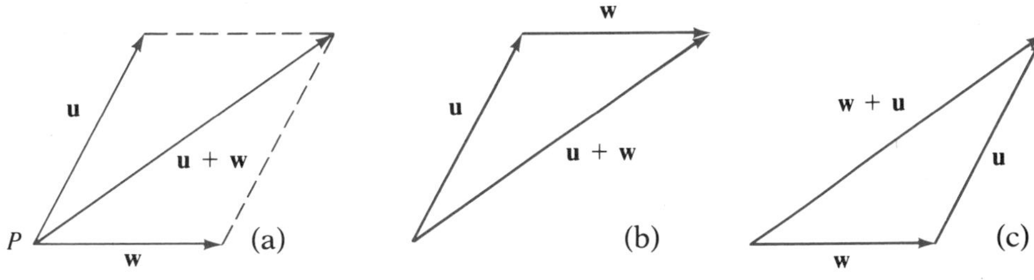

FIGURE 4.3

The usefulness of this vector addition is illustrated in the following two examples.

EXAMPLE 2 A wind is blowing from the south at 75 knots, and an airplane flies heading east at 100 knots. Find the resulting velocity of the airplane.

Solution Let $\mathbf{w}$ and $\mathbf{p}$ denote the wind velocity and the velocity of the airplane (relative to the air). It is a fact (which can be tested experimentally) that the resulting velocity of the airplane is $\mathbf{v} = \mathbf{w} + \mathbf{p}$, calculated by the parallelogram rule. Because $\|\mathbf{w}\| = 75$, $\|\mathbf{p}\| = 100$, and the parallelogram here is a rectangle, we obtain

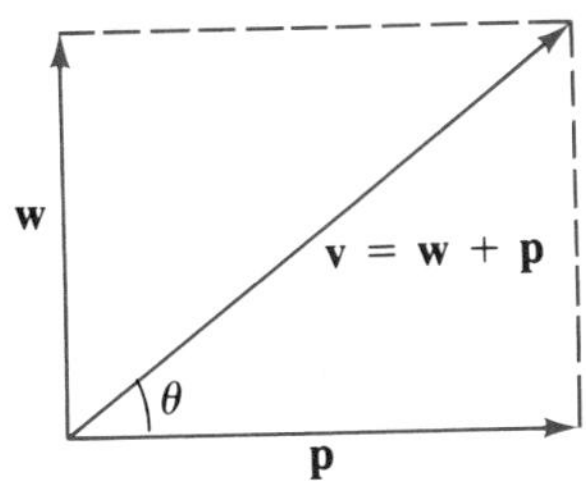

$$\|\mathbf{v}\| = \sqrt{75^2 + 100^2} = 25\sqrt{3^2 + 4^2} = 125$$

$$\cos\theta = \frac{100}{125} = 0.8$$

Hence the ground speed of the airplane is 125 knots, and its direction is θ radians north of east, where θ is obtained from $\cos\theta = 0.8$. The result (from tables or a calculator) is $\theta = 0.64$ (or $\theta = 37°$).

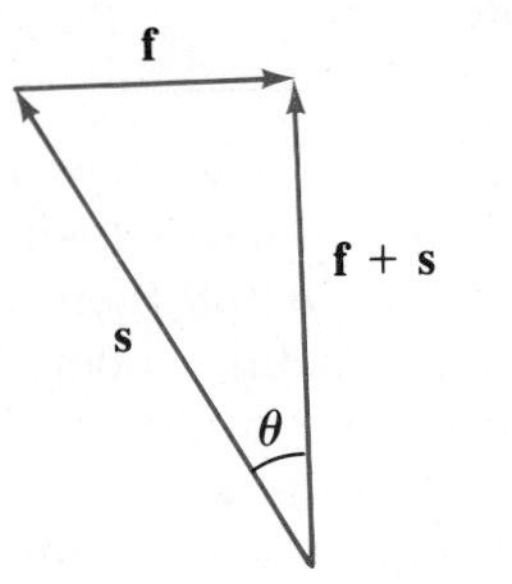

EXAMPLE 3 A river flows at 1 km per hour and a swimmer moves at 2 km per hour (relative to the water). At what angle must he swim to go straight across? What is his resulting speed?

Solution Let **f** denote the velocity of the river flow, and let **s** denote the swimmer's velocity. The resulting velocity is **f** + **s**, and we are to choose θ such that this has direction straight across the river (hence perpendicular to **f**). Because $\|\mathbf{f}\| = 1$ and $\|\mathbf{s}\| = 2$, we have $\sin \theta = \frac{1}{2}$ so $\theta = \frac{\pi}{6}$ (or 30°). Hence he must swim heading 30° upstream. His resulting speed is $\|\mathbf{f} + \mathbf{s}\| = \sqrt{2^2 - 1^2} = \sqrt{3} = 1.73$ km per hour.

The basic properties of vector addition are as follows:

1. $\mathbf{u} + \mathbf{v} = \mathbf{v} + \mathbf{u}$ for all vectors **u** and **v**
2. $\mathbf{u} + (\mathbf{v} + \mathbf{w}) = (\mathbf{u} + \mathbf{v}) + \mathbf{w}$ for all vectors **u**, **v**, and **w**
3. $\mathbf{v} + \mathbf{0} = \mathbf{v}$ for all vectors **v**
4. $\mathbf{v} + (-\mathbf{v}) = \mathbf{0}$ for all vectors **v**

The first of these has already been mentioned, and the last two are clear from the parallelogram rule. The second is demonstrated by Figure 4.4.

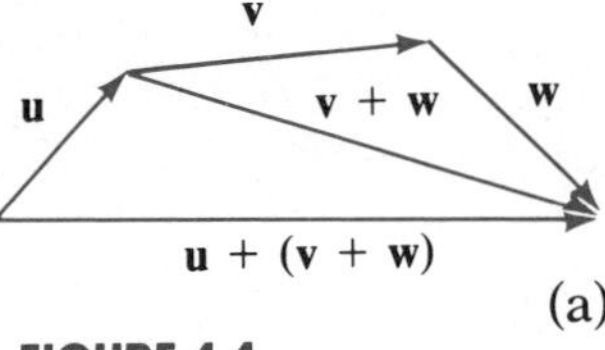

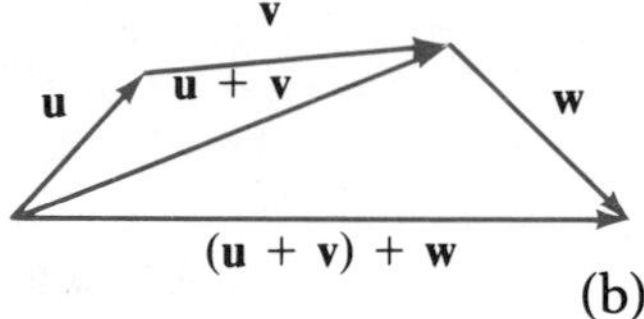

FIGURE 4.4

The two diagrams have **u**, **v**, and **w** in the same position. The vector across the bottom is $\mathbf{u} + (\mathbf{v} + \mathbf{w})$ in Figure 4.4(a) and $(\mathbf{u} + \mathbf{v}) + \mathbf{w}$ in Figure 4.4(b).

Because of the fact that $\mathbf{u} + (\mathbf{v} + \mathbf{w}) = (\mathbf{u} + \mathbf{v}) + \mathbf{w}$, we shall write this simply as $\mathbf{u} + \mathbf{v} + \mathbf{w}$. The foregoing discussion makes it clear that $\mathbf{u} + \mathbf{v} + \mathbf{w}$ can be regarded as the result of **u**, **v**, and **w** being placed end to end with the terminal point of each vector coinciding with the initial point of the next vector. The sum $\mathbf{u} + \mathbf{v} + \mathbf{w}$ then has the same initial point as **u** and the same terminal point as **w**. This is independent of the *order* in which the vectors are added (the reader is invited to draw $\mathbf{v} + \mathbf{u} + \mathbf{w}$ and $\mathbf{w} + \mathbf{v} + \mathbf{u}$ using the vectors in Figure 4.4).

It also works when more than three vectors are involved. For example, suppose forces $\mathbf{f}_1$, $\mathbf{f}_2$, $\mathbf{f}_3$, $\mathbf{f}_4$, and $\mathbf{f}_5$ act on a particle P as shown on the left in Figure 4.5, and it is desired to determine the vector force required to keep the particle from moving. The forces are re-drawn "end-to-end" in the right-hand diagram, and it is an experimentally verifiable fact that the resulting force on P is their vector sum $\mathbf{f} = \mathbf{f}_1 + \mathbf{f}_2 + \mathbf{f}_3 + \mathbf{f}_4 + \mathbf{f}_5$. Clearly the force necessary to restrain the particle is $-\mathbf{f}$, and this is depicted geometrically in the diagram.

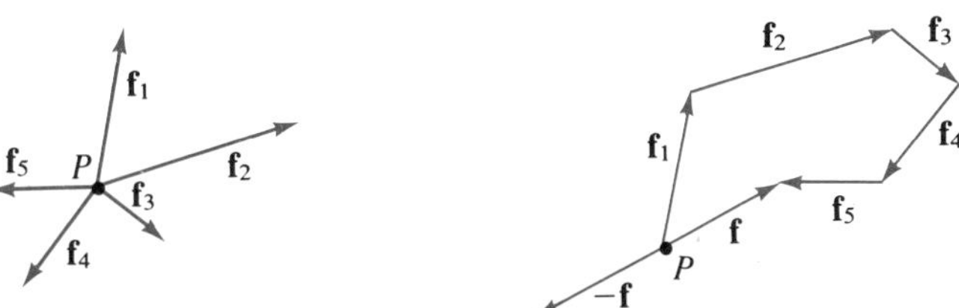

FIGURE 4.5

Vectors can be used effectively to give proofs of theorems in Euclidean geometry that make no use of coordinates. If A and B are two geometrical points, the vector from A to B is denoted as $\overrightarrow{AB}$.

EXAMPLE 4 Show that the diagonals of a parallelogram bisect each other.

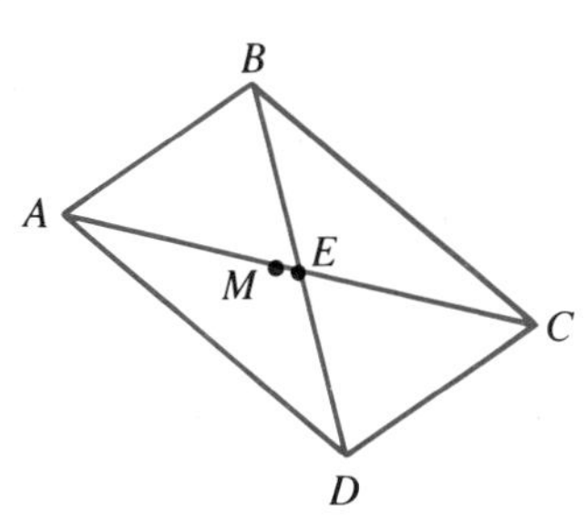

Solution Let the parallelogram have vertices A, B, C, and D, as shown; let E denote the intersection of the two diagonals; and let M denote the midpoint of diagonal AC. We must show that $M = E$ and that this is the midpoint of diagonal BD. This is accomplished by showing that $\overrightarrow{BM} = \overrightarrow{MD}$. (This suffices because the fact that $\overrightarrow{BM}$ and $\overrightarrow{MD}$ have the same direction means that $M = E$, and the fact that they have the same length means that $M = E$ is the midpoint of BD.) Now $\overrightarrow{AM} = \overrightarrow{MC}$ because M is the midpoint of AC, and $\overrightarrow{BA} = \overrightarrow{CD}$ because the figure is a parallelogram. Hence

$$\overrightarrow{BM} = \overrightarrow{BA} + \overrightarrow{AM} = \overrightarrow{CD} + \overrightarrow{MC} = \overrightarrow{MC} + \overrightarrow{CD} = \overrightarrow{MD}$$

where the first and last equalities use the parallelogram rule of vector addition.

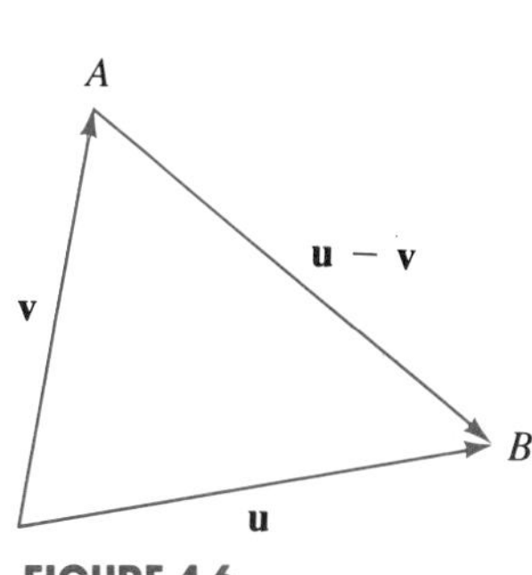

FIGURE 4.6

By analogy with numerical arithmetic, **vector subtraction** is defined as follows:

$$\mathbf{u} - \mathbf{v} = \mathbf{u} + (-\mathbf{v})$$

Like vector addition, subtraction has a geometric interpretation, as shown in Figure 4.6. Then $\mathbf{v} + \overrightarrow{AB} = \mathbf{u}$ by vector addition. Adding $-\mathbf{v}$ to each side gives

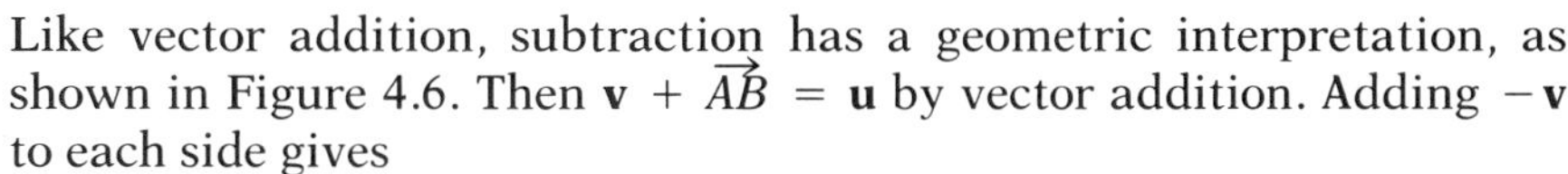

$$\begin{aligned} -\mathbf{v} + \mathbf{v} + \overrightarrow{AB} &= -\mathbf{v} + \mathbf{u} \\ \mathbf{0} + \overrightarrow{AB} &= \mathbf{u} + (-\mathbf{v}) \\ \overrightarrow{AB} &= \mathbf{u} - \mathbf{v} \end{aligned}$$

The situation in Figure 4.6 is best remembered by observing that

$$\mathbf{u} - \mathbf{v} \text{ is the vector that, when added to } \mathbf{v}, \text{ gives } \mathbf{u}$$

Now suppose a vector $\mathbf{v}$ is added to itself three times. The resulting vector $\mathbf{v} + \mathbf{v} + \mathbf{v}$ has the same direction as $\mathbf{v}$ and three times the magnitude, and it is natural to denote it as $3\mathbf{v}$. Similar definitions are in order for $2\mathbf{v}$, $4\mathbf{v}$, and so on. More generally, if $a > 0$ is any number, this suggests that we define $a\mathbf{v}$ to be a vector in the same direction as $\mathbf{v}$ but whose magnitude is $a\|\mathbf{v}\|$. This gives a way of multiplying any vector $\mathbf{v}$ by a *positive* scalar a. The general definition follows.

DEFINITION

Given any vector $\mathbf{v}$ and any scalar a, the **scalar product** of $\mathbf{v}$ by a is the vector $a\mathbf{v}$ defined by specifying its magnitude and direction as follows:

1. The magnitude of $a\mathbf{v}$ is $\|a\mathbf{v}\| = |a|\,\|\mathbf{v}\|$
2. The direction of $a\mathbf{v}$ is $\begin{cases} \text{the same as that of } \mathbf{v} \text{ if } a > 0 \text{ and } \mathbf{v} \neq \mathbf{0} \\ \text{unspecified if } a = 0 \text{ or } \mathbf{v} = \mathbf{0} \\ \text{opposite to that of } \mathbf{v} \text{ if } a < 0 \text{ and } \mathbf{v} \neq \mathbf{0} \end{cases}$

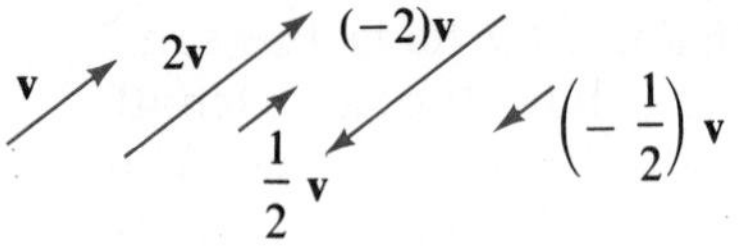

v 2v $(-2)\mathbf{v}$ $\frac{1}{2}\mathbf{v}$ $\left(-\frac{1}{2}\right)\mathbf{v}$

FIGURE 4.7

Because $|a| = a$ if a is positive, this formula for the magnitude of $a\mathbf{v}$ agrees with that given above. Some examples of scalar multiplication appear in Figure 4.7.

Taking $a = 0$ in the formula for $\|a\mathbf{v}\|$ yields $\|0\mathbf{v}\| = 0$. In other words, $0\mathbf{v}$ has magnitude zero so it is the zero vector; that is, $0\mathbf{v} = \mathbf{0}$. Similarly, $a\mathbf{0} = \mathbf{0}$ for each number a. Observe also that $1\mathbf{v} = \mathbf{v}$ because $1\mathbf{v}$ has the same magnitude and direction as $\mathbf{v}$. However, $(-1)\mathbf{v}$ has the same magnitude as $\mathbf{v}$ but the *opposite* direction, so $(-1)\mathbf{v} = -\mathbf{v}$. These properties of scalar multiplication are collected below for reference.

1. $\|a\mathbf{v}\| = |a|\,\|\mathbf{v}\|$ for all scalars a and vectors $\mathbf{v}$
2. $1\mathbf{v} = \mathbf{v}$ for all vectors $\mathbf{v}$
3. $(-1)\mathbf{v} = -\mathbf{v}$ for all vectors $\mathbf{v}$
4. $0\mathbf{v} = \mathbf{0}$ for all vectors $\mathbf{v}$
5. $a\mathbf{0} = \mathbf{0}$ for all scalars a

A vector is called a **unit vector** if its magnitude is 1.

EXAMPLE 5

If $\mathbf{v} \neq \mathbf{0}$, show that $\frac{1}{\|\mathbf{v}\|}\mathbf{v}$ is a unit vector in the same direction as $\mathbf{v}$.

Solution The vectors in the same direction as $\mathbf{v}$ are $a\mathbf{v}$, where $a > 0$. Because $\|a\mathbf{v}\| = a\|\mathbf{v}\|$, this is a unit vector when $a = 1/\|\mathbf{v}\|$.

Many properties of vector addition and scalar multiplication have been mentioned (and utilized) above. Several of these facts are consequences of the following eight fundamental properties.

THEOREM 1 Vector addition and scalar multiplication exhibit the following properties.

(1) $\mathbf{u} + \mathbf{v} = \mathbf{v} + \mathbf{u}$ for all vectors $\mathbf{u}$ and $\mathbf{v}$

(2) $\mathbf{u} + (\mathbf{v} + \mathbf{w}) = (\mathbf{u} + \mathbf{v}) + \mathbf{w}$ for all vectors $\mathbf{u}$, $\mathbf{v}$, and $\mathbf{w}$

(3) $\mathbf{u} + \mathbf{0} = \mathbf{u}$ for all vectors $\mathbf{u}$

(4) $\mathbf{u} + (-\mathbf{u}) = \mathbf{0}$ for all vectors $\mathbf{u}$

(5) $1\mathbf{u} = \mathbf{u}$ for all vectors $\mathbf{u}$

(6) $a(b\mathbf{u}) = (ab)\mathbf{u}$ for all vectors $\mathbf{u}$ and scalars a, b

(7) $(a + b)\mathbf{u} = a\mathbf{u} + b\mathbf{u}$ for all vectors $\mathbf{u}$ and scalars a, b

(8) $a(\mathbf{u} + \mathbf{v}) = a\mathbf{u} + a\mathbf{v}$ for all vectors $\mathbf{u}$, $\mathbf{v}$ and scalars a

Proof Only the last three remain to be verified. They follow easily from the definitions when the scalars a and b are positive or zero. The general case is not difficult (though it is a bit tedious), and the details are left as Exercise 14. ■

As for matrices, these properties enable us to carry out algebraic manipulations of vectors as though the vectors were variables.

EXAMPLE 6 (a) Simplify: $5(\mathbf{u} - 2\mathbf{v}) + 6(5\mathbf{u} + 2\mathbf{v}) - 2(\mathbf{v} - \mathbf{u})$.

(b) If $\mathbf{v} = 4\mathbf{w}$, show that $\mathbf{w} = \frac{1}{4}\mathbf{v}$.

Solution

(a) $5(\mathbf{u} - 2\mathbf{v}) + 6(5\mathbf{u} + 2\mathbf{v}) - 2(\mathbf{v} - \mathbf{u})$
$= 5\mathbf{u} - 10\mathbf{v} + 30\mathbf{u} + 12\mathbf{v} - 2\mathbf{v} + 2\mathbf{u}$
$= 37\mathbf{u}$

(b) $\frac{1}{4}\mathbf{v} = \frac{1}{4}(4\mathbf{w}) = (\frac{1}{4} \cdot 4)\mathbf{w} = 1\mathbf{w} = \mathbf{w}$

EXAMPLE 7 Show that the midpoints of the four sides of any quadrilateral are the vertices of a parallelogram. Here a quadrilateral is any figure with four vertices.

Solution Suppose the vertices of the quadrilateral are A, B, C, and D (in that order) and that E, F, G, and H are the midpoints of the sides as shown. It suffices to show $\overrightarrow{EF} = \overrightarrow{HG}$ (because then these two sides are

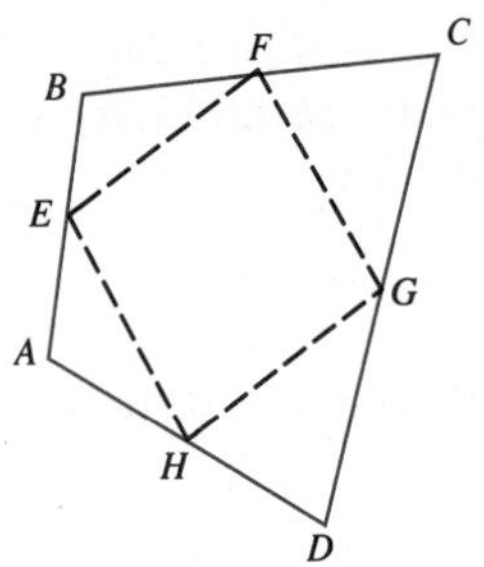

parallel and of equal length). Now the fact that E is the midpoint of AB means that $\vec{EB} = \frac{1}{2}\vec{AB}$. Similarly, $\vec{BF} = \frac{1}{2}\vec{BC}$, so

$$\vec{EF} = \vec{EB} + \vec{BF} = \frac{1}{2}\vec{AB} + \frac{1}{2}\vec{BC} = \frac{1}{2}(\vec{AB} + \vec{BC}) = \frac{1}{2}\vec{AC}$$

A similar argument shows that $\vec{HG} = \frac{1}{2}\vec{AC}$ too, so $\vec{EF} = \vec{HG}$ as required.

THEOREM 2

Suppose that **u** and **v** are nonzero vectors.

(1) **u** and **v** are parallel if and only if $\mathbf{u} = a\mathbf{v}$ for some $a \neq 0$.

(2) If **u** and **v** are not parallel, and $a\mathbf{u} + b\mathbf{v} = a_1\mathbf{u} + b_1\mathbf{v}$, then $a = a_1$ and $b = b_1$.

Proof

1. This follows from the definition of scalar multiplication.
2. We have $(a - a_1)\mathbf{u} = (b_1 - b)\mathbf{v}$. If $a \neq a_1$ then $\mathbf{u} = \left(\frac{b_1 - b}{a - a_1}\right)\mathbf{v}$, so **u** and **v** are parallel by (1). Hence $a = a_1$. Similarly, $b = b_1$. ■

The line through a vertex of a triangle and the midpoint of the opposite side is called a **median** of the triangle. It is well known that the three medians of any triangle are concurrent. (Lines are called **concurrent** if they all pass through a common point.) Theorem 2 can be used to show this (and more).

EXAMPLE 8

Show that the medians of a triangle meet at the point on each one third the way from the midpoint to the vertex (and so are concurrent).

Solution In the diagram let E and F be the midpoints of sides AB and AC, respectively. Then

$$\vec{EO} = s\vec{EC} \text{ and } \vec{FO} = t\vec{FB}$$

for *some* scalars s and t. We show $s = t = \frac{1}{3}$ by expressing $\vec{AO}$ two ways in the form $a\vec{AB} + b\vec{AC}$, and applying Theorem 2 (because $\vec{AB}$ and $\vec{AC}$ are nonparallel).

$$\begin{aligned}\vec{AO} = \vec{AE} + \vec{EO} = \vec{AE} + s\vec{EC} &= \vec{AE} + s(\vec{EA} + \vec{AC}) \\ &= (1 - s)\vec{AE} + s\vec{AC} \\ &= \frac{1}{2}(1 - s)\vec{AB} + s\vec{AC}\end{aligned}$$

Similarly, using $\overrightarrow{AO} = \overrightarrow{AF} + \overrightarrow{FO}$ gives $\overrightarrow{AO} = \frac{1}{2}(1 - t)\overrightarrow{AC} + t\overrightarrow{AB}$. Hence Theorem 2 gives $\frac{1}{2}(1 - s) = t$ and $\frac{1}{2}(1 - t) = s$. The only solution is $s = t = \frac{1}{3}$.

EXERCISES 4.1.1

1. Find $\|\mathbf{v}\|$ and θ in Example 1.

2. An airplane pilot flies at 300 km/hr in a direction 30° south of east. The wind is blowing from the south at 150 km/hr.

(a) Find the resulting direction and speed of the airplane.

(b) Find the speed of the airplane if the wind is from the west (at 150 km/hr).

3. A boat goes 12 knots heading north. The current is 5 knots from the west. In what direction does the boat actually move, and at what speed?

4. Use vectors to show that the line joining the midpoints of two sides of a triangle is parallel to the third side and half as long.

5. Let A, B, and C denote the three vertices of a triangle.

(a) If E is the midpoint of side BC, show that $\overrightarrow{AE} = \frac{1}{2}(\overrightarrow{AB} + \overrightarrow{AC})$.

(b) If F is the midpoint of side AC, show that $\overrightarrow{FE} = \frac{1}{2}\overrightarrow{AB}$.

6. Simplify the following where $\mathbf{u}$, $\mathbf{v}$, and $\mathbf{w}$ represent vectors.

(a) $6(\mathbf{u} + 3\mathbf{v} - \mathbf{w}) - 12(\mathbf{v} - 3\mathbf{u}) + 3(2\mathbf{w} - 2\mathbf{u} - 2\mathbf{v})$

(b) $8(2\mathbf{u} - \mathbf{v} + 3\mathbf{w}) + 3(5\mathbf{v} - 6\mathbf{w}) - 2(8\mathbf{u} + 3\mathbf{v} + 3\mathbf{w})$

7. Make a sketch as shown in Figure 4.4, illustrating each of the following ways of adding four vectors.

(a) $[(\mathbf{u} + \mathbf{v}) + \mathbf{w}] + \mathbf{z}$ **(b)** $(\mathbf{u} + \mathbf{v}) + (\mathbf{w} + \mathbf{z})$ **(c)** $\mathbf{u} + [\mathbf{v} + (\mathbf{w} + \mathbf{z})]$

8. Let A, B, C, D, E, and F be the vertices of a regular hexagon, taken in order. Show that $\overrightarrow{AB} + \overrightarrow{AC} + \overrightarrow{AD} + \overrightarrow{AE} + \overrightarrow{AF} = 3\overrightarrow{AD}$.

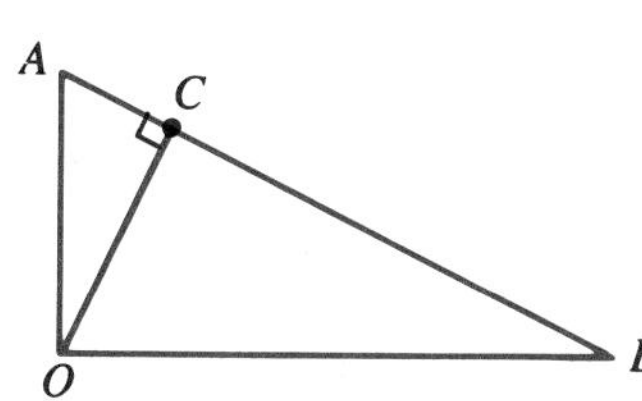

9. **(a)** Let P_1, P_2, P_3, P_4, P_5, and P_6 be six points equally spaced on a circle with center C. Show that $\overrightarrow{CP_1} + \overrightarrow{CP_2} + \overrightarrow{CP_3} + \overrightarrow{CP_4} + \overrightarrow{CP_5} + \overrightarrow{CP_6} = \mathbf{0}$.

(b) Show that the conclusion in part **(a)** holds for any *even* set of points evenly spaced on the circle.

(c) Show that the conclusion in part **(a)** holds for *three* points.

(d) Do you think it works for *any* finite set of points evenly spaced around the circle?

10. Let OAB be a right-angled triangle, as shown, with the right angle at O. If C is the foot of the perpendicular from O to the hypotenuse, show that

$$\overrightarrow{AC} = \frac{\|\overrightarrow{OA}\|^2}{\|\overrightarrow{AB}\|^2}\overrightarrow{AB}.$$

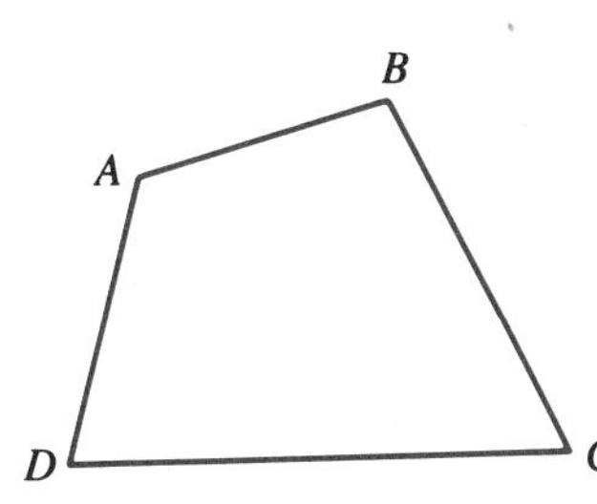

11. Consider a quadrilateral with vertices A, B, C, and D in order (as shown). If the diagonals AC and BD bisect each other, show that the quadrilateral is a parallelogram (this is the converse of Example 4). [*Hint:* Let E be the intersection of the diagonals. Show that $\overrightarrow{AB} = \overrightarrow{DC}$ by writing $\overrightarrow{AB} = \overrightarrow{AE} + \overrightarrow{EB}$.]

12. Use the technique of Example 8 to show that the diagonals of a parallelogram bisect each other.

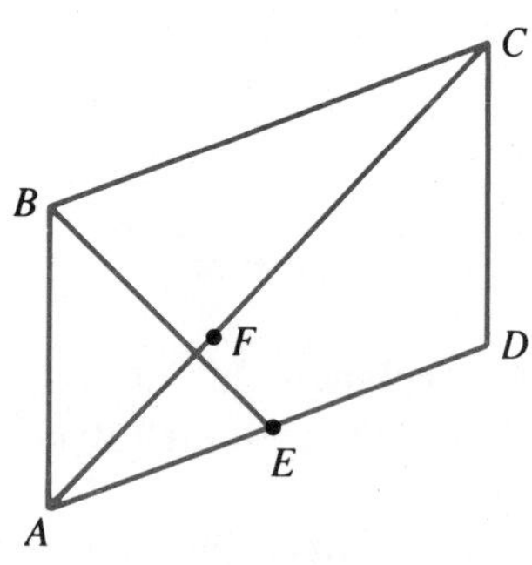

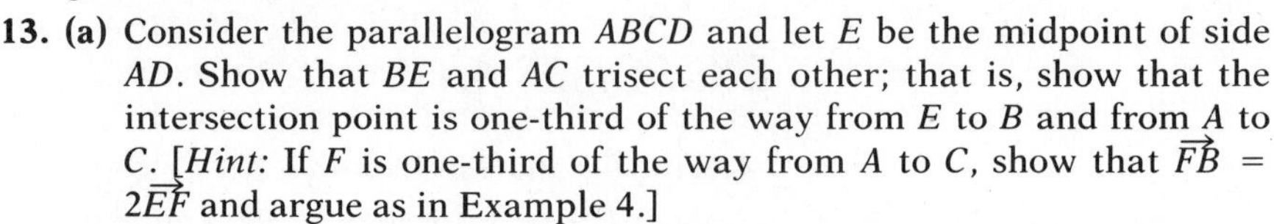

13. (a) Consider the parallelogram $ABCD$ and let E be the midpoint of side AD. Show that BE and AC trisect each other; that is, show that the intersection point is one-third of the way from E to B and from A to C. [*Hint:* If F is one-third of the way from A to C, show that $\overrightarrow{FB} = 2\overrightarrow{EF}$ and argue as in Example 4.]

    (b) Repeat (a) as in Example 8. [*Hint:* Let O be the intersection of AC and BE. If $\overrightarrow{AO} = s\overrightarrow{AC}$ and $\overrightarrow{EO} = t\overrightarrow{EB}$, use Theorem 2 to show that $s = t = \frac{1}{3}$.]

14. Prove the following parts of Theorem 1: **(a)** Part (6); **(b)** Part (7); **(c)** Part (8).

4.1.2 Coordinates and Lines

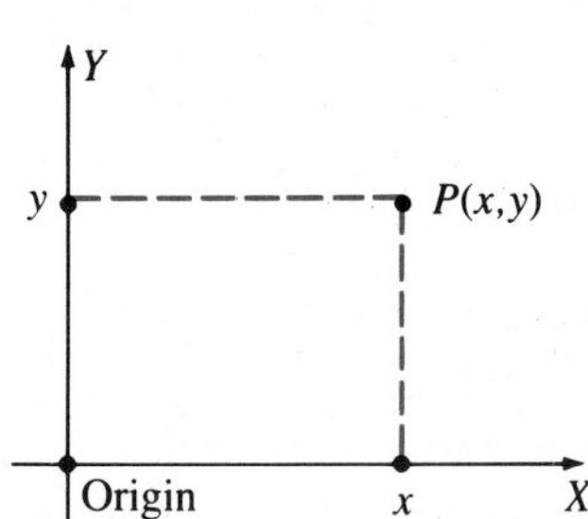

FIGURE 4.8

Examples 4, 7, and 8 use vectors to verify geometrical propositions without the benefit of coordinates. The fact that coordinates are unnecessary is not too surprising: the propositions give intrinsic properties of geometrical figures that do not depend on the choice of a coordinate system—that is on the way the figure is described. On the other hand, the introduction of coordinates gives us a convenient way of doing calculations such as the length of a vector or the angle between two vectors. These in turn can be applied to geometrical verifications. Consequently, our next task is to look for a way to assign coordinates to a vector.

Recall first how a plane is coordinatized. A point (called the **origin**) is selected, and two mutually perpendicular lines are chosen through the origin (called the **X-axis** and the **Y-axis**). Each of these axes is coordinatized by assigning a real number to every point in such a way that 0 is assigned to the origin. Then, given a point P in the plane, two numbers x and y (called the **coordinates** of P) are determined by dropping perpendiculars to the two axes (see Figure 4.8). When the coordinates of a point P are to be emphasized, we shall write $P = P(x,y)$. Hence the points in the plane correspond uniquely to ordered pairs (x,y) of real numbers.

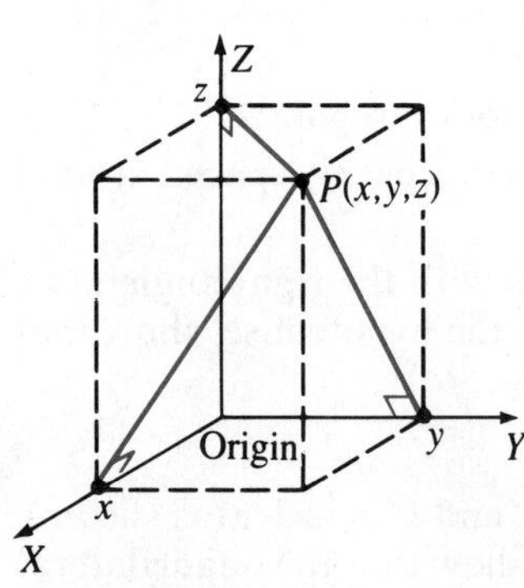

FIGURE 4.9

Coordinates are introduced into space in much the same way. A point (again called the **origin**) is selected, and three mutually perpendicular lines (called the **X-axis**, the **Y-axis**, and the **Z-axis**) are chosen through the origin. In Figure 4.9 the X-axis is to be visualized as coming out of the page, whereas the Y- and Z-axes lie in the plane of the page. These axes are coordinatized as in the plane and, given a point P in space, three numbers x, y, and z (called the **coordinates** of P) are determined by dropping perpendiculars to the three axes (see Figure 4.9). We write $P = P(x,y,z)$ so that points P correspond uniquely to ordered triples (x,y,z) of numbers.

The close analogy between the coordinatization of the plane and that of space suggests that many aspects of plane geometry are analogous to the corresponding situation in space. This is true. In fact, vectors provide a way of unifying the two studies. Consequently, we shall deal primarily with the situation in space, and all vectors will be assumed to be in space unless otherwise specified.

DEFINITION

Given a point $P(x, y, z)$, the **position vector** of P is defined to be the vector $\overrightarrow{OP}$ from the origin to P. It will be denoted

$$\mathbf{v} = (x,y,z)$$

and the numbers x, y, and z are called the X, Y- and Z-**components** of $\mathbf{v}$.

Every vector $\mathbf{v}$ is the position vector of a unique point $P(x,y,z)$ (if $\mathbf{v}$ is positioned with its initial point at the origin, then the terminal point is P). In particular, the zero vector is the position vector of the origin itself (because it has length 0), so

$$\mathbf{0} = (0,0,0)$$

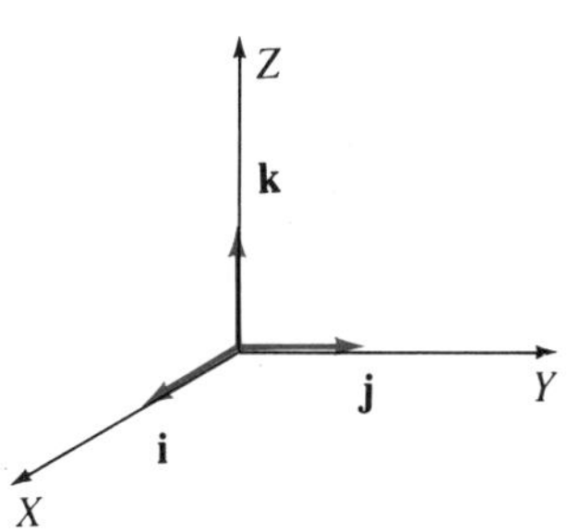

FIGURE 4.10

There are three other vectors that will be important as we continue our discussion: the so-called **coordinate vectors i, j,** and **k**. They are each of length 1 and point along the positive X-, Y-, and Z-axes, respectively (see Figure 4.10). In component form, they are

$$\mathbf{i} = (1,0,0)$$
$$\mathbf{j} = (0,1,0)$$
$$\mathbf{k} = (0,0,1)$$

Now consider a point $P_1(x,0,0)$ on the X-axis. Because P_1 is at a distance $|x|$ from the origin, it is clear that the position vector of P_1 is $x\mathbf{i}$ (the magnitude of $x\mathbf{i}$ is $\|x\mathbf{i}\| = |x|\ \|\mathbf{i}\| = |x|$, and the direction is along the positive X-axis if $x > 0$ and along the negative X-axis if $x < 0$). Hence, in component form,

$$x\mathbf{i} = (x,0,0)$$

Similarly,

$$y\mathbf{j} = (0,y,0)$$
$$z\mathbf{k} = (0,0,z)$$

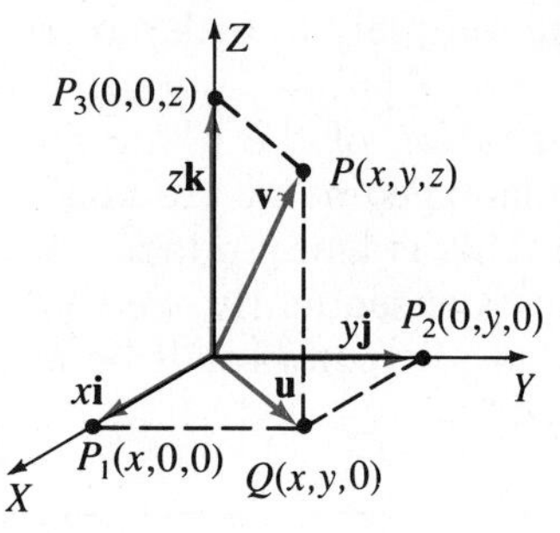

FIGURE 4.11

are the position vectors of $P_2(0,y,0)$ and $P_3(0,0,z)$ for any numbers y and z.

Now let $\mathbf{v} = (x,y,z)$ be *any* vector in component form. The components x, y, and z of $\mathbf{v}$ determine the vectors $x\mathbf{i}$, $y\mathbf{j}$, and $z\mathbf{k}$ along the X-, Y-, and Z-axes, respectively. Vector addition gives

$$\mathbf{u} = x\mathbf{i} + y\mathbf{j}$$

(see Figure 4.11), so

$$\begin{aligned} \mathbf{v} &= \mathbf{u} + z\mathbf{k} \\ &= x\mathbf{i} + y\mathbf{j} + z\mathbf{k} \end{aligned}$$

This proves Theorem 3.

THEOREM 3 If $\mathbf{v} = (x,y,z)$ is any vector in component form, then

$$\mathbf{v} = x\mathbf{i} + y\mathbf{j} + z\mathbf{k}$$

where $\mathbf{i}$, $\mathbf{j}$, and $\mathbf{k}$ are the coordinate vectors.

The most important reason for writing vectors in component form is that the vector operations defined above correspond precisely to matrix operations.

THEOREM 4 Let $\mathbf{u} = (x,y,z)$ and $\mathbf{u}_1 = (x_1, y_1, z_1)$ be two vectors in component form. Then:

(1) $\mathbf{u} = \mathbf{u}_1$ if and only if $x = x_1$, $y = y_1$, and $z = z_1$

(2) $\mathbf{u} + \mathbf{u}_1 = (x + x_1, y + y_1, z + z_1)$

(3) $a\mathbf{u} = (ax,ay,az)$ for any scalar a

(4) $\mathbf{u} - \mathbf{u}_1 = (x - x_1, y - y_1, z - z_1)$

Proof (1) The vectors $\mathbf{u} = (x,y,z)$ and $\mathbf{u}_1 = (x_1,y_1,z_1)$ are position vectors of points $P(x,y,z)$ and $P_1(x_1,y_1,z_1)$, respectively. Hence $\mathbf{u} = \mathbf{u}_1$ means $P = P_1$; that is, $x = x_1$, $y = y_1$, and $z = z_1$.

(2), (3), and (4). Theorem 3 shows that $\mathbf{u} = x\mathbf{i} + y\mathbf{j} + z\mathbf{k}$ and $\mathbf{u}_1 = x_1\mathbf{i} + y_1\mathbf{j} + z_1\mathbf{k}$. Hence Theorem 1 gives

$$\begin{aligned} \mathbf{u} + \mathbf{u}_1 &= (x\mathbf{i} + y\mathbf{j} + z\mathbf{k}) + (x_1\mathbf{i} + y_1\mathbf{j} + z_1\mathbf{k}) \\ &= (x + x_1)\mathbf{i} + (y + y_1)\mathbf{j} + (z + z_1)\mathbf{k} \\ &= (x + x_1, y + y_1, z + z_1) \end{aligned}$$

where the last step again utilizes Theorem 3. This gives (2); (3) and (4) are similar and are left as Exercise 18. ■

EXAMPLE 9 Given vectors $\mathbf{u} = (3,-1,2)$ and $\mathbf{v} = (1,0,-1)$, compute

$$\mathbf{u} + \mathbf{v},\ 3\mathbf{u},\text{ and } 2\mathbf{u} - 3\mathbf{v}$$

Solution These are just matrix manipulations.

$$\begin{aligned} \mathbf{u} + \mathbf{v} &= (3,-1,2) + (1,0,-1) = (4,-1,1) \\ 3\mathbf{u} &= 3(3,\ -1,2) = (9,-3,6) \\ 2\mathbf{u} - 3\mathbf{v} &= 2(3,-1,2) - 3(1,0,-1) = (3,-2,7) \end{aligned}$$

The following theorem is a particularly useful consequence of Theorem 4.

THEOREM 5 Given points $P_1(x_1,y_1,z_1)$ and $P_2(x_2,y_2,z_2)$, the vector from P_1 to P_2 is

$$\overrightarrow{P_1P_2} = (x_2 - x_1, y_2 - y_1, z_2 - z_1)$$

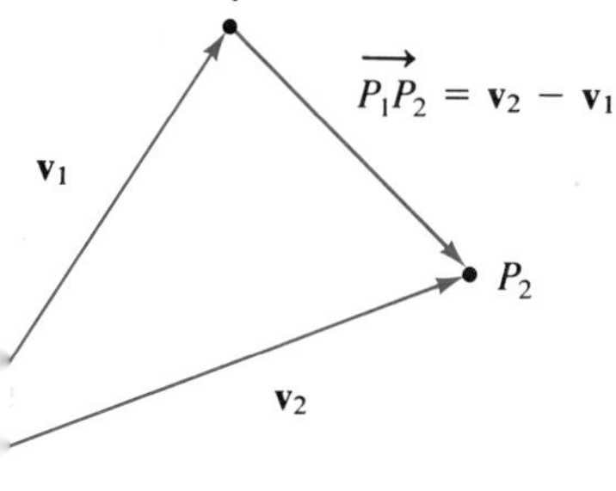

Proof Let $\mathbf{v_1} = (x_1,y_1,z_1)$ and $\mathbf{v_2} = (x_2,y_2,z_2)$ denote the position vectors of P_1 and P_2. Then

$$\overrightarrow{P_1P_2} = \mathbf{v_2} - \mathbf{v_1} = (x_2 - x_1, y_2 - y_1, z_2 - z_1)$$

using Theorem 4. (See the accompanying figure.) ■

These results all have natural analogs for vectors in the plane. The position vector $\mathbf{u}$ of a point $P = P(x,y)$ in the plane is defined to be the vector from the origin to P and is denoted $\mathbf{u} = (x,y)$. If $\mathbf{u_1} = (x_1,y_1)$ is another such vector, the analogs of Theorems 4 and 5 hold.

1. $\mathbf{u} = \mathbf{u_1}$ if and only if $x = x_1$, and $y = y_1$
2. $\mathbf{u} + \mathbf{u_1} = (x + x_1, y + y_1)$
3. $a\mathbf{u} = (ax,ay)$ for any scalar a
4. $\mathbf{u} - \mathbf{u_1} = (x - x_1, y - y_1)$
5. The vector from $P_1(x_1,y_1)$ to $P_2(x_2,y_2)$ is $\overrightarrow{P_1P_2} = (x_2 - x_1, y_2 - y_1)$.

The verifications are entirely analogous to those above and are omitted.

EXAMPLE 10 Find the coordinates of the point P one-third of the way from $P_1(3,-1,4)$ to $P_2(2,3,5)$.

Solution Write $P = P(x,y,z)$ and let $\mathbf{v_1} = (3,-1,4)$, $\mathbf{v_2} = (2,3,5)$, and $\mathbf{v} = (x,y,z)$ be the position vectors of P_1, P_2, and P, respectively. Hence

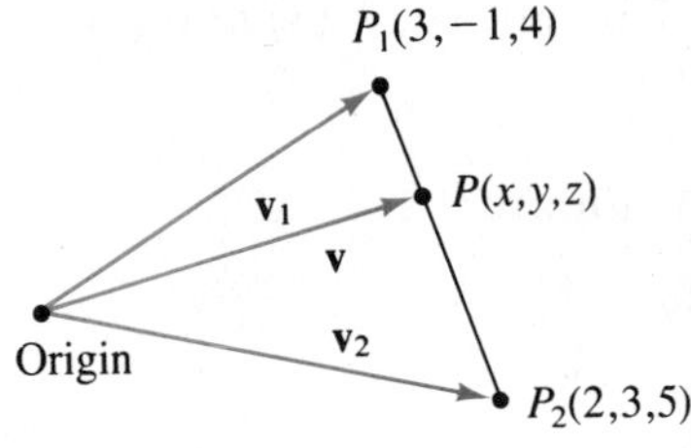

x, y, and z will be found if $\mathbf{v}$ can be determined in component form. But $\overrightarrow{P_1P} = \frac{1}{3}\overrightarrow{P_1P_2}$ by the choice of P, so

$$\overrightarrow{P_1P} = \frac{1}{3}[(2,3,5) - (3,-1,4)] = \frac{1}{3}(-1,4,1)$$

using Theorem 5. But then

$$\mathbf{v} = \mathbf{v_1} + \overrightarrow{P_1P} = (3,-1,4) + \frac{1}{3}(-1,4,1) = \left(\frac{8}{3},\frac{1}{3},\frac{13}{3}\right)$$

Thus $P = P\left(\frac{8}{3},\frac{1}{3},\frac{13}{3}\right)$ is the required point.

Given two arbitrary points $P_1(x_1,y_1,z_1)$ and $P_2(x_2,y_2,z_2)$, we define their **midpoint** $P = P(x,y,z)$ to be the point on the line segment between them halfway from one to the other. As in Example 10, let $\mathbf{v_1}$, $\mathbf{v_2}$, and $\mathbf{v}$ be the position vectors of P_1, P_2, and P, respectively. Then $\overrightarrow{P_1P} = \frac{1}{2}(\mathbf{v_2} - \mathbf{v_1})$ because P is the midpoint, so

$$\mathbf{v} = \mathbf{v_1} + \frac{1}{2}(\mathbf{v_2} - \mathbf{v_1}) = \frac{1}{2}(\mathbf{v_1} + \mathbf{v_2})$$

In other words the position vector of the midpoint of two points P_1 and P_2 is the "average" of the position vectors of P_1 and P_2. In component form, this gives the midpoint between $P_1(x_1,y_1,z_1)$ and $P_2(x_2,y_2,z_2)$ as

$$P\left(\frac{x_1 + x_2}{2}, \frac{y_1 + y_2}{2}, \frac{z_1 + z_2}{2}\right)$$

These vector techniques can be used to give a very simple way of describing straight lines in space. It is clear geometrically that there is exactly one line through a particular point in space that is parallel to a given nonzero vector.

DEFINITION Given a straight line, any nonzero vector that is parallel to the line is called a **direction vector** for the line.

Clearly every line has many direction vectors: in fact, any nonzero multiple of a direction vector will again serve as a direction vector.

Suppose $P_0 = P_0(x_0,y_0,z_0)$ is any point and $\mathbf{d} = (a,b,c)$ is any vector (assumed to be nonzero). Then there is a unique line through P_0 with direction vector $\mathbf{d}$, and we wish to give a condition that a point $P =$

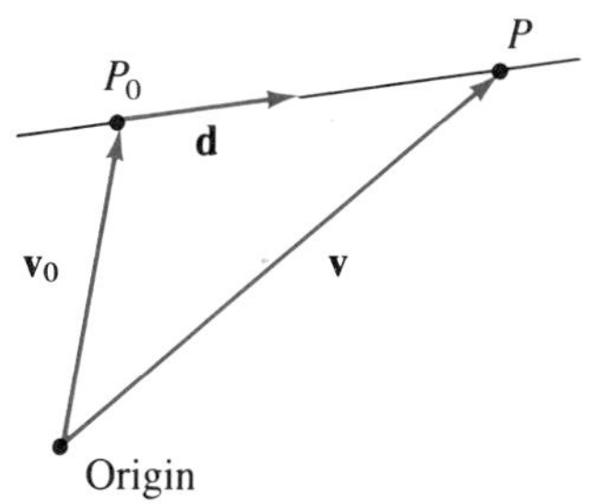

FIGURE 4.12

$P(x,y,z)$ will lie on that line. Let $\mathbf{v}_0 = (x_0,y_0,z_0)$ and $\mathbf{v} = (x,y,z)$ be the position vectors of P_0 and P, respectively (see Figure 4.12). Then $\mathbf{v} = \mathbf{v}_0 + \overrightarrow{P_0P}$, and P lies on the line if and only if $\overrightarrow{P_0P}$ is parallel to $\mathbf{d}$—that is, if and only if $\overrightarrow{P_0P} = t\mathbf{d}$ for some scalar t. Thus $\mathbf{v}$ is the position vector of a point on the line if and only if

$$\mathbf{v} = \mathbf{v_0} + t\mathbf{d} \qquad \text{for some scalar } t$$

This is called the **vector equation** of the line. In component form it becomes

$$(x,y,z) = (x_0,y_0,z_0) + t(a,b,c)$$

so that

$$\begin{aligned} x &= x_0 + ta \\ y &= y_0 + tb \qquad t \text{ any scalar} \\ z &= z_0 + tc \end{aligned}$$

These are called **parametric equations** for the line, and t is sometimes called a **parameter** for the line. Observe that t coordinatizes the line with P_0 corresponding to $t = 0$ and with the unit of distance equal to $\|\mathbf{d}\|$. For this reason it is sometimes convenient to choose $\mathbf{d}$ to be a unit vector.

EXAMPLE 11 Find parametric equations for the line through $P_0(2,1,3)$ with direction vector $\mathbf{d} = (2,0,-1)$.

Solution Here $\mathbf{v_0} = (2,1,3)$. If $\mathbf{v} = (x,y,z)$ is the position vector of an arbitrary point on the line, the vector equation is $\mathbf{v} = \mathbf{v_0} + t\mathbf{d}$. In component form this is

$$(x,y,z) = (2,1,3) + t(2,0,-1)$$

Hence the parametric equations are

$$\begin{aligned} x &= 2 + 2t \\ y &= 1 \qquad t \text{ a parameter} \\ z &= 3 - t \end{aligned}$$

EXAMPLE 12 Find the equations of the line through the points $P_0(2,0,1)$ and $P_1(4,-1,1)$.

Solution Let $\mathbf{d} = \overrightarrow{P_0P_1} = (4 - 2, -1 - 0, 1 - 1) = (2,-1,0)$ denote the vector from P_0 to P_1. Clearly $\mathbf{d}$ serves as a direction vector for the line. Using P_0 as the point on the line leads to the vector equation $(x,y,z) = (2,0,1) + t(2,-1,0)$. The parametric equations are

$$x = 2 + 2t$$
$$y = -t \qquad t \text{ a parameter}$$
$$z = 1$$

Note that if P_1 is used as the point on the line (rather than P_0), the equations are

$$x = 4 + 2s$$
$$y = -1 - s \qquad s \text{ a parameter}$$
$$z = 1$$

These are different from the preceding equations, but this is merely the result of a change in parameter. In fact, $s = t - 1$. Hence the equations of a line can "look" different but still describe the same set of points (the points on the line) as the parameter ranges over the real numbers.

Now suppose parametric equations

$$x = x_0 + ta$$
$$y = y_0 + tb$$
$$z = z_0 + tc$$

are given, where a, b, and c are not all zero. The set of all points $P(x,y,z)$ whose coordinates are given by these equations for some value of t is called the **graph** of the equations. Clearly $P_0(x_0,y_0,z_0)$ lies on this graph (it corresponds to $t = 0$) and the graph is just the line through P_0 with direction vector $\mathbf{d} = (a,b,c)$. Hence the direction vector $\mathbf{d} = (a,b,c)$ can be "read off" from the parametric equations. This and some other facts we have derived about lines are recorded in Theorem 6 for reference.

THEOREM 6 Let $\mathbf{d} = (a,b,c)$ be a nonzero vector and $P_0(x_0,y_0,z_0)$ a fixed point with position vector $\mathbf{v_0} = (x_0,y_0,z_0)$. The position vectors $\mathbf{v} = (x,y,z)$ of all points $P(x,y,z)$ on the line through P_0 with direction vector $\mathbf{d}$ are given by

$$\mathbf{v} = \mathbf{v_0} + t\mathbf{d}$$

where t is an arbitrary parameter. The points $P(x,y,z)$ on the line have coordinates

$$x = x_0 + ta$$
$$y = y_0 + tb \qquad t \text{ a parameter}$$
$$z = z_0 + tc$$

Conversely, the graph of these equations (where a, b, and c are not all zero) is the line through $P_0(x_0,y_0,z_0)$ with direction vector $\mathbf{d} = (a,b,c)$.

EXAMPLE 13 Find the equations of the line through $P_0(3,-1,2)$ parallel to the line with equations

$$\begin{aligned} x &= -1 + 2t \\ y &= 1 + t \\ z &= -3 + 4t \end{aligned}$$

Solution The given line has direction vector $\mathbf{d} = (2,1,4)$ by Theorem 6. Because the line we seek is parallel to this line, $\mathbf{d}$ serves as direction vector. Thus the parametric equations are

$$\begin{aligned} x &= 3 + 2t \\ y &= -1 + t \\ z &= 2 + 4t \end{aligned}$$

EXAMPLE 14 Determine whether the following lines intersect and, if so, find the point of intersection.

$$\begin{aligned} x &= 1 - 3t \qquad & x &= -1 + s \\ y &= 2 + 5t \qquad & y &= 3 - 4s \\ z &= 1 + t \qquad & z &= 1 - s \end{aligned}$$

Solution A typical point $P(x,y,z)$ on the first line has position vector $(x,y,z) = (1 - 3t,2 + 5t,1 + t)$ for some value of the parameter t. Similarly, a point on the second line has position vector $(x,y,z) = (-1 + s, 3 - 4s,1 - s)$ for some value of s. Hence if $P(x,y,z)$ lies on *both* lines, there must exist t and s such that

$$(1 - 3t,2 + 5t,1 + t) = (x,y,z) = (-1 + s,3 - 4s,1 - s)$$

This means that the three equations

$$\begin{aligned} 1 - 3t &= -1 + s \\ 2 + 5t &= 3 - 4s \\ 1 + t &= 1 - s \end{aligned}$$

must have a solution. Hence if there is *no* solution, the lines do not intersect. But in this case $t = 1$ and $s = -1$ satisfy all three equations. Thus the lines *do* intersect and the point of intersection is $(x,y,z) = (1 - 3t,2 + 5t,1 + t) = (-2,7,2)$ using $t = 1$. Of course this point can be found from $(x,y,z) = (-1 + s,3 - 4s,1 - s)$ using $s = -1$.

The analog of Theorem 6 holds for lines in the plane. If $\mathbf{d} = (a,b)$ is a nonzero vector and $P_0(x_0,y_0)$ a fixed point, then

$$\mathbf{v} = \mathbf{v}_0 + t\mathbf{d}$$

where t is an arbitrary parameter gives the position vectors $\mathbf{v} = (x,y)$ of all points $P = P(x,y)$ on the line through P_0 parallel to $\mathbf{d}$.

The description of (nonvertical) lines in the plane is usually done using the notion of slope. The next example derives the point–slope formula using vector techniques.

EXAMPLE 15 Show that the line through $P_0(x_0,y_0)$ with slope m has direction vector $\mathbf{d} = (1,m)$ and equation $y - y_0 = m(x - x_0)$.

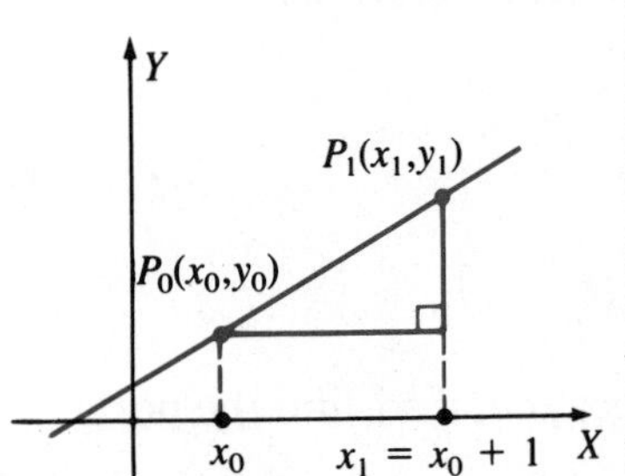

Solution Let $P_1(x_1,y_1)$ be the point on the line one unit to the right of P_0—that is, $x_1 = x_0 + 1$. Then $\mathbf{d} = \overrightarrow{P_0P_1}$ will serve as direction vector of the line, so

$$\mathbf{d} = (x_1 - x_0, y_1 - y_0) = (1, y_1 - y_0)$$

But the slope m can be computed as follows:

$$m = \frac{y_1 - y_0}{x_1 - x_0} = \frac{y_1 - y_0}{1} = y_1 - y_0$$

so $\mathbf{d} = (1,m)$. Hence the vector equation of the line is $(x,y) = (x_0,y_0) + t(1,m)$ and the scalar equations are $x = x_0 + t$, $y = y_0 + mt$. Eliminating t gives

$$y - y_0 = mt = m(x - x_0)$$

as asserted.

If we consider the line through $P_0(x_0,y_0)$ parallel to the Y-axis, then $\mathbf{d} = (0,1)$ is a direction vector that is not of the form $(1,m)$ for any m. This result confirms that the notion of slope makes no sense in this case. However, the vector method gives the vector equation of the line

$$(x,y) = (x_0,y_0) + t(0,1)$$

so the parametric equations for the line are

$$x = x_0$$
$$y = y_0 + t$$

Because y is clearly arbitrary, this is usually written simply as $x = x_0$.

EXERCISES 4.1.2

1. Let $\mathbf{u} = (-1,1,2)$, $\mathbf{v} = (2,0,3)$, and $\mathbf{w} = (-1,3,9)$. Compute the following in component form.

(a) $\mathbf{u} + 2\mathbf{v} - \mathbf{w}$ **(b)** $\frac{1}{3}(3\mathbf{u} - \mathbf{v} + 4\mathbf{w})$ **(c)** $6\mathbf{u} + 2\mathbf{v} - 2\mathbf{w}$

(d) $2\mathbf{v} - 3(\mathbf{u} + \mathbf{w})$ **(e)** $\frac{1}{3}(\mathbf{v} - 3\mathbf{u} + 2\mathbf{w})$ **(f)** $2(\mathbf{u} + \mathbf{v}) - (\mathbf{v} + \mathbf{w} - \mathbf{u})$

2. Determine whether **u** and **v** are parallel in each of the following cases.

(a) $\mathbf{u} = (1,2,-1)$; $\mathbf{v} = (2,1,0)$ **(b)** $\mathbf{u} = (3,-6,3)$; $\mathbf{v} = (-1,2,-1)$

(c) $\mathbf{u} = (1,0,1)$; $\mathbf{v} = (-1,0,1)$ **(d)** $\mathbf{u} = (2,0,-1)$; $\mathbf{v} = (-8,0,4)$

3. Let **u** and **v** be the position vectors of points P and Q, respectively, and let R be the point whose position vector is $\mathbf{u} + \mathbf{v}$. Express the following in terms of **u** and **v**.

(a) $\overrightarrow{QP}$ **(b)** $\overrightarrow{QR}$ **(c)** $\overrightarrow{RP}$ **(d)** $\overrightarrow{RO}$ where 0 is the origin

4. In each case, find $\overrightarrow{PQ}$ in component form.

(a) $P(1,-1,3)$, $Q(2,1,0)$ **(b)** $P(2,0,1)$, $Q(1,-1,6)$

(c) $P(0,0,1)$, $Q(1,0,-3)$ **(d)** $P(1,-1,2)$, $Q(1,-1,2)$

(e) $P(1,0,-3)$, $Q(-1,0,3)$ **(f)** $P(3,-1,6)$, $Q(1,1,4)$

5. In each case, find a point Q such that $\overrightarrow{PQ}$ has (i) the same direction as **v**; (ii) the opposite direction to **v**.

(a) $P(-1,2,2)$, $\mathbf{v} = (1,2,-1)$ **(b)** $P(3,0,-1)$, $\mathbf{v} = (2,-1,3)$

6. Let $\mathbf{u} = (3,-1,0)$, $\mathbf{v} = (4,0,1)$, and $\mathbf{w} = (1,1,3)$. In each case, find **x** such that:

(a) $3(2\mathbf{u} + \mathbf{x}) + \mathbf{w} = 2\mathbf{x} - \mathbf{v}$ **(b)** $2(3\mathbf{v} - \mathbf{x}) = 5\mathbf{w} + \mathbf{u} - 3\mathbf{x}$

7. Let $\mathbf{u} = (1,1,2)$, $\mathbf{v} = (0,1,2)$, and $\mathbf{w} = (1,0,-1)$. In each case, find numbers a, b, and c such that $\mathbf{x} = a\mathbf{u} + b\mathbf{v} + c\mathbf{w}$.

(a) $\mathbf{x} = (2,-1,6)$ **(b)** $\mathbf{x} = (1,3,0)$

8. Let $\mathbf{u} = (3,-1,0)$, $\mathbf{v} = (4,0,1)$, and $\mathbf{z} = (1,1,1)$. In each case, show that there are no numbers a, b, and c such that:

(a) $a\mathbf{u} + b\mathbf{v} + c\mathbf{z} = (1,2,1)$ **(b)** $a\mathbf{u} + b\mathbf{v} + c\mathbf{z} = (5,6,-1)$

9. Let $P_1 = P_1(2,1,-2)$ and $P_2 = P_2(1,-2,0)$. Find the coordinates of the point P:

(a) $\frac{1}{5}$ the way from P_1 to P_2 **(b)** $\frac{1}{4}$ the way from P_2 to P_1

10. Find the two points trisecting the segment between $P(2,3,5)$ and $Q(8,-6,2)$.

11. Let $P_1 = P_1(x_1,y_1,z_1)$ and $P_2 = P_2(x_2,y_2,z_2)$ be two points with position vectors $\mathbf{v_1}$ and $\mathbf{v_2}$, respectively. If r and s are positive integers, show that the point P lying $\frac{r}{r+s}$ the way from P_1 to P_2 has position vector

$$\mathbf{v} = \left[\frac{s}{r+s}\right]\mathbf{v_1} + \left[\frac{r}{r+s}\right]\mathbf{v_2}$$

12. Find the vector and parametric equations of the following lines.
    (a) The line parallel to $(2,-1,0)$ and passing through $P(1,-1,3)$
    (b) The line passing through $P(3,-1,4)$ and $Q(1,0,-1)$
    (c) The line passing through $P(3,-1,4)$ and $Q(3,-1,5)$
    (d) The line parallel to $(1,1,1)$ and passing through $P(1,1,1)$
    (e) The line passing through $P(1,0,-3)$ and parallel to the line with parametric equations $x = -1 + 2t$, $y = 2 - t$, and $z = 3 + 3t$.
    (f) The line passing through $P(2,-1,1)$ and parallel to the line with parametric equations $x = 2 - t$, $y = 1$, and $z = t$.
13. In each case, verify that the points P and Q lie on the line.
    (a) $x = 3 - 4t$, $y = 2 + t$, $z = 1 - t$ $\quad P(-1,3,0), Q(11,0,3)$
    (b) $x = 4 - t$, $y = 3$, $z = 1 - 2t$ $\quad P(2,3,-3), Q(-1,3,-9)$
14. Find the point of intersection (if any) of the following pairs of lines.
    (a) $x = 3 + t$, $y = 1 - 2t$, $z = 3 + 3t$ $\quad x = 4 + 2s$, $y = 6 + 3s$, $z = 1 + s$
    (b) $x = 1 - t$, $y = 2 + 2t$, $z = -1 + 3t$ $\quad x = 2s$, $y = 1 + s$, $z = 3$
    (c) $(x,y,z) = (3,-1,2) + t(1,1,-1)$ $\quad (x,y,z) = (1,1,-2) + s(2,0,3)$
    (d) $(x,y,z) = (4,-1,5) + t(1,0,1)$ $\quad (x,y,z) = (2,-7,12) + s(0,-2,3)$
15. Show that if a line passes through the origin, the position vectors of points on the line are all scalar multiples of some fixed nonzero vector.
16. Show that every line parallel to the Z-axis has parametric equations $x = x_0$, $y = y_0$, $z = t$ for some fixed numbers x_0 and y_0.
17. Let $\mathbf{d} = (a,b,c)$ be a vector where a, b, and c are *all* nonzero. Show that the equations of the line through $P_0(x_0,y_0,z_0)$ with direction vector $\mathbf{d}$ can be written in the form

$$\frac{x - x_0}{a} = \frac{y - y_0}{b} = \frac{z - z_0}{c}$$

    This is called the **symmetric form** of the equations.
18. (a) Prove (3) of Theorem 4. (b) Prove (4) of Theorem 4.
19. A cyclist riding north at 23 km/hr notices that the wind appears to be coming from the northwest. When he turns east and slows to 15 km/hr, the wind appears to be coming from the south. Find the wind vector $\mathbf{w} = (x,y)$ where the X-axis points east and the Y-axis points north. What is the wind speed?
20. The intersection of the three medians of a triangle is called the **centroid** of the triangle. If $\mathbf{u}$, $\mathbf{v}$, and $\mathbf{w}$ are the position vectors of the vertices of a triangle, show that the centroid has position vector $\frac{1}{3}(\mathbf{u} + \mathbf{v} + \mathbf{w})$.
21. Given four non-coplanar points in space, the figure with these points as vertices is called a **tetrahedron**. The line from a vertex through the cen-

troid (previous exercise) of the triangle formed by the remaining vertices is called a **median** of the tetrahedron. If **u**, **v**, **w**, and **x** are the position vectors of the four vertices, show that the point on a median one-fourth the way from the centroid to the vertex has position vector $\frac{1}{4}(\mathbf{u} + \mathbf{v} + \mathbf{w} + \mathbf{x})$. Conclude that the four medians are concurrrent.

SECTION 4.2 Orthogonality

4.2.1 Distance and the Dot Product

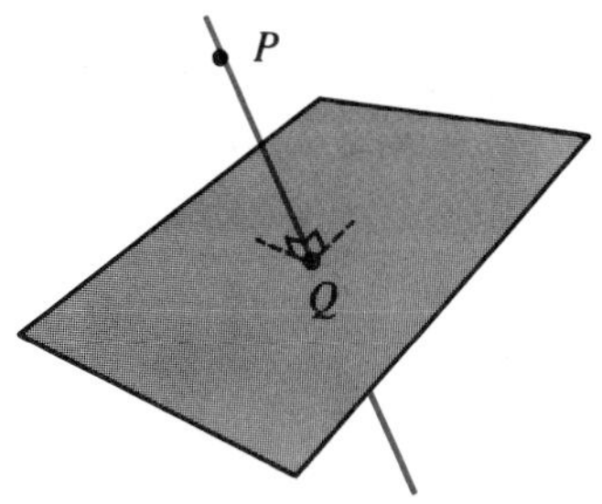

FIGURE 4.13

Any student of geometry soon realizes that the notion of perpendicular lines is fundamental. It comes up, for example, in minimization problems. As an illustration, suppose a point P and a plane are given and it is desired to find the point Q that lies in the plane and is closest to P, as shown in Figure 4.13. Clearly, what is required is to find the line through P that is perpendicular to the plane and then to obtain Q as the point of intersection of this line with the plane. Hence solving this *minimization* problem (finding the point in the plane *closest* to P) comes down to finding the line *perpendicular* to the plane, and this in turn requires a way to determine when two vectors are perpendicular. Surprisingly enough, this can be done by using the following formula for the length of a vector in terms of its components.

THEOREM 1

Let $\mathbf{v} = (x,y,z)$ be a vector. Then

$$\|\mathbf{v}\| = \sqrt{x^2 + y^2 + z^2}$$

where the notation $\sqrt{\ }$ indicates the positive square root.

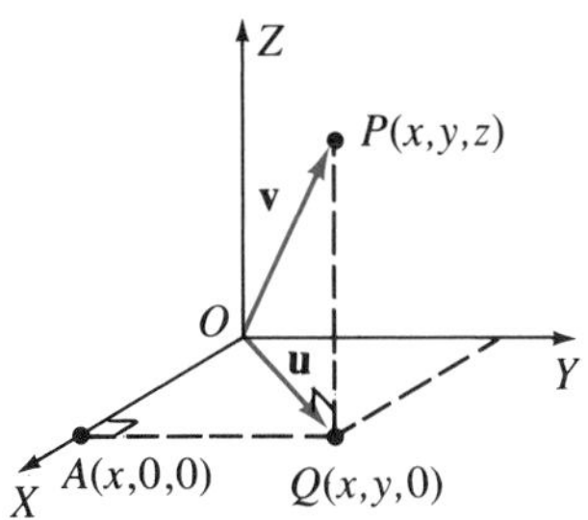

Proof This is an application of **Pythagoras' theorem:** Given a right-angled triangle, the square of the length of the hypotenuse equals the sum of the squares of the lengths of the other two sides. To apply this theorem, recall that **v** has its initial point at the origin and its terminal point at $P(x,y,z)$, as shown in the diagram, and let **u** be the position vector of $Q(x,y,0)$. If $A = A(x,0,0)$, the lengths of the line segments OA and AQ are $|x|$ and $|y|$, respectively, so $\|\mathbf{u}\|^2 = x^2 + y^2$ by Pythagoras' theorem. However, the length of PQ is $|z|$, and so, again by Pythagoras' theorem, $\|\mathbf{v}\|^2 = \|\mathbf{u}\|^2 + z^2 = (x^2 + y^2) + z^2$. This is the desired conclusion. ■

THEOREM 2
Distance Formula

The distance d between points $P_1(x_1,y_1,z_1)$ and $P_2(x_2,y_2,z_2)$ is given by

$$\sqrt{(x_2 - x_1)^2 + (y_2 - y_1)^2 + (z_2 - z_1)^2}$$

Proof Theorem 5 in Section 4.1 gives $\overrightarrow{P_1P_2} = (x_2 - x_1, y_2 - y_1, z_2 - z_1)$. So, by Theorem 1,

$$d = \|\overrightarrow{P_1P_2}\| = \sqrt{(x_2 - x_1)^2 + (y_2 - y_1)^2 + (z_2 - z_1)^2} \quad \blacksquare$$

EXAMPLE 1 Compute the distance between $P_1(3,-1,2)$ and $P_2(-1,1,0)$.

Solution The distance is

$$\sqrt{(-1 - 3)^2 + (1 - (-1))^2 + (0 - 2)^2} = \sqrt{16 + 4 + 4} = \sqrt{24} = 2\sqrt{6}.$$

If $\mathbf{v}$ is the velocity vector for some moving object, then its length $\|\mathbf{v}\|$ is called the **speed** of the object. Here is an example that uses the distance formula.

EXAMPLE 2 A rescue boat has a top speed of 13 knots. The captain wants to go due east as fast as possible in water with a current of 5 knots due south. Find the velocity vector $\mathbf{v} = (x,y)$ that he must achieve, assuming the X- and Y-axes point east and north, respectively, and find his resulting speed.

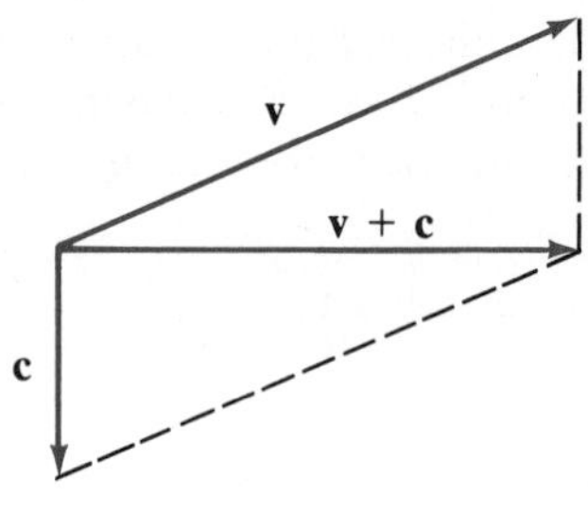

Solution Let $\mathbf{v}$ be a vector of length 13 (the speed of the boat) and in the direction he wants to go. The current $\mathbf{c}$ goes south at 5 knots, so $\mathbf{c} = (0,-5)$. Hence the actual motion of the boat is given by $\mathbf{v} + \mathbf{c}$. This must point east, so $\mathbf{v} + \mathbf{c} = (a,0)$ with $a > 0$. Hence $\mathbf{v} = (a,0) - \mathbf{c} = (a,5)$, and we determine a from the fact that $13 = \|\mathbf{v}\| = \sqrt{a^2 + 5^2}$. Hence $a = 12$ and $\mathbf{v} = (12,5)$ is the required vector. Because the actual motion of the boat is $\mathbf{v} + \mathbf{c} = (12,0)$, the resulting speed is 12 knots.

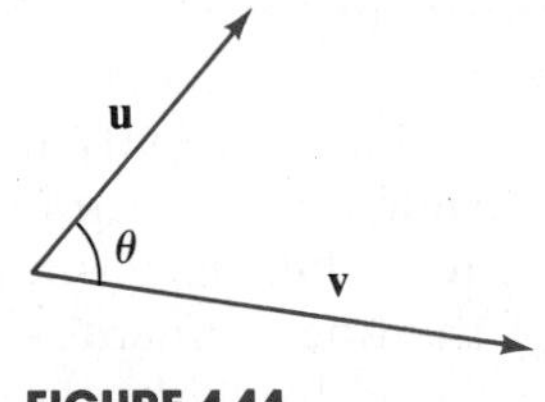

FIGURE 4.14

Let $\mathbf{u}$ and $\mathbf{v}$ be two nonzero vectors, and suppose they are positioned with a common initial point. Then they determine a unique angle θ in the range $0 \leq \theta \leq \pi$ (recall that π radians equals 180°). This angle θ will be referred to as **the angle between u and v** (see Figure 4.14). Clearly $\mathbf{u}$ and $\mathbf{v}$ are parallel if either $\theta = 0$ or $\theta = \pi$; they are said to be **orthogonal** if $\theta = \pi/2$—that is, if θ is a right angle. If one of $\mathbf{u}$ and $\mathbf{v}$ is $\mathbf{0}$, the angle between them is not defined.

DEFINITION The **dot product** $\mathbf{u} \cdot \mathbf{v}$ of two vectors $\mathbf{u}$ and $\mathbf{v}$ is defined as follows:

$$\mathbf{u} \cdot \mathbf{v} = \begin{cases} \|\mathbf{u}\|\,\|\mathbf{v}\| \cos\theta & \text{if } \mathbf{u} \neq \mathbf{0} \text{ and } \mathbf{v} \neq \mathbf{0} \\ 0 & \text{otherwise} \end{cases}$$

where θ is the angle between $\mathbf{u}$ and $\mathbf{v}$.

Note that $\mathbf{u} \cdot \mathbf{v}$ is a *number* (even though $\mathbf{u}$ and $\mathbf{v}$ are vectors). For this reason, $\mathbf{u} \cdot \mathbf{v}$ is sometimes called the **scalar product** of $\mathbf{u}$ and $\mathbf{v}$.

As defined above, the dot product of two vectors $\mathbf{u}$ and $\mathbf{v}$ is not easy to compute, because the angle between the vectors is not usually known. However, the following fundamental theorem provides a very easy way of calculating $\mathbf{u} \cdot \mathbf{v}$, provided that $\mathbf{u}$ and $\mathbf{v}$ are given in component form (and so leads to a method of computing the *angle*). Furthermore, it provides a rationale for our use of the term *dot product* at the beginning of Section 2.2. The proof depends on the **law of cosines** from trigonometry:

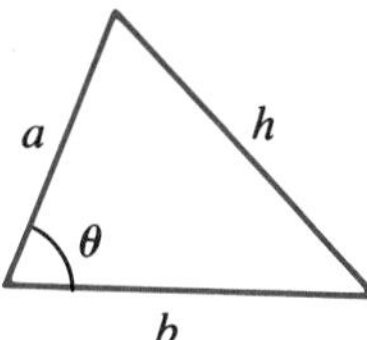

FIGURE 4.15

If a triangle has sides of lengths a, b, and h, and if θ is the angle between the sides of lengths a and b (as in Figure 4.15), then

$$h^2 = a^2 + b^2 - 2ab \cos \theta$$

Observe that this is a generalization of Pythagoras' theorem: If $\theta = \pi/2$ (that is, $\theta = 90°$), then $\cos \theta = 0$ and the result becomes $h^2 = a^2 + b^2$.

THEOREM 3

> Let $\mathbf{v_1} = (x_1,y_1,z_1)$ and $\mathbf{v_2} = (x_2,y_2,z_2)$ be two vectors given in component form. Then their dot product can be computed as follows:
>
> $$\mathbf{v_1} \cdot \mathbf{v_2} = x_1x_2 + y_1y_2 + z_1z_2$$
>
> In other words, $\mathbf{v_1} \cdot \mathbf{v_2}$ equals the dot product of the ordered triples (x_1,y_1,z_1) and (x_2,y_2,z_2).

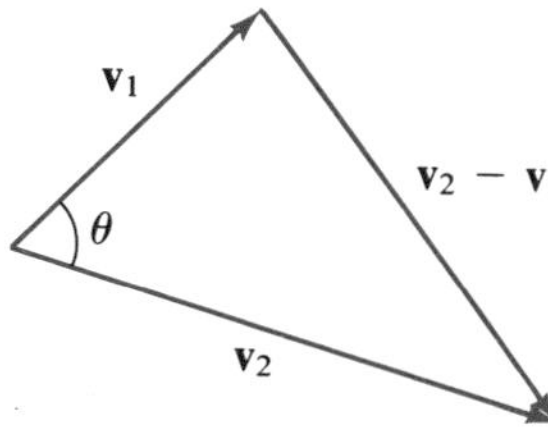

Proof Consider the triangle in the diagram and apply the law of cosines.

$$\begin{aligned}\|\mathbf{v_2} - \mathbf{v_1}\|^2 &= \|\mathbf{v_1}\|^2 + \|\mathbf{v_2}\|^2 - 2\|\mathbf{v_1}\|\,\|\mathbf{v_2}\| \cos \theta \\ &= \|\mathbf{v_1}\|^2 + \|\mathbf{v_2}\|^2 - 2\mathbf{v_1} \cdot \mathbf{v_2}\end{aligned}$$

It follows that $\mathbf{v_1} \cdot \mathbf{v_2} = \frac{1}{2}\{\|\mathbf{v_1}\|^2 + \|\mathbf{v_2}\|^2 - \|\mathbf{v_2} - \mathbf{v_1}\|^2\}$. Because $\mathbf{v_2} - \mathbf{v_1} = (x_2 - x_1, y_2 - y_1, z_2 - z_1)$ in component form, Theorem 1 gives

$$\mathbf{v_1} \cdot \mathbf{v_2} = \frac{1}{2}\{(x_1{}^2 + y_1{}^2 + z_1{}^2) + (x_2{}^2 + y_2{}^2 + z_2{}^2) - [(x_2 - x_1)^2 + (y_2 - y_1)^2 + (z_2 - z_1)^2]\}$$

The reader may verify that the right side reduces to $x_1x_2 + y_1y_2 + z_1z_2$. ■

EXAMPLE 3 Compute $\mathbf{u} \cdot \mathbf{v}$ when $\mathbf{u} = (2,-1,3)$ and $\mathbf{v} = (1,4,-1)$.

Solution $\mathbf{u} \cdot \mathbf{v} = 2 \cdot 1 + (-1) \cdot 4 + 3(-1) = -5$

The next theorem lists several basic properties of the dot product that we will use repeatedly.

THEOREM 4 Let **u**, **v**, and **w** denote arbitrary vectors.

(1) $\mathbf{u} \cdot \mathbf{v}$ is a number

(2) $\mathbf{u} \cdot \mathbf{v} = \mathbf{v} \cdot \mathbf{u}$

(3) $\mathbf{u} \cdot \mathbf{0} = 0 = \mathbf{0} \cdot \mathbf{u}$

(4) $\mathbf{u} \cdot \mathbf{u} = \|\mathbf{u}\|^2$

(5) $(k\mathbf{u}) \cdot \mathbf{v} = k(\mathbf{u} \cdot \mathbf{v}) = \mathbf{u} \cdot (k\mathbf{v})$ for any scalar k

(6) $\mathbf{u} \cdot (\mathbf{v} \pm \mathbf{w}) = \mathbf{u} \cdot \mathbf{v} \pm \mathbf{u} \cdot \mathbf{w}$

Proof (1), (2), and (3) are clear from the definition of $\mathbf{u} \cdot \mathbf{v}$.

(4) If $\mathbf{u} = (x,y,z)$, then $\mathbf{u} \cdot \mathbf{u} = x^2 + y^2 + z^2 = \|\mathbf{u}\|^2$ by Theorem 1.

(5) If $\mathbf{u} = (x,y,z)$ and $\mathbf{v} = (x_1,y_1,z_1)$, then, using Theorem 3,

$$\begin{aligned}(k\mathbf{u}) \cdot \mathbf{v} &= (kx,ky,kz) \cdot (x_1,y_1,z_1) \\ &= (kx)x_1 + (ky)y_1 + (kz)z_1 \\ &= k(xx_1 + yy_1 + zz_1) \\ &= k(\mathbf{u} \cdot \mathbf{v})\end{aligned}$$

The verification that $\mathbf{u} \cdot (k\mathbf{v}) = k(\mathbf{u} \cdot \mathbf{v})$ is analogous and is left to the reader.

(6) This is verified as in (5). It is left as Exercise 31. ■

Observe that (5) and (6) of Theorem 4 can be combined to allow such calculations as the following:

$$2\mathbf{u} \cdot (3\mathbf{v} - 2\mathbf{w} + 4\mathbf{z}) = 6(\mathbf{u} \cdot \mathbf{v}) - 4(\mathbf{u} \cdot \mathbf{w}) + 8(\mathbf{u} \cdot \mathbf{z})$$

Such calculations will be carried out without comment below. Here is an example.

EXAMPLE 4 Verify that $\|\mathbf{u}+\mathbf{v}\|^2 = \|\mathbf{u}\|^2 + \|\mathbf{v}\|^2 + 2(\mathbf{u}\cdot\mathbf{v})$.

Solution Apply Theorem 4.

$$\begin{aligned}\|\mathbf{u}+\mathbf{v}\|^2 &= (\mathbf{u}+\mathbf{v})\cdot(\mathbf{u}+\mathbf{v}) = (\mathbf{u}+\mathbf{v})\cdot\mathbf{u} + (\mathbf{u}+\mathbf{v})\cdot\mathbf{v}\\ &= \mathbf{u}\cdot\mathbf{u} + \mathbf{v}\cdot\mathbf{u} + \mathbf{u}\cdot\mathbf{v} + \mathbf{v}\cdot\mathbf{v}\\ &= \|\mathbf{u}\|^2 + \|\mathbf{v}\|^2 + 2(\mathbf{u}\cdot\mathbf{v})\end{aligned}$$

As we have mentioned, one important use of the dot product is in calculating the angle θ between two nonzero vectors $\mathbf{u}$ and $\mathbf{v}$. Because $\|\mathbf{u}\| \neq 0$ and $\|\mathbf{v}\| \neq 0$, the defining relationship $\mathbf{u}\cdot\mathbf{v} = \|\mathbf{u}\|\,\|\mathbf{v}\|\cos\theta$ can be solved for $\cos\theta$ to get

$$\cos\theta = \frac{\mathbf{u}\cdot\mathbf{v}}{\|\mathbf{u}\|\,\|\mathbf{v}\|}$$

This can be used to find θ. In this connection, it is worth noting that $\cos\theta$ has the same sign as $\mathbf{u}\cdot\mathbf{v}$, so $0 \le \theta < \pi/2$ if $\mathbf{u}\cdot\mathbf{v}$ is positive, whereas $\pi/2 < \theta \le \pi$ if $\mathbf{u}\cdot\mathbf{v}$ is negative. In particular, the vectors $\mathbf{u}$ and $\mathbf{v}$ are orthogonal (that is, $\theta = \pi/2$) if and only if $\mathbf{u}\cdot\mathbf{v} = 0$. We record this for future reference as Theorem 5.

THEOREM 5 Two nonzero vectors $\mathbf{u}$ and $\mathbf{v}$ are orthogonal if and only if $\mathbf{u}\cdot\mathbf{v} = 0$.

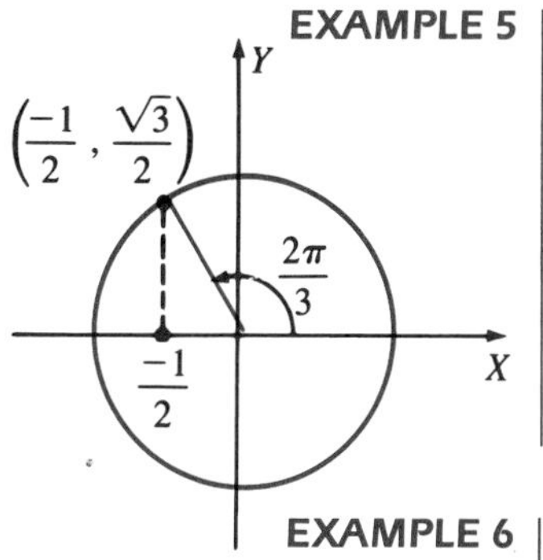

EXAMPLE 5 Compute the angle between $\mathbf{u} = (-1,1,2)$ and $\mathbf{v} = (2,1,-1)$.

Solution Compute $\cos\theta = \dfrac{\mathbf{u}\cdot\mathbf{v}}{\|\mathbf{u}\|\,\|\mathbf{v}\|} = \dfrac{-2 \quad +1 \quad -2}{\sqrt{6}\sqrt{6}} = \dfrac{-1}{2}$. Now recall that $\cos\theta$ and $\sin\theta$ are defined so that $(\cos\theta, \sin\theta)$ is the point on the unit circle determined by the angle θ (drawn counterclockwise, starting from the positive X-axis). In the present case, we know that $\cos\theta = -\frac{1}{2}$ and that $0 \le \theta \le \pi$. Because $\cos\pi/3 = \frac{1}{2}$, it follows that $\theta = 2\pi/3$ (see the diagram).

EXAMPLE 6 Show that the points $P(3,-1,1)$, $Q(4,1,4)$, and $R(6,0,4)$ are the vertices of a right triangle.

Solution The vectors along the sides of the triangle are

$$\overrightarrow{PQ} = (1,2,3),\ \overrightarrow{PR} = (3,1,3), \text{ and } \overrightarrow{QR} = (2,-1,0)$$

Evidently $\overrightarrow{PQ}\cdot\overrightarrow{QR} = (1,2,3)\cdot(2,-1,0) = 2 - 2 + 0 = 0$, so $\overrightarrow{PQ}$ and $\overrightarrow{QR}$ are orthogonal vectors. This means sides PQ and QR are perpendicular—that is, the angle at Q is a right angle.

The next example demonstrates how the dot product can be used to verify a geometrical theorem involving perpendicular lines.

EXAMPLE 7 A parallelogram with sides of equal length is called a **rhombus**. Show that the diagonals of a rhombus are perpendicular.

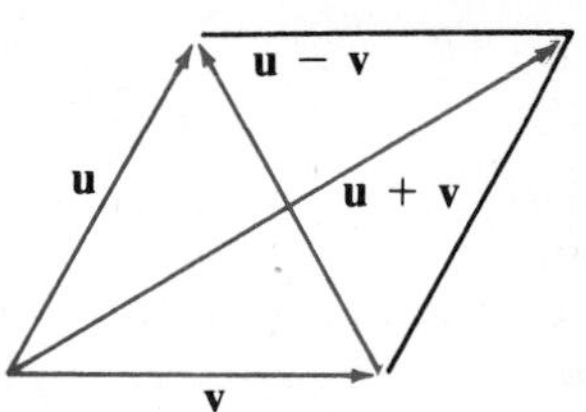

Solution Let **u** and **v** denote vectors along two adjacent sides of the rhombus, as shown in the diagram. Then the diagonals are $\mathbf{u} - \mathbf{v}$ and $\mathbf{u} + \mathbf{v}$, and we compute

$$\begin{aligned}(\mathbf{u} - \mathbf{v}) \cdot (\mathbf{u} + \mathbf{v}) &= \mathbf{u} \cdot (\mathbf{u} + \mathbf{v}) - \mathbf{v} \cdot (\mathbf{u} + \mathbf{v}) \\ &= \mathbf{u} \cdot \mathbf{u} + \mathbf{u} \cdot \mathbf{v} - \mathbf{v} \cdot \mathbf{u} - \mathbf{v} \cdot \mathbf{v} \\ &= \|\mathbf{u}\|^2 - \|\mathbf{v}\|^2 \\ &= 0\end{aligned}$$

because $\|\mathbf{u}\| = \|\mathbf{v}\|$ (it is a rhombus). Hence $\mathbf{u} - \mathbf{v}$ and $\mathbf{u} + \mathbf{v}$ are orthogonal.

If a nonzero vector **v** is given, it is often important to be able to write an arbitrary vector **u** as $\mathbf{u} = \mathbf{u}_1 + \mathbf{u}_2$, where $\mathbf{u}_1$ is parallel to **v** and $\mathbf{u}_2 = \mathbf{u} - \mathbf{u}_1$ is orthogonal to **v**. Suppose that **u** and $\mathbf{v} \neq \mathbf{0}$ emanate from a common initial point Q (see Figures 4.16 and 4.17). Let P be the terminal point of **u**, and let P_1 denote the foot of the perpendicular dropped from P to the line through Q parallel to **v**.

FIGURE 4.16

DEFINITION The vector $\mathbf{u}_1 = \overrightarrow{QP}_1$ is called **the projection of u on v**.

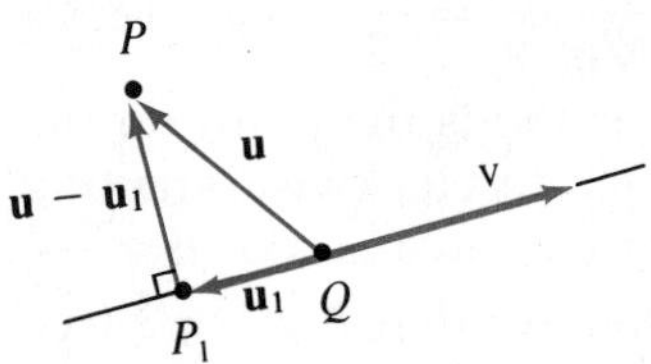

FIGURE 4.17

In Figure 4.16 the vector $\mathbf{u}_1$ has the same direction as **v**; however, it has the opposite direction from **v** if the angle between **u** and **v** is greater than $\pi/2$ (Figure 4.17). Note that the projection $\mathbf{u}_1$ is zero if and only if **u** and **v** are orthogonal.

Calculating the projection $\mathbf{u}_1$ of **u** on **v** is remarkably easy.

THEOREM 6 If **u** and $\mathbf{v} \neq \mathbf{0}$ are vectors, the projection $\mathbf{u}_1$ of **u** on **v** is given by

$$\mathbf{u}_1 = \frac{\mathbf{u} \cdot \mathbf{v}}{\|\mathbf{v}\|^2}\mathbf{v}$$

The vector $\mathbf{u} - \mathbf{u}_1$ is orthogonal to **v**.

Proof First observe that $\mathbf{u}_1$ is to be parallel to $\mathbf{v}$ and so has the form $\mathbf{u}_1 = t\mathbf{v}$ for some scalar t by Theorem 2 in Section 4.1. The vector $\mathbf{u}_1$ is chosen in such a way that $\mathbf{u} - \mathbf{u}_1$ and $\mathbf{v}$ are orthogonal (see Figures 4.16 and 4.17), and this requirement determines the scalar t. In fact it means that $(\mathbf{u} - \mathbf{u}_1) \cdot \mathbf{v} = 0$ by Theorem 5, and if $\mathbf{u}_1 = t\mathbf{v}$ is substituted here, the result is

$$0 = (\mathbf{u} - t\mathbf{v}) \cdot \mathbf{v} = \mathbf{u} \cdot \mathbf{v} - t(\mathbf{v} \cdot \mathbf{v}) = \mathbf{u} \cdot \mathbf{v} - t\|\mathbf{v}\|^2$$

It follows that $t = \dfrac{\mathbf{u} \cdot \mathbf{v}}{\|\mathbf{v}\|^2}$, where the assumption that $\mathbf{v} \neq \mathbf{0}$ guarantees that $\|\mathbf{v}\|^2 \neq 0$. ■

EXAMPLE 8 Find the projection of $\mathbf{u} = (2,-3,1)$ on $\mathbf{v} = (1,-1,3)$ and express $\mathbf{u} = \mathbf{u}_1 + \mathbf{u}_2$ where $\mathbf{u}_1$ is parallel to $\mathbf{v}$ and $\mathbf{u}_2$ is orthogonal to $\mathbf{v}$.

Solution The projection $\mathbf{u}_1$ of $\mathbf{u}$ on $\mathbf{v}$ is

$$\mathbf{u}_1 = \frac{\mathbf{u} \cdot \mathbf{v}}{\|\mathbf{v}\|^2}\mathbf{v} = \frac{2 + 3 + 3}{1^2 + (-1)^2 + 3^2}(1,-1,3) = \frac{8}{11}(1,-1,3)$$

The vector $\mathbf{u}_2 = \mathbf{u} - \mathbf{u}_1 = \frac{1}{11}(14,-25,-13)$ is orthogonal to $\mathbf{v}$ by Theorem 6 (alternatively, observe that $\mathbf{v} \cdot \mathbf{u}_2 = 0$), and $\mathbf{u} = \mathbf{u}_1 + \mathbf{u}_2$ as required.

EXAMPLE 9 Find the shortest distance from the point $P(1,3,-2)$ to the line through $P_0(2,0,-1)$ with direction vector $\mathbf{d} = (1,-1,0)$. Also find the point P_1 that lies on the line and is closest to P.

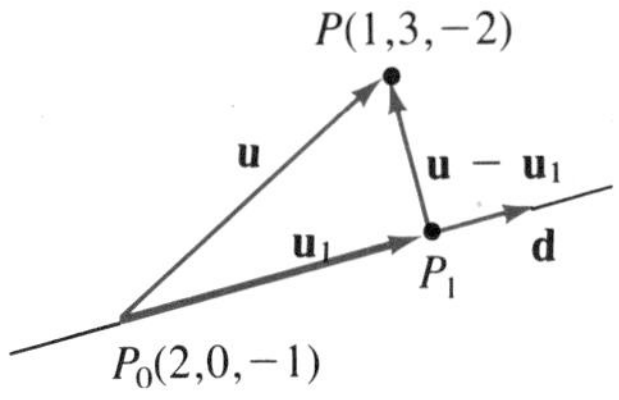

$P(1,3,-2)$, $\mathbf{u}$, $\mathbf{u} - \mathbf{u}_1$, $\mathbf{u}_1$, P_1, $\mathbf{d}$, $P_0(2,0,-1)$

Solution Let $\mathbf{u} = (1,3,-2) - (2,0,-1) = (-1,3,-1)$ denote the vector from P_0 to P, and let $\mathbf{u}_1$ denote the projection of $\mathbf{u}$ on $\mathbf{d}$. Thus

$$\mathbf{u}_1 = \frac{\mathbf{u} \cdot \mathbf{d}}{\|\mathbf{d}\|^2}\mathbf{d} = \frac{-1 - 3 + 0}{1^2 + (-1)^2 + 0^2}\mathbf{d} = -2\mathbf{d} = (-2,2,0)$$

by Theorem 6. The point P_1 on the line is closest to P, so the distance is

$$\|\overrightarrow{P_1P}\| = \|\mathbf{u} - \mathbf{u}_1\| = \|(1,1,-1)\| = \sqrt{3}$$

To find the coordinates of P_1, let $\mathbf{v}_0$ and $\mathbf{v}_1$ denote the position vectors of P_0 and P_1, respectively. Then $\mathbf{v}_0 = (2,0,-1)$ and $\mathbf{v}_1 = \mathbf{v}_0 + \mathbf{u}_1 = (0,2,-1)$. Hence $P_1 = P_1(0,2,-1)$ is the required point. It can be checked that the distance from P_1 to P is $\sqrt{3}$, as expected.

EXERCISES 4.2.1

1. Compute $\|\mathbf{v}\|$ if $\mathbf{v}$ equals:

(a) $(2,-1,1)$ **(b)** $(1,-1,2)$ **(c)** $(1,0,-1)$
(d) $(-1,0,2)$ **(e)** $2(1,-1,2)$ **(f)** $-3(1,1,2)$

2. Find a unit vector in the direction of:

(a) $(2,-2,1)$ **(b)** $(-2,-1,2)$

3. Find a unit vector in the direction from $(3,-1,4)$ to $(1,3,5)$.

4. Find the distance between the following pairs of points.

(a) $(3,-1,0)$ and $(2,0,1)$ **(b)** $(2,-1,2)$ and $(2,0,1)$
(c) $(-3,5,2)$ and $(-1,3,3)$ **(d)** $(4,0,-2)$ and $(3,2,0)$

5. Compute $\mathbf{u} \cdot \mathbf{v}$ where:

(a) $\mathbf{u} = (2,-1,3)$, $\mathbf{v} = (1,1,-2)$ **(b)** $\mathbf{u} = (1,2,-1)$, $\mathbf{v} = \mathbf{u}$
(c) $\mathbf{u} = (1,1,-3)$, $\mathbf{v} = 2(2,1,1)$ **(d)** $\mathbf{u} = (3,-1,5)$, $\mathbf{v} = (6,-7,-5)$
(e) $\mathbf{u} = (x,y,z)$, $\mathbf{v} = (a,b,c)$ **(f)** $\mathbf{u} = (a,b,c)$, $\mathbf{v} = \mathbf{0}$

6. Find the angle between the following pairs of vectors.

(a) $\mathbf{u} = (1,0,3)$, $\mathbf{v} = (2,0,1)$ **(b)** $\mathbf{u} = (3,-1,0)$, $\mathbf{v} = (-6,2,0)$
(c) $\mathbf{u} = (7,-1,3)$, $\mathbf{v} = (1,4,-1)$ **(d)** $\mathbf{u} = (2,1,-1)$, $\mathbf{v} = (3,6,3)$
(e) $\mathbf{u} = (1,-1,0)$, $\mathbf{v} = (0,1,1)$ **(f)** $\mathbf{u} = (0,3,4)$, $\mathbf{v} = (5\sqrt{2},-7,-1)$

7. Find all real numbers x such that:

(a) $(2,-1,3)$ and $(x,-2,1)$ are orthogonal

(b) $(2,-1,1)$ and $(1,x,2)$ are at an angle of $\frac{\pi}{3}$

8. Find all vectors $\mathbf{v} = (x,y,z)$ orthogonal to both:

(a) $\mathbf{u}_1 = (-1,-3,2)$ and $\mathbf{u}_2 = (0,1,1)$ **(b)** $\mathbf{u}_1 = (3,-1,2)$ and $\mathbf{u}_2 = (2,0,1)$
(c) $\mathbf{u}_1 = (2,0,-1)$ and $\mathbf{u}_2 = (-4,0,2)$ **(d)** $\mathbf{u}_1 = (2,-1,3)$ and $\mathbf{u}_2 = (0,0,0)$

9. Find two vectors $\mathbf{x}$ and $\mathbf{y}$ that are both orthogonal to $\mathbf{v} = (1,2,0)$ and such that $\mathbf{x}$ is orthogonal to $\mathbf{y}$.

10. Consider the triangle with vertices $P(2,0,-3)$, $Q(5,-2,1)$, and $R(7,5,3)$.

(a) Show that it is a right-angled triangle.

(b) Find the lengths of the three sides, and verify Pythagoras' theorem.

11. Show that the triangle with vertices $A(4,-7,9)$, $B(6,4,4)$, and $C(7,10,-6)$ is not a right-angled triangle.

12. Find the three internal angles of the triangle with vertices:

(a) $A(3,1,-2)$, $B(3,0,-1)$, and $C(5,2,-1)$

(b) $A(3,1,-2)$, $B(5,2,-1)$, and $C(4,3,-3)$.

13. Show that the line through $P_0(3,1,4)$ and $P_1(2,1,3)$ is perpendicular to the line through $P_2(1,-1,2)$ and $P_3(0,5,3)$.

14. In each case, compute the projection of $\mathbf{u}$ on $\mathbf{v}$.

(a) $\mathbf{u} = (5,7,1)$, $\mathbf{v} = (2,-1,3)$ **(b)** $\mathbf{u} = (3,-2,1)$, $\mathbf{v} = (4,1,1)$
(c) $\mathbf{u} = (1,-1,2)$, $\mathbf{v} = (3,-1,1)$ **(d)** $\mathbf{u} = (3,-2,-1)$, $\mathbf{v} = (-6,4,2)$

15. In each case, write $\mathbf{u} = \mathbf{u}_1 + \mathbf{u}_2$, where $\mathbf{u}_1$ is parallel to $\mathbf{v}$ and $\mathbf{u}_2$ is orthogonal to $\mathbf{v}$.
   (a) $\mathbf{u} = (2,-1,1)$, $\mathbf{v} = (1,-1,3)$
   (b) $\mathbf{u} = (3,1,0)$, $\mathbf{v} = (-2,1,4)$
   (c) $\mathbf{u} = (2,-1,0)$, $\mathbf{v} = (3,1,-1)$
   (d) $\mathbf{u} = (3,-2,1)$, $\mathbf{v} = (-6,4,-1)$
16. Calculate the distance from the point P to the line in each case, and find the point Q on the line closest to P.
   (a) $P(3,2,-1)$ line: $(x,y,z) = (2,1,3) + t(3,-1,-2)$
   (b) $P(1,-1,3)$ line: $(x,y,z) = (1,0,-1) + t(3,1,4)$
17. Show that two lines in the plane with slopes m_1 and m_2 are perpendicular if and only if $m_1m_2 = -1$. [*Hint:* Example 15 in Section 4.1.]
18. (a) Show that, of the four diagonals of a cube, no pair is perpendicular.
   (b) Consider a rectangular solid with sides of lengths a, b, and c. Show that it has two orthogonal diagonals if and only if the sum of two of a^2, b^2, and c^2 equals the third.
19. Let A, B, and $C(2, -1, 1)$ be the vertices of a triangle where $\overrightarrow{AB}$ is parallel to $(1, -1, 1)$, $\overrightarrow{AC}$ is parallel to $(2, 0, -1)$, and angle C is a right angle. Find the equation of the line through B and C.
20. Given $\mathbf{v} = (x,y,z)$ in component form, show that the projections of $\mathbf{v}$ on $\mathbf{i}$, $\mathbf{j}$, and $\mathbf{k}$ are $x\mathbf{i}$, $y\mathbf{j}$, and $z\mathbf{k}$, respectively.
21. Show that $(\mathbf{u} + \mathbf{v}) \cdot (\mathbf{u} - \mathbf{v}) = \|\mathbf{u}\|^2 - \|\mathbf{v}\|^2$ for any vectors $\mathbf{u}$ and $\mathbf{v}$.
22. (a) Show that $\|\mathbf{u} + \mathbf{v}\|^2 + \|\mathbf{u} - \mathbf{v}\|^2 = 2(\|\mathbf{u}\|^2 + \|\mathbf{v}\|^2)$ for any vectors $\mathbf{u}$ and $\mathbf{v}$.
   (b) What does this say about parallelograms?
23. Show that if the diagonals of a parallelogram are perpendicular, it is necessarily a rhombus. [*Hint:* Example 7.]

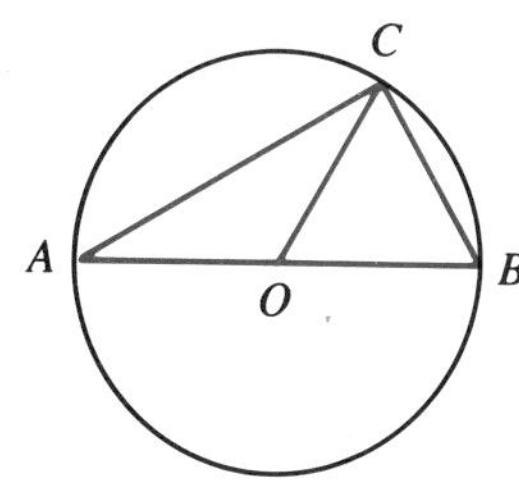

24. Let A and B be the end points of a diameter of a circle. If C is any point on the circle, show that AC and BC are perpendicular. [*Hint:* Express $\overrightarrow{AC}$ and $\overrightarrow{BC}$ in terms of $\mathbf{u} = \overrightarrow{OA}$ and $\mathbf{v} = \overrightarrow{OC}$, where O is the center.]
25. If $\mathbf{u}$ and $\mathbf{v}$ are orthogonal, show that $\|\mathbf{u} + \mathbf{v}\|^2 = \|\mathbf{u}\|^2 + \|\mathbf{v}\|^2$.
26. Let $\mathbf{u}$, $\mathbf{v}$, and $\mathbf{w}$ be pairwise orthogonal vectors.
   (a) Show that $\|\mathbf{u} + \mathbf{v} + \mathbf{w}\|^2 = \|\mathbf{u}\|^2 + \|\mathbf{v}\|^2 + \|\mathbf{w}\|^2$.
   (b) If $\mathbf{u}$, $\mathbf{v}$, and $\mathbf{w}$ are all the same length, show that they all make the same angle with $\mathbf{u} + \mathbf{v} + \mathbf{w}$.
27. Assume $\mathbf{u}$ and $\mathbf{v}$ are nonzero vectors that are not parallel. Show that $\mathbf{w} = \|\mathbf{u}\|\mathbf{v} + \|\mathbf{v}\|\mathbf{u}$ is a nonzero vector that bisects the angle between $\mathbf{u}$ and $\mathbf{v}$.
28. Let α, β, and γ be the angles a vector $\mathbf{v} \neq \mathbf{0}$ makes with the X-, Y-, and Z-axes, respectively. Then $\cos\alpha$, $\cos\beta$, and $\cos\gamma$ are called the **direction cosines** of the vector $\mathbf{v}$.
   (a) If $\mathbf{v} = (a,b,c)$, show that $\cos\alpha = \frac{a}{\|\mathbf{v}\|}$, $\cos\beta = \frac{b}{\|\mathbf{v}\|}$, and $\cos\gamma = \frac{c}{\|\mathbf{v}\|}$.
   (b) Show that $\cos^2\alpha + \cos^2\beta + \cos^2\gamma = 1$.
29. Let $\mathbf{v} \neq \mathbf{0}$ be any nonzero vector and suppose that a vector $\mathbf{u}$ can be written as $\mathbf{u} = \mathbf{p} + \mathbf{q}$, where $\mathbf{p}$ is parallel to $\mathbf{v}$ and $\mathbf{q}$ is orthogonal to $\mathbf{v}$.

Show that necessarily $\mathbf{p}$ is the projection of $\mathbf{u}$ on $\mathbf{v}$. [*Hint:* Argue as in the proof of Theorem 6.]

30. Use Theorem 4 in Section 4.1 and Theorem 1 to verify the formula $\|k\mathbf{v}\| = |k|\,\|\mathbf{v}\|$ for all scalars k and vectors $\mathbf{v}$.

31. Prove (6) of Theorem 4.

32. Let $\mathbf{v} \neq \mathbf{0}$ be a nonzero vector and let $a \neq 0$ be a scalar. If $\mathbf{u}$ is any vector, show that the projection of $\mathbf{u}$ on $\mathbf{v}$ equals the projection of $\mathbf{u}$ on $a\mathbf{v}$.

33. **(a)** Show that the **Cauchy–Schwarz inequality** $|\mathbf{u} \cdot \mathbf{v}| \leq \|\mathbf{u}\|\,\|\mathbf{v}\|$ holds for all vectors $\mathbf{u}$ and $\mathbf{v}$. [*Hint:* $|\cos\theta| \leq 1$ for all angles θ.]

(b) Show that $|\mathbf{u} \cdot \mathbf{v}| = \|\mathbf{u}\|\,\|\mathbf{v}\|$ if and only if $\mathbf{u}$ and $\mathbf{v}$ are parallel. [*Hint:* When is $\cos\theta = \pm 1$?]

(c) Show that $|x_1x_2 + y_1y_2 + z_1z_2| \leq \sqrt{x_1^2 + y_1^2 + z_1^2}\,\sqrt{x_2^2 + y_2^2 + z_2^2}$ holds for all numbers x_1, x_2, y_1, y_2, z_1, and z_2.

(d) Show that $|xy + yz + zx| \leq x^2 + y^2 + z^2$ for all x, y, and z.

(e) Show that $(x + y + z)^2 \leq 3(x^2 + y^2 + z^2)$ holds for all x, y, and z.

34. Prove that the **triangle inequality** $\|\mathbf{u} + \mathbf{v}\| \leq \|\mathbf{u}\| + \|\mathbf{v}\|$ holds for all vectors $\mathbf{u}$ and $\mathbf{v}$. [*Hint:* Consider the triangle with $\mathbf{u}$ and $-\mathbf{v}$ as two sides.]

35. Given a triangle, the line through a vertex that is perpendicular to the opposite side is called an **altitude** of the triangle. Show that the three altitudes of any triangle are concurrent. [*Hint:* In the diagram, show that $\overrightarrow{PC} \cdot \overrightarrow{AB} = 0$ by writing $\overrightarrow{AB} = \overrightarrow{AC} - \overrightarrow{BC}$ and using $\overrightarrow{PC} = \overrightarrow{PB} + \overrightarrow{BC}$ and $\overrightarrow{PC} = \overrightarrow{PA} + \overrightarrow{AC}$.]

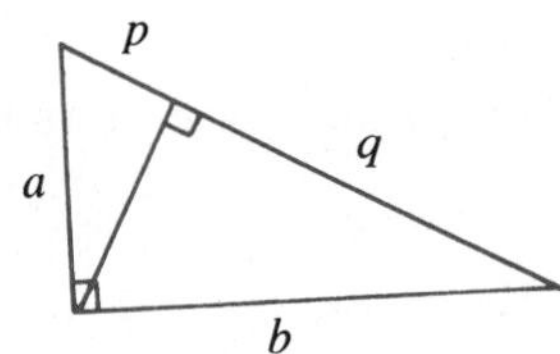

36. Prove Pythagoras' theorem. [*Hint:* In the diagram, let $h = p + q$. Then $\frac{a}{h} = \frac{p}{a}$ and $\frac{b}{h} = \frac{q}{b}$ using similar triangles.]

4.2.2 Planes and the Cross Product

It is evident geometrically that, among all planes that are perpendicular to a given straight line, there is exactly one containing a given point. This can be used to give a very simple description of a plane. In order to do this, it is necessary to introduce the following notion:

DEFINITION A nonzero vector $\mathbf{n}$ is called a **normal** to a plane if it is orthogonal to every vector in the plane.

For example, the coordinate vector $\mathbf{k}$ is a normal to the X-Y-plane.

Note that a normal to a given plane is by no means unique. Any nonzero multiple of a normal is again a normal, the point being that only the *direction* of the normal vector matters because its role is to determine the orientation of the plane.

Now suppose a point $P_0 = P_0(x_0, y_0, z_0)$ and a nonzero vector $\mathbf{n}$ are given, as shown in Figure 4.18. There is a unique plane that contains P_0

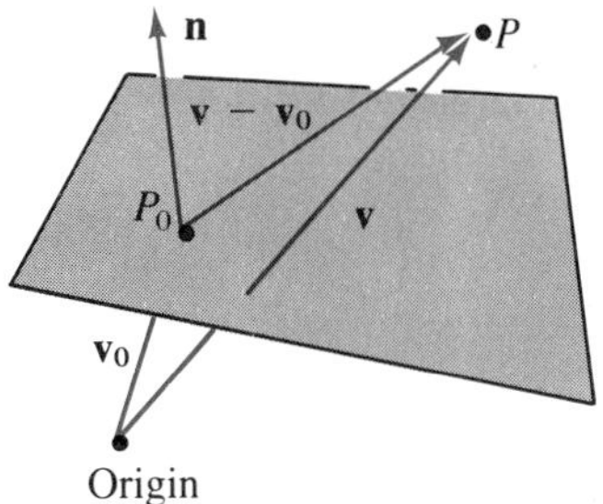

FIGURE 4.18

and has normal vector $\mathbf{n}$. To describe it, let $\mathbf{v}_0 = (x_0,y_0,z_0)$ be the position vector of P_0, and let $\mathbf{v} = (x,y,z)$ be the position vector of an arbitrary point $P = P(x,y,z)$. The vector from P_0 to P is $\mathbf{v} - \mathbf{v}_0$ and it is clear geometrically that P lies in the plane if and only if $\mathbf{v} - \mathbf{v}_0$ lies in the plane. Because $\mathbf{n}$ is a normal to the plane, this holds if and only if $\mathbf{v} - \mathbf{v}_0$ is orthogonal to $\mathbf{n}$—that is,

$$\mathbf{n} \cdot (\mathbf{v} - \mathbf{v}_0) = 0$$

This is called the **vector equation** of the plane. The vectors $\mathbf{n}$ and $\mathbf{v}_0$ are given, and *a vector* $\mathbf{v}$ *is the position vector of a point in the plane if and only if it satisfies this condition.* If $\mathbf{n} = (a,b,c)$ is given in component form, then, because $\mathbf{v} - \mathbf{v}_0 = (x - x_0, y - y_0, z - z_0)$, the vector equation becomes

$$a(x - x_0) + b(y - y_0) + c(z - z_0) = 0$$

This is the **scalar equation** of the plane through $P_0(x_0,y_0,z_0)$ with normal $\mathbf{n} = (a,b,c)$. It simplifies to

$$ax + by + cz = d$$

where $d = ax_0 + by_0 + cz_0$ is a constant.

EXAMPLE 10 Find the equation of the plane through $P_0(1,-1,3)$ with normal $\mathbf{n} = (3,-1,2)$.

Solution Here $\mathbf{v}_0 = (1,-1,3)$, so the general scalar equation becomes

$$3(x - 1) - (y + 1) + 2(z - 3) = 0$$

This simplifies to $3x - y + 2z = 10$.

Every plane with normal $\mathbf{n} = (a,b,c)$ has an equation of the form $ax + by + cz = d$. Conversely, suppose such an equation is given where a, b, and c are not all zero. The set of all points $P(x,y,z)$ whose coordinates x, y, and z satisfy the equation is called the **graph** of the equation, and it is clear that the graph of the equation $ax + by + cz = d$ is a plane with $\mathbf{n} = (a,b,c)$ as a normal vector (assuming that a, b, and c are not all zero). This observation is included in the following theorem, which collects the facts about planes for reference.

THEOREM 7 Let $P_0(x_0,y_0,z_0)$ be a fixed point with position vector $\mathbf{v}_0 = (x_0,y_0,z_0)$, and let $\mathbf{n} = (a,b,c)$ be a fixed nonzero vector. The position vectors $\mathbf{v} = (x,y,z)$ of all points $P(x,y,z)$ that lie on the plane through P_0 with normal $\mathbf{n}$ are given by the vector equation

$$\mathbf{n}\cdot(\mathbf{v} - \mathbf{v}_0) = 0$$

In component form this gives the scalar equation

$$a(x - x_0) + b(y - y_0) + c(z - z_0) = 0$$

of the plane in question. Conversely, the graph of any equation of the form

$$ax + by + cz = d$$

(where a, b, and c are not all zero) is a plane with normal $\mathbf{n} = (a,b,c)$.

EXAMPLE 11 Find the equation of the plane through $P_0(3,-1,2)$ that is parallel to the plane with equation $2x - 3y = 6$.

Solution The plane with equation $2x - 3y = 6$ has normal $\mathbf{n} = (2,-3,0)$ by Theorem 7, so $\mathbf{n}$ will serve as a normal to the plane we are looking for (the two planes are parallel). Because $P_0(3,-1,2)$ lies on the required plane, the equation is $2(x - 3) - 3(y + 1) + 0(z - 2) = 0$. This simplifies to $2x - 3y = 9$.

EXAMPLE 12 Find the shortest distance from the point $P_1(2,1,-3)$ to the plane with equation $3x - y + 4z = 1$. Also find the point on this plane closest to P_1.

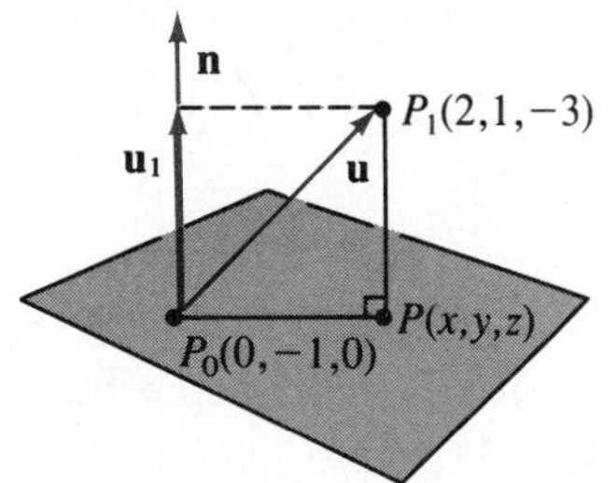

Solution 1 The plane in question has normal $\mathbf{n} = (3,-1,4)$. Choose any point P_0 in the plane—say $P_0(0,-1,0)$—and let $P(x,y,z)$ be the point in the plane closest to P_1 (see the diagram). The vector from P_0 to P_1 is $\mathbf{u} = (2,2,-3)$. Now erect $\mathbf{n}$ with its initial point at P_0. Then $\overrightarrow{PP_1} = \mathbf{u}_1$ is the projection of $\mathbf{u}$ on $\mathbf{n}$:

$$\mathbf{u}_1 = \frac{\mathbf{n}\cdot\mathbf{u}}{\|\mathbf{n}\|^2}\mathbf{n} = \frac{-8}{26}(3,-1,4) = \frac{-4}{13}(3,-1,4)$$

Hence the distance is $\|\overrightarrow{PP_1}\| = \|\mathbf{u}_1\| = \dfrac{4\sqrt{26}}{13}$. To calculate the point P, let $\mathbf{v} = (x,y,z)$ and $\mathbf{v}_0 = (0,-1,0)$ be the position vectors of P and P_0. Then $\mathbf{v} = \mathbf{v}_0 + \mathbf{u} - \mathbf{u}_1 = (0,-1,0) + (2,2,-3) + \frac{4}{13}(3,-1,4) = \left(\frac{38}{13}, \frac{9}{13}, \frac{-23}{13}\right)$

This gives the coordinates of P.

Solution 2 Let $\mathbf{v} = (x,y,z)$, and let $\mathbf{v}_1 = (2,1,-3)$ be the position vectors of P and P_1. Then P is on the line through P_1 with direction vector $\mathbf{n}$, so $\mathbf{v} = \mathbf{v}_1 + t\mathbf{n}$ for some value of t. In addition, P lies on the plane, so $\mathbf{n}\cdot\mathbf{v} = 3x - y + 4z = 1$. This determines t:

$$1 = \mathbf{n} \cdot \mathbf{v} = \mathbf{n} \cdot (\mathbf{v}_1 + t\mathbf{n}) = \mathbf{n} \cdot \mathbf{v}_1 + t\|\mathbf{n}\|^2 = -7 + t(26)$$

This gives $t = \frac{8}{26} = \frac{4}{13}$, so

$$(x,y,z) = \mathbf{v} = \mathbf{v}_1 + t\mathbf{n} = (2,1,-3) + \frac{4}{13}(3,-1,4) = \frac{1}{13}(38,9,-23)$$

as before. This determines P (in the diagram), and the reader can verify that the required distance is $\|\overrightarrow{PP_1}\| = \frac{4}{13}\sqrt{26}$, as before.

It is clear geometrically that if three distinct points P, Q, and R are given, there is a plane that contains all three, and, in fact, this plane is uniquely determined provided that P, Q, and R are not all on some line. In order to find the equation of this plane, it is necessary to find a normal to it. Now the vectors $\overrightarrow{PQ}$ and $\overrightarrow{PR}$ lie in the plane, so both must be orthogonal to the normal. In fact, *any* nonzero vector orthogonal to both $\overrightarrow{PQ}$ and $\overrightarrow{PR}$ will serve as a normal.

Hence the first thing that must be done is find a systematic way to discover a vector orthogonal to two given vectors.

DEFINITION

Given vectors $\mathbf{v}_1 = (x_1,y_1,z_1)$ and $\mathbf{v}_2 = (x_2,y_2,z_2)$, the **cross product** $\mathbf{v}_1 \times \mathbf{v}_2$ is defined by

$$\mathbf{v}_1 \times \mathbf{v}_2 = [(y_1z_2 - z_1y_2), -(x_1z_2 - z_1x_2), (x_1y_2 - y_1x_2)]$$

There is a useful way of remembering this definition. Recall that any vector $\mathbf{u} = (x,y,z)$ can be written as $\mathbf{u} = x\mathbf{i} + y\mathbf{j} + z\mathbf{k}$, where $\mathbf{i} = (1,0,0)$, $\mathbf{j} = (0,1,0)$, and $\mathbf{k} = (0,0,1)$ are coordinate vectors introduced in Section 4.1. Then the mnemonic is as follows:

MNEMONIC FOR THE CROSS PRODUCT

If $\mathbf{v}_1 = (x_1,y_1,z_1)$ and $\mathbf{v}_2 = (x_2,y_2,z_2)$ are two vectors, then

$$\begin{aligned}\mathbf{v}_1 \times \mathbf{v}_2 &= \det\begin{bmatrix} \mathbf{i} & \mathbf{j} & \mathbf{k} \\ x_1 & y_1 & z_1 \\ x_2 & y_2 & z_2 \end{bmatrix} \\ &= (y_1z_2 - z_1y_2)\mathbf{i} - (x_1z_2 - z_1x_2)\mathbf{j} + (x_1y_2 - y_1x_2)\mathbf{k}\end{aligned}$$

where the determinant is expanded along the first row by cofactors.

EXAMPLE 13 Find $\mathbf{u} \times \mathbf{v}$ if $\mathbf{u} = (2,-1,4)$ and $\mathbf{v} = (1,3,7)$.

Solution

$$\mathbf{u} \times \mathbf{v} = \det\begin{bmatrix} \mathbf{i} & \mathbf{j} & \mathbf{k} \\ 2 & -1 & 4 \\ 1 & 3 & 7 \end{bmatrix} = \mathbf{i}\det\begin{bmatrix} -1 & 4 \\ 3 & 7 \end{bmatrix} - \mathbf{j}\det\begin{bmatrix} 2 & 4 \\ 1 & 7 \end{bmatrix} + \mathbf{k}\det\begin{bmatrix} 2 & -1 \\ 1 & 3 \end{bmatrix}$$

$$= -19\mathbf{i} - 10\mathbf{j} + 7\mathbf{k}$$

$$= (-19,-10,7)$$

It is easily verified that $\mathbf{u} \times \mathbf{v}$ is orthogonal to both $\mathbf{u}$ and $\mathbf{v}$ in Example 13. This holds in general by the following result.

THEOREM 8 Let $\mathbf{w} = (x_1,y_1,z_1)$, $\mathbf{u} = (x_2,y_2,z_2)$, and $\mathbf{v} = (x_3,y_3,z_3)$. Then

$$\mathbf{w} \cdot (\mathbf{u} \times \mathbf{v}) = \det\begin{bmatrix} x_1 & y_1 & z_1 \\ x_2 & y_2 & z_2 \\ x_3 & y_3 & z_3 \end{bmatrix}$$

Proof Recall that $\mathbf{w} \cdot (\mathbf{u} \times \mathbf{v})$ is computed by multiplying corresponding components of $\mathbf{w}$ and $\mathbf{u} \times \mathbf{v}$ and then adding. The result follows by expanding the determinant along row 1. ■

Because of Theorem 8 and the mnemonic, several properties of the cross product follow easily from properties of determinants (they can also be verified directly).

THEOREM 9 Let $\mathbf{u}$, $\mathbf{v}$, and $\mathbf{w}$ denote arbitrary vectors.

(1) $\mathbf{u} \times \mathbf{v}$ is a vector
(2) $\mathbf{u} \times \mathbf{v}$ is orthogonal to both $\mathbf{u}$ and $\mathbf{v}$
(3) $\mathbf{u} \times \mathbf{0} = \mathbf{0} = \mathbf{0} \times \mathbf{u}$
(4) $\mathbf{u} \times \mathbf{u} = \mathbf{0}$
(5) $\mathbf{u} \times \mathbf{v} = -(\mathbf{v} \times \mathbf{u})$
(6) $(k\mathbf{u}) \times \mathbf{v} = k(\mathbf{u} \times \mathbf{v}) = \mathbf{u} \times (k\mathbf{v})$ for any scalar k
(7) $\mathbf{u} \times (\mathbf{v} + \mathbf{w}) = (\mathbf{u} \times \mathbf{v}) + (\mathbf{u} \times \mathbf{w})$
(8) $(\mathbf{v} + \mathbf{w}) \times \mathbf{u} = (\mathbf{v} \times \mathbf{u}) + (\mathbf{w} \times \mathbf{u})$

Proof (1) is clear; (2) follows from Theorem 8; and (3) and (4) follow because the determinant of a matrix is zero if either one row is zero or two rows are identical. If two rows are interchanged, the determinant changes sign, and this proves (5). The proof of (6), (7), and (8) is left as Exercise 22. ■

EXAMPLE 14 Find the equation of the plane through $P(1,3,-2)$, $Q(1,1,5)$, and $R(2,-2,3)$.

Solution The vectors $\overrightarrow{PQ} = (0,-2,7)$ and $\overrightarrow{PR} = (1,-5,5)$ lie in the plane, so

$$\overrightarrow{PQ} \times \overrightarrow{PR} = \det\begin{bmatrix} \mathbf{i} & \mathbf{j} & \mathbf{k} \\ 0 & -2 & 7 \\ 1 & -5 & 5 \end{bmatrix} = 25\mathbf{i} + 7\mathbf{j} + 2\mathbf{k} = (25,7,2)$$

is a normal to the plane (being orthogonal to both $\overrightarrow{PQ}$ and $\overrightarrow{PR}$). Because $P(1,3,-2)$ lies in the plane, the equation is $25(x - 1) + 7(y - 3) + 2(z + 2) = 0$. This simplifies to $25x + 7y + 2z = 42$. Incidentally, the reader may verify that this same equation is obtained if either Q or R is used as the point lying in the plane.

EXAMPLE 15 Find the shortest distance between the nonparallel lines

$$(x,y,z) = (1,0,-1) + t(2,0,1)$$
$$(x,y,z) = (3,1,0) + s(1,1,-1)$$

Then find the points A and B on the lines that are closest together.

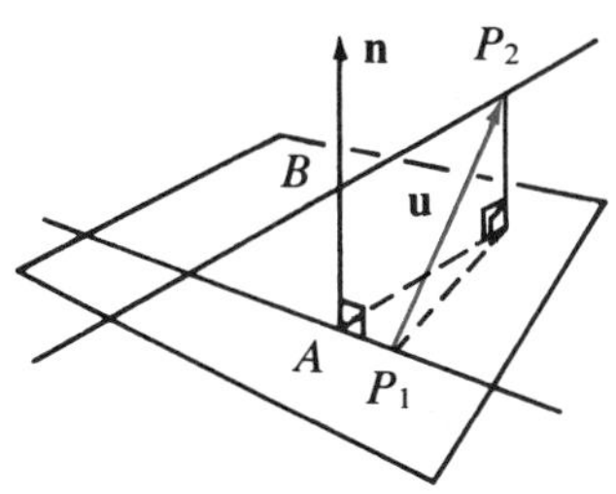

Solution Direction vectors for the two lines are $\mathbf{d}_1 = (2,0,1)$ and $\mathbf{d}_2 = (1,1,-1)$, so

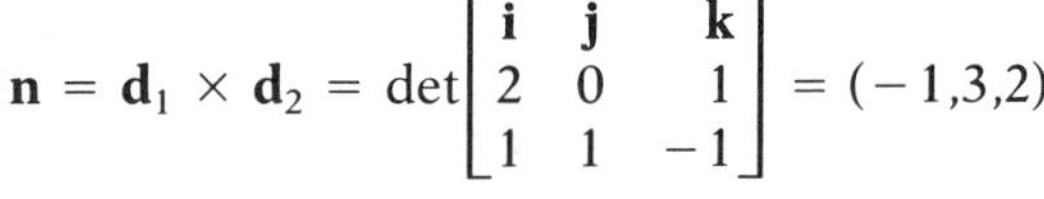

$$\mathbf{n} = \mathbf{d}_1 \times \mathbf{d}_2 = \det\begin{bmatrix} \mathbf{i} & \mathbf{j} & \mathbf{k} \\ 2 & 0 & 1 \\ 1 & 1 & -1 \end{bmatrix} = (-1,3,2)$$

is perpendicular to both lines. Consider the plane containing the first line with $\mathbf{n}$ as normal. This plane contains $P_1(1,0,-1)$ and is parallel to the second line. Hence the distance in question is just the shortest distance between $P_2(3,1,0)$ and this plane. The vector $\mathbf{u}$ from P_1 to P_2 is $\mathbf{u} = \overrightarrow{P_1P_2} = (2,1,1)$ and so, as in Example 12, the distance is the length of the projection of $\mathbf{u}$ on $\mathbf{n}$.

$$\text{Distance} = \left\| \frac{\mathbf{u} \cdot \mathbf{n}}{\|\mathbf{n}\|^2}\mathbf{n} \right\| = \frac{|\mathbf{u} \cdot \mathbf{n}|}{\|\mathbf{n}\|} = \frac{3}{\sqrt{14}} = \frac{3\sqrt{14}}{14}$$

Note that it is necessary that $\mathbf{n} = \mathbf{d}_1 \times \mathbf{d}_2$ not be zero for this calculation to be possible. As will be shown later (Theorem 11), this is guaranteed by the fact that $\mathbf{d}_1$ and $\mathbf{d}_2$ are *not* parallel.

The points A and B have coordinates $A(1 + 2t, 0, t - 1)$ and $B(3 + s, 1 + s, -s)$ for some s and t, so $\overrightarrow{AB} = (2 + s - 2t, 1 + s, 1 - s - t)$. This vector is orthogonal to $\mathbf{d}_1$ and $\mathbf{d}_2$, and the conditions $\overrightarrow{AB} \cdot \mathbf{d}_1 = 0$ and $\overrightarrow{AB} \cdot \mathbf{d}_2 = 0$ give equations $5t - s = 5$ and $t - 3s = 2$. The solution is

$s = \frac{-5}{14}$ and $t = \frac{13}{14}$, so the points are $A\left(\frac{40}{14}, 0, \frac{-1}{14}\right)$ and $B\left(\frac{37}{14}, \frac{9}{14}, \frac{5}{14}\right)$.
We have $\|\overrightarrow{AB}\| = \frac{3\sqrt{14}}{13}$, as before.

Recall that the dot product of two vectors **u** and **v** was defined by $\mathbf{u} \cdot \mathbf{v} = \|\mathbf{u}\|\,\|\mathbf{v}\| \cos \theta$, where θ is the angle between **u** and **v**. One virtue of this definition is that it does not depend on any coordinate system (although $\mathbf{u} \cdot \mathbf{v}$ can be *computed* using Theorem 3 when the components of **u** and **v** are given in some coordinate system). However, the cross product $\mathbf{u} \times \mathbf{v}$ has been defined in terms of components of **u** and **v**, and the question naturally arises whether $\mathbf{u} \times \mathbf{v}$ can be defined in terms of the vectors **u** and **v** themselves without any reference to coordinates. In other words, can the length and direction of $\mathbf{u} \times \mathbf{v}$ be given in terms of the length and direction of **u** and **v**? The answer is affirmative and is based to some extent on the following identity relating the dot product and the cross product.

THEOREM 10
Lagrange Identity

If **u** and **v** are any two vectors, then

$$\|\mathbf{u} \times \mathbf{v}\|^2 = \|\mathbf{u}\|^2\,\|\mathbf{v}\|^2 - (\mathbf{u} \cdot \mathbf{v})^2$$

Proof Given **u** and **v**, introduce a coordinate system and write $\mathbf{u} = (x_1, y_1, z_1)$ and $\mathbf{v} = (x_2, y_2, z_2)$ in component form. Then all the terms in the identity can be computed in terms of the components. The detailed proof is left as Exercise 21. ■

An expression for the magnitude of the vector $\mathbf{u} \times \mathbf{v}$ can be easily obtained from the Lagrange identity. If θ is the angle between **u** and **v**, substituting $\mathbf{u} \cdot \mathbf{v} = \|\mathbf{u}\|\,\|\mathbf{v}\| \cos \theta$ into the Lagrange identity gives

$$\|\mathbf{u} \times \mathbf{v}\|^2 = \|\mathbf{u}\|^2\,\|\mathbf{v}\|^2 - \|\mathbf{u}\|^2\,\|\mathbf{v}\|^2 \cos^2\theta = \|\mathbf{u}\|^2\,\|\mathbf{v}\|^2 \sin^2\theta$$

using the fact that $1 - \cos^2\theta = \sin^2\theta$. But $\sin \theta$ is non-negative on the range $0 \le \theta \le \pi$, so taking the positive square root of both sides gives

$$\|\mathbf{u} \times \mathbf{v}\| = \|\mathbf{u}\|\,\|\mathbf{v}\| \sin \theta$$

an expression for $\|\mathbf{u} \times \mathbf{v}\|$ that makes no reference to a coordinate system. Moreover, this expression has a nice geometrical interpretation. The parallelogram determined by the vectors **u** and **v** has base length $\|\mathbf{v}\|$ and altitude $\|\mathbf{u}\| \sin \theta$ (see Figure 4.19). Hence the area of the parallelogram formed by **u** and **v** is

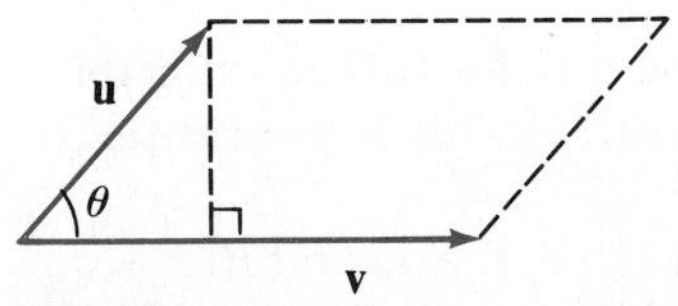

FIGURE 4.19

$$(\|\mathbf{u}\| \sin \theta)\|\mathbf{v}\| = \|\mathbf{u} \times \mathbf{v}\|$$

This proves the first part of Theorem 11.

THEOREM 11 If $\mathbf{u}$ and $\mathbf{v}$ are two vectors and θ is the angle between $\mathbf{u}$ and $\mathbf{v}$, then

(1) $\|\mathbf{u} \times \mathbf{v}\| = \|\mathbf{u}\|\,\|\mathbf{v}\| \sin\theta =$ area of the parallelogram determined by $\mathbf{u}$ and $\mathbf{v}$

(2) $\mathbf{u}$ and $\mathbf{v}$ are parallel if and only if $\mathbf{u} \times \mathbf{v} = \mathbf{0}$

Proof of (2) By (1), $\mathbf{u} \times \mathbf{v} = \mathbf{0}$ if and only if the area of the parallelogram is zero. But the area vanishes if and only if $\mathbf{u}$ and $\mathbf{v}$ have the same or opposite direction—that is, if they are parallel. ■

EXAMPLE 16 Find the area of the triangle with vertices $P(2,1,0)$, $Q(3,-1,1)$, and $R(1,0,1)$.

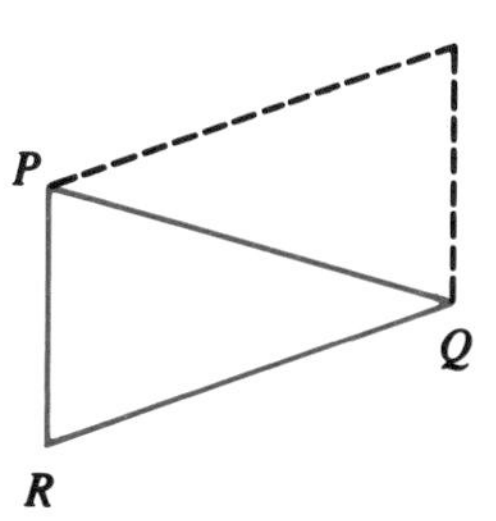

Solution First compute $\overrightarrow{RP} = (1,1,-1)$ and $\overrightarrow{RQ} = (2,-1,0)$. The area of the triangle is half the area of the parallelogram determined by these vectors, so it equals $\frac{1}{2}\|\overrightarrow{RP} \times \overrightarrow{RQ}\|$. Now

$$\overrightarrow{RP} \times \overrightarrow{RQ} = \det\begin{bmatrix} \mathbf{i} & \mathbf{j} & \mathbf{k} \\ 1 & 1 & -1 \\ 2 & -1 & 0 \end{bmatrix} = (-1,-2,-3)$$

so the area of the triangle is $\frac{1}{2}\|(-1,-2,-3)\| = \frac{1}{2}\sqrt{1+4+9} = \frac{1}{2}\sqrt{14}$.

Theorem 11 can also be used to find the shortest distance from a point to a line. The next example illustrates this by re-solving part of Example 9.

EXAMPLE 17 Find the shortest distance from the point $P(1,3,-2)$ to the line through $P_0(2,0,-1)$ with direction vector $\mathbf{d} = (1,-1,0)$.

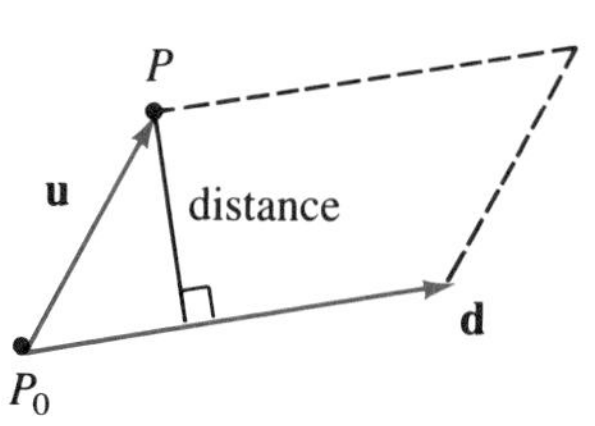

Solution Compute $\mathbf{u} = \overrightarrow{P_0P} = (-1,3,-1)$. Then

$$\mathbf{u} \times \mathbf{d} = \det\begin{bmatrix} \mathbf{i} & \mathbf{j} & \mathbf{k} \\ -1 & 3 & -1 \\ 1 & -1 & 0 \end{bmatrix} = (-1,-1,-2)$$

so the parallelogram determined by $\mathbf{u}$ and $\mathbf{d}$ has area $\|\mathbf{u} \times \mathbf{d}\| = \sqrt{1+1+4} = \sqrt{6}$. On the other hand, the area of this parallelogram equals $\|\mathbf{d}\|$ times the distance in question. Hence the distance from P to the line is

$$\frac{\|\mathbf{u} \times \mathbf{d}\|}{\|\mathbf{d}\|} = \frac{\sqrt{6}}{\sqrt{2}} = \sqrt{3}$$

Of course this agrees with Example 9, but the technique here is entirely different.

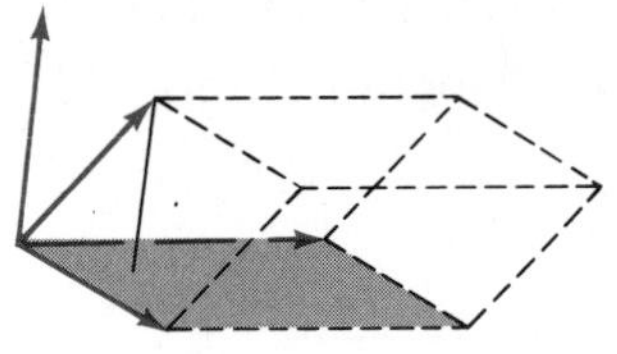

FIGURE 4.20

If three vectors **u**, **v**, and **w** are given, they determine a "squashed" rectangular solid called a **parallelepiped** (Figure 4.20), and it is often useful to be able to find the volume of such a solid. The base of the solid is the parallelogram determined by **u** and **v**, so it has area $A = \|\mathbf{u} \times \mathbf{v}\|$ by Theorem 11. The height of the solid is the length of the projection of **w** on $\mathbf{u} \times \mathbf{v}$. Hence

$$h = \left\| \frac{\mathbf{w} \cdot (\mathbf{u} \times \mathbf{v})}{\|\mathbf{u} \times \mathbf{v}\|^2} \right\| \|\mathbf{u} \times \mathbf{v}\| = \frac{|\mathbf{w} \cdot (\mathbf{u} \times \mathbf{v})|}{\|\mathbf{u} \times \mathbf{v}\|} = \frac{|\mathbf{w} \cdot (\mathbf{u} \times \mathbf{v})|}{A}$$

Thus the volume of the parallelepiped is $hA = |\mathbf{w} \cdot (\mathbf{u} \times \mathbf{v})|$.

THEOREM 12 The volume of the parallelepiped determined by **w**, **u**, and **v** (Figure 4.20) is given by $|\mathbf{w} \cdot (\mathbf{u} \times \mathbf{v})|$.

EXAMPLE 18 Find the volume of the parallelepiped determined by the vectors $\mathbf{w} = (1,2,-1)$, $\mathbf{u} = (1,1,0)$, and $\mathbf{v} = (-2,0,1)$.

Solution We use Theorem 8.

$$\mathbf{w} \cdot (\mathbf{u} \times \mathbf{v}) = \det \begin{bmatrix} 1 & 2 & -1 \\ 1 & 1 & 0 \\ -2 & 0 & 1 \end{bmatrix} = -3$$

Hence the volume is $|\mathbf{w} \cdot (\mathbf{u} \times \mathbf{v})| = |-3| = 3$.

We can now give a coordinate-free description of the cross product $\mathbf{u} \times \mathbf{v}$. In fact, the magnitude is given by $\|\mathbf{u} \times \mathbf{v}\| = \|\mathbf{u}\| \|\mathbf{v}\| \sin \theta$, and if $\mathbf{u} \times \mathbf{v} \neq \mathbf{0}$, its direction is very nearly determined by the fact that it is orthogonal to both **u** and **v** and so points along the line normal to the plane determined by **u** and **v**. It remains only to decide which of the two possible directions is correct.

Before this can be done, the basic issue of how coordinates themselves are assigned must be clarified. When coordinate axes are chosen in space, the procedure is as follows: An origin is selected and two perpendicular lines (the X- and Y-axes) are chosen through the origin, the positive direction on each of these axes being chosen quite arbitrarily. Then the line through the origin normal to this X-Y-plane is called the Z-axis, but there is a choice of which direction on this axis will be the positive one. The two possibilities are shown in Figure 4.21, and it is a standard convention that cartesian coordinates are always **right-**

hand coordinate systems. The reason for this terminology is that, in such a system, if the Z-axis is grasped in the right hand with the thumb pointing in the positive Z-direction, then the fingers will curl around from the positive X-axis to the positive Y-axis (through a right angle).

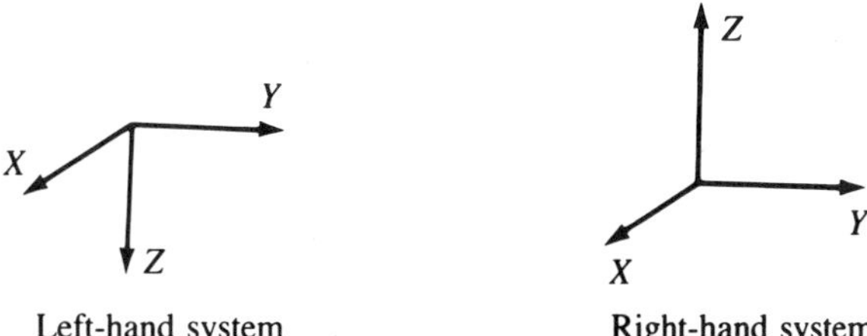

FIGURE 4.21

Suppose now that **u** and **v** are given and that θ is the angle between them (so $0 \leq \theta \leq \pi$). Then the direction of $\|\mathbf{u} \times \mathbf{v}\|$ is given by the so-called **right-hand rule**.

RIGHT-HAND RULE

> If the vector $\mathbf{u} \times \mathbf{v}$ is grasped in the right hand and the fingers curl around from **u** to **v** through the angle θ, the thumb points in the direction of $\mathbf{u} \times \mathbf{v}$.

To give an indication why this is true, introduce coordinates in such a way that the initial points of **u** and **v** are at the origin, **u** points along the positive X-axis, **v** lies in the X-Y-plane, and **v** and the positive Y-axis are on the same side of the X-axis. Then, in this system, **u** and **v** have component form $\mathbf{u} = (a,0,0)$ and $\mathbf{v} = (b,c,0)$, where $a > 0$ and $c > 0$. The situation is depicted in Figure 4.22. The right-hand rule asserts that $\mathbf{u} \times \mathbf{v}$ should point in the positive Z-direction. But our definition of $\mathbf{u} \times \mathbf{v}$ gives

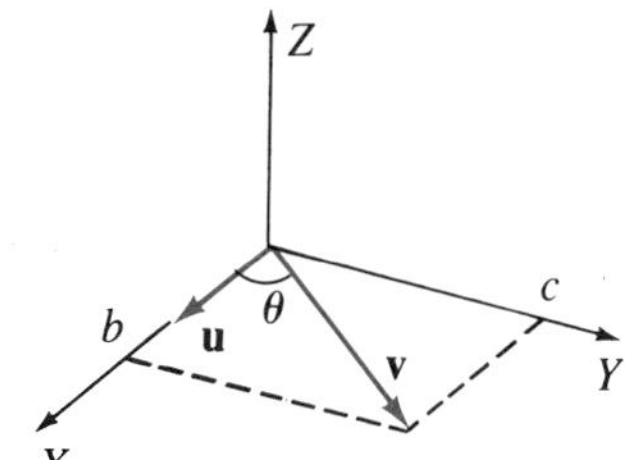

FIGURE 4.22

$$\mathbf{u} \times \mathbf{v} = \det\begin{bmatrix} \mathbf{i} & \mathbf{j} & \mathbf{k} \\ a & 0 & 0 \\ b & c & 0 \end{bmatrix} = (0,0,ac) = (ac)\mathbf{k}$$

and because $ac > 0$, this has the positive Z-direction.

EXERCISES 4.2.2

1. Compute $\mathbf{u} \times \mathbf{v}$ where:
   (a) $\mathbf{u} = (1,2,3)$, $\mathbf{v} = (1,0,-1)$
   (b) $\mathbf{u} = (3,-1,0)$, $\mathbf{v} = (-6,2,0)$
   (c) $\mathbf{u} = (3,-2,1)$, $\mathbf{v} = (1,1,1)$
   (d) $\mathbf{u} = (2,0,-1)$, $\mathbf{v} = (1,4,7)$

2. If **i**, **j**, and **k** are the coordinate vectors, verify that $\mathbf{i} \times \mathbf{j} = \mathbf{k}$, $\mathbf{j} \times \mathbf{k} = \mathbf{i}$, and $\mathbf{k} \times \mathbf{i} = \mathbf{j}$.
3. Show that $\mathbf{u} \times (\mathbf{v} \times \mathbf{w})$ need not equal $(\mathbf{u} \times \mathbf{v}) \times \mathbf{w}$ by calculating both when $\mathbf{u} = (1,1,1)$, $\mathbf{v} = (1,1,0)$, and $\mathbf{w} = (0,0,1)$.
4. Find two unit vectors orthogonal to both **u** and **v** if:
   (a) $\mathbf{u} = (1,2,2)$, $\mathbf{v} = (2,-1,2)$ **(b)** $\mathbf{u} = (1,2,-1)$, $\mathbf{v} = (3,1,2)$
5. Find the equation of each of the following planes.
   (a) Passing through $A(2,1,3)$, $B(3,-1,5)$, and $C(0,2,-4)$
   (b) Passing through $A(1,-1,6)$, $B(0,0,1)$, and $C(4,7,-11)$
   (c) Passing through $P(2,-1,4)$ and parallel to the plane with equation $3x - 2y - z = 0$
   (d) Passing through $P(3,0,-1)$ and parallel to the plane with equation $2x - y + z = 3$
   (e) Containing $P(3,0,-1)$ and the line $(x,y,z) = (0,0,2) + t(1,0,1)$
   (f) Containing $P(2,1,0)$ and the line $(x,y,z) = (3,-1,2) + t(1,0,-1)$
   (g) Containing the lines $(x,y,z) = (1,-1,2) + t(1,0,1)$ and $(x,y,z) = (0,0,2) + t(1,-1,0)$
   (h) Containing the lines $(x,y,z) = (3,1,0) + t(1,-1,3)$ and $(x,y,z) = (0,-2,5) + t(2,1,-1)$
   (i) Each point of which is equidistant from $P(2,-1,3)$ and $Q(1,1,-1)$
   (j) Each point of which is equidistant from $P(0,1,-1)$ and $Q(2,-1,-3)$
6. In each case, find the equation of the line.
   (a) Passing through $P(3,-1,4)$ and perpendicular to the plane $3x - 2y - z = 0$
   (b) Passing through $P(2,-1,3)$ and perpendicular to the plane $2x + y = 1$
   (c) Passing through $P(0,0,0)$ and perpendicular to the lines $(x,y,z) = (1,1,0) + t(2,0,-1)$ and $(x,y,z) = (2,1,-3) + t(1,-1,7)$
   (d) Passing through $P(1,1,-1)$ and perpendicular to the lines $(x,y,z) = (2,0,1) + t(1,1,-2)$ and $(x,y,z) = (5,5,-2) + t(1,2,-3)$
   (e) Passing through $P(2,1,-1)$, intersecting the line $(x,y,z) = (1,2,-1) + t(3,0,1)$ and perpendicular to that line
   (f) Passing through $P(1,1,2)$, intersecting the line $(x,y,z) = (2,1,0) + t(1,1,1)$, and perpendicular to that line
7. In each case, find the shortest distance from the point P to the plane, and find the point Q on the plane closest to P.
   (a) $P(2,3,0)$; plane with equation $5x - y + z = 1$
   (b) $P(3,1,-1)$; plane with equation $2x + y - z = 6$
8. **(a)** Does the line through $P(1,2,-3)$ with direction vector $\mathbf{d} = (1,2,-3)$ lie in the plane $2x - y - z = 3$? Explain.
   (b) Does the plane through $P(4,0,5)$, $Q(2,2,1)$, and $R(1,-1,2)$ pass through the origin? Explain.
9. Find the equations of the line of intersection of the following planes.
   (a) $2x - 3y + 2z = 5$ and $x + 2y - z = 4$

(b) $3x + y - 2z = 1$ and $x + y + z = 5$

10. Find the equations of *all* planes:

(a) Perpendicular to the line $(x,y,z) = (2,-1,3) + t(2,1,3)$

(b) Perpendicular to the line $(x,y,z) = (1,0,-1) + t(3,0,2)$

(c) Containing the origin

(d) Containing $P(3,2,-4)$

(e) Containing $P(1,1,-1)$ and $Q(0,1,1)$

(f) Containing $P(2,-1,1)$ and $Q(1,0,0)$

(g) Containing the line $(x,y,z) = (2,1,0) + t(1,-1,0)$

(h) Containing the line $(x,y,z) = (3,0,2) + t(1,2,-1)$

11. Find the shortest distance between the following parallel lines.

(a) $(x,y,z) = (2,-1,3) + t(1,-1,4)$
$(x,y,z) = (1,0,1) + t(1,-1,4)$

(b) $(x,y,z) = (3,0,2) + t(3,1,0)$
$(x,y,z) = (-1,2,2) + t(3,1,0)$

12. Find the shortest distance between the following nonparallel lines and the points on the lines that are closest together.

(a) $(x,y,z) = (3,0,1) + s(2,1,-3)$
$(x,y,z) = (1,1,-1) + t(1,0,1)$

(b) $(x,y,z) = (1,-1,0) + s(1,1,1)$
$(x,y,z) = (2,-1,3) + t(3,1,0)$

(c) $(x,y,z) = (3,1,-1) + s(1,1,-1)$
$(x,y,z) = (1,2,0) + t(1,0,2)$

(d) $(x,y,z) = (1,2,3) + s(2,0,-1)$
$(x,y,z) = (3,-1,0) + t(1,1,0)$

13. Find the area of the triangle with the following vertices.

(a) $A(3,-1,2)$, $B(1,1,0)$, and $C(1,2,-1)$

(b) $A(3,0,1)$, $B(5,1,0)$, and $C(7,2,-1)$

(c) $A(1,1,-1)$, $B(2,0,1)$, and $C(1,-1,3)$

(d) $A(3,-1,1)$, $B(4,1,0)$, and $C(2,-3,0)$

14. Find the volume of the parallelepiped determined by **w**, **u**, and **v** when:

(a) $\mathbf{w} = (2,1,1)$, $\mathbf{v} = (1,0,2)$, and $\mathbf{u} = (2,1,-1)$

(b) $\mathbf{w} = (1,0,3)$, $\mathbf{v} = (2,1,-3)$, and $\mathbf{u} = (1,1,1)$

15. Let P_0 be a point with position vector $\mathbf{v}_0$, and let $ax + by + cz = d$ be the equation of a plane with normal $\mathbf{n} = (a,b,c)$.

(a) Show that the point on the plane closest to P_0 has position vector $\mathbf{v} = \mathbf{v}_0 + \frac{d - (\mathbf{v}_0 \cdot \mathbf{n})}{\|\mathbf{n}\|^2}\mathbf{n}$. [*Hint:* $\mathbf{v} = \mathbf{v}_0 + t\mathbf{n}$ for some t, and $\mathbf{v} \cdot \mathbf{n} = d$.]

(b) Show that the shortest distance from P_0 to the plane is $\frac{|d - (\mathbf{v}_0 \cdot \mathbf{n})|}{\|\mathbf{n}\|}$.

(c) Let P_0' denote the reflection of P_0 in the plane—that is, the point on the opposite side of the plane such that the line through P_0 and P_0' is perpendicular to the plane. Show that $\mathbf{v}_0 + 2\frac{d - (\mathbf{v}_0 \cdot \mathbf{n})}{\|\mathbf{n}\|^2}\mathbf{n}$ is the position vector of P_0'.

16. Let A and B be points other than the origin, and let **a** and **b** be their position vectors. If **a** and **b** are not parallel, show that the plane through A, B, and the origin is given by $\{P(x,y,z) | (x,y,z) = s\mathbf{a} + t\mathbf{b}$ for some s and $t\}$.

17. Show that points A, B, and C are all on one line if and only if $\overrightarrow{AB} \times \overrightarrow{AC} = \mathbf{0}$.

18. Show that points A, B, C, and D are all on one plane if and only if $\overrightarrow{AB} \cdot (\overrightarrow{AC} \times \overrightarrow{AD}) = 0$.

19. Use Theorem 12 to confirm that, if $\mathbf{u}$, $\mathbf{v}$, and $\mathbf{w}$ are mutually perpendicular, the (rectangular) parallelepiped they determine has volume $\|\mathbf{u}\|\,\|\mathbf{v}\|\,\|\mathbf{w}\|$.

20. Show that the volume of the parallelepiped determined by $\mathbf{u}$, $\mathbf{v}$, and $\mathbf{u} \times \mathbf{v}$ is $\|\mathbf{u} \times \mathbf{v}\|^2$.

21. Complete the proof of Theorem 10.

22. Prove the following properties in Theorem 9.

(a) Property 6 **(b)** Property 7 **(c)** Property 8

23. **(a)** Show that $\mathbf{w} \cdot (\mathbf{u} \times \mathbf{v}) = \mathbf{u} \cdot (\mathbf{v} \times \mathbf{w}) = \mathbf{v} \cdot (\mathbf{w} \times \mathbf{u})$ holds for all vectors $\mathbf{w}$, $\mathbf{u}$, and $\mathbf{v}$. [*Hint:* Theorem 8.]

(b) Show that $\mathbf{v} - \mathbf{w}$ and $(\mathbf{u} \times \mathbf{v}) + (\mathbf{v} \times \mathbf{w}) + (\mathbf{w} \times \mathbf{u})$ are orthogonal.

24. Show that $\mathbf{u} \times (\mathbf{v} \times \mathbf{w}) = (\mathbf{u} \cdot \mathbf{w})\mathbf{v} - (\mathbf{u} \cdot \mathbf{v})\mathbf{w}$. [*Hint:* First do it for $\mathbf{u} = \mathbf{i}$, $\mathbf{j}$, and $\mathbf{k}$; then write $\mathbf{u} = x\mathbf{i} + y\mathbf{j} + z\mathbf{k}$ and use Theorem 9.]

25. Prove the **Jacobi identity:** $\mathbf{u} \times (\mathbf{v} \times \mathbf{w}) + \mathbf{v} \times (\mathbf{w} \times \mathbf{u}) + \mathbf{w} \times (\mathbf{u} \times \mathbf{v}) = \mathbf{0}$. [*Hint:* The previous exercise.]

26. Show that $(\mathbf{u} \times \mathbf{v}) \cdot (\mathbf{w} \times \mathbf{z}) = (\mathbf{u} \cdot \mathbf{w})(\mathbf{v} \cdot \mathbf{z}) - (\mathbf{u} \cdot \mathbf{z})(\mathbf{v} \cdot \mathbf{w})$. [*Hint:* Exercises 23 and 24.]

27. Let P, Q, R, and S be four points, not all on one plane. Show that the volume of the pyramid they determine is $\frac{1}{6}|\overrightarrow{PQ} \cdot (\overrightarrow{PR} \times \overrightarrow{PS})|$. [*Hint:* The volume of a "cone" with base area A and height h is $\frac{1}{3}Ah$.]

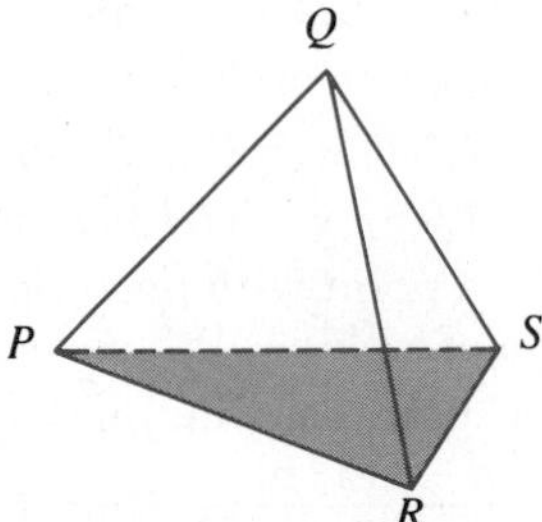

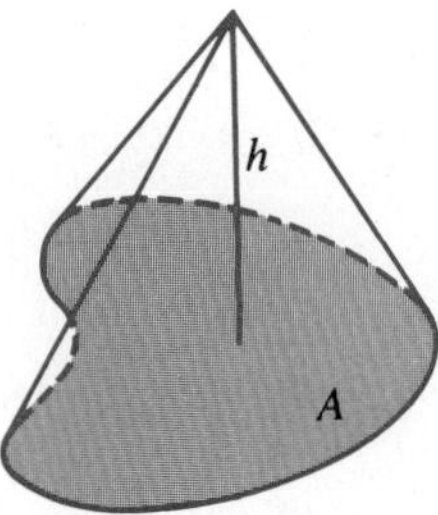

28. Consider a triangle with vertices A, B, and C. Let α, β, and γ denote the angles at A, B, and C, respectively, and let a, b, and c denote the lengths of the sides opposite A, B, and C, respectively. Write $\mathbf{u} = \overrightarrow{AB}$, $\mathbf{v} = \overrightarrow{BC}$, and $\mathbf{w} = \overrightarrow{CA}$.

(a) Deduce $\mathbf{u} + \mathbf{v} + \mathbf{w} = \mathbf{0}$.

(b) Show that $\mathbf{u} \times \mathbf{v} = \mathbf{w} \times \mathbf{u} = \mathbf{v} \times \mathbf{w}$. [*Hint:* Compute $\mathbf{u} \times (\mathbf{u} + \mathbf{v} + \mathbf{w})$ and $\mathbf{v} \times (\mathbf{u} + \mathbf{v} + \mathbf{w})$.]

(c) Deduce the **law of sines:**

$$\frac{\sin \alpha}{a} = \frac{\sin \beta}{b} = \frac{\sin \gamma}{c}$$

29. Let A be a 3×3 matrix. Given vectors $\mathbf{u}$, $\mathbf{v}$, and $\mathbf{w}$, show that the volume of the parallelepiped determined by $\mathbf{u}A$, $\mathbf{v}A$, and $\mathbf{w}A$ equals $|\det A|$ times the volume of the parallelepiped determined by $\mathbf{u}$, $\mathbf{v}$, and $\mathbf{w}$.

30. Let $\mathbf{n} = (a,b,c)$ be any nonzero vector.

(a) Show that every plane with $\mathbf{n}$ as normal has the vector equation $\mathbf{v} \cdot \mathbf{n} = d$ for some number d (where $\mathbf{v}$ denotes the position vector of an arbitrary point in the plane).

(b) Show that the (shortest) distance between two such planes $\mathbf{v} \cdot \mathbf{n} = d_1$ and $\mathbf{v} \cdot \mathbf{n} = d_2$ is $\dfrac{|d_2 - d_1|}{\|\mathbf{n}\|}$.

31. Given the cube with vertices $P(x,y,z)$, where each of x, y, and z is either 0 or 2, consider the plane perpendicular to the diagonal through $P(0,0,0)$ and $P(2,2,2)$ and bisecting it.

(a) Show that the plane meets six of the edges of the cube and bisects them.

(b) Show that the six points in **(a)** are the vertices of a regular hexagon.

SECTION 4.3 An Application to Least Squares Approximation (Optional)

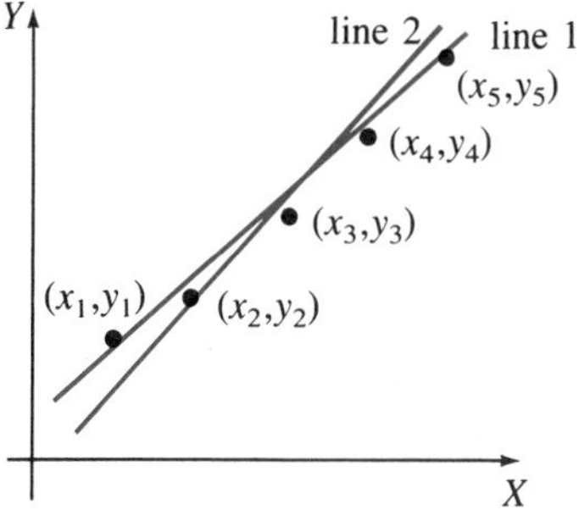

FIGURE 4.23

In many scientific investigations, data are collected that relate two variables. For example, if x is the number of dollars spent on advertising by a manufacturer and y is the value of sales in the region in question, the manufacturer could generate data by spending $x_1, x_2, \ldots, x_n$ dollars at different times and measuring the corresponding sales values $y_1, y_2, \ldots, y_n$. Suppose it is known that a linear relationship exists between the variables x and y — in other words, that $y = a + bx$ for some constants a and b. If the data are plotted, the points $(x_1,y_1), (x_2,y_2), \ldots, (x_n,y_n)$ may appear to lie on a straight line and we need to find the "best-fitting" line through these data points. For example, if five data points occur as shown in Figure 4.23, line 1 is clearly a better fit than line 2. In general, the problem is to find the values of the constants a and b such that the line $y = a + bx$ best approximates the data in question. Note that an *exact* fit would be obtained if a and b were such that $y_i = a + bx_i$ were true for each data point (x_i,y_i). But this is too much to expect. Experimental errors in measurement are bound to occur, so the choice of a and b should be made in such a way that the "errors" between the observed values y_i and the corresponding fitted values $a + bx_i$ are in some sense minimized.

The first thing we must do is explain exactly what we mean by the "best fit" of a line $y = a + bx$ to an observed set of data points (x_1,y_1), $(x_2,y_2), \ldots, (x_n,y_n)$. For convenience, write the linear function $a + bx$ as

$$f(x) = a + bx$$

so that the "fitted" points (on the line) have coordinates $(x_1,f(x_1)), \ldots,$

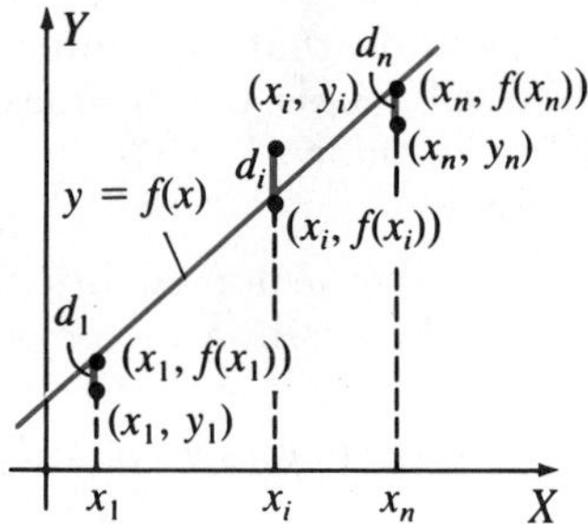

FIGURE 4.24

$(x_n,f(x_n))$. Figure 4.24 is a sketch of what the line $y = f(x)$ might look like. For each i the observed data point (x_i,y_i) and the fitted point $(x_i,f(x_i))$ need not be the same, and the distance d_i between them measures how far the line misses the observed point. For this reason d_i is often called the **error** at x_i, and a natural measure of how "close" the line $y = f(x)$ is to the observed data points is the sum $d_1 + d_2 + \cdots + d_n$ of all these errors. However, it turns out to be better to use the sum of squares

$$S = d_1^2 + d_2^2 + \cdots + d_n^2$$

as the measure of error, and the line $y = f(x)$ is to be chosen so as to make this sum as small as possible. This line is said to be the **least squares approximating line** for data points $(x_1,y_1), (x_2,y_2), \ldots, (x_n,y_n)$.

The square of the distance d_i is given by $d_i^2 = [y_i - f(x_i)]^2$ for each i, so the quantity S to be minimized is the sum

$$S = [y_1 - f(x_1)]^2 + [y_2 - f(x_2)]^2 + \cdots + [y_n - f(x_n)]^2$$

Note that all the numbers x_i and y_i are *given* here; what is required is that the *function* f be chosen in such a way as to minimize this expression. Because $f(x) = a + bx$, this amounts to choosing a and b so as to minimize S, and the problem can be solved using vector techniques. The following definitions simplify the notation.

$$\mathbf{y} = \begin{bmatrix} y_1 \\ y_2 \\ \vdots \\ y_n \end{bmatrix} \qquad M = \begin{bmatrix} 1 & x_1 \\ 1 & x_2 \\ \vdots & \vdots \\ 1 & x_n \end{bmatrix} \qquad \mathbf{z} = \begin{bmatrix} a \\ b \end{bmatrix}$$

Observe that

$$\mathbf{y} - M\mathbf{z} = \begin{bmatrix} y_1 - (a + bx_1) \\ y_2 - (a + bx_2) \\ \vdots \qquad \vdots \\ y_n - (a + bx_n) \end{bmatrix}$$

so the quantity S that is to be minimized is just the sum of the squares of the entries of this matrix.

Now suppose for a moment that $n = 3$. Then $\mathbf{y} - M\mathbf{z}$ is an ordered triple and so can be regarded as a vector (written as a column rather than as a row). Moreover, S is the square of the length of this vector.

$$S = \|\mathbf{y} - M\mathbf{z}\|^2$$

Here $\mathbf{y}$ and M are given, and we are asked to choose $\mathbf{z}$ such that the length of the vector $\mathbf{y} - M\mathbf{z}$ is as small as possible. To this end, consider the set P of *all* vectors $M\mathbf{z}$ where $\mathbf{z} = \begin{bmatrix} a \\ b \end{bmatrix}$ varies. Then P takes the form

$$P = \left\{ \begin{bmatrix} a + bx_1 \\ a + bx_2 \\ a + bx_3 \end{bmatrix} \middle| \; a \text{ and } b \text{ arbitrary} \right\}$$

If x_1, x_2, and x_3 are distinct, this is a plane through the origin (in fact, it contains

$$\mathbf{u} = \begin{bmatrix} x_1 \\ x_2 \\ x_3 \end{bmatrix} \text{ and } \mathbf{v} = \begin{bmatrix} 1 \\ 1 \\ 1 \end{bmatrix}, \text{ so } \mathbf{u} \times \mathbf{v} = \begin{bmatrix} x_2 - x_3 \\ x_3 - x_1 \\ x_1 - x_2 \end{bmatrix}$$

is a normal). Thus the task is to choose a point $M\mathbf{a}$ in P as close as possible to $\mathbf{y}$. It is clear geometrically (see Figure 4.25) that the vector $\mathbf{y} - M\mathbf{a}$ is orthogonal to every vector $M\mathbf{z}$ in the plane P. This means that

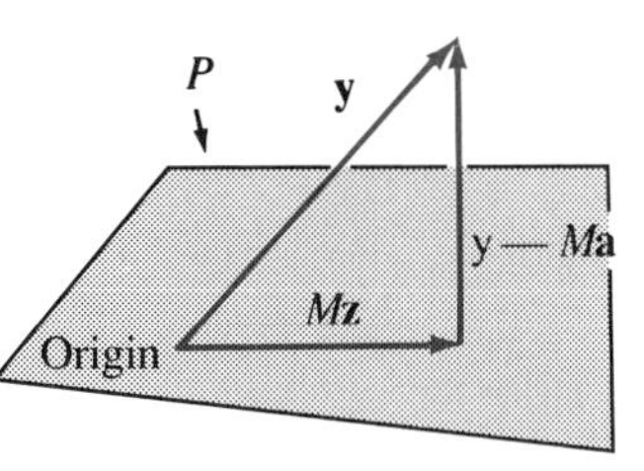

FIGURE 4.25

$$(M\mathbf{z}) \cdot (\mathbf{y} - M\mathbf{a}) = 0$$

for all $\mathbf{z}$, and this condition determines $\mathbf{a}$.

Now observe that the dot product of column vectors $\mathbf{u}$ and $\mathbf{v}$ can be written as $\mathbf{u} \cdot \mathbf{v} = \mathbf{u}^T\mathbf{v}$, where $\mathbf{u}$ and $\mathbf{v}$ are regarded as 3×1 matrices. Hence, for each $\mathbf{z}$,

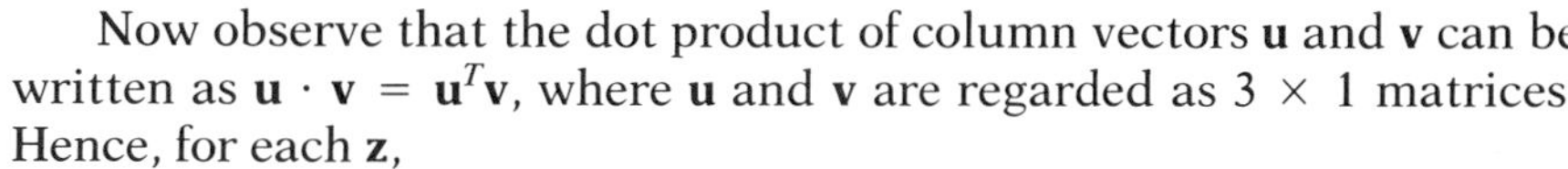

$$0 = (M\mathbf{z})^T(\mathbf{y} - M\mathbf{a}) = \mathbf{z}^T M^T(\mathbf{y} - M\mathbf{a}) = \mathbf{z} \cdot (M^T\mathbf{y} - M^TM\mathbf{a})$$

where the last dot product is in two dimensions. In other words, the vector $M^T\mathbf{y} - M^TM\mathbf{a}$ is orthogonal to *every* two-dimensional vector $\mathbf{z}$ and so must be zero (being orthogonal to itself!). This means that

$$M^TM\mathbf{a} = M^T\mathbf{y}$$

These are called the **normal equations** for $\mathbf{a}$ and can be solved using Gaussian elimination. Moreover, M^TM can be shown to be invertible when x_1, x_2, and x_3 are distinct (it is sufficient that at least two of x_1, x_2, and x_3 be distinct), so solving for $\mathbf{a}$ yields

$$\mathbf{a} = (M^TM)^{-1}M^T\mathbf{y}$$

This solves our problem (at least when $n = 3$) because if $\mathbf{a} = \begin{bmatrix} a_0 \\ b_0 \end{bmatrix}$, the best fitting line is $y = a_0 + b_0x$.

Of course this argument depends heavily on the fact that $n = 3$ so that we may avail ourselves of the theory of vectors in two and three dimensions and may use such notions as the length of a vector, orthogonality, and the dot product. However, all these notions extend to vectors in higher dimensions than two or three, and the entire argument goes through almost unaltered in the general context. This argument will be carried out in Chapter 6, and the result is the following useful theorem.

THEOREM 1 Suppose that n data points $(x_1,y_1), (x_2,y_2), \ldots, (x_n,y_n)$ are given, where at least two of $x_1, x_2, \ldots, x_n$ are distinct. Put

$$\mathbf{y} = \begin{bmatrix} y_1 \\ y_2 \\ \vdots \\ y_n \end{bmatrix} \qquad M = \begin{bmatrix} 1 & x_1 \\ 1 & x_2 \\ \vdots & \vdots \\ 1 & x_n \end{bmatrix}$$

Then the least squares approximating line for these data points has the equation

$$y = a_0 + b_0 x$$

where $\mathbf{a} = \begin{bmatrix} a_0 \\ b_0 \end{bmatrix}$ is given by the normal equations

$$(M^T M)\mathbf{a} = M^T \mathbf{y}$$

The condition that at least two of $x_1, x_2, \ldots, x_n$ are distinct ensures that $M^T M$ is an invertible matrix, so

$$\mathbf{a} = (M^T M)^{-1} M^T \mathbf{y}$$

EXAMPLE 1 Let data points $(x_1,y_1), (x_2,y_2), \ldots, (x_5,y_5)$ be given as in the accompanying table. Find the line that is the least squares approximation to these data.

x	y
1	1
3	2
4	3
6	4
7	5

Solution In this case we have

$$M^T M = \begin{bmatrix} 1 & 1 & \cdots & 1 \\ x_1 & x_2 & \cdots & x_5 \end{bmatrix} \begin{bmatrix} 1 & x_1 \\ 1 & x_2 \\ \vdots & \vdots \\ 1 & x_5 \end{bmatrix}$$

$$= \begin{bmatrix} 5 & x_1 + \cdots + x_5 \\ x_1 + \cdots + x_5 & x_1^2 + \cdots + x_5^2 \end{bmatrix} = \begin{bmatrix} 5 & 21 \\ 21 & 111 \end{bmatrix}$$

$$M^T \mathbf{y} = \begin{bmatrix} 1 & 1 & \cdots & 1 \\ x_1 & x_2 & \cdots & x_5 \end{bmatrix} \begin{bmatrix} y_1 \\ y_2 \\ \vdots \\ y_5 \end{bmatrix}$$

$$= \begin{bmatrix} y_1 + y_2 + \cdots + y_5 \\ x_1 y_1 + x_2 y_2 + \cdots + x_5 y_5 \end{bmatrix} = \begin{bmatrix} 15 \\ 78 \end{bmatrix}$$

so the normal equations $(M^T M)\mathbf{a} = M^T \mathbf{y}$ for $\mathbf{a} = \begin{bmatrix} a_0 \\ b_0 \end{bmatrix}$ become

$$\begin{bmatrix} 5 & 21 \\ 21 & 111 \end{bmatrix} \begin{bmatrix} a_0 \\ b_0 \end{bmatrix} = \begin{bmatrix} 15 \\ 78 \end{bmatrix}$$

The solution (using Gaussian elimination) is $\begin{bmatrix} a_0 \\ b_0 \end{bmatrix} = \begin{bmatrix} .24 \\ .66 \end{bmatrix}$, so the least squares approximating line for these data is $y = .24 + .66x$. Note that M^TM is indeed invertible here (the determinant is 114), and the inverse matrix is

$$(M^TM)^{-1} = \frac{1}{114}\begin{bmatrix} 111 & -21 \\ -21 & 5 \end{bmatrix}$$

Hence, as before,

$$\mathbf{a} = (M^TM)^{-1}M^T\mathbf{y} = \frac{1}{114}\begin{bmatrix} 27 \\ 75 \end{bmatrix} = \begin{bmatrix} .24 \\ .66 \end{bmatrix}$$

Suppose now that, rather than a straight line, we want to find the parabola $y = a + bx + cx^2$ that is the least squares approximation to the data points $(x_1,y_1), \ldots, (x_n,y_n)$. In the function $f(x) = a + bx + cx^2$, the *three* constants a, b, and c must be chosen to minimize the sum of squares of the errors:

$$S = [y_1 - f(x_1)]^2 + [y_2 - f(x_2)]^2 + \cdots + [y_n - f(x_n)]^2 \qquad (*)$$

Choosing a, b, and c amounts to choosing the (parabolic) function f that minimizes S. In general, there is a relationship $y = f(x)$ between the variables, and the range of candidate functions is limited—say, to all lines or to all parabolas. The task is to find, among the suitable candidates, that function that makes the quantity S as small as possible. The function that does so is called the least squares approximating function (of that type) for the data points.

As might be imagined, this is not always an easy task. However, if the functions $f(x)$ are restricted to polynomials of degree m,

$$f(x) = a_0 + a_1x + \cdots + a_mx^m$$

the analysis proceeds much as before (when $m = 1$). The problem is to choose the numbers $a_0, a_1, \ldots, a_m$ so as to minimize the sum

$$S = [y_1 - (a_0 + a_1x_1 + \cdots + a_mx_1{}^m)]^2 + \cdots$$
$$+ [y_n - (a_0 + a_1x_n + \cdots + a_mx_n{}^m)]^2$$

The resulting function $y = a_0 + a_1x + \cdots + a_mx^m$ is called the **least squares approximating polynomial of degree *m*** for the data $(x_1,y_1), \ldots, (x_n,y_n)$. By analogy with the above, define

$$\mathbf{y} = \begin{bmatrix} y_1 \\ y_2 \\ \vdots \\ y_n \end{bmatrix} \qquad M = \begin{bmatrix} 1 & x_1 & x_1^2 & \dots & x_1^m \\ 1 & x_2 & x_2^2 & \dots & x_2^m \\ \vdots & \vdots & \vdots & & \vdots \\ 1 & x_n & x_n^2 & \dots & x_n^m \end{bmatrix} \qquad \mathbf{a} = \begin{bmatrix} a_0 \\ a_1 \\ \vdots \\ a_m \end{bmatrix}$$

Then

$$\mathbf{y} - M\mathbf{a} = \begin{bmatrix} y_1 - (a_0 + a_1x_1 + \cdots + a_mx_1^m) \\ y_2 - (a_0 + a_1x_2 + \cdots + a_mx_2^m) \\ \vdots \\ y_n - (a_0 + a_1x_n + \cdots + a_mx_n^m) \end{bmatrix}$$

so S is the sum of the squares of the entries of $\mathbf{y} - M\mathbf{a}$. An analysis similar to that for Theorem 1 can be used to prove Theorem 2.

THEOREM 2 Suppose n data points $(x_1,y_1), (x_2,y_2), \ldots, (x_n,y_n)$ are given, where at least $m + 1$ of $x_1, x_2, \ldots, x_n$ are distinct (in particular $n \geq m + 1$). Put

$$\mathbf{y} = \begin{bmatrix} y_1 \\ y_2 \\ \vdots \\ y_n \end{bmatrix} \qquad M = \begin{bmatrix} 1 & x_1 & x_1^2 & \dots & x_1^m \\ 1 & x_2 & x_2^2 & \dots & x_2^m \\ \vdots & \vdots & \vdots & & \vdots \\ 1 & x_n & x_n^2 & \dots & x_n^m \end{bmatrix}$$

Then the least squares approximating polynomial of degree m for the data points has the equation

$$y = a_0 + a_1x + \cdots + a_mx^m$$

where $\mathbf{a} = \begin{bmatrix} a_0 \\ a_1 \\ \vdots \\ a_m \end{bmatrix}$ is given by the normal equations

$$(M^TM)\mathbf{a} = M^T\mathbf{y}$$

The condition that at least $m + 1$ of $x_1, x_2, \ldots, x_n$ be distinct ensures that the matrix MM^T is invertible, so

$$\mathbf{a} = (M^TM)^{-1}M^T\mathbf{y}$$

A proof of this theorem is given in Section 6.3.1.

EXAMPLE 2 Find the least squares approximating quadratic $y = a_0 + a_1x + a_2x^2$ for the following data points.

$$(-3,3), (-1,1), (0,1), (1,2), (3,4)$$

Solution This is an instance of Theorem 2 with $m = 2$. Here

$$\mathbf{y} = \begin{bmatrix} 3 \\ 1 \\ 1 \\ 2 \\ 4 \end{bmatrix} \qquad M = \begin{bmatrix} 1 & -3 & 9 \\ 1 & -1 & 1 \\ 1 & 0 & 0 \\ 1 & 1 & 1 \\ 1 & 3 & 9 \end{bmatrix}$$

Hence,

$$M^TM = \begin{bmatrix} 1 & 1 & 1 & 1 & 1 \\ -3 & -1 & 0 & 1 & 3 \\ 9 & 1 & 0 & 1 & 9 \end{bmatrix} \begin{bmatrix} 1 & -3 & 9 \\ 1 & -1 & 1 \\ 1 & 0 & 0 \\ 1 & 1 & 1 \\ 1 & 3 & 9 \end{bmatrix} = \begin{bmatrix} 5 & 0 & 20 \\ 0 & 20 & 0 \\ 20 & 0 & 164 \end{bmatrix}$$

$$M^T\mathbf{y} = \begin{bmatrix} 1 & 1 & 1 & 1 & 1 \\ -3 & -1 & 0 & 1 & 3 \\ 9 & 1 & 0 & 1 & 9 \end{bmatrix} \begin{bmatrix} 3 \\ 1 \\ 1 \\ 2 \\ 4 \end{bmatrix} = \begin{bmatrix} 11 \\ 4 \\ 66 \end{bmatrix}$$

The normal equations for $\mathbf{a}$ are

$$\begin{bmatrix} 5 & 0 & 20 \\ 0 & 20 & 0 \\ 20 & 0 & 164 \end{bmatrix} \mathbf{a} = \begin{bmatrix} 11 \\ 4 \\ 66 \end{bmatrix} \qquad \text{whence } \mathbf{a} = \begin{bmatrix} 1.15 \\ 0.20 \\ 0.26 \end{bmatrix}$$

This means that the least squares approximating quadratic for these data is $y = 1.15 + 0.20x + 0.26x^2$. Again the matrix M^TM is invertible, the inverse being $(M^TM)^{-1} = \frac{1}{420}\begin{bmatrix} 164 & 0 & -20 \\ 0 & 21 & 0 \\ -20 & 0 & 5 \end{bmatrix}$, so $\mathbf{a}$ can be calculated from $\mathbf{a} = (M^TM)^{-1}M^T\mathbf{y}$. However, this takes much more computation than Gaussian elimination.

Least squares approximation can be used to estimate physical constants, as is illustrated by the next example.

EXAMPLE 3 Hooke's law in mechanics asserts that the magnitude of the force f required to hold a spring is a linear function of the extension e of the spring (see the diagram on page 198). That is,

$$f = ke + e_0$$

where k and e_0 are constants depending only on the spring. The following data were collected for a particular spring.

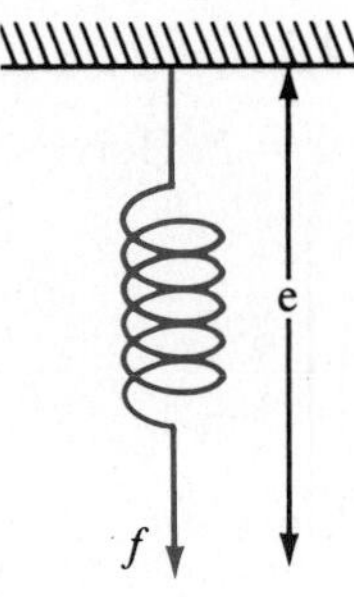

e	9	11	12	16	19
f	33	38	43	54	61

Find the least squares approximating line $f = a_0 + a_1e$ to these data, and use it to estimate k.

Solution Here f and e play the role of y and x in the general theory. We have

$$\mathbf{y} = \begin{bmatrix} 33 \\ 38 \\ 43 \\ 54 \\ 61 \end{bmatrix} \qquad M = \begin{bmatrix} 1 & 9 \\ 1 & 11 \\ 1 & 12 \\ 1 & 16 \\ 1 & 19 \end{bmatrix}$$

as in Theorem 1, so

$$M^TM = \begin{bmatrix} 1 & 1 & 1 & 1 & 1 \\ 9 & 11 & 12 & 16 & 19 \end{bmatrix} \begin{bmatrix} 1 & 9 \\ 1 & 11 \\ 1 & 12 \\ 1 & 16 \\ 1 & 19 \end{bmatrix} = \begin{bmatrix} 5 & 67 \\ 67 & 963 \end{bmatrix}$$

$$M^T\mathbf{y} = \begin{bmatrix} 1 & 1 & 1 & 1 & 1 \\ 9 & 11 & 12 & 16 & 19 \end{bmatrix} \begin{bmatrix} 33 \\ 38 \\ 43 \\ 54 \\ 61 \end{bmatrix} = \begin{bmatrix} 229 \\ 3254 \end{bmatrix}$$

Hence the normal equations for $\mathbf{a}$ are

$$\begin{bmatrix} 5 & 67 \\ 67 & 963 \end{bmatrix} \mathbf{a} = \begin{bmatrix} 229 \\ 3254 \end{bmatrix} \qquad \text{whence } \mathbf{a} = \begin{bmatrix} 7.70 \\ 2.84 \end{bmatrix}$$

The least squares approximating line is $f = 7.70 + 2.84e$, so the estimate for k is $k = 2.84$.

EXERCISES 4.3

1. Find the least squares approximating line $y = a_0 + a_1x$ to each of the following sets of data points.
   (a) (1,1), (3,2), (4,3), (6,4) **(b)** (2,4), (4,3), (7,2), (8,1)
   (c) $(-1,-1)$, (0,1), (1,2), (2,4), (3,6)
   (d) $(-2,3)$, $(-1,1)$, (0,0), $(1,-2)$, $(2,-4)$
2. Find the least squares approximating quadratic $y = a_0 + a_1x + a_2x^2$ for each of the following sets of data points.
   (a) (0,1), (2,2), (3,3), (4,5) **(b)** $(-2,1)$, (0,0), (3,2), (4,3)

3. If M is a square invertible matrix, show that $\mathbf{a} = M^{-1}\mathbf{y}$ (in the notation of Theorem 2).

4. Newton's laws of motion imply that an object dropped from rest at a height of 100 meters will be at a height $s = 100 - \frac{1}{2}gt^2$ meters t seconds later, where g is a constant called the acceleration of gravity. The values of s and t given in the table are observed. Write $x = t^2$, find the least squares approximating line $s = a + bx$ for these data, and use b to estimate g. Then find the least squares approximating quadratic $s = a_0 + a_1t + a_2t^2$, and use the value of a_2 to estimate g.

t	1	2	3
s	95	80	56

5. A naturalist measured the heights y_i (in meters) of several spruce trees with trunk diameters x_i (in centimeters). The data are as given in the table. Find the least squares approximating line for these data, and use it to estimate the height of a spruce tree with a trunk of diameter 10 cm.

x_i	5	7	8	12	13	16
y_i	2	3.3	4	7.3	7.9	10.1

5 Vector Spaces

SECTION 5.1 Examples and Basic Properties

Ordered pairs and ordered triples of real numbers arise naturally in Euclidean geometry as the Cartesian coordinates of points in the plane and in space. They have been in use since the sixteenth century (in fact, the name honors René Descartes, 1596–1650), but in more modern times it became clear that ordered n-tuples of real numbers are useful for values of n greater than 3. A good example is their use in Chapter 1 to describe solutions of systems of linear equations.

Our interest here is not so much with these n-tuples themselves as with the set of *all* n-tuples. If we denote the set of real numbers by $\mathbb{R}$, this set is described as follows:

DEFINITION Given $n \geq 1$, the set of all n-tuples with real entries is called **Euclidean *n*-space** and is denoted $\mathbb{R}^n$.

A word about notation is needed here. We shall continue to use both row and column notation for the n-tuples in $\mathbb{R}^n$ and, when it is clear from the context which notation is being used, we shall do so without comment.

The space $\mathbb{R}^3$ has a geometrical meaning when it is identified with the geometrical vectors discussed in Chapter 4. Thus a triple (a,b,c) in $\mathbb{R}^3$ is identified with the "arrow" from the origin to the point $P(a,b,c)$ with a, b, and c as coordinates. Then vector addition is given geometrically by the parallelogram law, and scalar multiplication takes on a geometrical meaning as well. The geometry is useful in that it often provides a "picture" of some aspect of the vector space that enhances our comprehension and even helps us understand the spaces $\mathbb{R}^n$ when $n > 3$. Of course $\mathbb{R}^2$ can be identified with the geometrical plane. And $\mathbb{R}^1$

is $\mathbb{R}$ itself — and so is identified with the points on a line. All these things are discussed in Chapter 4, and the reader who is totally unfamiliar with them would do well to read Section 4.1. However, having completed Chapter 4 is not required for understanding the present chapter, because our treatment here is algebraic in nature.

DEFINITION

Let $\mathbf{v} = (v_1, v_2, \ldots, v_n)$ and $\mathbf{u} = (u_1, u_2, \ldots, u_n)$ be n-tuples in $\mathbb{R}^n$.

(1) $\mathbf{u}$ and $\mathbf{v}$ are called **equal** (written $\mathbf{u} = \mathbf{v}$) if $u_1 = v_1$, $u_2 = v_2, \ldots, u_n = v_n$.

(2) The **sum** $\mathbf{u} + \mathbf{v}$ is defined by

$$\mathbf{u} + \mathbf{v} = (u_1 + v_1, u_2 + v_2, \ldots, u_n + v_n)$$

(3) If a is any real number, the **scalar product** $a\mathbf{v}$ is defined by

$$a\mathbf{v} = (av_1, av_2, \ldots, av_n)$$

(4) The **zero** n-tuple $\mathbf{0}$ in $\mathbb{R}^n$ is defined by

$$\mathbf{0} = (0, 0, \ldots, 0)$$

(5) The **negative** $-\mathbf{v}$ of the n-tuple $\mathbf{v}$ is defined by

$$-\mathbf{v} = (-v_1, -v_2, \ldots, -v_n)$$

(6) The **difference** $\mathbf{u} - \mathbf{v}$ is defined to be $\mathbf{u} + (-\mathbf{v})$. That is,

$$\mathbf{u} - \mathbf{v} = \mathbf{u} + (-\mathbf{v}) = (u_1 - v_1, u_2 - v_2, \ldots, u_n - v_n)$$

Of course, these definitions are consistent with the corresponding matrix operations (regarding the rows in $\mathbb{R}^n$ as $1 \times n$ matrices), so all the computations with matrices are available in $\mathbb{R}^n$. These definitions also conform to the operations introduced in Section 1.3 on the set of solutions (written as columns) to a homogeneous system of linear equations.

The following properties follow easily from the foregoing definitions (they also come from the corresponding properties of matrices).

PROPERTIES OF $\mathbb{R}^n$

Let $\mathbf{u}$, $\mathbf{v}$, and $\mathbf{w}$ denote n-tuples in $\mathbb{R}^n$, and let a and b be real numbers. Then

(1) $\mathbf{u} + \mathbf{v} = \mathbf{v} + \mathbf{u}$

(2) $\mathbf{u} + (\mathbf{v} + \mathbf{w}) = (\mathbf{u} + \mathbf{v}) + \mathbf{w}$

(3) $\mathbf{v} + \mathbf{0} = \mathbf{v}$

(4) $\mathbf{v} + (-\mathbf{v}) = \mathbf{0}$

(5) $a(\mathbf{v} + \mathbf{w}) = a\mathbf{v} + a\mathbf{w}$

(6) $(a + b)\mathbf{v} = a\mathbf{v} + b\mathbf{v}$

(7) $a(b\mathbf{v}) = (ab)\mathbf{v}$

(8) $1 \cdot \mathbf{v} = \mathbf{v}$

EXAMPLE 1 Let $\mathbf{u} = (3,-1,2,4)$ and $\mathbf{v} = (5,0,-6,1)$ in $\mathbb{R}^4$. Then

$$\mathbf{u} + \mathbf{v} = (8,-1,-4,5)$$
$$3\mathbf{u} - 5\mathbf{v} = (9,-3,6,12) - (25,0,-30,5) = (-16,-3,36,7)$$

EXAMPLE 2 Let $\mathbf{u} = (1,3,-1,0)$ $\mathbf{v} = (2,1,-5,1)$ and $\mathbf{w} = (1,1,1,-1)$ in $\mathbb{R}^4$.

(1) Show that the only numbers a, b, and c such that $a\mathbf{u} + b\mathbf{v} + c\mathbf{w} = \mathbf{0}$ are $a = b = c = 0$.

(2) Find numbers a, b, and c such that $(3,8,6,-4) = a\mathbf{u} + b\mathbf{v} + c\mathbf{w}$.

Solution

(1) The condition $a\mathbf{u} + b\mathbf{v} + c\mathbf{w} = \mathbf{0}$ gives

$$(a + 2b + c, 3a + b + c, -a - 5b + c, b - c) = (0,0,0,0).$$

Equating components gives the equations

$$\begin{aligned} a + 2b + c &= 0 \\ 3a + b + c &= 0 \\ -a - 5b + c &= 0 \\ b - c &= 0 \end{aligned}$$

for a, b, and c. The only solution is $a = b = c = 0$.

(2) As in (1), equating components gives the equations

$$\begin{aligned} a + 2b + c &= 3 \\ 3a + b + c &= 8 \\ -a - 5b + c &= 6 \\ b - c &= -4 \end{aligned}$$

for a, b, and c. The (unique) solution is $a = 2$, $b = -1$, and $c = 3$.

The foregoing properties of $\mathbb{R}^n$ are not exclusive to $\mathbb{R}^n$. For example, the set of all $m \times n$ matrices also has an addition and a scalar multiplication that satisfy these properties. It turns out that many other sets of mathematical objects have these properties, and the general study of such systems is the subject of this chapter.

DEFINITION A **vector space** consists of a nonempty set V of objects (called **vectors**) that can be added, that can be multiplied by a real number (called a **scalar** in this context), and for which certain axioms hold. If $\mathbf{v}$ and $\mathbf{w}$ are two vectors in V, their sum is expressed as $\mathbf{v} + \mathbf{w}$, and the scalar product of $\mathbf{v}$ by a real number a is denoted $a\mathbf{v}$. These operations are called **vector addition** and **scalar multiplication**, respectively, and the following axioms are assumed to hold.

Axioms for vector addition

A1. If $\mathbf{u}$ and $\mathbf{v}$ are in V, then $\mathbf{u} + \mathbf{v}$ is in V.

A2. $\mathbf{u} + \mathbf{v} = \mathbf{v} + \mathbf{u}$ for all $\mathbf{u}$ and $\mathbf{v}$ in V.

A3. $\mathbf{u} + (\mathbf{v} + \mathbf{w}) = (\mathbf{u} + \mathbf{v}) + \mathbf{w}$ for all $\mathbf{u}$, $\mathbf{v}$, and $\mathbf{w}$ in V.

A4. There exists an element $\mathbf{0}$ in V such that $\mathbf{v} + \mathbf{0} = \mathbf{v} = \mathbf{0} + \mathbf{v}$ for every $\mathbf{v}$ in V.

A5. For each $\mathbf{v}$ in V there exists an element $-\mathbf{v}$ in V such that $-\mathbf{v} + \mathbf{v} = \mathbf{0}$ and $\mathbf{v} + (-\mathbf{v}) = \mathbf{0}$.

Axioms for scalar multiplication

S1. If $\mathbf{v}$ is in V, then $a\mathbf{v}$ is in V for all a in $\mathbb{R}$.

S2. $a(\mathbf{v} + \mathbf{w}) = a\mathbf{v} + a\mathbf{w}$ for all $\mathbf{v}$ and $\mathbf{w}$ in V and all a in $\mathbb{R}$.

S3. $(a + b)\mathbf{v} = a\mathbf{v} + b\mathbf{v}$ for all $\mathbf{v}$ in V and all a and b in $\mathbb{R}$.

S4. $a(b\mathbf{v}) = (ab)\mathbf{v}$ for all $\mathbf{v}$ in V and all a and b in $\mathbb{R}$.

S5. $1\mathbf{v} = \mathbf{v}$ for all $\mathbf{v}$ in V.

The content of axioms A1 and S1 is described by saying that V is **closed** under vector addition and scalar multiplication. The element $\mathbf{0}$ in axiom A4 is called the **zero vector**, and the vector $-\mathbf{v}$ in axiom A5 is called the **negative** of $\mathbf{v}$. These are uniquely determined by the properties in axioms A4 and A5 (Theorem 3).

Note that the symbol + plays two different roles here. For example, axiom S3 reads $(a + b)\mathbf{v} = a\mathbf{v} + b\mathbf{v}$. Here the + on the left indicates ordinary addition of the numbers a and b, whereas the + on the right side indicates vector addition. Similarly, two kinds of multiplication are used, each indicated by juxtaposition. For example, the right side of axiom S4 is $(ab)\mathbf{v}$ in which the product ab is formed by ordinary numerical multiplication, whereas the product $(ab)\mathbf{v}$ of the number ab with $\mathbf{v}$ is a scalar multiplication. We shall continue to write vector expressions where the two meanings of addition and multiplication occur; it will always be clear from the context which is meant.

The properties of $\mathbb{R}^n$ that we have discussed give

EXAMPLE 3 | $\mathbb{R}^n$ is a vector space using the above addition and scalar multiplication.

The space $\mathbb{R}^n$ consists of special types of matrices. More generally, let $\mathbf{M}_{mn}$ denote the set of all $m \times n$ matrices with real entries. Then Theorem 1 in Section 2.1 gives the following information.

EXAMPLE 4 The set $\mathbf{M}_{mn}$ of all $m \times n$ matrices is a vector space using matrix addition and scalar multiplication. The zero element in this vector space is the zero matrix of size $m \times n$, and the vector-space negative of a matrix (required by axiom A5) is the same as that discussed in Section 2.1.

EXAMPLE 5 Show that

$$V = \{(x,x,y) \mid x \text{ and } y \text{ in } \mathbb{R}\}$$

is a vector space using the operations of $\mathbb{R}^3$.

Solution Axioms A2, A3, S2, S3, S4, and S5 all hold in $\mathbb{R}^3$ and so are satisfied in V. Hence we check the remaining axioms. Given a in $\mathbb{R}$, and $\mathbf{u} = (x,x,y)$ and $\mathbf{v} = (x_1,x_1,y_1)$ in V, we have

$$\mathbf{u} + \mathbf{v} = (x + x_1, x + x_1, y + y_1)$$
$$a\mathbf{v} = (ax_1,ax_1,ay_1)$$

so both lie in V (because the first components are equal). Hence axioms A1 and S1 are satisfied. The zero vector $\mathbf{0} = (0,0,0)$ of $\mathbb{R}^3$ lies in V, and $\mathbf{0} + \mathbf{v} = \mathbf{v}$ holds for all $\mathbf{v}$ in V (it holds for all $\mathbf{v}$ in $\mathbb{R}^3$). Thus $\mathbf{0}$ serves as the zero vector of V. This is axiom A4. Finally, if $\mathbf{v} = (x,x,y)$ is in V, then $-\mathbf{v} = (-x,-x,-y)$ is also in V and so serves as the negative of $\mathbf{v}$ in V. Hence axiom A5 is satisfied.

Another important source of examples of vector spaces is sets of polynomials, so we review some basic facts. A **polynomial** in an indeterminate x is an expression

$$p(x) = a_0 + a_1x + a_2x^2 + \cdots + a_nx^n$$

where $a_0, a_1, a_2, \ldots, a_n$ are real numbers called the **coefficients** of the polynomial. If all the coefficients are zero, the polynomial is called the **zero polynomial** and is denoted simply as 0. If $p(x) \neq 0$, the highest power of x with a nonzero coefficient is called the **degree** of $p(x)$ and is denoted as $\deg[p(x)]$. Hence $\deg(3 + 5x) = 1$, $\deg(1 + x + x^2) = 2$, and $\deg(4) = 0$. (The degree of the zero polynomial is not defined.)

Let $\mathbf{P}$ denote the set of all polynomials and suppose that

$$p(x) = a_0 + a_1x + a_2x^2 + \cdots$$
$$q(x) = b_0 + b_1x + b_2x^2 + \cdots$$

are two polynomials in $\mathbf{P}$ (possibly of different degrees). Then $p(x)$ and $q(x)$ are called **equal** [written $p(x) = q(x)$] if and only if all the corresponding coefficients agree — that is, $a_0 = b_0$, $a_1 = b_1$, $a_2 = b_2$, and so on. In particular, if $a_0 + a_1x + a_2x^2 + \cdots = 0$, this means that $a_0 = 0$,

$a_1 = 0, a_2 = 0, \ldots$, and this is the reason for calling x an **indeterminate.** The set **P** has an addition and scalar multiplication defined on it as follows: If $p(x)$ and $q(x)$ are as above and a is a real number,

$$p(x) + q(x) = (a_0 + b_0) + (a_1 + b_1)x + (a_2 + b_2)x^2 + \cdots$$
$$ap(x) = aa_0 + (aa_1)x + (aa_2)x^2 + \cdots$$

Evidently these are again polynomials, so **P** is closed under these operations. The other vector-space axioms are easily verified.

EXAMPLE 6 The set **P** of all polynomials is a vector space with the foregoing addition and scalar multiplication. The zero vector is the zero polynomial, and the negative of a polynomial $p(x) = a_0 + a_1x + a_2x^2 + \cdots$ is the polynomial $-p(x) = -a_0 - a_1x - a_2x^2 - \cdots$ obtained by negating all the coefficients.

There are other important examples of vector spaces consisting of polynomials. One of these is the set $\mathbf{P}_n$ of all polynomials of degree at most n, together with the zero polynomial. In other words, $\mathbf{P}_n$ consists of all polynomials

$$a_0 + a_1x + a_2x^2 + \cdots + a_nx^n$$

where $a_0, a_1, a_2, \ldots, a_n$ are real numbers, and so is closed under the addition and scalar multiplication in **P**. Moreover, the zero polynomial is included in $\mathbf{P}_n$, and the negative of a polynomial in $\mathbf{P}_n$ is again in $\mathbf{P}_n$. The other vector-space axioms are routinely verified.

EXAMPLE 7 For each $n \geq 0$, $\mathbf{P}_n$ is a vector space with the same vector addition and scalar multiplication as **P**. Again the zero polynomial serves as the zero vector, and the negative of a polynomial in $\mathbf{P}_n$ is just its negative in **P**.

If a and b are real numbers and $a \leq b$, the **interval** $[a, b]$ is defined to be the set of all real numbers x such that $a \leq x \leq b$. A (real-valued) **function** f on $[a, b]$ is a rule that associates every number x in $[a, b]$ with a real number denoted $f(x)$. The rule is frequently specified by giving a formula for $f(x)$ in terms of x. For example, $f(x) = 2^x$, $f(x) = \sin x$, and $f(x) = x^2 + 1$ are familiar functions. In fact, every polynomial $p(x)$ can be regarded as the formula for a function p. The set of all functions on $[a, b]$ is denoted $\mathbf{F}[a, b]$. Two functions f and g in $\mathbf{F}[a, b]$ are **equal** if $f(x) = g(x)$ for every x in $[a, b]$, and we describe this by saying that f and g have the **same action.** Note that two polynomials are equal (defined prior to Example 6) if and only if they are equal as functions.

If f and g are two functions in $\mathbf{F}[a, b]$, and r is a real number, define the sum $f + g$ and the scalar produce rf by

$$(f + g)(x) = f(x) + g(x) \qquad \text{for each } x \text{ in } [a, b]$$
$$(rf)(x) = rf(x) \qquad \text{for each } x \text{ in } [a, b]$$

In other words, the action of $f + g$ upon x is to associate it with the number $f(x) + g(x)$, and rf associates x with $rf(x)$.

These operations on $\mathbf{F}[a, b]$ are called **pointwise** addition and scalar multiplication of functions.

EXAMPLE 8 The set $\mathbf{F}[a, b]$ of all functions on the interval $[a, b]$ is a vector space if pointwise addition and scalar multiplication of functions are used. The zero function (in axiom A4) is denoted as 0 and has action defined by

$$0(x) = 0 \qquad \text{for all } x \text{ in } [a, b]$$

The negative of a function f is denoted $-f$ and has action defined by

$$(-f)(x) = -f(x) \qquad \text{for all } x \text{ in } [a, b]$$

Solution Axioms A1 and S1 are clearly satisfied because, if f and g are functions on $[a, b]$, then $f + g$ and rf are again such functions. The verification of the remaining axioms is left as Exercise 18.

Other examples of vector spaces will appear later, but these are sufficiently varied to indicate the scope of the concept and to illustrate the properties of vector spaces discussed below. With such a variety of examples, it may come as a surprise that a well-developed *theory* of vector spaces exists. That is, many properties can be shown to hold for *all* vector spaces and hence hold in every example. The following theorems collect several such properties, which will be used frequently below.

THEOREM 1
Cancellation

Let $\mathbf{u}$, $\mathbf{v}$, and $\mathbf{w}$ be vectors in a vector space V.

If $\mathbf{v} + \mathbf{u} = \mathbf{v} + \mathbf{w}$, then $\mathbf{u} = \mathbf{w}$.

Proof We are given $\mathbf{v} + \mathbf{u} = \mathbf{v} + \mathbf{w}$. If these were numbers instead of vectors, we would simply subtract $\mathbf{v}$ from both sides of the equation to obtain $\mathbf{u} = \mathbf{w}$. This can be accomplished with vectors by adding $-\mathbf{v}$ to both sides of the equation. The steps are as follows.

$$\mathbf{v} + \mathbf{u} = \mathbf{v} + \mathbf{w}$$
$$-\mathbf{v} + (\mathbf{v} + \mathbf{u}) = -\mathbf{v} + (\mathbf{v} + \mathbf{w}) \qquad \text{(axiom A5)}$$
$$(-\mathbf{v} + \mathbf{v}) + \mathbf{u} = (-\mathbf{v} + \mathbf{v}) + \mathbf{w} \qquad \text{(axiom A3)}$$

$$\mathbf{0} + \mathbf{u} = \mathbf{0} + \mathbf{w} \qquad \text{(axiom A5)}$$
$$\mathbf{u} = \mathbf{w} \qquad \text{(axiom A4)}$$

This is the desired conclusion. ∎

Note that axiom A2 permits variations of the cancellation property. For example, $\mathbf{u} + \mathbf{v} = \mathbf{w} + \mathbf{v}$ implies that $\mathbf{u} = \mathbf{w}$, because the condition can be written as $\mathbf{v} + \mathbf{u} = \mathbf{v} + \mathbf{w}$ by axiom A2.

As with many good mathematical theorems, the technique of the proof of Theorem 1 is at least as important as the theorem itself. The idea was to mimic the well-known process of numerical subtraction in a vector space V as follows: In order to subtract a vector $\mathbf{v}$ from both sides of a vector equation, we added $-\mathbf{v}$ to both sides. With this in mind, we define the **difference** $\mathbf{u} - \mathbf{v}$ of two vectors in V as

$$\mathbf{u} - \mathbf{v} = \mathbf{u} + (-\mathbf{v})$$

We shall say that this vector is the result of having **subtracted** $\mathbf{v}$ from $\mathbf{u}$, and, as in arithmetic, this operation has the property given in Theorem 2.

THEOREM 2 If $\mathbf{u}$ and $\mathbf{v}$ are vectors in a vector space V, the equation

$$\mathbf{x} + \mathbf{v} = \mathbf{u}$$

has one and only one solution $\mathbf{x}$ in V given by

$$\mathbf{x} = \mathbf{u} - \mathbf{v}$$

Proof The difference $\mathbf{x} = \mathbf{u} - \mathbf{v}$ is a solution to the equation because (using several axioms)

$$\mathbf{x} + \mathbf{v} = (\mathbf{u} - \mathbf{v}) + \mathbf{v} = [\mathbf{u} + (-\mathbf{v})] + \mathbf{v} = \mathbf{u} + (-\mathbf{v} + \mathbf{v}) = \mathbf{u} + \mathbf{0} = \mathbf{u}$$

To see that this is the only solution, suppose $\mathbf{x}_1$ is another so that $\mathbf{x}_1 + \mathbf{v} = \mathbf{u}$. Then $\mathbf{x} + \mathbf{v} = \mathbf{x}_1 + \mathbf{v}$ (they both equal $\mathbf{u}$), so $\mathbf{x} = \mathbf{x}_1$ by cancellation. ∎

THEOREM 3 Let V be a vector space.

(1) The zero vector $\mathbf{0}$ in V is uniquely determined by the property in axiom A4.

(2) Given a vector $\mathbf{v}$ in V, its negative $-\mathbf{v}$ is uniquely determined by the property in axiom A5.

Proof We leave the proof of (1) as Exercise 16(a). According to axiom A5, $-\mathbf{v} + \mathbf{v} = \mathbf{0}$. If $\mathbf{w}$ also has this property, then $\mathbf{w} + \mathbf{v} = \mathbf{0}$, so $\mathbf{w} + \mathbf{v} = -\mathbf{v} + \mathbf{v}$ (because both are equal to $\mathbf{0}$). Hence $\mathbf{w} = -\mathbf{v}$ by cancellation. ■

Because of this we are entitled to speak of *the* zero vector in a vector space and *the* negative of a vector. The next two theorems introduce some basic facts that will be used extensively below.

THEOREM 4 Let $\mathbf{v}$ denote a vector in a vector space V, and let a denote a real number.

(1) $0\mathbf{v} = \mathbf{0}$

(2) $a\mathbf{0} = \mathbf{0}$

(3) If $a\mathbf{v} = \mathbf{0}$, then either $a = 0$ or $\mathbf{v} = \mathbf{0}$.

(4) $(-1)\mathbf{v} = -\mathbf{v}$

(5) $(-a)\mathbf{v} = -(a\mathbf{v}) = a(-\mathbf{v})$ for all scalars a

Proof The proofs of (2) and (5) are left as Exercise 16.

(1) Observe that $0\mathbf{v} + 0\mathbf{v} = (0 + 0)\mathbf{v} = 0\mathbf{v} = 0\mathbf{v} + \mathbf{0}$ where the first equality is by axiom S3. It follows that $0\mathbf{v} = \mathbf{0}$ by cancellation.

(3) Assume $a\mathbf{v} = \mathbf{0}$; it suffices to show that if $a \neq 0$, then necessarily $\mathbf{v} = \mathbf{0}$. But $a \neq 0$ means we can scalar-multiply the given equation $a\mathbf{v} = \mathbf{0}$ by $\frac{1}{a}$ to obtain, by using (2) and axioms S4 and S5,

$$\mathbf{v} = 1\mathbf{v} = \left(\frac{1}{a}a\right)\mathbf{v} = \frac{1}{a}(a\mathbf{v}) = \frac{1}{a}\mathbf{0} = \mathbf{0}$$

(4) We have $-\mathbf{v} + \mathbf{v} = \mathbf{0}$ by axiom A5. On the other hand,

$$(-1)\mathbf{v} + \mathbf{v} = (-1)\mathbf{v} + 1\mathbf{v} = (-1 + 1)\mathbf{v} = 0\mathbf{v} = \mathbf{0}$$

using (1) and axioms S5 and S3. Hence $(-1)\mathbf{v} + \mathbf{v} = -\mathbf{v} + \mathbf{v}$ (because both are equal to $\mathbf{0}$), so $(-1)\mathbf{v} = -\mathbf{v}$ by cancellation. ■

Axioms S2 and S3 extend to more than two summands. For example,

$$a(\mathbf{u} + \mathbf{v} + \mathbf{w}) = a\mathbf{u} + a\mathbf{v} + a\mathbf{w} \quad \text{and} \quad (a + b + c)\mathbf{v} = a\mathbf{v} + b\mathbf{v} + c\mathbf{v}$$

hold for all values of the scalars and vectors involved. More generally,

$$a(\mathbf{v}_1 + \mathbf{v}_2 + \cdots + \mathbf{v}_n) = a\mathbf{v}_1 + a\mathbf{v}_2 + \cdots + a\mathbf{v}_n$$
$$(a_1 + a_2 + \cdots + a_n)\mathbf{v} = a_1\mathbf{v} + a_2\mathbf{v} + \cdots + a_n\mathbf{v}$$

hold for all $n \geq 1$, all numbers $a, a_1, \ldots, a_n$, and all vectors $\mathbf{v}, \mathbf{v}_1, \ldots, \mathbf{v}_n$. The verifications are by induction and are left to the reader (Exercise 17). These facts—together with the axioms, Theorems 4 and 5, and the definition of subtraction—enable us to simplify expressions involving sums of scalar multiples of vectors by collecting like terms, expanding, and taking out common factors. This has been discussed for the vector space of matrices in Section 2.1 (and for geometric vectors in Section 4.1); the manipulations in an arbitrary vector space are carried out in the same way. To illustrate, we rework Example 8 of Section 2.1 in the general context.

EXAMPLE 9 If $\mathbf{u}$, $\mathbf{v}$, and $\mathbf{w}$ are vectors in a vector space V, simplify

$$2(\mathbf{u} + 3\mathbf{w}) - 3(2\mathbf{w} - \mathbf{v}) - 3[2(2\mathbf{u} + \mathbf{v} - 4\mathbf{w}) - 4(\mathbf{u} - 2\mathbf{w})]$$

Solution The reduction proceeds as though $\mathbf{u}$, $\mathbf{v}$, and $\mathbf{w}$ were matrices or variables.

$$\begin{aligned}
&2(\mathbf{u} + 3\mathbf{w}) - 3(2\mathbf{w} - \mathbf{v}) - 3[2(2\mathbf{u} + \mathbf{v} - 4\mathbf{w}) - 4(\mathbf{u} - 2\mathbf{w})] \\
&\quad = 2\mathbf{u} + 6\mathbf{w} - 6\mathbf{w} + 3\mathbf{v} - 3[4\mathbf{u} + 2\mathbf{v} - 8\mathbf{w} - 4\mathbf{u} + 8\mathbf{w}] \\
&\quad = 2\mathbf{u} + 3\mathbf{v} - 3[2\mathbf{v}] \\
&\quad = 2\mathbf{u} + 3\mathbf{v} - 6\mathbf{v} \\
&\quad = 2\mathbf{u} - 3\mathbf{v}
\end{aligned}$$

The next example shows that the techniques for solving linear equations in Chapter 1 work for vector variables too.

EXAMPLE 10 Let $\mathbf{u}$ and $\mathbf{v}$ be vectors in a vector space V. Find vectors $\mathbf{x}$ and $\mathbf{y}$ in V such that

$$\begin{aligned}
\mathbf{x} - 4\mathbf{y} &= \mathbf{u} \\
2\mathbf{x} + 3\mathbf{y} &= \mathbf{v}
\end{aligned}$$

Solution The usual row operations on equations work here. Subtract twice the first equation from the second to obtain $11\mathbf{y} = \mathbf{v} - 2\mathbf{u}$. This gives $\mathbf{y} = \frac{1}{11}\mathbf{v} - \frac{2}{11}\mathbf{u}$. Substituting this in the first equation gives $\mathbf{x} = \mathbf{u} + 4\mathbf{y} = \frac{3}{11}\mathbf{u} + \frac{4}{11}\mathbf{v}$.

As an alternative solution, we could write the equations in matrix form:

$$\begin{bmatrix} 1 & -4 \\ 2 & 3 \end{bmatrix} \begin{bmatrix} \mathbf{x} \\ \mathbf{y} \end{bmatrix} = \begin{bmatrix} \mathbf{u} \\ \mathbf{v} \end{bmatrix}$$

as in Section 2.3.1, where the product of a matrix and a column of vectors is defined in the obvious way. But $\begin{bmatrix} 1 & -4 \\ 2 & 3 \end{bmatrix}^{-1} = \frac{1}{11}\begin{bmatrix} 3 & 4 \\ -2 & 1 \end{bmatrix}$, so

$$\begin{bmatrix} \mathbf{x} \\ \mathbf{y} \end{bmatrix} = \begin{bmatrix} 1 & -4 \\ 2 & 3 \end{bmatrix}^{-1} \begin{bmatrix} \mathbf{u} \\ \mathbf{v} \end{bmatrix} = \frac{1}{11}\begin{bmatrix} 3 & 4 \\ -2 & 1 \end{bmatrix} \begin{bmatrix} \mathbf{u} \\ \mathbf{v} \end{bmatrix} = \frac{1}{11}\begin{bmatrix} 3\mathbf{u} + 4\mathbf{v} \\ -2\mathbf{u} + \mathbf{v} \end{bmatrix}$$

Hence $\mathbf{x} = \frac{1}{11}(3\mathbf{u} + 4\mathbf{v})$ and $\mathbf{y} = \frac{1}{11}(-2\mathbf{u} + \mathbf{v})$, as before.

EXERCISES 5.1

1. Let $\mathbf{u} = (1, 2, -1, 0, 4)$ and $\mathbf{v} = (2, 7, 5, 3, -2)$. Compute:
(a) $\mathbf{u} + \mathbf{v}$ **(b)** $3\mathbf{u} - 2\mathbf{v}$ **(c)** $-2\mathbf{u} + \mathbf{v}$ **(d)** $-3(2\mathbf{u} - 3\mathbf{v})$

2. Vectors $\mathbf{u}$, $\mathbf{v}$, and $\mathbf{w}$ in $\mathbb{R}^n$ are called linearly independent if the only way $a\mathbf{u} + b\mathbf{v} + c\mathbf{w} = \mathbf{0}$ can hold is when $a = b = c = 0$. In each case, determine whether $\mathbf{u}$, $\mathbf{v}$, and $\mathbf{w}$ are linearly independent.
(a) $\mathbf{u} = (1, 2, -1, 1)$ $\mathbf{v} = (2, 1, 3, 0)$ $\mathbf{w} = (1, 0, 1, 2)$
(b) $\mathbf{u} = (1, 3, -1, 4)$ $\mathbf{v} = (2, 1, 1, -2)$ $\mathbf{w} = (4, -3, 5, -14)$
(c) $\mathbf{u} = (1, -1, 3, 2, -4)$ $\mathbf{v} = (2, 0, 1, 3, -5)$ $\mathbf{w} = (0, -2, 5, 1, -3)$
(d) $\mathbf{u} = (2, 1, 1, 3, 0)$ $\mathbf{v} = (1, 3, -1, 0, 4)$ $\mathbf{w} = (2, 1, 6, 8, 1)$

3. In each case determine scalars, a, b, and c (if they exist) such that the condition is satisfied:
(a) $a(1, 2, -1, 1) + b(2, 0, 1, 1) + c(1, 0, 2, 1) = (1, 4, -4, 1)$
(b) $a(1, 3, 0, 1) + b(2, -1, 1, 0) + c(3, 1, -1, 1) = (1, 4, -5, 2)$

4. In each case, show that V is a vector space using the operations of $\mathbb{R}^2$.
(a) $V = \{(x,0) \mid x \text{ in } \mathbb{R}\}$ **(b)** $V = \{(x,-x) \mid x \text{ in } \mathbb{R}\}$
(c) $V = \{(2x - y, x + y) \mid x \text{ and } y \text{ in } \mathbb{R}\}$
(d) $V = \{(3x - y, 2x + 5y) \mid x \text{ and } y \text{ in } \mathbb{R}\}$

5. Let V denote the set of ordered triples (x,y,z), and define addition on V as in $\mathbb{R}^3$. For each of the following definitions of scalar multiplication, decide whether V is a vector space.
(a) $a(x,y,z) = (ax,y,az)$ **(b)** $a(x,y,z) = (ax,0,az)$
(c) $a(x,y,z) = (0,0,0)$ **(d)** $a(x,y,z) = (2ax,2ay,2az)$

6. Are the following sets vector spaces with the indicated operations? If not, why not?
(a) The set of non-negative real numbers; ordinary addition and scalar multiplication.
(b) The set of all polynomials of degree ≥ 3, together with 0; operations of $\mathbf{P}$.

(c) The set of 2×2 matrices with zero determinant; usual matrix operations.

(d) The set of real numbers; usual operations.

(e) The set of complex numbers; usual addition and multiplication by a real number.

(f) A set $V = \{\mathbf{0}\}$ consisting of a single vector $\mathbf{0}$ and with $\mathbf{0} + \mathbf{0} = \mathbf{0}$ and $a\mathbf{0} = \mathbf{0}$ for all a in $\mathbb{R}$.

(g) The set V of all ordered pairs (x,y) with the addition of $\mathbb{R}^2$ but scalar multiplication $a(x,y) = (x,y)$ for all a in $\mathbb{R}$.

(h) The set V of all functions $f: \mathbb{R} \to \mathbb{R}$ with pointwise addition and scalar multiplication defined by $(af)(x) = f(ax)$.

(i) The set of all 2×2 matrices whose entries sum to 0; operations of $\mathbf{M}_{22}$.

(j) The set V of all 2×2 matrices with the addition of $\mathbf{M}_{22}$ but scalar multiplication $*$ defined by $a*X = aX^T$.

7. Let V be the set of positive real numbers, with vector addition being ordinary multiplication and scalar multiplication being $av = v^a$. Show that V is a vector space.

8. If V is the set of ordered pairs (x,y) of real numbers, show that it is a vector space if $(x,y) + (x_1,y_1) = (x + x_1, y + y_1 + 1)$ and $a(x,y) = (ax, ay + a - 1)$.

9. The line through the origin with slope m has the equation $y = mx$ and so consists of points $P(x,mx)$ with x in $\mathbb{R}$. Show that $V = \{(x,mx) \mid x \text{ in } \mathbb{R}\}$ is a vector space using the operations of $\mathbb{R}^2$.

10. Find $\mathbf{x}$ and $\mathbf{y}$ (in terms of $\mathbf{u}$ and $\mathbf{v}$) such that:

(a) $\begin{aligned} 2\mathbf{x} + \mathbf{y} &= \mathbf{u} \\ 5\mathbf{x} + 3\mathbf{y} &= \mathbf{v} \end{aligned}$

(b) $\begin{aligned} 3\mathbf{x} - 2\mathbf{y} &= \mathbf{u} \\ 4\mathbf{x} - 5\mathbf{y} &= \mathbf{v} \end{aligned}$

11. Find *all* vectors $\mathbf{x}$, $\mathbf{y}$, and $\mathbf{z}$ (in terms of $\mathbf{u}$, $\mathbf{v}$) such that:

(a) $\begin{aligned} \mathbf{x} - 2\mathbf{y} + \mathbf{z} &= 2\mathbf{u} - \mathbf{v} \\ 2\mathbf{x} - 3\mathbf{y} - \mathbf{z} &= \mathbf{v} - \mathbf{u} \\ -\mathbf{x} + 3\mathbf{y} - 4\mathbf{z} &= 4\mathbf{v} - 7\mathbf{u} \end{aligned}$

(b) $\begin{aligned} \mathbf{x} - \mathbf{y} + 2\mathbf{z} &= 3\mathbf{u} - \mathbf{v} \\ -3\mathbf{x} + 4\mathbf{y} - \mathbf{z} &= \mathbf{v} - \mathbf{u} \\ 5\mathbf{x} - 6\mathbf{y} + 5\mathbf{z} &= 7\mathbf{u} - 3\mathbf{v} \end{aligned}$

(c) $\begin{aligned} 3\mathbf{x} + 2\mathbf{y} + \mathbf{z} &= \mathbf{0} \\ 2\mathbf{x} + \mathbf{y} - 2\mathbf{z} &= \mathbf{0} \\ \mathbf{x} + \mathbf{y} + 3\mathbf{z} &= \mathbf{0} \end{aligned}$

(d) $\begin{aligned} 3\mathbf{x} - \mathbf{y} + 4\mathbf{z} &= \mathbf{0} \\ \mathbf{x} + \mathbf{y} - 5\mathbf{z} &= \mathbf{0} \\ \mathbf{x} - 3\mathbf{y} + 14\mathbf{z} &= \mathbf{0} \end{aligned}$

12. In each case, show that the condition $a\mathbf{u} + b\mathbf{v} + c\mathbf{w} = \mathbf{0}$ in V implies that $a = b = c = 0$.

(a) $V = \mathbb{R}^4$; $\mathbf{u} = (2,1,0,2)$, $\mathbf{v} = (1,1,-1,0)$, $\mathbf{w} = (0,1,2,1)$

(b) $V = \mathbf{M}_{22}$; $\mathbf{u} = \begin{bmatrix} 1 & 0 \\ 0 & 1 \end{bmatrix}$, $\mathbf{v} = \begin{bmatrix} 0 & 1 \\ 1 & 0 \end{bmatrix}$, $\mathbf{w} = \begin{bmatrix} 1 & 1 \\ 1 & -1 \end{bmatrix}$

(c) $V = \mathbf{P}_3$; $\mathbf{u} = x^3 + x$, $\mathbf{v} = x^2 + 1$, $\mathbf{w} = x^3 - x^2 + x + 1$

(d) $V = \mathbf{F}[0,\pi]$; $\mathbf{u} = \sin x$, $\mathbf{v} = \cos x$, $\mathbf{w} = 1$

13. Simplify each of the following.

(a) $3[2(\mathbf{u} - 2\mathbf{v} - \mathbf{w}) + 3(\mathbf{w} - \mathbf{v})] - 7(\mathbf{u} - 3\mathbf{v} - \mathbf{w})$

(b) $4(3\mathbf{u} - \mathbf{v} + \mathbf{w}) - 2[(3\mathbf{u} - 2\mathbf{v}) - 3(\mathbf{v} - \mathbf{w})] + 6(\mathbf{w} - \mathbf{u} - \mathbf{v})$

14. Show that $\mathbf{x} = \mathbf{v}$ is the only solution to the equation $\mathbf{x} + \mathbf{x} = 2\mathbf{v}$ in a vector space V. Cite all axioms used.

15. Show that $-\mathbf{0} = \mathbf{0}$ in any vector space. Cite all axioms used.

16. **(a)** Prove (1) of Theorem 3.

(b) Prove (2) of Theorem 4.

(c) Prove that $(-a)\mathbf{v} = -(a\mathbf{v})$ in Theorem 4 by first computing $(-a)\mathbf{v} + a\mathbf{v}$. Then do it using (4) of Theorem 4 and axiom S4.

(d) Prove that $a(-\mathbf{v}) = -(a\mathbf{v})$ in Theorem 4 in two ways, as in part **(c)**.

17. Let $\mathbf{v}, \mathbf{v}_1, \dots, \mathbf{v}_n$ denote vectors in a vector space V, and let $a, a_1, \dots, a_n$ denote numbers. Use induction on n to prove each of the following.

(a) $a(\mathbf{v}_1 + \mathbf{v}_2 + \cdots + \mathbf{v}_n) = a\mathbf{v}_1 + a\mathbf{v}_2 + \cdots + a\mathbf{v}_n$

(b) $(a_1 + a_2 + \cdots + a_n)\mathbf{v} = a_1\mathbf{v} + a_2\mathbf{v} + \cdots + a_n\mathbf{v}$

18. Verify axioms A2–A5 and S2 – S5 for the space $\mathbf{F}[a, b]$ of functions on $[a, b]$ (Example 8).

19. Prove each of the following for vectors $\mathbf{u}$ and $\mathbf{v}$ and scalars a and b.

(a) If $a\mathbf{v} = b\mathbf{v}$ and $\mathbf{v} \neq \mathbf{0}$, then $a = b$.

(b) If $a\mathbf{v} = a\mathbf{w}$ and $a \neq 0$, then $\mathbf{v} = \mathbf{w}$.

20. By calculating $(1 + 1)(\mathbf{v} + \mathbf{w})$ in two ways (using axioms S2 and S3), show that axiom A2 follows from the other axioms.

21. Let V be a vector space, and define V^n to be the set of all n-tuples $(\mathbf{v}_1, \mathbf{v}_2, \dots, \mathbf{v}_n)$ of n vectors $\mathbf{v}_i$, each belonging to V. Define addition and scalar multiplication on V^n as follows:

$$(\mathbf{u}_1, \mathbf{u}_2, \dots, \mathbf{u}_n) + (\mathbf{v}_1, \mathbf{v}_2, \dots, \mathbf{v}_n) = (\mathbf{u}_1 + \mathbf{v}_1, \mathbf{u}_2 + \mathbf{v}_2, \dots, \mathbf{u}_n + \mathbf{v}_n)$$
$$k(\mathbf{v}_1, \mathbf{v}_2, \dots, \mathbf{v}_n) = (k\mathbf{v}_1, k\mathbf{v}_2, \dots, k\mathbf{v}_n)$$

Show that V^n is a vector space.

22. Let V^n be the vector space of n-tuples from the preceding exercise, written as columns. If A is an $m \times n$ matrix, and X is in V^n, define AX in V^m by matrix multiplication. More precisely, if $A = [a_{ij}]$ and

$$X = \begin{bmatrix} \mathbf{v}_1 \\ \vdots \\ \mathbf{v}_n \end{bmatrix}, \text{ let } AX = \begin{bmatrix} \mathbf{u}_1 \\ \vdots \\ \mathbf{u}_n \end{bmatrix}, \text{ where } \mathbf{u}_2 = a_{i1}\mathbf{v}_1 + a_{i2}\mathbf{v}_2 + \cdots + a_{in}\mathbf{v}_n \text{ for each } i$$

Prove that:

(a) $B(AX) = (BA)X$

(b) $(A + A_1)X = AX + A_1X$

(c) $A(X + X_1) = AX + AX_1$

(d) $(kA)X = k(AX) = A(kX)$ if k is any number

(e) $IX = X$ if I is the $n \times n$ identity matrix

(f) Let E be an elementary matrix obtained by performing a row operation on (the rows of) I_n (see Section 2.3.2). Show that EX is the column resulting from performing that same row operation on the vectors (call them rows) of X. [*Hint:* Theorem 5 in Section 2.3.]

SECTION 5.2 Subspaces and Spanning Sets

Very often the most interesting vector spaces arise as parts of larger vector spaces. For example, consider the subset U of $\mathbb{R}^2$ defined as follows:

$$U = \{(x, -x) \mid x \text{ in } \mathbb{R}\}$$

If the vector addition and scalar multiplication of $\mathbb{R}^2$ are used, then the equations

$$(x, -x) + (y, -y) = ((x + y), -(x + y))$$
$$a(x, -x) = (ax, -(ax))$$

show that U is closed under addition (axiom A1) and scalar multiplication (axiom S1). The other eight axioms can also be verified directly, but because U is using the vector addition and scalar multiplication of a known vector space (in this case $\mathbb{R}^2$), these other axioms are automatically satisfied. This is a great relief (these axioms can be tedious to verify).

DEFINITION If V is a vector space, a subset U of V is called a **subspace** of V if U is itself a vector space using the vector addition and scalar multiplication of V.

If U is a subspace of V, it is clear (by axioms A1 and S1) that the sum of two vectors in U is again in U and that any scalar multiple of a vector in U is again in U—in short, that U is **closed** under the vector addition and scalar multiplication of V. The nice part is that the converse is also true: If U is closed under these operations, then all the other axioms are automatically satisfied. For example, axiom A2 asserts that $\mathbf{u} + \mathbf{u}_1 = \mathbf{u}_1 + \mathbf{u}$ holds for all vectors $\mathbf{u}$ and $\mathbf{u}_1$ in U. But this is clear because the equation is already true in V, and U uses the same addition as V. Similarly, axioms A3, S2, S3, S4, and S5 hold automatically in U because they are true in V. All that remains is to verify axioms A4 and A5.

THEOREM 1
Subspace Test

Let U be a subset of a vector space V. Then U is a subspace of V if and only if it satisfies the following three conditions.

(1) $\mathbf{0}$ lies in U where $\mathbf{0}$ is the zero vector of V.

(2) If $\mathbf{u}_1$ and $\mathbf{u}_2$ lie in U, then $\mathbf{u}_1 + \mathbf{u}_2$ lies in U.

(3) If $\mathbf{u}$ lies in U, then $a\mathbf{u}$ lies in U for all a in $\mathbb{R}$.

Proof If (1), (2) and (3) hold, then axiom A4 follows from (1), and axiom A5 follows from (3) (because $-\mathbf{u} = (-1)\mathbf{u}$ lies in U for all $\mathbf{u}$ in U). Hence by the discussion preceding the theorem, U is a subspace. Conversely, if U is a subspace, it is closed under addition and scalar multiplication, and this gives (2) and (3). If $\mathbf{z}$ denotes the zero vector of U, then $\mathbf{z} = 0\mathbf{z}$ by Theorem 4 in Section 5.1. But $0\mathbf{z} = \mathbf{0}$ in V by the same theorem, so $\mathbf{0} = \mathbf{z}$ lies in U. This proves (1). ■

If U is a subspace of V, the proof shows that U and V share the same zero vector. Also, if $\mathbf{u}$ lies in U, then $-\mathbf{u}$ lies in U; that is, the negative of a vector in U is the same as its negative in V.

This test provides an easy way of finding subspaces. Here are some examples.

EXAMPLE 1 If V is any vector space, then $\{\mathbf{0}\}$ and V are subspaces of V.

Solution $U = V$ clearly satisfies the conditions of the test. As to $U = \{\mathbf{0}\}$, it satisfies the conditions because $\mathbf{0} + \mathbf{0} = \mathbf{0}$ and $a\mathbf{0} = \mathbf{0}$ for all a in $\mathbb{R}$.

The vector space $\{\mathbf{0}\}$ is called the **zero subspace** of V. Because all zero subspaces look alike, we speak of the **zero vector space** and denote it by 0. It is the unique vector space containing just one vector.

EXAMPLE 2 Let $\mathbf{v}$ be a vector in a vector space V. Then the set

$$\mathbb{R}\mathbf{v} = \{a\mathbf{v} \mid a \text{ in } \mathbb{R}\}$$

of all scalar multiples of $\mathbf{v}$ is a subspace of V.

Solution Because $\mathbf{0} = 0\mathbf{v}$, it is clear that $\mathbf{0}$ lies in $\mathbb{R}\mathbf{v}$. Given two vectors $a\mathbf{v}$ and $a_1\mathbf{v}$ in $\mathbb{R}\mathbf{v}$, their sum $a\mathbf{v} + a_1\mathbf{v} = (a + a_1)\mathbf{v}$ is also a scalar multiple of $\mathbf{v}$ and so lies in $\mathbb{R}\mathbf{v}$. Hence $\mathbb{R}\mathbf{v}$ is closed under addition. Finally, given $a\mathbf{v}$, $r(a\mathbf{v}) = (ra)\mathbf{v}$ lies in $\mathbb{R}\mathbf{v}$, so $\mathbb{R}\mathbf{v}$ is closed under scalar multiplication.

EXAMPLE 3 Let A be an $m \times n$ matrix. Consider the set

$$U = \{A\mathbf{v} \mid \mathbf{v} \text{ lies in } \mathbb{R}^n, \mathbf{v} \text{ written as a column}\}$$

Then U is a subspace of $\mathbb{R}^m$ called the **image** or **range** of the matrix A.

Solution Note first that U is in fact a subset of $\mathbb{R}^m$ because A is $m \times n$. Each vector in U is of the form $A\mathbf{v}$ for some vector $\mathbf{v}$ in $\mathbb{R}^n$. To apply the subspace test, note that $\mathbf{0} = A\mathbf{0}$ has the required form, so $\mathbf{0}$ lies in U. Similarly, the equations $A\mathbf{v} + A\mathbf{v}_1 = A(\mathbf{v} + \mathbf{v}_1)$ and $r(A\mathbf{v}) = A(r\mathbf{v})$ show that sums and scalar multiples of vectors in U again have the required form. Hence U is a subspace of $\mathbb{R}^m$.

The next example gives another important subspace related to a matrix A. However, rather than specify the *form* of each vector in the subspace (as in Example 3), we describe it by specifying a *condition* that vectors must satisfy to be in the subspace.

EXAMPLE 4 Let A be an $m \times n$ matrix. The set

$$U = \{\mathbf{v} \text{ in } \mathbb{R}^n \mid A\mathbf{v} = \mathbf{0}\}$$

is a subspace of $\mathbb{R}^n$ called the **kernel** or **null space** of the matrix A. Note that U is the set of solutions to the homogeneous system of equations with A as coefficient matrix.

Solution Here U consists of all vectors $\mathbf{v}$ in $\mathbb{R}^n$ satisfying the condition that $A\mathbf{v} = \mathbf{0}$. Because $A\mathbf{0} = \mathbf{0}$, it is clear that $\mathbf{0}$ lies in U. If $\mathbf{v}$ and $\mathbf{v}_1$ both lie in U, then $A(\mathbf{v} + \mathbf{v}_1) = A\mathbf{v} + A\mathbf{v}_1 = \mathbf{0} + \mathbf{0} = \mathbf{0}$. This shows that $\mathbf{v} + \mathbf{v}_1$ qualifies for membership in U, so U is closed under addition. Similarly, $A(r\mathbf{v}) = r(A\mathbf{v}) = r\mathbf{0} = \mathbf{0}$, so $r\mathbf{v}$ lies in U. This means that U is closed under scalar multiplication and so is a subspace.

The next example describes a subset U of the space $\mathbf{M}_{22}$ by giving a *condition* that a matrix belongs to U, and then differently by giving the *form* of each matrix in U. Both characterizations of U are used to show that it is a subspace.

EXAMPLE 5 Let $A = \begin{bmatrix} 1 & 1 \\ 0 & 0 \end{bmatrix}$ be a fixed matrix in $\mathbf{M}_{22}$, and let

$$U = \{X \text{ in } \mathbf{M}_{22} \mid AX = XA\}$$

Then U is a subspace of $\mathbf{M}_{22}$.

Solution 1 If 0 is the 2×2 zero matrix, then $A0 = 0A$, so 0 satisfies the condition for membership in U. Next suppose that X and X_1 lie in U so that $AX = XA$ and $AX_1 = X_1A$. Then

$$A(X + X_1) = AX + AX_1 = XA + X_1A = (X + X_1)A$$
$$A(aX) = a(AX) = a(XA) = (aX)A \qquad \text{for all } a \text{ in } \mathbb{R}$$

so both $X + X_1$ and aX lie in U. Hence U is a subspace.

Solution 2 If X lies in U, write $X = \begin{bmatrix} x & y \\ z & w \end{bmatrix}$. Then the condition $AX = XA$ is

$$\begin{bmatrix} 1 & 1 \\ 0 & 0 \end{bmatrix}\begin{bmatrix} x & y \\ z & w \end{bmatrix} = \begin{bmatrix} x & y \\ z & w \end{bmatrix}\begin{bmatrix} 1 & 1 \\ 0 & 0 \end{bmatrix}$$

This means that $x + z = x$, $y + w = x$, and $z = 0$. Thus $w = x - y$, so X has the form $X = \begin{bmatrix} x & y \\ 0 & x - y \end{bmatrix}$ where x and y are arbitrary real numbers. Hence

$$U = \left\{ \begin{bmatrix} x & y \\ 0 & x - y \end{bmatrix} \;\middle|\; x \text{ and } y \text{ in } \mathbb{R} \right\}$$

specifies U by giving the *form* of all matrices in U. Now 0 clearly lies in U (when x and y are zero), and it is easily verified that sums and scalar multiples of matrices of this form are again of the same form and so lie in U. This shows again that U is a subspace of $\mathbf{M}_{22}$.

The two solutions here are quite different. The first has the advantage that it is brief and works in the same way for *any* square matrix A. The second is more tedious and would be different if another matrix A were used. However, the explicit form of the vectors (in this case, matrices) in a subspace is often needed in other contexts.

Suppose $p(x)$ is a polynomial and a is a number. Then the number $p(a)$ obtained by replacing x by a in the expression for $p(x)$ is called the **evaluation** of $p(x)$ at a. For example, if $p(x) = 5 - 6x + 2x^2$, then the evaluation of $p(x)$ at $a = 2$ is $p(2) = 1$. If $p(a) = 0$, the number a is called a **root** of $p(x)$.

EXAMPLE 6 Consider the set U of all polynomials in $\mathbf{P}$ that have 3 as a root:

$$U = \{p(x) \text{ in } \mathbf{P} \mid p(3) = 0\}$$

Then U is a subspace of $\mathbf{P}$.

Solution 1 Clearly the zero polynomial lies in U. Now let $p(x)$ and $q(x)$ lie in U so $p(3) = 0$ and $q(3) = 0$. If their sum is denoted $(p + q)(x)$, then it is easily verified that $(p + q)(a) = p(a) + q(a)$ holds for all numbers a. In particular, $(p + q)(3) = p(3) + q(3) = 0 + 0 = 0$, so U is closed under addition. The verification that U is closed under scalar multiplication is similar.

Solution 2 The form of all the polynomials in U follows from the factor theorem:†

$$U = \{(x - 3)q(x) \mid q(x) \text{ in } \mathbf{P}\}$$

†The factor theorem is given in Section 5.6.1.

The verification of this, and of the fact that this shows U to be a subspace of **P**, is left as Exercise 21.

The next example refers to the material in Chapter 4.

EXAMPLE 7 Regard $\mathbb{R}^3$ as the set of points in space. Then every plane through the origin is a subspace.

Solution By Theorem 7 of Section 4.2, every plane P through the origin has equation $ax + by + cz = 0$ for some numbers a, b, and c, not all zero. In other words, P is the following subset of $\mathbb{R}^3$.

$$P = \{(x,y,z) \text{ in } \mathbb{R}^3 \mid ax + by + cz = 0\}$$

It is clear from this that $\mathbf{0} = (0,0,0)$ lies in P; the verification that P is closed under the addition and scalar multiplication of $\mathbb{R}^3$ is left as Exercise 22.

The next example involves the notion of the derivative f' of a function f. (If the reader is not familiar with calculus, this example may be omitted.) A function f defined on the interval $[a, b]$ is called **differentiable** if the derivative function f' exists.

EXAMPLE 8 The subset $\mathbf{D}[a, b]$ of all **differentiable functions** on $[a, b]$ is a subspace of the vector space $\mathbf{F}[a, b]$ of all functions on $[a, b]$.

Solution The derivative of any constant function is the constant function 0; in particular, 0 itself is differentiable and so lies in $\mathbf{D}[a, b]$. If f and g both lie in $\mathbf{D}[a, b]$ (so that f' and g' exist), then it is a theorem of calculus that $f + g$ and af are both differentiable [in fact, $(f + g)' = f' + g'$ and $(af)' = af'$], so both lie in $\mathbf{D}[a, b]$. This shows that $\mathbf{D}[a, b]$ is a subspace of $\mathbf{F}[a, b]$.

EXAMPLE 9 Consider the two subsets P and Q of $\mathbb{R}^2$ defined by

$$P = \{(a,b) \text{ in } \mathbb{R}^2 \mid a \geq 0\} \qquad Q = \{(a,b) \text{ in } \mathbb{R}^2 \mid a^2 = b^2\}$$

Then P and Q both contain the zero vector, but they are not subspaces. In fact, P is closed under addition but not scalar multiplication, whereas Q is closed under scalar multiplication but not addition.

Solution It is clear that the zero vector (0,0) of $\mathbb{R}^2$ lies in both P and Q. If (a,b) and (a_1,b_1) both lie in P, then $a \geq 0$ and $a_1 \geq 0$. Hence $a + a_1 \geq 0$, so $(a,b) + (a_1,b_1) = (a + a_1, b + b_1)$ lies in P—that is, P is closed under addition. However, (1,0) lies in P but $(-1)(1,0) = (-1,0)$ is not in P, so P is not closed under scalar multiplication. On the other hand, if (a,b) lies in Q, then $a^2 = b^2$, so $k(a,b) = (ka,kb)$ also lies in Q [because

$(ka)^2 = k^2a^2 = k^2b^2 = (kb)^2$]. Thus Q is closed under scalar multiplication. However, Q is not closed under addition. For example, $(2,-2)$ and $(1,1)$ both lie in Q, but their sum $(2,-2) + (1,1) = (3,-1)$ does not lie in Q.

Theorem 2 in Section 1.3 can be described in our present terminology as follows: The set of solutions to a system of m homogeneous linear equations in n variables is closed under addition and scalar multiplication and so is a subspace of $\mathbb{R}^n$. Moreover, this theorem led to a convenient way of describing this subspace. Recall that in Example 4 in Section 1.3, the system of homogeneous linear equations

$$\begin{aligned} x_1 - 2x_2 + x_3 + x_4 &= 0 \\ -x_1 + 2x_2 + x_4 &= 0 \\ 2x_1 - 4x_2 + x_3 &= 0 \end{aligned}$$

was solved using elementary row operations, and the solution was found to be

$$\begin{bmatrix} x_1 \\ x_2 \\ x_3 \\ x_4 \end{bmatrix} = \begin{bmatrix} 2s + t \\ s \\ -2t \\ t \end{bmatrix} = s\begin{bmatrix} 2 \\ 1 \\ 0 \\ 0 \end{bmatrix} + t\begin{bmatrix} 1 \\ 0 \\ -2 \\ 1 \end{bmatrix}$$

where s and t are arbitrary parameters. Hence every solution can be found as a sum of scalar multiples of the two basic ones,

$$\begin{bmatrix} 2 \\ 1 \\ 0 \\ 0 \end{bmatrix} \text{ and } \begin{bmatrix} 1 \\ 0 \\ -2 \\ 1 \end{bmatrix}$$

Such descriptions often turn out to be the most convenient way to describe a subspace.

DEFINITION A vector $\mathbf{v}$ is called a **linear combination** of the vectors $\mathbf{v}_1, \mathbf{v}_2, \ldots, \mathbf{v}_n$ if it can be expressed in the form

$$\mathbf{v} = a_1\mathbf{v}_1 + a_2\mathbf{v}_2 + \cdots + a_n\mathbf{v}_n$$

where $a_1, a_2, \ldots, a_n$ are scalars called the **coefficients** of $\mathbf{v}_1, \mathbf{v}_2, \ldots, \mathbf{v}_n$.

EXAMPLE 10 Determine whether $(1,1,4)$ or $(1,5,1)$ is a linear combination of the vectors $\mathbf{v}_1 = (1,2,-1)$ and $\mathbf{v}_2 = (3,5,2)$ in $\mathbb{R}^3$.

Solution The question whether (1,1,4) is a linear combination of $\mathbf{v}_1$ and $\mathbf{v}_2$ comes down to whether numbers s and t can be found such that $(1,1,4) = s(1,2,-1) + t(3,5,2)$. Equating components gives three equations: $1 = s + 3t$, $1 = 2s + 5t$, and $4 = -s + 2t$. It is not difficult to verify that $s = -2$, $t = 1$ satisfy these equations, so (1,1,4) is indeed a linear combination of $\mathbf{v}_1$ and $\mathbf{v}_2$.

Turning to (1,5,1), the question is whether s and t can be found such that $(1,5,1) = s(1,2,-1) + t(3,5,2)$. This means that $1 = s + 3t$, $5 = 2s + 5t$, and $1 = -s + 2t$. The first and last of these equations have the unique solution $t = \frac{2}{5}$, $s = -\frac{1}{5}$, and these values do *not* satisfy the second equation. Hence no such s and t can be found, and (1,5,1) is *not* a linear combination of $\mathbf{v}_1$ and $\mathbf{v}_2$.

In the system of equations considered above, the solutions turned out to be just the set of *all* linear combinations of two particular solutions. More generally:

DEFINITION If $\{\mathbf{v}_1, \mathbf{v}_2, \ldots, \mathbf{v}_n\}$ is any set of vectors in a vector space V, the set of all linear combinations of these vectors is called their **span** and is denoted by

$$\text{span}\{\mathbf{v}_1, \mathbf{v}_2, \ldots, \mathbf{v}_n\}$$

If it happens that $V = \text{span}\{\mathbf{v}_1, \mathbf{v}_2, \ldots, \mathbf{v}_n\}$, then these vectors are called a **spanning set** for V.

For example, the span of two vectors $\mathbf{v}$ and $\mathbf{w}$ is the set

$$\text{span}\{\mathbf{v}, \mathbf{w}\} = \{s\mathbf{v} + t\mathbf{w} \mid s \text{ and } t \text{ in } \mathbb{R}\}$$

of all sums of scalar multiples of the vectors.

EXAMPLE 11 Describe span {(1,0,1), (0,1,1,)}.

Solution We have $s(1,0,1) + t(0,1,1) = (s, t, s + t)$ for all s and t in $\mathbb{R}$, so

$$\text{span}\,\{(1,0,1), (0,1,1)\} = \{(s, t, s + t) \mid s \text{ and } t \text{ in } \mathbb{R}\}$$

In the case of a single vector $\mathbf{v}$, the span is

$$\text{span}\{\mathbf{v}\} = \{s\mathbf{v} \mid s \text{ in } \mathbb{R}\} = \mathbb{R}\mathbf{v}$$

The notation $\mathbb{R}\mathbf{v}$ was introduced in Example 2 where it was verified that $\mathbb{R}\mathbf{v}$ is a subspace. It turns out that the span of any set of vectors is a subspace.

THEOREM 2 Let $U = \text{span}\{\mathbf{v}_1, \mathbf{v}_2, \ldots, \mathbf{v}_n\}$ in a vector space V. Then

(1) U is a subspace of V containing each of $\mathbf{v}_1, \mathbf{v}_2, \ldots, \mathbf{v}_n$.

(2) U is the "smallest" subspace containing these vectors in the sense that any subspace of V that contains each of $\mathbf{v}_1, \mathbf{v}_2, \ldots, \mathbf{v}_n$ must contain U.

Proof (1) Clearly $\mathbf{0} = 0\mathbf{v}_1 + \cdots + 0\mathbf{v}_n$ belongs to U. If $\mathbf{v} = a_1\mathbf{v}_1 + \cdots + a_n\mathbf{v}_n$ and $\mathbf{w} = b_1\mathbf{v}_1 + \cdots + b_n\mathbf{v}_n$ are two members of U, then

$$\mathbf{v} + \mathbf{w} = (a_1 + b_1)\mathbf{v}_1 + \cdots + (a_n + b_n)\mathbf{v}_n$$
$$a\mathbf{v} = (aa_1)\mathbf{v}_1 + \cdots + (aa_n)\mathbf{v}_n \qquad (\text{for } a \text{ in } \mathbb{R})$$

so both $\mathbf{v} + \mathbf{w}$ and $a\mathbf{v}$ lie in U. Hence U is a subspace of V. It contains each of $\mathbf{v}_1, \mathbf{v}_2, \ldots, \mathbf{v}_n$; for example, $\mathbf{v}_2 = 0\mathbf{v}_1 + 1\mathbf{v}_2 + 0\mathbf{v}_3 + \cdots + 0\mathbf{v}_n$. This proves (1).

(2) Let W be a subspace of V that contains each of $\mathbf{v}_1, \mathbf{v}_2, \ldots, \mathbf{v}_n$. Because W is closed under scalar multiplication, each of $a_1\mathbf{v}_1, a_2\mathbf{v}_2, \ldots, a_n\mathbf{v}_n$ lies in W for any choice of $a_1, a_2, \ldots$ in $\mathbb{R}$. But then $a_1\mathbf{v}_1 + a_2\mathbf{v}_2 + \cdots + a_n\mathbf{v}_n$ lies in W because W is closed under addition. This means that W contains every member of U, which proves (2). ■

EXAMPLE 12 Show that $\mathbb{R}^n = \text{span}\{\mathbf{e}_1, \mathbf{e}_2, \ldots, \mathbf{e}_n\}$, where

$$\begin{aligned} \mathbf{e}_1 &= (1, 0, 0, \ldots, 0) \\ \mathbf{e}_2 &= (0, 1, 0, \ldots, 0) \\ &\;\vdots \\ \mathbf{e}_n &= (0, 0, 0, \ldots, 1) \end{aligned}$$

Solution Because $\text{span}\{\mathbf{e}_1, \mathbf{e}_2, \ldots, \mathbf{e}_n\}$ is a subspace of $\mathbb{R}^n$, all we need to show is that every vector in $\mathbb{R}^n$ lies in $\text{span}\{\mathbf{e}_1, \mathbf{e}_2, \ldots, \mathbf{e}_n\}$. But if $\mathbf{v} = (a_1, a_2, \ldots, a_n)$ is any vector in $\mathbb{R}^n$, then

$$\mathbf{v} = (a_1, a_2, \ldots, a_n) = a_1\mathbf{e}_1 + a_2\mathbf{e}_2 + \cdots + a_n\mathbf{e}_n$$

Hence $\mathbf{v}$ lies in $\text{span}\{\mathbf{e}_1, \mathbf{e}_2, \ldots, \mathbf{e}_n\}$.

EXAMPLE 13 Show that $\mathbf{P}_n = \text{span}\{1, x, x^2, \ldots, x^n\}$.

Solution Because $\text{span}\{1, x, x^2, \ldots, x^n\}$ is a subspace of $\mathbf{P}_n$, it is required only to show that each polynomial $p(x)$ in $\mathbf{P}_n$ is a linear combination of $1, x, \ldots, x^n$. But this is clear because $p(x)$ has the form $p(x) = a_0 + a_1x + a_2x^2 + \cdots + a_nx^n$.

EXAMPLE 14 Show that $\mathbb{R}^3 = \text{span}\{(1,1,1), (1,1,0), (0,1,1)\}$.

Solution Write $\mathbf{v}_1 = (1,1,1)$, $\mathbf{v}_2 = (1,1,0)$, and $\mathbf{v}_3 = (0,1,1)$. Clearly $\text{span}\{\mathbf{v}_1, \mathbf{v}_2, \mathbf{v}_3\}$ is contained in $\mathbb{R}^3$. We have $\mathbb{R}^3 = \text{span}\{(1,0,0), (0,1,0), (0,0,1)\}$, so in order to prove that $\mathbb{R}^3$ is contained in $\text{span}\{\mathbf{v}_1, \mathbf{v}_2, \mathbf{v}_3\}$, it is enough to show that each of $(1,0,0)$, $(0,1,0)$, and $(0,0,1)$ lies in $\text{span}\{\mathbf{v}_1, \mathbf{v}_2, \mathbf{v}_3\}$ and to use Theorem 2. But they can be given explicitly as linear combinations of $\mathbf{v}_1$, $\mathbf{v}_2$, and $\mathbf{v}_3$:

$$(1,0,0) = (1,1,1) - (0,1,1) = \mathbf{v}_1 - \mathbf{v}_3$$
$$(0,0,1) = (1,1,1) - (1,1,0) = \mathbf{v}_1 - \mathbf{v}_2$$

and then, using the first of these,

$$(0,1,0) = (1,1,0) - (1,0,0) = \mathbf{v}_2 - (\mathbf{v}_1 - \mathbf{v}_3) = \mathbf{v}_2 - \mathbf{v}_1 + \mathbf{v}_3$$

EXAMPLE 15 Show that $\mathbf{P}_3 = \text{span}\{x^2 + x^3, x, 2x^2 + 1, 3\}$.

Solution Write $U = \text{span}\{x^2 + x^3, x, 2x^2 + 1, 3\}$. Then U is contained in $\mathbf{P}_3$, and we use the fact that $\mathbf{P}_3 = \text{span}\{1, x, x^2, x^3\}$ to show that $\mathbf{P}_3$ is contained in U. In fact, x and $1 = \frac{1}{3}3$ clearly lie in U. But then successively

$$x^2 = \frac{1}{2}[(2x^2 + 1) - 1]$$
$$x^3 = (x^2 + x^3) - x^2$$

also lie in U. Hence $\mathbf{P}_3$ is contained in U by Theorem 2.

The following notation is useful: If X and Y are two sets, then $X \subseteq Y$ means that X is **contained** in Y; that is, every member of X is a member of Y. Clearly $X = Y$ means that $X \subseteq Y$ and $Y \subseteq X$ are both true.

EXAMPLE 16 Let $\mathbf{u}$ and $\mathbf{v}$ be two vectors in a vector space V. Show that

$$\text{span}\{\mathbf{u},\mathbf{v}\} = \text{span}\{\mathbf{u} + \mathbf{v}, \mathbf{u} - \mathbf{v}\}$$

Solution We have $\text{span}\{\mathbf{u} + \mathbf{v}, \mathbf{u} - \mathbf{v}\} \subseteq \text{span}\{\mathbf{u}, \mathbf{v}\}$ because both $\mathbf{u} + \mathbf{v}$ and $\mathbf{u} - \mathbf{v}$ lie in $\text{span}\{\mathbf{u}, \mathbf{v}\}$. On the other hand,

$$\mathbf{u} = \frac{1}{2}(\mathbf{u} + \mathbf{v}) + \frac{1}{2}(\mathbf{u} - \mathbf{v})$$
$$\mathbf{v} = \frac{1}{2}(\mathbf{u} + \mathbf{v}) - \frac{1}{2}(\mathbf{u} - \mathbf{v})$$

so $\text{span}\{\mathbf{u}, \mathbf{v}\} \subseteq \text{span}\{\mathbf{u} + \mathbf{v}, \mathbf{u} - \mathbf{v}\}$ by Theorem 2.

EXERCISES 5.2

1. Which of the following are subspaces of $\mathbb{R}^3$? Support your answer.
   (a) $U = \{(a,b,1) \mid a \text{ and } b \text{ in } \mathbb{R}\}$
   (b) $U = \{(a,b,c) \mid a + 2b - c = 0; a, b, \text{ and } c \text{ in } \mathbb{R}\}$
   (c) $U = \{(0,0,c) \mid c \text{ in } \mathbb{R}\}$
   (d) $U = \{(a,b,0) \mid a^2 = b^2, a \text{ and } b \text{ in } \mathbb{R}\}$
   (e) $U = \{(a,a-1,c) \mid a \text{ and } c \text{ in } \mathbb{R}\}$
   (f) $U = \{(a,b,c) \mid a^2 + b^2 = c^2; a, b, \text{ and } c \text{ in } \mathbb{R}\}$
   (g) $U = \{(a,a,c) \mid a \text{ and } c \text{ in } \mathbb{R}\}$
2. Which of the following are subspaces of $\mathbf{P}_3$? Support your answer.
   (a) $\{f(x) \mid f(2) = 1\}$
   (b) $\{xg(x) \mid g(x) \text{ in } \mathbf{P}_2\}$
   (c) $\{xg(x) \mid g(x) \text{ in } \mathbf{P}_3\}$
   (d) $\{xg(x) + (1-x)h(x) \mid g(x) \text{ and } h(x) \text{ in } \mathbf{P}_2\}$
   (e) The set of all polynomials in $\mathbf{P}_3$ with constant term 0
   (f) $\{f(x) \mid \deg f(x) = 3\}$
3. Which of the following are subspaces of $\mathbf{M}_{22}$? Support your answer.
   (a) $\left\{\begin{bmatrix} a & b \\ 0 & c \end{bmatrix} \middle| \; a, b, \text{ and } c, \text{ in } \mathbb{R}\right\}$
   (b) $\left\{\begin{bmatrix} a & b \\ c & d \end{bmatrix} \middle| \; a + b = c + d; a, b, c, \text{ and } d \text{ in } \mathbb{R}\right\}$
   (c) $\{A \mid A \text{ in } \mathbf{M}_{22}, A = A^T\}$
   (d) $\{A \mid A \text{ in } \mathbf{M}_{22}, AB = 0\}$, B a fixed 2×2 matrix
   (e) $\{A \mid A \text{ in } \mathbf{M}_{22}, A^2 = A\}$
   (f) $\{A \mid A \text{ in } \mathbf{M}_{22}, A \text{ is not invertible}\}$
   (g) $\{A \mid A \text{ in } \mathbf{M}_{22}, BAC = CAB\}$, B and C fixed 2×2 matrices
4. Which of the following are subspaces of $\mathbf{F}[0,1]$? Support your answer.
   (a) $U = \{f \mid f(0) = 0\}$
   (b) $U = \{f \mid f(0) = 1\}$
   (c) $U = \{f \mid f(0) = f(1)\}$
   (d) $U = \{f \mid f(x) \geq 0 \text{ for all } x \text{ in } [0,1]\}$
   (e) $U = \{f \mid f(x) = f(y) \text{ for all } x \text{ and } y \text{ in } [0,1]\}$
   (f)† $U = \{f \mid \text{the derivative } f' \text{ exists in } [0,1]\}$
   (g)† $U = \left\{f \mid \int_0^1 f(x)dx = 0\right\}$
5. Let A be an $m \times n$ matrix. For which columns $\mathbf{b}$ in $\mathbb{R}^m$ is $U = \{\mathbf{x} \mid \mathbf{x} \text{ in } \mathbb{R}^n, A\mathbf{x} = \mathbf{b}\}$ a subspace of $\mathbb{R}^n$?

---

†This exercise requires calculus.

6. Let $\mathbf{v}$ be a vector in $\mathbb{R}^n$ (written as a column), and define $U = \{A\mathbf{v} \mid A \text{ in } \mathbf{M}_{mn}\}$.

(a) Show that U is a subspace of $\mathbb{R}^m$.

(b) Is U all of $\mathbb{R}^m$?

(c) Show that $U = \mathbb{R}^m$ if $\mathbf{v} \neq \mathbf{0}$.

7. Write each of the following as a linear combination of $x + 1$, $x^2 + x$, and $x^2 + 2$.

(a) $x^2 + 3x + 2$ **(b)** $2x^2 - 3x + 1$ **(c)** $x^2 + 1$ **(d)** x

8. Write each of the following as a linear combination of $(1,-1,1)$, $(1,0,1)$, and $(1,1,0)$.

(a) $(2,1,-1)$ **(b)** $(1,-7,5)$ **(c)** $\left(\frac{1}{2},2,\frac{1}{3}\right)$ **(d)** $(0,0,0)$

9. Determine whether $\mathbf{v}$ lies in span$\{\mathbf{u}, \mathbf{w}\}$ in each case.

(a) $\mathbf{v} = (1,-1,2)$; $\mathbf{u} = (1,1,1)$; $\mathbf{w} = (0,1,3)$

(b) $\mathbf{v} = (3,1,-3)$; $\mathbf{u} = (1,1,1)$; $\mathbf{w} = (0,1,3)$

(c) $\mathbf{v} = 3x^2 - 2x - 1$; $\mathbf{u} = x^2 + 1$; $\mathbf{w} = x + 2$

(d) $\mathbf{v} = x$; $\mathbf{u} = x^2 + 1$; $\mathbf{w} = x + 2$

(e) $\mathbf{v} = \begin{bmatrix} 1 & 3 \\ -1 & 1 \end{bmatrix}$; $\mathbf{u} = \begin{bmatrix} 1 & -1 \\ 2 & 1 \end{bmatrix}$; $\mathbf{w} = \begin{bmatrix} 2 & 1 \\ 1 & 0 \end{bmatrix}$

(f) $\mathbf{v} = \begin{bmatrix} 1 & -4 \\ 5 & 3 \end{bmatrix}$; $\mathbf{u} = \begin{bmatrix} 1 & -1 \\ 2 & 1 \end{bmatrix}$; $\mathbf{w} = \begin{bmatrix} 2 & 1 \\ 1 & 0 \end{bmatrix}$

10. Which of the following functions lie in span$\{\cos^2 x, \sin^2 x\}$? (Work in $\mathbf{F}[0, \pi]$.)

(a) $\cos 2x$ **(b)** 1 **(c)** x^2 **(d)** $1 + x^2$

11. (a) Show that $\mathbb{R}^3$ is spanned by $\{(1,0,1), (1,1,0), (0,1,1)\}$.

(b) Show that $\mathbf{P}_2$ is spanned by $\{1 + 2x^2, 3x, 1 + x\}$.

(c) Show that $\mathbf{M}_{22}$ is spanned by $\left\{\begin{bmatrix} 1 & 0 \\ 0 & 0 \end{bmatrix}, \begin{bmatrix} 1 & 0 \\ 0 & 1 \end{bmatrix}, \begin{bmatrix} 0 & 1 \\ 1 & 0 \end{bmatrix}, \begin{bmatrix} 1 & 1 \\ 0 & 1 \end{bmatrix}\right\}$.

12. If X and Y are two sets of vectors in a vector space V, and if $X \subseteq Y$, show that span $X \subseteq$ span Y.

13. Let $\mathbf{u}$, $\mathbf{v}$, and $\mathbf{w}$ denote vectors in a vector space V.

(a) Show that span$\{\mathbf{u}, \mathbf{v}, \mathbf{w}\}$ = span$\{\mathbf{u} + \mathbf{v}, \mathbf{u} + \mathbf{w}, \mathbf{v} + \mathbf{w}\}$

(b) Show that span$\{\mathbf{u}, \mathbf{v}, \mathbf{w}\}$ = span$\{\mathbf{u} - \mathbf{v}, \mathbf{u} + \mathbf{w}, \mathbf{w}\}$

14. Show that span$\{\mathbf{v}_1, \mathbf{v}_2, \ldots, \mathbf{v}_n, \mathbf{0}\}$ = span$\{\mathbf{v}_1, \mathbf{v}_2, \ldots, \mathbf{v}_n\}$ holds for any set of vectors $\{\mathbf{v}_1, \mathbf{v}_2, \ldots, \mathbf{v}_n\}$.

15. If X and Y are nonempty subsets of a vector space V such that span X = span $Y = V$, must there be a vector common to both X and Y? Justify your answer.

16. Is it possible that $\{(1,2,0), (1,1,1)\}$ can span the subspace $U = \{(a,b,0) \mid a \text{ and } b \text{ in } \mathbb{R}\}$?

17. Describe span$\{\mathbf{0}\}$.

18. Let $\mathbf{v}$ denote any vector in a vector space V. Show that span$\{\mathbf{v}\}$ = span$\{a\mathbf{v}\}$ for any $a \neq 0$.

19. If $\mathbb{R}^3$ is regarded as the set of points in space, show that span $\{\mathbf{v}\}$ is either 0 or a line through the origin for each $\mathbf{v}$ in $\mathbb{R}^3$.

20. Let U be a nonempty subset of a vector space V. Show that U is a subspace of V if and only if $\mathbf{u}_1 + a\mathbf{u}_2$ lies in U for all $\mathbf{u}_1$ and $\mathbf{u}_2$ in U and all a in $\mathbb{R}$.

21. Let U be the set in Example 6: $U = \{p(x) \text{ in } \mathbf{P} \mid p(3) = 0\}$. Use the factor theorem to show that U consists of multiples of $x - 3$; that is, show that $U = \{(x-3)q(x) \mid q(x) \text{ in } \mathbf{P}\}$. Use this to show that U is a subspace of $\mathbf{P}$.

22. Let P denote the set in Example 7. Show that P is a subspace of $\mathbb{R}^3$ by:

(a) Using the description $P = \{(x,y,z) \text{ in } \mathbb{R}^3 \mid ax + by + cz = 0\}$.

(b) Using Theorem 4 in Section 4.2 and the description $P = \{\mathbf{v} \text{ in } \mathbb{R}^3 \mid \mathbf{v} \cdot \mathbf{n} = 0\}$, $\mathbf{n} = (a,b,c)$.

23. Let $A_1, A_2, \ldots, A_m$ denote $n \times n$ matrices. If $\mathbf{b}$ is a nonzero column in $\mathbb{R}^n$ and $A_1\mathbf{b} = A_2\mathbf{b} = \cdots = A_m\mathbf{b} = \mathbf{0}$, show that $\{A_1, A_2, \ldots, A_m\}$ cannot span $\mathbf{M}_{nn}$.

24. Let $\{\mathbf{v}_1, \mathbf{v}_2, \ldots, \mathbf{v}_n\}$ and $\{\mathbf{u}_1, \ldots, \mathbf{u}_n\}$ be sets of vectors in a vector space; and let

$$X = \begin{bmatrix} \mathbf{v}_1 \\ \vdots \\ \mathbf{v}_n \end{bmatrix} \qquad Y = \begin{bmatrix} \mathbf{u}_1 \\ \vdots \\ \mathbf{u}_n \end{bmatrix}$$

as in Exercise 22 of Section 5.1.

(a) Show that $\text{span}\{\mathbf{v}_1, \mathbf{v}_2, \ldots, \mathbf{v}_n\} \subseteq \text{span}\{\mathbf{u}_1, \ldots, \mathbf{u}_n\}$ if and only if $AY = X$ for some $n \times n$ matrix A.

(b) If $X = AY$ where A is invertible, show that $\text{span}\{\mathbf{v}_1, \ldots, \mathbf{v}_n\} = \text{span}\{\mathbf{u}_1, \ldots, \mathbf{u}_n\}$.

25. If U and W are subspaces of a vector space V, let $U \cup W = \{\mathbf{v} \mid \mathbf{v} \text{ is in } U \text{ or } \mathbf{v} \text{ is in } W\}$. Show that $U \cup W$ is a subspace if and only if $U \subseteq W$ or $W \subseteq U$.

SECTION 5.3 Linear Independence and Dimension

5.3.1 Linear Independence

It is clear from the examples in Section 5.2 that a vector space V can have more than one spanning set, and the question arises as to which spanning sets are in some sense better than others. Consider the following spanning set for $\mathbb{R}^2$:

$$\mathbb{R}^2 = \text{span}\{(1,-1), (1,1), (2,1)\}$$

The vector (3,5) has two different representations as a linear combination of these spanning vectors.

$$(3,5) = -3(1,-1) - 2(1,1) + 4(2,1)$$
$$(3,5) = 2(1,-1) + 13(1,1) - 6(2,1)$$

On the other hand, the spanning set $\{(1,2), (2,3)\}$ for $\mathbb{R}^2$ has the property that a vector (x,y) in $\mathbb{R}^2$ has *exactly one* representation,

$$(x,y) = s(1,2) + t(2,3)$$

as a linear combination of $(1,2)$ and $(2,3)$; in fact $s = 2y - 3x$ and $t = 2x - y$. The present section is devoted to the study of spanning sets having the property that linear combinations have uniquely determined coefficients.

Suppose that $\{\mathbf{v}_1, \mathbf{v}_2, \ldots, \mathbf{v}_n\}$ is a set of vectors in a vector space V such that linear combinations of these vectors have uniquely determined coefficients. That is, if a vector $\mathbf{v}$ in V can be written as a linear combination of the vectors $\mathbf{v}_1, \mathbf{v}_2, \ldots, \mathbf{v}_n$ in two (ostensibly different) ways:

$$\mathbf{v} = s_1\mathbf{v}_1 + s_2\mathbf{v}_2 + \cdots + s_n\mathbf{v}_n \qquad (s_1, s_2, \ldots, s_n \text{ in } \mathbb{R})$$
$$\mathbf{v} = t_1\mathbf{v}_1 + t_2\mathbf{v}_2 + \cdots + t_n\mathbf{v}_n \qquad (t_1, t_2, \ldots, t_n \text{ in } \mathbb{R})$$

then $s_1 = t_1, s_2 = t_2, \ldots, s_n = t_n$. If this is applied to $\mathbf{v} = \mathbf{0}$, then because $\mathbf{0}$ can always be written as $\mathbf{0} = 0\mathbf{v}_1 + 0\mathbf{v}_2 + \cdots + 0\mathbf{v}_n$, the vectors $\mathbf{v}_1, \mathbf{v}_2, \ldots, \mathbf{v}_n$ exhibit the following weaker condition:

DEFINITION

A set of vectors $\{\mathbf{v}_1, \mathbf{v}_2, \ldots, \mathbf{v}_n\}$ is called **linearly independent** if it satisfies the following condition:

If $s_1\mathbf{v}_1 + s_2\mathbf{v}_2 + \cdots + s_n\mathbf{v}_n = \mathbf{0}$, then $s_1 = s_2 = \cdots = s_n = 0$.

Because this notion will be referred to frequently, it is worthwhile to formulate it slightly differently. The **trivial linear combination** of the vectors $\mathbf{v}_1, \mathbf{v}_2, \ldots, \mathbf{v}_n$ is the one with every coefficient zero.

$$0\mathbf{v}_1 + 0\mathbf{v}_2 + \cdots + 0\mathbf{v}_n$$

This is obviously one way of expressing $\mathbf{0}$ as a linear combination of these vectors, and they are linearly independent when it is the *only* way.

EXAMPLE 1 Show that $\{(1,0,-1), (2,1,2), (3,-2,0)\}$ is linearly independent in $\mathbb{R}^3$.

Solution Suppose that a linear combination of these vectors gives zero.

$$s_1(1,0,-1) + s_2(2,1,2) + s_3(3,-2,0) = (0,0,0)$$

We must show that it is the trivial combination — that is, that $s_1 = s_2 = s_3 = 0$. Equating components gives

$$\begin{aligned} s_1 + 2s_2 + 3s_3 &= 0 \\ s_2 - 2s_3 &= 0 \\ -s_1 + 2s_2 \qquad &= 0 \end{aligned}$$

It is easy to show that these equations have only the solution $s_1 = s_2 = s_3 = 0$.

EXAMPLE 2 Show that $\{1 + x, 3x + x^2, 2 + x - x^2\}$ is linearly independent in $\mathbf{P}_2$.

Solution Suppose a linear combination of these polynomials vanishes.

$$s_1(1 + x) + s_2(3x + x^2) + s_3(2 + x - x^2) = 0$$

Equating the coefficients of $1, x$, and x^2 gives a set of linear equations.

$$\begin{aligned} s_1 + \qquad 2s_3 &= 0 \\ s_1 + 3s_2 + s_3 &= 0 \\ s_2 - s_3 &= 0 \end{aligned}$$

Here again, the only solution is $s_1 = s_2 = s_3 = 0$, as the reader can verify.

EXAMPLE 3 Show that $\{\sin x, x, \cos x\}$ is linearly independent in the vector space $\mathbf{F}[0, 2\pi]$ of functions defined on the interval $[0, 2\pi]$.

Solution Suppose that a linear combination of these functions vanishes.

$$s_1(\sin x) + s_2 x + s_3(\cos x) = 0$$

This must hold for *all* values of x (by the definition of equality in $\mathbf{F}[0, 2\pi]$). Taking $x = 0$ yields $s_3 = 0$ (because $\sin 0 = 0$ and $\cos 0 = 1$). Hence

$$s_1(\sin x) + s_2 x = 0$$

and taking $x = \pi$ yields $s_2 \pi = 0$ (because $\sin \pi = 0$). Hence $s_2 = 0$, so

$$s_1(\sin x) = 0$$

Finally, $s_1 = 0$ follows from taking $x = \pi/2$ (because $\sin \pi/2 = 1$), so $s_1 = s_2 = s_3 = 0$.

EXAMPLE 4 Suppose that $\{\mathbf{u}, \mathbf{v}\}$ is a linearly independent set in a vector space V. Show that $\{\mathbf{u} + \mathbf{v}, \mathbf{u} - \mathbf{v}\}$ is also linearly independent.

Solution Suppose a linear combination of $\mathbf{u} + \mathbf{v}$ and $\mathbf{u} - \mathbf{v}$ vanishes.

$$s(\mathbf{u} + \mathbf{v}) + t(\mathbf{u} - \mathbf{v}) = \mathbf{0}$$

We must deduce that $s = t = 0$. Collecting coefficients of $\mathbf{u}$ and $\mathbf{v}$ gives

$$(s + t)\mathbf{u} + (s - t)\mathbf{v} = \mathbf{0}$$

Now this is a linear combination of $\mathbf{u}$ and $\mathbf{v}$ that vanishes, so because $\{\mathbf{u}, \mathbf{v}\}$ is linearly independent, all the coefficients must be zero. This yields linear equations $s + t = 0$ and $s - t = 0$, and the only solution is $s = t = 0$.

EXAMPLE 5 If $\mathbf{v} \neq \mathbf{0}$, the set $\{\mathbf{v}\}$ consisting of the single vector $\mathbf{v}$ is linearly independent.

Solution The linear combinations in this case are just the scalar multiples $s\mathbf{v}$, s in $\mathbb{R}$. If $s\mathbf{v} = \mathbf{0}$, then $s = 0$ by Theorem 4 in Section 5.1 (because $\mathbf{v} \neq \mathbf{0}$). This shows that $\{\mathbf{v}\}$ is linearly independent.

EXAMPLE 6 If $\{\mathbf{v}_1, \mathbf{v}_2, \ldots, \mathbf{v}_n\}$ is linearly independent in a vector space V, show that $\{a_1\mathbf{v}_1, a_2\mathbf{v}_2, \ldots, a_n\mathbf{v}_n\}$ is also linearly independent, provided that the numbers $a_1, a_2, \ldots, a_n$ are all nonzero.

Solution Suppose a linear combination of the new set vanishes.

$$s_1(a_1\mathbf{v}_1) + s_2(a_2\mathbf{v}_2) + \cdots + s_n(a_n\mathbf{v}_n) = \mathbf{0}$$

where $s_1, s_2, \ldots, s_n$ lie in $\mathbb{R}$. Then $s_1a_1 = s_2a_2 = \cdots = s_na_n = 0$ by the linear independence of $\{\mathbf{v}_1, \ldots, \mathbf{v}_n\}$. The fact that each a_i is nonzero now implies that $s_1 = s_2 = \cdots = s_n = 0$.

EXAMPLE 7 Show that no linearly independent set of vectors can contain the zero vector.

Solution Suppose that one of the vectors in $\{\mathbf{v}_1, \mathbf{v}_2, \ldots, \mathbf{v}_n\}$ is zero—say, $\mathbf{v}_1 = \mathbf{0}$. Then

$$1\mathbf{v}_1 + 0\mathbf{v}_2 + 0\mathbf{v}_3 + \cdots + 0\mathbf{v}_n = \mathbf{0}$$

is a nontrivial linear combination that vanishes, so $\{\mathbf{v}_1, \mathbf{v}_2, \ldots, \mathbf{v}_n\}$ cannot be linearly independent. The same conclusion can be reached if *any* $\mathbf{v}_i = \mathbf{0}$.

The notion of independence was motivated by the insistence that the set of vectors in question be such that linear combinations have uniquely determined coefficients. However, the definition of linear independence requires only that linear combinations *equaling zero* have uniquely determined coefficients (necessarily all zero). The first result about linearly independent sets asserts that they have the stronger uniqueness property.

THEOREM 1 Let $\{\mathbf{v}_1, \mathbf{v}_2, \ldots, \mathbf{v}_n\}$ be a linearly independent set of vectors in a vector space V. If a vector $\mathbf{v}$ has two (ostensibly different) representations

$$\mathbf{v} = s_1\mathbf{v}_1 + s_2\mathbf{v}_2 + \cdots + s_n\mathbf{v}_n$$
$$\mathbf{v} = t_1\mathbf{v}_1 + t_2\mathbf{v}_2 + \cdots + t_n\mathbf{v}_n$$

as linear combinations of these vectors, then $s_1 = t_1, s_2 = t_2, \ldots, s_n = t_n$.

Proof We have $s_1\mathbf{v}_1 + s_2\mathbf{v}_2 + \cdots + s_n\mathbf{v}_n = t_1\mathbf{v}_1 + t_2\mathbf{v}_2 + \cdots + t_n\mathbf{v}_n$ (because both sides equal $\mathbf{v}$), so, taking everything to the left side,

$$(s_1 - t_1)\mathbf{v}_1 + (s_2 - t_2)\mathbf{v}_2 + \cdots + (s_n - t_n)\mathbf{v}_n = \mathbf{0}$$

All the coefficients are zero by the independence of the $\mathbf{v}_i$, so $s_1 = t_1$, $s_2 = t_2, \ldots, s_n = t_n$, as required. ■

A set $\{\mathbf{v}_1, \mathbf{v}_2, \ldots, \mathbf{v}_n\}$ of vectors is called **linearly dependent** if it is not linearly independent. The following is a convenient test for linear dependence.

THEOREM 2 A set $\{\mathbf{v}_1, \mathbf{v}_2, \ldots, \mathbf{v}_n\}$ of vectors in a vector space V is linearly dependent if and only if some $\mathbf{v}_i$ is a linear combination of the others.

Proof Assume that $\{\mathbf{v}_1, \mathbf{v}_2, \ldots, \mathbf{v}_n\}$ is linearly dependent. Then some nontrivial linear combination vanishes — say, $a_1\mathbf{v}_1 + a_2\mathbf{v}_2 + \cdots + a_n\mathbf{v}_n = \mathbf{0}$ where some coefficient is not zero. Suppose $a_1 \neq 0$. Then $\mathbf{v}_1 = (-a_2/a_1)\mathbf{v}_2 + \cdots + (-a_n/a_1)\mathbf{v}_n$ gives $\mathbf{v}_1$ as a linear combination of the others. In general, if $a_i \neq 0$, then a similar argument expresses $\mathbf{v}_i$ as a linear combination of the others.

Conversely, suppose one of the vectors is a linear combination of the others—say, $\mathbf{v}_1 = a_2\mathbf{v}_2 + \cdots + a_n\mathbf{v}_n$. Then the nontrivial linear combination $1\mathbf{v}_1 - a_2\mathbf{v}_2 - \cdots - a_n\mathbf{v}_n$ equals zero, so the set $\{\mathbf{v}_1, \ldots, \mathbf{v}_n\}$ is not linearly independent; that is, it is linearly dependent. A similar argument works if *any* $\mathbf{v}_i$ is a linear combination of the others. ■

EXERCISES 5.3.1

1. Show that each of the following sets of vectors is linearly independent.
   (a) $\{(1,-1), (2,0)\}$ in $\mathbb{R}^2$
   (b) $\{(1,2), (-1,1)\}$ in $\mathbb{R}^2$
   (c) $\{(1,-1,0), (0,-1,2), (2,1,1)\}$ in $\mathbb{R}^3$
   (d) $\{(1,1,1), (0,1,1), (0,0,1)\}$ in $\mathbb{R}^3$

(e) $\{(1 + x, 1 - x, x + x^2\}$ in $\mathbf{P}_2$ (f) $\{x^2, x + 1, 1 - x - x^2\}$ in $\mathbf{P}_2$

(g) $\left\{\begin{bmatrix} 1 & 1 \\ 0 & 0 \end{bmatrix}, \begin{bmatrix} 1 & 0 \\ 1 & 0 \end{bmatrix}, \begin{bmatrix} 0 & 0 \\ 1 & -1 \end{bmatrix}, \begin{bmatrix} 0 & 1 \\ 0 & 1 \end{bmatrix}\right\}$ in $\mathbf{M}_{22}$

(h) $\left\{\begin{bmatrix} 1 & 1 \\ 1 & 0 \end{bmatrix}, \begin{bmatrix} 0 & 1 \\ 1 & 1 \end{bmatrix}, \begin{bmatrix} 1 & 0 \\ 1 & 1 \end{bmatrix}, \begin{bmatrix} 1 & 1 \\ 0 & 1 \end{bmatrix}\right\}$ in $\mathbf{M}_{22}$

2. Which of the following subsets of V are linearly independent?

(a) $V = \mathbb{R}^3$; $\{(1,-1,0), (3,2,-1), (3,5,-2)\}$

(b) $V = \mathbb{R}^3$; $\{(1,1,1), (1,-1,1), (0,0,1)\}$

(c) $V = \mathbb{R}^4$; $\{(1,-1,1,-1), (2,0,1,0), (0,-2,1,-2)\}$

(d) $V = \mathbb{R}^4$; $\{(1,1,0,0), (1,0,1,0), (0,0,1,1), (0,1,0,1)\}$

(e) $V = \mathbf{P}_2$; $\{x^2 + 1, x + 1, x\}$

(f) $V = \mathbf{P}_2$; $\{x^2 - x + 3, 2x^2 + x + 5, x^2 + 5x + 1\}$

(g) $V = \mathbf{M}_{22}$; $\left\{\begin{bmatrix} 1 & 1 \\ 0 & 1 \end{bmatrix}, \begin{bmatrix} 1 & 0 \\ 1 & 1 \end{bmatrix}, \begin{bmatrix} 1 & 0 \\ 0 & 1 \end{bmatrix}\right\}$

(h) $V = \mathbf{M}_{22}$; $\left\{\begin{bmatrix} -1 & 0 \\ 0 & -1 \end{bmatrix}, \begin{bmatrix} 1 & -1 \\ -1 & 1 \end{bmatrix}, \begin{bmatrix} 1 & 1 \\ 1 & 1 \end{bmatrix}, \begin{bmatrix} 0 & -1 \\ -1 & 0 \end{bmatrix}\right\}$

(i) $V = \mathbf{F}[1,2]$; $\left\{\frac{1}{x}, \frac{1}{x^2}, \frac{1}{x^3}\right\}$

(j) $V = \mathbf{F}[0,1]$; $\left\{\frac{1}{x^2 + x - 6}, \frac{1}{x^2 - 5x + 6}, \frac{1}{x^2 - 9}\right\}$

3. Which of the following are linearly independent in $\mathbf{F}[0, 2\pi]$?

(a) $\{\sin^2 x, \cos^2 x\}$

(b) $\{1, \sin^2 x, \cos^2 x\}$

(c) $\{x, \sin^2 x, \cos^2 x\}$

4. Find all values of x such that the following are linearly independent in $\mathbb{R}^3$.

(a) $\{(1,-1,0), (x,1,0), (0,2,3)\}$

(b) $\{(2,x,1), (1,0,1), (0,1,3)\}$

5. Let $A \neq 0$ and $B \neq 0$ be $n \times n$ matrices, and assume that A is symmetric and B is skew-symmetric (that is, $B^T = -B$). Show that $\{A, B\}$ is linearly independent.

6. Show that every nonempty subset of a linearly independent set of vectors is again linearly independent.

7. Show that every set of vectors containing a linearly dependent set is again linearly dependent.

8. Let f and g be functions on $[a, b]$, and assume that $f(a) = 1 = g(b)$ and $f(b) = 0 = g(a)$. Show that $\{f, g\}$ is linearly independent in $\mathbf{F}[a, b]$.

9. Let $\{A_1, A_2, \ldots, A_k\}$ be linearly independent in $\mathbf{M}_{mn}$, and suppose that U and V are invertible of size $m \times m$ and $n \times n$, respectively. Show that $\{UA_1V, UA_2V, \ldots, UA_kV\}$ is linearly independent.

10. Let $\{\mathbf{u}, \mathbf{v}\}$ be linearly independent in a vector space V.

(a) Show that $\{\mathbf{u}, \mathbf{u} + \mathbf{v}\}$ and $\{\mathbf{v}, \mathbf{u} + \mathbf{v}\}$ are both independent.

(b) Show that $\{\mathbf{u}, \mathbf{v}, \mathbf{u} + \mathbf{v}\}$ is *not* independent.

11. Show that the following statements are equivalent for two nonzero vectors $\mathbf{v}$ and $\mathbf{w}$ in a vector space V.

(a) $\{\mathbf{v}, \mathbf{w}\}$ is linearly independent.

(b) $\mathbf{v}$ is not in $\mathbb{R}\mathbf{w} = \text{span}\{\mathbf{w}\}$.

(c) $\mathbf{w}$ is not in $\mathbb{R}\mathbf{v} = \text{span}\{\mathbf{v}\}$.

(d) $\mathbb{R}\mathbf{v} \neq \mathbb{R}\mathbf{w}$

12. Regard $\mathbb{R}^3$ as the set of points in space. Show that $\mathbf{v}$ and $\mathbf{w}$ are linearly independent if and only if the lines through the origin with $\mathbf{v}$ and $\mathbf{w}$ as direction vectors are not parallel [see Chapter 4].

13. If $\mathbf{u} = (a,b)$ and $\mathbf{v} = (c,d)$, show that $\{\mathbf{u}, \mathbf{v}\}$ is linearly independent if and only if $\begin{bmatrix} a & b \\ c & d \end{bmatrix}$ is invertible. [*Hint:* Theorem 8, Section 2.3.2.]

14. Assume that $\{\mathbf{u}, \mathbf{v}\}$ is linearly independent in a vector space V. Write $\mathbf{u}' = a\mathbf{u} + b\mathbf{v}$ and $\mathbf{v}' = c\mathbf{u} + d\mathbf{v}$, where a, b, c, and d are numbers. Show that $\{\mathbf{u}', \mathbf{v}'\}$ is linearly independent if and only if the matrix $\begin{bmatrix} a & b \\ c & d \end{bmatrix}$ is invertible. [*Hint:* Theorem 8, Section 2.3.2.]

15. If $\{\mathbf{v}_1, \mathbf{v}_2, \ldots, \mathbf{v}_k\}$ is linearly independent and $\mathbf{w}$ is not in $\text{span}\{\mathbf{v}_1, \ldots, \mathbf{v}_k\}$, show that:

(a) $\{\mathbf{w}, \mathbf{v}_1, \mathbf{v}_2, \ldots, \mathbf{v}_k\}$ is linearly independent.

(b) $\{\mathbf{v}_1 + \mathbf{w}, \ldots, \mathbf{v}_k + \mathbf{w}\}$ is linearly independent.

16. If $\{\mathbf{v}_1, \mathbf{v}_2, \ldots, \mathbf{v}_k\}$ is linearly independent, show that $\{\mathbf{v}_1, \mathbf{v}_1 + \mathbf{v}_2, \ldots, \mathbf{v}_1 + \mathbf{v}_2 + \cdots + \mathbf{v}_k\}$ is also linearly independent.

17. Let $\{\mathbf{u}, \mathbf{v}, \mathbf{w}, \mathbf{z}\}$ be linearly independent. Which of the following are linearly dependent?

(a) $\{\mathbf{u} - \mathbf{v}, \mathbf{v} - \mathbf{w}, \mathbf{w} - \mathbf{u}\}$

(b) $\{\mathbf{u} + \mathbf{v}, \mathbf{v} + \mathbf{w}, \mathbf{w} + \mathbf{u}\}$

(c) $\{\mathbf{u} - \mathbf{v}, \mathbf{v} - \mathbf{w}, \mathbf{w} - \mathbf{z}, \mathbf{z} - \mathbf{u}\}$

(d) $\{\mathbf{u} + \mathbf{v}, \mathbf{v} + \mathbf{w}, \mathbf{w} + \mathbf{z}, \mathbf{z} + \mathbf{u}\}$

18.† Let V denote the space of all functions $f: \mathbb{R} \to \mathbb{R}$ for which the derivatives f' and f'' exist. Show that f_1, f_2, and f_3 in V are linearly independent provided that their **Wronskian** $w(x)$ is nonzero for some x, where

$$w(x) = \det \begin{bmatrix} f_1(x) & f_2(x) & f_3(x) \\ f_1'(x) & f_2'(x) & f_3'(x) \\ f_1''(x) & f_2''(x) & f_3''(x) \end{bmatrix}$$

19. Let $\{\mathbf{v}_1, \ldots, \mathbf{v}_n\}$ be linearly independent in a vector space V, and let A be an $n \times n$ matrix. Define $\mathbf{u}_1, \ldots, \mathbf{u}_n$ by

$$\begin{bmatrix} \mathbf{u}_1 \\ \vdots \\ \mathbf{u}_n \end{bmatrix} = A \begin{bmatrix} \mathbf{v}_1 \\ \vdots \\ \mathbf{v}_n \end{bmatrix}$$

(See Exercise 22 in Section 5.1.) Show that $\{\mathbf{u}_1, \ldots \mathbf{u}_n\}$ is linearly independent if and only if A is invertible.

---

†This exercise requires calculus.

20. Let $\{p, q\}$ be linearly independent polynomials. Show that $\{p, q, pq\}$ is linearly independent if and only if $\deg p \geq 1$ and $\deg q \geq 1$.

21. If z is a complex number, show that $\{z, z^2\}$ is linearly independent if and only if z is not real.

5.3.2 Basis and Dimension

The foregoing examples (particularly Examples 1 and 2) indicate that there is a connection between linear independence and solving systems of linear equations. The following theorem uses a simple fact about linear equations (Theorem 1 in Section 1.3) to prove a basic result about linear independence: The number of vectors in an independent set can never exceed the number in a spanning set.

THEOREM 3
Fundamental Theorem

> Suppose a vector space V can be spanned by n vectors. If any set of m vectors in V is linearly independent, then $m \leq n$.

Proof Let $V = \text{span}\{\mathbf{v}_1, \mathbf{v}_2, \ldots, \mathbf{v}_n\}$. We must show that every set $\{\mathbf{u}_1, \mathbf{u}_2, \ldots, \mathbf{u}_m\}$ of vectors in V with $m > n$ *fails* to be linearly independent. This is accomplished by showing that numbers $x_1, x_2, \ldots, x_m$ can be found, not all zero, such that

$$\sum_{j=1}^{m} x_j\mathbf{u}_j = x_1\mathbf{u}_1 + x_2\mathbf{u}_2 + \cdots + x_m\mathbf{u}_m = \mathbf{0}$$

Because V is spanned by vectors $\mathbf{v}_1, \mathbf{v}_2, \ldots, \mathbf{v}_n$, each vector $\mathbf{u}_j$ can be expressed as a linear combination

$$\mathbf{u}_j = a_{1j}\mathbf{v}_1 + a_{2j}\mathbf{v}_2 + \cdots + a_{nj}\mathbf{v}_n = \sum_{i=1}^{n} a_{ij}\mathbf{v}_i$$

Substituting these expressions into the preceding equation gives

$$\mathbf{0} = \sum_{j=1}^{m} x_j \left(\sum_{i=1}^{n} a_{ij}\mathbf{v}_i \right) = \sum_{i=1}^{n} \left(\sum_{j=1}^{m} a_{ij}x_j \right) \mathbf{v}_i$$

This will certainly be the case if each coefficient of $\mathbf{v}_i$ is zero—that is, if

$$\sum_{j=1}^{m} a_{ij}x_j = 0 \qquad \text{for } i = 1, 2, \ldots, n$$

But this is a system of n equations in the m variables $x_1, x_2, \ldots, x_m$, so because $m > n$, it has a nontrivial solution by Theorem 1 in Section 1.3. ■

We now come to a very important definition.

DEFINITION A set $\{\mathbf{e}_1, \mathbf{e}_2, \ldots, \mathbf{e}_n\}$ of vectors in a vector space V is called a **basis** of V if it satisfies the following two conditions:

(1) $\{\mathbf{e}_1, \mathbf{e}_2, \ldots, \mathbf{e}_n\}$ is linearly independent

(2) $V = \text{span}\{\mathbf{e}_1, \mathbf{e}_2, \ldots, \mathbf{e}_n\}$

Thus if a set of vectors $\{\mathbf{e}_1, \mathbf{e}_2, \ldots, \mathbf{e}_n\}$ is a basis, then *every* vector in V can be written as a linear combination of these vectors in a *unique* way (Theorem 1). But even more is true: Any two (finite) bases of V contain the same number of vectors.

THEOREM 4 Let $\{\mathbf{e}_1, \mathbf{e}_2, \ldots, \mathbf{e}_n\}$ and $\{\mathbf{f}_1, \mathbf{f}_2, \ldots, \mathbf{f}_m\}$ be two bases of a vector space V. Then $n = m$.

Proof Because $V = \text{span}\{\mathbf{e}_1, \mathbf{e}_2, \ldots, \mathbf{e}_n\}$, it follows from Theorem 3 that $m \leq n$. Similarly $n \leq m$, so $n = m$, as asserted. ■

Theorem 4 guarantees that, no matter which basis of V is chosen, it will contain the same number of vectors. Hence there is no ambiguity about the following definition.

DEFINITION If $\{\mathbf{e}_1, \mathbf{e}_2, \ldots, \mathbf{e}_n\}$ is a basis of the vector space V, the number n of vectors in the basis is called the **dimension** of V, and we write

$$\dim V = n$$

A vector space V is called **finite dimensional** if $V = 0$ or V has a finite basis.

In the discussion of bases to this point, we have tacitly assumed that a basis is nonempty and hence that the dimension of the space is at least 1. On the other hand, the zero space 0, consisting of the zero vector alone, has *no* (nonempty) basis (Example 7), and we define

$$\dim 0 = 0$$

This amounts to saying that the empty set of vectors is a basis of the zero space.

EXAMPLE 8 Show that $\dim \mathbb{R}^n = n$ and that $\{\mathbf{e}_1, \mathbf{e}_2, \ldots, \mathbf{e}_n\}$ is a basis where

$$\mathbf{e}_1 = (1, 0, 0, \ldots, 0), \mathbf{e}_2 = (0, 1, 0, \ldots, 0), \ldots, \mathbf{e}_n = (0, 0, 0, \ldots, 1)$$

Solution $\mathbb{R}^n = \text{span}\{\mathbf{e}_1, \mathbf{e}_2, \ldots, \mathbf{e}_n\}$ because

$$(a_1, a_2, \ldots, a_n) = a_1\mathbf{e}_1 + a_2\mathbf{e}_2 + \cdots + a_n\mathbf{e}_n$$

holds for all vectors $(a_1, a_2, \ldots, a_n)$ in $\mathbb{R}^n$. But this also shows that the vectors $\mathbf{e}_1, \mathbf{e}_2, \ldots, \mathbf{e}_n$ are linearly independent, because $a_1\mathbf{e}_1 + \cdots + a_n\mathbf{e}_n = \mathbf{0}$ implies that $a_1 = \cdots = a_n = 0$. Hence $\{\mathbf{e}_1, \mathbf{e}_2, \ldots, \mathbf{e}_n\}$ is a basis of $\mathbb{R}^n$, so $\dim \mathbb{R}^n = n$.

The basis of $\mathbb{R}^n$ in Example 8 will be called the **standard basis** for $\mathbb{R}^n$. Similar considerations apply to the space of all $m \times n$ matrices (see Exercise 5):

EXAMPLE 9 The space $\mathbf{M}_{mn}$ has dimension mn, and one basis consists of all $m \times n$ matrices with exactly one nonzero entry equal to 1.

EXAMPLE 10 Show that $\dim \mathbf{P}_n = n + 1$ and that $\{1, x, x^2, \ldots, x^n\}$ is a basis.

Solution Each polynomial $p(x) = a_0 + a_1x + \cdots + a_nx^n$ in $\mathbf{P}_n$ is clearly a linear combination of $1, x, \ldots, x^n$, so $\mathbf{P}_n = \text{span}\{1, x, \ldots, x^n\}$. On the other hand, if a linear combination of these vectors vanishes, $a_01 + a_1x + \cdots + a_nx^n = 0$, then $a_0 = a_1 = \cdots = a_n = 0$ because x is an indeterminate. So $\{1, x, \ldots, x^n\}$ is linearly independent and hence is a basis.

This result, together with Theorem 3, shows that the space $\mathbf{P}$ of *all* polynomials cannot be finite dimensional. In fact, if $\dim \mathbf{P} = n$, then $\mathbf{P}$ would be spanned by n polynomials. But the fact that $\{1, x, x^2, \ldots, x^n\}$ are $n + 1$ linearly independent vectors would then contradict the fundamental theorem (Theorem 3).

EXAMPLE 11 If $\mathbf{v} \neq \mathbf{0}$ is any nonzero vector in a vector space V, show that $\text{span}\{\mathbf{v}\} = \mathbb{R}\mathbf{v}$ has dimension 1.

Solution $\{\mathbf{v}\}$ clearly spans $\mathbb{R}\mathbf{v}$, and it is linearly independent by Example 5. Hence $\{\mathbf{v}\}$ is a basis of $\mathbb{R}\mathbf{v}$, and so $\dim \mathbb{R}\mathbf{v} = 1$.

EXAMPLE 12 As in Example 5 in Section 5.2, let $A = \begin{bmatrix} 1 & 1 \\ 0 & 0 \end{bmatrix}$ and consider the subspace

$$U = \{X \text{ in } \mathbf{M}_{22} \mid AX = XA\}$$

of $\mathbf{M}_{22}$. Show that $\dim U = 2$ and find a basis of U.

Solution It was shown in Example 5 of Section 5.2 that

$$U = \left\{ \begin{bmatrix} x & y \\ 0 & x - y \end{bmatrix} \,\middle|\, x \text{ and } y \text{ in } \mathbb{R} \right\}$$

Hence, each matrix X in U can be written

$$X = \begin{bmatrix} x & y \\ 0 & x - y \end{bmatrix} = x\begin{bmatrix} 1 & 0 \\ 0 & 1 \end{bmatrix} + y\begin{bmatrix} 0 & 1 \\ 0 & -1 \end{bmatrix}$$

so $U = \text{span}\left\{ \begin{bmatrix} 1 & 0 \\ 0 & 1 \end{bmatrix}, \begin{bmatrix} 0 & 1 \\ 0 & -1 \end{bmatrix} \right\}$. Moreover, the set $\left\{ \begin{bmatrix} 1 & 0 \\ 0 & 1 \end{bmatrix}, \begin{bmatrix} 0 & 1 \\ 0 & -1 \end{bmatrix} \right\}$ is linearly independent (verify this), so it is a basis of U, and $\dim U = 2$.

EXAMPLE 13 Show that the set V of all symmetric 2×2 matrices is a vector space, and find the dimension of V.

Solution A matrix A is symmetric if $A^T = A$. If A and B lie in V, then

$$(A + B)^T = A^T + B^T = A + B$$
$$(kA)^T = kA^T = kA$$

using Theorem 2 in Section 2.1, so $A + B$ and kA are also symmetric. This shows that V is a vector space (being a subspace of $\mathbf{M}_{22}$). Now a matrix A is symmetric when each entry equals the one directly across the main diagonal, so each 2×2 symmetric matrix has the form

$$\begin{bmatrix} a & c \\ c & b \end{bmatrix} = a\begin{bmatrix} 1 & 0 \\ 0 & 0 \end{bmatrix} + b\begin{bmatrix} 0 & 0 \\ 0 & 1 \end{bmatrix} + c\begin{bmatrix} 0 & 1 \\ 1 & 0 \end{bmatrix}$$

Hence the set $B = \left\{ \begin{bmatrix} 1 & 0 \\ 0 & 0 \end{bmatrix}, \begin{bmatrix} 0 & 0 \\ 0 & 1 \end{bmatrix}, \begin{bmatrix} 0 & 1 \\ 1 & 0 \end{bmatrix} \right\}$ spans V, and the reader can verify that B is linearly independent. Hence B is a basis of V, so $\dim V = 3$.

The fundamental theorem takes the following useful form when stated for vector spaces of dimension n.

THEOREM 5

Let V be a vector space and assume that $\dim V = n > 0$.

(1) No set of more than n vectors in V can be linearly independent.

(2) No set of fewer than n vectors can span V.

Proof V can be spanned by n vectors (any basis), so (1) restates the fundamental theorem. But the n basis vectors are also linearly independent, so no spanning set can have fewer than n vectors, again by Theorem 3. This gives (2). ■

Here is an application of (1) of Theorem 5.

EXAMPLE 14 Let A denote an $n \times n$ matrix. Then there exist $n^2 + 1$ real numbers a_0, $a_1, a_2, \ldots, a_{n^2}$, not all zero, such that

$$a_0 I + a_1 A + a_2 A^2 + \cdots + a_{n^2} A^{n^2} = 0$$

where I denotes the $n \times n$ identity matrix.

Solution The space $\mathbf{M}_{nn}$ of all $n \times n$ matrices has dimension n^2 by Example 9. Hence the $n^2 + 1$ matrices $I, A, \ldots, A^{n^2}$ cannot be linearly independent by property 1 of Theorem 5, so a nontrivial linear combination must vanish. This is the desired conclusion.

We note in passing that the result of Example 14 can be written as $f(A) = 0$ where $f(x) = a_0 + a_1 x + \cdots + a_{n^2} x^{n^2}$. In other words, A satisfies a polynomial of degree n^2. In fact, we know that A satisfies a polynomial of degree n (this is the Cayley–Hamilton theorem), but the brevity of the solution in Example 14 is an indication of the power of these methods.

EXERCISES 5.3.2

1. Show that the following are bases of the space V indicated.

(a) $\{(1,1,0), (1,0,1), (0,1,1)\}$; $V = \mathbb{R}^3$

(b) $\{(-1,1,1), (1,-1,1), (1,1,-1)\}$; $V = \mathbb{R}^3$

(c) $\left\{\begin{bmatrix} 1 & 0 \\ 0 & 1 \end{bmatrix}, \begin{bmatrix} 0 & 1 \\ 1 & 0 \end{bmatrix}, \begin{bmatrix} 1 & 1 \\ 0 & 1 \end{bmatrix}, \begin{bmatrix} 1 & 0 \\ 0 & 0 \end{bmatrix}\right\}$; $V = \mathbf{M}_{22}$

(d) $\{1 + x, x + x^2, x^2 + x^3, x^3\}$; $V = \mathbf{P}_3$

2. Exhibit a basis and calculate the dimension of each of the following subspaces of $\mathbb{R}^4$.

(a) $\{(a, a + b, a - b, b) \mid a \text{ and } b \text{ in } \mathbb{R}\}$

(b) $\{(a + b, a - b, a, b) \mid a \text{ and } b \text{ in } \mathbb{R}\}$

(c) $\{(a, b, c + a, c) \mid a, b, \text{ and } c \text{ in } \mathbb{R}\}$

(d) $\{(a - b, b + c, a, b + c) \mid a, b, \text{ and } c \text{ in } \mathbb{R}\}$

(e) $\{(a, b, c, d) \mid a + b + c + d = 0\}$ **(f)** $\{(a, b, c, d) \mid a + b = c + d\}$

3. Exhibit a basis and calculate the dimension of each of the following subspaces of $\mathbf{P}_2$.

(a) $\{a(1 + x) + b(x + x^2) \mid a \text{ and } b \text{ in } \mathbb{R}\}$

(b) $\{a + b(x + x^2) \mid a \text{ and } b \text{ in } \mathbb{R}\}$

(c) $\{p(x) \mid p(1) = 0\}$ **(d)** $\{p(x) \mid p(x) = p(-x)\}$

4. Exhibit a basis and calculate the dimension of each of the following subspaces of $\mathbf{M}_{22}$.
   (a) $\{A \mid A^T = -A\}$
   (b) $\left\{A \,\middle|\, A\begin{bmatrix} 1 & 1 \\ -1 & 0 \end{bmatrix} = \begin{bmatrix} 1 & 1 \\ -1 & 0 \end{bmatrix} A\right\}$
   (c) $\left\{A \,\middle|\, A\begin{bmatrix} 1 & 0 \\ -1 & 0 \end{bmatrix} = \begin{bmatrix} 0 & 0 \\ 0 & 0 \end{bmatrix}\right\}$
   (d) $\left\{A \,\middle|\, A\begin{bmatrix} 1 & 1 \\ -1 & 0 \end{bmatrix} = \begin{bmatrix} 0 & 1 \\ -1 & 1 \end{bmatrix} A\right\}$
5. (a) Show that $\left\{\begin{bmatrix} 1 & 0 \\ 0 & 0 \end{bmatrix}, \begin{bmatrix} 0 & 1 \\ 0 & 0 \end{bmatrix}, \begin{bmatrix} 0 & 0 \\ 1 & 0 \end{bmatrix}, \begin{bmatrix} 0 & 0 \\ 0 & 1 \end{bmatrix}\right\}$ is a basis of $\mathbf{M}_{22}$.
   (b) Complete the solution to Example 9.
6. Let $A = \begin{bmatrix} 1 & 1 \\ 0 & 0 \end{bmatrix}$ and define $U = \{X \mid X \text{ is in } \mathbf{M}_{22} \text{ and } AX = X\}$.
   (a) Find a basis of U containing A.
   (b) Find a basis of U not containing A.
7. In each case, find a basis of the subspace U.
   (a) $\left\{\begin{bmatrix} x \\ y \\ z \end{bmatrix} \,\middle|\, \begin{matrix} 2x - y + z = 0 \\ x + 2y - z = 0 \\ x - 3y + 2z = 0 \end{matrix}\right\}$
   (b) $\left\{\begin{bmatrix} x \\ y \\ z \end{bmatrix} \,\middle|\, \begin{matrix} 3x - y + z = 0 \\ x + 3y - 2z = 0 \\ 2x - 4y + 3z = 0 \end{matrix}\right\}$
   (c) $\left\{\begin{bmatrix} r \\ s \\ t \\ u \end{bmatrix} \,\middle|\, \begin{matrix} 3r - s + 2u = 0 \\ r + s + t = 0 \\ 2r - 2s - t + 2u = 0 \\ r - 3s - 2t + 2u = 0 \end{matrix}\right\}$
   (d) $\left\{\begin{bmatrix} r \\ s \\ t \\ u \end{bmatrix} \,\middle|\, \begin{matrix} r + s - t + 2u = 0 \\ 3r - s + 2t - u = 0 \\ r - 3s + 4t - 5u = 0 \\ 5r - 3s + 5t - 4u = 0 \end{matrix}\right\}$
8. (a) Let V denote the set of all 2×2 matrices whose column sums are all equal. Show that V is a subspace of $\mathbf{M}_{22}$, and compute dim V.
   (b) Repeat part **(a)** for 3×3 matrices.
   (c) Repeat part **(a)** for $n \times n$ matrices.
9. (a) Let $V = \{(x^2 + x + 1)p(x) \mid p(x) \text{ in } \mathbf{P}_2\}$. Show that V is a subspace of $\mathbf{P}_4$ and find dim V. [*Hint:* If $f(x)\, g(x) = 0$ in $\mathbf{P}_2$, and $f(x) \neq 0$, then $g(x) = 0$.]
   (b) Repeat with $V = \{(x^2 - x)p(x) \mid p(x) \text{ in } \mathbf{P}_3\}$, a subset of $\mathbf{P}_5$.
   (c) Generalize.
10. If $V = \mathbf{F}[a, b]$ as in Example 8 in Section 5.1, show that the set of constant functions is a subspace of dimension 1 (f is **constant** if there is a number c such that $f(x) = c$ for all x).
11. (a) If U is an invertible $n \times n$ matrix and $\{A_1, A_2, \ldots\}$ is a basis for $\mathbf{M}_{mn}$, show that $\{A_1U, A_2U, \ldots\}$ is also a basis.
    (b) Show that part **(a)** fails if U is not invertible. [*Hint:* $U\mathbf{x} = \mathbf{0}$ for some $\mathbf{x} \neq \mathbf{0}$ in $\mathbb{R}^n$.]
12. Show that $\{(a,b), (a_1,b_1)\}$ is a basis of $\mathbb{R}^2$ if and only if $\{a + bx, a_1 + b_1x\}$ is a basis of $\mathbf{P}_1$.
13. Let $\mathbf{D}_n$ denote the set of all functions f from the set $\{1, 2, \ldots, n\}$ to $\mathbb{R}$.

(a) Show that $\mathbf{D}_n$ is a vector space with pointwise addition and scalar multiplication.

(b) Show that $\{S_1, S_2, \ldots, S_n\}$ is a basis of $\mathbf{D}_n$ where, for each $k = 1, 2, \ldots, n$, the function S_k is defined by $S_k(k) = 1$, whereas $S_k(j) = 0$ if $j \neq k$.

14. A polynomial $p(x)$ is **even** if $p(-x) = p(x)$ and **odd** if $p(-x) = -p(x)$. Let E_n and O_n denote the sets of even and odd polynomials in $\mathbf{P}_n$.

(a) Show that E_n is a subspace of $\mathbf{P}_n$ and find $\dim E_n$.

(b) Show that O_n is a subspace of $\mathbf{P}_n$ and find $\dim O_n$.

15. Suppose that $\{\mathbf{v}_1, \mathbf{v}_2, \ldots, \mathbf{v}_n\}$ is a maximal linearly independent set for a vector space V. That is, $\{\mathbf{v}_1, \ldots, \mathbf{v}_n\}$ is linearly independent and no set of more than n vectors is linearly independent. Show that $\{\mathbf{v}_1, \ldots, \mathbf{v}_n\}$ is a basis of V.

16. Suppose that $\{\mathbf{v}_1, \mathbf{v}_2, \ldots, \mathbf{v}_n\}$ is a minimal spanning set for a vector space V. That is, $V = \text{span}\{\mathbf{v}_1, \ldots, \mathbf{v}_n\}$ and V cannot be spanned by fewer than n vectors. Show that $\{\mathbf{v}_1, \ldots, \mathbf{v}_n\}$ is a basis of V.

17. If V is any vector space, let V^n be the space of n-tuples from V defined in Exercise 22 in Section 5.1. If V is finite dimensional, show that $\dim(V^n) = n(\dim V)$.

5.3.3 Existence of Bases

Up to this point, we have had no guarantee that an arbitrary vector space *has* a basis — and hence no guarantee that one can speak *at all* of the dimension of V. However, Theorem 7 below shows that any space that is spanned by a finite set of vectors has a (finite) basis. Theorem 6, which we will examine first, provides a useful connection between linear independence and spanning sets.

THEOREM 6 Let $\{\mathbf{v}_1, \mathbf{v}_2, \ldots, \mathbf{v}_n\}$ be a linearly independent set of vectors in a vector space V. The following conditions are equivalent for a vector $\mathbf{v}$ in V:

(1) $\{\mathbf{v}, \mathbf{v}_1, \mathbf{v}_2, \ldots, \mathbf{v}_n\}$ is linearly independent.

(2) $\mathbf{v}$ does not lie in $\text{span}\{\mathbf{v}_1, \mathbf{v}_2, \ldots, \mathbf{v}_n\}$.

Proof Assume (1) is true and suppose, if possible, that $\mathbf{v}$ lies in $\text{span}\{\mathbf{v}_1, \ldots, \mathbf{v}_n\}$ — say, $\mathbf{v} = a_1\mathbf{v}_1 + \cdots + a_n\mathbf{v}_n$. Then $\mathbf{v} - a_1\mathbf{v}_1 - \cdots - a_n\mathbf{v}_n = \mathbf{0}$ is a nontrivial linear combination, contrary to (1). So (1) implies (2). Conversely, assume that (2) holds and suppose that $a\mathbf{v} + a_1\mathbf{v}_1 + \cdots + a_n\mathbf{v}_n = \mathbf{0}$. If $a \neq 0$, then $\mathbf{v} = (-a^{-1}a_1)\mathbf{v}_1 + \cdots + (-a^{-1}a_n)\mathbf{v}_n$, contrary to (2). So $a = 0$ and hence $a_1\mathbf{v}_1 + \cdots + a_n\mathbf{v}_n = \mathbf{0}$. This implies that $a_1 = \cdots = a_n = 0$ because $\{\mathbf{v}_1, \ldots, \mathbf{v}_n\}$ is linearly independent. This proves that (2) implies (1). ∎

THEOREM 7

Let $V \neq 0$ be a vector space spanned by n vectors.

(1) Each set of linearly independent vectors is part of a basis.

(2) Each spanning set of nonzero vectors contains a basis.

(3) V has a basis, and $\dim V \leq n$.

Proof

Property (1) Let $\{\mathbf{v}_1, \mathbf{v}_2, \ldots, \mathbf{v}_k\}$ be linearly independent. If it spans V, there is nothing to prove. If not, choose $\mathbf{v}_{k+1}$ outside $\text{span}\{\mathbf{v}_1, \ldots, \mathbf{v}_k\}$. Then $\{\mathbf{v}_1, \ldots, \mathbf{v}_k, \mathbf{v}_{k+1}\}$ is linearly independent by Theorem 6. If this spans V, we are finished. If not, choose $\mathbf{v}_{k+2}$ outside $\text{span}\{\mathbf{v}_1, \ldots, \mathbf{v}_{k+1}\}$ so that $\{\mathbf{v}_1, \ldots, \mathbf{v}_{k+2}\}$ is linearly independent. Continue this process. Either a basis is reached at some stage or, if not, arbitrarily large independent sets will be found in V. But this latter possibility cannot occur, by Theorem 3, because V is spanned by a finite number of vectors.

Property (2) Let $V = \text{span}\{\mathbf{v}_1, \mathbf{v}_2, \ldots, \mathbf{v}_m\}$, where each $\mathbf{v}_i \neq \mathbf{0}$. If $\{\mathbf{v}_1, \ldots, \mathbf{v}_m\}$ is linearly independent, we are finished. If not, one of these vectors lies in the span of the others (Theorem 2). Relabeling if necessary, assume that $\mathbf{v}_1$ lies in $\text{span}\{\mathbf{v}_2, \ldots, \mathbf{v}_m\}$ so that $V = \text{span}\{\mathbf{v}_2, \ldots, \mathbf{v}_m\}$. Now repeat the argument. If $\{\mathbf{v}_2, \ldots, \mathbf{v}_m\}$ is linearly independent, we are finished. If not, we have (after possible relabeling) $V = \text{span}\{\mathbf{v}_3, \ldots, \mathbf{v}_m\}$. Continue this process. If a basis is encountered at some stage, we are finished. If not, we ultimately reach $V = \text{span}\{\mathbf{v}_m\}$. But then $\{\mathbf{v}_m\}$ is a basis.

Property (3) V has a spanning set of n vectors, one of which is nonzero because $V \neq 0$. Hence property (3) follows from property (2). ■

Recall that a vector space is called finite dimensional if it has a finite basis. A corollary to Theorem 7 improves on this.

COROLLARY

A nonzero vector space is finite dimensional if and only if it can be spanned by finitely many vectors.

Theorem 7 is very useful, and two important consequences are collected below.

THEOREM 8

Let V be a vector space, and assume that $\dim V = n > 0$.

(1) Any set of n linearly independent vectors in V is a basis (that is, it necessarily spans V).

(2) Any spanning set of n nonzero vectors in V is a basis (that is, it is necessarily linearly independent).

Proof

Property (1) If the n independent vectors do not span V, they are part of a basis of more than n vectors by property (1) of Theorem 7. This contradicts Theorem 4.

Property (2) If the n vectors in a spanning set are not linearly independent, they contain a basis of fewer than n vectors by property (2) of Theorem 7, contradicting Theorem 4. ■

Theorem 8 saves time when we are verifying that a set of n vectors is a basis of a space known to have dimension n. Here are some examples.

EXAMPLE 15 Show that $\{(1,0,0,0), (1,1,0,0), (1,1,1,0), (1,1,1,1)\}$ is a basis of $\mathbb{R}^4$.

Solution Because $\dim(\mathbb{R}^4) = 4$, it suffices (by Theorem 8) to show either that these vectors are linearly independent or that they span $\mathbb{R}^4$. But subtracting successive vectors shows that the standard basis $\{(1,0,0,0), (0,1,0,0), (0,0,1,0), (0,0,0,1)\}$ is contained in the span of these vectors, so they span $\mathbb{R}^4$.

EXAMPLE 16 Let V denote the space of all symmetric 2×2 matrices. Find a basis of V consisting of invertible matrices.

Solution It was established in Example 13 that $\dim V = 3$, so what is needed is a set of three invertible, symmetric matrices that are linearly independent (or, alternatively, span V). Clearly $\left\{\begin{bmatrix}1 & 0\\0 & 1\end{bmatrix}, \begin{bmatrix}1 & 0\\0 & -1\end{bmatrix}, \begin{bmatrix}0 & 1\\1 & 0\end{bmatrix}\right\}$ is such a set, and hence it is a basis of the required type.

The next theorem collects some very useful information on the dimension of subspaces of finite-dimensional spaces.

THEOREM 9 Let V be a vector space of dimension n, and let U and W denote subspaces of V. Then:

(1) U is finite dimensional and $\dim U \leq n$.

(2) Any basis of U is part of a basis for V.

(3) If $U \subseteq W$ and $\dim U = \dim W$, then $U = W$.

Proof Surprisingly enough, the hard part is showing that U is finite dimensional. If $U = 0$, this is clear, so assume $U \neq 0$ and choose $\mathbf{u}_1 \neq \mathbf{0}$ in U. If $U = \text{span}\{\mathbf{u}_1\}$, then U is finite dimensional. If $U \neq \text{span}\{\mathbf{u}_1\}$, choose $\mathbf{u}_2$ in U outside $\text{span}\{\mathbf{u}_1\}$. Then $\{\mathbf{u}_1, \mathbf{u}_2\}$ is linearly independent by Theorem 6. If $U = \text{span}\{\mathbf{u}_1, \mathbf{u}_2\}$, then U is finite dimensional. If not, continue the process to find $\mathbf{u}_3$ in U such that $\{\mathbf{u}_1, \mathbf{u}_2, \mathbf{u}_3\}$ is linearly independent. At each stage, either we have found a finite basis for U (and so proved that U is finite dimensional) or the process continues. But the process cannot continue indefinitely, because the space V (having dimension n) cannot contain more than n independent vectors. Hence the process must terminate, so U must be finite dimensional.

The rest is easy. U has a finite basis of (say) $m = \dim U$ vectors, and because these are linearly independent, $m \leq n$ by Theorem 5. This proves (1) of Theorem 9, and (2) then follows from Theorem 7. Finally, let $\dim U = \dim W = m$. Then any basis $\{\mathbf{u}_1, \ldots, \mathbf{u}_m\}$ of U is an independent set of m vectors in W and so is a basis of W by Theorem 8. In particular, $\{\mathbf{u}_1, \ldots, \mathbf{u}_m\}$ spans W, so because it also spans U, $W = \text{span}\{\mathbf{u}_1, \ldots, \mathbf{u}_m\} = U$. This proves property (3). ■

EXAMPLE 17 If a is a number, let W denote the subspace of all polynomials in $\mathbf{P}_n$ with a as a root.

$$W = \{p(x) \mid p(x) \text{ is in } \mathbf{P}_n \text{ and } p(a) = 0\}$$

Show that $\{(x - a), (x - a)^2, \ldots, (x - a)^n\}$ is a basis of W.

Solution Observe first that $(x - a), (x - a)^2, (x - a)^3, \ldots, (x - a)^n$ are members of W and are linearly independent (Exercise 16). Write

$$U = \text{span}\{(x - a), (x - a)^2, (x - a)^3, \ldots, (x - a)^n\}$$

Then we have $U \subseteq W \subseteq \mathbf{P}_n$, $\dim U = n$, and $\dim \mathbf{P}_n = n + 1$. Hence $n \leq \dim W \leq n + 1$ by property (1) of Theorem 9, so $\dim W = n$ or $\dim W = n + 1$. But then $W = U$ or $W = \mathbf{P}_n$ by property (3) of Theorem 9. Because $W \neq \mathbf{P}_n$, it follows that $W = U$.

The next example uses ideas from Chapter 4.

EXAMPLE 18 The only subspaces of three-dimensional Euclidean space $\mathbb{R}^3$ are 0, $\mathbb{R}^3$, lines through the origin, and planes through the origin.

Solution Let U be such a subspace. Because $\dim \mathbb{R}^3 = 3$, the only possibilities for $\dim U$ are 0, 1, 2, and 3. If $\dim U = 0$ or 3, then $U = 0$ or $\mathbb{R}^3$. If $\dim U = 1$, let $\{\mathbf{d}\}$ be a basis so that $U = \mathbb{R}\mathbf{d}$. Then each vector in U has the form $\mathbf{u} = t\mathbf{d}$, t in $\mathbb{R}$, and this is just the line through the origin with direction vector $\mathbf{d}$. Finally, if $\dim U = 2$, let $\{\mathbf{u}_1, \mathbf{u}_2\}$ be a basis of U. If $\mathbf{n} = \mathbf{u}_1 \times \mathbf{u}_2$, we claim that U is the plane P through the origin with normal $\mathbf{n}$. The plane P can be described by

$$P = \{\mathbf{v} \mid \mathbf{v} \cdot \mathbf{n} = 0\}$$

This is a subspace (as the reader can verify), and because $\mathbf{u}_1$ and $\mathbf{u}_2$ are orthogonal to $\mathbf{n} = \mathbf{u}_1 \times \mathbf{u}_2$, we have $U \subseteq P \subseteq \mathbb{R}^3$. Because $\dim U = 2$ and $\dim \mathbb{R}^3 = 3$, this means that $\dim P = 2$ or 3, so $P = U$ or $P = \mathbb{R}^3$. Clearly $P \neq \mathbb{R}^3$, so $P = U$, as required.

As a last application of the notion of dimension, we give a useful condition that a matrix is invertible.

THEOREM 10

The following conditions are equivalent for an $n \times n$ matrix A.

(1) The rows of A are linearly independent in $\mathbb{R}^n$.

(2) The rows of A span $\mathbb{R}^n$.

(3) The columns of A are linearly independent in $\mathbb{R}^n$.

(4) The columns of A span $\mathbb{R}^n$.

(5) A is invertible.

Proof We show that (1), (2), and (5) are equivalent. The equivalence of (3), (4), and (5) is analogous and is left as Exercise 14.

Property (1) implies property (2). This follows from Theorem 8 because $\dim \mathbb{R}^n = n$.

Property (2) implies property (5). By Theorem 9 of Section 2.3, it suffices to find a matrix B such that $BA = I$. Let $\mathbf{r}_1, \mathbf{r}_2, \ldots, \mathbf{r}_n$ denote the rows of A, and write row i of B as a matrix $\mathbf{k} = [k_1\ k_2\ \ldots\ k_n]$. Then $I = BA$ gives (using block multiplication)

$$\text{row } i \text{ of } I = \text{row } i \text{ of } BA = \mathbf{k}A = [k_1\ k_2 \ldots k_n]\begin{bmatrix}\mathbf{r}_1\\ \mathbf{r}_2\\ \vdots\\ \mathbf{r}_n\end{bmatrix}$$

$$= k_1\mathbf{r}_1 + k_2\mathbf{r}_2 + \cdots + k_n\mathbf{r}_n$$

But property (2) implies that such k_i exist, so each row of B can be found.

Property (5) implies property (1). Let $k_1\mathbf{r}_1 + k_2\mathbf{r}_2 + \cdots + k_n\mathbf{r}_n = \mathbf{0}$, where $\mathbf{r}_i$ denotes the ith row of A. Then property (1) follows if this implies that $k_1 = k_2 = \cdots = k_n = 0$. If we write $\mathbf{k} = [k_1\ k_2 \ldots k_n]$, then, again using block multiplication,

$$\mathbf{k}A = [k_1\ k_2 \ldots k_n]\begin{bmatrix}\mathbf{r}_1\\ \mathbf{r}_2\\ \vdots\\ \mathbf{r}_n\end{bmatrix} = k_1\mathbf{r}_1 + k_2\mathbf{r}_2 + \cdots + k_n\mathbf{r}_n = \mathbf{0}$$

Right multiplication by A^{-1} [which exists by property (5)] gives $\mathbf{k} = \mathbf{0}$, so $k_1 = k_2 = \cdots = k_n = 0$, as required. ■

Hence, for example, to verify that $A = \begin{bmatrix} 3 & -1 & 0\\ 1 & 0 & 4\\ 0 & 1 & 3\end{bmatrix}$ is an invertible matrix, it suffices to check that the rows $(3,-1,0)$, $(1,0,4)$, and $(0,1,3)$ are linearly independent in $\mathbb{R}^3$ (or, equivalently, that they span $\mathbb{R}^3$). This holds in reverse too.

EXAMPLE 19 Show that $(1,2,-1)$, $(3,1,-4)$, and $(1,1,7)$ are a basis of $\mathbb{R}^3$.

Solution Let $A = \begin{bmatrix} 1 & 2 & -1\\ 3 & 1 & -4\\ 1 & 1 & 7\end{bmatrix}$ be the matrix with these vectors as rows. It suffices by Theorem 10 to check that this matrix is invertible. This is easily verified (for example, $\det A = -41$, so A is invertible by Theorem 2 in Section 3.2).

EXAMPLE 20 If A is a square matrix, show that A is invertible if and only if the transpose A^T is invertible.

Solution This is clear from Theorem 10 because the rows of A are the columns of A^T.

EXERCISES 5.3.3

1. In each case, find a basis for V that includes the vector $\mathbf{v}$.
   (a) $V = \mathbb{R}^3$, $\mathbf{v} = (1,-1,1)$
   (b) $V = \mathbb{R}^3$, $\mathbf{v} = (0,1,1)$
   (c) $V = \mathbf{M}_{22}$, $\mathbf{v} = \begin{bmatrix} 1 & 1\\ 1 & 1\end{bmatrix}$
   (d) $V = \mathbf{P}_3$, $\mathbf{v} = x^2 - x + 1$
2. In each case, find a basis for V among the given vectors.
   (a) $V = \mathbb{R}^3$, $\{(1,1,-1), (2,0,1), (-1,1,-2), (1,2,1)\}$
   (b) $V = \mathbf{P}_2$, $\{x^2 + 3, x + 2, x^2 - 2x - 1, x^2 + x\}$
3. In each case, find a basis of V containing $\mathbf{v}$ and $\mathbf{w}$.

(a) $V = \mathbb{R}^4$, $\mathbf{v} = (1,-1,1,-1)$, $\mathbf{w} = (0,1,0,1)$
(b) $V = \mathbb{R}^4$, $\mathbf{v} = (0,0,1,1)$, $\mathbf{w} = (1,1,1,1)$
(c) $V = \mathbf{M}_{22}$, $\mathbf{v} = \begin{bmatrix} 1 & 0 \\ 0 & 1 \end{bmatrix}$, $\mathbf{w} = \begin{bmatrix} 0 & 1 \\ 1 & 0 \end{bmatrix}$
(d) $V = \mathbf{P}_3$, $\mathbf{v} = x^2 + 1$, $\mathbf{w} = x^2 + x$

4. **(a)** If z is not a real number, show that $\{z, z^2\}$ is a basis of the space $\mathbb{C}$ of all complex numbers..
   (b) If z is neither real nor pure imaginary, show that $\{z, \bar{z}\}$ is a basis of $\mathbb{C}$.
5. Find a basis of $\mathbf{M}_{22}$ consisting of matrices with the property $A^2 = A$.
6. **(a)** Find a basis of $\mathbf{P}_3$ consisting of polynomials whose coefficients sum to 4.
   (b) What if they sum to 0?
7. If $\{\mathbf{u}, \mathbf{v}, \mathbf{w}\}$ is a basis of V, determine which of the following are bases.
   (a) $\{\mathbf{u} + \mathbf{v}, \mathbf{u} + \mathbf{w}, \mathbf{v} + \mathbf{w}\}$ **(b)** $\{2\mathbf{u} + \mathbf{v} + 3\mathbf{w}, 3\mathbf{u} + \mathbf{v} - \mathbf{w}, \mathbf{u} - 4\mathbf{w}\}$
   (c) $\{\mathbf{u}, \mathbf{u} + \mathbf{v} + \mathbf{w}\}$ **(d)** $\{\mathbf{u}, \mathbf{u} + \mathbf{w}, \mathbf{u} - \mathbf{w}, \mathbf{v} + \mathbf{w}\}$
8. **(a)** Can two vectors span $\mathbb{R}^3$? Can they be linearly independent? Explain.
   (b) Can four vectors span $\mathbb{R}^3$? Can they be linearly independent? Explain.
9. Show that any nonzero vector in a finite dimensional vector space is part of a basis.
10. Given $\mathbf{v}_1, \mathbf{v}_2, \mathbf{v}_3, \ldots, \mathbf{v}_k$ and $\mathbf{v}$, let $U = \text{span}\{\mathbf{v}_1, \mathbf{v}_2, \ldots, \mathbf{v}_k\}$ and $W = \text{span}\{\mathbf{v}_1, \ldots, \mathbf{v}_k, \mathbf{v}\}$. Show that either $\dim W = \dim U$ or $\dim W = 1 + \dim U$.
11. Use Theorem 10 to test whether each of the following matrices is invertible.
    (a) $\begin{bmatrix} 1 & -1 & 2 \\ 0 & 1 & 5 \\ 2 & 1 & 19 \end{bmatrix}$ **(b)** $\begin{bmatrix} 1 & 2 & -1 \\ 3 & 1 & 1 \\ 5 & 3 & -1 \end{bmatrix}$ **(c)** $\begin{bmatrix} 2 & 1 & -2 \\ 1 & 0 & 3 \\ 4 & 1 & 4 \end{bmatrix}$
12. Use Theorem 10 to determine whether each of the following sets of vectors is a basis of $\mathbb{R}^4$.
    (a) $\{(1,-1,1,0), (1,0,1,-1), (1,1,1,1), (0,1,2,3)\}$
    (b) $\{(1,1,0,0), (1,0,1,0), (0,1,0,1), (0,0,1,1)\}$
13. Use Theorem 10 to show that a triangular matrix (Section 3.1) is invertible provided that the entries on the main diagonal are nonzero.
14. Complete the proof of Theorem 10.
15. If A is an $n \times n$ matrix, prove that $\det A = 0$ if and only if some row of A is a linear combination of the others.
16. Complete Example 17 by showing that $\{(x - a), (x - a)^2, \ldots, (x - a)^n\}$ is linearly independent in $\mathbf{P}_n$. [*Hint:* They have degrees $1, 2, \ldots, n$.]
17. Suppose U is a subspace of $\mathbf{P}_1$ and $U \neq 0$, $U \neq \mathbf{P}_1$. Show that either $U = \mathbb{R}$ or $U = \mathbb{R}(a + x)$ for some a in $\mathbb{R}$.
18. **(a)** Let $p(x)$ and $q(x)$ lie in $\mathbf{P}_1$, and suppose that $p(1) \neq 0$, $q(2) \neq 0$, and $p(2) = 0 = q(1)$. Show that $\{p(x), q(x)\}$ is a basis of $\mathbf{P}_1$. [*Hint:* If $rp(x) + sq(x) = 0$, evaluate at $x = 1, x = 2$.]

(b) Let $B = \{p_0(x), p_1(x), \ldots, p_n(x)\}$ be a set of polynomials in $\mathbf{P}_n$. Assume that there exist numbers $a_0, a_1, \ldots, a_n$ such that $p_i(a_i) \neq 0$ for each i but $p_i(a_j) = 0$ if i is different from j. Show that B is a basis of $\mathbf{P}_n$.

19. Let $\{\mathbf{v}_1, \mathbf{v}_2, \ldots, \mathbf{v}_n\}$ be a basis of $\mathbb{R}^n$ (written as columns), and let A be an $n \times n$ matrix.

(a) If A is invertible, show that $\{A\mathbf{v}_1, A\mathbf{v}_2, \ldots, A\mathbf{v}_n\}$ is a basis of $\mathbb{R}^n$.

(b) If $\{A\mathbf{v}_1, A\mathbf{v}_2, \ldots, A\mathbf{v}_n\}$ is a basis of $\mathbb{R}^n$, show that A is invertible.

20. Let V be the set of all infinite sequences $(a_0, a_1, a_2, \ldots)$ of real numbers. Define addition and scalar multiplication by $(a_0, a_1, \ldots) + (b_0, b_1, \ldots) = (a_0 + b_0, a_1 + b_1, \ldots)$ and $r(a_0, a_1, \ldots) = (ra_0, ra_1, \ldots)$.

(a) Show that V is a vector space.

(b) Show that V is not finite dimensional.

(c) [For those with some calculus.] Show that the set of convergent sequences (that is, $\lim_{n\to\infty} a_n$ exists) is a subspace, also of infinite dimension.

21. If U and W are subspaces of V, we define their **sum** $U + W$ and their **intersection** $U \cap W$ as follows:

$$U + W = \{\mathbf{u} + \mathbf{w} \mid \mathbf{u} \text{ in } U \text{ and } \mathbf{w} \text{ in } W\}$$
$$U \cap W = \{\mathbf{v} \mid \mathbf{v} \text{ in both } U \text{ and } W\}$$

(a) Show that $U + W$ and $U \cap W$ are subspaces of V.

(b) In each case, determine the subspaces $U + W$ and $U \cap W$ of V.

(i) $V = \mathbb{R}^5$, $U = \{(a, b, 0,0,0) \mid a \text{ and } b \text{ in } \mathbb{R}\}$ and $W = \{(0,0, c, d, e) \mid c, d, \text{ and } e \text{ in } \mathbb{R}\}$

(ii) $V = \mathbb{R}^3$, $U = \{(a, b, a) \mid a \text{ and } b \text{ in } \mathbb{R}\}$ and $W = \{(a, a, a) \mid a \text{ in } \mathbb{R}\}$

(iii) $V = \mathbb{R}^3$, $U = \{(a, b, a - b) \mid a \text{ and } b \text{ in } \mathbb{R}\}$ and $W = \{(a, a + b, b) \mid a \text{ and } b \text{ in } \mathbb{R}\}$

(iv) $V = \mathbb{R}^3$, $U = \{(a, a + b, b) \mid a \text{ and } b \text{ in } \mathbb{R}\}$ and $W = \{(a, b, 2a - b) \mid a \text{ and } b \text{ in } \mathbb{R}\}$

(c) **(i)** Show that $\text{span}\{\mathbf{u}, \mathbf{w}\} = \mathbb{R}\mathbf{u} + \mathbb{R}\mathbf{w}$ for any vectors $\mathbf{u}$ and $\mathbf{w}$.

(ii) Show that $\text{span}\{\mathbf{u}_1, \ldots, \mathbf{u}_m, \mathbf{w}_1, \ldots, \mathbf{w}_n\} = \text{span}\{\mathbf{u}_1, \ldots, \mathbf{u}_m\} + \text{span}\{\mathbf{w}_1, \ldots, \mathbf{w}_n\}$ for any vectors $\mathbf{u}_i$ and $\mathbf{w}_j$.

(d) Let U and W be subspaces of V. Show that:

(i) $U + W$ contains U and W and is contained in any subspace that contains U and W.

(ii) $U \cap W$ is contained in U and W and contains any subspace that is contained in U and W.

(e) Show that $U \cap W = 0$ if and only if $\{\mathbf{u}, \mathbf{w}\}$ is linearly independent for any nonzero vectors $\mathbf{u}$ in U and $\mathbf{w}$ in W.

(f) If U and W are finite dimensional, show that $U + W$ is finite dimensional and that $\dim(U + W) \leq \dim U + \dim W$. [*Hint:* A basis of U together with a basis of W spans $U + W$.]

(g) If U and W are finite dimensional, and if $U \cap W = 0$, show that $\dim(U + W) = \dim U + \dim W$. [*Hint:* See parts **(e)** and **(f)**.]

SECTION 5.4 Rank of a Matrix

Theorem 10 in Section 5.3 shows that the notions of linear independence and spanning sets are useful in discussing the invertibility of square matrices. The question arises whether these ideas shed any light on nonsquare matrices (which, of course, cannot be invertible). Before this can be discussed, two subspaces related to a matrix must be introduced.

DEFINITION If A is an $m \times n$ matrix, the rows of A are vectors in $\mathbb{R}^n$, and the subspace of $\mathbb{R}^n$ spanned by these rows is called the **row space** of A and is denoted as row A. Similarly, the space spanned by the columns of A is called the **column space** of A and is denoted as col A.

The following properties of these spaces will be needed.

THEOREM 1 Let A, U, and V be matrices of sizes $m \times n$, $m \times m$, and $n \times n$, respectively.

(1) $\text{row}(UA) \subseteq \text{row } A$ with equality if U is invertible.

(2) $\text{col}(AV) \subseteq \text{col } A$ with equality if V is invertible.

Proof Let $\mathbf{u}_i$ denote row i of the matrix U, and write $\mathbf{u}_i = [u_{i1}\ u_{i2} \dots u_{im}]$. If $\mathbf{r}_1, \mathbf{r}_2, \dots, \mathbf{r}_m$ denote the rows of A, then block multiplication gives

$$\text{row } i \text{ of } UA = \mathbf{u}_i A = [u_{i1}\ u_{i2} \dots u_{im}] \begin{bmatrix} \mathbf{r}_1 \\ \mathbf{r}_2 \\ \vdots \\ \mathbf{r}_m \end{bmatrix}$$

$$= u_{i1}\mathbf{r}_1 + u_{i2}\mathbf{r}_2 + \cdots + u_{im}\mathbf{r}_m$$

Thus each row of UA lies in the row space of A, so $\text{row}(UA) \subseteq \text{row } A$, as asserted. If U is invertible, then $A = U^{-1}(UA)$, so, using what we have just proved,

$$\text{row } A = \text{row}[U^{-1}(UA)] \subseteq \text{row}(UA)$$

Thus $\text{row } A = \text{row}(UA)$ if U is invertible. This proves (1), and (2) follows in the same way. [Alternatively, apply (1) to $(AV)^T = V^T A^T$.] ■

This leads to important results in matrix theory. The first is the surprising fact that row A and col A have the *same dimension* for any matrix A.

To see this, consider first the case of a matrix R in row-echelon form (Section 1.2). Recall that in each nonzero row of R, the first nonzero entry (from the left) is a 1, called the leading 1 for that row. Moreover, these leading 1's occupy distinct columns, and these columns have only zero entries below the leading 1. This implies (Exercise 15) that

(a) These columns are a basis of col R.

(b) The rows of R are linearly independent (and so are a basis of row R).

These facts lead to the following fundamental theorem.

THEOREM 2
Rank Theorem

Let A denote any $m \times n$ matrix. Then

$$\dim(\text{row } A) = \dim(\text{col } A)$$

Moreover, suppose A can be carried to a matrix R in row-echelon form by a series of elementary row operations. If r denotes the number of nonzero rows in R, then

(1) The rows of R are a basis of row A.

(2) If the leading 1's lie in columns $j_1, j_2, \ldots, j_r$ of R, then the corresponding columns $j_1, j_2, \ldots, j_r$ of A are a basis of col A.

Proof Theorem 7 of Section 2.3 asserts that $R = UA$ for some invertible matrix U. Hence row A = row R by Theorem 1, and (1) follows from (b) above.

To prove (2), let $\mathbf{c}_1, \mathbf{c}_2, \ldots, \mathbf{c}_n$ denote the columns of A. Then $A = [\mathbf{c}_1\ \mathbf{c}_2 \ldots \mathbf{c}_n]$ in block form, and

$$R = UA = U[\mathbf{c}_1\ \mathbf{c}_2 \ldots \mathbf{c}_n] = [U\mathbf{c}_1\ U\mathbf{c}_2 \ldots U\mathbf{c}_n].$$

Hence, in the notation of (2), the set $B = \{U\mathbf{c}_{j_1}, U\mathbf{c}_{j_2}, \ldots, U\mathbf{c}_{j_r}\}$ consists of the columns of R that contain a leading 1, so B is a basis of col R by (a) above. But then the fact that U is invertible implies that $\{\mathbf{c}_{j_1}, \mathbf{c}_{j_2}, \ldots, \mathbf{c}_{j_r}\}$ is linearly independent. Furthermore, if $\mathbf{c}_j$ is any column of A, then $U\mathbf{c}_j$ is a linear combination of the columns in B. Again, the invertibility of U implies that $\mathbf{c}_j$ is a linear combination of $\mathbf{c}_{j_1}, \mathbf{c}_{j_2}, \ldots, \mathbf{c}_{j_r}$. This proves (2).

Finally, $\dim(\text{row } A) = r = \dim(\text{col } A)$ by (1) and (2). ■

DEFINITION

The common dimension of the row and column spaces of an $m \times n$ matrix A is called the **rank** of A and is denoted rank A.

Recall that in Section 1.2.2 it was asserted (without proof) that no matter how a matrix A is reduced (by row operations) to a matrix R in row-echelon form, the number r of nonzero rows of R is always the same and that this number r was called the "rank" of A. Part (1) of the rank theorem not only shows that rank determined in this way is the same as rank determined in the present way but also verifies that rank is independent of how the matrix A is carried to row-echelon form. We record this for reference.

COROLLARY 1

Suppose a matrix A can be carried to a matrix R in row-echelon form by a series of elementary row operations. Then the rank of A is equal to the number of nonzero rows of R.

The next corollary follows from Theorem 2, because the dimension of a vector space cannot exceed the number of vectors in any spanning set.

COROLLARY 2

If A is an $m \times n$ matrix, then rank $A \leq m$ and rank $A \leq n$.

The fact that the rows of A are just the columns of the transpose matrix A^T gives

COROLLARY 3

If A is any matrix, rank A = rank A^T.

Theorem 1 immediately yields

COROLLARY 4

Let A be an $m \times n$ matrix, and let U and V be invertible matrices of size $m \times m$ and $n \times n$, respectively. Then:

$$\text{rank } A = \text{rank}(UA) = \text{rank}(AV)$$

If A is an $n \times n$ matrix, then the row space is spanned by the n rows and so has dimension n if and only if those rows are linearly independent. Combining this with Theorem 10 in Section 5.3 gives

COROLLARY 5

An $n \times n$ matrix A is invertible if and only if rank $A = n$.

EXAMPLE 1

Compute the rank of the matrix $A = \begin{bmatrix} 1 & 2 & 2 & -1 \\ 3 & 6 & 5 & 0 \\ 1 & 2 & 1 & 2 \end{bmatrix}$ and find bases for the row space and the column space of A.

Solution The reduction of A to row-echelon form is as follows:

$$\begin{bmatrix} 1 & 2 & 2 & -1 \\ 3 & 6 & 5 & 0 \\ 1 & 2 & 1 & 2 \end{bmatrix} \rightarrow \begin{bmatrix} 1 & 2 & 2 & -1 \\ 0 & 0 & -1 & 3 \\ 0 & 0 & -1 & 3 \end{bmatrix} \rightarrow \begin{bmatrix} 1 & 2 & 2 & -1 \\ 0 & 0 & 1 & -3 \\ 0 & 0 & 0 & 0 \end{bmatrix}$$

Hence rank$(A) = 2$, and $\{(1,2,2,-1), (0,0,1,-3)\}$ is a basis of the row space of A. Moreover, the leading 1's are in columns 1 and 3 of the row-echelon matrix, so Theorem 2 shows that columns 1 and 3 of A are a basis $\left\{\begin{bmatrix} 1 \\ 3 \\ 1 \end{bmatrix}, \begin{bmatrix} 2 \\ 5 \\ 1 \end{bmatrix}\right\}$ of col A. It is worth noting that a basis of col A can be obtained from the fact that the columns of A are the rows of A^T. Hence we reduce A^T to row-echelon form.

$$\begin{bmatrix} 1 & 3 & 1 \\ 2 & 6 & 2 \\ 2 & 5 & 1 \\ -1 & 0 & 2 \end{bmatrix} \rightarrow \begin{bmatrix} 1 & 3 & 1 \\ 0 & 0 & 0 \\ 0 & -1 & -1 \\ 0 & 3 & 3 \end{bmatrix} \rightarrow \begin{bmatrix} 1 & 3 & 1 \\ 0 & 1 & 1 \\ 0 & 0 & 0 \\ 0 & 0 & 0 \end{bmatrix}$$

Thus another basis of the column space of A is $\left\{\begin{bmatrix} 1 \\ 3 \\ 1 \end{bmatrix}, \begin{bmatrix} 0 \\ 1 \\ 1 \end{bmatrix}\right\}$. Of course this reconfirms that the rank of A is 2.

The rank theorem can be used to find bases of subspaces in $\mathbb{R}^n$.

EXAMPLE 2 Find a basis for the following subspace of $\mathbb{R}^4$.

$$U = \text{span}\{(1,1,2,3), (2,3,1,0), (1,3,-4,-9)\}$$

Solution U is just the row space of $\begin{bmatrix} 1 & 1 & 2 & 3 \\ 2 & 3 & 1 & 0 \\ 1 & 3 & -4 & -9 \end{bmatrix}$, so we reduce this to row-echelon form.

$$\begin{bmatrix} 1 & 1 & 2 & 3 \\ 2 & 3 & 1 & 0 \\ 1 & 3 & -4 & -9 \end{bmatrix} \rightarrow \begin{bmatrix} 1 & 1 & 2 & 3 \\ 0 & 1 & -3 & -6 \\ 0 & 2 & -6 & -12 \end{bmatrix} \rightarrow \begin{bmatrix} 1 & 1 & 2 & 3 \\ 0 & 1 & -3 & -6 \\ 0 & 0 & 0 & 0 \end{bmatrix}$$

The required basis is $\{(1,1,2,3), (0,1,-3,-6)\}$.

Let m, n, and r be positive integers, and assume that $r \leq n$ and $r \leq m$. Recall that the block $m \times n$ matrix $\begin{bmatrix} I_r & 0 \\ 0 & 0 \end{bmatrix}$ denotes the $m \times n$ matrix with the $r \times r$ identity matrix in the upper left corner and zeros elsewhere. Such matrices were discussed in Section 2.2. The reason for recalling them here is the following theorem.

THEOREM 3 Let A denote an $m \times n$ matrix. Then A has rank r if and only if there exist invertible matrices U and V (of sizes $m \times m$ and $n \times n$, respectively), such that

$$UAV = \begin{bmatrix} I_r & 0 \\ 0 & 0 \end{bmatrix}$$

in block form, where I_r is the $r \times r$ identity matrix.

Proof If such U and V exist, then rank $A = r$ by Corollary 4 of Theorem 2. Conversely, by Theorem 7 in Section 2.3, there is a matrix R in reduced row-echelon form such that $R = UA$, where U is $m \times m$ and invertible. If rank $A = r$, then only the first r rows of R are nonzero and these rows contain every column of I_r. Hence elementary *column* operations will reduce R to the required block form. This can be accomplished by doing the corresponding *row* operations to R^T to obtain its row-echelon form $U_1R^T = \begin{bmatrix} I_r & 0 \\ 0 & 0 \end{bmatrix}$, where U_1 is $n \times n$ and invertible. Transposition gives $RU_1{}^T = \begin{bmatrix} I_r & 0 \\ 0 & 0 \end{bmatrix}^T = \begin{bmatrix} I_r & 0 \\ 0 & 0 \end{bmatrix}$, so we take $V = U_1{}^T$. ■

If A is an $m \times n$ matrix, this theorem not only proves the existence of the matrices U and V but also gives us a way to find them. If R is the reduced row-echelon form of A, the row operations that carry A to R also carry I_m to an invertible matrix U such that $UA = R$ (Theorem 7 in Section 2.3). Hence they carry the block matrix $[AI_m]$ to $[RU]$. Similarly, the reduced row-echelon form of R^T is $\begin{bmatrix} I_r & 0 \\ 0 & 0 \end{bmatrix}$, and the corresponding row operations carry I_n to a matrix V^T such that $V^TR^T = \begin{bmatrix} I_r & 0 \\ 0 & 0 \end{bmatrix}$. Hence

$$UAV = RV = (V^TR^T)^T = \begin{bmatrix} I_r & 0 \\ 0 & 0 \end{bmatrix}^T = \begin{bmatrix} I_r & 0 \\ 0 & 0 \end{bmatrix}$$

so U and V are the required matrices. To summarize:

(1) Find U from $[AI_m] \to [RU]$.

(2) Find V from $[R^T I_n] \to \left[\begin{bmatrix} I_r & 0 \\ 0 & 0 \end{bmatrix} V^T\right]$.

Here is an example.

EXAMPLE 3

Given $A = \begin{bmatrix} 1 & -1 & 1 & 2 \\ 2 & -2 & 1 & -1 \\ -1 & 1 & 0 & 3 \end{bmatrix}$, find invertible matrices U and V such that $UAV = \begin{bmatrix} I_r & 0 \\ 0 & 0 \end{bmatrix}$, where $r = \text{rank } A$.

Solution The matrix U and the reduced row-echelon form R of A are computed by the row reduction $[AI_3] \to [RU]$.

$$\left[\begin{array}{rrrr|rrr} 1 & -1 & 1 & 2 & 1 & 0 & 0 \\ 2 & -2 & 1 & -1 & 0 & 1 & 0 \\ -1 & 1 & 0 & 3 & 0 & 0 & 1 \end{array}\right] \to \left[\begin{array}{rrrr|rrr} 1 & -1 & 0 & -3 & -1 & 1 & 0 \\ 0 & 0 & 1 & 5 & 2 & -1 & 0 \\ 0 & 0 & 0 & 0 & -1 & 1 & 1 \end{array}\right]$$

Hence

$$R = \begin{bmatrix} 1 & -1 & 0 & -3 \\ 0 & 0 & 1 & 5 \\ 0 & 0 & 0 & 0 \end{bmatrix} \quad \text{and} \quad U = \begin{bmatrix} -1 & 1 & 0 \\ 2 & -1 & 0 \\ -1 & 1 & 1 \end{bmatrix}$$

In particular, $r = \text{rank } R = 2$. Now row-reduce $[R^T I_4] \to \left[\begin{bmatrix} I_r & 0 \\ 0 & 0 \end{bmatrix} V^T\right]$:

$$\left[\begin{array}{rrr|rrrr} 1 & 0 & 0 & 1 & 0 & 0 & 0 \\ -1 & 0 & 0 & 0 & 1 & 0 & 0 \\ 0 & 1 & 0 & 0 & 0 & 1 & 0 \\ -3 & 5 & 0 & 0 & 0 & 0 & 1 \end{array}\right] \to \left[\begin{array}{rrr|rrrr} 1 & 0 & 0 & 1 & 0 & 0 & 0 \\ 0 & 1 & 0 & 0 & 0 & 1 & 0 \\ 0 & 0 & 0 & 1 & 1 & 0 & 0 \\ 0 & 0 & 0 & 3 & 0 & -5 & 1 \end{array}\right]$$

whence

$$V^T = \begin{bmatrix} 1 & 0 & 0 & 0 \\ 0 & 0 & 1 & 0 \\ 1 & 1 & 0 & 0 \\ 3 & 0 & -5 & 1 \end{bmatrix} \qquad V = \begin{bmatrix} 1 & 0 & 1 & 3 \\ 0 & 0 & 1 & 0 \\ 0 & 1 & 0 & -5 \\ 0 & 0 & 0 & 1 \end{bmatrix}$$

Then $UAV = \begin{bmatrix} I_2 & 0 \\ 0 & 0 \end{bmatrix}$, as is easily verified.

Let A be an $m \times n$ matrix, and consider the associated homogeneous system of linear equations

$$A\mathbf{x} = \mathbf{0}$$

The set of solutions to this system is a subspace of $\mathbb{R}^n$ (Theorem 2 in Section 1.3) called the **null space** of the matrix A. The dimension of this space is clearly of interest, and it can be found by using the methods of Chapter 1 in any particular case.

EXAMPLE 4

Find a basis for the null space of $A = \begin{bmatrix} 1 & -2 & 1 & 1 \\ -1 & 2 & 0 & 1 \\ 2 & -4 & 1 & 0 \end{bmatrix}$.

Solution The corresponding system of linear equations is

$$\begin{aligned} x_1 - 2x_2 + x_3 + x_4 &= 0 \\ -x_1 + 2x_2 + x_4 &= 0 \\ 2x_1 - 4x_2 + x_3 &= 0 \end{aligned}$$

The reduction of the augmented matrix to reduced row-echelon form was carried out in Example 4 in Section 1.3. The result is

$$\begin{bmatrix} 1 & -2 & 1 & 1 & 0 \\ -1 & 2 & 0 & 1 & 0 \\ 2 & -4 & 1 & 0 & 0 \end{bmatrix} \to \begin{bmatrix} 1 & -2 & 0 & -1 & 0 \\ 0 & 0 & 1 & 2 & 0 \\ 0 & 0 & 0 & 0 & 0 \end{bmatrix}$$

The leading variables are x_1 and x_3, so the nonleading variables are assigned parameters $x_2 = s$, $x_4 = t$. Then the equations are solved for the leading variables, and the result is

$$\begin{bmatrix} x_1 \\ x_2 \\ x_3 \\ x_4 \end{bmatrix} = s\begin{bmatrix} 2 \\ 1 \\ 0 \\ 0 \end{bmatrix} + t\begin{bmatrix} 1 \\ 0 \\ -2 \\ 1 \end{bmatrix}$$

It follows that the null space of A is the span of $\left\{\begin{bmatrix} 2 \\ 1 \\ 0 \\ 0 \end{bmatrix}, \begin{bmatrix} 1 \\ 0 \\ -2 \\ 1 \end{bmatrix}\right\}$. The reader can check that these vectors are linearly independent and so are a basis.

Now suppose A is any $m \times n$ matrix, and denote rank A as r. Then A can be carried to a matrix R in reduced row-echelon form, and R has the same null space (by Theorem 1) and has exactly r nonzero rows. Hence there are r leading variables, so because there are n variables in all, there are $n - r$ nonleading variables. As in Example 4, these variables are assigned parameter values. This gives every solution as a linear combination of $n - r$ particular solutions (the parameters are the coefficients), and so these $n - r$ solutions span the null space. In fact, they are necessarily a basis, because as the next theorem shows, the null space has dimension $n - r$.

THEOREM 4

Let A be an $m \times n$ matrix and denote $r = \text{rank } A$. Then the null space of A is a subspace of $\mathbb{R}^n$ of dimension $n - r$.

Proof Let $N(A) = \{\mathbf{x} \mid A\mathbf{x} = \mathbf{0}\}$ denote the null space. By Theorem 3, let $UAV = \begin{bmatrix} I_r & 0 \\ 0 & 0 \end{bmatrix}$, where U and V are invertible. Then $N(A) = N(UA)$ because U is invertible, and $\dim N(UA) = \dim N(UAV)$ because V is invertible (Exercise 20). Hence $\dim N(A) = \dim N(UAV)$, and this equals $n - r$ because $UAV = \begin{bmatrix} I_r & 0 \\ 0 & 0 \end{bmatrix}$ is $m \times n$, so $N(UAV)$ consists of all columns in $\mathbb{R}^n$ with the first r entries zero. ■

Corollary 5 of Theorem 2 asserts that an $n \times n$ matrix A is invertible if and only if rank $A = n$. This can be rephrased as follows: A is invertible if and only if rank A is as large as possible, and in this form, it makes sense for nonsquare matrices. If A is $m \times n$, then the largest possible rank is the smaller of m and n. The next theorem shows that $m \times n$ matrices with rank m do retain some of the properties of invertible matrices. (Of course, an analogous theorem holds if the rank is n.)

THEOREM 5

The following conditions are equivalent for an $m \times n$ matrix.

(1) rank $A = m$.

(2) The rows of A are linearly independent.

(3) If $\mathbf{x}A = \mathbf{0}$ with $\mathbf{x}$ in $\mathbb{R}^m$, then necessarily $\mathbf{x} = \mathbf{0}$.

(4) AA^T is invertible.

Proof (Optional)

(1) implies (2). This is by Theorem 8 in Section 5.3 because the rows span row A, and row A has dimension m by (1).

(2) implies (3). If $\mathbf{x}A = \mathbf{0}$, write $\mathbf{x} = [x_1\ x_2 \dots x_m]$, and denote the rows of A by $\mathbf{v}_1, \mathbf{v}_2, \dots, \mathbf{v}_m$. Then block multiplication gives $x_1\mathbf{v}_1 + x_2\mathbf{v}_2 + \cdots + x_m\mathbf{v}_m = \mathbf{x}A = \mathbf{0}$, so (2) implies $x_1 = x_2 = \cdots = x_m = 0$. Hence $\mathbf{x} = \mathbf{0}$.

(3) implies (4). By Theorem 8 in Section 2.3, it suffices to show that, if $\mathbf{x}(AA^T) = \mathbf{0}$ with $\mathbf{x}$ in $\mathbb{R}^m$, then $\mathbf{x} = \mathbf{0}$. Now $\mathbf{x}A$ is a row in $\mathbb{R}^n$, and we compute

$$(\mathbf{x}A)(\mathbf{x}A)^T = \mathbf{x}AA^T\mathbf{x}^T = \mathbf{0}\mathbf{x}^T = 0$$

This means that $\mathbf{x}A = \mathbf{0}$ (if $\mathbf{y}\mathbf{y}^T = 0$ with $\mathbf{y}$ in $\mathbb{R}^n$, then $\mathbf{y} = \mathbf{0}$), so $\mathbf{x} = \mathbf{0}$ by (3).

(4) implies (1). If $\mathbf{v}_i$ is the ith row of A, it suffices to show that $\{\mathbf{v}_1, \ldots, \mathbf{v}_m\}$ is linearly independent. Let $x_1\mathbf{v}_1 + x_2\mathbf{v}_2 + \cdots + x_m\mathbf{v}_m = \mathbf{0}$, where the x_i lie in $\mathbb{R}$. If $\mathbf{x} = [x_1\ x_2\ \ldots\ x_m]$, this means $\mathbf{x}A = \mathbf{0}$, whence $\mathbf{x}AA^T = \mathbf{0}$. But then $\mathbf{x} = \mathbf{0}$ by (4), so $x_1 = x_2 = \cdots = x_m = 0$, as required. ■

EXAMPLE 5 If at least two of the numbers x, y, and z are distinct, show that $S = \begin{bmatrix} 3 & x+y+z \\ x+y+z & x^2+y^2+z^2 \end{bmatrix}$ is invertible.

Solution If $A = \begin{bmatrix} 1 & 1 & 1 \\ x & y & z \end{bmatrix}$, then $S = AA^T$, so it is necessary only to verify that the rows of A are linearly independent. This is left to the reader.

EXERCISES 5.4

1. In each case, find bases for the row and column spaces of A and determine the rank of A.

(a) $A = \begin{bmatrix} 2 & -4 & 6 & 8 \\ 2 & -1 & 3 & 2 \\ 4 & -5 & 9 & 10 \\ 0 & -1 & 1 & 2 \end{bmatrix}$ **(b)** $A = \begin{bmatrix} 2 & 1 & 1 \\ 1 & 0 & 1 \\ 2 & -1 & 1 \\ 1 & 1 & 1 \end{bmatrix}$

(c) $A = \begin{bmatrix} 1 & -1 & 5 & -2 & 2 \\ 2 & 3 & -2 & 5 & 1 \\ 0 & 5 & -12 & 9 & -3 \\ 1 & 4 & -7 & 7 & -1 \end{bmatrix}$ **(d)** $A = \begin{bmatrix} 1 & 2 & -1 & 3 \\ 0 & 3 & 1 & 1 \end{bmatrix}$

2. In each case, find a basis of the subspace U.

(a) $U = \operatorname{span}\{(1,-1,0,3), (2,1,5,1), (4,-2,5,7)\}$

(b) $U = \operatorname{span}\{(1,-1,2,5,1), (3,1,4,2,7), (1,1,0,0,0), (5,1,6,7,8)\}$

(c) $U = \operatorname{span}\left\{\begin{bmatrix} 1 \\ 1 \\ 0 \\ 0 \end{bmatrix}, \begin{bmatrix} 0 \\ 0 \\ 1 \\ 1 \end{bmatrix}, \begin{bmatrix} 1 \\ 0 \\ 1 \\ 0 \end{bmatrix}, \begin{bmatrix} 0 \\ 1 \\ 0 \\ 1 \end{bmatrix}\right\}$

(d) $U = \text{span}\left\{\begin{bmatrix} 1 \\ 5 \\ -6 \end{bmatrix}, \begin{bmatrix} 2 \\ 6 \\ -8 \end{bmatrix}, \begin{bmatrix} 3 \\ 7 \\ -10 \end{bmatrix}, \begin{bmatrix} 4 \\ 8 \\ 12 \end{bmatrix}\right\}$

3. **(a)** Can a 3×4 matrix have independent columns? Independent rows? Explain.

(b) If A is 4×3 and rank $A = 2$, can A have independent columns? Independent rows? Explain.

(c) If A is an $m \times n$ matrix and rank $A = m$, show that $m \leq n$.

(d) Can a nonsquare matrix have its rows independent and its columns independent? Explain.

4. **(a)** Show that rank $UA \leq$ rank A, with equality if U is invertible.

(b) Show that rank $AV \leq$ rank A, with equality if V is invertible.

5. Show that rank $(AB) \leq$ rank A and that rank $(AB) \leq$ rank B.

6. Prove (2) of Theorem 1 by applying (1) to $(AV)^T = V^T A^T$.

7. Show that the rank does not change when an elementary row or column operation is performed on a matrix.

8. Show that every $m \times n$ matrix A with m linearly independent rows can be obtained from some $n \times n$ invertible matrix by deleting the last $n - m$ rows.

9. In each case, find a basis of the null space of A. Then compute rank A and verify Theorem 4.

(a) $A = \begin{bmatrix} 3 & 1 & 1 \\ 2 & 0 & 1 \\ 4 & 2 & 1 \\ 1 & -1 & 1 \end{bmatrix}$ (b) $A = \begin{bmatrix} 3 & 5 & 5 & 2 & 0 \\ 1 & 0 & 2 & 2 & 1 \\ 1 & 1 & 1 & -2 & -2 \\ -2 & 0 & -4 & -4 & -2 \end{bmatrix}$

10. Show that the column space of an $m \times n$ matrix A is $V = \{A\mathbf{x} \mid \mathbf{x} \text{ in } \mathbb{R}^n\}$.

11. Let A be an $m \times n$ matrix. Show that $A\mathbf{x} = \mathbf{b}$ has a solution for every $\mathbf{b}$ in $\mathbb{R}^m$ if and only if rank $A = m$.

12. If A is $m \times n$ and $\mathbf{b}$ is $m \times 1$, show that $\mathbf{b}$ lies in the column space of A if and only if rank $[A\ \mathbf{b}] =$ rank A.

13. **(a)** Show that $A\mathbf{x} = \mathbf{b}$ has a solution if and only if rank $A =$ rank $[A\ \mathbf{b}]$. [*Hint:* Exercise 11.]

(b) If $A\mathbf{x} = \mathbf{b}$ has no solution, show that rank $[A\ \mathbf{b}] = 1 +$ rank A.

14. Formulate and prove the analog of Theorem 5 for $m \times n$ matrices with independent columns.

15. Let R denote a matrix in row-echelon form.

(a) Show that the rows of R are linearly independent and so are a basis of row R.

(b) Show that the columns of R containing leading 1's are a basis of col R.

16. In each case, find invertible U and V such that $UAV = \begin{bmatrix} I_r & 0 \\ 0 & 0 \end{bmatrix}$, where $r =$ rank A.

(a) $A = \begin{bmatrix} 1 & 1 & -1 \\ -2 & -2 & 4 \end{bmatrix}$ (b) $A = \begin{bmatrix} 3 & 2 \\ 2 & 1 \end{bmatrix}$

(c) $A = \begin{bmatrix} 1 & -1 & 2 & 1 \\ 2 & -1 & 0 & 3 \\ 0 & 1 & -4 & 1 \end{bmatrix}$ (d) $A = \begin{bmatrix} 1 & 1 & 0 & -1 \\ 3 & 2 & 1 & 1 \\ 1 & 0 & 1 & 3 \end{bmatrix}$

17. Let X be an $m \times k$ matrix. If I is the $m \times m$ identity matrix, show that $I + XX^T$ is invertible. [*Hint:* $I + XX^T = [I\ X]\begin{bmatrix} I \\ X^T \end{bmatrix}$ in block form. Use Theorem 5.]

18. Show that an $m \times n$ matrix A has rank m if and only if there is an $n \times m$ matrix B such that $AB = I_m$.

19. If A and B are $m \times n$ matrices, show that rank $(A + B) \leq$ rank A + rank B. [*Hint:* If U and V are the column spaces of A and B, respectively, show that the column space of $A + B$ is contained in $U + V$ and that $\dim(U + V) \leq \dim U + \dim V$. (See item 21 in Exercises 5.3.3.)]

20. Let $N(A) = \{\mathbf{x} \mid A\mathbf{x} = \mathbf{0}\}$ denote the null space of the $m \times n$ matrix A.

(a) Show that $N(A) = N(UA)$ for any invertible $m \times m$ matrix U.

(b) Show that $\dim N(A) = \dim N(AV)$ for any invertible $n \times n$ matrix V. [*Hint:* If $\{\mathbf{x}_1, \ldots, \mathbf{x}_k\}$ is a basis of $N(A)$, consider $\{V^{-1}\mathbf{x}_1, \ldots, V^{-1}\mathbf{x}_k\}$.

21. Two $m \times n$ matrices A and B are called **equivalent** (written $A \overset{e}{\sim} B$) if there exist invertible matrices U and V (sizes $m \times m$ and $n \times n$) such that $A = UBV$.

(a) Prove the following properties of equivalence.

(i) $A \overset{e}{\sim} A$ for all $m \times n$ matrices A.

(ii) If $A \overset{e}{\sim} B$, then $B \overset{e}{\sim} A$.

(iii) If $A \overset{e}{\sim} B$ and $B \overset{e}{\sim} C$, then $A \overset{e}{\sim} C$.

(b) Prove that two $m \times n$ matrices are equivalent if and only if they have the same rank. [*Hint:* Use part **(a)** and Theorem 3.]

22. Let A be an $m \times n$ matrix of rank r.

(a) Show that A can be factored as $A = PQ$, where P is $m \times r$ and has r independent columns and Q is $r \times n$ and has r independent rows. [*Hint:* Let $UAV = \begin{bmatrix} I_r & 0 \\ 0 & 0 \end{bmatrix}$ as in Theorem 3, and write U^{-1} and V^{-1} in block form as follows: $U^{-1} = \begin{bmatrix} U_1 & U_2 \\ U_3 & U_4 \end{bmatrix}$ and $V^{-1} = \begin{bmatrix} V_1 & V_2 \\ V_3 & V_4 \end{bmatrix}$, where U_1 and V_1 are $r \times r$.]

(b) For each matrix A in Exercise 16, write $A = PQ$ as in part **(a)**.

23. Suppose that R and S are two $m \times n$ reduced row-echelon matrices that are row-equivalent. Prove that $R = S$. [*Hint:* Let rank R = rank $S = r$. Use induction on m. If $m > 1$, write $R = US$, where U is an $m \times m$ invertible matrix. Show that the first column of U equals the first column of I_m by examining the first nonzero column of R and of US. Write $U =$

$\left[\begin{array}{c|c} 1 & X \\ \hline 0 & V \end{array}\right]$, $R = \left[\begin{array}{c|c} r & P \\ \hline 0 & R' \end{array}\right]$, and $S = \left[\begin{array}{c|c} s & Q \\ \hline 0 & S' \end{array}\right]$ in block form, and conclude $R' = S'$ by induction. Then show that the first r columns of U equal the first r columns of I_m by examining the columns in $R = US$ that contain leading 1's. Finally, write $U = \left[\begin{array}{c|c} I_r & Y \\ \hline 0 & Z \end{array}\right]$, $R = \left[\begin{array}{c|c} R_1 & R_2 \\ \hline 0 & 0 \end{array}\right]$, and $S = \left[\begin{array}{c|c} S_1 & S_2 \\ \hline 0 & 0 \end{array}\right]$, where R_1 and S_1 are $r \times r$. Conclude that $R = S$.]

SECTION 5.5 Coordinates

In geometry it is customary to label each point in space by an ordered triple of numbers called the coordinates of that point (for plane geometry, two coordinates suffice). In this way geometric space becomes identified with the vector space $\mathbb{R}^3$, so the operations of vector addition and scalar multiplication in $\mathbb{R}^3$ take on a geometric interpretation. Sometimes it is desirable to choose a different coordinate system that is more convenient for a particular geometrical purpose. The best way to look at this is to view a coordinate system as the result of choosing a basis for $\mathbb{R}^3$ (or $\mathbb{R}^2$) and then to realize that selecting a more appropriate coordinate system amounts to finding a better basis. In the present section, coordinates are defined relative to any basis in a finite-dimensional vector space, and then the way that changes in the basis affect the coordinates is described in detail.

A point $P(a,b,c)$ in space has coordinates $x = a$, $y = b$, and $z = c$ and is associated with the vector (a,b,c) in $\mathbb{R}^3$. On the other hand, given a vector in $\mathbb{R}^3$, one way to recover the coordinates is to express the vector in terms of the standard basis $\mathbf{e}_1 = (1,0,0)$, $\mathbf{e}_2 = (0,1,0)$, and $\mathbf{e}_3 = (0,0,1)$.

$$(a,b,c) = a\mathbf{e}_1 + b\mathbf{e}_2 + c\mathbf{e}_3$$

Then the x-, y-, and z-coordinates of the vector may be "read off" as the coefficients of $\mathbf{e}_1$, $\mathbf{e}_2$, and $\mathbf{e}_3$, respectively. A similar observation is true in $\mathbb{R}^n$.

This procedure can be carried out in any finite dimensional vector space V with one small change: Up to now the *order* of the vectors in a basis has been of no importance. However, in this section we shall speak of an **ordered basis** $\{\mathbf{b}_1, \mathbf{b}_2, \ldots, \mathbf{b}_n\}$, which is just a basis where the order in which the vectors are listed is taken into account. Hence $\{\mathbf{b}_2, \mathbf{b}_1, \mathbf{b}_3\}$ would be a different *ordered* basis from $\{\mathbf{b}_1, \mathbf{b}_2, \mathbf{b}_3\}$.

DEFINITION If $B = \{\mathbf{b}_1, \mathbf{b}_2, \ldots, \mathbf{b}_n\}$ is an ordered basis in a vector space V, and if

$$\mathbf{v} = v_1\mathbf{b}_1 + v_2\mathbf{b}_2 + \cdots + v_n\mathbf{b}_n$$

is a vector in V, then the (uniquely determined) numbers $v_1, v_2, \ldots,$ v_n are called the **coordinates** of **v** with respect to the basis B. The **coordinate vector** of **v** with respect to B is defined to be

$$C_B(\mathbf{v}) = \begin{bmatrix} v_1 \\ v_2 \\ \vdots \\ v_n \end{bmatrix}$$

The reason for writing $C_B(\mathbf{v})$ as a column instead of a row will become clear later.

EXAMPLE 1 Find the coordinate vector for $\mathbf{v} = (2,1,3)$ with respect to the ordered basis $B = \{(1,1,0), (1,0,1), (0,1,1)\}$ of $\mathbb{R}^3$.

Solution Express **v** as a linear combination of the basis vectors.

$$\mathbf{v} = (2,1,3) = 0(1,1,0) + 2(1,0,1) + 1(0,1,1)$$

Hence the coordinate vector is $C_B(\mathbf{v}) = \begin{bmatrix} 0 \\ 2 \\ 1 \end{bmatrix}$.

EXAMPLE 2 Let $B = \{(1, 0, \ldots, 0), (0, 1, \ldots, 0), \ldots, (0, 0, \ldots, 1)\}$ denote the standard ordered basis of $\mathbb{R}^n$. Then

$$C_B(\mathbf{v}) = \mathbf{v}^T \qquad \text{for all } \mathbf{v} \text{ in } \mathbb{R}^n$$

Solution This follows because $\mathbf{v} = (v_1, v_2, \ldots, v_n) = v_1\mathbf{e}_1 + v_2\mathbf{e}_2 + \cdots + v_n\mathbf{e}_n$, where the standard basis vectors are denoted $\mathbf{e}_1, \mathbf{e}_2, \ldots, \mathbf{e}_n$, respectively.

An often-encountered problem in geometry is the following: An equation is given using variables x and y, and it is desired to simplify the form of the equation by changing variables. Here is an example.

EXAMPLE 3 Simplify $9x^2 + 10xy + 2y^2 = 9$ by substituting $x = x_1 - y_1$, $y = x_1 + 2y_1$.

Solution Substitution in the equation gives

$$9(x_1^2 - 2x_1y_1 + y_1^2) + 10(x_1^2 + x_1y_1 - 2y_1^2) + 2(x_1^2 + 4x_1y_1 + 4y_1^2) = 9$$

This reduces to $21x_1^2 - 3y_1^2 = 9$, or $7x_1^2 - y_1^2 = 3$. Hence the equation assumes a simpler form with these new variables x_1 and y_1.

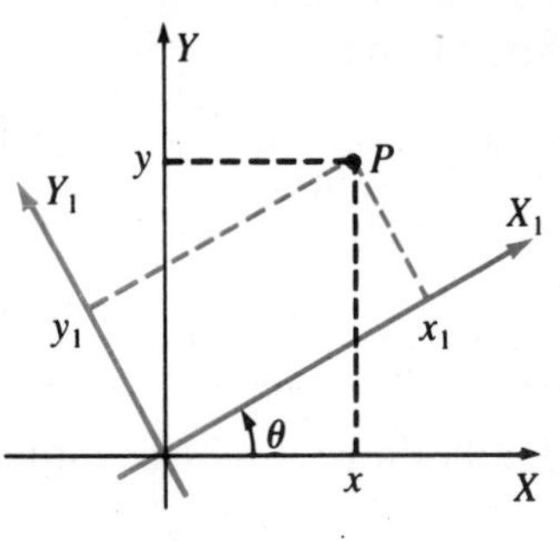

FIGURE 5.1

It turns out that a change of variables can always be made so that the equation, when expressed in terms of the new variables x_1 and y_1, has no x_1y_1-term. Moreover, the new variables arise as the coordinates of (x,y) with respect to an appropriate new basis of $\mathbb{R}^2$. In fact, the new basis is obtained from the standard basis by a rotation.

In Figure 5.1, new X_1-Y_1-axes are obtained by rotating the X-Y-axes counterclockwise through an angle θ. Then a point P will have two sets of coordinates: (x,y) with respect to the X-Y-axes and (x_1,y_1) with respect to the X_1-Y_1-axes. Now the X-Y-coordinates are determined by the vectors $\mathbf{e}_1 = (1,0)$ and $\mathbf{e}_2 = (0,1)$—that is, by the standard basis $B = \{\mathbf{e}_1, \mathbf{e}_2\}$. Similarly, the X_1-Y_1-coordinates are determined by the basis $D = \{\mathbf{f}_1, \mathbf{f}_2\}$ (see Figure 5.2). This basis and the relationship of the new coordinates and the old ones are recorded in Theorem 1.

THEOREM 1 Let B denote the standard basis in $\mathbb{R}^2$, and let $D = \{\mathbf{f}_1, \mathbf{f}_2\}$ denote the new basis obtained from B by rotating the coordinate system around the origin counterclockwise through an angle θ. Then

$$\mathbf{f}_1 = (\cos\theta, \sin\theta)$$
$$\mathbf{f}_2 = (-\sin\theta, \cos\theta)$$

If $\mathbf{v} = (x,y)$ is any vector in $\mathbb{R}^2$, then $C_B(\mathbf{v}) = \begin{bmatrix} x \\ y \end{bmatrix}$ and $C_D(\mathbf{v}) = \begin{bmatrix} x_1 \\ y_1 \end{bmatrix}$ are related by the matrix equation

$$C_D(\mathbf{v}) = \begin{bmatrix} \cos\theta & \sin\theta \\ -\sin\theta & \cos\theta \end{bmatrix} C_B(\mathbf{v}) \quad \text{or} \quad C_B(\mathbf{v}) = \begin{bmatrix} \cos\theta & -\sin\theta \\ \sin\theta & \cos\theta \end{bmatrix} C_D(\mathbf{v})$$

In equation form, this is

$$\begin{aligned} x_1 &= x\cos\theta + y\sin\theta \\ y_1 &= -x\sin\theta + y\cos\theta \end{aligned} \quad \text{or} \quad \begin{aligned} x &= x_1\cos\theta - y_1\sin\theta \\ y &= x_1\sin\theta + y_1\cos\theta \end{aligned}$$

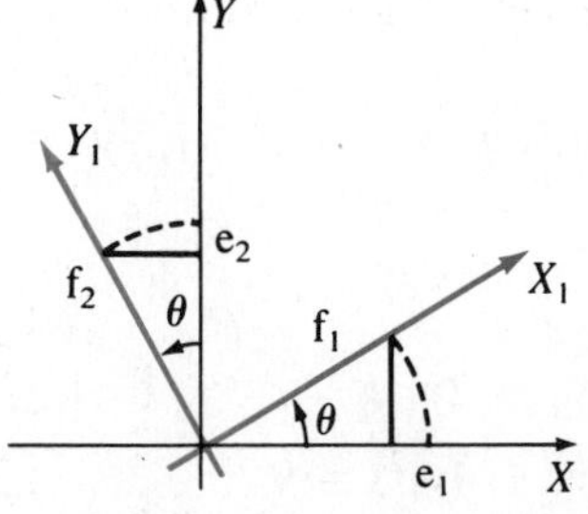

FIGURE 5.2

Proof Simple trigonometry gives the new basis (see Figure 5.2), as in the theorem. Then

$$(x,y) = \mathbf{v} = x_1\mathbf{f}_1 + y_1\mathbf{f}_2 = x_1(\cos\theta, \sin\theta) + y_1(-\sin\theta, \cos\theta)$$

so comparing entries yields

$$x = x_1\cos\theta - y_1\sin\theta$$
$$y = x_1\sin\theta + y_1\cos\theta$$

This can be written $\begin{bmatrix} x \\ y \end{bmatrix} = \begin{bmatrix} \cos\theta & -\sin\theta \\ \sin\theta & \cos\theta \end{bmatrix} \begin{bmatrix} x_1 \\ y_1 \end{bmatrix}$, and the rest is because $\begin{bmatrix} \cos\theta & -\sin\theta \\ \sin\theta & \cos\theta \end{bmatrix}^{-1} = \begin{bmatrix} \cos\theta & \sin\theta \\ -\sin\theta & \cos\theta \end{bmatrix}$. ■

It should be observed that the matrix $\begin{bmatrix} \cos\theta & \sin\theta \\ -\sin\theta & \cos\theta \end{bmatrix}$ is independent of $\mathbf{v}$ and depends only on the two bases B and D. We shall return to this below. For the present we give an illustration of how this result is used.

EXAMPLE 4 Consider the equation $x^2 + xy + y^2 = 1$. Assume that the axes are rotated through an angle θ. Determine θ so that, when the equation is expressed in terms of the new variables x_1 and y_1, there is no x_1y_1-term.

Solution Because (see the proof of Theorem 1) $x = x_1 \cos\theta - y_1 \sin\theta$ and $y = x_1 \sin\theta + y_1 \cos\theta$, the equation becomes

$$[x_1\cos\theta - y_1\sin\theta]^2 + [x_1\cos\theta - y_1\sin\theta]\,[x_1\sin\theta + y_1\cos\theta] + [x_1\sin\theta + y_1\cos\theta]^2 = 1$$

The coefficient of x_1y_1 on the left side is

$$-2\sin\theta\cos\theta + \cos^2\theta - \sin^2\theta + 2\sin\theta\cos\theta = \cos^2\theta - \sin^2\theta$$

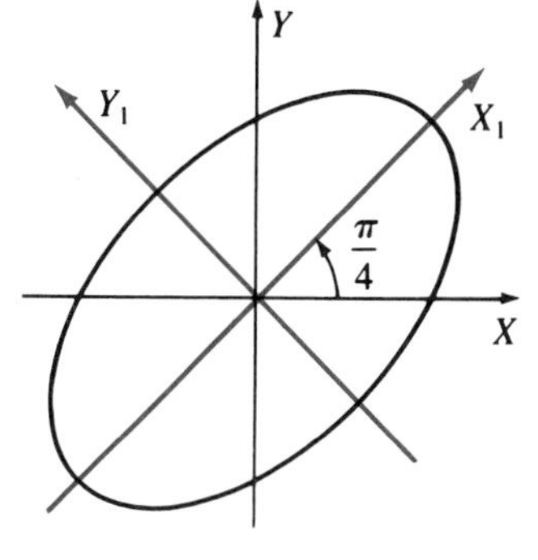

This vanishes if $\cos^2\theta = \sin^2\theta$, and one such angle is $\theta = \pi/4$ (when $\cos\theta = \sin\theta = 1/\sqrt{2}$). Hence the new basis is $\{\mathbf{f}_1, \mathbf{f}_2\}$, where

$$\mathbf{f}_1 = \left[\frac{1}{\sqrt{2}}, \frac{1}{\sqrt{2}}\right] \quad \text{and} \quad \mathbf{f}_2 = \left[\frac{-1}{\sqrt{2}}, \frac{1}{\sqrt{2}}\right]$$

and the new variables are

$$x_1 = \frac{1}{\sqrt{2}}(x + y) \quad \text{and} \quad y_1 = \frac{1}{\sqrt{2}}(-x + y)$$

Because $x = \frac{1}{\sqrt{2}}(x_1 - y_1)$ and $y = \frac{1}{\sqrt{2}}(x_1 + y_1)$, the equation becomes $3x_1^2 + y_1^2 = 2$. In this form the graph of the equation is seen to be an ellipse, and the angle θ has been chosen such that the new X_1-Y_1-axes are the axes of symmetry of the ellipse (see the diagram). It turns out that it is always possible to eliminate the xy-term in an equation $ax^2 + bxy + cy^2 = d$ by rotating the axes through an appropriate angle θ (Exercise 10).

Suppose two ordered bases B and D are given in an n-dimensional vector space V. Then any vector $\mathbf{v}$ in V will have coordinates $C_B(\mathbf{v})$ and $C_D(\mathbf{v})$ with respect to these bases. And, as in Theorem 1, it turns out that there is an $n \times n$ matrix P such that

$$C_D(\mathbf{v}) = P\,C_B(\mathbf{v})$$

for every vector $\mathbf{v}$. To see this, consider first the case where V has dimension $n = 2$. If $B = \{\mathbf{b}_1, \mathbf{b}_2\}$ and $D = \{\mathbf{d}_1, \mathbf{d}_2\}$ express the vectors $\mathbf{b}_1$ and $\mathbf{b}_2$ in terms of $\mathbf{d}_1$ and $\mathbf{d}_2$. Then

$$\mathbf{b}_1 = r\mathbf{d}_1 + s\mathbf{d}_2 \qquad \text{that is, } C_D(\mathbf{b}_1) = \begin{bmatrix} r \\ s \end{bmatrix}$$

$$\mathbf{b}_2 = t\mathbf{d}_1 + u\mathbf{d}_2 \qquad \text{that is, } C_D(\mathbf{b}_2) = \begin{bmatrix} t \\ u \end{bmatrix}$$

Given any vector $\mathbf{v}$ in V, write it as a linear combination of $\mathbf{b}_1$ and $\mathbf{b}_2$.

$$\mathbf{v} = v_1\mathbf{b}_1 + v_2\mathbf{b}_2 \qquad \text{that is, } C_B(\mathbf{v}) = \begin{bmatrix} v_1 \\ v_2 \end{bmatrix}$$

Substitution gives

$$\mathbf{v} = v_1(r\mathbf{d}_1 + s\mathbf{d}_2) + v_2(t\mathbf{d}_1 + u\mathbf{d}_2) = (rv_1 + tv_2)\mathbf{d}_1 + (sv_1 + uv_2)\mathbf{d}_2$$

Hence

$$C_D(\mathbf{v}) = \begin{bmatrix} rv_1 + tv_2 \\ sv_1 + uv_2 \end{bmatrix} = \begin{bmatrix} r & t \\ s & u \end{bmatrix}\begin{bmatrix} v_1 \\ v_2 \end{bmatrix} = P\,C_B(\mathbf{v})$$

where $P = \begin{bmatrix} r & t \\ s & u \end{bmatrix}$ is a matrix depending only on B and D and not on $\mathbf{v}$. Indeed, the columns of P are just $C_D(\mathbf{b}_1) = \begin{bmatrix} r \\ s \end{bmatrix}$ and $C_D(\mathbf{b}_2) = \begin{bmatrix} t \\ u \end{bmatrix}$, so we can represent P in block form by listing its columns in order:

$$P = [C_D(\mathbf{b}_1)\; C_D(\mathbf{b}_2)]$$

P is called the transition matrix from B to D, and when it is necessary to emphasize the bases, we shall write $P = P_{D \leftarrow B}$. This process works in general.

DEFINITION If $B = \{\mathbf{b}_1, \mathbf{b}_2, \ldots, \mathbf{b}_n\}$ and D are any two ordered bases of V, define the **transition matrix** $P_{D \leftarrow B}$ from B to D in block form by listing its columns:

$$P_{D \leftarrow B} = [C_D(\mathbf{b}_1)\; C_D(\mathbf{b}_2) \ldots C_D(\mathbf{b}_n)]$$

> The reason for the form of this notation will be apparent in Theorems 2 and 3.

This means that the first column of $P_{D \leftarrow B}$ is $C_D(\mathbf{b}_1)$, the second column is $C_D(\mathbf{b}_2)$, and so on. With this, the above argument proves the case $n = 2$ of the following theorem. The general case is analogous and is left as Exercise 12.

THEOREM 2 Let $B = \{\mathbf{b}_1, \mathbf{b}_2, \ldots, \mathbf{b}_n\}$ and $D = \{\mathbf{d}_1, \mathbf{d}_2, \ldots, \mathbf{d}_n\}$ be two ordered bases of a vector space V of dimension n. If $P_{D \leftarrow B}$ is defined as above, then

$$C_D(\mathbf{v}) = P_{D \leftarrow B}\, C_B(\mathbf{v})$$

holds for every vector $\mathbf{v}$ in V.

EXAMPLE 5 Find the transition matrix $P_{D \leftarrow B}$ for the following ordered bases of $\mathbb{R}^2$.

$$B = \{\mathbf{b}_1 = (1,0), \mathbf{b}_2 = (1,1)\}, \text{ and } D = \{\mathbf{d}_1 = (1,-1), \mathbf{d}_2 = (1,1)\}$$

Then verify Theorem 2 for $\mathbf{v} = (5,7)$.

Solution Express $\mathbf{b}_1$ and $\mathbf{b}_2$ in terms of $\mathbf{d}_1$ and $\mathbf{d}_2$, and read off $P_{D \leftarrow B}$:

$$\mathbf{b}_1 = \frac{1}{2}\mathbf{d}_1 + \frac{1}{2}\mathbf{d}_2 \qquad \text{and} \qquad \mathbf{b}_2 = 0\mathbf{d}_1 + 1\mathbf{d}_2$$

so $P_{D \leftarrow B} = [C_D(\mathbf{b}_1)\; C_D(\mathbf{b}_2)] = \begin{bmatrix} \frac{1}{2} & 0 \\ \frac{1}{2} & 1 \end{bmatrix}$. If $\mathbf{v} = (5,7)$, then

$$\mathbf{v} = -2\mathbf{b}_1 + 7\mathbf{b}_2 \qquad \text{so } C_B(\mathbf{v}) = \begin{bmatrix} -2 \\ 7 \end{bmatrix}$$

$$\mathbf{v} = -\mathbf{d}_1 + 6\mathbf{d}_2 \qquad \text{so } C_D(\mathbf{v}) = \begin{bmatrix} -1 \\ 6 \end{bmatrix}$$

Then $P_{D \leftarrow B}\, C_B(\mathbf{v}) = \begin{bmatrix} \frac{1}{2} & 0 \\ \frac{1}{2} & 1 \end{bmatrix} \begin{bmatrix} -2 \\ 7 \end{bmatrix} = \begin{bmatrix} -1 \\ 6 \end{bmatrix} = C_D(\mathbf{v})$, as expected.

EXAMPLE 6 In $\mathbf{P}_2$ find $P_{D \leftarrow B}$ if $B = \{1, x, x^2\}$ and $D = \{1, (1-x), (1-x)^2\}$. Then use this to express $p = p(x) = a + bx + cx^2$ as a polynomial in powers of $(1 - x)$.

Solution The transition matrix $P_{D \leftarrow B}$ is computed as follows:

$$\begin{aligned} 1 &= 1 + 0(1 - x) + 0(1 - x)^2 \\ x &= 1 - \ \ (1 - x) + 0(1 - x)^2 \\ x^2 &= 1 - 2(1 - x) + 1(1 - x)^2 \end{aligned}$$

so $P_{D \leftarrow B} = [C_D(1)\ C_D(x)\ C_D(x^2)] = \begin{bmatrix} 1 & 1 & 1 \\ 0 & -1 & -2 \\ 0 & 0 & 1 \end{bmatrix}$. We have $C_B(p) = \begin{bmatrix} a \\ b \\ c \end{bmatrix}$,

so

$$C_D(p) = P_{D \leftarrow B}\, C_B(p) = \begin{bmatrix} 1 & 1 & 1 \\ 0 & -1 & -2 \\ 0 & 0 & 1 \end{bmatrix} \begin{bmatrix} a \\ b \\ c \end{bmatrix} = \begin{bmatrix} a + b + c \\ -b - 2c \\ c \end{bmatrix}$$

Hence $p(x) = (a + b + c) - (b + 2c)(1 - x) + c(1 - x)^2$.

We conclude this section with some important properties of these transition matrices $P_{D \leftarrow B}$. The following property of coordinate vectors will be needed. Suppose that V is an n-dimensional vector space and that $B = \{\mathbf{b}_1, \ldots, \mathbf{b}_n\}$ is an ordered basis of V. If A is an $n \times n$ matrix, then

$$A[C_B(\mathbf{v})] = \mathbf{0} \text{ for all } \mathbf{v} \text{ in } V \text{ implies } A = 0$$

In fact, $A[C_B(\mathbf{b}_i)]$ is the ith column of A, because $C_B(\mathbf{b}_i)$ is the ith column of the $n \times n$ identity matrix.

THEOREM 3 Let B, D, and E be three ordered bases of an n-dimensional vector space.

(1) $P_{B \leftarrow B} = I$

(2) $P_{D \leftarrow B}$ is invertible and $(P_{D \leftarrow B})^{-1} = P_{B \leftarrow D}$

(3) $P_{E \leftarrow D} P_{D \leftarrow B} = P_{E \leftarrow B}$

Proof The proof of (1) is left as Exercise 3. If (1) and (3) are assumed to be true, (2) follows because, taking $E = B$ in (3), $P_{B \leftarrow D} P_{D \leftarrow B} = P_{B \leftarrow B} = I$ by (1). Similarly $P_{D \leftarrow B} P_{B \leftarrow D} = I$, and (2) follows.

It remains to verify (3). For all $\mathbf{v}$ in V, we have $C_D(\mathbf{v}) = P_{D \leftarrow B} C_B(\mathbf{v})$ and $C_E(\mathbf{v}) = P_{E \leftarrow D} C_D(\mathbf{v})$ by Theorem 2. Eliminating $C_D(\mathbf{v})$ gives

$$C_E(\mathbf{v}) = P_{E \leftarrow D} C_D(\mathbf{v}) = P_{E \leftarrow D} P_{D \leftarrow B} C_B(\mathbf{v})$$

But also $C_E(\mathbf{v}) = P_{E\leftarrow B}C_B(\mathbf{v})$, again by Theorem 2, so

$$(P_{E\leftarrow D}P_{D\leftarrow B} - P_{E\leftarrow B})C_B(\mathbf{v}) = \mathbf{0}$$

for all $\mathbf{v}$ in V. Now (3) follows by the remark preceding this theorem. ■

Suppose that $B = \{\mathbf{b}_1, \mathbf{b}_2, \ldots, \mathbf{b}_n\}$ is an ordered basis of a space V. Given $\mathbf{v}$ and $\mathbf{w}$ in V, write them as linear combinations of the basis vectors.

$$\mathbf{v} = v_1\mathbf{b}_1 + v_2\mathbf{b}_2 + \cdots + v_n\mathbf{b}_n$$
$$\mathbf{w} = w_1\mathbf{b}_1 + w_2\mathbf{b}_2 + \cdots + w_n\mathbf{b}_n$$

Adding gives

$$\mathbf{v} + \mathbf{w} = (v_1 + w_1)\mathbf{b}_1 + (v_2 + w_2)\mathbf{b}_2 + \cdots + (v_n + w_n)\mathbf{b}_n$$

and, in terms of components, this asserts that

$$C_B(\mathbf{v} + \mathbf{w}) = \begin{bmatrix} v_1 + w_1 \\ v_2 + w_2 \\ \vdots \\ v_n + w_n \end{bmatrix} = \begin{bmatrix} v_1 \\ v_2 \\ \vdots \\ v_n \end{bmatrix} + \begin{bmatrix} w_1 \\ w_2 \\ \vdots \\ w_n \end{bmatrix} = C_B(\mathbf{v}) + C_B(\mathbf{w})$$

This proves the first part of the following theorem; the other part is left as Exercise 15.

THEOREM 4 Let B be any ordered basis of a vector space V.

(1) $C_B(\mathbf{v} + \mathbf{w}) = C_B(\mathbf{v}) + C_B(\mathbf{w})$ for all $\mathbf{v}$ and $\mathbf{w}$ in V

(2) $C_B(r\mathbf{v}) = r\,C_B(\mathbf{v})$ for all r in $\mathbb{R}$ and $\mathbf{v}$ in V

By virtue of the properties listed in Theorem 4, C_B is called a linear transformation. Linear transformations will be studied extensively in Chapter 8.

EXERCISES 5.5

1. Find an angle θ, and a new basis obtained from the standard basis of $\mathbb{R}^2$ by a counterclockwise rotation through the angle θ, such that the following equations have no x_1y_1-term, where x_1 and y_1 are new variables obtained.

(a) $2x^2 + xy + y^2 = 3$ (b) $x^2 + xy + 2y^2 = 4$
(c) $4x^2 + \sqrt{3}xy + y^2 = 1$ (d) $3x^2 + \sqrt{3}xy + 2y^2 = 4$

2. In each case, find the coordinates of $\mathbf{v}$ with respect to the basis B of the vector space V.

(a) $V = \mathbf{P}_2, \mathbf{v} = 2x^2 + x - 1, B = \{x + 1, x^2, 3\}$
(b) $V = \mathbf{P}_2, \mathbf{v} = ax^2 + bx + c, B = \{x^2, x + 1, x + 2\}$
(c) $V = \mathbb{R}^3, \mathbf{v} = (1,-1,2), B = \{(1,-1,0), (1,1,1), (0,1,1)\}$
(d) $V = \mathbb{R}^3, \mathbf{v} = (a,b,c), B = \{(1,-1,2), (1,1,-1), (0,0,1)\}$
(e) $V = \mathbf{M}_{22}, \mathbf{v} = \begin{bmatrix} 1 & 2 \\ -1 & 0 \end{bmatrix}, B = \left\{\begin{bmatrix} 1 & 1 \\ 0 & 0 \end{bmatrix}, \begin{bmatrix} 1 & 0 \\ 1 & 0 \end{bmatrix}, \begin{bmatrix} 0 & 0 \\ 1 & 1 \end{bmatrix}, \begin{bmatrix} 1 & 0 \\ 0 & 1 \end{bmatrix}\right\}$

3. If $B = \{\mathbf{b}_1, \mathbf{b}_2, \ldots, \mathbf{b}_n\}$ is an ordered basis of a vector space V, describe $C_B(\mathbf{b}_1), C_B(\mathbf{b}_2), \ldots, C_B(\mathbf{b}_n)$ and prove property (1) of Theorem 3.

4. In each case find $P_{D\leftarrow B}$, where B and D are ordered bases of V. Then verify $C_D(\mathbf{v}) = P_{D\leftarrow B}C_B(\mathbf{v})$.

(a) $V = \mathbb{R}^2, B = \{(0,-1), (2,1)\}, D = \{(0,1), (1,1)\}, \mathbf{v} = (3,-5)$
(b) $V = \mathbf{P}_2, B = \{x, 1 + x, x^2\}, D = \{2, x + 3, x^2 - 1\}, \mathbf{v} = 1 + x + x^2$
(c) $V = \mathbf{M}_{22}, B = \left\{\begin{bmatrix} 1 & 0 \\ 0 & 0 \end{bmatrix}, \begin{bmatrix} 0 & 1 \\ 0 & 0 \end{bmatrix}, \begin{bmatrix} 0 & 0 \\ 0 & 1 \end{bmatrix}, \begin{bmatrix} 0 & 0 \\ 1 & 0 \end{bmatrix}\right\},$

$D = \left\{\begin{bmatrix} 1 & 1 \\ 0 & 0 \end{bmatrix}, \begin{bmatrix} 1 & 0 \\ 1 & 0 \end{bmatrix}, \begin{bmatrix} 1 & 0 \\ 0 & 1 \end{bmatrix}, \begin{bmatrix} 0 & 1 \\ 1 & 0 \end{bmatrix}\right\}, \mathbf{v} = \begin{bmatrix} 3 & -1 \\ 1 & 4 \end{bmatrix}$

5. In $\mathbb{R}^3$ find $P_{D\leftarrow B}$, where $B = \{(1,0,0), (1,1,0), (1,1,1)\}$ and $D = \{(1,0,1), (1,0,-1), (0,1,0)\}$. If $\mathbf{v} = (a,b,c)$, show that $C_D(\mathbf{v}) = \frac{1}{2}\begin{bmatrix} a + c \\ a - c \\ 2b \end{bmatrix}$ and $C_B(\mathbf{v}) = \begin{bmatrix} a - b \\ b - c \\ c \end{bmatrix}$, and verify Theorem 2.

6. In $\mathbf{P}_3$ find $P_{D\leftarrow B}$ if $B = \{1, x, x^2, x^3\}$ and $D = \{1, (1 - x), (1 - x)^2, (1 - x)^3\}$. Then express $p = a + bx + cx^2 + dx^3$ as a polynomial in powers of $(1 - x)$.

7. For the situation of Example 5, let $\mathbf{v} = (a, b)$ be *any* vector in $\mathbb{R}^2$ and show that $C_B(\mathbf{v}) = \begin{bmatrix} a - b \\ b \end{bmatrix}$ and $C_D(\mathbf{v}) = \begin{bmatrix} \frac{1}{2}(a - b) \\ \frac{1}{2}(a + b) \end{bmatrix}$. Then verify that $C_D(\mathbf{v}) = P_{D\leftarrow B}C_B(\mathbf{v})$.

8. In each case verify that $P_{D\leftarrow B}$ is the inverse of $P_{B\leftarrow D}$ and that $P_{E\leftarrow D}P_{D\leftarrow B} = P_{E\leftarrow B}$, where B, D, and E are ordered bases of V.

(a) $V = \mathbb{R}^3, B = \{(1,1,1), (1,-2,1), (1,0,-1)\}, D = \{(1,0,0), (0,1,0), (0,0,1)\}, E = \{(1,1,1), (1,-1,0), (-1,0,1)\}$
(b) $V = \mathbf{P}_2, B = \{1, x, x^2\}, D = \{1 + x + x^2, 1 - x, -1 + x^2\}, E = \{x^2, x, 1\}$

9. Use property (2) of Theorem 3, with D the standard basis of $\mathbb{R}^n$, to find the inverse of:

(a) $A = \begin{bmatrix} 1 & 1 & 0 \\ 1 & 0 & 1 \\ 0 & 1 & 1 \end{bmatrix}$ (b) $A = \begin{bmatrix} 1 & 2 & 1 \\ 2 & 3 & 0 \\ -1 & 0 & 2 \end{bmatrix}$

10. Consider the equation $ax^2 + bxy + cy^2 = d$, where $b \neq 0$. Introduce new variables x_1 and y_1 by rotating the axes counterclockwise through an angle θ. Show that the resulting equation has no x_1y_1-term if θ is given by

$$\cos 2\theta = \frac{a - c}{\sqrt{b^2 + (a - c)^2}}, \quad \sin 2\theta = \frac{b}{\sqrt{b^2 + (a - c)^2}}$$

[*Hint:* $\sin 2\theta = 2 \sin \theta \cos \theta$, and $\cos 2\theta = \cos^2\theta - \sin^2\theta$.]

11. Let $B = \{\mathbf{b}_1, \mathbf{b}_2, \ldots, \mathbf{b}_n\}$ be any ordered basis of $\mathbb{R}^n$, written as columns. If $Q = [\mathbf{b}_1\ \mathbf{b}_2\ \ldots\ \mathbf{b}_n]$ is the matrix with the $\mathbf{b}_i$ as columns, show that $QC_B(\mathbf{v}) = \mathbf{v}$ for all $\mathbf{v}$ in $\mathbb{R}^n$.

12. Prove Theorem 2 by expressing each $\mathbf{b}_j$ as follows:

$$\mathbf{b}_j = p_{1j}\mathbf{d}_1 + p_{2j}\mathbf{d}_2 + \cdots + p_{nj}\mathbf{d}_n = \sum_{i=1}^{n} p_{ij}\mathbf{d}_i$$

Show that, if $P = [p_{ij}]$, then $P = [C_D(\mathbf{b}_1) \ldots C_D(\mathbf{b}_n)]$ and $C_D(\mathbf{v}) = PC_B(\mathbf{v})$ for all $\mathbf{v}$ in V.

13. Let $B = \{\mathbf{e}_1, \ldots, \mathbf{e}_n\}$ be the standard ordered basis of $\mathbb{R}^n$, written as columns. If $D = \{\mathbf{d}_1, \ldots, \mathbf{d}_n\}$ is any ordered basis, show that $P_{B \leftarrow D} = [\mathbf{d}_1 \ldots \mathbf{d}_n]$.

14. Find $P_{D \leftarrow B}$ if $B = \{\mathbf{b}_1, \mathbf{b}_2, \mathbf{b}_3, \mathbf{b}_4\}$ and $D = \{\mathbf{b}_2, \mathbf{b}_3, \mathbf{b}_1, \mathbf{b}_4\}$. Matrices arising when the bases differ only in the *order* of the vectors are called **permutation matrices.**

15. Let B be an ordered basis of a vector space V. Let $\mathbf{v}$ and $\mathbf{w}$ be vectors in V.

(a) Prove Theorem 4.

(b) Show that $C_B(a\mathbf{v} + b\mathbf{w}) = aC_B(\mathbf{v}) + bC_B(\mathbf{w})$ for all numbers a and b.

(c) If $C_B(\mathbf{v}) = C_B(\mathbf{w})$, show that $\mathbf{v} = \mathbf{w}$.

(d) Show that $\{\mathbf{v}, \mathbf{w}\}$ is linearly independent in V if and only if $\{C_B(\mathbf{v}), C_B(\mathbf{w})\}$ is linearly independent. [*Hint:* See parts **(b)** and **(c)**.]

(e) Extend part **(d)** to m vectors in V.

SECTION 5.6 Applications of Vector Spaces (Optional)[†]

5.6.1 Vector Spaces of Polynomials

The vector space of all polynomials of degree at most n is denoted $\mathbf{P}_n$, and it was established in Section 5.3 that $\mathbf{P}_n$ has dimension $n + 1$; in fact, $\{1, x, x^2, \ldots, x^n\}$ is a basis. The next theorem shows that any $n + 1$ polynomials of distinct degrees will form a basis.

[†]The two applications in this section are independent and may be taken in any order.

THEOREM 1

Let $p_0(x), p_1(x), p_2(x), \ldots, p_n(x)$ be polynomials in $\mathbf{P}_n$ of degrees 0, 1, 2, ..., n, respectively. Then $\{p_0(x), \ldots, p_n(x)\}$ is a basis of $\mathbf{P}_n$.

Proof Because $\dim \mathbf{P}_n = n + 1$, it suffices by Theorem 8 in Section 5.3 to show that $\{p_0(x), p_1(x), \ldots, p_n(x)\}$ is linearly independent. So suppose some linear combination vanishes:

$$a_0p_0(x) + a_1p_1(x) + \cdots + a_np_n(x) = 0$$

The aim is to show that $a_0 = a_1 = \cdots = a_n = 0$. Because $p_n(x)$ has degree n and every other polynomial $p_k(x)$ has degree *lower* than n, the term involving x^n on the left side is a_nax^n, where $a \neq 0$ is the leading coefficient of $p_n(x)$. It follows that $a_na = 0$, so $a_n = 0$. Hence the foregoing linear combination becomes

$$a_0p_0(x) + a_1p_1(x) + \cdots + a_{n-1}p_{n-1}(x) = 0$$

and a similar argument shows that $a_{n-1} = 0$. Continue in this way to obtain $a_k = 0$ for all k. ■

An immediate consequence is that $\{1, (x - a), (x - a)^2, \ldots, (x - a)^n\}$ is a basis of $\mathbf{P}_n$ for any number a. Hence

COROLLARY 1

If a is any number, every polynomial $f(x)$ of degree at most n has an **expansion in powers** of $(x - a)$.

$$f(x) = a_0 + a_1(x - a) + a_2(x - a)^2 + \cdots + a_n(x - a)^n \qquad (*)$$

If $f(x)$ is evaluated at $x = a$, then (*) becomes

$$f(a) = a_0 + a_1(a - a) + \cdots + a_n(a - a)^n = a_0$$

Hence $a_0 = f(a)$, and (*) can be written $f(x) = f(a) + (x - a)g(x)$, where $g(x)$ is a polynomial of degree $n - 1$ (this assumes that $n \geq 1$). If it happens that $f(a) = 0$, then it is clear that $f(x)$ has the form $f(x) = (x - a)g(x)$. Conversely, every such polynomial certainly satisfies $f(a) = 0$, and we obtain

COROLLARY 2

Remainder Theorem

Let $f(x)$ be a polynomial of degree $n \geq 1$ and let a be any number. Then

(1) $f(x) = f(a) + (x - a)\,g(x)$ for some polynomial $g(x)$ of degree $n - 1$.

Factor Theorem

(2) $f(a) = 0$ if and only if $f(x) = (x - a)g(x)$ for some polynomial $g(x)$.

The polynomial $g(x)$ can be easily computed by using "long division" to divide $f(x)$ by $(x - a)$.

All the coefficients in the expansion (*) of $f(x)$ in powers of $(x - a)$ can be determined in terms of the derivatives[†] of $f(x)$. (These will be familiar to students of calculus.) Let $f^{(n)}(x)$ denote the nth derivative of the polynomial $f(x)$, and write $f^{(0)}(x) = f(x)$. Then, if

$$f(x) = a_0 + a_1(x - a) + a_2(x - a)^2 + \cdots + a_n(x - a)^n$$

it is clear that $a_0 = f(a) = f^{(0)}(a)$. Differentiation gives

$$f^{(1)}(x) = a_1 + 2a_2(x - a) + 3a_3(x - a)^2 + \cdots + na_n(x - a)^{n-1}$$

and substituting $x = a$ yields $a_1 = f^{(1)}(a)$. This process continues to give $a_2 = \dfrac{f^{(2)}(a)}{2!}$, $a_3 = \dfrac{f^{(3)}(a)}{3!}$, and so on, where $k! = k(k - 1) \ldots 2 \cdot 1$. Hence

COROLLARY 3
Taylor's Theorem

If $f(x)$ is a polynomial of degree n, then

$$f(x) = f(a) + \frac{f^{(1)}(a)}{1!}(x - a) + \frac{f^{(2)}(a)}{2!}(x - a)^2 + \cdots + \frac{f^{(n)}(a)}{n!}(x - a)^n$$

EXAMPLE 1

Expand $f(x) = 5x^3 + 10x + 2$ as a polynomial in powers of $x - 1$.

Solution The derivatives are $f^{(1)}(x) = 15x^2 + 10$, $f^{(2)}(x) = 30x$, and $f^{(3)}(x) = 30$. Hence the Taylor expansion is

$$\begin{aligned} f(x) &= f(1) + \frac{f^{(1)}(1)}{1!}(x - 1) + \frac{f^{(2)}(1)}{2!}(x - 1)^2 + \frac{f^{(3)}(1)}{3!}(x - 1)^3 \\ &= 17 + 25(x - 1) + 15(x - 1)^2 + 5(x - 1)^3 \end{aligned}$$

Taylor's theorem is useful in that it provides a formula for the coefficients in the expansion. It is dealt with in calculus texts and will not be pursued here.

Theorem 1 produces bases of $\mathbf{P}_n$ consisting of polynomials of distinct degrees. A different criterion is involved in the next theorem.

[†]The discussion of Taylor's theorem may be omitted with no loss of continuity.

THEOREM 2 Let $f_0(x), f_1(x), \ldots, f_n(x)$ be polynomials in $\mathbf{P}_n$. Assume that numbers $a_0, a_1, \ldots, a_n$ exist such that

$$f_i(a_i) \neq 0 \quad \text{for each } i$$
$$f_i(a_j) = 0 \quad \text{if } i \neq j$$

Then

(1) $\{f_0(x), \ldots, f_n(x)\}$ is a basis of $\mathbf{P}_n$.

(2) If $f(x)$ is any polynomial in $\mathbf{P}_n$, its expansion as a linear combination of these basis vectors is

$$f(x) = \frac{f(a_0)}{f_0(a_0)}f_0(x) + \frac{f(a_1)}{f_1(a_1)}f_1(x) + \cdots + \frac{f(a_n)}{f_n(a_n)}f_n(x)$$

Proof (1) It suffices to show that it is linearly independent (because $\dim \mathbf{P}_n = n + 1$). Suppose that

$$r_0 f_0(x) + r_1 f_1(x) + \cdots + r_n f_n(x) = 0 \qquad r_i \text{ in } \mathbb{R}$$

Because $f_i(a_0) = 0$ for all $i > 0$, taking $x = a_0$ gives $r_0 f_0(a_0) = 0$. But then the fact that $f_0(a_0) \neq 0$ shows that $r_0 = 0$. The proof that $r_i = 0$ for $i > 0$ is analogous.

(2) By (1), $f(x) = r_0 f_0(x) + \cdots + r_n f_n(x)$ for *some* numbers r_i. Again, evaluating at a_0 gives $f(a_0) = r_0 f_0(a_0)$, so $r_0 = f(a_0)/f_0(a_0)$. Similarly, $r_i = f(a_i)/f_i(a_i)$ for each i. ■

EXAMPLE 2 Show that $\{x^2 - x, x^2 - 2x, x^2 - 3x + 2\}$ is a basis of $\mathbf{P}_2$.

Solution Write $f_0(x) = x^2 - x = x(x - 1)$, $f_1(x) = x^2 - 2x = x(x - 2)$, and $f_2(x) = x^2 - 3x + 2 = (x - 1)(x - 2)$. Then the conditions of the theorem are satisfied with $a_0 = 2$, $a_1 = 1$, and $a_2 = 0$.

If $n + 1$ distinct numbers are given, there is a natural choice of the polynomials $f_i(x)$ in Theorem 2. To illustrate, let a_0, a_1, and a_2 be distinct and write

$$f_0(x) = \frac{(x - a_1)(x - a_2)}{(a_0 - a_1)(a_0 - a_2)} \qquad f_1(x) = \frac{(x - a_0)(x - a_2)}{(a_1 - a_0)(a_1 - a_2)}$$

$$f_2(x) = \frac{(x - a_0)(x - a_1)}{(a_2 - a_0)(a_2 - a_1)}$$

Then $f_0(a_0) = f_1(a_1) = f_2(a_2) = 1$, whereas $f_i(a_j) = 0$ for $i \neq j$. Hence Theorem 2 applies, and because $f_i(a_i) = 1$ for each i, the formula for expanding any polynomial is simplified.

In fact, this can be generalized with no extra effort. If $a_0, a_1, \ldots, a_n$ are distinct numbers, define the **Lagrange polynomials** $c_0(x), c_1(x), \ldots, c_n(x)$ relative to these numbers as follows:

$$c_k(x) = \frac{\prod_{i \neq k} (x - a_i)}{\prod_{i \neq k} (a_k - a_i)} \qquad k = 0, 1, 2, \ldots, n$$

Here the numerator is the product of all the terms $(x - a_0)$, $(x - a_1), \ldots, (x - a_n)$ with $(x - a_k)$ omitted, and a similar remark applies to the denominator. If $n = 3$, these are just the polynomials in the preceding paragraph. If $n = 4$, the polynomial $c_1(x)$ takes the form

$$c_1(x) = \frac{(x - a_0)(x - a_2)(x - a_3)}{(a_1 - a_0)(a_1 - a_2)(a_1 - a_3)}$$

In the general case, it is clear that $c_i(a_i) = 1$ for each i and that $c_i(a_j) = 0$ if $i \neq j$. Hence Theorem 2 specializes as given in Theorem 3.

THEOREM 3 Lagrange Interpolation Expansion

Let $a_0, a_1, \ldots, a_n$ be distinct numbers. The corresponding set

$$\{c_0(x), c_1(x), \ldots, c_n(x)\}$$

of Lagrange polynomials is a basis of $\mathbf{P}_n$, and any polynomial $f(x)$ in $\mathbf{P}_n$ has the following unique expansion as a linear combination of these polynomials.

$$f(x) = f(a_0)c_0(x) + f(a_1)c_1(x) + \cdots + f(a_n)c_n(x)$$

EXAMPLE 3 Find the Lagrange interpolation expansion for $f(x) = x^2 - 2x + 1$ relative to $a_0 = -1$, $a_1 = 0$, and $a_2 = 1$.

Solution The Lagrange polynomials are

$$c_0(x) = \frac{(x - 0)(x - 1)}{(-1 - 0)(-1 - 1)} = \frac{1}{2}(x^2 - x)$$

$$c_1(x) = \frac{(x + 1)(x - 1)}{(0 + 1)(0 - 1)} = -(x^2 - 1)$$

$$c_2(x) = \frac{(x + 1)(x - 0)}{(1 + 1)(1 - 0)} = \frac{1}{2}(x^2 + x)$$

Because $f(-1) = 4$, $f(0) = 1$, and $f(1) = 0$, the expansion is $f(x) = 2(x^2 - x) - (x^2 - 1)$.

The Lagrange interpolation expansion gives an easy proof of the following important fact.

COROLLARY Let $f(x)$ be a polynomial in $\mathbf{P}_n$, and let $a_0, a_1, \ldots, a_n$ denote distinct numbers. If $f(a_i) = 0$ for all i, then $f(x)$ is the zero polynomial (that is, all coefficients are zero).

Proof All the coefficients in the Lagrange expansion of $f(x)$ are zero. ■

EXERCISES 5.6.1

1. Show that any set of polynomials of distinct degrees is linearly independent.

†2. Expand each of the following as a polynomial in powers of $x - 1$.
   (a) $f(x) = x^3 - 2x^2 + x - 1$ **(b)** $f(x) = x^3 + x + 1$
   (c) $f(x) = x^4$ **(d)** $f(x) = x^3 - 3x^2 + 3x$

†3. Prove Taylor's theorem for polynomials.

†4. Use Taylor's theorem to derive the **binomial theorem:**

$$(1 + x)^n = \binom{n}{0} + \binom{n}{1}x + \binom{n}{2}x^2 + \cdots + \binom{n}{n}x^n$$

Here the **binomial coefficients** $\binom{n}{r}$ are defined by $\binom{n}{r} = \frac{n!}{r!(n-r)!}$ where $n! = n(n-1)\cdots 2 \cdot 1$ if $n \geq 1$ and $0! = 1$.

†5. Let $f(x)$ be a polynomial of degree n. Show that, given any polynomial $g(x)$ in $\mathbf{P}_n$, there exist numbers $b_0, b_1, \ldots, b_n$ such that

$$g(x) = b_0 f(x) + b_1 f^{(1)}(x) + \cdots + b_n f^{(n)}(x)$$

where $f^{(k)}(x)$ denotes the kth derivative of $f(x)$.

6. Use Theorem 2 to show that the following are bases of $\mathbf{P}_2$.
   (a) $\{x^2 - 2x, x^2 + 2x, x^2 - 4\}$
   (b) $\{x^2 - 3x + 2, x^2 - 4x + 3, x^2 - 5x + 6\}$

7. Find the Lagrange interpolation expansion of $f(x)$ relative to $a_0 = 1$, $a_1 = 2$, and $a_2 = 3$.
   (a) $f(x) = x^2 + 1$ **(b)** $f(x) = x^2 + x + 1$

8. Let $a_0, a_1, \ldots, a_n$ be distinct numbers. If $f(x)$ and $g(x)$ in $\mathbf{P}_n$ satisfy $f(a_i) = g(a_i)$ for all i, show that $f(x) = g(x)$. [*Hint:* See the corollary to Theorem 3.]

---

†This exercise requires polynomial differentiation.

9. Let $a_0, a_1, \ldots, a_n$ be distinct numbers. If $f(x)$ in $\mathbf{P}_{n+1}$ satisfies $f(a_i) = 0$ for each $i = 0, 1, \ldots, n$, show that $f(x) = r(x - a_0)(x - a_1) \ldots (x - a_n)$ for some r in $\mathbb{R}$. [*Hint:* r is the coefficient of x^{n+1} in $f(x)$. Consider $f(x) - r(x - a_0) \ldots (x - a_n)$ and use the corollary to Theorem 3.]

10. Let a and b denote distinct numbers.

(a) Show that $\{(x - a), (x - b)\}$ is a basis of $\mathbf{P}_1$.

(b) Show that $\{(x - a)^2, (x - a)(x - b), (x - b)^2\}$ is a basis of $\mathbf{P}_2$.

(c) Show that $\{(x - a)^n, (x - a)^{n-1}(x - b), \ldots, (x - a)(x - b)^{n-1}, (x - b)^n\}$ is a basis of $\mathbf{P}_n$. [*Hint* for part **(c)**: If a linear combination vanishes, evaluate at $x = a$ and $x = b$. Then reduce to the case $n - 2$ by using the fact that if $p(x)q(x) = 0$ in $\mathbf{P}$, then either $p(x) = 0$ or $q(x) = 0$.]

11. Let a and b be two distinct numbers. Assume that $n \geq 2$ and let

$$U_n = \{f(x) \text{ in } \mathbf{P}_n \mid f(a) = 0 = f(b)\}$$

(a) Show that $U_n = \{(x - a)(x - b)p(x) \mid p(x) \text{ in } \mathbf{P}_{n-2}\}$.

(b) Show that $\dim U_n = n - 1$. [*Hint:* If $p(x)q(x) = 0$ in $\mathbf{P}$, then either $p(x) = 0$ or $q(x) = 0$.]

(c) Show that $\{(x - a)^{n-1}(x - b), (x - a)^{n-2}(x - b)^2, \ldots, (x - a)^2(x - b)^{n-2}, (x - a)(x - b)^{n-1}\}$ is a basis of U_n. [*Hint:* Exercise 10.]

5.6.2 Differential Equations of First and Second Order†

Let f be a function of a variable x, and let f' and f'' denote the first and second derivatives of f. Equations of the form

$$f' + 3f = 0$$
$$f'' + 2f' + f = 0$$

are called **differential equations,** and solving many practical problems comes down to finding functions f satisfying such an equation. The study of differential equations is a very large theory, and the present book gives a short introduction to how linear algebra aids in the solution of these equations (we return to this subject in Section 7.4.2). Of course, an acquaintance with calculus is required.

The simplest example is the **first-order** equation

$$f' + af = 0$$

where a is a number. It is easily verified that $f(x) = e^{-ax}$ is one solution, and this equation is simple enough for us to find *all* solutions. In fact, suppose f is *any* solution so that $f'(x) + af(x) = 0$. Then consider the new function given by $g(x) = f(x)e^{ax}$. The chain rule of differentiation gives

†This section requires differential calculus.

$$\begin{aligned} g'(x) &= f(x)[ae^{ax}] + f'(x)e^{ax} \\ &= af(x)e^{ax} - af(x)e^{ax} \\ &= 0 \end{aligned}$$

Hence the function $g(x)$ has zero derivative and so must be a constant—say, $g(x) = c$. But then $f(x)e^{ax} = c$, so

$$f(x) = ce^{-ax}$$

In other words, every solution $f(x)$ is just a multiple of the "basic" solution e^{-ax}.

At this point we can see where linear algebra comes into play. The aim is to describe *all* solutions of the equation $f' + af = 0$—that is, to describe the set

$$U = \{f \mid f' \text{ exists and } f' + af = 0\}$$

But this set U is a vector space. In fact, if f and f_1 both lie in U (so $f' + af = 0$ and $f_1' + af_1 = 0$), then given a number c, the basic theory of differentiation shows that $(f + f_1)' = f' + f'_1$ and $(cf)' = cf'$ both exist and that $f + f_1$ and cf lie in U:

$$\begin{aligned} (f + f_1)' + a(f + f_1) &= (f' + af) + (f_1' + af_1) = 0 \\ (cf)' + a(cf) &= c(f' + af) = 0 \end{aligned}$$

Hence U is a vector space (in fact, it is a subspace of the space of all real-valued functions), and the previous paragraph shows that e^{-ax} lies in U and *every* member of U is a scalar multiple of e^{-ax}. This can be expressed as in Theorem 4.

THEOREM 4

The set of solutions of the first-order differential equation

$$f' + af = 0$$

is a one-dimensional vector space, and $\{e^{-ax}\}$ is a basis.

EXAMPLE 4

Assume that the number $n(t)$ of bacteria in a culture at time t has the property that the rate of change of n is proportional to n itself. If there are n_0 bacteria present when $t = 0$, find the number at time t.

Solution Let k denote the proportionality constant. The rate of change of $n(t)$ is its time-derivative $n'(t)$, so the given relationship is $n'(t) = kn(t)$. Thus $n' - kn = 0$, and Theorem 5 shows that all solutions n are given by $n(t) = ce^{kt}$, where c is a constant. In this case, the constant c is determined by the requirement that there be n_0 bacteria present when $t = 0$. Hence $n_0 = n(0) = ce^{k \cdot 0} = c$, so

$$n(t) = n_0 e^{kt}$$

gives the number at time t. Of course the constant k depends on the strain of bacteria.

The condition that $n(0) = n_0$ in Example 4 is called an **initial condition** or a **boundary condition** and serves to select one solution from the available ones. Only one initial condition is needed here, because the space of solutions is one-dimensional.

Now consider **second-order** differential equations of the form

$$f'' + af' + bf = 0$$

where a and b are constants. Again the set

$$U = \{f \mid f'' + af' + bf = 0\}$$

is a vector space, and here $\dim U = 2$ (we omit the proof). In order to find a basis for U, it is necessary to introduce the **characteristic polynomial**

$$x^2 + ax + b$$

of the equation. Suppose that λ is a real root of this polynomial—that is, $\lambda^2 + a\lambda + b = 0$. Then the function

$$g(x) = e^{\lambda x}$$

is a solution to the differential equation:

$$\begin{aligned} g''(x) + ag'(x) + bg(x) &= \lambda^2 e^{\lambda x} + a\lambda e^{\lambda x} + be^{\lambda x} \\ &= (\lambda^2 + a\lambda + b)e^{\lambda x} \\ &= 0 \end{aligned}$$

Hence if λ and μ are two distinct real roots of the characteristic polynomial, then $e^{\lambda x}$ and $e^{\mu x}$ are solutions to the differential equation and so lie in U. Moreover, they are linearly independent, because if $re^{\lambda x} + se^{\mu x} = 0$ for numbers r and s, and if $r \neq 0$, then $e^{(\lambda - \mu)x} = \dfrac{-s}{r}$, so $e^{(\lambda - \mu)x}$ is constant. This is impossible if $\lambda \neq \mu$, so the assumption that $r \neq 0$ is invalid. Thus $r = 0$, and similarly $s = 0$. Hence $\{e^{\lambda x}, e^{\mu x}\}$ is a linearly independent set in U and so, because $\dim U = 2$, it is a basis. This establishes the first part of Theorem 5.

THEOREM 5 Let U denote the space of solutions of the second-order differential equation

$$f'' + af' + bf = 0$$

Assume that λ and μ are real roots of the characteristic polynomial $x^2 + ax + b$. Then

(1) If $\lambda \neq \mu$, then $\{e^{\lambda x}, e^{\mu x}\}$ is a basis of U.

(2) If $\lambda = \mu$, then $\{e^{\lambda x}, xe^{\lambda x}\}$ is a basis of U.

Proof Except for the fact that $\dim U = 2$, (1) has been proved above. If $\lambda = \mu$, the verification that $xe^{\lambda x}$ is a solution and that $\{e^{\lambda x}, xe^{\lambda x}\}$ is linearly independent is left as Exercise 4. Then (2) follows, because $\dim U = 2$. ■

EXAMPLE 5 Find all solutions f of $f'' - f' - 6f = 0$.

Solution The characteristic polynomial is $x^2 - x - 6 = (x - 3) \cdot (x + 2)$. The roots are 3 and -2, so $\{e^{3x}, e^{-2x}\}$ is a basis for the space of solutions. Hence every solution has the form

$$f(x) = ce^{3x} + de^{-2x}$$

where c and d are constants.

The function $f(x) = ce^{3x} + de^{-2x}$ in Example 5 is sometimes referred to as the **general solution** of the differential equation. The constants c and d are determined by two boundary conditions.

EXAMPLE 6 Find the solution of $f'' + 4f' + 4f = 0$ that satisfies the boundary conditions $f(0) = 1, f(1) = -1$.

Solution The characteristic polynomial is $x^2 + 4x + 4 = (x + 2)^2$, so -2 is a double root. Hence $\{e^{-2x}, xe^{-2x}\}$ is a basis for the space of solutions, and the general solution takes the form $f(x) = ce^{-2x} + dxe^{-2x}$. Applying the boundary conditions gives $1 = f(0) = c$ and $-1 = f(1) = (c + d)e^{-2}$. Hence $c = 1$ and $d = -(1 + e^2)$, so the required solution is

$$f(x) = e^{-2x} - (1 + e^2)xe^{-2x}$$

One further question remains: What happens if the roots of the characteristic polynomial are not real? To answer this, we must first state precisely what $e^{\lambda x}$ means when λ is not real. If q is a real number, define

$$e^{iq} = \cos q + i \sin q$$

where $i^2 = -1$. Then the relationship $e^{iq}e^{iq_1} = e^{i(q+q_1)}$ holds for all real q and q_1, as is easily verified. If $\lambda = p + iq$, where p and q are real numbers, we define

$$e^\lambda = e^p e^{iq} = e^p(\cos q + i \sin q)$$

Then it is a routine exercise to show that

(1) $e^\lambda e^\mu = e^{\lambda+\mu}$

(2) $e^\lambda = 1$ if and only if $\lambda = 0$

(3) $(e^{\lambda x})' = \lambda e^{\lambda x}$

These imply easily that $f(x) = e^{\lambda x}$ is a solution to $f'' + af' + bf = 0$ if λ is a (possibly complex) root of the characteristic polynomial $x^2 + ax + b$. Now write $\lambda = p + iq$ so that

$$f(x) = e^{\lambda x} = e^{px}\cos(qx) + ie^{px}\sin(qx)$$

For convenience, denote the real and imaginary parts of $f(x)$ as $u(x) = e^{px}\cos(qx)$ and $v(x) = e^{px}\sin(qx)$. Then the fact that $f(x)$ satisfies the differential equation gives

$$0 = f'' + af' + bf = (u'' + au' + bu) + i(v'' + av' + bv)$$

Equating real and imaginary parts shows that $u(x)$ and $v(x)$ are both solutions to the differential equation. This proves part of Theorem 6.

THEOREM 6 Let U denote the space of solutions of the second-order differential equation

$$f'' + af' + bf = 0$$

where a and b are real. Suppose λ is a nonreal root of the characteristic polynomial $x^2 + ax + b$. If $\lambda = p + iq$, where p and q are real, then

$$\{e^{px}\cos(qx), e^{px}\sin(qx)\}$$

is a basis of U.

Proof The foregoing discussion shows that these functions lie in U. Because $\dim U = 2$ (a fact we have not proved), it suffices to show that they are linearly independent. But if

$$re^{px}\cos(qx) + se^{px}\sin(qx) = 0$$

for all x, then $r\cos(qx) + s\sin(qx) = 0$ for all x (because $e^{px} \neq 0$). Taking $x = 0$ gives $r = 0$, and taking $x = \frac{\pi}{2q}$ gives $s = 0$ ($q \neq 0$ because λ is not real). This is what we wanted. ■

EXAMPLE 7 Find the solution $f(x)$ to $f'' - 2f' + 2f = 0$ that satisfies $f(0) = 2$ and $f\left(\frac{\pi}{2}\right) = 0$.

Solution The characteristic polynomial $x^2 - 2x + 2$ has roots $1 + i$ and $1 - i$. Taking $\lambda = 1 + i$ (quite arbitrarily) gives $p = q = 1$ in the notation of Theorem 6, so $\{e^x\cos x, e^x\sin x\}$ is a basis for the space of solutions. The general solution is thus $f(x) = e^x(r\cos x + s\sin x)$. The boundary conditions yield $2 = f(0) = r$ and $0 = f\left(\frac{\pi}{2}\right) = e^{\pi/2}s$. Thus $r = 2$ and $s = 0$, and the required solution is $f(x) = 2e^x\cos x$.

The following is an important special case of Theorem 6.

THEOREM 7 If $q \neq 0$ is a real number, the space of solutions to the differential equation $f'' + q^2f = 0$ has basis $\{\cos(qx), \sin(qx)\}$.

Proof The characteristic polynomial $x^2 + q^2$ has roots qi and $-qi$, so Theorem 6 applies with $p = 0$. ■

In many situations, the displacement $s(t)$ of some object at time t turns out to have an oscillating form $s(t) = c\sin(at) + d\cos(at)$. These are called **simple harmonic motions**; an example follows.

EXAMPLE 8 A weight is attached to an extension spring. If it is pulled from the equilibrium position and released, it is observed to oscillate up and down. Let $e(t)$ denote the distance of the weight below the equilibrium position t seconds later. It is known **(Hooke's Law)** that the acceleration $e''(t)$ of the weight is proportional to the displacement $e(t)$ and in the opposite direction. That is,

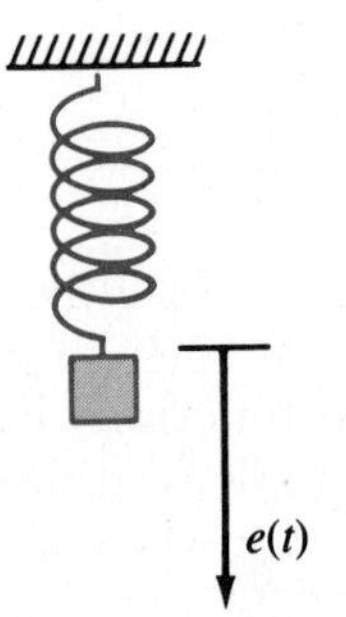

$$e''(t) = -ke(t)$$

where $k > 0$ is called the **spring constant.** Find $e(t)$ if the maximum extension is 10 cm below the equilibrium position, and find the **period** of the oscillation (time taken for the weight to make a full oscillation).

Solution It follows from Theorem 7 (with $q^2 = k$) that

$$e(t) = c\sin(\sqrt{k}t) + d\cos(\sqrt{k}t)$$

where c and d are constants. The condition $e(0) = 0$ gives $d = 0$, so $e(t) = c\sin(\sqrt{k}t)$. Now the maximum value of the function $\sin x$ is 1 (when $x = \pi/2$), so $c = 10$ (when $t = \frac{\pi}{2\sqrt{k}}$). Hence

$$e(t) = 10\sin(\sqrt{k}t)$$

Finally the weight goes through a full oscillation as $\sqrt{k}t$ increases from 0 to 2π. The time taken is $t = \dfrac{2\pi}{\sqrt{k}}$, the period of the oscillation.

EXERCISES 5.6.2

1. Find a solution f to each of the following differential equations satisfying the given boundary conditions.

(a) $f' - 3f = 0$; $f(1) = 2$

(b) $f' + f = 0$; $f(1) = 1$

(c) $f'' + 2f' - 15f = 0$; $f(1) = f(0) = 0$

(d) $f'' + f' - 6f = 0$; $f(0) = 0, f(1) = 1$

(e) $f'' - 2f' + f = 0$; $f(1) = f(0) = 1$

(f) $f'' - 4f' + 4f = 0$; $f(0) = 2, f(-1) = 0$

(g) $f'' - 3af' + 2a^2f = 0, a \neq 0$; $f(0) = 0, f(1) = 1 - e^a$

(h) $f'' - a^2f = 0, a \neq 0$; $f(0) = 1, f(1) = 0$

(i) $f'' - 2f' + 5f = 0$; $f(0) = 1, f\left(\frac{\pi}{4}\right) = 0$

(j) $f'' + 4f' + 5f = 0$; $f(0) = 0, f\left(\frac{\pi}{2}\right) = 1$

2. Show that the solution to $f' + af = 0$ satisfying $f(x_0) = k$ is $f(x) = ke^{a(x_0-x)}$.

3. If the characteristic polynomial of $f'' + af' + bf = 0$ has real roots, show that $f = 0$ is the only solution satisfying $f(0) = 0 = f(1)$.

4. Complete the proof of Theorem 5. [*Hint:* If λ is a double root of $x^2 + ax + b$, then $a = -2\lambda$ and $b = \lambda^2$.]

5. **(a)** Given the equation $f' + af = b$, $(a \neq 0)$, make the substitution $f(x) = g(x) + b/a$ and obtain a differential equation for g. Thence derive the general solution for $f' + af = b$.

(b) Find the general solution to $f' + f = 2$.

6. Consider the differential equation $f'' + af' + bf = g$, where g is some fixed function. Assume that f_0 is one solution of this equation.

(a) Show that the general solution is $cf_1 + df_2 + f_0$, where c and d are constants and $\{f_1, f_2\}$ is any basis for the solutions to $f'' + af' + bf = 0$.

(b) Find a solution to $f'' + f' - 6f = 2x^3 - x^2 - 2x$. [*Hint:* try $f(x) = \frac{-1}{3}x^3$.]

7. A radioactive element decays at a rate proportional to the amount present. Suppose an initial mass of 10 g decays to 8 g in 3 hours.

(a) Find the mass t hours later.

(b) Find the "half-life" of the element—the time it takes to decay to half its mass.

8. The population $N(t)$ of a region at time t increases at a rate proportional to the population. If the population doubles in 5 years and is 3,000,000 initially, find $N(t)$.

9. Consider a spring, as in Example 8. If the period of the oscillation is 30 sec, find the spring constant k.

10. As a pendulum swings (see the diagram), let t measure time since it was vertical. The angle $\theta = \theta(t)$ from the vertical can be shown to satisfy the equation $\theta'' + k\theta = 0$, provided that θ is small. If the maximal angle is $\theta = .05$ radians, find $\theta(t)$ in terms of k. If the period is .5 sec, find k. [Assume that $\theta = 0$ when $t = 0$.]

6 Inner Product Spaces

SECTION 6.1 Inner Products and Norms

If a physicist were asked to describe a vector, his or her response would probably be that it is a quantity with both magnitude and direction, and such things as displacement, velocity, and force would be cited as examples. However, the discussion of vectors in Chapter 5 makes no reference to the magnitude or length of a vector or to the angle between two vectors. These are geometrical ideas, whereas the treatment there is algebraic and focuses on the addition and scalar multiplication of vectors and on the related notions of spanning set, linear independence, and basis. The plan in this chapter is to define an *inner product* on an arbitrary vector space V (of which the dot product is an example in $\mathbb{R}^n$) and use it to introduce the notions of length and orthogonality in V.

DEFINITION

An **inner product** on a vector space V is a function that assigns a number $\langle \mathbf{v},\mathbf{w} \rangle$ to every pair $\mathbf{v}$, $\mathbf{w}$ of vectors in V in such a way that the following axioms are satisfied.

P1. $\langle \mathbf{v},\mathbf{w} \rangle$ is a real number for all $\mathbf{v}$ and $\mathbf{w}$ in V.

P2. $\langle \mathbf{v},\mathbf{w} \rangle = \langle \mathbf{w},\mathbf{v} \rangle$ for all $\mathbf{v}$ and $\mathbf{w}$ in V.

P3. $\langle \mathbf{v} + \mathbf{w}, \mathbf{u} \rangle = \langle \mathbf{v},\mathbf{u} \rangle + \langle \mathbf{w},\mathbf{u} \rangle$ for all $\mathbf{u}$, $\mathbf{v}$, and $\mathbf{w}$ in V.

P4. $\langle r\mathbf{v},\mathbf{w} \rangle = r\langle \mathbf{v},\mathbf{w} \rangle$ for all $\mathbf{v}$ and $\mathbf{w}$ in V and all r in $\mathbb{R}$.

P5. $\langle \mathbf{v},\mathbf{v} \rangle > 0$ for all $\mathbf{v} \neq \mathbf{0}$ in V.

A vector space V with an inner product $\langle\ ,\ \rangle$ will be called an **inner product space.**

The **dot product** $\mathbf{v} \cdot \mathbf{w}$ of two n-tuples $\mathbf{v} = (v_1, v_2, \ldots, v_n)$ and $\mathbf{w} =$

$(w_1, w_2, \ldots, w_n)$ in Euclidean space $\mathbb{R}^n$ was defined in Section 2.2 as follows:

$$\mathbf{v} \cdot \mathbf{w} = v_1w_1 + v_2w_2 + \cdots + v_nw_n$$

This was useful in describing matrix multiplication and was given a geometrical interpretation in Chapter 4.

EXAMPLE 1 Show that $\mathbb{R}^n$ is an inner product space with the dot product as inner product.

Solution We verify Axiom P5, leaving the rest as Exercise 2. If $\mathbf{v} = (v_1, v_2, \ldots, v_n)$ then $\mathbf{v} \cdot \mathbf{v} = v_1^2 + v_2^2 + \cdots + v_n^2$ is nonnegative, and is zero if and only if $v_1 = v_2 = \cdots = v_n = 0$.

EXAMPLE 2 Let $B = \{\mathbf{b}_1, \mathbf{b}_2, \ldots, \mathbf{b}_n\}$ be any basis of a vector space V, and suppose that $\mathbf{v} = v_1\mathbf{b}_1 + \cdots + v_n\mathbf{b}_n$ and $\mathbf{w} = w_1\mathbf{b}_1 + \cdots + w_n\mathbf{b}_n$ are vectors in V, where v_i and w_i are real numbers. Show that

$$\langle \mathbf{v},\mathbf{w} \rangle = v_1w_1 + v_2w_2 + \cdots + v_nw_n$$

defines an inner product on V.

Solution The verification can be carried out directly. Alternatively, observe that $\langle \mathbf{v},\mathbf{w} \rangle = C_B(\mathbf{v}) \cdot C_B(\mathbf{w})$ and use Theorem 4 in Section 5.5. The details are omitted.

The next two examples require some knowledge of calculus.

EXAMPLE 3[†] Given polynomials $p = p(x)$ and $q = q(x)$ in $\mathbf{P}_n$, define

$$\langle p,q \rangle = \int_0^1 p(x)q(x)\,dx$$

Show that this is an inner product on $\mathbf{P}_n$.

Solution Axioms P1 and P2 are clear. As to axiom P4,

$$\langle rp,q \rangle = \int_0^1 rp(x)q(x)\,dx = r\int_0^1 p(x)q(x)\,dx = r\langle p,q \rangle$$

Axiom P3 is similar. Finally, $\langle p,p \rangle = \int_0^1 p(x)^2\,dx$, and it is a theorem of calculus that this is zero if and only if $p(x)$ is the zero polynomial. This gives axiom P5.

[†]This example (and others below referring to it) may be omitted with no loss of continuity by students with no calculus background.

Other functions besides polynomials can be integrated. The next example is verified in exactly the same way as Example 3.

EXAMPLE 4 Let $\mathbf{C}[a, b]$ denote the vector space of **continuous functions** from $[a, b]$ to $\mathbb{R}$, a subspace of $\mathbf{F}[a, b]$. Then

$$\langle f,g \rangle = \int_a^b f(x)g(x)\,dx$$

defines an inner product on $\mathbf{C}[a, b]$.

If $\mathbf{v}$ is any vector, then, using axiom P3,

$$\langle \mathbf{0},\mathbf{v} \rangle = \langle \mathbf{0} + \mathbf{0},\mathbf{v} \rangle = \langle \mathbf{0},\mathbf{v} \rangle + \langle \mathbf{0},\mathbf{v} \rangle$$

and it follows that the number $\langle \mathbf{0},\mathbf{v} \rangle$ must be zero. This observation is recorded below for reference; the other proofs are left as Exercise 9.

THEOREM 1 Let $\langle\ ,\ \rangle$ be an inner product on a space V, let $\mathbf{v}$, $\mathbf{u}$, and $\mathbf{w}$ denote vectors in V, and let r denote a real number.

(1) $\langle \mathbf{u},\mathbf{v} + \mathbf{w} \rangle = \langle \mathbf{u},\mathbf{v} \rangle + \langle \mathbf{u},\mathbf{w} \rangle$

(2) $\langle \mathbf{w}, r\mathbf{v} \rangle = r\langle \mathbf{w},\mathbf{v} \rangle$

(3) $\langle \mathbf{v},\mathbf{0} \rangle = 0 = \langle \mathbf{0},\mathbf{v} \rangle$

(4) $\langle \mathbf{v},\mathbf{v} \rangle = 0$ if and only if $\mathbf{v} = \mathbf{0}$

If $\langle\ ,\ \rangle$ is an inner product on a space V, then, given $\mathbf{u}$, $\mathbf{v}$, and $\mathbf{w}$ in V,

$$\langle r\mathbf{u} + s\mathbf{v},\mathbf{w} \rangle = \langle r\mathbf{u},\mathbf{w} \rangle + \langle s\mathbf{v},\mathbf{w} \rangle = r\langle \mathbf{u},\mathbf{w} \rangle + s\langle \mathbf{v},\mathbf{w} \rangle$$

by Axioms P3 and P4. Moreover, there is nothing special about the fact that there are two terms in the linear combination or that it is in the first component. Again, the proof is left as Exercise 9.

THEOREM 2 Let $\langle\ ,\ \rangle$ be an inner product in a vector space V.

(1) $\langle r_1\mathbf{v}_1 + r_2\mathbf{v}_2 + \cdots + r_n\mathbf{v}_n,\mathbf{w} \rangle = r_1\langle \mathbf{v}_1,\mathbf{w} \rangle + r_2\langle \mathbf{v}_2,\mathbf{w} \rangle + \cdots + r_n\langle \mathbf{v}_n,\mathbf{w} \rangle$

(2) $\langle \mathbf{v}, s_1\mathbf{w}_1 + s_2\mathbf{w}_2 + \cdots + s_m\mathbf{w}_m \rangle = s_1\langle \mathbf{v},\mathbf{w}_1 \rangle + s_2\langle \mathbf{v},\mathbf{w}_2 \rangle + \cdots + s_m\langle \mathbf{v},\mathbf{w}_m \rangle$

hold for all r_i and s_j in $\mathbb{R}$ and all $\mathbf{v}$, $\mathbf{w}$, $\mathbf{v}_i$, and $\mathbf{w}_j$ in V.

This result could be described by saying that inner products "preserve" linear combinations. Moreover, two applications of Theorem 2 give

$$\begin{aligned}\langle \mathbf{u} + \mathbf{v}, \mathbf{w} + \mathbf{p}\rangle &= \langle \mathbf{u}, \mathbf{w} + \mathbf{p}\rangle + \langle \mathbf{v}, \mathbf{w} + \mathbf{p}\rangle \\ &= \langle \mathbf{u}, \mathbf{w}\rangle + \langle \mathbf{u}, \mathbf{p}\rangle + \langle \mathbf{v}, \mathbf{w}\rangle + \langle \mathbf{v}, \mathbf{p}\rangle\end{aligned}$$

In other words: To compute $\langle \mathbf{u} + \mathbf{v}, \mathbf{w} + \mathbf{p}\rangle$ take the inner product of each term in the first component with each term in the second component, and add the results. (The analog in ordinary algebra is $(x + y) \cdot (z + t) = xz + xt + yz + yt$.) Thus $\langle \mathbf{u} + \mathbf{v}, \mathbf{w} + \mathbf{p} + \mathbf{y}\rangle$ would expand to a sum of six inner products.

EXAMPLE 5 If $\mathbf{u}$ and $\mathbf{v}$ are vectors in an inner product space, expand $\langle 2\mathbf{u} - \mathbf{v}, 3\mathbf{u} + 2\mathbf{v}\rangle$.

Solution

$$\begin{aligned}\langle 2\mathbf{u} - \mathbf{v}, 3\mathbf{u} + 2\mathbf{v}\rangle &= \langle 2\mathbf{u}, 3\mathbf{u}\rangle + \langle 2\mathbf{u}, 2\mathbf{v}\rangle + \langle -\mathbf{v}, 3\mathbf{u}\rangle + \langle -\mathbf{v}, 2\mathbf{v}\rangle \\ &= 6\langle \mathbf{u}, \mathbf{u}\rangle + 4\langle \mathbf{u}, \mathbf{v}\rangle - 3\langle \mathbf{u}, \mathbf{v}\rangle - 2\langle \mathbf{v}, \mathbf{v}\rangle \\ &= 6\langle \mathbf{u}, \mathbf{u}\rangle + \langle \mathbf{u}, \mathbf{v}\rangle - 2\langle \mathbf{v}, \mathbf{v}\rangle\end{aligned}$$

If $P(x,y,z)$ is a point in space, the vector $\mathbf{v} = (x,y,z)$ in $\mathbb{R}^3$ is called the position vector of P and is thought of geometrically as the "arrow" from the origin to P (see Chapter 4). By Pythagoras' theorem, $\mathbf{v}$ has length $\sqrt{x^2 + y^2 + z^2} = \sqrt{\mathbf{v} \cdot \mathbf{v}}$, and this suggests the way to define the "length" of a vector in any inner product space. The word *norm* is also used where it is desirable to avoid the geometric connotations of the word *length*.

DEFINITION If $\langle\ ,\ \rangle$ is an inner product on a space V, the **norm** or **length** $\|\mathbf{v}\|$ of a vector $\mathbf{v}$ in V is defined by

$$\|\mathbf{v}\| = \sqrt{\langle \mathbf{v}, \mathbf{v}\rangle}$$

Here it is the positive square root that is taken, and axiom P5 guarantees that $\langle \mathbf{v}, \mathbf{v}\rangle \geq 0$ so $\|\mathbf{v}\|$ is a real number.

EXAMPLE 6 The length of $\mathbf{v} = (v_1, v_2, \ldots, v_n)$ in $\mathbb{R}^n$ is

$$\|\mathbf{v}\| = \sqrt{v_1{}^2 + v_2{}^2 + \cdots + v_n{}^2}$$

where the dot product is used.

EXAMPLE 7 The norm of a polynomial $p = p(x)$ in $\mathbf{P}_n$ (with the inner product from Example 3) is given by

$$\|p\| = \sqrt{\int_0^1 p(x)^2\, dx}$$

A vector $\mathbf{v}$ is called a **unit vector** if $\|\mathbf{v}\| = 1$. The following result is useful.

THEOREM 3 If $\mathbf{v} \neq \mathbf{0}$ is any vector in an inner product space V, then

$$\hat{\mathbf{v}} = \frac{1}{\|\mathbf{v}\|}\mathbf{v}$$

is the unique unit vector that is a positive multiple of $\mathbf{v}$.

Proof If $\hat{\mathbf{v}} = t\mathbf{v}$ with $t > 0$, the requirement that $\|\hat{\mathbf{v}}\| = 1$ implies that

$$1 = \|\hat{\mathbf{v}}\|^2 = \langle t\mathbf{v}, t\mathbf{v}\rangle = t^2\langle \mathbf{v},\mathbf{v}\rangle = t^2\|\mathbf{v}\|^2$$

Because t is positive, this gives $t = \dfrac{1}{\|\mathbf{v}\|}$. ■

EXAMPLE 8 Find a unit vector $\hat{\mathbf{v}}$ that is a positive multiple of $\mathbf{v} = (1, -1, 2, 0)$ in $\mathbb{R}^4$ using the dot product.

Solution $\|\mathbf{v}\|^2 = 1^2 + (-1)^2 + 2^2 + 0^2 = 6$, so $\hat{\mathbf{v}} = \dfrac{1}{\sqrt{6}}\mathbf{v}$.

The next theorem reveals an important and useful fact about the relationship between norms and inner products.

THEOREM 4 Schwarz Inequality If $\mathbf{v}$ and $\mathbf{w}$ are two vectors in an inner product space V, then

$$\langle \mathbf{v},\mathbf{w}\rangle^2 \leq \|\mathbf{v}\|^2\,\|\mathbf{w}\|^2$$

Moreover, equality occurs if and only if one of $\mathbf{v}$ and $\mathbf{w}$ is a scalar multiple of the other.

Proof If either $\mathbf{v} = \mathbf{0}$ or $\mathbf{w} = \mathbf{0}$, the inequality holds (in fact, it is an equality). Otherwise let $\hat{\mathbf{v}} = \dfrac{1}{\|\mathbf{v}\|}\mathbf{v}$ and $\hat{\mathbf{w}} = \dfrac{1}{\|\mathbf{w}\|}\mathbf{w}$. Then $\|\hat{\mathbf{v}}\| = 1 = \|\hat{\mathbf{w}}\|$ and

$$\|\hat{\mathbf{v}} - \hat{\mathbf{w}}\|^2 = \langle \hat{\mathbf{v}} - \hat{\mathbf{w}}, \hat{\mathbf{v}} - \hat{\mathbf{w}}\rangle = \|\hat{\mathbf{v}}\|^2 - 2\langle \hat{\mathbf{v}},\hat{\mathbf{w}}\rangle + \|\hat{\mathbf{w}}\|^2 = 2(1 - \langle \hat{\mathbf{v}},\hat{\mathbf{w}}\rangle)$$

A similar calculation gives $\|\hat{\mathbf{v}} + \hat{\mathbf{w}}\|^2 = 2(1 + \langle \hat{\mathbf{v}},\hat{\mathbf{w}}\rangle)$. These imply that $-1 \leq \langle \hat{\mathbf{v}},\hat{\mathbf{w}}\rangle \leq 1$, and the inequality follows because $\langle \hat{\mathbf{v}},\hat{\mathbf{w}}\rangle = \dfrac{\langle \mathbf{v},\mathbf{w}\rangle}{\|\mathbf{v}\|\,\|\mathbf{w}\|}$.

Now suppose $\mathbf{v}$ and $\mathbf{w}$ are such that equality holds: $\langle\mathbf{v},\mathbf{w}\rangle^2 = \|\mathbf{v}\|^2\,\|\mathbf{w}\|^2$. We are to show that one is a multiple of the other. If either is zero, this is clear. Otherwise $\langle\hat{\mathbf{v}},\hat{\mathbf{w}}\rangle^2 = 1$, so either $\|\hat{\mathbf{v}} - \hat{\mathbf{w}}\| = 0$ or $\|\hat{\mathbf{v}} + \hat{\mathbf{w}}\| = 0$ by the above equations. But then $\hat{\mathbf{v}} = \pm\hat{\mathbf{w}}$, so $\mathbf{v} = \pm\dfrac{\|\mathbf{v}\|}{\|\mathbf{w}\|}\mathbf{w}$, as required. ■

The Schwarz inequality is one of the most useful inequalities in mathematics. Here are two important special cases.

EXAMPLE 9
Cauchy Inequality

If $v_1, v_2, \ldots, v_n$ and $w_1, w_2, \ldots, w_n$ are any two sequences of real numbers, then

$$(v_1w_1 + v_2w_2 + \cdots + v_nw_n)^2 \leq (v_1^2 + v_2^2 + \cdots + v_n^2)(w_1^2 + w_2^2 + \cdots + w_n^2)$$

Solution If $\mathbf{v} = (v_1, v_2, \ldots, v_n)$ and $\mathbf{w} = (w_1, w_2, \ldots, w_n)$ in $\mathbb{R}^n$, then this is just the Schwarz inequality (using the dot product).

If the Schwarz inequality is applied to the space $\mathbf{C}[a, b]$ of continuous functions on the interval $[a, b]$ with the norm defined in Example 4, the result is Example 10.

EXAMPLE 10

If f and g are continuous functions on the interval $[a, b]$, then

$$\left\{\int_a^b f(x)g(x)\,dx\right\}^2 \leq \int_a^b f(x)^2\,dx \int_a^b g(x)^2\,dx$$

Another famous inequality, the so-called **triangle inequality,** also comes from the Schwarz inequality. It is included in the following list of basic properties of the norm of a vector.

THEOREM 5

If V is an inner product space, the norm $\|\cdot\|$ has the following properties.

(1) $\|\mathbf{v}\| \geq 0$ for every vector $\mathbf{v}$ in V

(2) $\|\mathbf{v}\| = 0$ if and only if $\mathbf{v} = \mathbf{0}$

(3) $\|r\mathbf{v}\| = |r|\,\|\mathbf{v}\|$ for every $\mathbf{v}$ in V and every r in $\mathbb{R}$

(4) $\|\mathbf{v} + \mathbf{w}\| \leq \|\mathbf{v}\| + \|\mathbf{w}\|$ for all $\mathbf{v}$ and $\mathbf{w}$ in V (Triangle inequality)

Proof Because $\|\mathbf{v}\| = \sqrt{\langle\mathbf{v},\mathbf{v}\rangle}$, properties (1) and (2) follow immediately from (3) and (4) of Theorem 1. As to (3), compute

$$\|r\mathbf{v}\|^2 = \langle r\mathbf{v}, r\mathbf{v}\rangle = r^2\langle \mathbf{v},\mathbf{v}\rangle = r^2\|\mathbf{v}\|^2$$

Hence (3) follows by taking positive square roots. Finally, the fact that $\langle \mathbf{v}, \mathbf{w}\rangle \le \|\mathbf{v}\|\,\|\mathbf{w}\|$ by the Schwarz inequality gives

$$\begin{aligned}\|\mathbf{v} + \mathbf{w}\|^2 &= \langle \mathbf{v} + \mathbf{w}, \mathbf{v} + \mathbf{w}\rangle = \|\mathbf{v}\|^2 + 2\langle \mathbf{v},\mathbf{w}\rangle + \|\mathbf{w}\|^2 \\ &\le \|\mathbf{v}\|^2 + 2\|\mathbf{v}\|\,\|\mathbf{w}\| + \|\mathbf{w}\|^2 \\ &= (\|\mathbf{v}\| + \|\mathbf{w}\|)^2\end{aligned}$$

Hence (4) follows by taking positive square roots. ■

It is worth noting that the usual triangle inequality for absolute values, $|r + s| \le |r| + |s|$ for all real numbers r and s, is a special case of (4) where $V = \mathbb{R} = \mathbb{R}^1$ and the dot product $\langle r,s\rangle = rs$ is used.

In many calculations in an inner product space, it is required to show that some vector $\mathbf{v}$ is zero. This is often accomplished most easily by showing that its length $\|\mathbf{v}\|$ is zero. Here is an example.

EXAMPLE 11 Let $\{\mathbf{v}_1, \ldots, \mathbf{v}_n\}$ be a spanning set for an inner product space V. If $\mathbf{v}$ in V satisfies $\langle \mathbf{v},\mathbf{v}_i\rangle = 0$ for each $i = 1, 2, \ldots, n$, show that $\mathbf{v} = \mathbf{0}$.

Solution Write $\mathbf{v} = r_1\mathbf{v}_1 + \cdots + r_n\mathbf{v}_n$, r_i in $\mathbb{R}$. To show that $\mathbf{v} = \mathbf{0}$, we show that $\|\mathbf{v}\|^2 = \langle \mathbf{v},\mathbf{v}\rangle = 0$. Using Theorem 2, we have

$$\langle \mathbf{v},\mathbf{v}\rangle = \langle \mathbf{v}, r_1\mathbf{v}_1 + \cdots + r_n\mathbf{v}_n\rangle = r_1\langle \mathbf{v},\mathbf{v}_1\rangle + \cdots + r_n\langle \mathbf{v},\mathbf{v}_n\rangle = 0$$

by hypothesis, and the result follows.

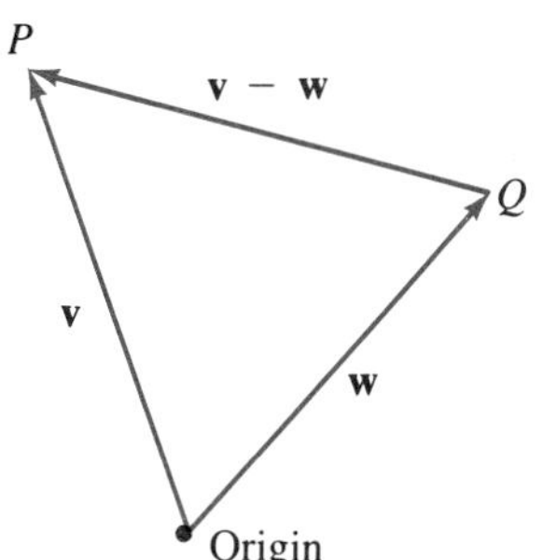

FIGURE 6.1

If $P(x,y,z)$ and $Q(x_1,y_1,z_1)$ are two points in space, let $\mathbf{v} = (x,y,z)$ and $\mathbf{w} = (x_1,y_1,z_1)$ denote their position vectors (see Chapter 4). Then $\mathbf{v} - \mathbf{w} = (x - x_1, y - y_1, z - z_1)$ is the vector from Q to P (see Figure 6.1), so $\|\mathbf{v} - \mathbf{w}\|$ is the distance between P and Q. It is sometimes referred to as the distance between $\mathbf{v}$ and $\mathbf{w}$, and so it can be generalized as follows:

DEFINITION Define the **distance** between vectors $\mathbf{v}$ and $\mathbf{w}$ in an inner product space V to be

$$d(\mathbf{v},\mathbf{w}) = \|\mathbf{v} - \mathbf{w}\|$$

This then reduces to the usual geometric distance in $\mathbb{R}^3$ (when the dot product is employed), and it has the following properties.

THEOREM 6 Let V be an inner product space.

(1) $d(\mathbf{v},\mathbf{w}) \ge 0$ for all $\mathbf{v}$, $\mathbf{w}$ in V

(2) $d(\mathbf{v},\mathbf{w}) = 0$ if and only if $\mathbf{v} = \mathbf{w}$

(3) $d(\mathbf{v},\mathbf{w}) = d(\mathbf{w},\mathbf{v})$ for all $\mathbf{v}$ and $\mathbf{w}$ in V

(4) $d(\mathbf{v},\mathbf{w}) \leq d(\mathbf{v},\mathbf{u}) + d(\mathbf{u},\mathbf{w})$ for all $\mathbf{v}$, $\mathbf{u}$, and $\mathbf{w}$ in V

Proof (1), (2), and (3) are left as Exercise 20. The proof of (4) uses the triangle inequality as follows:

$$d(\mathbf{v},\mathbf{w}) = \|\mathbf{v} - \mathbf{w}\| = \|(\mathbf{v} - \mathbf{u}) + (\mathbf{u} - \mathbf{w})\| \leq \|\mathbf{v} - \mathbf{u}\| + \|\mathbf{u} - \mathbf{w}\|$$
$$= d(\mathbf{v},\mathbf{u}) + d(\mathbf{u},\mathbf{w}) \quad \blacksquare$$

We conclude this section with a description of all inner products on $\mathbb{R}^n$ (written as columns). The next example of an inner product on $\mathbb{R}^2$ turns out to be a prototype.

EXAMPLE 12

Let $A = \begin{bmatrix} 2 & 1 \\ 1 & 2 \end{bmatrix}$. Given $\mathbf{v}$ and $\mathbf{w}$ in $\mathbb{R}^2$, define $\langle \mathbf{v},\mathbf{w} \rangle = \mathbf{v}^T A \mathbf{w}$. Show that this is an inner product on $\mathbb{R}^2$.

Solution We verify axioms P2 and P5 and leave the rest to the reader (see Exercise 31). To obtain axiom P2, observe that A is symmetric; that is, $A = A^T$. Then

$$\langle \mathbf{w},\mathbf{v} \rangle = \mathbf{w}^T A \mathbf{v} = \mathbf{w}^T A^T \mathbf{v} = (\mathbf{v}^T A \mathbf{w})^T = \langle \mathbf{v},\mathbf{w} \rangle^T = \langle \mathbf{v},\mathbf{w} \rangle$$

where $\langle \mathbf{v},\mathbf{w} \rangle^T = \langle \mathbf{v},\mathbf{w} \rangle$ because $\langle \mathbf{v},\mathbf{w} \rangle$ is a 1×1 matrix (a number) and so is symmetric. Now consider axiom P5. If $\mathbf{v} = \begin{bmatrix} v_1 \\ v_2 \end{bmatrix}$, then $\langle \mathbf{v},\mathbf{v} \rangle = 2v_1^2 + 2v_1v_2 + 2v_2^2$. Complete the square in the first two terms.

$$\langle \mathbf{v},\mathbf{v} \rangle = 2\left[v_1^2 + v_1v_2 + \left(\frac{1}{2}v_2\right)^2\right] - \frac{1}{2}v_2^2 + 2v_2^2$$
$$= 2\left(v_1 + \frac{1}{2}v_2\right)^2 + \frac{3}{2}v_2^2$$
$$= \frac{1}{2}[(2v_1 + v_2)^2 + 3v_2^2]$$

This can vanish only if $2v_1 + v_2 = 0$ and $v_2 = 0$—that is, if $\mathbf{v} = \begin{bmatrix} 0 \\ 0 \end{bmatrix}$. This is P5. Note that P5 also follows from $\langle \mathbf{v},\mathbf{v} \rangle = v_1^2 + (v_1 + v_2)^2 + v_2^2$.

If A is *any* $n \times n$ symmetric matrix, it is easy to verify (Exercise 31) that axioms P1, P2, P3, and P4 are satisfied if $\langle \mathbf{v},\mathbf{w} \rangle = \mathbf{v}^T A \mathbf{w}$ for all $\mathbf{v}$ and $\mathbf{w}$ in $\mathbb{R}^n$. Such products are called **symmetric bilinear forms** on V. However, axiom A5 need *not* be satisfied, as will be shown in the proof of Theorem 7.

DEFINITION

An $n \times n$ matrix A is called **positive definite** if $\langle \mathbf{v},\mathbf{w} \rangle = \mathbf{v}^T A \mathbf{w}$ is an inner product on $\mathbb{R}^n$. In other words, an A is positive definite if

(1) A is symmetric.

(2) $\mathbf{v}^T A \mathbf{v} > 0$ for all $\mathbf{v} \neq \mathbf{0}$ in $\mathbb{R}^n$.

Note that the identity matrix I is positive definite by virtue of the fact that $\mathbf{v}^T I \mathbf{w} = \mathbf{v} \cdot \mathbf{w}$ is an inner product on $\mathbb{R}^n$.

THEOREM 7

Every inner product on $\mathbb{R}^n$ (written as columns) is given as follows:

$$\langle \mathbf{v},\mathbf{w} \rangle = \mathbf{v}^T A \mathbf{w}$$

for some $n \times n$ positive definite matrix A. However, not every symmetric matrix A yields an inner product in this way.

Proof Let $\{\mathbf{e}_1, \mathbf{e}_2, \ldots, \mathbf{e}_n\}$ be the standard basis of $\mathbb{R}^n$. Given two vectors $\mathbf{v} = \sum_{i=1}^{n} v_i \mathbf{e}_i$ and $\mathbf{w} = \sum_{j=1}^{n} w_j \mathbf{e}_j$, compute $\langle \mathbf{v},\mathbf{w} \rangle$ by taking the inner product of each term $v_i \mathbf{e}_i$ with each term $w_j \mathbf{e}_j$. The result is a double sum.

$$\langle \mathbf{v},\mathbf{w} \rangle = \sum_{i=1}^{n} \sum_{j=1}^{n} \langle v_i \mathbf{e}_i, w_j \mathbf{e}_j \rangle = \sum_{i=1}^{n} \sum_{j=1}^{n} v_i \langle \mathbf{e}_i,\mathbf{e}_j \rangle w_j$$

It is easy to check that this is a matrix product.

$$\langle \mathbf{v},\mathbf{w} \rangle = [v_1\ v_2\ \ldots\ v_n] \begin{bmatrix} \langle \mathbf{e}_1,\mathbf{e}_1 \rangle & \langle \mathbf{e}_1,\mathbf{e}_2 \rangle & \ldots & \langle \mathbf{e}_1,\mathbf{e}_n \rangle \\ \langle \mathbf{e}_2,\mathbf{e}_1 \rangle & \langle \mathbf{e}_2,\mathbf{e}_2 \rangle & \ldots & \langle \mathbf{e}_2,\mathbf{e}_n \rangle \\ \vdots & \vdots & & \vdots \\ \langle \mathbf{e}_n,\mathbf{e}_1 \rangle & \langle \mathbf{e}_n,\mathbf{e}_2 \rangle & \ldots & \langle \mathbf{e}_n,\mathbf{e}_n \rangle \end{bmatrix} \begin{bmatrix} w_1 \\ w_2 \\ \vdots \\ w_n \end{bmatrix}$$

Hence $\langle \mathbf{v},\mathbf{w} \rangle = \mathbf{v}^T A \mathbf{w}$, where A is the $n \times n$ matrix whose (i,j)-entry is $\langle \mathbf{e}_i,\mathbf{e}_j \rangle$. The fact that $\langle \mathbf{e}_i,\mathbf{e}_j \rangle = \langle \mathbf{e}_j,\mathbf{e}_i \rangle$ shows that A is symmetric.

However, not every symmetric matrix yields an inner product. For example, $A = \begin{bmatrix} 1 & 0 \\ 0 & -1 \end{bmatrix}$ is symmetric and $[xy] \begin{bmatrix} 1 & 0 \\ 0 & -1 \end{bmatrix} \begin{bmatrix} x_1 \\ y_1 \end{bmatrix} = xx_1 - yy_1$ for all $\begin{bmatrix} x \\ y \end{bmatrix}$ and $\begin{bmatrix} x_1 \\ y_1 \end{bmatrix}$ in $\mathbb{R}^2$. In particular $\left\langle \begin{bmatrix} 1 \\ 1 \end{bmatrix}, \begin{bmatrix} 1 \\ 1 \end{bmatrix} \right\rangle = 0$, so axiom P5 fails. ■

EXAMPLE 13 Let the inner product $\langle\ ,\ \rangle$ be defined on $\mathbb{R}^2$ by

$$\left\langle \begin{bmatrix} v_1 \\ v_2 \end{bmatrix}, \begin{bmatrix} w_1 \\ w_2 \end{bmatrix} \right\rangle = 2v_1w_1 - v_1w_2 - v_2w_1 + v_2w_2$$

Find a symmetric 2×2 matrix A such that $\langle \mathbf{v},\mathbf{w} \rangle = \mathbf{v}^T A\mathbf{w}$ for all $\mathbf{v}$, $\mathbf{w}$ in $\mathbb{R}^2$.

Solution The (i,j)-entry of the matrix A is the coefficient of v_iw_j in the expression, so $A = \begin{bmatrix} 2 & -1 \\ -1 & 1 \end{bmatrix}$. The reader can verify that $\langle \mathbf{v},\mathbf{w} \rangle = \mathbf{v}^T A\mathbf{w}$, where $\mathbf{v} = \begin{bmatrix} v_1 \\ v_2 \end{bmatrix}$ and $\mathbf{w} = \begin{bmatrix} w_1 \\ w_2 \end{bmatrix}$.

EXAMPLE 14 If A and B are positive definite $n \times n$ matrices, show that $A + B$ is also positive definite.

Solution $A + B$ is symmetric because $(A + B)^T = A^T + B^T = A + B$ (A and B are symmetric). Given $\mathbf{v} \neq \mathbf{0}$ in $\mathbb{R}^n$, $\mathbf{v}^T(A + B)\mathbf{v} = \mathbf{v}^T A\mathbf{v} + \mathbf{v}^T B\mathbf{v}$ is positive, because $\mathbf{v}^T A\mathbf{v}$ and $\mathbf{v}^T B\mathbf{v}$ are both positive.

EXERCISES 6.1

1. In each case, determine which of axioms P1–P5 fail to hold.
- **(a)** $V = \mathbb{R}^2$, $\langle (x_1,y_1), (x_2,y_2) \rangle = x_1y_1x_2y_2$
- **(b)** $V = \mathbb{R}^3$, $\langle (x_1,x_2,x_3), (y_1,y_2,y_3) \rangle = x_1y_1 - x_2y_2 + x_3y_3$
- **(c)** $V = \mathbb{C}$, $\langle z,w \rangle = z\overline{w}$, where $\overline{w}$ is complex conjugation
- **(d)** $V = \mathbf{P}_3$, $\langle p(x),q(x) \rangle = p(1)q(1)$
- **(e)** $V = \mathbf{M}_{22}$, $\langle A,B \rangle = \det(AB)$
- **(f)** $V = \mathbf{F}[0,1]$, $\langle f,g \rangle = f(1)g(0) + f(0)g(1)$

2. **(a)** Verify that the dot product on $\mathbb{R}^n$ satisfies axioms P1–P5.
- **(b)** Complete Example 2 and show that it yields the dot product on $\mathbb{R}^n$ with appropriate choice of the basis.

3. In each case, find a scalar multiple of $\mathbf{v}$ that is a unit vector.
- **(a)** $\mathbf{v} = (1,-1,2,0)$ in $\mathbb{R}^4$ (dot product)
- **(b)** $\mathbf{v} = (2,-3,1,1)$ in $\mathbb{R}^4$ (dot product)
- **(c)** $\mathbf{v} = \begin{bmatrix} 1 \\ 3 \end{bmatrix}$ in $\mathbb{R}^2$ $\left(\langle \mathbf{v},\mathbf{w} \rangle = \mathbf{v}^T \begin{bmatrix} 1 & 1 \\ 1 & 2 \end{bmatrix} \mathbf{w} \right)$
- **(d)** $\mathbf{v} = \begin{bmatrix} 3 \\ -1 \end{bmatrix}$ in $\mathbb{R}^2$ $\left(\langle \mathbf{v},\mathbf{w} \rangle = \mathbf{v}^T \begin{bmatrix} 1 & -1 \\ -1 & 2 \end{bmatrix} \mathbf{w} \right)$

4. In each case, find the distance between $\mathbf{u}$ and $\mathbf{v}$.
- **(a)** $\mathbf{u} = (1,2,2)$, $\mathbf{v} = (2,-2,1)$
- **(b)** $\mathbf{u} = (1,1,0)$, $\mathbf{v} = (0,1,-1)$
- **(c)** $\mathbf{u} = (3,-1,2,0)$, $\mathbf{v} = (1,1,1,3)$
- **(d)** $\mathbf{u} = (1,2,-1,2)$, $\mathbf{v} = (2,1,-1,3)$

5. If $p = p(x)$ and $q = q(x)$ are polynomials in $\mathbf{P}_n$, define

$$\langle p,q\rangle = p(0)q(0) + p(1)q(1) + \cdots + p(n)q(n)$$

Show that this is an inner product on $\mathbf{P}_n$. [*Hint* for P5: If $p(0) = p(1) = \cdots = p(n) = 0$, then $p = 0$—Corollary of Theorem 3 in Section 5.6.]

6. Let $\mathbf{D}_n$ denote the space of all functions from the set $\{1, 2, 3, \ldots, n\}$ to $\mathbb{R}$ with pointwise addition and scalar multiplication (see Exercise 13 in Section 5.3.2). Show that $\langle\ ,\ \rangle$ is an inner product on $\mathbf{D}_n$ if $\langle f,g\rangle = f(1)g(1) + f(2)g(2) + \cdots + f(n)g(n)$.

7. If $V = \mathbf{M}_{mn}$, define $\langle A,B\rangle = \text{tr}(AB^T)$, where the trace tr$X$ of a square matrix X is the sum of the entries on the main diagonal. Show this is an inner product on $\mathbf{M}_{mn}$. [*Hint:* Exercise 25 in Section 2.2.]

8. Let re(z) denote the real part of the complex number z. Show that $\langle\ ,\ \rangle$ is an inner product on $\mathbb{C}$ if $\langle z,w\rangle = \text{re}(z\overline{w})$.

9. **(a)** Prove Theorem 1. **(b)** Prove Theorem 2.

10. Let **u** and **v** be vectors in an inner product space V.
    (a) Expand $\langle 2\mathbf{u} - 7\mathbf{v}, 3\mathbf{u} + 5\mathbf{v}\rangle$. **(b)** Expand $\langle 3\mathbf{u} - 4\mathbf{v}, 5\mathbf{u} + \mathbf{v}\rangle$.
    (c) Show that $\|\mathbf{u} + \mathbf{v}\|^2 = \|\mathbf{u}\|^2 + 2\langle\mathbf{u},\mathbf{v}\rangle + \|\mathbf{v}\|^2$.
    (d) Show that $\|\mathbf{u} - \mathbf{v}\|^2 = \|\mathbf{u}\|^2 - 2\langle\mathbf{u},\mathbf{v}\rangle + \|\mathbf{v}\|^2$.
    (e) Show that $\langle\mathbf{u} + \mathbf{v}, \mathbf{u} - \mathbf{v}\rangle = \|\mathbf{u}\|^2 - \|\mathbf{v}\|^2$.

11. Show that $\|\mathbf{v}\|^2 + \|\mathbf{w}\|^2 = \frac{1}{2}\{\|\mathbf{v} + \mathbf{w}\|^2 + \|\mathbf{v} - \mathbf{w}\|^2\}$ for any **v** and **w** in an inner product space.

12. Let $\langle\ ,\ \rangle$ be an inner product on a vector space V. Show that the corresponding distance function is translation invariant. That is, show that $d(\mathbf{v},\mathbf{w}) = d(\mathbf{v} + \mathbf{u}, \mathbf{w} + \mathbf{u})$ for all **v**, **w**, and **u** in V.

13. **(a)** Show that $\langle\mathbf{u},\mathbf{v}\rangle = \frac{1}{4}[\|\mathbf{u} + \mathbf{v}\|^2 - \|\mathbf{u} - \mathbf{v}\|^2]$ for all **u**, **v** in an inner product space V.
    (b) If $\langle\ ,\ \rangle$ and $\langle\ ,\ \rangle'$ are two inner products on V that have equal associated norm functions, show that $\langle\mathbf{u},\mathbf{v}\rangle = \langle\mathbf{u},\mathbf{v}\rangle'$ holds for all **u** and **v**.

14. Let **v** denote a vector in an inner product space V.
    (a) Show that $W = \{\mathbf{w} \mid \mathbf{w} \text{ in } V, \langle\mathbf{v},\mathbf{w}\rangle = 0\}$ is a subspace of V.
    (b) If $V = \mathbb{R}^3$ with the dot product, and if $\mathbf{v} = (1,-1,2)$, find a basis for W.

15. Given vectors $\mathbf{w}_1, \mathbf{w}_2, \ldots, \mathbf{w}_n$ and **v**, assume that $\langle\mathbf{v},\mathbf{w}_i\rangle = 0$ for each i. Show that $\langle\mathbf{v},\mathbf{w}\rangle = 0$ for all **w** in span$\{\mathbf{w}_1, \mathbf{w}_2, \ldots, \mathbf{w}_n\}$.

16. If $V = \text{span}\{\mathbf{v}_1, \mathbf{v}_2, \ldots, \mathbf{v}_n\}$ and $\langle\mathbf{v},\mathbf{v}_i\rangle = \langle\mathbf{w},\mathbf{v}_i\rangle$ holds for each i, show that $\mathbf{v} = \mathbf{w}$.

17. Use the Cauchy inequality to prove that:
    (a) $(r_1 + r_2 + \cdots + r_n)^2 \le n(r_1{}^2 + r_2{}^2 + \cdots + r_n{}^2)$ for all r_i in $\mathbb{R}$.
    (b) $r_1r_2 + r_1r_3 + r_2r_3 \le r_1{}^2 + r_2{}^2 + r_3{}^2$ for all r_1, r_2, r_3. [*Hint:* See part **(a)**.]

18. Use the Schwarz inequality in an inner product space to show that:
    (a) If $\|\mathbf{u}\| \le 1$, then $\langle\mathbf{u},\mathbf{v}\rangle^2 \le \|\mathbf{v}\|^2$ for all **v** in V.
    (b) $(x \cos\theta + y \sin\theta)^2 \le x^2 + y^2$ for all real x, y, and θ.

(c) $\|r_1\mathbf{v}_1 + \cdots + r_n\mathbf{v}_n\|^2 \le \{r_1\|\mathbf{v}_1\| + \cdots + r_n\|\mathbf{v}_n\|\}^2$ for all vectors $\mathbf{v}_i$, and all $r_i > 0$ in $\mathbb{R}$.

19. If A is a $2 \times n$ matrix, let $\mathbf{u}$ and $\mathbf{v}$ denote the rows of A.

(a) Show that $AA^T = \begin{bmatrix} \|\mathbf{u}\|^2 & \mathbf{u}\cdot\mathbf{v} \\ \mathbf{u}\cdot\mathbf{v} & \|\mathbf{v}\|^2 \end{bmatrix}$.

(b) Show that $\det(AA^T) \ge 0$.

20. Prove properties (1), (2), and (3) of Theorem 6.

21. If $V = \mathbb{R}^2$, define $\|(x,y)\| = |x| + |y|$.

(a) Show that $\|\cdot\|$ satisfies the conditions in Theorem 5.

(b) Show that $\|\cdot\|$ doesn't arise from an inner product on $\mathbb{R}^2$. [*Hint:* If it did, use Theorem 7 to find numbers a, b, and c such that $\|(x,y)\|^2 = ax^2 + bxy + cy^2$ for all x and y.]

22. Prove the Schwarz inequality as follows: Show that $\|t\mathbf{v} + \mathbf{w}\|^2 = at^2 + bt + c$, where $a = \|\mathbf{v}\|^2$, $b = 2\langle\mathbf{v},\mathbf{w}\rangle$, and $c = \|\mathbf{w}\|^2$. Conclude that $b^2 - 4ac \le 0$ from the quadratic formula. For equality, argue that $at^2 + bt + c$ has one real root t_0, so $\mathbf{w} = -t_0\mathbf{v}$.

23. Show that the sum of two inner products on V is again an inner product.

24. In each case, show that $\langle\mathbf{v},\mathbf{w}\rangle = \mathbf{v}^TA\mathbf{w}$ defines an inner product on $\mathbb{R}^2$ and hence show that A is positive definite.

(a) $A = \begin{bmatrix} 2 & 1 \\ 1 & 1 \end{bmatrix}$ **(b)** $A = \begin{bmatrix} 5 & -3 \\ -3 & 2 \end{bmatrix}$ **(c)** $A = \begin{bmatrix} 3 & 2 \\ 2 & 3 \end{bmatrix}$ **(d)** $A = \begin{bmatrix} 3 & 4 \\ 4 & 6 \end{bmatrix}$

25. In each case, find a symmetric matrix A such that $\langle\mathbf{v},\mathbf{w}\rangle = \mathbf{v}^TA\mathbf{w}$.

(a) $\left\langle \begin{bmatrix} v_1 \\ v_2 \end{bmatrix}, \begin{bmatrix} w_1 \\ w_2 \end{bmatrix} \right\rangle = v_1w_1 + 2v_1w_2 + 2v_2w_1 + 5v_2w_2$

(b) $\left\langle \begin{bmatrix} v_1 \\ v_2 \end{bmatrix}, \begin{bmatrix} w_1 \\ w_2 \end{bmatrix} \right\rangle = v_1w_1 - v_1w_2 - v_2w_1 + 2v_2w_2$

(c) $\left\langle \begin{bmatrix} v_1 \\ v_2 \\ v_3 \end{bmatrix}, \begin{bmatrix} w_1 \\ w_2 \\ w_3 \end{bmatrix} \right\rangle = 2v_1w_1 + v_2w_2 + v_3w_3 - v_1w_2 - v_2w_1 + v_2w_3 + v_3w_2$

(d) $\left\langle \begin{bmatrix} v_1 \\ v_2 \\ v_3 \end{bmatrix}, \begin{bmatrix} w_1 \\ w_2 \\ w_3 \end{bmatrix} \right\rangle = v_1w_1 + 2v_2w_2 + 5v_3w_3 - 2v_1w_3 - 2v_3w_1$

26. If A is positive definite, show that A is invertible. [*Hint:* Theorem 8 in Section 2.3.]

27. If A is positive definite, show that each of the following are positive definite.

(a) A^{-1} **(b)** kA, $k > 0$ **(c)** A^2 **(d)** A^3

28. If A and B are positive definite matrices, show that the block matrix $\begin{bmatrix} A & 0 \\ 0 & B \end{bmatrix}$ is also positive definite.

29. (a) Show that $A = UU^T$ is positive definite if U is an $n \times m$ matrix of rank n. [*Hint:* Theorem 5 in Section 5.4.]

(b) If A is invertible and symmetric, show that A^2 is positive definite.

30. Show that $A = \begin{bmatrix} a & b \\ b & c \end{bmatrix}$ is positive definite if and only if $a > 0$ and $\det A > 0$.

31. Let A be a symmetric $n \times n$ matrix and define $\langle \mathbf{v},\mathbf{w} \rangle = \mathbf{v}^T A \mathbf{w}$ for all $\mathbf{v}$, $\mathbf{w}$ in $\mathbb{R}^n$.
    (a) Show that $\langle\ ,\ \rangle$ satisfies P1–P4.
    (b) Show that part (a) fails if A is not symmetric. [*Hint:* Consider $A = \begin{bmatrix} 0 & 1 \\ -1 & 0 \end{bmatrix}$.]
    (c) If A satisfies $\mathbf{x}^T A \mathbf{x} = 0$ for all $\mathbf{x}$ in $\mathbb{R}^n$, show that $A = 0$. [*Hint:* Compute $\langle \mathbf{x} + \mathbf{y}, \mathbf{x} + \mathbf{y} \rangle$, and take $\mathbf{x}$ and $\mathbf{y}$ to be columns of the identity matrix.]
    (d) If A and B are symmetric and $\mathbf{x}^T A \mathbf{x} = \mathbf{x}^T B \mathbf{x}$ for all $\mathbf{x}$ in $\mathbb{R}^n$, show that $A = B$.

SECTION 6.2 Orthogonality

6.2.1 Orthogonal Sets of Vectors

The idea that two lines can be perpendicular is fundamental in geometry, and this section is devoted to introducing this notion into a general inner product space V. To motivate the definition, we first use the Schwarz inequality to define the angle between two nonzero vectors in V. If $\mathbf{v} \neq \mathbf{0}$ and $\mathbf{w} \neq \mathbf{0}$, the inequality can be written

$$\left[\frac{\langle \mathbf{v},\mathbf{w} \rangle}{\|\mathbf{v}\|\,\|\mathbf{w}\|}\right]^2 \leq 1 \qquad \text{so that} \qquad -1 \leq \frac{\langle \mathbf{v},\mathbf{w} \rangle}{\|\mathbf{v}\|\,\|\mathbf{w}\|} \leq 1$$

Now it is a fact that given a number between -1 and 1, there is exactly one angle θ in the range 0 to π whose cosine equals that number. Hence let θ be the unique angle such that:

(1) $0 \leq \theta \leq \pi$

(2) $\cos \theta = \dfrac{\langle \mathbf{v},\mathbf{w} \rangle}{\|\mathbf{v}\|\,\|\mathbf{w}\|}$

This angle θ is called the **angle between v and w**. The terminology is justified by the geometric interpretation of $\mathbb{R}^3$ in Chapter 4.

We shall not refer further to the angle between two vectors in the sequel. Our main reason for mentioning it at all is that it suggests the following: Let θ be the angle between two vectors $\mathbf{v} \neq \mathbf{0}$ and $\mathbf{w} \neq \mathbf{0}$. Then θ is a right angle if and only if $\cos \theta = 0$—that is, if $\langle \mathbf{v},\mathbf{w} \rangle = 0$.

DEFINITION

Two vectors $\mathbf{v}$ and $\mathbf{w}$ in an inner product space V are said to be **orthogonal** if

$$\langle \mathbf{v},\mathbf{w} \rangle = 0$$

A set $\{\mathbf{e}_1, \mathbf{e}_2, \ldots, \mathbf{e}_n\}$ of vectors is called an **orthogonal set of vectors** if

(1) Each $\mathbf{e}_i \neq \mathbf{0}$.

(2) $\langle \mathbf{e}_i,\mathbf{e}_j \rangle = 0$ for all $i \neq j$.

If, in addition, $\|\mathbf{e}_i\| = 1$ for each i, the set $\{\mathbf{e}_1, \mathbf{e}_2, \ldots, \mathbf{e}_n\}$ is called an **orthonormal** set.

Clearly $\mathbf{0}$ is orthogonal to every vector, and $\mathbf{0}$ is the only vector that is orthogonal to itself. The condition in property 2 is expressed by saying that the vectors $\mathbf{e}_1, \ldots, \mathbf{e}_n$ are **pairwise orthogonal.** The reason for excluding the zero vector from orthogonal sets of vectors is that the orthogonal sets of interest (for example, orthogonal bases) consist of nonzero vectors.

EXAMPLE 1 The standard basis of $\mathbb{R}^n$ is orthonormal (using the dot product).

The second part of the next result shows that it is easy to convert an orthogonal set into an orthonormal one.

THEOREM 1 Let $\{\mathbf{e}_1, \mathbf{e}_2, \ldots, \mathbf{e}_n\}$ be an orthogonal set of vectors.

(1) $\{r_1\mathbf{e}_1, r_2\mathbf{e}_2, \ldots, r_n\mathbf{e}_n\}$ is also orthogonal for any choice of $r_i \neq 0$ in $\mathbb{R}$.

(2) $\left\{\frac{1}{\|\mathbf{e}_1\|}\mathbf{e}_1, \frac{1}{\|\mathbf{e}_2\|}\mathbf{e}_2, \ldots, \frac{1}{\|\mathbf{e}_n\|}\mathbf{e}_n\right\}$ is an orthonormal set.

The proof is left as Exercise 11(a). This process of passing from an orthogonal set to an orthonormal one is called **normalizing** the orthogonal set.

EXAMPLE 2 Show that

$$\{\mathbf{e}_1 = (1,1,1,-1), \mathbf{e}_2 = (1,0,1,2), \mathbf{e}_3 = (-1,0,1,0), \mathbf{e}_4 = (-1,3,-1,1)\}$$

is orthogonal in $\mathbb{R}^4$ (with the dot product), and find the corresponding normalized set.

Solution $\langle \mathbf{e}_1,\mathbf{e}_2 \rangle = 1 + 0 + 1 - 2 = 0$, and we leave it to the reader to verify that $\langle \mathbf{e}_i,\mathbf{e}_j \rangle = 0$ for all $i \neq j$. Because each $\mathbf{e}_i \neq \mathbf{0}$, the set is orthogonal. We have $\|\mathbf{e}_1\| = 2$, $\|\mathbf{e}_2\| = \sqrt{6}$, $\|\mathbf{e}_3\| = \sqrt{2}$, $\|\mathbf{e}_4\| = 2\sqrt{3}$. Hence normalization gives

$$\left\{\frac{1}{2}\mathbf{e}_1, \frac{1}{\sqrt{6}}\mathbf{e}_2, \frac{1}{\sqrt{2}}\mathbf{e}_3, \frac{1}{2\sqrt{3}}\mathbf{e}_4\right\}$$

as the corresponding orthonormal set.

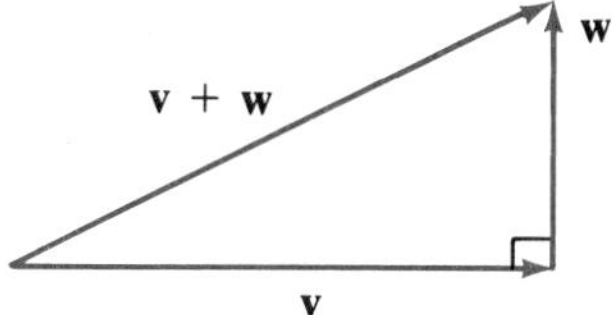

FIGURE 6.2

Given a right-angled triangle, let $\mathbf{v}$ and $\mathbf{w}$ be the geometric vectors along the two perpendicular sides, as shown in Figure 6.2. Then the vector along the hypotenuse is $\mathbf{v} + \mathbf{w}$ by the parallelogram law, and Pythagoras' theorem may be expressed as $\|\mathbf{v} + \mathbf{w}\|^2 = \|\mathbf{v}\|^2 + \|\mathbf{w}\|^2$. The reason is that $\{\mathbf{v},\mathbf{w}\}$ is an orthogonal set, and this suggests the following general form of Pythagoras' theorem.

THEOREM 2
Pythagoras' Theorem

If $\{\mathbf{e}_1, \mathbf{e}_2, \ldots, \mathbf{e}_n\}$ is an orthogonal set of vectors, then

$$\|\mathbf{e}_1 + \mathbf{e}_2 + \cdots + \mathbf{e}_n\|^2 = \|\mathbf{e}_1\|^2 + \|\mathbf{e}_2\|^2 + \cdots + \|\mathbf{e}_n\|^2$$

Proof Use Theorem 2 in Section 6.1 to compute

$$\begin{aligned}\|\mathbf{e}_1 + \mathbf{e}_2 + \cdots + \mathbf{e}_n\|^2 &= \langle \mathbf{e}_1 + \mathbf{e}_2 + \cdots + \mathbf{e}_n, \mathbf{e}_1 + \mathbf{e}_2 + \cdots + \mathbf{e}_n\rangle \\ &= \langle \mathbf{e}_1,\mathbf{e}_1\rangle + \langle \mathbf{e}_2,\mathbf{e}_2\rangle + \cdots + \langle \mathbf{e}_n,\mathbf{e}_n\rangle + \sum_{i \neq j} \langle \mathbf{e}_i,\mathbf{e}_j\rangle\end{aligned}$$

The result follows because $\langle \mathbf{e}_i,\mathbf{e}_i\rangle = \|\mathbf{e}_i\|^2$ for each i, whereas $\langle \mathbf{e}_i,\mathbf{e}_j\rangle = 0$ if $i \neq j$. ■

The next theorem gives a connection (perhaps unexpected) between orthogonality and linear independence. It will be used extensively below.

THEOREM 3

Every orthogonal set of vectors is linearly independent.

Proof If $\{\mathbf{e}_1, \mathbf{e}_2, \ldots, \mathbf{e}_n\}$ is orthogonal, suppose a linear combination vanishes.

$$r_1\mathbf{e}_1 + r_2\mathbf{e}_2 + \cdots + r_n\mathbf{e}_n = \mathbf{0}$$

If we compute the inner product of this with $\mathbf{e}_1$, the result is

$$\begin{aligned}0 = \langle \mathbf{0},\mathbf{e}_1\rangle &= \langle r_1\mathbf{e}_1 + r_2\mathbf{e}_2 + \cdots + r_n\mathbf{e}_n,\mathbf{e}_1\rangle \\ &= r_1\langle \mathbf{e}_1,\mathbf{e}_1\rangle + r_2\langle \mathbf{e}_2,\mathbf{e}_1\rangle + \cdots + r_n\langle \mathbf{e}_n,\mathbf{e}_1\rangle \\ &= r_1\|\mathbf{e}_1\|^2 + 0 + \cdots + 0\end{aligned}$$

Recall that orthogonal sets consist of *nonzero* vectors so that $\|\mathbf{e}_1\|^2 \neq 0$. Hence $r_1 = 0$, and similarly $r_2 = r_3 = \cdots = r_n = 0$, as required. ■

This leads immediately to the notion of an **orthogonal basis** of an inner product space V—that is, an orthogonal set that spans V (and so is a basis by Theorem 3). We shall show in the following section (Theorem 8) that every finite dimensional inner product space *has* an orthogonal basis.

EXAMPLE 3 Show that $\left\{\begin{bmatrix} 2 \\ -1 \\ 0 \end{bmatrix}, \begin{bmatrix} 0 \\ 1 \\ 1 \end{bmatrix}, \begin{bmatrix} 0 \\ -1 \\ 2 \end{bmatrix}\right\}$ is an orthogonal basis of $\mathbb{R}^3$ with inner product $\langle \mathbf{v},\mathbf{w} \rangle = \mathbf{v}^T A \mathbf{w}$, where $A = \begin{bmatrix} 1 & 1 & 0 \\ 1 & 2 & 0 \\ 0 & 0 & 1 \end{bmatrix}$.

Solution We have

$$\left\langle \begin{bmatrix} 2 \\ -1 \\ 0 \end{bmatrix}, \begin{bmatrix} 0 \\ 1 \\ 1 \end{bmatrix} \right\rangle = [2\ -1\ 0] \begin{bmatrix} 1 & 1 & 0 \\ 1 & 2 & 0 \\ 0 & 0 & 1 \end{bmatrix} \begin{bmatrix} 0 \\ 1 \\ 1 \end{bmatrix} = [1\ 0\ 0] \begin{bmatrix} 0 \\ 1 \\ 1 \end{bmatrix} = 0$$

and the reader can verify that the other pairs are orthogonal too. Hence the set is orthogonal, so it is linearly independent by Theorem 3. But because $\dim \mathbb{R}^3 = 3$, it is a basis by Theorem 8 in Section 5.3.

When we are working with a basis of a vector space V, it is usually a tedious calculation to express a vector in V as a linear combination of the basis vectors. However, in an inner product space this problem is very much simplified when an orthogonal basis is used, because explicit formulas exist for the coefficients.

THEOREM 4 Expansion Theorem

Let $\{\mathbf{e}_1, \mathbf{e}_2, \ldots, \mathbf{e}_n\}$ be an orthogonal basis of an inner product space V. If $\mathbf{v}$ is any vector in V, then

$$\mathbf{v} = \frac{\langle \mathbf{v},\mathbf{e}_1 \rangle}{\|\mathbf{e}_1\|^2}\mathbf{e}_1 + \frac{\langle \mathbf{v},\mathbf{e}_2 \rangle}{\|\mathbf{e}_2\|^2}\mathbf{e}_2 + \cdots + \frac{\langle \mathbf{v},\mathbf{e}_n \rangle}{\|\mathbf{e}_n\|^2}\mathbf{e}_n$$

is the expansion of $\mathbf{v}$ as a linear combination of the basis vectors.

Proof Because the $\mathbf{e}_i$ are a basis, $\mathbf{v} = r_1\mathbf{e}_1 + r_2\mathbf{e}_2 + \cdots + r_n\mathbf{e}_n$ holds for *some* r_i in $\mathbb{R}$. The problem is to show that $r_i = \frac{\langle \mathbf{v},\mathbf{e}_i \rangle}{\|\mathbf{e}_i\|^2}$ holds for each i. The orthogonality of the $\mathbf{e}_i$ gives

$$\begin{aligned}\langle \mathbf{v},\mathbf{e}_i\rangle &= \langle r_1\mathbf{e}_1 + \cdots + r_n\mathbf{e}_n, \mathbf{e}_i\rangle \\ &= r_1\langle \mathbf{e}_1,\mathbf{e}_i\rangle + \cdots + r_i\langle \mathbf{e}_i,\mathbf{e}_i\rangle + \cdots + r_n\langle \mathbf{e}_n,\mathbf{e}_i\rangle \\ &= 0 + \cdots + r_i\|\mathbf{e}_i\|^2 + \cdots + 0 \\ &= r_i\|\mathbf{e}_i\|^2\end{aligned}$$

The result follows because $\|\mathbf{e}_i\|^2 \neq 0$. ■

Thus the fact that the basis is orthogonal means that the expansion of $\mathbf{v}$ as a linear combination of basis vectors is reduced to evaluating inner products. It is even easier if the basis is orthonormal; then the denominators $\|\mathbf{e}_i\|^2$ all equal 1.

The coefficients $\frac{\langle \mathbf{v},\mathbf{e}_1\rangle}{\|\mathbf{e}_1\|^2}, \frac{\langle \mathbf{v},\mathbf{e}_2\rangle}{\|\mathbf{e}_2\|^2}, \ldots, \frac{\langle \mathbf{v},\mathbf{e}_n\rangle}{\|\mathbf{e}_n\|^2}$ in the Expansion Theorem are sometimes called the **Fourier coefficients** of $\mathbf{v}$ with respect to the orthogonal basis $\{\mathbf{e}_1, \mathbf{e}_2, \ldots, \mathbf{e}_n\}$. This is in honor of the French mathematician J. B. J. Fourier (1768–1830). His original work was with a particular orthogonal set in the space $C[a,b]$, and we will have more to say about that in Section 6.3.2.

EXAMPLE 4 In $\mathbb{R}^4$ (with the dot product), expand $\mathbf{v} = (a,b,c,d)$ as a linear combination of the basis

$$\{\mathbf{e}_1 = (1,1,1,-1), \mathbf{e}_2 = (1,0,1,2), \mathbf{e}_3 = (-1,0,1,0), \mathbf{e}_4 = (-1,3,-1,1)\}$$

Solution The task is to find r_1, r_2, r_3, and r_4 such that $\mathbf{v} = r_1\mathbf{e}_1 + r_2\mathbf{e}_2 + r_3\mathbf{e}_3 + r_4\mathbf{e}_4$. Because the basis is orthogonal (see Example 2), Theorem 4 gives

$$r_1 = \frac{\langle \mathbf{v},\mathbf{e}_1\rangle}{\|\mathbf{e}_1\|^2} = \frac{a + b + c - d}{4}$$

$$r_2 = \frac{\langle \mathbf{v},\mathbf{e}_2\rangle}{\|\mathbf{e}_2\|^2} = \frac{a + c + 2d}{6}$$

$$r_3 = \frac{\langle \mathbf{v},\mathbf{e}_3\rangle}{\|\mathbf{e}_3\|^2} = \frac{-a + c}{2}$$

$$r_4 = \frac{\langle \mathbf{v},\mathbf{e}_4\rangle}{\|\mathbf{e}_4\|^2} = \frac{-a + 3b - c + d}{12}$$

The reader may verify that $\mathbf{v} = r_1\mathbf{e}_1 + r_2\mathbf{e}_2 + r_3\mathbf{e}_3 + r_4\mathbf{e}_4$ does indeed hold.

The next theorem identifies a class of matrices that will simplify the relationship between orthogonal bases of an inner product space and will play a fundamental role in the next two chapters.

THEOREM 5 The following conditions are equivalent for an $n \times n$ matrix P.

(1) P is invertible and $P^{-1} = P^T$.

(2) The rows of P are orthonormal (with respect to the dot product).

(3) The columns of P are orthonormal (with respect to the dot product).

Proof First recall that condition (1) is equivalent to $PP^T = I$ by Theorem 9 in Section 2.3. Let $\mathbf{r}_1, \mathbf{r}_2, \ldots, \mathbf{r}_n$ denote the rows of P. Then $\mathbf{r}_j^T$ is the jth column of P^T, so the (i,j)-entry of PP^T is $\mathbf{r}_i \cdot \mathbf{r}_j$. Thus $PP^T = I$ means that $\mathbf{r}_i \cdot \mathbf{r}_j = 0$ if $i \neq j$ and $\mathbf{r}_i \cdot \mathbf{r}_j = 1$ if $i = j$, so condition (1) is equivalent to (2). The equivalence of (1) and (3) is similar. ■

DEFINITION An $n \times n$ matrix P is called an **orthogonal matrix** if it satisfies one (and hence all) of the conditions in Theorem 5.

The matrices in the following example arose from rotations in the plane (see Theorem 1 in Section 5.5).

EXAMPLE 5 The matrix $\begin{bmatrix} \cos\theta & \sin\theta \\ -\sin\theta & \cos\theta \end{bmatrix}$ is orthogonal for any angle θ.

It is not enough that the rows of a matrix A are merely orthogonal in order for A to be an orthogonal matrix.

EXAMPLE 6 The matrix $\begin{bmatrix} 2 & 1 & 1 \\ -1 & 1 & 1 \\ 0 & -1 & 1 \end{bmatrix}$ has orthogonal rows but is not an orthogonal matrix. However, if the rows are normalized, the resulting matrix

$$\begin{bmatrix} \frac{2}{\sqrt{6}} & \frac{1}{\sqrt{6}} & \frac{1}{\sqrt{6}} \\ \frac{-1}{\sqrt{3}} & \frac{1}{\sqrt{3}} & \frac{1}{\sqrt{3}} \\ 0 & \frac{-1}{\sqrt{2}} & \frac{1}{\sqrt{2}} \end{bmatrix}$$ is orthogonal.

EXAMPLE 7 If P and Q are orthogonal matrices, then PQ is also orthogonal, as is $P^{-1} = P^T$.

Solution P and Q are invertible, so PQ is also invertible and $(PQ)^{-1} =$

$Q^{-1}P^{-1} = Q^TP^T = (PQ)^T$. Hence PQ is orthogonal. Similarly, $(P^{-1})^{-1} = P = (P^T)^T = (P^{-1})^T$ shows that P^{-1} is orthogonal.

If V is an inner product space, the next result reveals a close connection between inner products of vectors in V and dot products of the associated coordinate vectors discussed in Section 5.5.

THEOREM 6 Let $E = \{\mathbf{e}_1, \mathbf{e}_2, \ldots, \mathbf{e}_n\}$ be an ordered orthonormal basis of an inner product space V.

(1) $\langle \mathbf{v},\mathbf{w}\rangle = C_E(\mathbf{v}) \cdot C_E(\mathbf{w})$

(2) $\|\mathbf{v}\| = \|C_E(\mathbf{v})\|$

(3) A basis B of V is an orthonormal basis if and only if the transition matrix $P_{E\leftarrow B}$ is an orthogonal matrix.

Proof The proofs of (1) and (2) are left as Exercise 11(b). Write $B = \{\mathbf{b}_1, \mathbf{b}_2, \ldots, \mathbf{b}_n\}$, and recall (Section 5.5) that

$$P_{E\leftarrow B} = [C_E(\mathbf{b}_1)\; C_E(\mathbf{b}_2) \ldots C_E(\mathbf{b}_n)]$$

Then Theorem 5 asserts that $P_{E\leftarrow B}$ is an orthogonal matrix if and only if its columns $\{C_E(\mathbf{b}_1), \ldots, C_E(\mathbf{b}_n)\}$ form an orthonormal set with respect to the dot product. But (1) and (2) show that this happens if and only if $\{\mathbf{b}_1, \ldots, \mathbf{b}_n\}$ is an orthonormal set in V. ■

EXAMPLE 8 Show that $E = \left\{\mathbf{e}_1 = \begin{bmatrix}1\\0\\0\end{bmatrix}, \mathbf{e}_2 = \begin{bmatrix}1\\-1\\0\end{bmatrix}, \mathbf{e}_3 = \begin{bmatrix}0\\0\\1\end{bmatrix}\right\}$ is an orthonormal basis for $\mathbb{R}^3$ with respect to the inner product $\langle \mathbf{v},\mathbf{w}\rangle = \mathbf{v}^TA\mathbf{w}$, where $A = \begin{bmatrix}1 & 1 & 0\\1 & 2 & 0\\0 & 0 & 1\end{bmatrix}$, and verify (1) and (2) of Theorem 6 for $\mathbf{v} = \begin{bmatrix}1\\2\\-1\end{bmatrix}$ and $\mathbf{w} = \begin{bmatrix}2\\1\\1\end{bmatrix}$.

Solution The verifications that E is orthonormal, that $\langle \mathbf{v},\mathbf{w}\rangle = 10$, and that $\|\mathbf{v}\| = \sqrt{14}$ are left to the reader. Theorem 4 gives the expansions of $\mathbf{v}$ and $\mathbf{w}$.

$$\mathbf{v} = \langle \mathbf{v},\mathbf{e}_1\rangle\mathbf{e}_1 + \langle \mathbf{v},\mathbf{e}_2\rangle\mathbf{e}_2 + \langle \mathbf{v},\mathbf{e}_3\rangle\mathbf{e}_3 = 3\mathbf{e}_1 - 2\mathbf{e}_2 - \mathbf{e}_3$$

$$\mathbf{w} = \langle \mathbf{w},\mathbf{e}_1\rangle\mathbf{e}_1 + \langle \mathbf{w},\mathbf{e}_2\rangle\mathbf{e}_2 + \langle \mathbf{w},\mathbf{e}_3\rangle\mathbf{e}_3 = 3\mathbf{e}_1 - \mathbf{e}_2 + \mathbf{e}_3$$

Hence $C_E(\mathbf{v}) = \begin{bmatrix} 3 \\ -2 \\ -1 \end{bmatrix}$ and $C_E(\mathbf{w}) = \begin{bmatrix} 3 \\ -1 \\ 1 \end{bmatrix}$, so $C_E(\mathbf{v}) \cdot C_E(\mathbf{w}) = 10$ and $\|C_E(\mathbf{v})\| = \sqrt{14}$.

EXERCISES 6.2.1

1. Obtain an orthonormal basis of $\mathbb{R}^3$ by normalizing the following.
   (a) $\{(1,-1,2), (0,2,1), (5,1,-2)\}$ **(b)** $\{(1,1,1), (4,1,-5), (2,-3,1)\}$
2. In each case, show that the set of vectors is orthogonal in the space V.
   (a) $\{(1,-1,2,5), (4,1,1,-1), (-7,28,5,5)\}$, $V = \mathbb{R}^4$ with the dot product
   (b) $\{(2,-1,4,5), (0,-1,1,-1), (0,3,2,-1)\}$, $V = \mathbb{R}^4$ with the dot product
   (c) $\left\{\begin{bmatrix} 1 \\ -1 \end{bmatrix}, \begin{bmatrix} 1 \\ 0 \end{bmatrix}\right\}$, $V = \mathbb{R}^2$, $\langle \mathbf{v},\mathbf{w} \rangle = \mathbf{v}^T A \mathbf{w}$ where $A = \begin{bmatrix} 2 & 2 \\ 2 & 5 \end{bmatrix}$
   (d) $\left\{\begin{bmatrix} 1 \\ 1 \\ 1 \end{bmatrix}, \begin{bmatrix} -1 \\ 0 \\ 1 \end{bmatrix}, \begin{bmatrix} 1 \\ -6 \\ 1 \end{bmatrix}\right\}$, $V = \mathbb{R}^3$, $\langle \mathbf{v},\mathbf{w} \rangle = \mathbf{v}^T A \mathbf{w}$ where $A = \begin{bmatrix} 2 & 0 & 1 \\ 0 & 1 & 0 \\ 1 & 0 & 2 \end{bmatrix}$
3. In each case, show that B is an orthogonal basis of $\mathbb{R}^3$, and use Theorem 4 to expand $\mathbf{v} = (a,b,c)$ as a linear combination of the basis vectors. Use the dot product.
   (a) $\{(1,-1,3), (-2,1,1), (4,7,1)\}$ **(b)** $\{(1,0,-1), (1,4,1), (2,-1,2)\}$
   (c) $\{(1,2,3), (-1,-1,1), (5,-4,1)\}$ **(d)** $\{(1,1,1), (1,-1,0), (1,1,-2)\}$
4. In each case, write $\mathbf{v}$ as a linear combination of the orthogonal basis of the subspace U. Use the dot product.
   (a) $\mathbf{v} = (13,-20,15)$; $U = \text{span}\{(1,-2,3), (-1,1,1)\}$
   (b) $\mathbf{v} = (14,1,-8,5)$; $U = \text{span}\{(2,-1,0,3), (2,1,-2,-1)\}$
5. In each case, find the cosine of the angle between $\mathbf{u}$ and $\mathbf{v}$.
   (a) $\mathbf{u} = (1,2,2)$, $\mathbf{v} = (2,-2,1)$ **(b)** $\mathbf{u} = (1,1,0)$, $\mathbf{v} = (0,1,-1)$
   (c) $\mathbf{u} = (3,-1,2,0)$, $\mathbf{v} = (1,1,1,3)$ **(d)** $\mathbf{u} = (1,2,-1,2)$, $\mathbf{v} = (2,1,-1,3)$
6. **(a)** Show that $\{\mathbf{u},\mathbf{v}\}$ is orthogonal if and only if $\|\mathbf{u} + \mathbf{v}\|^2 = \|\mathbf{u}\|^2 + \|\mathbf{v}\|^2$.
   (b) If $\mathbf{u} = \mathbf{v} = (1,1)$ and $\mathbf{w} = (-1,0)$, show that $\|\mathbf{u} + \mathbf{v} + \mathbf{w}\|^2 = \|\mathbf{u}\|^2 + \|\mathbf{v}\|^2 + \|\mathbf{w}\|^2$ but $\{\mathbf{u},\mathbf{v},\mathbf{w}\}$ is *not* orthogonal. Hence the converse to Pythagoras' theorem need not hold for more than two vectors.
7. Let $\mathbf{v}$ and $\mathbf{w}$ be vectors in an inner product space V. Show that:
   (a) $\mathbf{v}$ is orthogonal to $\mathbf{w}$ if and only if $\|\mathbf{v} + \mathbf{w}\| = \|\mathbf{v} - \mathbf{w}\|$.
   (b) $\mathbf{v} + \mathbf{w}$ and $\mathbf{v} - \mathbf{w}$ are orthogonal if and only if $\|\mathbf{v}\| = \|\mathbf{w}\|$.
8. Normalize the rows to make each of the following matrices orthogonal.
   (a) $\begin{bmatrix} 1 & 1 \\ -1 & 1 \end{bmatrix}$ **(b)** $\begin{bmatrix} 3 & -4 \\ 4 & 3 \end{bmatrix}$
   (c) $\begin{bmatrix} 1 & 2 \\ -4 & 2 \end{bmatrix}$ **(d)** $\begin{bmatrix} a & b \\ -b & a \end{bmatrix}$, $(a,b) \neq (0,0)$

(e) $\begin{bmatrix} \cos\theta & \sin\theta & 0 \\ -\sin\theta & \cos\theta & 0 \\ 0 & 0 & 2 \end{bmatrix}$ (f) $\begin{bmatrix} 2 & 1 & -1 \\ 1 & -1 & 1 \\ 0 & 1 & 1 \end{bmatrix}$

(g) $\begin{bmatrix} -1 & 2 & 2 \\ 2 & -1 & 2 \\ 2 & 2 & -1 \end{bmatrix}$ (h) $\begin{bmatrix} 2 & 6 & -3 \\ 3 & 2 & 6 \\ -6 & 3 & 2 \end{bmatrix}$

9. If P is a triangular orthogonal matrix, show that P is diagonal and that all diagonal entries are 1 or -1.

10. If P is orthogonal, show that kP is orthogonal if and only if $k = 1$ or $k = -1$.

11. **(a)** Prove Theorem 1.

(b) Prove (1) and (2) of Theorem 6.

12. Let P be an orthogonal matrix.

(a) Show that $\det P = 1$ or $\det P = -1$.

(b) Give 2×2 examples of P such that $\det P = 1$ and $\det P = -1$.

(c) If $\det P = -1$, show that $I + P$ has no inverse. [*Hint:* $P^T(I + P) = (I + P)^T$.]

(d) If P is $n \times n$ and $\det P \neq (-1)^n$, show that $I - P$ has no inverse. [*Hint:* $P^T(I - P) = -(I - P)^T$.]

13. If the first two rows of an orthogonal matrix are $\left[\frac{1}{3}, \frac{2}{3}, \frac{2}{3}\right]$ and $\left[\frac{2}{3}, \frac{1}{3}, \frac{-2}{3}\right]$, find all possible third rows.

14. If $B = \{\mathbf{b}_1, \ldots, \mathbf{b}_n\}$ and $E = \{\mathbf{e}_1, \ldots, \mathbf{e}_n\}$ are ordered bases of an inner product space V, and if E is orthonormal, show that the transition matrix P from B to E is $P_{E \leftarrow B} = [\langle \mathbf{e}_i, \mathbf{b}_j \rangle]$.

15. We call a square matrix E a **projection matrix** if $E^2 = E = E^T$.

(a) If E is a projection matrix, show that $P = I - 2E$ is orthogonal and symmetric.

(b) If P is orthogonal and symmetric, show that $E = \frac{1}{2}(I - P)$ is a projection matrix.

(c) If U is $m \times n$ and $U^TU = I$ (for example, a unit row in $\mathbb{R}^n$), show that $E = UU^T$ is a projection matrix.

16. [Requires calculus]. Show that $\left\{1, x - \frac{1}{2}, x^2 - x + \frac{1}{6}\right\}$ is an orthogonal basis of $\mathbf{P}_2$ with the inner product $\langle p,q \rangle = \int_0^1 p(x)q(x)\,dx$, and find the corresponding orthonormal basis.

17. A matrix that we obtain from the identity matrix by writing its rows in a different order is called a **permutation matrix.** Show that every permutation matrix is orthogonal. Generalize.

18. Let $\mathbf{r}_1, \mathbf{r}_2, \ldots, \mathbf{r}_n$ denote the rows of the $n \times n$ matrix $A = [a_{ij}]$.

(a) Show that the (i,j)-entry of AA^T is $\mathbf{r}_i \cdot \mathbf{r}_j$.

(b) If the rows of A are orthogonal, show that the (i,j)-entry of A^{-1} is $\dfrac{a_{ji}}{\|\mathbf{r}_j\|^2}$.

19. **(a)** Let A be an $m \times n$ matrix. Show that the following are equivalent.

(i) A has orthogonal rows.

(ii) A can be factored as $A = DP$, where D is invertible and diagonal and P has orthonormal rows.

(iii) AA^T is an invertible, diagonal matrix.

(b) Show that an $n \times n$ matrix A has orthogonal rows if and only if A can be factored as $A = DP$, where P is orthogonal and D is diagonal and invertible.

(c) Formulate and prove the analog of parts **(a)** and **(b)** for matrices with orthogonal columns.

20. Let A be a skew-symmetric matrix; that is, $A^T = -A$. Assume that A is an $n \times n$ matrix.

(a) Show that $I + A$ is invertible. [*Hint:* By Theorem 8 in Section 2.3, it suffices to show that $\mathbf{x}(I + A) = \mathbf{0}$, $\mathbf{x}$ in $\mathbb{R}^n$, implies $\mathbf{x} = \mathbf{0}$. Compute $\mathbf{x} \cdot \mathbf{x} = \mathbf{x}\mathbf{x}^T$, and use the fact that $\mathbf{x}A = -\mathbf{x}$ and $\mathbf{x}A^2 = \mathbf{x}$.]

(b) Show that $P = (I - A)(I + A)^{-1}$ is orthogonal.

(c) Show that every orthogonal matrix P such that $I + P$ is invertible arises as in part **(b)** from some skew-symmetric matrix A. [*Hint:* Solve $P = (I - A)(I + A)^{-1}$ for A.]

21. Show that the following are equivalent for an $n \times n$ matrix P.

(a) P is orthogonal.

(b) $\|\mathbf{v}P\| = \|\mathbf{v}\|$ for all rows $\mathbf{v}$ in $\mathbb{R}^n$.

(c) $\|\mathbf{v}P - \mathbf{w}P\| = \|\mathbf{v} - \mathbf{w}\|$ for all rows $\mathbf{v}$ and $\mathbf{w}$ in $\mathbb{R}^n$.

(d) $(\mathbf{v}P) \cdot (\mathbf{w}P) = \mathbf{v} \cdot \mathbf{w}$ for all rows $\mathbf{v}$ and $\mathbf{w}$ in $\mathbb{R}^n$.

[*Hints:* For part **(c)** implies part **(d)**: Exercise 13 in Section 6.1. For part **(d)** implies part **(a)**: Show that row i of P equals $\mathbf{e}_iP$, where $\mathbf{e}_i$ is row i of the identity matrix.]

22. Let $\{\mathbf{e}_1, \mathbf{e}_2, \ldots, \mathbf{e}_n\}$ be an orthogonal basis of V, and let θ_i be the angle between $\mathbf{v} \neq \mathbf{0}$ and $\mathbf{e}_i$ for each i. Show that $\cos^2\theta_1 + \cos^2\theta_2 + \ldots + \cos^2\theta_n = 1$.

23. Let $\{\mathbf{e}_1, \mathbf{e}_2, \ldots, \mathbf{e}_n\}$ be an orthonormal basis of V. If $P = [p_{ij}]$ is an $n \times n$ matrix, define $\mathbf{b}_i = p_{i1}\mathbf{e}_1 + \cdots + p_{in}\mathbf{e}_n$ for each i. Show that $B = \{\mathbf{b}_1, \mathbf{b}_2, \ldots, \mathbf{b}_n\}$ is an orthonormal basis if and only if P is an orthogonal matrix. [*Hint:* Theorem 6.]

24. Show that every 2×2 orthogonal matrix has the form $\begin{bmatrix} \cos\theta & \sin\theta \\ -\sin\theta & \cos\theta \end{bmatrix}$ or $\begin{bmatrix} \cos\theta & \sin\theta \\ \sin\theta & -\cos\theta \end{bmatrix}$ for some angle θ. [*Hint:* If $a^2 + b^2 = 1$, then $a = \cos\theta$ and $b = \sin\theta$ for some angle θ.]

25. Let $\mathbf{n} \neq \mathbf{0}$ and $\mathbf{w} \neq \mathbf{0}$ be nonparallel vectors in $\mathbb{R}^3$ (as in Chapter 4).

(a) Show that $\left\{\mathbf{n}, \mathbf{n} \times \mathbf{w}, \mathbf{w} - \frac{\mathbf{n} \cdot \mathbf{w}}{\|\mathbf{n}\|^2}\mathbf{n}\right\}$ is an orthogonal basis of $\mathbb{R}^3$.

(b) Show that span $\left\{\mathbf{n} \times \mathbf{w}, \mathbf{w} - \frac{\mathbf{n} \cdot \mathbf{w}}{\|\mathbf{n}\|^2}\mathbf{n}\right\}$ is the plane through the origin with normal $\mathbf{n}$.

26. [Refers to Section 5.6.] Show that the Lagrange polynomials $c_0(x)$, $c_1(x), \ldots, c_n(x)$ defined in Section 5.6 are an orthonormal basis of $\mathbf{P}_n$ with respect to the inner product

$$\langle p(x), q(x)\rangle = p(a_0)q(a_0) + p(a_1)q(a_1) + \cdots + p(a_n)q(a_n)$$

(see Exercise 5 in Section 6.1), where $a_0, a_1, \ldots, a_n$ are distinct numbers. Then deduce Theorem 3 in Section 5.6 from the expansion theorem.

6.2.2 Projections and the Gram–Schmidt Algorithm

It is clear from the expansion theorem (Theorem 4) that an orthogonal basis is preferred if it is desired to expand vectors easily as linear combinations of the basis vectors. However, we have yet to establish that every finite dimensional space *has* an orthogonal basis. The following theorem fills this gap.

THEOREM 7 Every finite dimensional inner product space has an orthogonal basis.

Proof Let V be an inner product space with $\dim V = n$. We proceed by induction on n. If $n = 1$, then any basis $\{\mathbf{v}_1\}$ is orthogonal. So assume inductively that every inner product space of dimension n has an orthogonal basis. If $\dim V = n + 1$, let $\{\mathbf{v}_1, \ldots, \mathbf{v}_n, \mathbf{v}_{n+1}\}$ be any basis and write $U = \operatorname{span}\{\mathbf{v}_1, \ldots, \mathbf{v}_n\}$. Then $\dim U = n$, so let $\{\mathbf{e}_1, \ldots, \mathbf{e}_n\}$ be an orthogonal basis of U. It suffices to find any nonzero vector $\mathbf{v}$ such that $\langle \mathbf{v},\mathbf{e}_i\rangle = 0$ for each i (then $\{\mathbf{e}_1, \ldots, \mathbf{e}_n, \mathbf{v}\}$ will be an orthogonal basis by Theorem 3 together with Theorem 8 of Section 5.3). Consider

$$\mathbf{v} = \mathbf{v}_{n+1} - t_1\mathbf{e}_1 - t_2\mathbf{e}_2 - \cdots - t_n\mathbf{e}_n$$

where the t_i are numbers to be determined. Then $\mathbf{v} \neq \mathbf{0}$ for all choices of t_i, because $\mathbf{v}_{n+1}$ is not in U. Hence we are finished if, for each i, we can choose t_i such that $\langle \mathbf{v},\mathbf{e}_i\rangle = 0$. But the orthogonality of $\{\mathbf{e}_1, \mathbf{e}_2, \ldots, \mathbf{e}_n\}$ gives

$$\begin{aligned}\langle \mathbf{v},\mathbf{e}_i\rangle &= \langle \mathbf{v}_{n+1},\mathbf{e}_i\rangle - t_1\langle \mathbf{e}_1,\mathbf{e}_i\rangle - \cdots - t_i\langle \mathbf{e}_i,\mathbf{e}_i\rangle - \cdots - t_n\langle \mathbf{e}_n,\mathbf{e}_i\rangle \\ &= \langle \mathbf{v}_{n+1},\mathbf{e}_i\rangle - 0 - \cdots - t_i\|\mathbf{e}_i\|^2 - \cdots - 0 \\ &= \langle \mathbf{v}_{n+1},\mathbf{e}_i\rangle - t_i\|\mathbf{e}_i\|^2.\end{aligned}$$

Hence $\langle \mathbf{v},\mathbf{e}_i\rangle = 0$ if we choose $t_i = \dfrac{\langle \mathbf{v}_{n+1},\mathbf{e}_i\rangle}{\|\mathbf{e}_i\|^2}$ for each i. ■

Theorem 7 guarantees the existence of an orthogonal basis in any finite dimensional inner product space V. However, the proof yields much more: It gives a systematic procedure by which *any* basis of V can be modified in a systematic way to yield an orthogonal basis. This procedure is called the Gram–Schmidt orthogonalization algorithm. It is based on the following fundamental fact, the proof of which is contained in the proof of Theorem 7.

ORTHOGONAL LEMMA Let $\{\mathbf{e}_1, \mathbf{e}_2, \ldots, \mathbf{e}_m\}$ be an orthogonal set of vectors in an inner product space V, and let $\mathbf{v}$ be any vector *not* in $\text{span}\{\mathbf{e}_1, \mathbf{e}_2, \ldots, \mathbf{e}_m\}$. Define

$$\mathbf{e}_{m+1} = \mathbf{v} - \frac{\langle \mathbf{v},\mathbf{e}_1\rangle}{\|\mathbf{e}_1\|^2}\mathbf{e}_1 - \frac{\langle \mathbf{v},\mathbf{e}_2\rangle}{\|\mathbf{e}_2\|^2}\mathbf{e}_2 - \cdots - \frac{\langle \mathbf{v},\mathbf{e}_m\rangle}{\|\mathbf{e}_m\|^2}\mathbf{e}_m$$

Then $\{\mathbf{e}_1,\mathbf{e}_2, \ldots, \mathbf{e}_m, \mathbf{e}_{m+1}\}$ is an orthogonal set of vectors.

This result will be used repeatedly in this section.

EXAMPLE 9 Given the orthogonal set $\{(1,0,-2), (2,3,1)\}$ in $\mathbb{R}^3$ (with the dot product), find an orthogonal basis of $\mathbb{R}^3$ containing these vectors.

Solution For convenience, write $\mathbf{e}_1 = (1,0,-2)$ and $\mathbf{e}_2 = (2,3,1)$. Then $\{\mathbf{e}_1,\mathbf{e}_2\}$ is orthogonal, and $\mathbf{v} = (1,0,0)$ lies outside $\text{span}\{\mathbf{e}_1,\mathbf{e}_2\}$ because $\{\mathbf{e}_1,\mathbf{e}_2, \mathbf{v}\}$ is linearly independent. By the Orthogonal Lemma, let

$$\begin{aligned}
\mathbf{e}_3 &= \mathbf{v} - \frac{\langle \mathbf{v},\mathbf{e}_1\rangle}{\|\mathbf{e}_1\|^2}\mathbf{e}_1 - \frac{\langle \mathbf{v},\mathbf{e}_2\rangle}{\|\mathbf{e}_2\|^2}\mathbf{e}_2 \\
&= (1,0,0) - \frac{1}{5}(1,0,-2) - \frac{2}{14}(2,3,1) \\
&= \frac{1}{35}[(35,0,0) - (7,0,-14) - (10,15,5)] \\
&= \frac{1}{35}(18,-15,9) \\
&= \frac{3}{35}(6,-5,3).
\end{aligned}$$

It is easily verified that $\mathbf{e}_3$ is orthogonal to both $\mathbf{e}_1$ and $\mathbf{e}_2$. Hence $\{\mathbf{e}_1,\mathbf{e}_2,\mathbf{e}_3\}$ is orthogonal (as the lemma asserts) and so is a basis of $\mathbb{R}^3$ (by Theorem 3, because $\dim \mathbb{R}^3 = 3$). Note that it may be more convenient to use $\mathbf{e}_3{}' = (6,-5,3)$ in place of $\mathbf{e}_3$ to avoid fractions. Then $\{\mathbf{e}_1,\mathbf{e}_2,\mathbf{e}_3{}'\}$ is also an orthogonal basis of $\mathbb{R}^3$ (using Theorem 1).

The orthogonal lemma shows how any vector outside the span of an orthogonal set of vectors can be used to enlarge the orthogonal set. It can be used repeatedly to construct an orthogonal basis for an inner product space from any given (possibly nonorthogonal) basis. More precisely, suppose $\{\mathbf{v}_1,\mathbf{v}_2, \ldots, \mathbf{v}_n\}$ is a basis of an inner product space V. The idea is to use the $\mathbf{v}_i$ to construct an orthogonal basis $\{\mathbf{e}_1,\mathbf{e}_2, \ldots, \mathbf{e}_n\}$ one vector at a time. To start the process, take

$$\mathbf{e}_1 = \mathbf{v}_1$$

With an eye on the orthogonal lemma, we use $\mathbf{v}_2$ to define $\mathbf{e}_2$:

$$\mathbf{e}_2 = \mathbf{v}_2 - \frac{\langle \mathbf{v}_2, \mathbf{e}_1 \rangle}{\|\mathbf{e}_1\|^2}\mathbf{e}_1$$

The fact that $\{\mathbf{v}_1, \mathbf{v}_2\}$ is linearly independent shows that $\mathbf{v}_2$ does not lie in span$\{\mathbf{e}_1\}$, so the lemma guarantees that $\{\mathbf{e}_1, \mathbf{e}_2\}$ is orthogonal. Moreover,

$$\text{span}\{\mathbf{e}_1, \mathbf{e}_2\} = \text{span}\{\mathbf{v}_1, \mathbf{v}_2\}$$

as may be readily verified. This completes stage 2 of the process. Next we utilize $\mathbf{v}_3$ to find a vector $\mathbf{e}_3$ orthogonal to both $\mathbf{e}_1$ and $\mathbf{e}_2$. Again the lemma suggests defining

$$\mathbf{e}_3 = \mathbf{v}_3 - \frac{\langle \mathbf{v}_3, \mathbf{e}_1 \rangle}{\|\mathbf{e}_1\|^2}\mathbf{e}_1 - \frac{\langle \mathbf{v}_3, \mathbf{e}_2 \rangle}{\|\mathbf{e}_2\|^2}\mathbf{e}_2$$

As before, the fact that $\{\mathbf{v}_1, \mathbf{v}_2, \mathbf{v}_3\}$ is linearly independent means $\mathbf{v}_3$ does not lie in span$\{\mathbf{v}_1, \mathbf{v}_2\}$ = span$\{\mathbf{e}_1, \mathbf{e}_2\}$ so the lemma shows that $\{\mathbf{e}_1, \mathbf{e}_2, \mathbf{e}_3\}$ is orthogonal. Again, the reader can verify that

$$\text{span}\{\mathbf{e}_1, \mathbf{e}_2, \mathbf{e}_3\} = \text{span}\{\mathbf{v}_1, \mathbf{v}_2, \mathbf{v}_3\}$$

This completes stage 3 of the construction. The process continues in the same way, and in stage n we construct an orthogonal set $\{\mathbf{e}_1, \mathbf{e}_2, \ldots, \mathbf{e}_n\}$ such that

$$\text{span}\{\mathbf{e}_1, \mathbf{e}_2, \ldots, \mathbf{e}_n\} = \text{span}\{\mathbf{v}_1, \mathbf{v}_2, \ldots, \mathbf{v}_n\} = V$$

In other words, $\{\mathbf{e}_1, \mathbf{e}_2, \ldots, \mathbf{e}_n\}$ is an orthogonal basis for V. The procedure can be summarized as follows.

THEOREM 8 Gram–Schmidt Orthogonalization Algorithm

Let V be an inner product space, and let $\{\mathbf{v}_1, \mathbf{v}_2, \ldots, \mathbf{v}_n\}$ be any basis of V. Define vectors $\mathbf{e}_1, \mathbf{e}_2, \ldots, \mathbf{e}_n$ in V successively as follows:

$$\begin{aligned}
\mathbf{e}_1 &= \mathbf{v}_1 \\
\mathbf{e}_2 &= \mathbf{v}_2 - \frac{\langle \mathbf{v}_2, \mathbf{e}_1 \rangle}{\|\mathbf{e}_1\|^2}\mathbf{e}_1 \\
\mathbf{e}_3 &= \mathbf{v}_3 - \frac{\langle \mathbf{v}_3, \mathbf{e}_1 \rangle}{\|\mathbf{e}_1\|^2}\mathbf{e}_1 - \frac{\langle \mathbf{v}_3, \mathbf{e}_2 \rangle}{\|\mathbf{e}_2\|^2}\mathbf{e}_2 \\
&\vdots \quad \vdots
\end{aligned}$$

$$\mathbf{e}_k = \mathbf{v}_k - \frac{\langle \mathbf{v}_k, \mathbf{e}_1 \rangle}{\|\mathbf{e}_1\|^2}\mathbf{e}_1 - \frac{\langle \mathbf{v}_k, \mathbf{e}_2 \rangle}{\|\mathbf{e}_2\|^2}\mathbf{e}_2 - \cdots - \frac{\langle \mathbf{v}_k, \mathbf{e}_{k-1} \rangle}{\|\mathbf{e}_{k-1}\|^2}\mathbf{e}_{k-1}$$

for each $k = 2, 3, \ldots, n$. Then

(1) $\{\mathbf{e}_1, \mathbf{e}_2, \ldots, \mathbf{e}_n\}$ is an orthogonal basis of V.

(2) $\operatorname{span}\{\mathbf{e}_1, \mathbf{e}_2, \ldots, \mathbf{e}_k\} = \operatorname{span}\{\mathbf{v}_1, \mathbf{v}_2, \ldots, \mathbf{v}_k\}$ holds for each $k = 1, 2, \ldots, n$.

Recall (Theorem 7 in Section 5.3) that any linearly independent set of vectors in a vector space V of dimension n is part of a basis of V. A slight modification of the Gram–Schmidt algorithm shows that if V is an inner product space, then any orthogonal set of vectors is part of an orthogonal basis of V. The details are left as Exercise 9(a).

EXAMPLE 10 Use the Gram–Schmidt algorithm to find an orthogonal basis for

$$U = \operatorname{span}\{\mathbf{v}_1 = (1,1,-1,-1), \mathbf{v}_2 = (3,2,0,1), \mathbf{v}_3 = (1,0,1,0)\}$$

Solution Take $\mathbf{e}_1 = \mathbf{v}_1 = (1,1,-1,-1)$. The algorithm gives

$$\mathbf{e}_2 = \mathbf{v}_2 - \frac{\langle \mathbf{v}_2, \mathbf{e}_1 \rangle}{\|\mathbf{e}_1\|^2}\mathbf{e}_1 = (3,2,0,1) - \frac{4}{4}(1,1,-1,-1) = (2,1,1,2)$$

$$\mathbf{e}_3 = \mathbf{v}_3 - \frac{\langle \mathbf{v}_3, \mathbf{e}_1 \rangle}{\|\mathbf{e}_1\|^2}\mathbf{e}_1 - \frac{\langle \mathbf{v}_3, \mathbf{e}_2 \rangle}{\|\mathbf{e}_2\|^2}\mathbf{e}_2 = \mathbf{v}_3 - \frac{0}{4}\mathbf{e}_1 - \frac{3}{10}\mathbf{e}_2 = \frac{1}{10}(4,-3,7,-6)$$

Hence $\{(1,1,-1,-1), (2,1,1,2), \frac{1}{10}(4,-3,7,-6)\}$ is the orthogonal basis provided by the algorithm. In hand calculations it may be convenient to eliminate fractions, so $\{(1,1,-1,-1), (2,1,1,2), (4,-3,7,-6)\}$ is also an orthogonal basis for U.

The next example involves polynomial calculus and may be omitted without loss of continuity.

EXAMPLE 11 Consider $V = \mathbf{P}_3$ with the inner product $\langle p,q \rangle = \int_{-1}^{1} p(x)q(x)\,dx$. If the Gram–Schmidt algorithm is applied to the basis $\{1, x, x^2, x^3\}$, show that the result is the orthogonal basis

$$\left\{1, x, \frac{1}{3}(3x^2 - 1), \frac{1}{5}(5x^3 - 3x)\right\}$$

Solution Take $\mathbf{e}_1 = 1$. Then the algorithm gives

$$\mathbf{e}_2 = x - \frac{\langle x, \mathbf{e}_1 \rangle}{\|\mathbf{e}_1\|^2}\mathbf{e}_1 = x - \frac{0}{2}\mathbf{e}_1 = x$$

$$\begin{aligned}\mathbf{e}_3 &= x^2 - \frac{\langle x^2,\mathbf{e}_1\rangle}{\|\mathbf{e}_1\|^2}\mathbf{e}_1 - \frac{\langle x^2,\mathbf{e}_2\rangle}{\|\mathbf{e}_2\|^2}\mathbf{e}_2 \\ &= x^2 - \frac{2/3}{2}1 - \frac{0}{2/3}x \\ &= \frac{1}{3}(3x^2 - 1)\end{aligned}$$

The verification that $\mathbf{e}_4 = \frac{1}{5}(5x^3 - 3x)$ is omitted.

The polynomials in Example 11 are such that the leading coefficient is 1 in each case. In other contexts (the study of differential equations, for example) it is customary to take multiples $p(x)$ of these polynomials such that $p(1) = 1$. The resulting orthogonal basis of $\mathbf{P}_3$ is

$$\left\{1, x, \frac{1}{2}(3x^2 - 1), \frac{1}{2}(5x^3 - 3x^2)\right\}$$

and these are the first four **Legendre polynomials,** so called to honor the French mathematician A. M. Legendre (1752–1833). They are important in the study of differential equations.

Before proceeding, it is useful to introduce a new notion. Let U be a subspace of an inner product space V. The **orthogonal complement** $U^\perp$ of U in V is defined to be the set of all vectors $\mathbf{v}$ in V that are orthogonal to every vector in U. More formally,

$$U^\perp = \{\mathbf{v} \mid \mathbf{v} \text{ in } V, \langle \mathbf{v},\mathbf{u}\rangle = 0 \text{ for all } \mathbf{u} \text{ in } U\}$$

(Many people refer verbally to $U^\perp$ as "U-perp.") It is an easy exercise (Exercise 19) to see that $U^\perp$ is also a subspace of V and, if $U = \text{span}\{\mathbf{u}_1, \mathbf{u}_2, \ldots, \mathbf{u}_m\}$, that

$$U^\perp = \{\mathbf{v} \mid \mathbf{v} \text{ in } V, \langle \mathbf{v},\mathbf{u}_i\rangle = 0 \text{ for } i = 1, 2, \ldots, m\}$$

EXAMPLE 12 Given the subspace $U = \text{span}\{(1,-1,0,1), (2,0,1,-1)\}$ of $\mathbb{R}^4$, calculate $U^\perp$ (using the dot product).

Solution $U^\perp$ consists of all vectors $\mathbf{v} = (x,y,z,w)$ orthogonal to both $\mathbf{u}_1 = (1,-1,0,1)$ and $\mathbf{u}_2 = (2,0,1,-1)$. The conditions are

$$\begin{aligned}\mathbf{v}\cdot\mathbf{u}_1 &= x - y + w = 0 \\ \mathbf{v}\cdot\mathbf{u}_2 &= 2x + z - w = 0\end{aligned}$$

The solutions are $\mathbf{v} = (s,s + t,-2s + t,t) = s(1,1,-2,0) + t(0,1,1,1)$, where s and t are parameters, so

$$U^\perp = \text{span}\{(1,1,-2,0), (0,1,1,1)\}$$

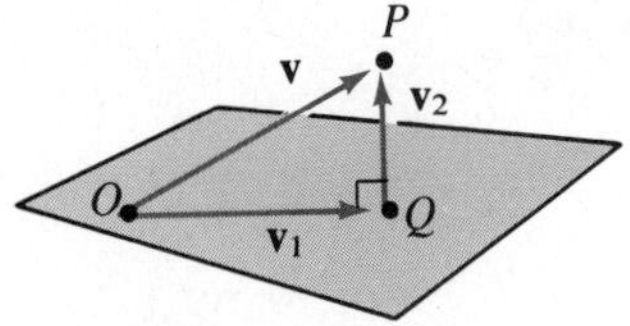

FIGURE 6.3

The orthogonal complement is related to the fundamental notion of a projection, which is so important in geometry and physics. A simple geometrical example will illustrate the situation. Suppose a point P and a plane through the origin O are given, and we want to find the point Q in the plane that is closest to P. Our geometric intuition assures us that such a point Q exists. If we let $\mathbf{v}$ be the position vector of P (that is, the vector from O to P), then what is required is to find the position vector $\mathbf{v}_1$ of Q (see Figure 6.3). Again, our geometric insight assures us that, if we can write $\mathbf{v} = \mathbf{v}_1 + \mathbf{v}_2$ in such a way that $\mathbf{v}_1$ lies in the plane and $\mathbf{v}_2$ is *perpendicular* to the plane, then $\mathbf{v}_1$ will be the vector we want.

Now we make two observations: first, that the set U of position vectors of points in the plane is a *subspace* of $\mathbb{R}^3$ (because the plane contains the origin) and, second, that the condition that $\mathbf{v}_2$ is perpendicular to the plane means that $\mathbf{v}_2$ is *orthogonal* to every vector in U. Hence the problem can be reformulated as follows: Given a vector $\mathbf{v}$ and a subspace U of $\mathbb{R}^3$, is it possible to write $\mathbf{v}$ as a sum $\mathbf{v} = \mathbf{v}_1 + \mathbf{v}_2$, where $\mathbf{v}_1$ lies in U and $\mathbf{v}_2$ lies in $U^\perp$? In this form, the question makes sense in any inner product space. Moreover, the answer is yes, and the vectors $\mathbf{v}_1$ and $\mathbf{v}_2$ are uniquely determined by $\mathbf{v}$ and U.

In general, let U be any subspace of an inner product space V, and let $\{\mathbf{e}_1, \mathbf{e}_2, \dots, \mathbf{e}_m\}$ be an orthogonal basis of U. Given any vector $\mathbf{v}$ in V, we wish to write $\mathbf{v}$ in the form

$$\mathbf{v} = \mathbf{v}_1 + \mathbf{v}_2 \qquad \mathbf{v}_1 \in U, \mathbf{v}_2 \in U^\perp$$

It is clear that the vector

$$\mathbf{v}_1 = \frac{\langle \mathbf{v}_1 \mathbf{e}_1 \rangle}{\|\mathbf{e}_1\|^2}\mathbf{e}_1 + \frac{\langle \mathbf{v}, \mathbf{e}_2 \rangle}{\|\mathbf{e}_2\|^2}\mathbf{e}_2 + \cdots + \frac{\langle \mathbf{v}, \mathbf{e}_m \rangle}{\|\mathbf{e}_m\|^2}\mathbf{e}_m$$

lies in U. Because $U = \operatorname{span}\{\mathbf{e}_1, \dots, \mathbf{e}_m\}$, the orthogonal lemma shows that

$$\mathbf{v}_2 = \mathbf{v} - \mathbf{v}_1 = \mathbf{v} - \frac{\langle \mathbf{v}, \mathbf{e}_1 \rangle}{\|\mathbf{e}_1\|^2}\mathbf{e}_1 - \frac{\langle \mathbf{v}, \mathbf{e}_2 \rangle}{\|\mathbf{e}_2\|^2}\mathbf{e}_2 - \cdots - \frac{\langle \mathbf{v}, \mathbf{e}_m \rangle}{\|\mathbf{e}_m\|^2}\mathbf{e}_m$$

lies in $U^\perp$ (being orthogonal to each of $\mathbf{e}_1, \mathbf{e}_2, \dots, \mathbf{e}_m$). Hence, $\mathbf{v} = \mathbf{v}_1 + \mathbf{v}_2$ is a decomposition of $\mathbf{v}$ of the desired type.

But even more is true: This is the *only* way that $\mathbf{v}$ can be written as the sum of a vector in U and one in $U^\perp$. More precisely, if $\mathbf{v} = \mathbf{w}_1 + \mathbf{w}_2$ where $\mathbf{w}_1$ lies in U and $\mathbf{w}_2$ lies in $U^\perp$, then necessarily $\mathbf{w}_1 = \mathbf{v}_1$ and $\mathbf{w}_2 = \mathbf{v}_2$. Indeed, $\mathbf{w}_1 + \mathbf{w}_2 = \mathbf{v}_1 + \mathbf{v}_2$ (both equal $\mathbf{v}$), and hence

$$\mathbf{w}_1 - \mathbf{v}_1 = \mathbf{v}_2 - \mathbf{w}_2$$

Now this vector lies in U (because $\mathbf{w}_1 - \mathbf{v}_1$ is in U) and also in $U^\perp$ (because $\mathbf{v}_2 - \mathbf{w}_2$ is in $U^\perp$), so it must be $\mathbf{0}$ (it is orthogonal to itself!). Hence $\mathbf{w}_1 = \mathbf{v}_1$ and $\mathbf{w}_2 = \mathbf{v}_2$, as asserted.

Thus each vector $\mathbf{v}$ in V can be *uniquely* represented as a sum $\mathbf{v} = \mathbf{v}_1 + \mathbf{v}_2$ where $\mathbf{v}_1$ lies in U and $\mathbf{v}_2$ lies in $U^\perp$. The vectors $\mathbf{v}_1$ and $\mathbf{v}_2$ depend only on $\mathbf{v}$ and U, not on the choice of orthogonal basis $\{\mathbf{e}_1, \ldots, \mathbf{e}_m\}$ of U used to compute them. Hence we can make the following definition.

DEFINITION

If $\mathbf{v} = \mathbf{v}_1 + \mathbf{v}_2$, $\mathbf{v}_1$ in U, $\mathbf{v}_2$ in $U^\perp$, the vector $\mathbf{v}_1$ in U is called the **projection of v on** U and is denoted by

$$\text{proj}_U(\mathbf{v})$$

The vector $\mathbf{v}_2$ in $U^\perp$ is called the **component of v orthogonal to** U and is given by

$$\mathbf{v} - \text{proj}_U(\mathbf{v})$$

This discussion is summarized in the following fundamental theorem.

THEOREM 9
Projection Theorem

Let V be an inner product space and let U be a finite dimensional subspace of V. Then every vector $\mathbf{v}$ in V can be written uniquely as the sum of a vector $\text{proj}_U(\mathbf{v})$ in U and a vector $\mathbf{v} - \text{proj}_U(\mathbf{v})$ in $U^\perp$:

$$\mathbf{v} = \text{proj}_U(\mathbf{v}) + [\mathbf{v} - \text{proj}_U(\mathbf{v})]$$

Furthermore, if $\{\mathbf{e}_1, \mathbf{e}_2, \ldots, \mathbf{e}_m\}$ is any orthogonal basis of U, then $\text{proj}_U(\mathbf{v})$ can be computed as follows:

$$\text{proj}_U(\mathbf{v}) = \frac{\langle \mathbf{v},\mathbf{e}_1\rangle}{\|\mathbf{e}_1\|^2}\mathbf{e}_1 + \frac{\langle \mathbf{v},\mathbf{e}_2\rangle}{\|\mathbf{e}_2\|^2}\mathbf{e}_2 + \cdots + \frac{\langle \mathbf{v},\mathbf{e}_m\rangle}{\|\mathbf{e}_m\|^2}\mathbf{e}_m$$

EXAMPLE 13

In $\mathbb{R}^4$ write $\mathbf{v} = (3,-2,5,1)$ as the sum of a vector $\mathbf{v}_1$ in $U = \text{span}\{(3,-1,2,0), (1,1,-1,3)\}$ and a vector $\mathbf{v}_2$ orthogonal to every vector in U. Use the dot product.

Solution The vectors $\mathbf{e}_1 = (3,-1,2,0)$ and $\mathbf{e}_2 = (1,1,-1,3)$ are orthogonal, so

$$\begin{aligned}\mathbf{v}_1 = \text{proj}_U(\mathbf{v}) &= \frac{\langle \mathbf{v},\mathbf{e}_1\rangle}{\|\mathbf{e}_1\|^2}\mathbf{e}_1 + \frac{\langle \mathbf{v},\mathbf{e}_2\rangle}{\|\mathbf{e}_2\|^2}\mathbf{e}_2 \\ &= \frac{21}{14}\mathbf{e}_1 + \frac{-1}{12}\mathbf{e}_2 = \frac{1}{12}(53,-19,37,-3)\end{aligned}$$

by Theorem 9. Hence $\mathbf{v}_2 = \mathbf{v} - \mathbf{v}_1 = \frac{1}{12}(-17,-5,23,15)$. It is clear that $\mathbf{v}_1$ lies in U and that $\mathbf{v} = \mathbf{v}_1 + \mathbf{v}_2$; the reader may verify that $\mathbf{v}_2$ is orthogonal to $\mathbf{e}_1$ and $\mathbf{e}_2$ and hence to every vector in $U = \text{span}\{\mathbf{e}_1,\mathbf{e}_2\}$.

EXAMPLE 14 In $\mathbb{R}^4$ let $U = \{(r, r + s, s, r + 2s) \mid r, s \text{ in } \mathbb{R}\}$. Find the projection of $\mathbf{v} = (3,-1,0,2)$ onto U, and express $\mathbf{v}$ as the sum of a vector in U and a vector in $U^\perp$. Use the dot product.

Solution We have $U = \text{span}\{(1,1,0,1), (0,1,1,2)\}$. These vectors are independent but not orthogonal, so the Gram–Schmidt process gives an orthogonal basis $\{\mathbf{e}_1, \mathbf{e}_2\}$ of U:

$$\mathbf{e}_1 = (1,1,0,1)$$

$$\mathbf{e}_2 = (0,1,1,2) - \frac{(0,1,1,2)\cdot(1,1,0,1)}{\|(1,1,0,1)\|^2}(1,1,0,1) = (-1,0,1,1)$$

Hence

$$\text{proj}_U(\mathbf{v}) = \frac{\mathbf{v}\cdot\mathbf{e}_1}{\|\mathbf{e}_1\|^2}\mathbf{e}_1 + \frac{\mathbf{v}\cdot\mathbf{e}_2}{\|\mathbf{e}_2\|^2}\mathbf{e}_2 = \frac{4}{3}\mathbf{e}_1 + \frac{-1}{3}\mathbf{e}_2 = \frac{1}{3}(5,4,-1,3)$$

Finally, the component of $\mathbf{v}$ orthogonal to U is $\mathbf{v} - \text{proj}_U(\mathbf{v}) = \frac{1}{3}(4,-7,1,3)$, so the required decomposition is

$$\mathbf{v} = \frac{1}{3}(5,4,-1,3) + \frac{1}{3}(4,-7,1,3)$$

In the projection theorem, the subspace U of the inner product space V is required to be finite dimensional, but no such restriction is placed on V. If $\dim V$ is finite, however, we have the following useful result.

THEOREM 10 If U is a subspace of a finite dimensional inner product space V, then

$$\dim V = \dim U + \dim U^\perp$$

Proof Let $\{\mathbf{u}_1, \ldots, \mathbf{u}_m\}$ and $\{\mathbf{w}_1, \ldots, \mathbf{w}_k\}$ be orthogonal bases of U and $U^\perp$, respectively. Then $\{\mathbf{u}_1, \ldots, \mathbf{u}_m, \mathbf{w}_1, \ldots, \mathbf{w}_k\}$ is orthogonal (by the definition of $U^\perp$) and so is linearly independent. Hence it is sufficient to show that these vectors span V. But this follows from the projection theorem. ■

Our discussion of projections was motivated by the problem of finding the point in a plane through the origin that was closest to a given point. After the usual identification of points with their position vectors, this became the task of finding the vector in a subspace closest to a given vector. The answer (based to some extent on geometric intuition) was the projection of the given vector on the subspace. The next theorem shows that this is true in general.

THEOREM 11
Approximation Theorem

Let U be a finite dimensional subspace of an inner product space V. If $\mathbf{v}$ is any vector in V, then $\text{proj}_U(\mathbf{v})$ is the vector in U that is closest to $\mathbf{v}$. Here *closest* means that

$$\|\mathbf{v} - \text{proj}_U(\mathbf{v})\| \leq \|\mathbf{v} - \mathbf{u}\|$$

for all $\mathbf{u}$ in U.

Proof For any $\mathbf{u}$ in U, Write $\mathbf{v} - \mathbf{u}$ as follows:

$$\mathbf{v} - \mathbf{u} = [\mathbf{v} - \text{proj}_U(\mathbf{v})] + [\text{proj}_U(\mathbf{v}) - \mathbf{u}]$$

The first member lies in $U^\perp$ (by the projection theorem), whereas the second lies in U (because $\mathbf{u}$ is in U). Hence the two are orthogonal. So, by Pythagoras' theorem,

$$\|\mathbf{v} - \mathbf{u}\|^2 = \|\mathbf{v} - \text{proj}_U(\mathbf{v})\|^2 + \|\text{proj}_U(\mathbf{v}) - \mathbf{u}\|^2 \geq \|\mathbf{v} - \text{proj}_U(\mathbf{v})\|^2$$

The result follows. ■

EXAMPLE 15

Consider the plane through the origin with equation $2x + y - z = 0$. Find the point in this plane closest to $P_0(2,-1,-3)$.

Solution The plane is the subspace U whose points (x,y,z) satisfy $z = 2x + y$. Thus

$$U = \{(s, t, 2s + t) \mid s, t \text{ in } \mathbb{R}\} = \text{span}\{(0,1,1,), (1,0,2)\}$$

By the Gram–Schmidt process, $\{\mathbf{e}_1 = (0,1,1), \mathbf{e}_2 = (1,-1,1)\}$ is an orthogonal basis of U. Hence the vector in U closest to $\mathbf{v} = (2,-1,-3)$ is

$$\text{proj}_U(\mathbf{v}) = \frac{\langle \mathbf{v},\mathbf{e}_1\rangle}{\|\mathbf{e}_1\|^2}\mathbf{e}_1 + \frac{\langle \mathbf{v},\mathbf{e}_2\rangle}{\|\mathbf{e}_2\|^2}\mathbf{e}_2 = -2\mathbf{e}_1 + 0\mathbf{e}_2 = (0,-2,-2)$$

Hence the point in U closest to $P_0(2,-1,-3)$ is $Q(0,-2,-2)$.

We conclude with another example involving calculus; it can be omitted with no loss of continuity.

EXAMPLE 16

Consider the space $\mathbf{C}[-1,1]$ of real-valued continuous functions on the interval $[-1,1]$ with inner product $\langle f,g\rangle = \int_{-1}^{1} f(x)g(x)\,dx$. Find the polynomial $p = p(x)$ of degree at most 2 that best approximates the absolute-value function f given by $f(x) = |x|$.

Solution Here we want the vector p in the subspace $U = \mathbf{P}_2$ that is closest to f. In Example 11 the Gram–Schmidt algorithm was applied to give an orthogonal basis $\{\mathbf{e}_1 = 1, \mathbf{e}_2 = x, \mathbf{e}_3 = 3x^2 - 1\}$ of $\mathbf{P}_2$ (where, for convenience, we have changed $\mathbf{e}_3$ by a numerical factor). Hence the required polynomial is

$$\begin{aligned} p &= \text{proj}_{\mathbf{P}_2}(f) \\ &= \frac{\langle f,\mathbf{e}_1\rangle}{\|\mathbf{e}_1\|^2}\mathbf{e}_1 + \frac{\langle f,\mathbf{e}_2\rangle}{\|\mathbf{e}_2\|^2}\mathbf{e}_2 + \frac{\langle f,\mathbf{e}_3\rangle}{\|\mathbf{e}_3\|^2}\mathbf{e}_3 \\ &= \frac{1}{2}\mathbf{e}_1 + 0\mathbf{e}_2 + \frac{1/2}{8/5}\mathbf{e}_3 \\ &= \frac{3}{16}(5x^2 + 1) \end{aligned}$$

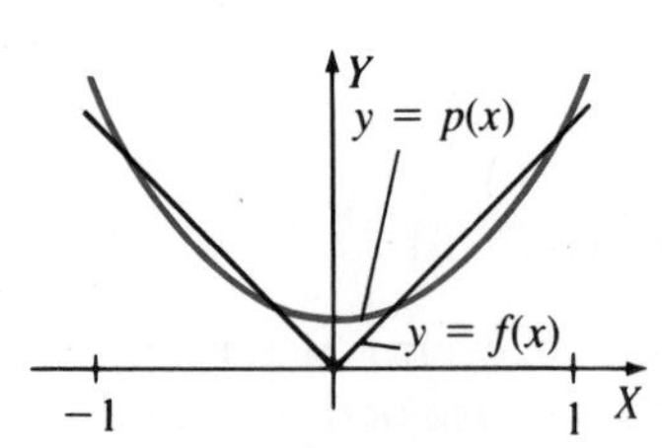

The graphs of $p(x)$ and $f(x)$ are given in the diagram.

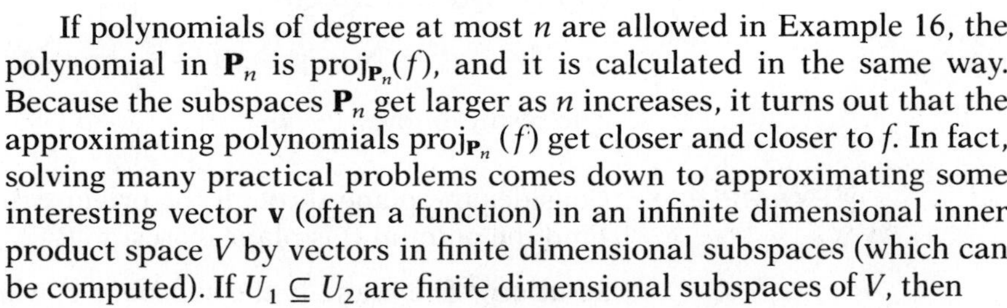
If polynomials of degree at most n are allowed in Example 16, the polynomial in $\mathbf{P}_n$ is $\text{proj}_{\mathbf{P}_n}(f)$, and it is calculated in the same way. Because the subspaces $\mathbf{P}_n$ get larger as n increases, it turns out that the approximating polynomials $\text{proj}_{\mathbf{P}_n}(f)$ get closer and closer to f. In fact, solving many practical problems comes down to approximating some interesting vector $\mathbf{v}$ (often a function) in an infinite dimensional inner product space V by vectors in finite dimensional subspaces (which can be computed). If $U_1 \subseteq U_2$ are finite dimensional subspaces of V, then

$$\|\mathbf{v} - \text{proj}_{U_2}(\mathbf{v})\| \le \|\mathbf{v} - \text{proj}_{U_1}(\mathbf{v})\|$$

by Theorem 11 (because $\text{proj}_{U_1}(\mathbf{v})$ lies in U_1 and hence in U_2). Thus $\text{proj}_{U_2}(\mathbf{v})$ is a better approximation to $\mathbf{v}$ than $\text{proj}_{U_1}(\mathbf{v})$. Hence a general method in approximation theory might be described as follows: Given $\mathbf{v}$, use it to constuct a sequence of finite dimensional subspaces

$$U_1 \subseteq U_2 \subseteq U_3 \subseteq \cdots$$

of V in such a way that $\|\mathbf{v} - \text{proj}_{U_k}(\mathbf{v})\|$ approaches zero as k increases. Then $\text{proj}_{U_k}(\mathbf{v})$ is a suitable approximation to $\mathbf{v}$ if k is large enough. For more information, the interested reader may wish to consult *Interpolation and Approximation* by Philip J. Davis (New York: Blaisdell, 1963).

EXERCISES 6.2.2

Use the dot product in $\mathbb{R}^n$ unless otherwise instructed.

1. In each case, use the Gram–Schmidt algorithm to convert the given basis B of V into an orthogonal basis.

(a) $V = \mathbb{R}^2, B = \{(1,-1), (2,1)\}$

(b) $V = \mathbb{R}^2, B = \{(2,1), (1,2)\}$

(c) $V = \mathbb{R}^3, B = \{(1,-1,1), (1,0,1), (1,1,2)\}$
(d) $V = \mathbb{R}^3, B = \{(0,1,1), (1,1,1), (1,-2,2)\}$

2. In each case, write $\mathbf{v}$ as the sum of a vector in U and a vector in $U^\perp$.
(a) $\mathbf{v} = (1,5,7), U = \text{span}\{(1,-2,3), (-1,1,1)\}$
(b) $\mathbf{v} = (2,1,6), U = \text{span}\{(3,-1,2), (2,0,-3)\}$
(c) $\mathbf{v} = (3,1,5,9), U = \text{span}\{(1,0,1,1), (0,1,-1,1), (-2,0,1,1)\}$
(d) $\mathbf{v} = (2,0,1,6), U = \text{span}\{(1,1,1,1), (1,1,-1,-1), (1,-1,1,-1)\}$
(e) $\mathbf{v} = (a,b,c,d), U = \text{span}\{(1,0,0,0), (0,1,0,0), (0,0,1,0)\}$
(f) $\mathbf{v} = (a,b,c,d), U = \text{span}\{(1,-1,2,0), (-1,1,1,1)\}$

3. Let $\mathbf{v} = (1,-2,1,6)$ in $\mathbb{R}^4$, and let $U = \text{span}\{(2,1,3,-4), (1,2,0,1)\}$.
(a) Compute $\text{proj}_U(\mathbf{v})$.
(b) Show that $\{(1,0,2,-3), (4,7,1,2)\}$ is another orthogonal basis of U.
(c) Use the basis in part **(b)** to compute $\text{proj}_U(\mathbf{v})$.

4. In each case, use the Gram–Schmidt algorithm to find an orthogonal basis of the subspace U of V, and find the vector in U closest to $\mathbf{v}$.
(a) $V = \mathbb{R}^3; U = \text{span}\{(1,1,1), (0,1,1)\}; \mathbf{v} = (-1,2,1)$
(b) $V = \mathbb{R}^3; U = \text{span}\{(1,-1,0), (-1,0,1)\}; \mathbf{v} = (2,1,0)$
(c) $V = \mathbb{R}^4; U = \text{span}\{(1,0,1,0), (1,1,1,0), (1,1,0,0)\}; \mathbf{v} = (2,0,-1,3)$
(d) $V = \mathbb{R}^4; U = \text{span}\{(1,-1,0,1), (1,1,0,0), (1,1,0,1)\}; \mathbf{v} = (2,0,3,1)$

5. In each case, use the Gram–Schmidt process to convert the basis $B = \{1, x, x^2\}$ into an orthogonal basis of $\mathbf{P}_2$.
(a) $\langle p,q \rangle = p(0)q(0) + p(1)q(1) + p(2)q(2)$
(b)† $\langle p,q \rangle = \int_0^2 p(x)q(x)\,dx.$

6.† Using the inner product $\langle p,q \rangle = \int_0^1 p(x)q(x)\,dx$ on $\mathbf{P}_2$, write $\mathbf{v}$ as the sum of a vector in U and a vector in $U^\perp$.
(a) $\mathbf{v} = x^2, U = \text{span}\{x + 1, 9x - 5\}$ (b) $\mathbf{v} = x^2 + 1, U = \text{span}\{1, 2x - 1\}$

7. Let $U = \text{span}\{\mathbf{v}_1, \mathbf{v}_2, \ldots, \mathbf{v}_k\}$, $\mathbf{v}_i$ in $\mathbb{R}^n$, and let A be the $k \times n$ matrix with the $\mathbf{v}_i$ as rows.
(a) Show that $U^\perp = \{\mathbf{x} \mid \mathbf{x} \text{ in } \mathbb{R}^n, A\mathbf{x}^T = \mathbf{0}\}$.
(b) Use part **(a)** to find $U^\perp$ if $U = \text{span}\{(1,-1,2,1), (1,0,-1,1)\}$.

8. Let U be a subspace of an inner product space V, and let $\dim U = m$ and $\dim V = n$.
(a) Show that there is an orthogonal basis $\{\mathbf{e}_1, \mathbf{e}_2, \ldots, \mathbf{e}_n\}$ of V such that $\{\mathbf{e}_1, \mathbf{e}_2, \ldots, \mathbf{e}_m\}$ and $\{\mathbf{e}_{m+1}, \ldots, \mathbf{e}_n\}$ are orthogonal bases of U and $U^\perp$.
(b) If $\{\mathbf{e}_1, \ldots, \mathbf{e}_m, \ldots, \mathbf{e}_n\}$ is any orthogonal basis of V such that $\{\mathbf{e}_1, \ldots, \mathbf{e}_m\}$ is an orthogonal basis of U, show that $\{\mathbf{e}_{m+1}, \ldots, \mathbf{e}_n\}$ is an orthogonal basis of $U^\perp$.

9. (a) Show that every orthogonal set in a vector space of finite dimension is part of an orthogonal basis.

†Requires calculus.

(b) If $\{\mathbf{e}_1, \mathbf{e}_2, \ldots, \mathbf{e}_{n-1}\}$ is orthonormal in an inner product space of dimension n, prove that there are exactly two vectors $\mathbf{e}_n$ such that $\{\mathbf{e}_1, \mathbf{e}_2, \ldots, \mathbf{e}_{n-1}, \mathbf{e}_n\}$ is an orthonormal basis.

10. Let $\mathbf{e}_1$ denote any unit vector in $\mathbb{R}^n$. Show that there exists an orthogonal matrix P with $\mathbf{e}_1$ as its first row. [*Hint:* Gram–Schmidt.]

11. Let $B = \{\mathbf{v}_1, \mathbf{v}_2, \ldots, \mathbf{v}_n\}$ be an ordered basis of an inner product space V, and let $E = \{\mathbf{e}_1, \mathbf{e}_2, \ldots, \mathbf{e}_n\}$ be the ordered orthogonal basis produced by the Gram–Schmidt algorithm. Show that $P_{E\leftarrow B}$ is an upper triangular matrix with 1's on the main diagonal.

12. Verify that $\text{span}\{\mathbf{e}_1, \mathbf{e}_2, \ldots, \mathbf{e}_k\} = \text{span}\{\mathbf{v}_1, \mathbf{v}_2, \ldots, \mathbf{v}_k\}$ holds for each $k = 1, 2, \ldots, m$ in Theorem 8. [*Hint:* Use induction on k.]

13. If the Gram–Schmidt process is used on an orthogonal basis $\{\mathbf{v}_1, \ldots, \mathbf{v}_n\}$ of a subspace U, show that $\mathbf{e}_k = \mathbf{v}_k$ holds for each $k = 1, 2, \ldots, n$. That is, show that the algorithm reproduces the same basis.

14. Let U be a finite dimensional subspace of an inner product space V.

(a) Show that $\mathbf{v}$ lies in U if and only if $\mathbf{v} = \text{proj}_U(\mathbf{v})$.

(b) If $V = \mathbb{R}^3$, show that $(-5,4,-3)$ lies in $\text{span}\{(3,-2,5), (-1,1,1)\}$ but that $(-1,0,2)$ does not.

15. Let U be a finite dimensional subspace of an inner product space V.

(a) Show that $\text{proj}_U(\mathbf{v} + \mathbf{w}) = \text{proj}_U(\mathbf{v}) + \text{proj}_U(\mathbf{w})$ for all $\mathbf{v}$ and $\mathbf{w}$ in V.

(b) Show that $\text{proj}_U(r\mathbf{v}) = r\,\text{proj}_U(\mathbf{v})$ for all $\mathbf{v}$ in V and all r in $\mathbb{R}$.

16. Let V be a finite dimensional inner product space. If U is any subspace of V, show that $\mathbf{v} = \text{proj}_U(\mathbf{v}) + \text{proj}_{U^\perp}(\mathbf{v})$ holds for all $\mathbf{v}$ in V.

17. Let V be an inner product space of dimension n. If $\mathbf{w} \neq \mathbf{0}$ in V, show that:

(a) $(\mathbb{R}\mathbf{w})^\perp = \{\mathbf{v} \mid \mathbf{v} \text{ in } V, \langle \mathbf{v},\mathbf{w}\rangle = 0\}$

(b) $\dim(\mathbb{R}\mathbf{w})^\perp = n - 1$. [*Hint:* Theorem 10.]

18. Let U be a subspace of a finite dimensional inner product space V. Show that $U = (U^\perp)^\perp$. [*Hint:* $U \subseteq U^{\perp\perp}$ is easy; use Theorem 10 twice.]

19. If X is any set of vectors in an inner product space V, define $X^\perp = \{\mathbf{v} \mid \mathbf{v} \text{ in } V, \langle \mathbf{v},\mathbf{x}\rangle = 0 \text{ for all } \mathbf{x} \text{ in } X\}$.

(a) Show that $X^\perp$ is a subspace of V.

(b) If $U = \text{span}\{\mathbf{u}_1, \mathbf{u}_2, \ldots, \mathbf{u}_m\}$, show that $U^\perp = \{\mathbf{u}_1, \ldots, \mathbf{u}_m\}^\perp$.

(c) If $X \subseteq Y$, show that $Y^\perp \subseteq X^\perp$.

(d) Show that $X^\perp \cap Y^\perp = (X \cup Y)^\perp$.

20. **(a)** Let E be an $n \times n$ matrix, and let $U = \{\mathbf{v}E \mid \mathbf{v} \text{ in } \mathbb{R}^n\}$. Show that the following are equivalent.

(i) $E^2 = E = E^T$ (E is a **projection matrix**)

(ii) $(\mathbf{v} - \mathbf{v}E) \cdot (\mathbf{u}E) = 0$ for all $\mathbf{u}$ and $\mathbf{v}$ in $\mathbb{R}^n$

(iii) $\text{proj}_U(\mathbf{v}) = \mathbf{v}E$ for all $\mathbf{v}$ in $\mathbb{R}^n$

[*Hint:* Part **(ii)** implies part **(iii)**: Write $\mathbf{v} = \mathbf{v}E + (\mathbf{v} - \mathbf{v}E)$ and use the uniqueness in Theorem 9. Part **(iii)** implies part **(i)**: $\mathbf{v} - \mathbf{v}E$ lies in $U^\perp$ for all $\mathbf{v}$ in $\mathbb{R}^n$.]

(b) If E is a projection matrix, show that $I - E$ is also.

(c) If $EF = 0 = FE$ and E and F are projection matrices, show that $E + F$ is also.

(d) If A is $m \times n$ and AA^T is invertible, show that $E = A^T(AA^T)^{-1}A$ is a projection matrix.

21. Let $\mathbf{v}$ be a vector in an inner product space V.

(a) Show that $\|\mathbf{v}\| \geq \|\text{proj}_U(\mathbf{v})\|$ holds for all finite dimensional subspaces U. [*Hint:* Pythagoras' theorem.]

(b) If $\{\mathbf{e}_1, \mathbf{e}_2, \ldots, \mathbf{e}_m\}$ is any orthogonal set in V, prove **Bessel's inequality**:

$$\frac{\langle \mathbf{v}, \mathbf{e}_1 \rangle^2}{\|\mathbf{e}_1\|^2} + \cdots + \frac{\langle \mathbf{v}, \mathbf{e}_m \rangle^2}{\|\mathbf{e}_m\|^2} \leq \|\mathbf{v}\|^2$$

22. (a) Let S denote a set of vectors in a finite dimensional inner product space V, and suppose that $\langle \mathbf{u}, \mathbf{v} \rangle = 0$ for all $\mathbf{u}$ in S implies $\mathbf{v} = \mathbf{0}$. Show that $V = \text{span } S$. [*Hint:* Write $U = \text{span } S$ and use Theorem 10.]

(b) Let $A_1, A_2, \ldots, A_k$ be $n \times n$ matrices. Show that the following are equivalent.

(i) If $A_i\mathbf{b} = \mathbf{0}$ for all i (where $\mathbf{b}$ is a column in $\mathbb{R}^n$), then $\mathbf{b} = \mathbf{0}$.

(ii) The set of all rows of the matrices A_i spans $\mathbb{R}^n$.

23. Let A be an $n \times n$ matrix of rank r. Show that there is an invertible $n \times n$ matrix U such that UA is a row-echelon matrix with the property that the first r rows are orthogonal. [*Hint:* Let R be the row-echelon form of A, and use the Gram–Schmidt process on the nonzero rows of R from the bottom up. Use Theorem 5 in Section 2.3.]

24. Let U be a subspace of $\mathbb{R}^n$. Show that there exists an $n \times n$ matrix A such that $U = \{\mathbf{x} \mid A\mathbf{x} = \mathbf{0}\}$

25. Let $\mathbf{v}_1, \mathbf{v}_2, \ldots, \mathbf{v}_{n-1}$ be vectors in $\mathbb{R}^n$. Consider the $(n - 1) \times n$ matrix A with the $\mathbf{v}_i$ as rows, and let A_i denote the $(n - 1) \times (n - 1)$ matrix obtained from A by deleting column i. Define the vector $\mathbf{c}$ in $\mathbb{R}^n$ by

$$\mathbf{c} = [\det A_1, -\det A_2, \det A_3, \ldots, (-1)^{n+1}\det A_n]$$

Show that:

(a) $\mathbf{v}_i \cdot \mathbf{c} = 0$ for all $i = 1, 2, \ldots, n - 1$.

(b) $\mathbf{c} \neq \mathbf{0}$ if and only if $\{\mathbf{v}_1, \mathbf{v}_2, \ldots, \mathbf{v}_{n-1}\}$ is linearly independent.

(c) If $\{\mathbf{v}_1, \mathbf{v}_2, \ldots, \mathbf{v}_{n-1}\}$ is linearly independent, use Theorem 10 to show that all solutions to the system of $n - 1$ homogeneous equations

$$A\mathbf{x} = \mathbf{0}$$

are given by $t\mathbf{c}$, t a parameter.

[*Hints:* **(a)** Write $B_i = \begin{bmatrix} \mathbf{v}_i \\ A \end{bmatrix}$ and show that $\det B_i = 0$. **(b)** If some $\det A_i \neq 0$, the rows of A_i are linearly independent. Conversely, if the $\mathbf{v}_i$ are independent, consider $A = UR$ where R is in reduced row-echelon form.]

6.2.3 LP-Factorization (Optional)

The main virtue of orthogonal matrices is that they can be inverted easily: Simply take the transpose. This fact combines with the following theorem to give a useful way of simplifying many matrix calculations (for example, in least squares approximation; see Section 6.3.1). The result concerns matrices with independent rows (or columns), and the idea is to factor such a matrix as the product of an invertible lower (respectively, upper) triangular matrix and a matrix with *orthonormal* rows (respectively, columns).

Suppose A is an $m \times n$ matrix with linearly independent rows $\mathbf{r}_1, \mathbf{r}_2, \ldots, \mathbf{r}_m$. Then the Gram–Schmidt process can be applied to these rows to produce orthogonal rows $\mathbf{e}_1, \mathbf{e}_2, \ldots, \mathbf{e}_m$, where $\mathbf{e}_1 = \mathbf{r}_1$ and

$$\mathbf{e}_k = \mathbf{r}_k - \frac{\mathbf{r}_k \cdot \mathbf{e}_1}{\|\mathbf{e}_1\|^2}\mathbf{e}_1 - \frac{\mathbf{r}_k \cdot \mathbf{e}_2}{\|\mathbf{e}_2\|^2}\mathbf{e}_2 - \cdots - \frac{\mathbf{r}_k \cdot \mathbf{e}_{k-1}}{\|\mathbf{e}_{k-1}\|^2}\mathbf{e}_{k-1}$$

for each $k = 2, 3, \ldots, m$. Now let $\mathbf{p}_i = \dfrac{1}{\|\mathbf{e}_i\|}\mathbf{e}_i$ for each i. Then $\mathbf{p}_1, \mathbf{p}_2, \ldots, \mathbf{p}_m$ are orthonormal rows, and the foregoing equation becomes

$$\|\mathbf{e}_k\|\,\mathbf{p}_k = \mathbf{r}_k - (\mathbf{r}_k \cdot \mathbf{p}_1)\mathbf{p}_1 - (\mathbf{r}_k \cdot \mathbf{p}_2)\,\mathbf{p}_2 - \cdots - (\mathbf{r}_k \cdot \mathbf{p}_{k-1})\,\mathbf{p}_{k-1}$$

If, for each k, this is solved for $\mathbf{r}_k$ in terms of $\mathbf{p}_1, \mathbf{p}_2, \ldots, \mathbf{p}_k$, we get

$$\begin{aligned}
\mathbf{r}_1 &= \|\mathbf{e}_1\|\,\mathbf{p}_1\\
\mathbf{r}_2 &= (\mathbf{r}_2 \cdot \mathbf{p}_1)\,\mathbf{p}_1 + \|\mathbf{e}_2\|\,\mathbf{p}_2\\
\mathbf{r}_3 &= (\mathbf{r}_3 \cdot \mathbf{p}_1)\,\mathbf{p}_1 + (\mathbf{r}_3 \cdot \mathbf{p}_2)\,\mathbf{p}_2 + \|\mathbf{e}_3\|\,\mathbf{p}_3\\
\vdots\ & \quad \vdots\\
\mathbf{r}_m &= (\mathbf{r}_m \cdot \mathbf{p}_1)\,\mathbf{p}_1 + (\mathbf{r}_m \cdot \mathbf{p}_2)\,\mathbf{p}_2 + (\mathbf{r}_m \cdot \mathbf{p}_3)\,\mathbf{p}_3 + \ldots + \|\mathbf{e}_m\|\,\mathbf{p}_m
\end{aligned}$$

In matrix form, these equations give the required factorization of A:

$$A = \begin{bmatrix} \mathbf{r}_1 \\ \mathbf{r}_2 \\ \mathbf{r}_3 \\ \vdots \\ \mathbf{r}_m \end{bmatrix} = \begin{bmatrix} \|\mathbf{e}_1\| & 0 & 0 & \cdots & 0 \\ \mathbf{r}_2 \cdot \mathbf{p}_1 & \|\mathbf{e}_2\| & 0 & \cdots & 0 \\ \mathbf{r}_3 \cdot \mathbf{p}_1 & \mathbf{r}_3 \cdot \mathbf{p}_2 & \|\mathbf{e}_3\| & \cdots & 0 \\ \vdots & \vdots & \vdots & & \vdots \\ \mathbf{r}_m \cdot \mathbf{p}_1 & \mathbf{r}_m \cdot \mathbf{p}_2 & \mathbf{r}_m \cdot \mathbf{p}_3 & \cdots & \|\mathbf{e}_m\| \end{bmatrix} \cdot \begin{bmatrix} \mathbf{p}_1 \\ \mathbf{p}_2 \\ \mathbf{p}_3 \\ \vdots \\ \mathbf{p}_m \end{bmatrix} = LP \qquad (*)$$

Note that the first factor L is invertible because $\|\mathbf{e}_i\| \neq 0$ for each i, and the second factor P has orthonormal rows $\mathbf{p}_1, \mathbf{p}_2, \ldots, \mathbf{p}_m$. Moreover, an analogous procedure works if A has linearly independent columns (alternatively, apply the foregoing to A^T). The result is Theorem 12.

THEOREM 12 Let A denote an $m \times n$ matrix.

(1) If the rows of A are linearly independent, then A can be factored as

$$A = LP$$

where L is an $m \times m$ lower triangular invertible matrix and P is an $m \times n$ matrix with orthonormal rows. In particular, $PP^T = I_m$.

(2) If the columns of A are linearly independent, then A can be factored as

$$A = QR$$

where R is an $n \times n$ invertible upper triangular matrix and Q is an $m \times n$ matrix with orthonormal columns. In particular, $Q^TQ = I_n$.

For a square matrix, having independent rows (or columns) is equivalent to being invertible, whereas having orthonormal rows (or columns) is equivalent to being orthogonal. Hence:

THEOREM 13 If A is square and invertible, then A has factorizations

$$A = LP \qquad \text{and} \qquad A = QR$$

where P and Q are orthogonal matrices, L is invertible and lower triangular, and R is invertible and upper triangular.

The argument leading to Theorem 12 gives formulas, in equation (*), for L and P. Here is an example.

EXAMPLE 17 Express $A = \begin{bmatrix} 1 & -1 & 0 & 0 \\ 1 & 0 & 1 & 0 \\ 0 & 1 & 1 & 1 \end{bmatrix}$ in the form $A = LP$, where L is lower triangular and invertible, and P has orthonormal rows.

Solution Write $\mathbf{r}_1 = (1,-1,0,0)$, $\mathbf{r}_2 = (1,0,1,0)$, $\mathbf{r}_3 = (0,1,1,1)$ for the rows of A. Apply the Gram–Schmidt algorithm to obtain orthogonal rows $\mathbf{e}_1$, $\mathbf{e}_2$, and $\mathbf{e}_3$.

$$\begin{aligned} \mathbf{e}_1 &= \mathbf{r}_1 &&= (1,-1,0,0) \\ \mathbf{e}_2 &= \mathbf{r}_2 - \frac{1}{2}\mathbf{e}_1 &&= \left(\frac{1}{2},\frac{1}{2},1,0\right) \\ \mathbf{e}_3 &= \mathbf{r}_3 + \frac{1}{2}\mathbf{e}_1 - \mathbf{e}_2 &&= (0,0,0,1) \end{aligned}$$

Hence, using the notation preceding Theorem 12, we get

$$P = \begin{bmatrix} \mathbf{p}_1 \\ \mathbf{p}_2 \\ \mathbf{p}_3 \end{bmatrix} = \begin{bmatrix} \frac{1}{\sqrt{2}} & \frac{-1}{\sqrt{2}} & 0 & 0 \\ \frac{1}{\sqrt{6}} & \frac{1}{\sqrt{6}} & \frac{2}{\sqrt{6}} & 0 \\ 0 & 0 & 0 & 1 \end{bmatrix} = \frac{1}{\sqrt{6}} \begin{bmatrix} \sqrt{3} & -\sqrt{3} & 0 & 0 \\ 1 & 1 & 2 & 0 \\ 0 & 0 & 0 & \sqrt{6} \end{bmatrix}$$

$$L = \begin{bmatrix} \|\mathbf{e}_1\| & 0 & 0 \\ \mathbf{r}_2 \cdot \mathbf{p}_1 & \|\mathbf{e}_2\| & 0 \\ \mathbf{r}_3 \cdot \mathbf{p}_1 & \mathbf{r}_3 \cdot \mathbf{p}_2 & \|\mathbf{e}_3\| \end{bmatrix} = \frac{1}{2} \begin{bmatrix} 2\sqrt{2} & 0 & 0 \\ \sqrt{2} & \sqrt{6} & 0 \\ -\sqrt{2} & \sqrt{6} & 2 \end{bmatrix}$$

The reader can confirm that $A = LP$.

Theorem 12 is useful in computations in the following way: If A has independent rows, then AA^T is invertible (Theorem 5 in Section 5.4), and in practice it is often necessary to invert AA^T (see, for example, Section 6.3.1). But if $A = LP$ as in (1) of Theorem 12, then $AA^T = LPP^TL^T = LL^T$, so

$$(AA^T)^{-1} = (L^{-1})^T L^{-1}$$

The inverse of L is particularly easy to find because L is triangular. In fact, the main reason for the importance of Theorem 12 is that AA^T may be difficult to invert. This is because the inversion process can be numerically unstable so that errors in the computation become too large. The processes of factoring $A = LP$ and finding L^{-1} are essentially back substitutions and so are less unstable.

EXAMPLE 18 Use Theorem 12 to invert AA^T, where A is the matrix in Example 17.

Solution The factorization $A = LP$ was found in Example 17. We have

$$L = \frac{1}{2} \begin{bmatrix} 2\sqrt{2} & 0 & 0 \\ \sqrt{2} & \sqrt{6} & 0 \\ -\sqrt{2} & \sqrt{6} & 2 \end{bmatrix}, \quad L^{-1} = \frac{1}{6} \begin{bmatrix} 3\sqrt{2} & 0 & 0 \\ -\sqrt{6} & 2\sqrt{6} & 0 \\ 6 & -6 & 6 \end{bmatrix}$$

Hence

$$(AA^T)^{-1} = (L^{-1})^T L^{-1} = \frac{1}{3} \begin{bmatrix} 5 & -4 & 3 \\ -4 & 5 & -3 \\ 3 & -3 & 3 \end{bmatrix}$$

The reader can verify that this is the inverse of AA^T.

EXERCISES 6.2.3

1. In each case, factor A as $A = LP$, where L is invertible and lower triangular and P has orthonormal rows. Then compute $(AA^T)^{-1}$ as in Example 18.

   (a) $A = \begin{bmatrix} 1 & -1 \\ -1 & 0 \end{bmatrix}$ (b) $A = \begin{bmatrix} 2 & 1 \\ 1 & 1 \end{bmatrix}$

   (c) $A = \begin{bmatrix} 1 & 1 & 1 & 0 \\ 1 & 1 & 0 & 0 \\ 1 & 0 & 0 & 0 \end{bmatrix}$ (d) $A = \begin{bmatrix} 1 & -1 & 0 & 1 \\ 1 & 0 & 1 & -1 \\ 0 & 1 & 1 & 0 \end{bmatrix}$

2. If $A = LP = L_1P_1$ are two LP-factorizations of the invertible matrix A, show that $L_1 = LD$ and $P_1 = DP$ for some diagonal matrix D with diagonal entries ± 1. [*Hint:* Consider $L_1^{-1}L = P_1P^{-1}$.]

SECTION 6.3 Applications of Inner Products (Optional)†

6.3.1 Best Approximation and Least Squares

A system of linear equations need not have a solution. However, even when no solution exists, it is often desirable to find a "best approximation" to a solution. This section gives one definition of what "best approximation" means. Then it will be shown that such an approximation always exists, and a method for finding it will be described. The result will then be applied to least squares approximation of data, a subject introduced in Section 4.3.

Suppose A is an $m \times n$ matrix and $\mathbf{b}$ is a column in $\mathbb{R}^m$, and consider the system

$$A\mathbf{x} = \mathbf{b}$$

of m linear equations in n variables. This need not have a solution. However, given any column $\mathbf{u}$ in $\mathbb{R}^n$, the distance $\|\mathbf{b} - A\mathbf{u}\|$ is a measure of how far $A\mathbf{u}$ is from $\mathbf{b}$ (where we use the dot product in $\mathbb{R}^m$). Hence it is natural to ask whether there is a column $\mathbf{u}$ in $\mathbb{R}^n$ that is as close as possible to a solution in the sense that

$$\|\mathbf{b} - A\mathbf{u}\|^2$$

is the minimum value of $\|\mathbf{b} - A\mathbf{x}\|^2$ as $\mathbf{x}$ ranges over all columns in $\mathbb{R}^n$.

The approximation theorem (Theorem 11 in Section 6.2) answers this question in the affirmative. To see how, define

$$U = \{A\mathbf{x} \mid \mathbf{x} \text{ lies in } \mathbb{R}^n\}$$

†The two applications in this section are independent and may be taken in any order.

Then U is a subspace of $\mathbb{R}^m$, so we are to find $A\mathbf{x}$ in U as close as possible to $\mathbf{b}$. The approximation theorem guarantees a solution—call it $A\mathbf{u}$—and, in fact

$$A\mathbf{u} = \text{proj}_U(\mathbf{b}).$$

However, there are two computational problems involved here. First, we need an orthogonal basis of U to compute $\text{proj}_U(\mathbf{b})$. Second, we end up with $A\mathbf{u}$ rather than $\mathbf{u}$ itself. So it is useful to find a way to compute $\mathbf{u}$ directly. The key observation is that $\mathbf{b} - A\mathbf{u}$ lies in $U^\perp$ by the projection theorem (Theorem 9 in Section 6.2) and so is orthogonal to every vector $A\mathbf{x}$ in U. Thus,

$$0 = (A\mathbf{x}) \cdot (\mathbf{b} - A\mathbf{u}) = (A\mathbf{x})^T(\mathbf{b} - A\mathbf{u}) = \mathbf{x}^T A^T(\mathbf{b} - A\mathbf{u}) = \mathbf{x} \cdot [A^T(\mathbf{b} - A\mathbf{u})]$$

for all $\mathbf{x}$ in $\mathbb{R}^n$. In other words, $A^T(\mathbf{b} - A\mathbf{u})$ is orthogonal to *every* vector in $\mathbb{R}^n$, and so must be zero. Hence $\mathbf{u}$ satisfies

$$(A^TA)\mathbf{u} = A^T\mathbf{b}$$

This is a system of linear equations called the **normal equations** for $\mathbf{u}$. Note that this system may have more than one solution (see Exercise 4). However, the $n \times n$ matrix A^TA is invertible if (and only if) the columns of A are linearly independent (Theorem 5, Section 5.4), so in this case, $\mathbf{u}$ is uniquely determined and is given explicitly by

$$\mathbf{u} = (A^TA)^{-1}A^T\mathbf{b}$$

However, the most efficient way to find $\mathbf{u}$ is to apply Gaussian elimination to the normal equations. This discussion is summarized in the following theorem.

THEOREM 1 Best Approximation Theorem

Let A be an $m \times n$ matrix, let $\mathbf{b}$ be any column in $\mathbb{R}^m$, and consider the system

$$A\mathbf{x} = \mathbf{b}$$

of m equations in n variables. Any solution $\mathbf{u}$ to the normal equations

$$(A^TA)\mathbf{u} = A^T\mathbf{b}$$

is a best approximation to a solution to $A\mathbf{x} = \mathbf{b}$ in the sense that $\|\mathbf{b} = A\mathbf{u}\|$ is the minimum value of $\|\mathbf{b} - A\mathbf{x}\|$ as $\mathbf{x}$ ranges over all columns in $\mathbb{R}^n$.

If the columns of A are linearly independent, then A^TA is invertible and $\mathbf{u}$ is given uniquely by $\mathbf{u} = (A^TA)^{-1}A^T\mathbf{b}$.

Note that if A is $n \times n$ and invertible, then

$$\mathbf{u} = (A^TA)^{-1}A^T\mathbf{b} = A^{-1}\mathbf{b}$$

is the solution to the system of equations, and $\|\mathbf{b} - A\mathbf{u}\| = 0$. Hence $(A^TA)^{-1}A^T$ is playing the role of the inverse of the non-square matrix A in the case where A has linearly independent columns. The matrix $A^T(AA^T)^{-1}$ plays a similar role when the rows of A are linearly independent. These are both special cases of the so-called **generalized inverse** of a matrix A. However, we shall not pursue this topic here.

EXAMPLE 1 The equations

$$\begin{aligned} 3x - y &= 4 \\ x + 2y &= 0 \\ 2x + y &= 1 \end{aligned}$$

have no solution. Find the vector $\mathbf{u} = \begin{bmatrix} x_0 \\ y_0 \end{bmatrix}$ that best approximates a solution.

Solution In this case,

$$A = \begin{bmatrix} 3 & -1 \\ 1 & 2 \\ 2 & 1 \end{bmatrix} \quad \text{so} \quad A^TA = \begin{bmatrix} 3 & 1 & 2 \\ -1 & 2 & 1 \end{bmatrix}\begin{bmatrix} 3 & -1 \\ 1 & 2 \\ 2 & 1 \end{bmatrix} = \begin{bmatrix} 14 & 1 \\ 1 & 6 \end{bmatrix}$$

is invertible. The normal equations $A^TA\mathbf{u} = A^T\mathbf{b}$ are

$$\begin{bmatrix} 14 & 1 \\ 1 & 6 \end{bmatrix}\mathbf{u} = \begin{bmatrix} 14 \\ -3 \end{bmatrix}, \quad \text{so} \quad \mathbf{u} = \frac{1}{83}\begin{bmatrix} 87 \\ -56 \end{bmatrix}$$

Thus $x_0 = 87/83$ and $y_0 = -56/83$. With these values of x and y, the left sides of the equations are

$$\begin{aligned} 3x_0 - y_0 &= 317/83 = 3.82 \\ x_0 + 2y_0 &= -25/83 = -0.30 \\ 2x_0 + y_0 &= 118/83 = 1.42 \end{aligned}$$

This is as close as possible to a solution.

EXAMPLE 2 The average number g of goals per game scored by a hockey player seems to be related linearly to two factors: the number x_1 of years of experience and the number x_2 of goals in the preceding 10 games. The accompanying data were collected on four players. Find the linear function $g = a_0 + a_1x_1 + a_2x_2$ that best fits these data.

Solution If the relationship is given by $g = r_0 + r_1x_1 + r_2x_2$, then the data can be described as follows:

g	x_1	x_2
.8	5	3
.8	3	4
.6	1	5
.4	2	1

$$\begin{bmatrix} 1 & 5 & 3 \\ 1 & 3 & 4 \\ 1 & 1 & 5 \\ 1 & 2 & 1 \end{bmatrix} \begin{bmatrix} r_0 \\ r_1 \\ r_2 \end{bmatrix} = \begin{bmatrix} .8 \\ .8 \\ .6 \\ .4 \end{bmatrix}$$

Using the notation in Theorem 1,

$$\mathbf{u} = (A^T A)^{-1} A^T \mathbf{b}$$

$$= \frac{1}{294} \begin{bmatrix} 833 & -119 & -133 \\ -119 & 35 & 7 \\ -133 & 7 & 35 \end{bmatrix} \begin{bmatrix} 1 & 1 & 1 & 1 \\ 5 & 3 & 1 & 2 \\ 3 & 4 & 5 & 1 \end{bmatrix} \begin{bmatrix} .8 \\ .8 \\ .6 \\ .4 \end{bmatrix} = \begin{bmatrix} .14 \\ .09 \\ .08 \end{bmatrix}$$

Hence the best-fitting function is $g = .14 + .09x_1 + .08x_2$. The computation would have been reduced if the normal equations had been constructed and then solved by Gaussian elimination.

Theorem 1 applies directly to least squares approximation. This was treated in Section 4.3, though it was not possible to prove the main theorem (Theorem 2 in Section 4.3) with the techniques then available.

Suppose that data are available giving pairs of corresponding values of the two variables x and y:

$$(x_1, y_1), (x_2, y_2), \ldots, (x_n, y_n)$$

Given such data pairs, assume for the moment that the variables x and y are related by a polynomial of degree m.

$$y = p(x) = r_0 + r_1 x + \cdots + r_m x^m$$

Then for each x_i we have *two* values of the variable y, the observed value y_i and the computed value $p(x_i)$. The question now is this: Is it possible to choose the coefficients $r_0, r_1, \ldots, r_m$ in such a way that the $p(x_i)$ are "as close as possible" to the corresponding y_i? To apply Theorem 1, the following notation is convenient:

$$\mathbf{y} = \begin{bmatrix} y_1 \\ y_2 \\ \vdots \\ y_n \end{bmatrix} \qquad p(\mathbf{x}) = \begin{bmatrix} p(x_1) \\ p(x_2) \\ \vdots \\ p(x_n) \end{bmatrix}$$

Then the problem takes the following form: Choose $r_0, r_1, \ldots, r_n$ such that

$$\|\mathbf{y} - p(\mathbf{x})\|^2 = [y_1 - p(x_1)]^2 + [y_2 - p(x_2)]^2 + \cdots + [y_n - p(x_n)]^2$$

is as small as possible. A polynomial $p(x)$ satisfying this condition is called a **least squares approximating polynomial** of degree m for the data pairs given. Now write

$$\mathbf{r} = \begin{bmatrix} r_0 \\ r_1 \\ \vdots \\ r_m \end{bmatrix} \qquad M = \begin{bmatrix} 1 & x_1 & x_1^2 & \dots & x_1^m \\ 1 & x_2 & x_2^2 & \dots & x_2^m \\ \vdots & \vdots & \vdots & & \vdots \\ 1 & x_n & x_n^2 & \dots & x_n^m \end{bmatrix}$$

Then $p(\mathbf{x})$ can be written

$$p(\mathbf{x}) = \begin{bmatrix} p(x_1) \\ p(x_2) \\ \vdots \\ p(x_n) \end{bmatrix} = \begin{bmatrix} r_0 + r_1x_1 + \cdots + r_mx_1^m \\ r_0 + r_1x_2 + \cdots + r_mx_2^m \\ \vdots \\ r_0 + r_1x_n + \cdots + r_mx_n^m \end{bmatrix} = M\mathbf{r}$$

so we are to find $\mathbf{r}$ in $\mathbb{R}^{m+1}$ such that $\|\mathbf{y} - M\mathbf{r}\|^2$ is as small as possible. In this form, Theorem 1 applies directly and gives the first part of Theorem 2.

THEOREM 2 Let n data pairs $(x_1,y_1), (x_2,y_2), \ldots, (x_n,y_n)$ be given, and write

$$\mathbf{y} = \begin{bmatrix} y_1 \\ y_2 \\ \vdots \\ y_n \end{bmatrix} \qquad M = \begin{bmatrix} 1 & x_1 & x_1^2 & \dots & x_1^m \\ 1 & x_2 & x_2^2 & \dots & x_2^m \\ \vdots & \vdots & \vdots & & \vdots \\ 1 & x_n & x_n^2 & \dots & x_n^m \end{bmatrix}$$

(1) If $\mathbf{u} = \begin{bmatrix} u_0 \\ u_1 \\ \vdots \\ u_m \end{bmatrix}$ is any solution to the normal equations

$$(M^TM)\mathbf{u} = M^T\mathbf{y}$$

then the polynomial

$$\overline{p}(x) = u_0 + u_1x + u_2x^2 + \cdots + u_mx^m$$

is a least squares approximating polynomial of degree m for the given data pairs.

(2) If at least $m + 1$ of the numbers $x_1, x_2, \ldots, x_n$ are distinct (so $n \geq m + 1$), the matrix M^TM is invertible and $\mathbf{u}$ is uniquely determined by

$$\mathbf{u} = (M^TM)^{-1}M^T\mathbf{y}$$

Proof It remains to prove that the columns of M are linearly independent in (2). Suppose a linear combination of the columns vanishes:

$$r_0\begin{bmatrix}1\\1\\\vdots\\1\end{bmatrix} + r_1\begin{bmatrix}x_1\\x_2\\\vdots\\x_n\end{bmatrix} + \cdots + r_m\begin{bmatrix}x_1{}^m\\x_2{}^m\\\vdots\\x_n{}^m\end{bmatrix} = \begin{bmatrix}0\\0\\\vdots\\0\end{bmatrix}$$

If we write $p(x) = r_0 + r_1x + \cdots + r_mx^m$, equating coefficients shows that $p(x_1) = p(x_2) = \cdots = p(x_n) = 0$. But then the hypothesis asserts that the polynomial $p(x)$ has at least $m + 1$ distinct roots, so because $p(x)$ has degree m, it must be the zero polynomial. Thus $r_0 = r_1 = \cdots = r_m = 0$, as required. ■

Several examples illustrating the use of this theorem were given in Section 4.3. The interested reader is referred to them.

There is an extension of Theorem 2 that should be mentioned. Given the data pairs $(x_1,y_1), \ldots, (x_n,y_n)$, that theorem shows how to find a polynomial

$$p(x) = r_0 + r_1x + \cdots + r_mx^m$$

such that $\|\mathbf{y} - p(\mathbf{x})\|^2$ is as small as possible, where $\mathbf{y}$ and $p(\mathbf{x})$ are as above. Choosing the appropriate polynomial $p(x)$ amounts to choosing the coefficients $r_0, r_1, \ldots, r_m$, and the theorem gives a formula for the optimal choices. Now $p(x)$ is a linear combination of the functions $1, x, \ldots, x^m$, where the r_i are the coefficients, and this suggests applying the method to linear combinations of other functions. If $f_0(x), f_1(x), \ldots, f_m(x)$ are given functions, write

$$f(x) = r_0 f_0(x) + r_1 f_1(x) + \cdots + r_m f_m(x)$$

where $r_0, r_1, \ldots, r_m$ are real numbers. Then the more general question is whether $r_0, r_1, \ldots, r_m$ can be found such that $\|\mathbf{y} - f(\mathbf{x})\|^2$ is as small as possible, where now

$$f(\mathbf{x}) = \begin{bmatrix}f(x_1)\\f(x_2)\\\vdots\\f(x_n)\end{bmatrix}$$

The theorem follows.

THEOREM 3 Let n data pairs $(x_1,y_1), (x_2,y_2), \ldots, (x_n,y_n)$ be given, and suppose that $m + 1$ functions $f_0(x), f_1(x), \ldots, f_m(x)$ are specified. Write

$$\mathbf{y} = \begin{bmatrix}y_1\\y_2\\\vdots\\y_n\end{bmatrix} \qquad M = \begin{bmatrix}f_0(x_1) & f_1(x_1) & \ldots & f_m(x_1)\\f_0(x_2) & f_1(x_2) & \ldots & f_m(x_2)\\\vdots & \vdots & & \vdots\\f_0(x_n) & f_1(x_n) & \ldots & f_m(x_n)\end{bmatrix}$$

(1) If $\mathbf{u} = \begin{bmatrix} u_0 \\ u_1 \\ \vdots \\ u_m \end{bmatrix}$ is any solution to the normal equations

$$(M^T M)\mathbf{u} = M^T\mathbf{y}$$

then

$$\bar{f}(x) = u_0 f_0(x) + u_1 f_1(x) + \cdots + u_m f_m(x)$$

is the best approximation for these data among all functions $f(x)$ of the form

$$f(x) = r_0 f_0(x) + r_1 f_1(x) + \cdots + r_m f_m(x) \qquad r_i \text{ in } \mathbb{R}$$

in the sense that $\|\mathbf{y} - \bar{f}(\mathbf{x})\| \le \|\mathbf{y} - f(\mathbf{x})\|$ holds for all choices of the r_i

(2) If $M^T M$ is invertible (that is, rank $M = m + 1$), then $\mathbf{u}$ is uniquely determined by

$$\mathbf{u} = (M^T M)^{-1} M^T \mathbf{y}$$

Proof Observe that $f(\mathbf{x}) = M\mathbf{r}$, where $\mathbf{r} = \begin{bmatrix} r_0 \\ r_1 \\ \vdots \\ r_m \end{bmatrix}$, so we are asked to choose $\mathbf{r}$ to minimize $\|\mathbf{y} - M\mathbf{r}\|^2$. Theorem 1 applies as before. ■

The function $\bar{f}(x) = u_0 f_0(x) + u_1 f_1(x) + \cdots + u_m f_m(x)$ is called a **least squares approximating function** of the form $r_0 f_0(x) + \cdots + r_m f_m(x)$. This theorem contains Theorem 2 as a special case ($f_i(x) = x^i$ for each i), but there is no guarantee that $M^T M$ will be invertible in the general case if $m + 1$ of the x_i are distinct. Conditions for this to hold depend on the choice of the functions $f_0(x), f_1(x), \ldots, f_m(x)$.

EXAMPLE 3 Given the data pairs $(-1,0)$, $(0,1)$, and $(1,4)$, find the least squares approximating function of the form $r_0 x + r_1 2^x$.

Solution The functions are $f_0(x) = x$ and $f_1(x) = 2^x$, so the matrix M is

$$M = \begin{bmatrix} f_0(x_1) & f_1(x_1) \\ f_0(x_2) & f_1(x_2) \\ f_0(x_3) & f_1(x_3) \end{bmatrix} = \begin{bmatrix} -1 & 2^{-1} \\ 0 & 2^0 \\ 1 & 2^1 \end{bmatrix} = \frac{1}{2}\begin{bmatrix} -2 & 1 \\ 0 & 2 \\ 2 & 4 \end{bmatrix}$$

In this case $M^TM = \frac{1}{4}\begin{bmatrix} 8 & 6 \\ 6 & 21 \end{bmatrix}$ is invertible, so the normal equations

$$\frac{1}{4}\begin{bmatrix} 8 & 6 \\ 6 & 21 \end{bmatrix}\mathbf{u} = \begin{bmatrix} 4 \\ 9 \end{bmatrix} \quad \text{have solution} \quad \mathbf{u} = \frac{1}{11}\begin{bmatrix} 10 \\ 16 \end{bmatrix}$$

Hence the best-fitting function of the form $r_0x + r_12^x$ is $\bar{f}(x) = \frac{10}{11}x + \frac{16}{11}2^x$. Note that $\bar{f}(\mathbf{x}) = \begin{bmatrix} \bar{f}(-1) \\ \bar{f}(0) \\ \bar{f}(1) \end{bmatrix} = \begin{bmatrix} -2/11 \\ 16/11 \\ 42/11 \end{bmatrix}$, compared with $\mathbf{y} = \begin{bmatrix} 0 \\ 1 \\ 4 \end{bmatrix}$.

EXERCISES 6.3.1

1. Find the best approximation to a solution of each of the following systems of equations.

(a)
$$\begin{aligned} x + y - z &= 5 \\ 2x - y + 6z &= 1 \\ 3x + 2y - z &= 6 \\ -x + 4y + z &= 0 \end{aligned}$$

(b)
$$\begin{aligned} 3x + y + z &= 6 \\ 2x + 3y - z &= 1 \\ 2x - y + z &= 0 \\ 3x - 3y + 3z &= 8 \end{aligned}$$

2. Find a least squares approximating function of the form $r_0x + r_1x^2 + r_2(-1)^x$ for each of the following sets of data pairs.

(a) $(-1,1), (0,3), (1,1), (2,0)$

(b) $(0,1), (1,1), (2,5), (3,10)$

3. Find the least squares approximating function of the form $r_0x^2 + r_1 \sin \frac{\pi x}{2}$ for each of the following sets of data pairs.

(a) $(0,0), (1,2), (2,3)$

(b) $\left(-1, \frac{1}{2}\right), (0,0), (1,4)$

4. Let A be any $m \times n$ matrix and write $K = \{\mathbf{k} \mid A^TA\mathbf{k} = \mathbf{0}\}$. Let $\mathbf{b}$ be an m-column. Show that, if there is an n-column $\mathbf{u}$ such that $\|\mathbf{b} - A\mathbf{u}\|$ is minimal, then *all* such vectors have the form $\mathbf{u} + \mathbf{k}$ for some $\mathbf{k}$ in K. [*Hint:* $\|\mathbf{b} - A\mathbf{w}\|$ is minimal if and only if $A^TA\mathbf{w} = A^T\mathbf{b}$.]

5. The yield y of wheat in bushels per acre appears to be a linear function of the number of days x_1 of sunshine, the number of inches x_2 of rain, and the number of pounds x_3 of fertilizer applied per acre. Find the best fit to the data (on page 325) of an equation of the form $y = r_0 + r_1x_1 + r_2x_2 + r_3x_3$. [*Hint:* If a calculator for inverting A^TA is not available, the inverse is given in the answer.]

y	x_1	x_2	x_3
28	50	18	10
30	40	20	16
21	35	14	10
23	40	12	12
23	30	16	14

6. Given the situation in Theorem 3, write

$$f(x) = r_0 f_0(x) + r_1 f_1(x) + \cdots + r_m f_m(x)$$

Suppose that $f(x)$ has at most k roots for any choice of the coefficients $r_0, r_1, \ldots, r_m$.

(a) Show that $M^T M$ is invertible if at least $k + 1$ of the x_i are distinct.

(b) If at least two of the x_i are distinct, show that there is always a best approximation of the form $r_0 + r_1 e^x$.

(c) If at least three of the x_i are distinct, show that there is always a best approximation of the form $r_0 + r_1 x + r_2 e^x$. [Calculus is needed.]

7. If A is an $m \times n$ matrix, it can be proved that there exists a unique $n \times m$ matrix $A^{\#}$ satisfying the following four conditions: $AA^{\#}A = A$; $A^{\#}AA^{\#} = A^{\#}$; $AA^{\#}$ and $A^{\#}A$ are symmetric. The matrix $A^{\#}$ is called the **generalized inverse** of A, or the **Moore–Penrose** inverse.

(a) If A is square and invertible, show that $A^{\#} = A^{-1}$.

(b) If rank $A = m$, show that $A^{\#} = A^T(AA^T)^{-1}$.

(c) If rank $A = n$, show that $A^{\#} = (A^TA)^{-1}A^T$.

6.3.2 Introduction to Fourier Approximation

In this section we shall investigate an important orthogonal set in the space $\mathbf{C}[-\pi, \pi]$ of continuous functions on the interval $[-\pi, \pi]$, using the inner product

$$\langle f, g \rangle = \int_{-\pi}^{\pi} f(x)g(x)\, dx$$

Of course calculus will be needed. The orthogonal set in question is

$$\{1, \sin x, \cos x, \sin 2x, \cos 2x, \sin 3x, \cos 3x, \ldots\}$$

and the first such investigation was carried out by Jean Baptiste Joseph Fourier (1768–1830), who used these functions in 1822 to study the conduction of heat in solids.

Standard techniques of integration give

$$\|1\|^2 = \int_{-\pi}^{-\pi} 1^2 \, dx = 2\pi$$

$$\|\sin kx\|^2 = \int_{-\pi}^{\pi} \sin^2(kx) \, dx = \pi \qquad \text{for any } k = 1, 2, 3, \ldots$$

$$\|\cos kx\|^2 = \int_{-\pi}^{\pi} \cos^2(kx) \, dx = \pi \qquad \text{for any } k = 1, 2, 3, \ldots$$

We leave the verifications to the reader, together with the task of showing that these functions are orthogonal: $\langle \sin(kx), \sin(mx) \rangle = 0 = \langle \cos(kx), \cos(mx) \rangle$ if $k \neq m$, and

$$\langle \sin(kx), \cos(mx) \rangle = \int_{-\pi}^{\pi} \sin(kx)\cos(mx) \, dx = 0 \qquad k, m = 0, 1, \ldots$$

(Note that $1 = \cos(0x)$, so the function 1 is included.)

Now define the following subspace of $\mathbf{C}[-\pi, \pi]$:

$$T_n = \text{span}\{1, \sin x, \cos x, \sin 2x, \cos 2x, \ldots, \sin nx, \cos nx\}$$

The aim is to use the approximation theorem (Theorem 11 in Section 6.2) so, given a function f in $\mathbf{C}[-\pi, \pi]$, define the **Fourier coefficients** of f by

$$a_0 = \frac{\langle f(x), 1 \rangle}{\|1^2\|} = \frac{1}{2\pi} \int_{-\pi}^{\pi} f(x) \, dx$$

$$a_k = \frac{\langle f(x), \cos(kx) \rangle}{\|\cos(kx)\|^2} = \frac{1}{\pi} \int_{-\pi}^{\pi} f(x) \cos(kx) \, dx \qquad k = 1, 2, \ldots$$

$$b_k = \frac{\langle f(x), \sin(kx) \rangle}{\|\sin(kx)\|^2} = \frac{1}{\pi} \int_{-\pi}^{\pi} f(x) \sin(kx) \, dx \qquad k = 1, 2, \ldots$$

Then the approximation theorem gives Theorem 4.

THEOREM 4 Let f be any continuous real-valued function defined on the interval $[-\pi, \pi]$. If $a_0, a_1, \ldots$ and $b_0, b_1, \ldots$ are the Fourier coefficients of f, then

$$t_{nf}(x) = a_0 + a_1 \cos x + b_1 \sin x + a_2 \cos 2x + b_2 \sin 2x + \cdots$$
$$\cdots + a_n \cos nx + b_n \sin nx$$

is a function in T_n that is closest to f in the sense that

$$\|f - t_{nf}\| \leq \|f - t\|$$

holds for all functions t in T_n.

The function t_{nf} is called the nth **Fourier approximation** to the function f.

EXAMPLE 4 Find the fifth Fourier approximation to the function $f(x)$ defined on $[-\pi, \pi]$ as follows:

$$f(x) = \begin{cases} \pi + x & -\pi \leq x < 0 \\ \pi - x & 0 \leq x \leq \pi \end{cases}$$

Y
π
$-\pi$ 0 π X

Solution The graph of $y = f(x)$ appears in the diagram. The Fourier coefficients are computed as follows. The details of the integrations (usually by parts) are omitted.

$$a_0 = \frac{1}{2\pi}\int_{-\pi}^{\pi} f(x)\,dx = \frac{\pi}{2}$$

$$a_k = \frac{1}{\pi}\int_{-\pi}^{\pi} f(x)\cos(kx)\,dx = \frac{2}{\pi k^2}[1 - \cos(k\pi)] = \begin{cases} 0 & \text{if } k \text{ is even} \\ \dfrac{4}{\pi k^2} & \text{if } k \text{ is odd} \end{cases}$$

$$b_k = \frac{1}{\pi}\int_{-\pi}^{\pi} f(x)\sin(kx)\,dx = 0 \qquad \text{for all } k = 1, 2, \ldots$$

Hence the fifth Fourier approximation is

$$t_{5f}(x) = \frac{\pi}{2} + \frac{4}{\pi}\left\{\cos x + \frac{1}{3^2}\cos(3x) + \frac{1}{5^2}\cos(5x)\right\}$$

EXAMPLE 5 Find the fourth Fourier approximation for the function $f(x) = x$ on the interval $[-\pi, \pi]$.

Solution We have $a_k = 0$ for all $k \geq 0$ in this case, whereas

$$b_k = \frac{1}{\pi}\int_{-\pi}^{\pi} x\sin(kx)\,dx = \frac{2}{k\pi}[-\pi\cos(k\pi)]$$

$$= \frac{2}{k}(-1)^{k+1} \quad \text{for all } k = 1, 2, 3, \ldots$$

Again, we omit the details of the integration by parts. Hence the

Fourier approximation is

$$t_{4f}(x) = 2\left\{\sin x - \frac{\sin(2x)}{2} + \frac{\sin(3x)}{3} - \frac{\sin(4x)}{4}\right\}$$

We say that a function f is an **even function** if $f(x) = f(-x)$ holds for all x; f is called an **odd function** if $f(-x) = -f(x)$ holds for all x. Examples of even functions are constant functions, the even powers x^2, $x^4, \ldots,$ and $\cos(kx)$; and these functions are characterized by the fact that the graph of $y = f(x)$ is symmetric about the Y-axis. Examples of odd functions are the odd powers $x, x^3, \ldots$ and $\sin(kx)$ where $k > 0$, and the graph of $y = f(x)$ is symmetric about the origin if f is odd. The usefulness of these functions stems from the fact that

$$\int_{-\pi}^{\pi} f(x)\,dx = 0 \qquad \text{if } f \text{ is odd}$$

$$\int_{-\pi}^{\pi} f(x)\,dx = 2\int_{0}^{\pi} f(x)\,dx \qquad \text{if } f \text{ is even}$$

These facts often simplify the computations of the Fourier coefficients. For example:

1. The Fourier sine coefficients b_k all vanish if f is even.
2. The Fourier cosine coefficients a_k all vanish if f is odd.

This is because $f(x)\sin(kx)$ is odd in the first case and $f(x)\cos(kx)$ is odd in the second case. These observations are illustrated in the two foregoing examples.

The functions 1, $\cos(kx)$, and $\sin(kx)$ that occur in the Fourier approximation for $f(x)$ are all easy to generate as an electrical voltage (when x is time). By summing these signals (with the amplitudes given by the Fourier coefficients), it is possible to produce an electrical signal with (the approximation to) $f(x)$ as the voltage. Hence these Fourier approximations play a fundamental role in electronics.

Finally, the Fourier approximations $t_{1f}, t_{2f}, \ldots$ of a function f get better and better as n increases. The reason is that the subspaces T_n increase

$$T_1 \subseteq T_2 \subseteq T_3 \subseteq \cdots \subseteq T_n \subseteq \cdots$$

so, because $t_{nf} = \text{proj}_{T_n}(f)$, we get (see the discussion following Example 16 in Section 6.2)

$$\|f - t_{1f}\| \geq \|f - t_{2f}\| \geq \cdots \geq \|f - t_{nf}\| \geq \cdots$$

This draws our attention to the infinite series

$$a_0 + a_1 \cos x + b_1 \sin x + a_2 \cos 2x + b_2 \sin 2x + \cdots \qquad (*)$$

where the a_k and b_k are the Fourier coefficients of f. This is called the **Fourier series** for $f(x)$, and the question arises immediately whether such an infinite sum makes any sense at all. This leads to the notion of *convergence,* which will not be dealt with here. However, whether the series (*) makes sense or not, two observations about it can be made: First, the sum of the first $2n + 1$ terms is just t_{nf}, so these partial sums get closer and closer to f as n increases. Second, if f happens to lie in T_n for some n, then $a_k = b_k = 0$ for all $k > n$, so the sum is actually finite and it adds to f. It turns out that (*) converges to f for every function f in $\mathbf{C}[-\pi, \pi]$ such that $f(-\pi) = f(\pi)$. The proof of this is given in books on Fourier analysis. [For example, R. V. Churchill, *Fourier Series and Boundary Value Problems* (New York: McGraw-Hill, 1941).] This subject not only had great historical impact on the development of mathematics but has become one of the standard tools in science and engineering.

EXERCISES 6.3.2

1. In each case, find the Fourier approximation t_{5f} of the given function in $\mathbf{C}[-\pi, \pi]$.

(a) $f(x) = \pi - x$

(b) $f(x) = |x| = \begin{cases} x & \text{if } 0 \le x \le \pi \\ -x & \text{if } -\pi \le x < 0 \end{cases}$

(c) $f(x) = x^2$

(d) $f(x) = \begin{cases} 0 & \text{if } -\pi \le x < 0 \\ x & \text{if } 0 \le x \le \pi \end{cases}$

2. **(a)** Find t_{5f} for the even function on $[-\pi, \pi]$ satisfying $f(x) = x$ for $0 \le x \le \pi$.

(b) Find t_{6f} for the even function on $[-\pi, \pi]$ satisfying $f(x) = \sin x$ for $0 \le x \le \pi$.

[*Hint:* If $k > 1$, $\displaystyle\int \sin x \cos(kx)\, dx = \frac{1}{2}\left[\frac{\cos[(k-1)x]}{k-1} - \frac{\cos[(k+1)x]}{k+1}\right]$.]

3. **(a)** Prove that $\displaystyle\int_{-\pi}^{\pi} f(x)\, dx = 0$ if f is odd and that $\displaystyle\int_{-\pi}^{\pi} f(x)\, dx = 2\int_0^{\pi} f(x)\, dx$ if f is even.

(b) Prove that $\frac{1}{2}[f(x) + f(-x)]$ is even and that $\frac{1}{2}[f(x) - f(-x)]$ is odd for any function f.

4. Show that $\{1, \cos x, \cos(2x), \cos(3x), \ldots\}$ is an orthogonal set in $\mathbf{C}[0, \pi]$ with respect to the inner product $\langle f, g\rangle = \displaystyle\int_0^{\pi} f(x)g(x)\, dx$.

7 Eigenvalues and Diagonalization

SECTION 7.1 Eigenvalues and Eigenvectors

7.1.1 Eigenvalues and the Characteristic Polynomial

This section deals with certain numbers associated with a square matrix called the *eigenvalues* of the matrix. These numbers arise in the study of quadratic forms (Section 7.4.1), in the study of differential equations (Section 7.4.2), and in many physical applications (as natural frequencies and as energy levels of electrons, to name only two).

DEFINITION

If A is an $n \times n$ matrix, a number λ is called an **eigenvalue** of A if

$$A\mathbf{p} = \lambda\mathbf{p}$$

for some nonzero n-column $\mathbf{p}$. In this case the (nonzero) vector $\mathbf{p}$ is called an **eigenvector** of A corresponding to λ. Eigenvalues and eigenvectors are often called **characteristic values** and **characteristic vectors,** respectively. If λ is an eigenvalue of A, the set

$$E_\lambda = E_\lambda(A) = \{\mathbf{p} \mid \mathbf{p} \text{ in } \mathbb{R}^n, A\mathbf{p} = \lambda\mathbf{p}\}$$

is a vector space (subspace of $\mathbb{R}^n$) called the **eigenspace** associated with λ.

Hence E_λ consists of all eigenvectors corresponding to λ, together with the zero vector. The condition $A\mathbf{p} = \lambda\mathbf{p}$ can be written

$$(\lambda I - A)\mathbf{p} = \mathbf{0}$$

where I is the $n \times n$ identity matrix. Hence E_λ consists of all solutions to this system of linear equations, and λ is an eigenvalue if it happens that E_λ contains at least one nonzero vector.

EXAMPLE 1

Show that $\lambda = -3$ is an eigenvalue of $A = \begin{bmatrix} 5 & 8 & 16 \\ 4 & 1 & 8 \\ -4 & -4 & -11 \end{bmatrix}$, and find the corresponding eigenspace E_{-3}.

Solution If $\lambda = -3$, the condition that $(\lambda I - A)\,\mathbf{p} = \mathbf{0}$ becomes

$$\begin{aligned} -8x_1 - 8x_2 - 16x_3 &= 0 \\ -4x_1 - 4x_2 - 8x_3 &= 0 \\ 4x_1 + 4x_2 + 8x_3 &= 0 \end{aligned} \qquad \text{where} \qquad \mathbf{p} = \begin{bmatrix} x_1 \\ x_2 \\ x_3 \end{bmatrix}$$

The solution is $\mathbf{p} = s\mathbf{p}_1 + t\mathbf{p}_2$, where $\mathbf{p}_1 = \begin{bmatrix} -1 \\ 1 \\ 0 \end{bmatrix}$ and $\mathbf{p}_2 = \begin{bmatrix} -2 \\ 0 \\ 1 \end{bmatrix}$. Hence $E_{-3} = \text{span}\{\mathbf{p}_1, \mathbf{p}_2\}$.

The technique in Example 1 leads to a way of determining all the possible eigenvalues. Let A be any $n \times n$ matrix. A number λ is an eigenvalue of A if and only if

$$A\mathbf{p} = \lambda\mathbf{p} \text{ for some } \mathbf{p} \neq \mathbf{0}$$

As above, this condition can be written

$$(\lambda I - A)\mathbf{p} = \mathbf{0} \text{ for some } \mathbf{p} \neq \mathbf{0}$$

where I is the $n \times n$ identity matrix. Now recall (Theorem 10 in Section 2.3) that a matrix U is invertible if and only if $U\mathbf{p} = \mathbf{0}$ implies that $\mathbf{p} = \mathbf{0}$. Hence $\mathbf{p}$ is an eigenvalue of A if and only if $\lambda I - A$ is *not* invertible, and this in turn means that

$$\det(\lambda I - A) = 0$$

by Theorem 2 in Section 3.2. This is a very useful test for the eigenvalues of A, and it suggests the following definition:

DEFINITION

The **characteristic polynomial** of the $n \times n$ matrix A is defined to be

$$c_A(x) = \det(xI - A)$$

Theorem 1 summarizes this discussion.

THEOREM 1

Let A be an $n \times n$ matrix. The eigenvalues of A are the roots of the characteristic polynomial of A. That is, they are the numbers λ

satisfying

$$c_A(\lambda) = \det(\lambda I - A) = 0$$

where I is the $n \times n$ identity matrix. In this case, the eigenspace E_λ corresponding to λ is given by

$$E_\lambda = \{\mathbf{p} \mid (\lambda I - A)\mathbf{p} = \mathbf{0}\}$$

and so consists of all solutions to a system of n linear equations in n variables. The eigenvectors corresponding to λ are the nonzero vectors in the eigenspace E_λ.

Hence determining the eigenspaces of a matrix is reduced to two problems: First find the eigenvalues, and then find the eigenspaces as sets of solutions to linear homogeneous equations. This latter problem has been treated in this book. The problem of determining the eigenvalues is more difficult, and we will not spend a great deal of time on it. Hence the examples and exercises will be so constructed that the roots of the characteristic polynomials encountered are relatively easy to find (usually integers). The reader should not be misled by this into thinking that eigenvalues are so easily obtained for the matrices that occur in practical applications! In fact, the eigenvalues are *not* usually found as roots of the characteristic polynomial, and two other techniques are briefly described in Section 7.1.3 below.

EXAMPLE 2 Find the characteristic polynomial of the matrix $A = \begin{bmatrix} 5 & 8 & 16 \\ 4 & 1 & 8 \\ -4 & -4 & -11 \end{bmatrix}$ mentioned in Example 1, and use it to find all eigenvalues and eigenspaces of A.

Solution The characteristic polynomial is

$$c_A(x) = \det(xI - A) = \det\begin{bmatrix} x-5 & -8 & -16 \\ -4 & x-1 & -8 \\ 4 & 4 & x+11 \end{bmatrix}$$

Adding the last row to the second row gives

$$\begin{aligned} c_A(x) &= \det\begin{bmatrix} x-5 & -8 & -16 \\ 0 & x+3 & x+3 \\ 4 & 4 & x+11 \end{bmatrix} \\ &= (x+3)(x^2+2x-3) = (x+3)^2(x-1) \end{aligned}$$

so the eigenvalues are $\lambda = -3$ and $\lambda = 1$. The case $\lambda = -3$ was dealt with in Example 1, where it was shown that the corresponding eigenspace is $E_{-3} = \text{span}\{\mathbf{p}_1, \mathbf{p}_2\}$ where $\mathbf{p}_1 = \begin{bmatrix} -1 \\ 1 \\ 0 \end{bmatrix}$ and $\mathbf{p}_2 = \begin{bmatrix} -2 \\ 0 \\ 1 \end{bmatrix}$. However, the characteristic polynomial has revealed a new eigenvalue $\lambda = 1$. The corresponding eigenspace consists of all $\mathbf{p} = \begin{bmatrix} x_1 \\ x_2 \\ x_3 \end{bmatrix}$ satisfying $(I - A)\mathbf{p} = \mathbf{0}$. This gives the equations

$$\begin{bmatrix} -4 & -8 & -16 \\ -4 & 0 & -8 \\ 4 & 4 & 12 \end{bmatrix} \begin{bmatrix} x_1 \\ x_2 \\ x_3 \end{bmatrix} = \begin{bmatrix} 0 \\ 0 \\ 0 \end{bmatrix} \quad \text{whence} \quad \mathbf{p} = t\mathbf{p}_3, \text{ where } \mathbf{p}_3 = \begin{bmatrix} 2 \\ 1 \\ -1 \end{bmatrix}$$

Hence the eigenspace corresponding to $\lambda = 1$ is $E_1 = \text{span}\{\mathbf{p}_3\}$.

The characteristic polynomial in Example 2 is $(x + 3)^2(x - 1)$ and the eigenspaces E_{-3} and E_1 have dimensions 2 and 1, respectively. These dimensions are equal to the multiplicities of $\lambda = -3$ and $\lambda = 1$ as roots of the characteristic polynomial, and the reader might be justified in asking whether this is always true. The next example shows that this inference cannot be drawn in general (it is true for symmetric matrices, however, as we shall see).

EXAMPLE 3 Find the characteristic polynomial, eigenvalues, and eigenspaces for $A = \begin{bmatrix} 2 & 1 & 1 \\ 2 & 1 & -2 \\ -1 & 0 & -2 \end{bmatrix}$.

Solution The characteristic polynomial is

$$\begin{aligned} c_A(x) = \det(xI - A) &= \det \begin{bmatrix} x - 2 & -1 & -1 \\ -2 & x - 1 & 2 \\ 1 & 0 & x + 2 \end{bmatrix} \\ &= x^3 - x^2 - 5x - 3 = (x + 1)^2(x - 3) \end{aligned}$$

so the eigenvalues are $\lambda = -1, 3$. For $\lambda = -1$, the eigenspace consists of all $\mathbf{p} = \begin{bmatrix} x_1 \\ x_2 \\ x_3 \end{bmatrix}$ such that $(-I - A)\mathbf{p} = \mathbf{0}$. This gives

$$\begin{bmatrix} -3 & -1 & -1 \\ -2 & -2 & 2 \\ 1 & 0 & 1 \end{bmatrix} \begin{bmatrix} x_1 \\ x_2 \\ x_3 \end{bmatrix} = \begin{bmatrix} 0 \\ 0 \\ 0 \end{bmatrix} \quad \text{so that } \mathbf{p} = t\mathbf{p}_1, \text{ where } \mathbf{p}_1 = \begin{bmatrix} -1 \\ 2 \\ 1 \end{bmatrix}$$

Hence the eigenspace is $E_{-1} = \mathbb{R}\mathbf{p}_1$, and this has dimension 1 even though the eigenvalue $\lambda = -1$ is a *double* root of $c_A(x)$. For $\lambda = 3$, the eigenspace is given by $E_3 = \mathbb{R}\mathbf{p}_2$, where $\mathbf{p}_2 = \begin{bmatrix} 5 \\ 6 \\ -1 \end{bmatrix}$.

The following example gives one situation wherein the characteristic polynomial and eigenvalues are easy to determine.

EXAMPLE 4 If A is a triangular matrix, show that the eigenvalues of A are the entries on the main diagonal.

Solution If the main diagonal entries of A are $a_{11}, a_{22}, \ldots, a_{nn}$, then those of $xI - A$ are $x - a_{11}, x - a_{22}, \ldots, x - a_{nn}$. Hence Theorem 3 in Section 3.1 gives

$$c_A(x) = \det(xI - A) = (x - a_{11})(x - a_{22}) \ldots (x - a_{nn})$$

The result follows.

EXAMPLE 5 Show that A and A^T have the same characteristic polynomial and hence the same eigenvalues.

Solution We use the fact (Theorem 3 in Section 3.2) that a matrix and its transpose have the same determinant. Hence

$$c_{A^T}(x) = \det[xI - A^T] = \det[(xI - A)^T] = \det(xI - A) = c_A(x)$$

The result follows from Theorem 1.

There are many examples of matrices with real entries that have eigenvalues that are not real (for example, $\begin{bmatrix} 0 & 1 \\ -1 & 0 \end{bmatrix}$ has eigenvalues i and $-i$, where $i^2 = -1$). We will have more to say about such matrices in Section 7.3, where the following important result will be proved.

THEOREM 2 If A is a real symmetric matrix, each root of the characteristic polynomial $c_A(x)$ is real.

EXAMPLE 6 Confirm Theorem 2 for any symmetric 2×2 matrix A.

Solution Write $A = \begin{bmatrix} a & b \\ b & c \end{bmatrix}$. The characteristic polynomial is

$$c_A(x) = \det\begin{bmatrix} x - a & -b \\ -b & x - c \end{bmatrix} = x^2 - (a + c)x + (ac - b^2)$$

This has real roots because the discriminant $(a + c)^2 - 4(ac - b^2) = (a - c)^2 + 4b^2$ is nonnegative.

EXERCISES 7.1.1

1. Find the characteristic polynomial, eigenvalues, and a basis for each eigenspace for A if:

(a) $A = \begin{bmatrix} 1 & 2 \\ 3 & 2 \end{bmatrix}$ **(b)** $A = \begin{bmatrix} 2 & -4 \\ -1 & -1 \end{bmatrix}$ **(c)** $A = \begin{bmatrix} 7 & 0 & -4 \\ 0 & 5 & 0 \\ 5 & 0 & -2 \end{bmatrix}$

(d) $A = \begin{bmatrix} 1 & 1 & -3 \\ 2 & 0 & 6 \\ 1 & -1 & 5 \end{bmatrix}$ **(e)** $A = \begin{bmatrix} 1 & -2 & 3 \\ 2 & 6 & -6 \\ 1 & 2 & -1 \end{bmatrix}$ **(f)** $A = \begin{bmatrix} 0 & 1 & 1 \\ 1 & 0 & 1 \\ 1 & 1 & 0 \end{bmatrix}$

(g) $A = \begin{bmatrix} 3 & 1 & 1 \\ -4 & -2 & -5 \\ 2 & 2 & 5 \end{bmatrix}$ **(h)** $A = \begin{bmatrix} 2 & 1 & 1 \\ 0 & 1 & 0 \\ 1 & -1 & 2 \end{bmatrix}$

(i) $A = \begin{bmatrix} \lambda & 0 & 0 \\ 0 & \lambda & 0 \\ 0 & 0 & \mu \end{bmatrix}, \lambda \neq \mu$

2. If $A = \begin{bmatrix} \lambda & a & b \\ 0 & \lambda & c \\ 0 & 0 & \lambda \end{bmatrix}$, show that λ is the only eigenvalue of A and that

$$E_\lambda = \left\{ \begin{bmatrix} r \\ s_1 \\ s_2 \end{bmatrix} \middle| \begin{bmatrix} a & b \\ 0 & c \end{bmatrix} \begin{bmatrix} s_1 \\ s_2 \end{bmatrix} = \begin{bmatrix} 0 \\ 0 \end{bmatrix} \right\}$$

3. Show that A has $\lambda = 0$ as an eigenvalue if and only if A is not invertible.

4. Let A denote an $n \times n$ matrix and put $A_1 = A - \alpha I$, α in $\mathbb{R}$. Show that λ is an eigenvalue of A if and only if $\lambda - \alpha$ is an eigenvalue of A_1. How do the eigenvectors compare? (Hence the eigenvalues of A_1 are just those of A "shifted" by α.)

5. Identify the vector $\mathbf{v} = \begin{bmatrix} x \\ y \end{bmatrix}$ with the point $P(x,y)$ in the plane. Let A be a 2×2 matrix.
   (a) If $\mathbf{v} \neq \mathbf{0}$, show that $\mathbb{R}\mathbf{v}$ is the line through $\mathbf{0}$ and $\mathbf{v}$.
   (b) Show that, if L is any line through the origin, so is $AL = \{A\mathbf{v} \mid \mathbf{v} \text{ in } L\}$ (provided that $AL \neq \{\mathbf{0}\}$).
   (c) If $A\mathbf{v} \neq \mathbf{0}$, show that $\mathbf{v}$ is an eigenvector of A if and only if the line $L = \mathbb{R}\mathbf{v}$ is *fixed* by A—that is, $AL = L$.

6. Show that the eigenvalues of $\begin{bmatrix} \cos\theta & \sin\theta \\ -\sin\theta & \cos\theta \end{bmatrix}$ are $e^{i\theta}$ and $e^{-i\theta}$. (See Appendix A.)

7. Find the characteristic polynomial of the $n \times n$ identity matrix I. Show that I has exactly one eigenvalue, and find the eigenspace.

8. Given $A = \begin{bmatrix} a & b \\ c & d \end{bmatrix}$, show that:

   (a) $c_A(x) = x^2 - (a + d)x + \det A$

   (b) The eigenvalues are $\frac{1}{2}[(a + d) \pm \sqrt{(a - d)^2 + 4bc}]$.

9. Let A be an $n \times n$ matrix, and suppose that the characteristic polynomial has the form

$$c_A(x) = a_n x^n + a_{n-1}x^{n-1} + \cdots + a_1 x + a_0$$

   (a) Show that $a_0 = c_A(0) = (-1)^n \det A$.

   (b) Show that $a_n = 1$.

10. Let A be any $n \times n$ matrix and $r \neq 0$ a real number.

   (a) Show that the eigenvalues of rA are precisely the numbers $r\lambda$, where λ is an eigenvalue of A.

   (b) Show that $c_{rA}(x) = r^n c_A\left(\frac{x}{r}\right)$.

11. (a) If each row of A has the same sum s, show that s is an eigenvalue.

   (b) If each column of A has the same sum s, show that s is an eigenvalue.

12. Let A be an invertible $n \times n$ matrix.

   (a) Show that the eigenvalues of A are nonzero.

   (b) Show that the eigenvalues of A^{-1} are precisely the numbers $\frac{1}{\lambda}$, where λ is an eigenvalue of A.

   (c) Show that $c_{A^{-1}}(x) = \frac{(-x)^n}{\det A} c_A\left(\frac{1}{x}\right)$.

13. Suppose λ is an eigenvalue of a square matrix A with eigenvector $\mathbf{v} \neq \mathbf{0}$.

   (a) Show that λ^2 is an eigenvalue of A^2 (with the same $\mathbf{v}$).

   (b) Show that $\lambda^3 - 2\lambda + 3$ is an eigenvalue of $A^3 - 2A + 3I$.

   (c) Show that $p(\lambda)$ is an eigenvalue of $p(A)$ for any nonzero polynomial $p(x)$.

14. If A is an $n \times n$ matrix, show that $c_{A^2}(x^2) = (-1)^n c_A(x) c_A(-x)$.

15. An $n \times n$ matrix A is called **nilpotent** if $A^k = 0$ for some $k \geq 1$.

   (a) Show that each triangular matrix with zeros on the main diagonal is nilpotent.

   (b) Show that $\lambda = 0$ is the only eigenvalue (even complex) of A. [*Hint:* Exercise 13.]

   (c) Deduce that $c_A(x) = x^n$.

16. (a) Given $p(x) = a_0 + a_1 x + a_2 x^2 + a_3 x^3 + x^4$, show that

$$A = \begin{bmatrix} 0 & 1 & 0 & 0 \\ 0 & 0 & 1 & 0 \\ 0 & 0 & 0 & 1 \\ -a_0 & -a_1 & -a_2 & -a_3 \end{bmatrix}$$

   is a matrix whose characteristic polynomial equals $p(x)$. [A is called the **companion matrix** for $p(x)$.]

(b) Generalize to polynomials of degree n with the coefficient of x^n being 1.

17. Let $A = \begin{bmatrix} 0 & a & b \\ a & 0 & c \\ b & c & 0 \end{bmatrix}$ and $B = \begin{bmatrix} c & a & b \\ a & b & c \\ b & c & a \end{bmatrix}$.

(a) Show that $x^3 - (a^2 + b^2 + c^2)x - 2abc$ has real roots by considering A.

(b) Show that $a^2 + b^2 + c^2 \geq ab + ac + bc$ by considering B.

18. Let $A = \begin{bmatrix} B & 0 \\ 0 & C \end{bmatrix}$, where B and C are square matrices.

(a) Show that $c_A(x) = c_B(x)\, c_C(x)$.

(b) If $\mathbf{p}$ and $\mathbf{q}$ are eigenvectors of B and C, respectively, show that $\begin{bmatrix} \mathbf{p} \\ \mathbf{0} \end{bmatrix}$ and $\begin{bmatrix} \mathbf{0} \\ \mathbf{q} \end{bmatrix}$ are eigenvectors of A, and show how every eigenvector of A arises from such eigenvectors.

7.1.2 Similar Matrices

In this brief section, a relationship (called similarity) for square matrices is introduced that will be fundamental for the rest of this chapter and will recur in Section 9.1.

DEFINITION

Two $n \times n$ matrices A and B are called **similar** (denoted $A \sim B$) if

$$B = P^{-1}AP$$

holds for some invertible matrix P.

The matrix P is not unique (if $A = B = I$, *any* invertible P will do), and it is sometimes difficult to decide whether A and B are similar.

EXAMPLE 7

Let $A = \begin{bmatrix} 2 & 1 \\ -1 & -1 \end{bmatrix}$, $B = \begin{bmatrix} -2 & 5 \\ -1 & 3 \end{bmatrix}$, and $B_1 = \begin{bmatrix} 5 & 2 \\ 4 & 1 \end{bmatrix}$. Show that A and B are similar but A and B_1 are not.

Solution If $P = \begin{bmatrix} -1 & 3 \\ 1 & -2 \end{bmatrix}$, the reader can verify that $P^{-1}AP = B$ (rather than computing P^{-1}, it is easier to check that P is invertible and $AP = PB$). Hence A and B are similar. On the other hand, it is easy to verify that similar matrices have equal determinants, and so A and B_1 are not similar because $\det A = -1$ whereas $\det B_1 = -3$.

The fact (utilized in Example 7) that similar matrices have equal determinants is sometimes expressed by saying that the determinant of a matrix is a **similarity invariant.** Other examples are given in the following theorem. Define the **trace** of a square matrix A (denoted tr A) to be the sum of the entries on the main diagonal of A.

THEOREM 3 If A and B are similar matrices, they have the same determinant, the same rank, the same trace, the same characteristic polynomial, and the same eigenvalues.

Proof Let $B = P^{-1}AP$, where P is invertible. Then Theorems 1 and 2 in Section 3.2 give

$$\det B = \det P^{-1} \det A \det P = \frac{1}{\det P} \det A \det P = \det A$$

Next, Corollary 4 of Theorem 2 in Section 5.4 gives

$$\text{rank } B = \text{rank}(P^{-1}AP) = \text{rank}(AP) = \text{rank } A$$

As to the trace, it suffices to show $\text{tr}(XA) = \text{tr}(AX)$ holds for any matrix X, because then

$$\text{tr } B = \text{tr}(P^{-1}AP) = \text{tr}[(AP)P^{-1}] = \text{tr } A$$

But if $X = [x_{ij}]$ and $A = [a_{ij}]$, then the (i,j)-entry of AX is $\sum_{k=1}^{n} a_{ik}x_{kj}$. Hence

$$\text{tr}(AX) = \sum_{m=1}^{n} \left[\sum_{k=1}^{n} a_{mk}x_{km} \right] = \sum_{k=1}^{n} \left[\sum_{m=1}^{n} x_{km}a_{mk} \right] = \text{tr}(XA)$$

Turning to the characteristic polynomial, the fact that $xI = P^{-1}(xI)P$ gives

$$c_B(x) = \det(xI - B) = \det(P^{-1}(xI - A)P) = \det(xI - A) = c_A(x)$$

Hence A and B share the same characteristic polynomial and so have the same eigenvalues by Theorem 1. ■

Note that it is possible for two nonsimilar matrices to share all the attributes in Theorem 3 (see Exercise 2).

Similarity can be used to simplify a variety of matrix calculations, particularly when it can be shown that the matrix in question is similar to a diagonal matrix. Here is an example.

EXAMPLE 8 Let $B = \begin{bmatrix} 5 & 4 \\ -2 & -1 \end{bmatrix}$. Find the characteristic polynomial of B and compute

$B^3 - 5B^2$ by using the fact that $B = P^{-1}AP$, where $A = \begin{bmatrix} 1 & 0 \\ 0 & 3 \end{bmatrix}$ and $P = \begin{bmatrix} 1 & 2 \\ 1 & 1 \end{bmatrix}$.

Solution We leave it to the reader to verify that $B = P^{-1}AP$ (equivalently, $PB = AP$). Hence B and A are similar, so, using Theorem 3,

$$c_B(x) = c_A(x) = \det\begin{bmatrix} x-1 & 0 \\ 0 & x-3 \end{bmatrix} = (x-1)(x-3)$$

This computation is much easier than calculating $c_B(x)$ directly. Direct calculation of $B^3 - 5B^2$ would also be tedious, but the fact that $B = P^{-1}AP$ allows it to be carried out with A in place of B. In fact, $B^2 = (P^{-1}AP)(P^{-1}AP) = P^{-1}A^2P$, and similarly $B^3 = P^{-1}A^3P$. It follows that $B^3 - 5B^2 = P^{-1}(A^3 - 5A^2)P$, and the matrix $A^3 - 5A^2$ is easy to compute because A is diagonal. The result is

$$\begin{aligned} B^3 - 5B^2 &= P^{-1}(A^3 - 5A^2)P \\ &= \begin{bmatrix} -1 & 2 \\ 1 & -1 \end{bmatrix}\left[\begin{pmatrix} 1 & 0 \\ 0 & 27 \end{pmatrix} - 5\begin{pmatrix} 1 & 0 \\ 0 & 9 \end{pmatrix}\right]\begin{bmatrix} 1 & 2 \\ 1 & 1 \end{bmatrix} \\ &= \begin{bmatrix} -32 & -28 \\ 14 & 10 \end{bmatrix} \end{aligned}$$

The type of simplification in this example occurs for any polynomial in B in place of $B^3 - 5B^2$ (See Exercise 6).

One further comment on similarity is in order. Suppose $A \sim B$, say $B = P^{-1}AP$ for some invertible matrix P. This implies that $A = PBP^{-1}$, and this, in turn, can be written in the form $A = Q^{-1}BQ$ where $Q = P^{-1}$. Hence $B \sim A$ and we have verified the second of the following three properties of similarity (the others are left as Exercise 12).

THEOREM 4 Let A, B, and C denote $n \times n$ matrices.

(1) $A \sim A$ for all A.

(2) If $A \sim B$, then $B \sim A$.

(3) If $A \sim B$ and $B \sim C$, then $A \sim C$.

The properties of similarity in Theorem 4 are useful in the following way. If we want to prove that A and B are similar, it is sometimes more convenient to show that both are similar to some "nice" matrix D (possibly diagonal): $A \sim D$ and $B \sim D$. Then $D \sim B$ by (2) of Theorem 4, so $A \sim D$ and $D \sim B$ imply that $A \sim B$ by (3).

By virtue of Theorem 4, similarity is an example of an equivalence relation. The general concept is as follows: If S is any set, a relation $\equiv$ on S is called an **equivalence relation** on S if it satisfies the following conditions:

(1) $s \equiv s$ for all s in S.

(2) If $s \equiv t$ with s and t in S, then $t \equiv s$.

(3) If $r \equiv s$ and $s \equiv t$ with r, s, and t in S, then $r \equiv t$.

Equality is the prototype equivalence relation.

In matrix theory, we have already encountered two equivalence relations of note. The first is the following: Two $m \times n$ matrices A and B are called **row-equivalent** if A can be carried to B by a sequence of elementary row operations (equivalently, if $B = UA$ for some invertible matrix U). This is analyzed in detail in Exercise 20 of Section 2.3.2, and the most important theorem in Chapter 1 is that every $m \times n$ matrix is row-equivalent to a reduced row-echelon matrix.

The second equivalence relation we have encountered is the following: Two $m \times n$ matrices A and B are called **equivalent** if they have the same rank (equivalently, $B = UAV$ for some invertible matrices U and V), and Theorem 3 in Section 5.4 shows that every $m \times n$ matrix of rank r is equivalent to an $m \times n$ matrix of the form $\begin{bmatrix} I_r & 0 \\ 0 & 0 \end{bmatrix}$.

Similarity is also an equivalence relation on the set of $n \times n$ matrices, so it is natural to look for "nice" matrices to which every matrix is similar. It turns out that there are various choices, called **canonical forms** in this context. A complete discussion of canonical forms lies outside the scope of this book. However, for many purposes, diagonal matrices are an appropriate choice as the "nice" matrices under this new relationship of similarity. Unfortunately, not every square matrix is similar to a diagonal matrix, but every symmetric matrix does have this property—a fact that is useful in applications. Methods for deciding when a matrix is "diagonalizable" and for finding the corresponding diagonal matrix will be given in the following section.

EXERCISES 7.1.2

1. By computing the trace, determinant, and rank, show that A and B are *not* similar in each case.

(a) $A = \begin{bmatrix} 1 & 2 \\ 2 & 1 \end{bmatrix}, B = \begin{bmatrix} 1 & 1 \\ -1 & 1 \end{bmatrix}$ **(b)** $A = \begin{bmatrix} 3 & 1 \\ 2 & -1 \end{bmatrix}, B = \begin{bmatrix} 1 & 1 \\ 2 & 1 \end{bmatrix}$

(c) $A = \begin{bmatrix} 2 & 1 \\ 1 & -1 \end{bmatrix}, B = \begin{bmatrix} 3 & 0 \\ 1 & -1 \end{bmatrix}$ **(d)** $A = \begin{bmatrix} 3 & 1 \\ -1 & 2 \end{bmatrix}, B = \begin{bmatrix} 2 & -1 \\ 3 & 2 \end{bmatrix}$

(e) $A = \begin{bmatrix} 2 & 1 & 1 \\ 1 & 0 & 1 \\ 1 & 1 & 0 \end{bmatrix}$, $B = \begin{bmatrix} 1 & -2 & 1 \\ -2 & 4 & -2 \\ -3 & 6 & -3 \end{bmatrix}$

(f) $A = \begin{bmatrix} 1 & 2 & -3 \\ 1 & -1 & 2 \\ 0 & 3 & -5 \end{bmatrix}$, $B = \begin{bmatrix} -2 & 1 & 3 \\ 6 & -3 & -9 \\ 0 & 0 & 0 \end{bmatrix}$

2. Show that I is *not* similar to $A = \begin{bmatrix} 1 & 2 \\ 0 & 1 \end{bmatrix}$, even though they have the same determinant, the same rank, the same trace, and the same characteristic polynomial.

3. Show that $\begin{bmatrix} 1 & 2 & -1 & 0 \\ 2 & 0 & 1 & 1 \\ 1 & 1 & 0 & -1 \\ 4 & 3 & 0 & 0 \end{bmatrix}$ and $\begin{bmatrix} 1 & -1 & 3 & 0 \\ -1 & 0 & 1 & 1 \\ 0 & -1 & 4 & 1 \\ 5 & -1 & -1 & -4 \end{bmatrix}$ are *not* similar.

4. If $A \sim B$, show that:

(a) $A^T \sim B^T$ **(b)** $A^{-1} \sim B^{-1}$

(c) $rA \sim rB$ for r in $\mathbb{R}$ **(d)** $A^n \sim B^n$ for $n \geq 1$.

5. In each case, find $P^{-1}AP$ and then compute A^n.

(a) $A = \begin{bmatrix} 6 & -5 \\ 2 & -1 \end{bmatrix}$, $P = \begin{bmatrix} 1 & 5 \\ 1 & 2 \end{bmatrix}$

(b) $A = \begin{bmatrix} -7 & -12 \\ 6 & 10 \end{bmatrix}$, $P = \begin{bmatrix} -3 & 4 \\ 2 & -3 \end{bmatrix}$

6. Given a polynomial $p(x) = r_0 + r_1x + \cdots + r_nx^n$ and a square matrix A, the matrix $p(A) = r_0I + r_1A + \cdots + r_nA^n$ is called the **evaluation** of $p(x)$ at A. Let $B = P^{-1}AP$.

(a) Show that $p(B) = P^{-1}p(A)P$ for all polynomials $p(x)$.

(b) Find $p(B)$ if $p(x) = x^3 + 3x^2 + x - 1$, $A = \begin{bmatrix} -1 & 0 \\ 0 & 2 \end{bmatrix}$, and $P = \begin{bmatrix} 3 & 2 \\ 2 & 1 \end{bmatrix}$.

7. If A is invertible, show that AB is similar to BA for all B.

8. (a) Show that the only matrix similar to a scalar matrix $A = rI$, r in $\mathbb{R}$, is A itself.

(b) If A has the property that the only matrix similar to it is A itself, show that $A = rI$ for some r in $\mathbb{R}$. [*Hint:* See Exercise 30 of Section 2.2.]

9. Let λ be an eigenvalue of A with corresponding eigenvector $\mathbf{p}$. If $B = P^{-1}AP$ is similar to A, show that $P^{-1}\mathbf{p}$ is an eigenvector of B corresponding to λ.

10. Suppose a function f is defined that assigns to every $n \times n$ matrix A a number $f(A)$. Assume that $f(PQ) = f(QP)$ holds for all $n \times n$ matrices P and Q. Show that f is a similarity invariant; that is, if A is similar to B, then $f(A) = f(B)$.

11. Suppose $\lambda_1, \lambda_2, \ldots, \lambda_n$ are the eigenvalues of an $n \times n$ matrix A, including repetitions. Show that $\det A = \lambda_1\lambda_2 \ldots \lambda_n$. [*Hint:* Look at the constant coefficient in $c_A(x)$. Use Theorem 1 to write $c_A(x) = (x - \lambda_1)(x - \lambda_2) \cdots (x - \lambda_n)$.]

12. Complete the proof of Theorem 4.
13. Let P be an invertible $n \times n$ matrix. If A is any $n \times n$ matrix, write $T_P(A) = P^{-1}AP$. Verify that:
    (a) $T_P(I) = I$
    (b) $T_P(AB) = T_P(A)\,T_P(B)$
    (c) $T_P(A + B) = T_P(A) + T_P(B)$
    (d) $T_P(rB) = rT_P(A)$
    (e) $T_P(A^k) = [T_P(A)]^k$ for $k \geq 1$
    (f) If A is invertible, $T_P(A^{-1}) = [T_P(A)]^{-1}$.
    (g) If Q is invertible, $T_Q[T_P(A)] = T_{PQ}(A)$.
14. Assume the 2×2 matrix A is similar to an upper triangular matrix. If $\operatorname{tr} A = 0 = \operatorname{tr} A^2$, show that $A^2 = 0$.
15. Show that A is similar to A^T for all 2×2 matrices A. [*Hint:* Let $A = \begin{bmatrix} a & b \\ c & d \end{bmatrix}$. If $c = 0$, treat the cases $b = 0$ and $b \neq 0$ separately. If $c \neq 0$, reduce to the case $c = 1$ using Exercise 4 **(c)**.]

7.1.3 Computing Eigenvalues (Optional)

In practice, the problem of finding eigenvalues and eigenvectors of a matrix is virtually never solved by finding the roots of the characteristic polynomial. Iterative methods are much better. Two of these will be described briefly in this section.

An eigenvalue λ of an $n \times n$ matrix A is said to be a **dominant eigenvalue** if

$$|\lambda| > |\mu| \qquad \text{for all eigenvalues } \mu \neq \lambda$$

Any corresponding eigenvector is called a **dominant eigenvector** of A. When such an eigenvalue exists, one technique for finding it is as follows: Let $\mathbf{v}_0$ in $\mathbb{R}^n$ be a first approximation to a dominant eigenvector, and compute successive approximations $\mathbf{v}_1, \mathbf{v}_2, \ldots$ by

$$\mathbf{v}_1 = A\mathbf{v}_0, \qquad \mathbf{v}_2 = A\mathbf{v}_1 \qquad \mathbf{v}_3 = A\mathbf{v}_2, \ldots$$

In general, we define

$$\mathbf{v}_{k+1} = A\mathbf{v}_k \qquad \text{for each } k \geq 0$$

If the first estimate $\mathbf{v}_0$ is good enough (see below), these vectors $\mathbf{v}_n$ will approximate dominant eigenvectors of A. This technique is called the **power method**. Moreover, it can be used to approximate the dominant eigenvalue λ. Observe that if $\mathbf{p}$ is any dominant eigenvector, then

$$\frac{\mathbf{p} \cdot (A\mathbf{p})}{\|\mathbf{p}\|^2} = \frac{\mathbf{p} \cdot (\lambda\mathbf{p})}{\|\mathbf{p}^2\|} = \lambda$$

Because the vectors $\mathbf{v}_1, \mathbf{v}_2, \ldots, \mathbf{v}_n, \ldots$ approximate dominant eigenvectors, we define the **Rayleigh quotients** as follows:

$$r_k = \frac{\mathbf{v}_k \cdot A\mathbf{v}_k}{\|\mathbf{v}_k\|^2} = \frac{\mathbf{v}_k \cdot \mathbf{v}_{k+1}}{\|\mathbf{v}_k\|^2} \qquad \text{for } k \geq 1$$

Then the numbers r_k approximate the dominant eigenvalue λ.

EXAMPLE 9 Use the power method to approximate a dominant eigenvector and eigenvalue of $A = \begin{bmatrix} 1 & 1 \\ 2 & 0 \end{bmatrix}$.

Solution The eigenvalues of A are 2 and -1, with eigenvectors $\begin{bmatrix} 1 \\ 1 \end{bmatrix}$ and $\begin{bmatrix} 1 \\ -2 \end{bmatrix}$. Take $\mathbf{v}_0 = \begin{bmatrix} 1 \\ 0 \end{bmatrix}$ as the first approximation and compute $\mathbf{v}_1$, $\mathbf{v}_2, \ldots,$ successively, from $\mathbf{v}_1 = A\mathbf{v}_0$, $\mathbf{v}_2 = A\mathbf{v}_1, \ldots$. The result is

$$\mathbf{v}_1 = \begin{bmatrix} 1 \\ 2 \end{bmatrix}, \mathbf{v}_2 = \begin{bmatrix} 3 \\ 2 \end{bmatrix}, \mathbf{v}_3 = \begin{bmatrix} 5 \\ 6 \end{bmatrix}, \mathbf{v}_4 = \begin{bmatrix} 11 \\ 10 \end{bmatrix}, \mathbf{v}_5 = \begin{bmatrix} 21 \\ 22 \end{bmatrix}, \ldots$$

These vectors are approaching scalar multiples of the dominant eigenvector $\begin{bmatrix} 1 \\ 1 \end{bmatrix}$. Moreover, the Rayleigh quotients are

$$r_1 = \frac{7}{5}, r_2 = \frac{27}{13}, r_3 = \frac{115}{61}, r_4 = \frac{451}{221}, \ldots$$

and these are approaching the dominant eigenvalue 2.

To see why the power method works, let $\lambda_1, \lambda_2, \ldots, \lambda_m$ be eigenvalues of A with λ_1 dominant, and let $\mathbf{u}_1, \mathbf{u}_2, \ldots, \mathbf{u}_m$ be corresponding eigenvectors. What is required is that the first approximation $\mathbf{v}_0$ be a linear combination of these eigenvectors:

$$\mathbf{v}_0 = a_1\mathbf{u}_1 + a_2\mathbf{u}_2 + \cdots + a_m\mathbf{u}_m \qquad \text{with } a_1 \neq 0$$

If $k \geq 1$, the fact that $A^k\mathbf{u}_i = \lambda_i^k\mathbf{u}_i$ for each i gives

$$\mathbf{v}_k = a_1\lambda_1^k\mathbf{u}_1 + a_2\lambda_2^k\mathbf{u}_2 + \cdots + a_m\lambda_m^k\mathbf{u}_m \qquad \text{for each } k$$

Hence

$$\frac{1}{\lambda_1^k}\mathbf{v}_k = a_1\mathbf{u}_1 + a_2\left(\frac{\lambda_2}{\lambda_1}\right)^k \mathbf{u}_2 + \cdots + a_m\left(\frac{\lambda_m}{\lambda_1}\right)^k \mathbf{u}_m$$

The right side approaches $a_1\mathbf{u}_1$ as k increases because λ_1 is dominant

$\left(\left|\frac{\lambda_i}{\lambda_1}\right| < 1 \text{ for each } i > 1\right)$. Because $a_1 \neq 0$, this means that $\mathbf{v}_k$ approximates the dominant eigenvector $a_1\lambda_1{}^k\mathbf{u}_1$.

The power method requires that the first approximation $\mathbf{v}_0$ be a linear combination of eigenvectors. (In Example 9 the eigenvectors form a basis of $\mathbb{R}^2$.) But even in this case, the method fails if $a_1 = 0$, where a_1 is the coefficient of the dominant eigenvector (try $\mathbf{v}_0 = \begin{bmatrix} -1 \\ 2 \end{bmatrix}$ in Example 9). In general, the rate of convergence is quite slow if any of the ratios $\left|\frac{\lambda_i}{\lambda_1}\right|$ is near 1. Also, because the method requires repeated multiplications by A, it is not recommended unless these multiplications are easy to carry out (for example, if most of the entries of A are zero).

A much better method depends on the factorization (using the Gram–Schmidt algorithm) of an invertible matrix A in the form

$$A = QR$$

where Q is orthogonal and R is invertible and upper triangular. The ***QR*-algorithm** uses this repeatedly to create a sequence of matrices $A_1 = A, A_2, A_3, \ldots$, as follows:

1. Define $A_1 = A$ and factor it as $A_1 = Q_1R_1$.
2. Define $A_2 = R_1Q_1$ and factor it as $A_2 = Q_2R_2$.
3. Define $A_3 = R_2Q_2$ and factor it as $A_3 = Q_3R_3$.

$$\vdots \qquad\qquad \vdots$$

In general, A_k is factored as $A_k = Q_kR_k$ and we define $A_{k+1} = R_kQ_k$. Then A_{k+1} is similar to A_k (in fact, $A_{k+1} = R_kQ_k = (Q_k^{-1}A_k)Q_k$), and hence each A_k has the same eigenvalues as A. If the eigenvalues of A are real, the remarkable thing is that the sequence of matrices $A_1, A_2, A_3, \ldots$ converges to an upper triangular matrix with these eigenvalues on the main diagonal. [See below for the case of complex eigenvalues.]

EXAMPLE 10

If $A = \begin{bmatrix} 1 & 1 \\ 2 & 0 \end{bmatrix}$ as in Example 9, use the QR-algorithm to approximate the eigenvalues.

Solution The matrices A_1, A_2, and A_3 are as follows:

$$A_1 = \begin{bmatrix} 1 & 1 \\ 2 & 0 \end{bmatrix} \text{ where } Q_1 = \frac{1}{\sqrt{5}}\begin{bmatrix} 1 & -2 \\ 2 & 1 \end{bmatrix} \text{ and } R_1 = \frac{1}{\sqrt{5}}\begin{bmatrix} 5 & 1 \\ 0 & -2 \end{bmatrix}$$

$$A_2 = \frac{1}{5}\begin{bmatrix} 7 & -9 \\ -4 & -2 \end{bmatrix} = \begin{bmatrix} 1.4 & -1.8 \\ -0.8 & -0.4 \end{bmatrix} \text{ where } Q_2 = \frac{1}{\sqrt{65}}\begin{bmatrix} 7 & 4 \\ -4 & 7 \end{bmatrix}$$

$$\text{and } R_2 = \frac{1}{\sqrt{65}}\begin{bmatrix} 13 & -11 \\ 0 & -10 \end{bmatrix}$$

$$A_3 = \frac{1}{13}\begin{bmatrix} 27 & -5 \\ 8 & -14 \end{bmatrix} = \begin{bmatrix} 2.08 & -0.38 \\ 0.62 & -1.08 \end{bmatrix}$$

This is converging to $\begin{bmatrix} 2 & * \\ 0 & -1 \end{bmatrix}$ and so is approximating the eigenvalues 2 and -1 on the main diagonal.

It is beyond the scope of this book to pursue a detailed discussion of these methods, and the reader is referred to J. M. Wilkinson, *The Algebraic Eigenvalue Problem* (Oxford, England: Oxford University Press, 1965) or G. W. Stewart, *Introduction to Matrix Computations* (New York: Academic Press, 1973). We conclude with some remarks on the QR-algorithm.

Shifting Convergence is accelerated if, at stage k of the algorithm, a number s_k is chosen and $A_k - s_kI$ is factored in the form Q_kR_k, rather than A_k itself. Then

$$Q_k^{-1}A_kQ_k = Q_k^{-1}(Q_kR_k + s_kI)Q_k = R_kQ_k + s_kI$$

so we take $A_{k+1} = R_kQ_k + s_kI$. If the shifts s_k are carefully chosen, convergence can be greatly improved.

Preliminary Preparation A matrix of the form illustrated in the diagram

$$\begin{bmatrix} * & * & * & * & * \\ * & * & * & * & * \\ 0 & * & * & * & * \\ 0 & 0 & * & * & * \\ 0 & 0 & 0 & * & * \end{bmatrix}$$

is said to be in **upper Hessenberg** form, and the QR-factorizations of such matrices are greatly simplified. A series of orthogonal matrices H_1, $H_2, \ldots, H_m$ (called **Householder matrices**) can be easily constructed such that

$$B = H_m^T \cdots H_1^T AH_1 \cdots H_m$$

is in upper Hessenberg form. Then the QR-algorithm can be efficiently applied to B and, because B is similar to A, produces the eigenvalues of A.

Complex Eigenvalues If some of the eigenvalues of a real matrix A are not real, the QR-algorithm converges to a block upper triangular matrix where the diagonal blocks are either 1×1 (the real eigenvalues) or 2×2 (each providing a pair of conjugate complex eigenvalues of A).

EXERCISES 7.1.3

1. In each case, find the exact eigenvalues and determine corresponding eigenvectors. Then start with $\mathbf{v}_0 = \begin{bmatrix} 1 \\ 1 \end{bmatrix}$ and compute $\mathbf{v}_4$ and r_3 using the power method.

   (a) $\begin{bmatrix} 2 & -4 \\ -3 & 3 \end{bmatrix}$ **(b)** $\begin{bmatrix} 5 & 2 \\ -3 & -2 \end{bmatrix}$ **(c)** $\begin{bmatrix} 1 & 2 \\ 2 & 1 \end{bmatrix}$ **(d)** $\begin{bmatrix} 3 & 1 \\ 1 & 0 \end{bmatrix}$

2. In each case, find the exact eigenvalues and then approximate them using the QR-algorithm.

   (a) $\begin{bmatrix} 1 & 1 \\ 1 & 0 \end{bmatrix}$ **(b)** $\begin{bmatrix} 3 & 1 \\ 1 & 0 \end{bmatrix}$

3. Apply the power method to $A = \begin{bmatrix} 0 & 1 \\ -1 & 0 \end{bmatrix}$, starting at $\mathbf{v}_0 = \begin{bmatrix} 1 \\ 1 \end{bmatrix}$. Does it converge? Explain.

4. If A is symmetric, show that each matrix A_k in the QR-algorithm is also symmetric. Deduce that they converge to a diagonal matrix.

5. Given a matrix A, let A_k, Q_k and R_k, $k \geq 1$, be the matrices constructed in the QR-algorithm. Show that $A^k = (Q_1Q_2 \cdots Q_k)(R_k \cdots R_2R_1)$ for each $k \geq 1$ and hence that this is a QR-factorization of A^k. [*Hint:* Show that $Q_kR_k = R_{k-1}Q_{k-1}$ for each $k \geq 2$, and use this equality to compute $(Q_1Q_2 \cdots Q_k)(R_k \cdots R_2R_1)$ "from the center out." Use the fact that $(AB)^{n+1} = A(BA)^nB$ for any square matrices A and B.]

SECTION 7.2 Diagonalization

7.2.1 Independent Eigenvectors

In this section we will be concerned with the following class of matrices.

DEFINITION An $n \times n$ matrix A is said to be **diagonalizable** if it is similar to a real diagonal matrix—that is, if $P^{-1}AP$ is diagonal for some invertible real matrix P.

A condition on A will be given below that will determine whether such a matrix P exists, and if P does exist, an effective method of finding P will be outlined. Then the more difficult question of whether the matrix P can be chosen to be orthogonal will be considered. This turns out to be the case precisely when A is symmetric.

Let A denote an $n \times n$ matrix, and suppose that an invertible matrix P exists that diagonalizes A, say

$$P^{-1}AP = D = \begin{bmatrix} \lambda_1 & 0 & 0 & \dots & 0 \\ 0 & \lambda_2 & 0 & \dots & 0 \\ 0 & 0 & \lambda_3 & \dots & 0 \\ \vdots & \vdots & \vdots & & \vdots \\ 0 & 0 & 0 & \dots & \lambda_n \end{bmatrix}$$

where $\lambda_1, \lambda_2, \dots, \lambda_n$ are real numbers. It follows from Theorem 3 in Section 7.1 that these numbers λ_i are precisely the eigenvalues of A, but the following argument will rediscover this fact and lead to a way of finding P (if it exists). The idea is to determine P by finding each of its columns. So let $\mathbf{p}_1, \mathbf{p}_2, \dots, \mathbf{p}_n$ denote the columns of P, and write P in block form as follows:

$$P = [\mathbf{p}_1 \; \mathbf{p}_2 \dots \mathbf{p}_n]$$

The condition $P^{-1}AP = D$ can be rewritten as $AP = PD$:

$$A[\mathbf{p}_1 \; \mathbf{p}_2 \dots \mathbf{p}_n] = [\mathbf{p}_1 \; \mathbf{p}_2 \dots \mathbf{p}_n] \begin{bmatrix} \lambda_1 & 0 & \dots & 0 \\ 0 & \lambda_2 & \dots & 0 \\ \vdots & \vdots & & \vdots \\ 0 & 0 & \dots & \lambda_n \end{bmatrix}$$

If both sides are written in terms of their columns, the result is

$$[A\mathbf{p}_1 \; A\mathbf{p}_2 \dots A\mathbf{p}_n] = [\lambda_1\mathbf{p}_1 \; \lambda_2\mathbf{p}_2 \dots \lambda_n\mathbf{p}_n]$$

Equating columns gives

$$A\mathbf{p}_i = \lambda_i\mathbf{p}_i \qquad i = 1, 2, \dots, n$$

so that the entries of D are eigenvalues of A and the corresponding columns of P are eigenvectors. The fact that P is invertible shows that the eigenvectors used must be linearly independent by Theorem 10 in Section 5.3. This proves that if A is diagonalizable, then it has n linearly independent eigenvectors. The converse of this statement is also true. If $\mathbf{p}_1, \mathbf{p}_2, \dots, \mathbf{p}_n$ are linearly independent eigenvectors, then $P = [\mathbf{p}_1 \; \mathbf{p}_2 \dots \mathbf{p}_n]$ is invertible and the above computation shows $AP = PD$ where D is diagonal. Hence $P^{-1}AP = D$, so the following theorem has been proved.

THEOREM 1

> An $n \times n$ matrix is diagonalizable if and only if it has n linearly independent eigenvectors (which are then a basis of $\mathbb{R}^n$).

The foregoing argument actually gives the following method for finding P:

DIAGONALIZATION ALGORITHM

> Given an $n \times n$ matrix A:
>
> (1) Find the eigenvalues of A.
>
> (2) Find (if possible) n linearly independent eigenvectors $\mathbf{p}_1, \mathbf{p}_2, \ldots, \mathbf{p}_n$.
>
> (3) Form $P = [\mathbf{p}_1\ \mathbf{p}_2 \ldots \mathbf{p}_n]$—the matrix with the $\mathbf{p}_i$ as columns.
>
> (4) Then $P^{-1}AP$ is diagonal, the diagonal entries being the eigenvalues corresponding to $\mathbf{p}_1, \mathbf{p}_2, \ldots, \mathbf{p}_n$, respectively.

EXAMPLE 1

Diagonalize the matrix $A = \begin{bmatrix} 5 & 8 & 16 \\ 4 & 1 & 8 \\ -4 & -4 & -11 \end{bmatrix}$ mentioned in Example 2 in Section 7.1.

Solution In that example, the eigenvalues were found to be $\lambda = -3$ and $\lambda = 1$, with eigenspaces $E_{-3} = \text{span}\{\mathbf{p}_1, \mathbf{p}_2\}$ and $E_1 = \text{span}\{\mathbf{p}_3\}$, where

$$\mathbf{p}_1 = \begin{bmatrix} -1 \\ 1 \\ 0 \end{bmatrix} \quad \mathbf{p}_2 = \begin{bmatrix} -2 \\ 0 \\ 1 \end{bmatrix} \quad \mathbf{p}_3 = \begin{bmatrix} 2 \\ 1 \\ -1 \end{bmatrix}$$

These vectors are linearly independent, so

$$P = [\mathbf{p}_1\ \mathbf{p}_2\ \mathbf{p}_3] = \begin{bmatrix} -1 & -2 & 2 \\ 1 & 0 & 1 \\ 0 & 1 & -1 \end{bmatrix}$$

is invertible and will diagonalize A. In fact,

$$P^{-1} = \begin{bmatrix} -1 & 0 & -2 \\ 1 & 1 & 3 \\ 1 & 1 & 2 \end{bmatrix} \quad \text{and} \quad P^{-1}AP = \begin{bmatrix} -3 & 0 & 0 \\ 0 & -3 & 0 \\ 0 & 0 & 1 \end{bmatrix}$$

Note that the diagonal entries $-3, -3, 1$ of $P^{-1}AP$ are (in order) the eigenvalues corresponding to the columns $\mathbf{p}_1, \mathbf{p}_2$, and $\mathbf{p}_3$ of P.

Of course, it may happen that a matrix is *not* diagonalizable. Then the process will fail at step 2: It will be impossible to find n independent eigenvectors.

EXAMPLE 2

Show that the matrix $A = \begin{bmatrix} 2 & 1 & 1 \\ 2 & 1 & -2 \\ -1 & 0 & -2 \end{bmatrix}$ mentioned in Example 3 of Section 7.1 is not diagonalizable.

Solution The eigenvalues were determined in this example to be $\lambda = -1$ and $\lambda = 3$, with eigenspaces $E_{-1} = \text{span}\{\mathbf{p}_1\}$ and $E_3 = \text{span}\{\mathbf{p}_2\}$, where

$$\mathbf{p}_1 = \begin{bmatrix} -1 \\ 2 \\ 1 \end{bmatrix} \quad \text{and} \quad \mathbf{p}_2 = \begin{bmatrix} 5 \\ 6 \\ -1 \end{bmatrix}$$

These are independent, but there are not enough of them. All the eigenvectors corresponding to $\lambda = -1$ are multiples of $\mathbf{p}_1$, so none can lie in an independent set containing $\mathbf{p}_1$. Similarly, no eigenvector corresponding to $\lambda = 3$ is independent of $\mathbf{p}_2$. Because these are the only eigenvalues of A, it is impossible to find three independent eigenvectors. Hence A is not diagonalizable by Theorem 1.

Observe that the matrices in Examples 1 and 2 were both of size 3×3 but each had only *two* eigenvalues. The reason why the matrix in Example 1 was diagonalizable is that the eigenspaces had dimensions 2 and 1, respectively, so two independent eigenvectors could be found in the first space and one in the second, making up the required independent set of three. The situation in Example 2 was different: Both eigenspaces had dimension 1, so there was no possibility of finding *three* independent eigenvectors.

This discussion makes it clear that the dimensions of the eigenspaces play a vital role in determining whether there exist n linearly independent eigenvectors—and hence whether the matrix is diagonalizable. We can clearly choose independent eigenvectors in any eigenspace, and the question arises whether vectors from *different* eigenspaces (corresponding to *distinct* eigenvalues) will be linearly independent. The answer is yes.

THEOREM 2

Let $\lambda_1, \lambda_2, \ldots, \lambda_k$ be distinct eigenvalues of a matrix A. If $\mathbf{p}_1, \mathbf{p}_2, \ldots, \mathbf{p}_k$ are eigenvectors corresponding to $\lambda_1, \lambda_2, \ldots, \lambda_k$, respectively, then $\{\mathbf{p}_1, \mathbf{p}_2, \ldots, \mathbf{p}_k\}$ is a linearly independent set.

Proof Use induction on k. If $k = 1$, then $\{\mathbf{p}_1\}$ is linearly independent because $\mathbf{p}_1 \neq \mathbf{0}$. Now assume that $k > 1$ and that the theorem holds for any $k - 1$ eigenvalues. Suppose a linear combination of the $\mathbf{p}_i$ vanishes:

$$r_1\mathbf{p}_1 + r_2\mathbf{p}_2 + r_3\mathbf{p}_3 + \cdots + r_k\mathbf{p}_k = \mathbf{0} \qquad (*)$$

It is required to show that $r_1 = r_2 = \cdots = r_k = 0$. Left-multiplication by A, together with the fact that $A\mathbf{p}_i = \lambda_i\mathbf{p}_i$ holds for each i, yields

$$r_1\lambda_1\mathbf{p}_1 + r_2\lambda_2\mathbf{p}_2 + r_3\lambda_3\mathbf{p}_3 + \cdots + r_k\lambda_k\mathbf{p}_k = \mathbf{0}$$

Now multiply (*) by λ_1 and subtract it from this to obtain

$$r_2(\lambda_2 - \lambda_1)\mathbf{p}_2 + r_3(\lambda_3 - \lambda_1)\mathbf{p}_3 + \cdots + r_k(\lambda_k - \lambda_1)\mathbf{p}_k = \mathbf{0}$$

But $\{\mathbf{p}_2, \ldots, \mathbf{p}_k\}$ is linearly independent by the induction hypothesis, so $r_2(\lambda_2 - \lambda_1) = r_3(\lambda_3 - \lambda_1) = \cdots = r_k(\lambda_k - \lambda_1) = 0$. The fact that the λ_i are distinct now implies that $r_2 = r_3 = \cdots = r_k = 0$, so (*) becomes $r_1\mathbf{p}_1 = \mathbf{0}$. Because $\mathbf{p}_1 \neq \mathbf{0}$, this implies that $r_1 = 0$ and completes the proof. ■

This theorem combines with Theorem 1 to yield the following useful fact.

THEOREM 3 An $n \times n$ matrix with n distinct eigenvalues is diagonalizable.

EXAMPLE 3 Show that $A = \begin{bmatrix} 1 & 0 & 0 \\ 1 & 2 & -3 \\ 1 & -1 & 0 \end{bmatrix}$ is diagonalizable.

Solution The characteristic polynomial is

$$c_A(x) = \det(xI - A) = \det\begin{bmatrix} x-1 & 0 & 0 \\ -1 & x-2 & 3 \\ -1 & 1 & x \end{bmatrix} = (x-1)(x-3)(x+1)$$

Hence the eigenvalues are 1, 3, and -1, so A is diagonalizable by Theorem 3.

It is important to note that the converse of Theorem 3 is false: There exist diagonalizable $n \times n$ matrices A that do not have n distinct eigenvalues. For example, the 3×3 matrix in Example 1 is diagonalizable, but $\lambda = -3$ and $\lambda = 1$ are the only eigenvalues.

An eigenvalue λ of a matrix A is said to have **multiplicity** m if it is repeated m times as a root of the characteristic polynomial—that is, if

$$c_A(x) = (x - \lambda)^m q(x) \qquad \text{where } q(\lambda) \neq 0$$

This is illustrated for the matrices in Examples 1, 2, and 3 in the accompanying table.

Matrix	*Characteristic Polynomial*	*Eigenvalues*	*Multiplicity*	*Dimension of Eigenspace*
$\begin{bmatrix} 5 & 8 & 16 \\ 4 & 1 & 8 \\ -4 & -4 & -11 \end{bmatrix}$	$(x+3)^2(x-1)$	-3 1	2 1	2 1
$\begin{bmatrix} 2 & 1 & 1 \\ 2 & 1 & -2 \\ -1 & 0 & -2 \end{bmatrix}$	$(x+1)^2(x-3)$	-1 3	2 1	1 1
$\begin{bmatrix} 1 & 0 & 0 \\ 1 & 2 & -3 \\ 1 & -1 & 0 \end{bmatrix}$	$(x-1)(x-3)(x+1)$	1 3 -1	1 1 1	1 1 1

The first and last matrices here are diagonalizable because the dimensions of the eigenspaces add up to $3 = n$ in both cases; however, this is not true for the second matrix, which is not diagonalizable (see the discussion following Example 2). Moreover, in both diagonalizable cases, the multiplicity of each eigenvalue equals the dimension of the corresponding eigenspace, whereas this *fails* for the eigenvalue $\lambda = -1$ of the second (nondiagonalizable) matrix. These observations are all made precise in Theorem 5 below. The proof requires the following fact, the proof of which is deferred to Section 9.2 (Theorem 9), where more powerful methods will be available.

THEOREM 4 If λ is an eigenvalue of A of multiplicity m, then $\dim(E_\lambda) \le m$.

THEOREM 5 Let A be an $n \times n$ matrix, and let

$$c_A(x) = (x - \lambda_1)^{m_1}(x - \lambda_2)^{m_2} \cdots (x - \lambda_k)^{m_k}$$

where $\lambda_1, \lambda_2, \ldots, \lambda_k$ denote the distinct real eigenvalues of A (with multiplicities $m_1, m_2, \ldots, m_k$). Write $d_i = \dim(E_{\lambda_i})$ for each i. Then the following statements are equivalent:

(1) A is diagonalizable.
(2) $d_1 + d_2 + \cdots + d_k = n$
(3) $d_i = m_i$ for each i.

Proof

(1) implies (2). Let B be a set of n independent eigenvectors. Then each vector in B corresponds to exactly one λ_i; say t_i of them lie in E_{λ_i}. These vectors are linearly independent, so $t_i \leq \dim(E_{\lambda_i}) = d_i$ holds for each i. Hence

$$n = t_1 + \cdots + t_k \leq d_1 + \cdots + d_k$$

But $d_1 + \cdots + d_k \leq n$ always holds (Exercise 7), so (2) follows.

(2) implies (3). Because m_i is the multiplicity of λ_i in $c_A(x)$, we have $d_i \leq m_i$ for each i by Theorem 4. Hence (2) gives

$$n = d_1 + \cdots + d_k \leq m_1 + \cdots + m_k = \deg [c_A(x)] = n$$

This means that $d_1 + \cdots + d_k = m_1 + \cdots + m_k$, so because $d_i \leq m_i$ for each i, we must have $d_i = m_i$. This proves (3).

(3) implies (1). Let B_i denote a basis of E_{λ_i} for each i, and let B consist of all vectors belonging to at least one of the B_i. Then Exercise 7 shows that B is linearly independent and contains $d_1 + \cdots + d_k$ vectors. But (3) implies that $d_1 + \cdots + d_k = m_1 + \cdots + m_k = \deg[c_A(x)] = n$, and (1) follows. ■

The proof that (3) implies (1) makes the process of diagonalization clear: Find the eigenspaces corresponding to the various eigenvectors of A and choose a basis of each. Then the collection of all vectors in these bases is an independent set of eigenvectors, and A is diagonalizable (by Theorem 1) if this set contains n vectors.

We conclude this section with an application of diagonalization to linear recurrence relations. This is treated in more generality in Section 8.3, but diagonalization techniques give a solution in certain cases. Here is an example.

EXAMPLE 4 Suppose a sequence of numbers $x_0, x_1, x_2, \ldots$ is determined by the condition that $x_0 = x_1 = 1$, and each successive x_n is given by

$$x_{n+2} = 6x_n + x_{n+1} \qquad n \geq 0$$

Find a formula for x_n in terms of n.

Solution It is clear that the sequence is completely determined by these conditions. In fact, the next few numbers are

$$x_2 = 6x_0 + x_1 = 7$$
$$x_3 = 6x_1 + x_2 = 13$$
$$x_4 = 6x_2 + x_3 = 55$$
$$\vdots$$

A general pattern for these numbers is not apparent, but a clever device transforms the problem into a simpler recurrence involving matrices. The idea is to compute the vector

$$\mathbf{v}_n = \begin{bmatrix} x_n \\ x_{n+1} \end{bmatrix}$$

for each $n \geq 0$ rather than x_n itself. The recurrence $x_{n+2} = 6x_n + x_{n+1}$ gives

$$\mathbf{v}_{n+1} = \begin{bmatrix} x_{n+1} \\ x_{n+2} \end{bmatrix} = \begin{bmatrix} 0 & 1 \\ 6 & 1 \end{bmatrix} \begin{bmatrix} x_n \\ x_{n+1} \end{bmatrix} = A\mathbf{v}_n \qquad \text{where } A = \begin{bmatrix} 0 & 1 \\ 6 & 1 \end{bmatrix}$$

Hence $\mathbf{v}_1 = A\mathbf{v}_0$, $\mathbf{v}_2 = A\mathbf{v}_1 = A^2\mathbf{v}_0$, $\mathbf{v}_3 = A\mathbf{v}_2 = A^3\mathbf{v}_0, \ldots$, and clearly

$$\mathbf{v}_n = A^n\mathbf{v}_0 \qquad n = 0, 1, 2, \ldots$$

Moreover, $\mathbf{v}_0 = \begin{bmatrix} 1 \\ 1 \end{bmatrix}$ is prescribed, so the problem comes down to computing A^n. But this calculation is simplified if A is diagonalized.

The characteristic polynomial for A is $c_A(x) = (x - 3)(x + 2)$. Hence the eigenvectors are $\mathbf{p}_1 = \begin{bmatrix} 1 \\ 3 \end{bmatrix}$ and $\mathbf{p}_2 = \begin{bmatrix} 1 \\ -2 \end{bmatrix}$, so $P = \begin{bmatrix} 1 & 1 \\ 3 & -2 \end{bmatrix}$ is invertible and $P^{-1}AP = \begin{bmatrix} 3 & 0 \\ 0 & -2 \end{bmatrix} = D$. Finally, $P^{-1} = \frac{1}{5}\begin{bmatrix} 2 & 1 \\ 3 & -1 \end{bmatrix}$, so

$$\begin{aligned} \begin{bmatrix} x_n \\ x_{n+1} \end{bmatrix} &= \mathbf{v}_n = A^n\mathbf{v}_0 = PD^nP^{-1}\mathbf{v}_0 \\ &= \frac{1}{5}\begin{bmatrix} 1 & 1 \\ 3 & -2 \end{bmatrix} \begin{bmatrix} 3^n & 0 \\ 0 & (-2)^n \end{bmatrix} \begin{bmatrix} 2 & 1 \\ 3 & -1 \end{bmatrix} \begin{bmatrix} 1 \\ 1 \end{bmatrix} \\ &= \frac{1}{5}\begin{bmatrix} 3^{n+1} - (-2)^{n+1} \\ 3^{n+2} + (-2)^{n+2} \end{bmatrix} \end{aligned}$$

It follows that $x_n = \frac{1}{5}(3^{n+1} - (-2)^{n+1})$ for each $n = 0, 1, 2, \ldots$. It is easy to verify that this formula gives $x_0 = 1$, $x_1 = 1$ and satisfies $x_{n+2} = 6x_n + x_{n+1}$ for any $n \geq 0$, as required.

This approach works more generally—but only if the corresponding matrix is diagonalizable. This turns out to be the case precisely when the eigenvalues of A are distinct (Exercise 14). The general case is treated by other methods in Section 8.3.

EXERCISES 7.2.1

1. In each case, decide whether the matrix A is diagonalizable. If so, find P such that $P^{-1}AP$ is diagonal.

(a) $\begin{bmatrix} 1 & 4 \\ 3 & 2 \end{bmatrix}$ **(b)** $\begin{bmatrix} 2 & -4 \\ -1 & 2 \end{bmatrix}$ **(c)** $\begin{bmatrix} 1 & 0 & 0 \\ 1 & 2 & 1 \\ 0 & 0 & 1 \end{bmatrix}$

(d) $\begin{bmatrix} 3 & 0 & 6 \\ 0 & -3 & 0 \\ 5 & 0 & 2 \end{bmatrix}$ **(e)** $\begin{bmatrix} 3 & 1 & 6 \\ 2 & 1 & 0 \\ -1 & 0 & -3 \end{bmatrix}$ **(f)** $\begin{bmatrix} 0 & 0 & 1 \\ 1 & 0 & -1 \\ 0 & 1 & 1 \end{bmatrix}$

2. If $A = \begin{bmatrix} 6 & 2 \\ 4 & -1 \end{bmatrix}$ and $B = \begin{bmatrix} 1 & 0 \\ 4 & -14 \end{bmatrix}$, verify that A and B are diagonalizable but AB is not.

3. If A is an $n \times n$ matrix, show that A is diagonalizable if and only if A^T is diagonalizable.

4. If A and B are similar matrices, show that A is diagonalizable if and only if B is diagonalizable. (Hence, being diagonalizable is a similarity invariant.)

5. If A is diagonalizable, show that each of the following is also diagonalizable.

(a) A^n, $n \geq 1$ **(b)** kA **(c)** A^{-1} if it exists

(d) $p(A)$, $p(x)$ any polynomial (see Exercise 6 in Section 7.1.2)

6. Give an example of two diagonalizable matrices A and B whose sum $A + B$ is not diagonalizable.

7. Let $\lambda_1, \lambda_2, \ldots, \lambda_k$ be distinct eigenvalues of an $n \times n$ matrix A, and let B_i be a linearly independent subset of E_{λ_i} containing d_i vectors (for each i). If B consists of all vectors in at least one of the B_i, show that B is a linearly independent set of $d_1 + d_2 + \cdots + d_k$ vectors. [*Hint:* Theorem 2.]

8. Show that the only diagonalizable matrix A that has only one eigenvalue λ is the scalar matrix $A = \lambda I$. [*Hint:* The diagonalization algorithm.]

9. (a) Show that two diagonalizable matrices are similar if and only if they have the same eigenvalues with the same multiplicities.

(b) Show that $I = \begin{bmatrix} 1 & 0 \\ 0 & 1 \end{bmatrix}$ and $A = \begin{bmatrix} 1 & 1 \\ 0 & 1 \end{bmatrix}$ have the same characteristic polynomial (and hence the same eigenvalues) but are not similar.

10. Let A denote an $n \times n$ upper triangular matrix.

(a) If all the main diagonal entries of A are distinct, show that A is diagonalizable.

(b) If all the main diagonal entries are equal, show that A can be diagonalizable only if it is *already* diagonal. [*Hint:* See Exercise 8.]

(c) Show that $\begin{bmatrix} 1 & 0 & 1 \\ 0 & 1 & 0 \\ 0 & 0 & 2 \end{bmatrix}$ is diagonalizable but that $\begin{bmatrix} 1 & 1 & 0 \\ 0 & 1 & 0 \\ 0 & 0 & 2 \end{bmatrix}$ is not.

11. Let A be a diagonalizable $n \times n$ matrix with eigenvalues $\lambda_1, \lambda_2, \ldots, \lambda_n$ (including multiplicities). Show that $\det A = \lambda_1 \lambda_2 \cdots \lambda_n$ and $\operatorname{tr} A = \lambda_1 + \lambda_2 + \cdots + \lambda_n$.

12. Let $A = \begin{bmatrix} B & 0 \\ 0 & C \end{bmatrix}$ where B and C are square matrices.

(a) If B and C are diagonalizable via Q and R (that is, $Q^{-1}BQ$ and $R^{-1}CR$ are diagonal), show that A is diagonalizable via $\begin{bmatrix} Q & 0 \\ 0 & R \end{bmatrix}$.

(b) Use **(a)** to diagonalize A if $B = \begin{bmatrix} 5 & 3 \\ 3 & 5 \end{bmatrix}$ and $C = \begin{bmatrix} 7 & -1 \\ -1 & 7 \end{bmatrix}$.

(c) If A is diagonalizable, show that both B and C are diagonalizable.

13. Solve each of the following linear recurrences by the method given in Example 4.

(a) $x_{n+2} = 3x_n + 2x_{n+1}, x_0 = 1, x_1 = 1$

(b) $x_{n+2} = 2x_n - x_{n+1}, x_0 = 1, x_1 = 2$

14. Generalize Example 4 as follows: Assume that the sequence $x_0, x_1, x_2, \ldots$ satisfies

$$x_{n+k} = r_0x_n + r_1x_{n+1} + \cdots + r_{k-1}x_{n+k-1}$$

for all $n \geq 0$. Define

$$A = \begin{bmatrix} 0 & 1 & 0 & \cdots & 0 \\ 0 & 0 & 1 & \cdots & 0 \\ \vdots & \vdots & \vdots & & \vdots \\ 0 & 0 & 0 & \cdots & 1 \\ r_0 & r_1 & r_2 & \cdots & r_{k-1} \end{bmatrix}, \qquad \mathbf{v}_n = \begin{bmatrix} x_n \\ x_{n+1} \\ \vdots \\ x_{n+k-1} \end{bmatrix}$$

Then show that:

(a) $\mathbf{v}_n = A^n\mathbf{v}_0$ for all n

(b) $c_A(x) = x^k - r_{k-1}x^{k-1} - \cdots - r_1x - r_0$

(c) If λ is an eigenvalue of A, the eigenspace E_λ has dimension 1, and $\mathbf{p} = (1, \lambda, \lambda^2, \ldots, \lambda^{k-1})^T$ is an eigenvector. [*Hint:* Use $c_A(\lambda) = 0$ to show that $E_\lambda = \mathbb{R}\mathbf{p}$.]

(d) A is diagonalizable if and only if the eigenvalues of A are distinct. [*Hint:* See part **(c)** and Theorem 5.]

(e) If $\lambda_1, \lambda_2, \ldots, \lambda_k$ are distinct real eigenvalues, there exist constants $t_1, t_2, \ldots, t_k$ such that $x_n = t_1\lambda_1{}^n + \cdots + t_k\lambda_k{}^n$ holds for all n. [*Hint:* If D is diagonal with $\lambda_1, \lambda_2, \ldots, \lambda_k$ as the main diagonal entries, show that $A^n = PD^nP^{-1}$ has entries that are linear combinations of $\lambda_1{}^n, \lambda_2{}^n, \ldots, \lambda_k{}^n$.]

7.2.2 Orthogonal Diagonalization of Symmetric Matrices

An $n \times n$ matrix A is diagonalizable if and only if it has n linearly independent eigenvectors $\mathbf{p}_1, \mathbf{p}_2, \ldots, \mathbf{p}_n$, and then the $n \times n$ matrix

$$P = [\mathbf{p}_1\ \mathbf{p}_2 \cdots \mathbf{p}_n]$$

is invertible and $P^{-1}AP$ is diagonal. All this was discussed at length above. Moreover, $\{\mathbf{p}_1, \mathbf{p}_2, \ldots, \mathbf{p}_n\}$ is a basis of $\mathbb{R}^n$ (written as columns), and because it became clear in Chapter 6 that the really "nice" bases of an inner product space are the orthonormal ones, the question of when the eigenvectors are orthonormal arises.

DEFINITION

An $n \times n$ matrix A is said to be **orthogonally diagonalizable** when an orthogonal matrix P can be found such that $P^{-1}AP$ is diagonal.

This condition turns out to characterize the symmetric matrices.

THEOREM 6 Principal Axes Theorem

The following conditions are equivalent for an $n \times n$ matrix A.

(1) A has an orthonormal set of n eigenvectors.

(2) A is orthogonally diagonalizable.

(3) A is symmetric.

Proof

(1) *is equivalent to* (2). This follows from the following observations: If $P = [\mathbf{p}_1 \cdots \mathbf{p}_n]$ is an $n \times n$ matrix, then P is orthogonal if and only if $\{\mathbf{p}_1, \ldots, \mathbf{p}_n\}$ is an orthonormal set in $\mathbb{R}^n$; and $P^{-1}AP$ is diagonal if and only if $\{\mathbf{p}_1, \ldots, \mathbf{p}_n\}$ consists of eigenvectors of A (see the proof of Theorem 1).

(2) *implies* (3). If $P^TAP = D$ is diagonal, where $P^{-1} = P^T$, then $A = PDP^T$. Because $D^T = D$, this gives $A^T = PD^TP^T = A$.

(3) *implies* (2). If A is an $n \times n$ symmetric matrix, we proceed by induction on n. If $n = 1$, A is already diagonal. If $n > 1$, assume that (3) implies (2) for $(n-1) \times (n-1)$ symmetric matrices. By Theorem 2, in Section 7.1, let λ_1 be a (real) eigenvalue of A, and let $A\mathbf{p}_1 = \lambda\mathbf{p}_1$, where $\|\mathbf{p}_1\| = 1$. Use the Gram–Schmidt algorithm to find an orthonormal basis $\{\mathbf{p}_1, \mathbf{p}_2, \ldots, \mathbf{p}_n\}$ for $\mathbb{R}^n$, and

let $P_1 = [\mathbf{p}_1\ \mathbf{p}_2\ \cdots\ \mathbf{p}_n]$. Then P_1 is orthogonal and $P_1^TAP_1 = \begin{bmatrix} \lambda_1 & X \\ 0 & A_1 \end{bmatrix}$ in block form. Because A is symmetric, it follows that $X = 0$ and A_1 is symmetric. Then, by induction, there exists an $(n-1) \times (n-1)$ orthogonal matrix Q such that $Q^TA_1Q = D_1$ is diagonal. Hence $P_2 = \begin{bmatrix} 1 & 0 \\ 0 & Q \end{bmatrix}$ is orthogonal and

$$\begin{aligned}(P_1P_2)^T A (P_1P_2) &= P_2^T(P_1^TAP_1)P_2 \\ &= \begin{bmatrix} 1 & 0 \\ 0 & Q^T \end{bmatrix} \cdot \begin{bmatrix} \lambda_1 & 0 \\ 0 & A_1 \end{bmatrix} \cdot \begin{bmatrix} 1 & 0 \\ 0 & Q \end{bmatrix} \\ &= \begin{bmatrix} \lambda_1 & 0 \\ 0 & D_1 \end{bmatrix}\end{aligned}$$

is diagonal. Because P_1P_2 is orthogonal, this proves (2). ■

A set of orthonormal eigenvectors of a symmetric matrix A is called a set of **principal axes** for A. The name comes from geometry, and this is discussed in Section 7.4.1. Theorem 6 is also called the **real spectral theorem**, and the set of distinct eigenvalues is called the **spectrum** of the matrix. In full generality, the spectral theorem is a similar result for matrices with complex entries (Theorem 8 in the next section).

EXAMPLE 5 Find an orthogonal matrix P such that $P^{-1}AP$ is diagonal where $A = \begin{bmatrix} 1 & 0 & -1 \\ 0 & 1 & 2 \\ -1 & 2 & 5 \end{bmatrix}$.

Solution The characteristic polynomial of A is

$$c_A(x) = \det\begin{bmatrix} x-1 & 0 & 1 \\ 0 & x-1 & -2 \\ 1 & -2 & x-5 \end{bmatrix} = x(x-1)(x-6)$$

Thus the eigenvalues are $\lambda = 0$, 1, and 6, and corresponding eigenvectors are

$$\mathbf{p}_1 = \begin{bmatrix} 1 \\ -2 \\ 1 \end{bmatrix} \quad \mathbf{p}_2 = \begin{bmatrix} 2 \\ 1 \\ 0 \end{bmatrix} \quad \mathbf{p}_3 = \begin{bmatrix} -1 \\ 2 \\ 5 \end{bmatrix}$$

respectively. Moreover, by what appears to be remarkably good luck, these eigenvectors are *orthogonal*. We have $\|\mathbf{p}_1\|^2 = 6$, $\|\mathbf{p}_2\|^2 = 5$, and $\|\mathbf{p}_3\|^2 = 30$, so

$$P = \left[\frac{1}{\sqrt{6}}\mathbf{p}_1 \ \frac{1}{\sqrt{5}}\mathbf{p}_2 \ \frac{1}{\sqrt{30}}\mathbf{p}_3\right] = \frac{1}{\sqrt{30}}\begin{bmatrix} \sqrt{5} & 2\sqrt{6} & -1 \\ -2\sqrt{5} & \sqrt{6} & 2 \\ \sqrt{5} & 0 & 5 \end{bmatrix}$$

is orthogonal (thus $P^{-1} = P^T$), and

$$P^TAP = \begin{bmatrix} 0 & 0 & 0 \\ 0 & 1 & 0 \\ 0 & 0 & 6 \end{bmatrix}$$

by the diagonalization algorithm.

Actually, the fact that the eigenvectors in Example 5 are orthogonal was no coincidence. Theorem 2 guarantees they are linearly independent (they correspond to distinct eigenvalues); the fact that the matrix is symmetric implies that they are orthogonal. In order to prove this, we need the following useful fact about symmetric matrices.

THEOREM 7

If A is an $n \times n$ symmetric matrix, then

$$(A\mathbf{u}) \cdot \mathbf{v} = \mathbf{u} \cdot (A\mathbf{v})$$

for all columns $\mathbf{u}$ and $\mathbf{v}$.

Proof Recall that $\mathbf{u} \cdot \mathbf{v} = \mathbf{u}^T\mathbf{v}$ for all columns $\mathbf{u}$ and $\mathbf{v}$. Because $A^T = A$, we get

$$(A\mathbf{u}) \cdot \mathbf{v} = (A\mathbf{u})^T\mathbf{v} = \mathbf{u}^TA^T\mathbf{v} = \mathbf{u}^TA\mathbf{v} = \mathbf{u} \cdot (A\mathbf{v}) \quad \blacksquare$$

THEOREM 8

If A is a symmetric matrix, then eigenvectors of A corresponding to distinct eigenvalues are orthogonal.

Proof Let $A\mathbf{p} = \lambda\mathbf{p}$ and $A\mathbf{q} = \mu\mathbf{q}$, where $\lambda \neq \mu$. Using Theorem 7, we compute

$$\lambda(\mathbf{p} \cdot \mathbf{q}) = (\lambda\mathbf{p}) \cdot \mathbf{q} = (A\mathbf{p}) \cdot \mathbf{q} = \mathbf{p} \cdot (A\mathbf{q}) = \mathbf{p} \cdot (\mu\mathbf{q}) = \mu(\mathbf{p} \cdot \mathbf{q})$$

Hence $(\lambda - \mu)(\mathbf{p} \cdot \mathbf{q}) = 0$, and so $\mathbf{p} \cdot \mathbf{q} = 0$ because $\lambda \neq \mu$. $\blacksquare$

Now the procedure for diagonalizing a symmetric $n \times n$ matrix is clear. Find the distinct eigenvalues (all real by Theorem 2 in Section 7.1), and find orthonormal bases for each eigenspace (the Gram–Schmidt algorithm may be needed). Then the set of all these basis vectors is orthonormal (by Theorem 8) and contains n vectors.

EXAMPLE 6 Diagonalize $A = \begin{bmatrix} 8 & -2 & 2 \\ -2 & 5 & 4 \\ 2 & 4 & 5 \end{bmatrix}$.

Solution The characteristic polynomial is

$$c_A(x) = \det\begin{bmatrix} x-8 & 2 & -2 \\ 2 & x-5 & -4 \\ -2 & -4 & x-5 \end{bmatrix} = x(x-9)^2$$

Hence the distinct eigenvalues are $\lambda = 0$ and 9 of multiplicities 1 and 2, respectively, so $\dim(E_0) = 1$ and $\dim(E_9) = 2$ by Theorem 5 (A is diagonalizable, being symmetric). One eigenvector for $\lambda = 0$ is $\mathbf{p}_1 = \begin{bmatrix} 1 \\ 2 \\ -2 \end{bmatrix}$, so $E_0 = \text{span}\{\mathbf{p}_1\}$. Gaussian elimination gives

$$E_9 = \text{span}\left\{\begin{bmatrix} -2 \\ 1 \\ 0 \end{bmatrix}, \begin{bmatrix} 2 \\ 0 \\ 1 \end{bmatrix}\right\}$$

and these eigenvectors are both orthogonal to $\mathbf{p}_1$ (as Theorem 8 guarantees) but not to each other. The Gram–Schmidt process gives an orthogonal basis

$$\left\{\mathbf{p}_2 = \begin{bmatrix} -2 \\ 1 \\ 0 \end{bmatrix}, \mathbf{p}_3 = \begin{bmatrix} 2 \\ 4 \\ 5 \end{bmatrix}\right\}$$

of E_9. Normalizing gives orthonormal eigenvectors

$$\left\{\frac{1}{3}\mathbf{p}_1, \frac{1}{\sqrt{5}}\mathbf{p}_2, \frac{1}{3\sqrt{5}}\mathbf{p}_3\right\},$$

so

$$P = \begin{bmatrix} \frac{1}{3}\mathbf{p}_1 & \frac{1}{\sqrt{5}}\mathbf{p}_2 & \frac{1}{3\sqrt{5}}\mathbf{p}_3 \end{bmatrix} = \frac{1}{3\sqrt{5}}\begin{bmatrix} \sqrt{5} & -6 & 2 \\ 2\sqrt{5} & 3 & 4 \\ -2\sqrt{5} & 0 & 5 \end{bmatrix}$$

is an orthogonal matrix such that $P^{-1}AP$ is diagonal.

It is worth noting that other, more convenient matrices P exist. For example,

$$\mathbf{q}_2 = \begin{bmatrix} 2 \\ 1 \\ 2 \end{bmatrix} \quad \text{and} \quad \mathbf{q}_3 = \begin{bmatrix} -2 \\ 2 \\ 1 \end{bmatrix}$$

lie in E_9 and they are orthogonal. Moreover, they both have norm 3 (as does $\mathbf{p}_1$), so

$$Q = \begin{bmatrix} \frac{1}{3}\mathbf{p}_1 & \frac{1}{3}\mathbf{q}_2 & \frac{1}{3}\mathbf{q}_3 \end{bmatrix} = \frac{1}{3}\begin{bmatrix} 1 & 2 & -2 \\ 2 & 1 & 2 \\ -2 & 2 & 1 \end{bmatrix}$$

is a nicer orthogonal matrix with the property that $Q^{-1}AQ$ is diagonal.

Theorem 7 in Section 6.1 implies that every inner product $\langle\ ,\ \rangle$ on the space of n-columns is given by $\langle \mathbf{u},\mathbf{v}\rangle = \mathbf{u}^T A\mathbf{v}$ for some symmetric matrix A. However, not every symmetric matrix A yields an inner product in this way. The problem is that axiom P5,

$$\langle \mathbf{v},\mathbf{v}\rangle = \mathbf{v}^T A\mathbf{v} > 0 \qquad \text{for all } \mathbf{v} \neq \mathbf{0}$$

may fail. A symmetric matrix A is called **positive definite** if it enjoys this property, and the following theorem characterizes such matrices in terms of their eigenvalues.

THEOREM 9 A symmetric matrix A is positive definite if and only if all its eigenvalues are positive.

Proof Because A is symmetric, let $P^TAP = D = \begin{bmatrix} \lambda_1 & 0 & \dots & 0 \\ 0 & \lambda_2 & \dots & 0 \\ \vdots & \vdots & & \vdots \\ 0 & 0 & \dots & \lambda_n \end{bmatrix}$, where $\lambda_1, \lambda_2, \dots, \lambda_n$ are all eigenvalues of A and $P^T = P^{-1}$. This gives $A = PDP^T$, so given any n-column $\mathbf{v} \neq \mathbf{0}$, $\mathbf{v}^TA\mathbf{v} = \mathbf{v}^TPDP^T\mathbf{v} = (P^T\mathbf{v})^TD(P^T\mathbf{v})$. Now $P^T\mathbf{v} \neq \mathbf{0}$ because P^T is invertible, so if $P^T\mathbf{v} = \begin{bmatrix} r_1 \\ \vdots \\ r_n \end{bmatrix}$, this gives

$$\mathbf{v}^TA\mathbf{v} = \lambda_1 r_1^{\,2} + \lambda_2 r_2^{\,2} + \cdots + \lambda_n r_n^{\,2}$$

This is positive if each $\lambda_i > 0$, because some $r_i \neq 0$. Hence A is positive definite if each λ_i is positive. Conversely, if A is positive definite, then $\lambda_1 r_1^{\,2} + \lambda_2 r_2^{\,2} + \cdots + \lambda_n r_n^{\,2} = \mathbf{v}^TA\mathbf{v} > 0$ for any choice of $r_1, r_2, \dots, r_n$, where $\mathbf{v} = P\begin{bmatrix} r_1 \\ \vdots \\ r_n \end{bmatrix} \neq \mathbf{0}$. This implies that each $\lambda_i > 0$. ■

EXAMPLE 7 Show that $A = \begin{bmatrix} 2 & 2 \\ 2 & 5 \end{bmatrix}$ is positive definite.

Solution The characteristic polynomial is

$$c_A(x) = \det\begin{bmatrix} x-2 & -2 \\ -2 & x-5 \end{bmatrix} = x^2 - 7x + 6 = (x-6)(x-1)$$

Hence the eigenvalues are 6 and 1. Because they are both positive, A is positive definite. In this case, the condition for positive definiteness can be checked directly: If $\mathbf{v} = \begin{bmatrix} x \\ y \end{bmatrix}$, then

$$\mathbf{v}^T A\mathbf{v} = [x\ y]\begin{bmatrix} 2 & 2 \\ 2 & 5 \end{bmatrix}\begin{bmatrix} x \\ y \end{bmatrix} = 2x^2 + 4xy + 5y^2 = 2(x+y)^2 + 3y^2,$$

and this is clearly positive if $\mathbf{v} \neq \mathbf{0}$.

EXERCISES 7.2.2

1. For each matrix A, find an orthogonal matrix P such that $P^{-1}AP$ is diagonal.

(a) $A = \begin{bmatrix} 3 & 0 & 0 \\ 0 & 2 & 2 \\ 0 & 2 & 5 \end{bmatrix}$ **(b)** $A = \begin{bmatrix} 3 & 0 & 7 \\ 0 & 5 & 0 \\ 7 & 0 & 3 \end{bmatrix}$

(c) $A = \begin{bmatrix} 1 & 1 & 0 \\ 1 & 1 & 0 \\ 0 & 0 & 2 \end{bmatrix}$ **(d)** $A = \begin{bmatrix} 5 & -2 & -4 \\ -2 & 8 & -2 \\ -4 & -2 & 5 \end{bmatrix}$

(e) $A = \begin{bmatrix} 5 & 3 & 0 & 0 \\ 3 & 5 & 0 & 0 \\ 0 & 0 & 7 & 1 \\ 0 & 0 & 1 & 7 \end{bmatrix}$ **(f)** $A = \begin{bmatrix} 3 & 5 & -1 & 1 \\ 5 & 3 & 1 & -1 \\ -1 & 1 & 3 & 5 \\ 1 & -1 & 5 & 3 \end{bmatrix}$

2. Consider $A = \begin{bmatrix} 0 & a & 0 \\ a & 0 & c \\ 0 & c & 0 \end{bmatrix}$ where one of $a, c \neq 0$. Show that $c_A(x) = x(x-k)(x+k)$, where $k = \sqrt{a^2 + c^2}$ and find an orthogonal matrix P such that $P^{-1}AP$ is diagonal.

3. Consider $A = \begin{bmatrix} 0 & 0 & a \\ 0 & b & 0 \\ a & 0 & 0 \end{bmatrix}$. Show that $c_A(x) = (x-b)(x-a)(x+a)$ and find an orthogonal matrix P such that $P^{-1}AP$ is diagonal.

4. Given $A = \begin{bmatrix} b & a \\ a & b \end{bmatrix}$, show that $c_A(x) = (x-a-b)(x+a-b)$ and find an orthogonal matrix P such that $P^{-1}AP$ is diagonal.

5. Consider $A = \begin{bmatrix} b & 0 & a \\ 0 & b & 0 \\ a & 0 & b \end{bmatrix}$. Show that $c_A(x) = (x-b)(x-b-a) \cdot (x-b+a)$ and find an orthogonal matrix P such that $P^{-1}AP$ is diagonal.

6. Show that the following are equivalent for a symmetric matrix A.
   (a) A is orthogonal. (b) $A^2 = I$ (c) All eigenvalues of A are ± 1.
   [*Hint:* For part (**b**) if and only if part (**c**), use Theorem 6.]
7. If $A = A^T$, show that $\det A$ and $\operatorname{tr} A$ are, respectively, the product and sum of the eigenvalues of A (including multiplicities).
8. Show that a symmetric matrix is invertible if and only if all its eigenvalues are nonzero.
9. (a) If A is symmetric and every eigenvalue of a is nonnegative, show that there is a symmetric matrix B such that $B^2 = A$. [*Hint:* If $A = P^TDP$ with D diagonal, show that $D = E^2$ for some (diagonal) matrix E.]
   (b) Show that an $n \times n$ matrix A is positive definite if and only if $A = B^2$, where B is symmetric and invertible. [*Hint:* See Exercise 29(b) in Section 6.1.]
10. Let A be any $n \times n$ matrix. Show that:
    (a) A^TA is symmetric.
    (b) Every eigenvalue of A^TA is nonnegative. [*Hint:* If $A^TA\mathbf{p} = \lambda\mathbf{p}$, $\mathbf{p} \neq \mathbf{0}$, compute $A\mathbf{p} \cdot A\mathbf{p}$.]
    (c) A^TA is positive definite if and only if A is invertible.
11. We call matrices A and B **orthogonally similar** (and write $A \overset{\circ}{\sim} B$) if $B = P^TAP$ for an orthogonal matrix P.
    (a) Show that
        (i) $A \overset{\circ}{\sim} A$ for all A.
        (ii) $A \overset{\circ}{\sim} B$ implies that $B \overset{\circ}{\sim} A$.
        (iii) $A \overset{\circ}{\sim} B$ and $B \overset{\circ}{\sim} C$ imply that $A \overset{\circ}{\sim} C$.
    (b) Show that the following are equivalent for two symmetric matrices A and B.
        (i) A and B are similar.
        (ii) A and B are orthogonally similar.
        (iii) A and B have the same eigenvalues.

    [*Hint:* For part (**i**) implies part (**ii**), use Theorem 3 in Section 7.1 and Theorem 6.]

    Remark: Part (**a**) shows that orthogonal similarity is an equivalence relation on the set of symmetric $n \times n$ matrices, and part (**b**) implies that each such matrix is orthogonally similar to exactly one diagonal matrix, apart from the order of the diagonal entries. Hence these classes of diagonal matrices are an appropriate choice of canonical form for symmetric matrices under this relationship.
12. Assume that A and B are orthogonally similar (Exercise 11).
    (a) Show that A^{-1} and B^{-1} are orthogonally similar.
    (b) Show that A^2 and B^2 are orthogonally similar.
    (c) Show that, if A is symmetric, so is B.
13. Prove the converse of Theorem 7. If $(A\mathbf{u}) \cdot \mathbf{v} = \mathbf{u} \cdot (A\mathbf{v})$ for all n-columns $\mathbf{u}$ and $\mathbf{v}$, then A is symmetric.

SECTION 7.3 Complex Matrices (Optional)

If A is an $n \times n$ matrix, the characteristic polynomial $c_A(x)$ is a polynomial of degree n and the eigenvalues of A are just the roots of $c_A(x)$. In each of our examples these roots have been *real* numbers (in fact, the examples have been carefully chosen so this will be the case!) but it need not happen, even though the characteristic polynomial has real coefficients. For example, if $A \begin{bmatrix} 0 & 1 \\ -1 & 0 \end{bmatrix}$, then $c_A(x) = x^2 + 1$ has roots i and $-i$, where i is the complex number satisfying $i^2 = -1$. Therefore, we have to deal with the possibility that the eigenvalues of a (real) square matrix might be complex numbers.

In fact, nearly everything in this book would remain true if the phrase *real number* were replaced by *complex number* wherever it occurs. Then we would deal with matrices with complex entries, systems of linear equations with complex coefficients (and complex solutions), determinants of complex matrices, and vector spaces with scalar multiplication by any complex number allowed. Moreover, the proofs of most theorems about (the real version of) these concepts extend easily to the complex case. It is not our intention here to give a full treatment of complex linear algebra. However, we will carry the theory far enough to prove that the eigenvalues of a symmetric matrix A are real (Theorem 2 in Section 7.1) and to prove the Spectral Theorem (an extension of the principal axes theorem (Theorem 6 in Section 7.2)).

The set of complex numbers is denoted $\mathbb{C}$. We require only the most basic properties of these numbers (mainly conjugation and absolute values), and the reader can find this material in Appendix A.

If $n \geq 1$, we denote the set of all n-tuples of complex numbers by $\mathbb{C}^n$. As for $\mathbb{R}^n$, these n-tuples will be written either as rows or as columns and will be referred to as vectors. We define vector operations on $\mathbb{C}^n$ as follows:

$$(v_1, v_2, \ldots, v_n) + (w_1, w_2, \ldots, w_n) = (v_1 + w_1, v_2 + w_2, \ldots, v_n + w_n)$$
$$u(v_1, v_2, \ldots, v_n) = (uv_1, uv_2, \ldots, uv_n) \text{ for } u \text{ in } \mathbb{C}$$

With these definitions, $\mathbb{C}^n$ satisfies the axioms for a vector space (with complex scalars) given in Chapter 5. Thus we can speak of spanning sets for $\mathbb{C}^n$, of linearly independent subsets, and of bases. In all cases, the definitions are identical with the real case, except that the scalars are allowed to be complex numbers. In particular, the standard basis of $\mathbb{R}^n$ remains a basis of $\mathbb{C}^n$, called the **standard basis** of $\mathbb{C}^n$.

There is a natural generalization to $\mathbb{C}^n$ of the dot product in $\mathbb{R}^n$. Given $\mathbf{z} = (z_1, z_2, \ldots, z_n)$ and $\mathbf{w} = (w_1, w_2, \ldots, w_n)$ in $\mathbb{C}^n$, define their

standard inner product $\langle \mathbf{z},\mathbf{w} \rangle$ by

$$\langle \mathbf{z},\mathbf{w} \rangle = z_1\overline{w}_1 + z_2\overline{w}_2 + \cdots + z_n\overline{w}_n$$

Clearly, if z and w actually lie in $\mathbb{R}^n$, then $\langle \mathbf{z},\mathbf{w} \rangle = \mathbf{z} \cdot \mathbf{w}$ is the usual dot product.

EXAMPLE 1 If $\mathbf{z} = (2, 1 - i, 2i, 3 - i)$ and $\mathbf{w} = (1 - i, -1, -i, 3 + 2i)$, then

$$\langle \mathbf{z},\mathbf{w} \rangle = 2(1 + i) + (1 - i)(-1) + (2i)(i) + (3 - i)(3 - 2i) = 6 - 6i$$
$$\langle \mathbf{z},\mathbf{z} \rangle = 2 \cdot 2 + (1 - i)(1 + i) + (2i)(-2i) + (3 - i)(3 + i) = 20$$

Note that $\langle \mathbf{z},\mathbf{w} \rangle$ is a complex number in general. However, if $\mathbf{w} = \mathbf{z} = (z_1, z_2, \ldots, z_n)$, the definition gives $\langle \mathbf{z},\mathbf{z} \rangle = |z_1|^2 + \cdots + |z_n|^2$, which is a nonnegative real number and equals 0 if and only if $\mathbf{z} = \mathbf{0}$. This explains the conjugation in the definition of $\langle \mathbf{z},\mathbf{w} \rangle$, and it gives (4) of the following theorem.

THEOREM 1 Let $\mathbf{z}$, $\mathbf{z}_1$, $\mathbf{w}$, and $\mathbf{w}_1$ denote n-tuples in $\mathbb{C}^n$, and let λ denote a complex number.

(1) $\langle \mathbf{z} + \mathbf{z}_1, \mathbf{w} \rangle = \langle \mathbf{z},\mathbf{w} \rangle + \langle \mathbf{z}_1,\mathbf{w} \rangle$ and $\langle \mathbf{z},\mathbf{w} + \mathbf{w}_1 \rangle = \langle \mathbf{z},\mathbf{w} \rangle + \langle \mathbf{z},\mathbf{w}_1 \rangle$

(2) $\langle \lambda\mathbf{z},\mathbf{w} \rangle = \lambda\langle \mathbf{z},\mathbf{w} \rangle$ and $\langle \mathbf{z},\lambda\mathbf{w} \rangle = \overline{\lambda}\langle \mathbf{z},\mathbf{w} \rangle$

(3) $\langle \mathbf{z},\mathbf{w} \rangle = \overline{\langle \mathbf{w},\mathbf{z} \rangle}$

(4) $\langle \mathbf{z},\mathbf{z} \rangle \geq 0$, and $\langle \mathbf{z},\mathbf{z} \rangle = 0$ if and only if $\mathbf{z} = \mathbf{0}$

Proof We leave (1) and (2) to the reader (Exercise 8), and (4) is proved above. To prove (3), write $\mathbf{z} = (z_1, z_2, \ldots, z_n)$ and $\mathbf{w} = (w_1, w_2, \ldots, w_n)$. Then

$$\overline{\langle \mathbf{w},\mathbf{z} \rangle} = \overline{(w_1\overline{z}_1 + \cdots + w_n\overline{z}_n)} = \overline{w}_1\,\overline{\overline{z}}_1 + \cdots + \overline{w}_n\,\overline{\overline{z}}_n$$
$$= z_1\,\overline{w}_1 + \cdots + z_n\overline{w}_n = \langle \mathbf{z},\mathbf{w} \rangle \qquad \blacksquare$$

As for the dot product on $\mathbb{R}^n$ (and, indeed, for any real inner product), property (4) enables us to define the **norm** or **length** $\|\mathbf{z}\|$ of a vector $\mathbf{z} = (z_1, z_2, \ldots, z_n)$ in $\mathbb{C}^n$:

$$\|\mathbf{z}\| = \sqrt{\langle \mathbf{z},\mathbf{z} \rangle} = \sqrt{|z_1|^2 + |z_2|^2 + \cdots + |z_n|^2}$$

This satisfies the properties of a norm function set out in Theorem 5 of Section 6.1. The only properties we shall need are the following (the proof is left to the reader).

THEOREM 2

If $\mathbf{z}$ is any vector in $\mathbb{C}^n$, then

(1) $\|\mathbf{z}\| \geq 0$ and $\|\mathbf{z}\| = 0$ if and only if $\mathbf{z} = \mathbf{0}$

(2) $\|\lambda\mathbf{z}\| = |\lambda|\,\|\mathbf{z}\|$ for all complex numbers λ

A vector $\mathbf{u}$ in $\mathbb{C}^n$ is called a **unit vector** if $\|\mathbf{u}\| = 1$. Property (2) then shows that if $\mathbf{z} \neq \mathbf{0}$ is any nonzero vector in $\mathbb{C}^n$, then

$$\mathbf{u} = \frac{1}{\|\mathbf{z}\|}\mathbf{z}$$

is a unit vector.

EXAMPLE 2 In $\mathbb{C}^4$, find a unit vector $\mathbf{u}$ that is a positive real multiple of $\mathbf{z} = (1 - i, i, 2, 3 + 4i)$.

Solution $\|\mathbf{z}\| = \sqrt{2 + 1 + 4 + 25} = \sqrt{32}$, so take $\mathbf{u} = \frac{1}{4\sqrt{2}}\mathbf{z}$.

A matrix $Z = [z_{ij}]$ is called a **complex matrix** if each entry z_{ij} is a complex number. The notion of conjugation for complex numbers extends to matrices as follows: Define the **conjugate** of Z to be the matrix

$$\overline{Z} = [\overline{z}_{ij}]$$

obtained from Z by conjugating every entry. Then

$$\overline{Z + W} = \overline{Z} + \overline{W} \qquad \text{and} \qquad \overline{ZW} = \overline{Z}\,\overline{W}$$

holds for all (complex) matrices of appropriate size (using properties C_1 and C_2 of Appendix A).

Transposition of complex matrices is defined just as in the real case. The following notion is fundamental in the study of complex matrices.

DEFINITION

The **conjugate transpose** Z^* of a complex matrix Z is defined by

$$Z^* = (\overline{Z})^T = \overline{(Z^T)}$$

Observe that $Z^* = Z^T$ when Z is real.

EXAMPLE 3

$$\begin{bmatrix} 3 & 1 - i & 2 + i \\ 2i & 5 + 2i & -i \end{bmatrix}^* = \begin{bmatrix} 3 & -2i \\ 1 + i & 5 - 2i \\ 2 - i & i \end{bmatrix}$$

The following properties of Z^* follow easily from the rules for transposition of real matrices (Exercise 8) and extend these rules to complex matrices. Note the conjugate in property (3).

THEOREM 3 Let Z and W denote the complex matrices, and let λ be a complex number.

(1) $(Z^*)^* = Z$

(2) $(Z + W)^* = Z^* + W^*$

(3) $(\lambda Z)^* = \overline{\lambda} Z^*$

(4) $(ZW)^* = W^* Z^*$

If A is a real symmetric matrix, it is clear that $A^* = A$. The complex matrices that satisfy this condition turn out to be the most natural generalization of the real symmetric matrices.

DEFINITION A square complex matrix H is called **Hermitian** if $H^* = H$.

The name honors Charles Hermite (1822–1901). These matrices are easy to recognize because the "reflection" of each entry in the main diagonal must be the conjugate of that entry, and the entries on the main diagonal must be real.

EXAMPLE 4 $\begin{bmatrix} 3 & i & 2+i \\ -i & -2 & -7 \\ 2-i & -7 & 1 \end{bmatrix}$ is Hermitian, whereas $\begin{bmatrix} 1 & i \\ i & -2 \end{bmatrix}$ and $\begin{bmatrix} 1 & i \\ -i & i \end{bmatrix}$ are not.

The following gives a very useful characterization of Hermitian matrices in terms of the standard inner product in $\mathbb{C}^n$.

THEOREM 4 An $n \times n$ complex matrix H is Hermitian if and only if

$$\langle H\mathbf{v}, \mathbf{w}\rangle = \langle \mathbf{v}, H\mathbf{w}\rangle$$

for all columns $\mathbf{v}$ and $\mathbf{w}$ in $\mathbb{C}^n$.

Proof If H is Hermitian, we have $H = H^* = (\overline{H})^T$, so taking transposes gives $H^T = \overline{H}$. If $\mathbf{v}$ and $\mathbf{w}$ are columns in $\mathbb{C}^n$, then $\langle \mathbf{v}, \mathbf{w}\rangle = \mathbf{v}^T\overline{\mathbf{w}}$, so

$$\langle H\mathbf{v}, \mathbf{w}\rangle = (H\mathbf{v})^T\overline{\mathbf{w}} = \mathbf{v}^T H^T \overline{\mathbf{w}} = \mathbf{v}^T \overline{H}\overline{\mathbf{w}} = \mathbf{v}^T\overline{H\mathbf{w}} = \langle \mathbf{v}, H\mathbf{w}\rangle$$

To prove the converse, let $\mathbf{e}_j$ denote column j of the identity matrix. If $H = [h_{ij}]$ we have

$$\overline{h}_{ij} = \langle \mathbf{e}_i, H\mathbf{e}_j \rangle = \langle H\mathbf{e}_i, \mathbf{e}_j \rangle = h_{ji}$$

Hence $\overline{H} = H^T$, so $H^* = H$. ■

Let Z be an $n \times n$ complex matrix. As in the real case, a complex number λ is called an **eigenvalue** of Z if $Z\mathbf{v} = \lambda\mathbf{v}$ holds for some $\mathbf{v} \neq \mathbf{0}$ in $\mathbb{C}^n$. In this case $\mathbf{v}$ is called an **eigenvector** of Z corresponding to λ. The **characteristic polynomial** $c_Z(x)$ is defined by

$$c_Z(x) = \det(xI - Z)$$

This polynomial has complex coefficients (possibly nonreal). However, the proof of Theorem 1 in Section 7.1 goes through to show that the eigenvalues of Z are the roots (possibly complex) of $c_Z(x)$. It is at this point that the advantage of working with complex numbers becomes apparent. The real numbers are incomplete in the sense that the characteristic polynomial of a real matrix may fail to have all its roots real. However, this difficulty does not occur for the complex numbers. The so-called fundamental theorem of algebra ensures that *every* polynomial of positive degree with complex coefficients has a complex root. Hence every square complex matrix has a (complex) eigenvalue. Indeed (Appendix A), $c_Z(x)$ factors completely as follows:

$$c_Z(x) = (x - \lambda_1)(x - \lambda_2) \cdots (x - \lambda_n)$$

where $\lambda_1, \lambda_2, \ldots \lambda_n$ are the eigenvalues of Z (with possible repetitions due to multiple roots).

The next result shows that, for Hermitian matrices, the eigenvalues are actually real. Because symmetric real matrices are Hermitian, it gives a proof of Theorem 2 in Section 7.1. It also extends Theorem 8 in Section 7.2, which asserts that eigenvectors of a symmetric real matrix that correspond to distinct eigenvalues are actually orthogonal. In the complex context, two vectors $\mathbf{v}$ and $\mathbf{w}$ in $\mathbb{C}^n$ are said to be **orthogonal** if $\langle \mathbf{v}, \mathbf{w} \rangle = 0$.

THEOREM 5 Let H denote a Hermitian matrix.

(1) The eigenvalues of H are real.

(2) Eigenvectors of H corresponding to distinct eigenvalues are orthogonal.

Proof Let λ and μ be eigenvalues of H with eigenvectors $\mathbf{v}$ and $\mathbf{w}$. Then $H\mathbf{v} = \lambda\mathbf{v}$ and $H\mathbf{w} = \mu\mathbf{w}$, so Theorem 4 gives

$$\lambda\langle \mathbf{v},\mathbf{w}\rangle = \langle \lambda\mathbf{v},\mathbf{w}\rangle = \langle H\mathbf{v},\mathbf{w}\rangle = \langle \mathbf{v},H\mathbf{w}\rangle = \langle \mathbf{v},\mu\mathbf{w}\rangle = \overline{\mu}\langle \mathbf{v},\mathbf{w}\rangle \qquad (*)$$

If $\mu = \lambda$ and $\mathbf{w} = \mathbf{v}$, this becomes $\lambda\langle \mathbf{v},\mathbf{v}\rangle = \overline{\lambda}\langle \mathbf{v},\mathbf{v}\rangle$. Because $\langle \mathbf{v},\mathbf{v}\rangle = \|\mathbf{v}\|^2 \neq 0$, this implies $\lambda = \overline{\lambda}$. Thus λ is real, proving (1). Similarly, μ is real, so (*) gives $\lambda\langle \mathbf{v},\mathbf{w}\rangle = \mu\langle \mathbf{v},\mathbf{w}\rangle$. If $\lambda \neq \mu$, this implies $\langle \mathbf{v},\mathbf{w}\rangle = 0$, proving (2). ■

The principal axes theorem (Theorem 6 in Section 7.2) asserts that every real symmetric matrix A is orthogonally diagonalizable—that is, P^TAP is diagonal where P is an orthogonal matrix ($P^{-1} = P^T$). The next theorem identifies the complex analogs of these orthogonal real matrices. As in the real case, a set of nonzero vectors $\{\mathbf{v}_1, \mathbf{v}_2, \ldots, \mathbf{v}_m\}$ in $\mathbb{C}^n$ is called **orthogonal** if $\langle \mathbf{v}_i,\mathbf{v}_j\rangle = 0$ whenever $i \neq j$, and it is **orthonormal** if, in addition, $\|\mathbf{v}_i\| = 1$ for each i.

THEOREM 6

The following are equivalent for an $n \times n$ complex matrix U.

(1) $U^{-1} = U^*$.

(2) The rows of U are an orthonormal set in $\mathbb{C}^n$.

(3) The columns of U are an orthonormal set in $\mathbb{C}^n$.

The proof is a direct adaptation of the proof of Theorem 5 in Section 6.2.

DEFINITION

A square complex matrix U is called **unitary** if it satisfies the conditions in Theorem 6.

Thus a real matrix is unitary if and only if it is orthogonal.

EXAMPLE 5

The matrix $Z = \begin{bmatrix} 1+i & 1 \\ 1-i & i \end{bmatrix}$ has orthogonal columns, but the rows are not orthogonal. Normalizing gives the unitary matrix $\frac{1}{2}\begin{bmatrix} 1+i & \sqrt{2} \\ 1-i & \sqrt{2}i \end{bmatrix}$.

Given a real symmetric matrix A, the diagonalization algorithm in Section 7.2 is a procedure for finding an orthogonal matrix P such that P^TAP is diagonal. The following example illustrates Theorem 5 and shows that the algorithm works for complex matrices.

EXAMPLE 6

Consider the Hermitian matrix $H = \begin{bmatrix} 3 & 2+i \\ 2-i & 7 \end{bmatrix}$. Find the eigenvalues of H, find two orthonormal eigenvectors, and so find a unitary matrix U such that U^*HU is diagonal.

Solution The characteristic polynomial of H is

$$c_H(x) = \det(xI - H) = \det\begin{bmatrix} x-3 & -2-i \\ -2+i & x-7 \end{bmatrix} = (x-2)(x-8)$$

Hence the eigenvalues are $\lambda = 2,8$ (both real as expected), and corresponding eigenvectors are $\begin{bmatrix} 2+i \\ -1 \end{bmatrix}$ and $\begin{bmatrix} 1 \\ 2-i \end{bmatrix}$ (orthogonal as expected). They each have length $\sqrt{6}$, so, as in the (real) diagonalization algorithm, let $U = \frac{1}{\sqrt{6}}\begin{bmatrix} 2+i & 1 \\ -1 & 2-i \end{bmatrix}$ be the unitary matrix with the normalized eigenvectors as columns. Then $U^*HU = \begin{bmatrix} 2 & 0 \\ 0 & 8 \end{bmatrix}$ is diagonal.

An $n \times n$ complex matrix Z is called **unitarily diagonalizable** if U^*ZU is diagonal for some unitary matrix U. As Example 6 suggests, we are going to prove that every Hermitian matrix is unitarily diagonalizable. However, with only a little extra effort, we can get a very important theorem that has this result as an easy consequence.

A complex matrix is called **upper triangular** if each entry below the main diagonal is zero. We owe the following theorem to Issai Schur (1875–1941).

THEOREM 7
Schur's Theorem

If Z is any $n \times n$ complex matrix, there exists a unitary matrix U such that

$$U^*ZU = T \qquad \text{is upper triangular}$$

Moreover, the entries on the main diagonal of T are the eigenvalues $\lambda_1, \lambda_2, \ldots, \lambda_n$ of Z (including multiplicities).

Proof We use induction on n. If $n = 1$, Z is already upper triangular. If $n > 1$, assume the theorem is valid for $(n-1) \times (n-1)$ complex matrices. Let λ_1 be an eigenvalue of Z, and let $\mathbf{u}_1$ be an eigenvector with $\|\mathbf{u}_1\| = 1$. Then $\mathbf{u}_1$ is part of a basis of $\mathbb{C}^n$ (by the analogue of Theorem 7 in Section 5.3), so the (complex analog of the) Gram–Schmidt process provides $\mathbf{u}_2, \ldots, \mathbf{u}_n$ such that $\{\mathbf{u}_1, \ldots, \mathbf{u}_n\}$ is an orthonormal basis of $\mathbb{C}^n$.

If $U_1 = [\mathbf{u}_1\ \mathbf{u}_2\ \ldots\ \mathbf{u}_n]$ is the matrix with these vectors as its columns, then

$$U_1{}^*ZU_1 = \begin{bmatrix} \lambda_1 & X_1 \\ 0 & Z_1 \end{bmatrix}$$

in block form. Now apply induction to find a unitary $(n - 1) \times (n - 1)$ matrix W_1 such that $W_1{}^*Z_1W_1 = T_1$ is upper triangular. Then $U_2 = \begin{bmatrix} 1 & 0 \\ 0 & W_1 \end{bmatrix}$ is a unitary $n \times n$ matrix. Hence $U = U_1U_2$ is unitary (using Theorem 6), and

$$\begin{aligned} U^*ZU &= U_2{}^*(U_1{}^*ZU_1)U_2 \\ &= \begin{bmatrix} 1 & 0 \\ 0 & W_1{}^* \end{bmatrix} \begin{bmatrix} \lambda_1 & X_1 \\ 0 & Z_1 \end{bmatrix} \begin{bmatrix} 1 & 0 \\ 0 & W_1 \end{bmatrix} = \begin{bmatrix} \lambda_1 & X_1W_1 \\ 0 & T_1 \end{bmatrix} \end{aligned}$$

is upper triangular. Finally, Z and $U^*ZU = T$ have the same eigenvalues by (the complex version of) Theorem 3 in Section 7.1, and they are the diagonal entries of T because T is upper triangular. ■

The fact that similar matrices have the same traces and determinants gives

COROLLARY

Let Z be an $n \times n$ complex matrix, and let $\lambda_1, \lambda_2, \ldots, \lambda_n$ denote the eigenvalues of Z, including multiplicities. Then

$$\det Z = \lambda_1\lambda_2 \cdots \lambda_n \quad \text{and} \quad \operatorname{tr} Z = \lambda_1 + \lambda_2 + \cdots + \lambda_n.$$

Schur's theorem asserts that every complex matrix can be "unitarily triangularized." However, we cannot substitute "unitarily diagonalized" here. In fact, if $Z = \begin{bmatrix} 1 & 1 \\ 0 & 1 \end{bmatrix}$ there is no invertible complex matrix U at all such that $U^{-1}ZU$ is diagonal. However, the situation is much better for Hermitian matrices.

THEOREM 8
Spectral Theorem

If H is Hermitian, there is a unitary matrix U such that U^*HU is diagonal.

Proof By Schur's theorem, let $U^*HU = T$ be upper triangular where U is unitary. Because $U^{-1} = U^*$, this gives $H = UTU^*$. But H is Hermitian, so

$$UTU^* = H = H^* = (UTU^*)^* = UT^*U^*$$

It follows by cancellation that $T = T^*$, so T is both upper and lower triangular. Hence T is actually diagonal. ■

The principal axes theorem asserts that a real matrix A is symmetric if and only if it is orthogonally diagonalizable (that is, P^TAP is diagonal for some real orthogonal matrix P). Theorem 8 is the complex analog of half of this result; however, the converse is false for complex matrices: There exist unitarily diagonalizable matrices that are not Hermitian.

EXAMPLE 7

Show that the non-Hermitian matrix $Z = \begin{bmatrix} 0 & 1 \\ -1 & 0 \end{bmatrix}$ is unitarily diagonalizable.

Solution The characteristic polynomial is $c_Z(x) = x^2 + 1$. Hence the eigenvalues are i and $-i$, and it is easy to verify that $\begin{bmatrix} i \\ -1 \end{bmatrix}$ and $\begin{bmatrix} -1 \\ i \end{bmatrix}$ are corresponding eigenvectors. Moreover, these eigenvectors are orthogonal and both have length $\sqrt{2}$, so $U = \frac{1}{\sqrt{2}}\begin{bmatrix} i & -1 \\ -1 & i \end{bmatrix}$ is a unitary matrix such that $U^*ZU = \begin{bmatrix} i & 0 \\ 0 & -i \end{bmatrix}$ is diagonal.

There is a very simple way to characterize those complex matrices that are unitarily diagonalizable.

DEFINITION

An $n \times n$ complex matrix N is called **normal** if $NN^* = N^*N$.

It is clear that every Hermitian matrix is normal, as is the matrix $\begin{bmatrix} 0 & 1 \\ -1 & 0 \end{bmatrix}$ in Example 7. We conclude with the following result.

THEOREM 9

An $n \times n$ complex matrix Z is unitarily diagonalizable if and only if Z is normal.

Proof Assume first that $U^*ZU = D$, where U is unitary and D is diagonal. Then $DD^* = D^*D$ as is easily verified. Because $DD^* = U^*(ZZ^*)U$

and $D^*D = U^*(Z^*Z)U$, it follows by cancellation that $ZZ^* = Z^*Z$. Conversely, assume Z is normal—that is, $ZZ^* = Z^*Z$. By Schur's theorem, let $U^*ZU = T$, where T is upper triangular and U is unitary. Moreover T is normal too:

$$TT^* = U^*(ZZ^*)U = U^*(Z^*Z)U = T^*T$$

Hence it suffices to show that a normal $n \times n$ upper triangular matrix T must be diagonal. If $T = [t_{ij}]$, then equating (1,1)-entries in TT^* and T^*T gives

$$|t_{11}|^2 + |t_{12}|^2 + \cdots + |t_{1n}|^2 = |t_{11}|^2$$

This implies $t_{12} = t_{13} = \cdots = t_{1n} = 0$, so $T = \begin{bmatrix} t_{11} & 0 \\ 0 & T_1 \end{bmatrix}$ in block form. Hence $T^* = \begin{bmatrix} \bar{t}_{11} & 0 \\ 0 & T_1{}^* \end{bmatrix}$, so $TT^* = T^*T$ implies $T_1T_1{}^* = T_1{}^*T_1$. Thus T_1 is diagonal by induction, and the proof is complete. ■

EXERCISES 7.3

1. In each case, compute the norm of the complex vector.

(a) $(1, 1 - i, -2, i)$ **(b)** $(1 - i, 1 + i, 1, -1)$

(c) $(2 + i, 1 - i, 2, 0, -i)$ **(d)** $(-2, -i, 1 + i, 1 - i, 2i)$.

2. In each case, determine whether the two vectors are orthogonal.

(a) $(4, -3i, 2 + i), (i, 2, 2 - 4i)$ **(b)** $(i, -i, 2 + i), (i, i, 2 - i)$

(c) $(1, 1, i, i), (1, i, -i, 1)$

(d) $(4 + 4i, 2 + i, 2i), (-1 + i, 2, 3 - 2i)$.

3. A subset U of $\mathbb{C}^n$ is called a **complex subspace** of $\mathbb{C}^n$ if it contains $\mathbf{0}$ and if, given $\mathbf{v}$ and $\mathbf{w}$ in U, both $\mathbf{v} + \mathbf{w}$ and $z\mathbf{v}$ lie in U (z any complex number). In each case, determine whether U is a subspace of $\mathbb{C}^3$.

(a) $U = \{(w, \overline{w}, 0) \mid w \text{ in } \mathbb{C}\}$ **(b)** $U = \{(w, 2w, a) \mid w \text{ in } \mathbb{C}, a \text{ in } \mathbb{R}\}$

(c) $U = \mathbb{R}^3$

(d) $U = \{(v + w, v - 2w, v) \mid v, w \text{ in } \mathbb{C}\}$

4. In each case, find a basis over $\mathbb{C}$, and determine the dimension of the complex subspace U of $\mathbb{C}^3$ (see the previous exercise).

(a) $U = \{(w, v + w, v - iw) \mid v, w \text{ in } \mathbb{C}\}$

(b) $U = \{(iv + w, 0, 2v - w) \mid v, w \text{ in } \mathbb{C}\}$

(c) $U = \{(u, v, w) \mid iu - 3v + (1 - i)w = 0;\ u, v, w \text{ in } \mathbb{C}\}$

(d) $U = \{(u, v, w) \mid 2u + (1 + i)v - iw = 0;\ u, v, w \text{ in } \mathbb{C}\}$.

5. In each case, determine whether the given matrix is Hermitian, unitary, or normal.

(a) $\begin{bmatrix} 1 & -i \\ i & i \end{bmatrix}$ **(b)** $\begin{bmatrix} 2 & 3 \\ -3 & 2 \end{bmatrix}$ **(c)** $\begin{bmatrix} 1 & i \\ -i & 2 \end{bmatrix}$ **(d)** $\begin{bmatrix} 1 & -i \\ i & -1 \end{bmatrix}$

(e) $\frac{1}{\sqrt{2}} \cdot \begin{bmatrix} 1 & -1 \\ 1 & 1 \end{bmatrix}$ **(f)** $\begin{bmatrix} 1 & 1+i \\ 1+i & -i \end{bmatrix}$

(g) $\begin{bmatrix} 2i & -1+i \\ 1+i & i \end{bmatrix}$ **(h)** $\frac{1}{\sqrt{2}|z|} \cdot \begin{bmatrix} z & z \\ \bar{z} & -\bar{z} \end{bmatrix}, z \neq 0$

6. In each case, find a unitary matrix U such that U^*ZU is diagonal.

(a) $Z = \begin{bmatrix} 1 & i \\ -i & 1 \end{bmatrix}$ **(b)** $Z = \begin{bmatrix} 4 & 3-i \\ 3+i & 1 \end{bmatrix}$

(c) $Z = \begin{bmatrix} a & b \\ -b & a \end{bmatrix}$; a, b real **(d)** $Z = \begin{bmatrix} 2 & 1+i \\ 1-i & 3 \end{bmatrix}$

(e) $Z = \begin{bmatrix} 1 & 0 & 1+i \\ 0 & 2 & 0 \\ 1-i & 0 & 0 \end{bmatrix}$ **(f)** $Z = \begin{bmatrix} 1 & 0 & 0 \\ 0 & 1 & 1+i \\ 0 & 1-i & 2 \end{bmatrix}$

7. Show that $\langle Z\mathbf{v},\mathbf{w}\rangle = \langle \mathbf{v},Z^*\mathbf{w}\rangle$ holds for all $n \times n$ matrices Z and for all columns $\mathbf{v}$ and $\mathbf{w}$ in $\mathbb{C}^n$.

8. **(a)** Prove (1) and (2) of Theorem 1.

(b) Prove Theorem 2.

(c) Prove Theorem 3.

9. **(a)** Show that Z is Hermitian if and only if $\bar{Z} = Z^T$.

(b) Show that the diagonal entries of any Hermitian matrix are real.

10. **(a)** Show that every complex matrix Z can be written uniquely in the form $Z = A + iB$, where A and B are real matrices.

(b) If $Z = A + iB$ as in **(a)**, show that Z is Hermitian if and only if A is symmetric and B is skew-symmetric (that is, $B^T = -B$).

11. If Z is any complex $n \times n$ matrix, show that ZZ^* and $Z + Z^*$ are Hermitian.

12. A complex matrix S is called **skew-Hermitian** if $S^* = -S$.

(a) Show that $Z - Z^*$ is skew-Hermitian for any square complex matrix Z.

(b) If S is skew-Hermitian, show that S^2 and iS are Hermitian.

(c) If S is skew-Hermitian, show that the eigenvalues of S are pure imaginary ($i\lambda$ for real λ).

(d) Show that every $n \times n$ complex matrix Z can be written uniquely as $Z = H + S$, where H is Hermitian and S is skew-Hermitian.

13. Let U be a unitary matrix. Show that:

(a) $\|U\mathbf{v}\| = \|\mathbf{v}\|$ for all columns $\mathbf{v}$ in $\mathbb{C}^n$.

(b) $|\lambda| = 1$ for every eigenvalue λ of U.

14. **(a)** If Z is an invertible complex matrix, show that Z^* is invertible and that $(Z^*)^{-1} = (Z^{-1})^*$

(b) Show that the inverse of a unitary matrix is again unitary.

(c) If U is unitary, show that U^* is unitary.

15. Let Z be an $m \times n$ matrix such that $Z^*Z = I_m$ (for example, Z is a unit row in $\mathbb{C}^n$).

(a) Show that $H = ZZ^*$ is Hermitian and satisfies $H^2 = H$.

(b) Show that $U = I - 2ZZ^*$ is both unitary and Hermitian (so $U^{-1} = U^* = U$).

16. (a) If N is normal, show that zN is also normal for all complex numbers z.

(b) Show that (a) fails if "normal" is replaced by "Hermitian."

17. Show that a real 2×2 normal matrix is either symmetric or has the form $\begin{bmatrix} a & b \\ -b & a \end{bmatrix}$.

18. If H is Hermitian, show that all the coefficients of $c_H(x)$ are real numbers.

19. (a) If $A = \begin{bmatrix} 1 & 1 \\ 0 & 1 \end{bmatrix}$, show that $U^{-1}AU$ is not diagonal for any invertible complex matrix U.

(b) If $A = \begin{bmatrix} 0 & 1 \\ -1 & 0 \end{bmatrix}$, show that $U^{-1}AU$ is not upper triangular for any *real* invertible matrix U.

20. If Z is any $n \times n$ matrix, show that U^*ZU is lower triangular for some unitary matrix U.

21. If Z is a 3×3 matrix, show that $Z^2 = 0$ if and only if there exists a unitary matrix U such that U^*ZU has either the form $\begin{bmatrix} 0 & 0 & u \\ 0 & 0 & v \\ 0 & 0 & 0 \end{bmatrix}$ or the form $\begin{bmatrix} 0 & u & v \\ 0 & 0 & 0 \\ 0 & 0 & 0 \end{bmatrix}$.

22. If $Z^2 = Z$, show that rank Z = tr Z. [*Hint:* Schur's theorem.]

SECTION 7.4 Applications of Diagonalization (Optional)[†]

7.4.1 Quadratic Forms

An expression like $q = 3x^2 + 5xy - 7y^2$ is called a quadratic form in the variables x and y. Such forms were analyzed in Section 5.5, where we found that if new coordinate axes X' and Y' are obtained by rotating the X-Y-system, the rotation angle could be chosen so that, when q is expressed in terms of the new coordinates x' and y', it has the form $q = \lambda x'^2 + \mu y'^2$. In other words, the "cross term" could be eliminated by rotating the coordinate system. This can be seen geometrically by observing that, if a quadratic form is set equal to a constant, the resulting

[†]The two applications in the section are independent and may be taken in any order.

equation has the form

$$ax^2 + bxy + cy^2 = d$$

where a, b, c, and d are real numbers, and the graph of such an equation is a circle, ellipse, hyperbola, or some degenerate form (for example, $x^2 - y^2 = 0$ represents two straight lines). These curves are studied in analytic geometry, where it is shown that if the new coordinate axes are chosen along the axes of symmetry of the curve, the equation (in the new variables) involves no cross term. This geometric approach works with three variables too, where the form of various quadratic surfaces is studied. However, the geometry becomes more intricate in this case, and purely algebraic methods are a great help. When forms of four or more variables are considered, the algebraic techniques are all but indispensable.

The purpose of this section is to show how our diagonalization techniques can be used to advantage. This has far-reaching applications; quadratic forms arise in such diverse areas as statistics, physics, theory of functions of several variables, and number theory, as well as in geometry.

A **quadratic form** $q = q(\mathbf{x})$ in the n variables $x_1, x_2, \dots, x_n$ is a linear combination of terms $x_1^2, x_2^2, \dots, x_n^2$ and cross terms x_1x_2, x_1x_3, and so on:

$$q = a_{11}x_1^2 + a_{22}x_2^2 + \cdots + a_{nn}x_n^2 + a_{12}x_1x_2 + a_{13}x_1x_3 + \cdots$$

In summation notation, this is a double sum

$$q = q(\mathbf{x}) = \sum_{i=1}^{n}\sum_{j=1}^{n} a_{ij}x_ix_j = \mathbf{x}^TA\mathbf{x}$$

where $\mathbf{x} = \begin{bmatrix} x_1 \\ \vdots \\ x_n \end{bmatrix}$ and $A = [a_{ij}]$ is an $n \times n$ matrix. Note that if $i \neq j$, two separate terms $a_{ij}x_ix_j$ and $a_{ji}x_jx_i$ are listed, each of which involves x_ix_j, and they can (rather cleverly) be replaced by

$$\frac{1}{2}(a_{ij} + a_{ji})x_ix_j \qquad \text{and} \qquad \frac{1}{2}(a_{ij} + a_{ji})\,x_jx_i$$

respectively, *without altering the sum*. Hence there is no loss of generality in assuming that x_ix_j and x_jx_i have the same coefficient in the sum for q. In other words, *we may assume that A is symmetric*.

EXAMPLE 1 Write $q = x_1^2 + 3x_3^2 + 2x_1x_2 - x_1x_3$ in the form $q(\mathbf{x}) = \mathbf{x}^TA\mathbf{x}$, where A is a symmetric 3×3 matrix.

Solution The cross terms are $2x_1x_2 = x_1x_2 + x_2x_1$ and $-x_1x_3 = -\frac{1}{2}x_1x_3 - \frac{1}{2}x_3x_1$. Of course, x_2x_3 and x_3x_2 both have coefficient zero, as does x_2^2. Hence

$$q(\mathbf{x}) = [x_1x_2x_3]\begin{bmatrix} 1 & 1 & \frac{-1}{2} \\ 1 & 0 & 0 \\ \frac{-1}{2} & 0 & 3 \end{bmatrix}\begin{bmatrix} x_1 \\ x_2 \\ x_3 \end{bmatrix}$$

is the required form.

We shall assume from now on that all quadratic forms are given by $q(\mathbf{x}) = \mathbf{x}^TA\mathbf{x}$, where A is symmetric. Given such a form, the problem is to find new variables $y_1, y_2, \ldots, y_n$, related to $x_1, \ldots, x_n$, with the property that when q is expressed in terms of $y_1, y_2, \ldots, y_n$, there are no cross terms. If we write

$$\mathbf{y} = \begin{bmatrix} y_1 \\ y_2 \\ \vdots \\ y_n \end{bmatrix}$$

this amounts to asking that $q = \mathbf{y}^TD\mathbf{y}$ where D is diagonal. It turns out that this can always be accomplished and, not surprisingly, that D is the matrix obtained when the symmetric matrix A is diagonalized. In fact, as Theorem 6 in Section 7.2 shows, a matrix P can be found that is orthogonal (that is, $P^{-1} = P^T$) and diagonalizes A:

$$P^TAP = D = \begin{bmatrix} \lambda_1 & 0 & \ldots & 0 \\ 0 & \lambda_2 & \ldots & 0 \\ \vdots & \vdots & & \vdots \\ 0 & 0 & \ldots & \lambda_n \end{bmatrix}$$

The diagonal entries $\lambda_1, \lambda_2, \ldots, \lambda_n$ are the (not necessarily distinct) eigenvalues of A, repeated according to their multiplicities in $c_A(x)$, and the columns $\mathbf{p}_1, \mathbf{p}_2, \ldots, \mathbf{p}_n$ of P are the corresponding eigenvectors of A. Now define new variables $\mathbf{y}$ by the equations

$$\mathbf{x} = P\mathbf{y} \quad \text{equivalently,} \quad \mathbf{y} = P^T\mathbf{x}$$

Then substitution in $q(\mathbf{x}) = \mathbf{x}^TA\mathbf{x}$ gives

$$q = (P\mathbf{y})^TA(P\mathbf{y}) = \mathbf{y}^T(P^TAP)\mathbf{y} = \mathbf{y}^TD\mathbf{y} = \lambda_1y_1^2 + \lambda_2y_2^2 + \cdots + \lambda_ny_n^2$$

Hence this change of variables produces the desired simplification in q. Furthermore, the fact that P is orthogonal means that the eigenvectors

$\mathbf{p}_i$ were chosen so that $B = \{\mathbf{p}_1, \mathbf{p}_2, \ldots, \mathbf{p}_n\}$ is an orthonormal basis of $\mathbb{R}^n$ (Theorem 5 in Section 6.2). If P is written in block form, $P = [\mathbf{p}_1\ \mathbf{p}_2 \ldots \mathbf{p}_n]$, then

$$\mathbf{x} = P\mathbf{y} = [\mathbf{p}_1\ \mathbf{p}_2 \ldots \mathbf{p}_n]\begin{bmatrix} y_1 \\ y_2 \\ \vdots \\ y_n \end{bmatrix} = y_1\mathbf{p}_1 + y_2\mathbf{p}_2 + \cdots + y_n\mathbf{p}_n$$

so the coordinate vector of $\mathbf{x}$ with respect to B is $C_B(\mathbf{x}) = \mathbf{y}$. This proves the following important theorem.

THEOREM 1 Diagonalization Theorem

Let $q = \mathbf{x}^TA\mathbf{x}$ be a quadratic form in the variables $x_1, x_2, \ldots, x_n$, where $\mathbf{x} = \begin{bmatrix} x_1 \\ x_2 \\ \vdots \\ x_n \end{bmatrix}$ and A is a symmetric $n \times n$ matrix. Let P be an orthogonal matrix such that P^TAP is diagonal, and define new variables $\mathbf{y} = \begin{bmatrix} y_1 \\ y_2 \\ \vdots \\ y_n \end{bmatrix}$ by

$$\mathbf{x} = P\mathbf{y} \qquad \text{equivalently,} \quad \mathbf{y} = P^T\mathbf{x}$$

If q is expressed in terms of these new variables $y_1, y_2, \ldots, y_n$, the result is

$$q = \lambda_1 y_1^2 + \lambda_2 y_2^2 + \cdots + \lambda_n y_n^2$$

where $\lambda_1, \lambda_2, \ldots, \lambda_n$ are the eigenvalues of A repeated according to their multiplicities. Finally, the columns $\mathbf{p}_1, \mathbf{p}_2, \ldots, \mathbf{p}_n$ of P are the corresponding eigenvectors; $B = \{\mathbf{p}_1, \mathbf{p}_2, \ldots, \mathbf{p}_n\}$ is an orthonormal basis of $\mathbb{R}^n$; and

$$\mathbf{x} = y_1\mathbf{p}_1 + y_2\mathbf{p}_2 + \cdots + y_n\mathbf{p}_n$$

Hence the vector $\mathbf{y} = C_B(\mathbf{x})$ is the coordinate vector of $\mathbf{x}$ with respect to B.

The orthonormal eigenvectors $\mathbf{p}_1, \cdots, \mathbf{p}_n$ are called **principal axes** for the quadratic form q.

Thus the computation of $q = q(\mathbf{x}) = \mathbf{x}^TA\mathbf{x}$ at a specific vector $\mathbf{x}$ is a routine matter when the eigenvalues $\lambda_1, \ldots, \lambda_n$ and the corresponding principal axes $\mathbf{p}_1, \mathbf{p}_2, \ldots, \mathbf{p}_n$ are known. The coefficients y_i in the expansion $\mathbf{x} = y_1\mathbf{p}_1 + \cdots + y_n\mathbf{p}_n$ are given by $y_i = \mathbf{x} \cdot \mathbf{p}_i$ by the expansion theorem (Theorem 4 in Section 6.2), so

$$q = q(\mathbf{x}) = \lambda_1(\mathbf{x} \cdot \mathbf{p}_1)^2 + \cdots + \lambda_n(\mathbf{x} \cdot \mathbf{p}_n)^2$$

EXAMPLE 2 Find new variables y_1, y_2, y_3, and y_4 such that

$$q = 3(x_1^2 + x_2^2 + x_3^2 + x_4^2) + 2x_1x_2 - 10x_1x_3 + 10x_1x_4 + 10x_2x_3 - 10x_2x_4 + 2x_3x_4$$

has diagonal form, and find the corresponding principal axes.

Solution The form can be written as $q = \mathbf{x}^TA\mathbf{x}$, where

$$\mathbf{x} = \begin{bmatrix} x_1 \\ x_2 \\ x_3 \\ x_4 \end{bmatrix} \qquad A = \begin{bmatrix} 3 & 1 & -5 & 5 \\ 1 & 3 & 5 & -5 \\ -5 & 5 & 3 & 1 \\ 5 & -5 & 1 & 3 \end{bmatrix}$$

A routine calculation yields

$$c_A(x) = \det(xI - A) = (x - 12)(x + 8)(x - 4)^2$$

so the eigenvalues are $\lambda_1 = 12$, $\lambda_2 = -8$, and $\lambda_3 = \lambda_4 = 4$. The corresponding eigenvectors are

$$\mathbf{p}_1 = \begin{bmatrix} 1 \\ -1 \\ -1 \\ 1 \end{bmatrix} \qquad \mathbf{p}_2 = \begin{bmatrix} 1 \\ -1 \\ 1 \\ -1 \end{bmatrix} \qquad \mathbf{p}_3 = \begin{bmatrix} 1 \\ 1 \\ 1 \\ 1 \end{bmatrix} \qquad \mathbf{p}_4 = \begin{bmatrix} 1 \\ 1 \\ -1 \\ -1 \end{bmatrix}$$

These are orthogonal and all have length 2, so $\frac{1}{2}\mathbf{p}_1, \frac{1}{2}\mathbf{p}_2, \frac{1}{2}\mathbf{p}_3, \frac{1}{2}\mathbf{p}_4$ are the required principal axes. The matrix

$$P = \begin{bmatrix} \frac{1}{2}\mathbf{p}_1 & \frac{1}{2}\mathbf{p}_2 & \frac{1}{2}\mathbf{p}_3 & \frac{1}{2}\mathbf{p}_4 \end{bmatrix} = \frac{1}{2}\begin{bmatrix} 1 & 1 & 1 & 1 \\ -1 & -1 & 1 & 1 \\ -1 & 1 & 1 & -1 \\ 1 & -1 & 1 & -1 \end{bmatrix}$$

is thus orthogonal, and $P^{-1}AP = P^TAP$ is diagonal. Hence the new variables $\mathbf{y}$ and the old variables $\mathbf{x}$ are related by $\mathbf{x} = P\mathbf{y}$ and $\mathbf{y} = P^T\mathbf{x}$. Explicitly,

$$y_1 = \frac{1}{2}(x_1 - x_2 - x_3 + x_4) \qquad x_1 = \frac{1}{2}(y_1 + y_2 + y_3 + y_4)$$

$$y_2 = \frac{1}{2}(x_1 - x_2 + x_3 - x_4) \qquad x_2 = \frac{1}{2}(-y_1 - y_2 + y_3 + y_4)$$

$$y_3 = \frac{1}{2}(x_1 + x_2 + x_3 + x_4) \qquad x_3 = \frac{1}{2}(-y_1 + y_2 + y_3 - y_4)$$

$$y_4 = \frac{1}{2}(x_1 + x_2 - x_3 - x_4) \qquad x_4 = \frac{1}{2}(y_1 - y_2 + y_3 - y_4)$$

If these are substituted in the expression for q, the result is

$$q = 12{y_1}^2 - 8{y_2}^2 + 4{y_3}^2 + 4{y_4}^2$$

This is the required diagonal form.

Consider the graph of the equation $rx^2 + sy^2 = t$. Call the graph an **ellipse** if $rs > 0$ and a **hyperbola** if $rs < 0$. (We regard the point $x^2 + y^2 = 0$ as a degenerate ellipse and the lines $x^2 - y^2 = 0$ as a degenerate hyperbola.) Our theory asserts that every equation $ax^2 + bxy + cy^2 = d$ can be transformed into such a diagonal form by a change of coordinates, and the following application of Theorem 1 gives a simple way of deciding which conic it is.

COROLLARY If a, b, and c are not all zero, the graph of $ax^2 + bxy + cy^2 = d$ is an ellipse if $b^2 - 4ac < 0$ and a hyperbola if $b^2 - 4ac > 0$.

Proof If $q = ax^2 + bxy + cy^2$, the matrix of q is $A = \begin{bmatrix} a & \frac{1}{2}b \\ \frac{1}{2}b & c \end{bmatrix}$.

Theorem 1 shows that new variables x_1 and y_1 can be found so that $q = \lambda_1{x_1}^2 + \lambda_2{y_1}^2$. Moreover, there is an orthogonal matrix P such that $P^{-1}AP = \begin{bmatrix} \lambda_1 & 0 \\ 0 & \lambda_2 \end{bmatrix}$, so

$$\lambda_1\lambda_2 = \det\begin{bmatrix} \lambda_1 & 0 \\ 0 & \lambda_2 \end{bmatrix} = \det(P^{-1}AP) = \det A = -\frac{1}{4}(b^2 - 4ac)$$

The graph $q = \lambda_1{x_1}^2 + \lambda_2{y_1}^2 = d$ is an ellipse if $\lambda_1\lambda_2 > 0$ (that is, $b^2 - 4ac < 0$) and a hyperbola if $\lambda_1\lambda_2 < 0$ ($b^2 - 4ac > 0$). ■

It was shown in Exercise 10 in Section 5.5 that, given an equation $ax^2 + bxy + cy^2 = d$, the coordinate axes can always be rotated through an angle θ so that the equation, when expressed in terms of the new variables x_1 and y_1, will have no x_1y_1-term.

EXAMPLE 3 Consider the equation $x^2 + xy + y^2 = 1$ dealt with in Example 4 in Section 5.5. Use Theorem 1 to find new variables x_1 and y_1 so that the equation takes the form $rx_1^2 + sy_1^2 = 1$, and show that this change of coordinates can be achieved by a rotation.

Solution Using the notation of Example 4 in Section 5.5, write $\mathbf{x} = \begin{bmatrix} x \\ y \end{bmatrix}$ and $q = q(\mathbf{x}) = x^2 + xy + y^2$ so that the equation is $q(\mathbf{x}) = 1$. Now $q(\mathbf{x}) = \mathbf{x}^T A\mathbf{x}$, where $A = \begin{bmatrix} 1 & \frac{1}{2} \\ \frac{1}{2} & 1 \end{bmatrix}$. The eigenvalues of A are $\lambda_1 = \frac{3}{2}$ and $\lambda_2 = \frac{1}{2}$, with eigenvectors $\mathbf{p}_1 = \begin{bmatrix} 1 \\ 1 \end{bmatrix}$ and $\mathbf{p}_2 = \begin{bmatrix} -1 \\ 1 \end{bmatrix}$. Hence $P = \frac{1}{\sqrt{2}}\begin{bmatrix} 1 & -1 \\ 1 & 1 \end{bmatrix}$ is orthogonal and

$$P^{-1}AP = P^T AP = \begin{bmatrix} \frac{3}{2} & 0 \\ 0 & \frac{1}{2} \end{bmatrix}$$

as in Theorem 1. The new variables $\mathbf{x}_1 = \begin{bmatrix} x_1 \\ y_1 \end{bmatrix}$ are related to $\mathbf{x}$ by

$$\begin{bmatrix} x_1 \\ y_1 \end{bmatrix} = P^T \begin{bmatrix} x \\ y \end{bmatrix} = \frac{1}{\sqrt{2}} \begin{bmatrix} x + y \\ -x + y \end{bmatrix}$$

Hence $x_1 = \frac{1}{\sqrt{2}}(x + y)$ and $x_2 = \frac{1}{\sqrt{2}}(-x + y)$, as before, and

$$q = q(\mathbf{x}_1) = \lambda_1 x_1^2 + \lambda_2 y_1^2 = \frac{3}{2}x_1^2 + \frac{1}{2}y_1^2$$

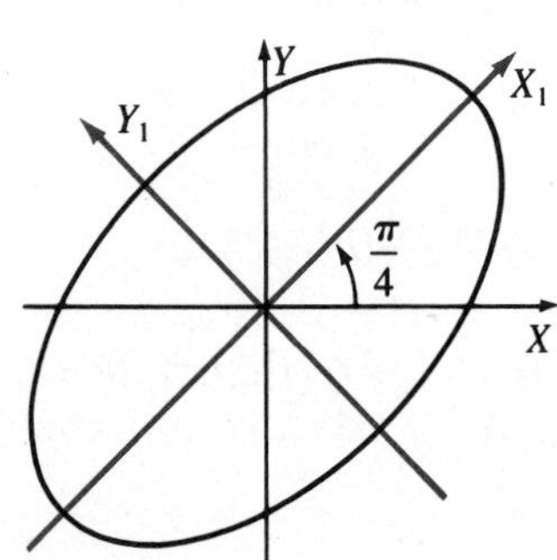

again by Theorem 1. Thus the equation $q = 1$ becomes $3x_1^2 + y_1^2 = 2$, as in Example 4 in Section 5.5. Finally, if the X-Y-axes are rotated counterclockwise through an angle θ (see the diagram), Theorem 1 in Section 5.5 shows that the coordinates relative to the new X_1-Y_1-axes are related to the X-Y-coordinates by

$$\begin{bmatrix} x_1 \\ y_1 \end{bmatrix} = \begin{bmatrix} \cos\theta & \sin\theta \\ -\sin\theta & \cos\theta \end{bmatrix} \begin{bmatrix} x \\ y \end{bmatrix}$$

Comparing this to $\mathbf{x}_1 = P^T\mathbf{x}$ gives $\cos\theta = \frac{1}{\sqrt{2}} = \sin\theta$, so the change of variables can be achieved by a rotation of $\theta = \pi/4$.

If the eigenvalues λ_1 and λ_2 had been interchanged in Example 3, the orthogonal matrix would have been $P_1 = \frac{1}{\sqrt{2}}\begin{bmatrix} -1 & 1 \\ 1 & 1 \end{bmatrix}$, and, although the resulting change of variables would have diagonalized q, it could *not* have been achieved by a rotation. In fact, det $P_1 = -1$ whereas det $P = 1$, and it can be shown that orthogonal 2×2 matrices with determinant 1 correspond to rotations, whereas those with determinant -1 correspond to reflections through a line through the origin. (Note that det $P = \pm 1$ for every orthogonal matrix P.)

If the columns of P are not required to be orthonormal in Theorem 1 (but merely orthogonal), then the reduction of q can be pushed further. The eigenvectors $\mathbf{p}_1, \ldots, \mathbf{p}_n$ of A may be put in any desired order, so P can be chosen in such a way that the positive eigenvalues of A come first (say, $\lambda_1, \lambda_2, \ldots, \lambda_k$), then the negative ones ($\lambda_{k+1}, \lambda_{k+2}, \ldots, \lambda_r$), and finally the zero eigenvalues ($\lambda_{r+1} = \cdots = \lambda_n = 0$). Of course, there may be no eigenvalues in any of these categories. Now define new variables

$$\mathbf{z} = \begin{bmatrix} z_1 \\ z_2 \\ \vdots \\ z_n \end{bmatrix} \quad \text{by} \quad \begin{aligned} z_i &= \sqrt{\lambda_i}\,y_i && \text{if } i = 1, 2, \ldots, k \\ z_i &= \sqrt{-\lambda_i}\,y_i && \text{if } i = k+1, k+2, \ldots, r \\ z_i &= y_i && \text{if } i = r+1, r+2, \ldots, n \end{aligned}$$

If these are substituted in $q = \lambda_1 y_1^2 + \lambda_2 y_2^2 + \cdots + \lambda_n y_n^2$, the result is

$$q = z_1^2 + \cdots + z_k^2 - z_{k+1}^2 - \cdots - z_r^2$$

This can be put in matrix form. Let E denote the diagonal matrix whose diagonal entries (in order) are

$$\sqrt{\lambda_1}, \sqrt{\lambda_2}, \ldots, \sqrt{\lambda_k}, \sqrt{-\lambda_{k+1}}, \ldots, \sqrt{-\lambda_r}, 1, \ldots, 1$$

Then E is invertible and $\mathbf{z} = E\mathbf{y}$. If we define $Q = PE^{-1}$, then Q is invertible, and the fact that $\mathbf{x} = P\mathbf{y}$ (from Theorem 1) relates the new variables:

$$\mathbf{x} = Q\mathbf{z} \qquad \text{or} \qquad \mathbf{z} = Q^{-1}\mathbf{x} = EP^T\mathbf{x}$$

Moreover, the columns of $Q = PE^{-1}$ are multiples of the columns of P (because E^{-1} is diagonal) and so are an orthogonal basis of $\mathbb{R}^n$ (but not necessarily orthonormal).

When a quadratic form q is given by $q = z_1^2 + \cdots + z_k^2 - z_{k+1}^2 - \cdots - z_r^2$, it is said to be **completely diagonalized.** This discussion shows that, once the orthogonal matrix P is found (as in Theorem 1) that renders q in the form $q = \lambda_1 y_1^2 + \cdots + \lambda_n y_n^2$, the matrix Q that completely diagonalizes q is easily obtained.

EXAMPLE 4 If q is the quadratic form of Example 2,

$$\begin{aligned} q = {} & 3(x_1^2 + x_2^2 + x_3^2 + x_4^2) \\ & + 2x_1x_2 - 10x_1x_3 + 10x_1x_4 + 10x_2x_3 - 10x_2x_4 + 2x_3x_4 \end{aligned}$$

find variables z_1, z_2, z_3, and z_4 such that q is completely diagonalized.

Solution In Example 2 the eigenvalues of A were found to be 12, -8, 4, and 4. If we list them with the positive ones first—12, 4, 4, -8—the corresponding orthonormal eigenvectors (using different notation from Example 2) are

$$\mathbf{p}_1 = \frac{1}{2}\begin{bmatrix} 1 \\ -1 \\ -1 \\ 1 \end{bmatrix} \quad \mathbf{p}_2 = \frac{1}{2}\begin{bmatrix} 1 \\ 1 \\ 1 \\ 1 \end{bmatrix} \quad \mathbf{p}_3 = \frac{1}{2}\begin{bmatrix} 1 \\ 1 \\ -1 \\ -1 \end{bmatrix} \quad \mathbf{p}_4 = \frac{1}{2}\begin{bmatrix} 1 \\ -1 \\ 1 \\ -1 \end{bmatrix}$$

Hence Theorem 1 gives

$$P^TAP = \begin{bmatrix} 12 & 0 & 0 & 0 \\ 0 & 4 & 0 & 0 \\ 0 & 0 & 4 & 0 \\ 0 & 0 & 0 & -8 \end{bmatrix} \quad \text{where} \quad P = \frac{1}{2}\begin{bmatrix} 1 & 1 & 1 & 1 \\ -1 & 1 & 1 & -1 \\ -1 & 1 & -1 & 1 \\ 1 & 1 & -1 & -1 \end{bmatrix}$$

Note that this P is *distinct* from the P used in Example 2; the columns have been permuted to get the eigenvalues in the desired order.

With this choice of P, the variables z_1, z_2, z_3, and z_4 are given by $\mathbf{z} = EP^T\mathbf{x}$, where $\mathbf{E} = \begin{bmatrix} \sqrt{12} & 0 & 0 & 0 \\ 0 & \sqrt{4} & 0 & 0 \\ 0 & 0 & \sqrt{4} & 0 \\ 0 & 0 & 0 & \sqrt{8} \end{bmatrix} = 2\begin{bmatrix} \sqrt{3} & 0 & 0 & 0 \\ 0 & 1 & 0 & 0 \\ 0 & 0 & 1 & 0 \\ 0 & 0 & 0 & \sqrt{2} \end{bmatrix}$. The result is

$$\begin{aligned} z_1 &= \sqrt{3}(x_1 - x_2 - x_3 + x_4) \\ z_2 &= x_1 + x_2 + x_3 + x_4 \\ z_3 &= x_1 + x_2 - x_3 - x_4 \\ z_4 &= \sqrt{2}(x_1 - x_2 + x_3 - x_4) \end{aligned}$$

This discussion focuses attention on the number of positive and negative terms appearing when a quadratic form q is completely diagonalized. The next theorem shows that these numbers are uniquely determined by q, and the result is sometimes called **Sylvester's law of inertia.**

THEOREM 2 Sylvester's Law of Inertia

Let $q = q(\mathbf{x}) = \mathbf{x}^TA\mathbf{x}$ be a quadratic form. Suppose Q is any invertible matrix such that the change of variables $\mathbf{x} = Q\mathbf{z}$ completely diagonalizes q.

$$q(\mathbf{x}) = q(Q\mathbf{z}) = z_1^2 + z_2^2 + \cdots + z_k^2 - z_{k+1}^2 - \cdots - z_r^2$$

The numbers k and r are uniquely determined by A. In fact, r is the rank of A, and k is the number of positive eigenvalues of A.

Proof The foregoing discussion shows that such a matrix Q exists with r and k as in the theorem. Suppose Q' is another such matrix and that

$$q(\mathbf{x}) = q(Q'\mathbf{z}) = z_1^2 + z_2^2 + \cdots + z_{k'}^2 - z_{k'+1}^2 - \cdots - z_{r'}^2$$

Then we are to show that $k = k'$ and $r = r'$. It follows (see Exercise 31 in Section 6.1) that Q'^TAQ' is diagonal with r' nonzero entries on the main diagonal (k' ones and $r' - k'$ minus ones). Hence $r' = \text{rank}(Q'^TAQ') = \text{rank } A$, using Corollary 4 of Theorem 2 in Section 5.4. Similarly, $r = \text{rank } A$ so $r' = r$.

To prove $k = k'$, let $\mathbf{q}_1, \mathbf{q}_2, \ldots, \mathbf{q}_n$ denote the columns of Q. They are a basis of $\mathbb{R}^n$ and

$$\mathbf{x} = Q\mathbf{z} = [\mathbf{q}_1\ \mathbf{q}_2 \ldots \mathbf{q}_n]\begin{bmatrix} z_1 \\ z_2 \\ \vdots \\ z_n \end{bmatrix} = z_1\mathbf{q}_1 + z_2\mathbf{q}_2 + \cdots + z_n\mathbf{q}_n$$

Now let $U = \text{span}\{\mathbf{q}_1, \mathbf{q}_2, \ldots, \mathbf{q}_k, \mathbf{q}_{r+1}, \ldots, \mathbf{q}_n\}$. Given $\mathbf{u}$ in U, write

$$\mathbf{u} = u_1\mathbf{q}_1 + \cdots + u_k\mathbf{q}_k + 0\mathbf{q}_{k+1} + \cdots + 0\mathbf{q}_r + u_{r+1}\mathbf{q}_{r+1} + \cdots + u_n\mathbf{q}_n$$

Then $q(\mathbf{u}) = u_1^2 + \cdots + u_k^2 \geq 0$. On the other hand, if $\mathbf{q}_1', \mathbf{q}_2', \ldots, \mathbf{q}_n'$ denote the columns of Q' and $V = \text{span}\{\mathbf{q}'_{k'+1}, \mathbf{q}'_{k'+2}, \ldots, \mathbf{q}'_{r'}\}$, a similar argument shows that $q(\mathbf{v}) < 0$ for all $\mathbf{v}$ in V, $\mathbf{v} \neq \mathbf{0}$. Hence U and V contain no common vector except $\mathbf{0}$. But then any basis of U, together with a basis of V, is a basis of the subspace $U + V = \{\mathbf{u} + \mathbf{v} \mid \mathbf{u} \text{ in } U, \mathbf{v} \text{ in } V\}$ (Exercise 21 in Section 5.3.3), so

$$\dim(U + V) = \dim U + \dim V = (k + n - r) + (r' - k') = k + n - k'$$

because $r = r'$. But $\dim(U + V) \leq n$ because $U + V$ is a subspace of $\mathbb{R}^n$, so $k + n - k' \leq n$. Hence $k \leq k'$, and a similar argument shows that $k' \leq k$. ■

The numbers k and r are called, respectively, the **index** and **rank** of the quadratic form q. The index can be described as the number of positive eigenvalues of the matrix A defining the quadratic form (also called the index of A). The number $2k - r$ of positive eigenvalues minus the number of negative ones is called the **signature** of q. The rank of q is just the rank of A. Clearly $k \leq r$, and the case $k = r$ occurs just when A has no negative eigenvalues. This means that $q(\mathbf{x}) \geq 0$ for all $\mathbf{x}$, and q is said to be a **positive** quadratic form in this case. The case $k = r = n$ occurs just when all eigenvalues are positive, and q is called **positive definite** in this case (as is the matrix A).

EXAMPLE 5 Find the index, rank, and signature of the quadratic form

$$q = x_2^2 + x_3^2 + 4x_1x_2 - 4x_1x_3$$

Solution We have $q(\mathbf{x}) = \mathbf{x}^T A\mathbf{x}$, where $A = \begin{bmatrix} 0 & 2 & -2 \\ 2 & 1 & 0 \\ -2 & 0 & 1 \end{bmatrix}$, and the characteristic polynomial of A is

$$c_A(x) = (x - 1)(x^2 - x - 8) = (x - 1)(x - \lambda_2)(x - \lambda_3)$$

where $\lambda_2 = \frac{1}{2}(\sqrt{33} + 1)$ and $\lambda_3 = -\frac{1}{2}(\sqrt{33} - 1)$. Hence the diagonalized form of q is

$$q = y_1^2 + \frac{1}{2}(\sqrt{33} + 1)y_2^2 - \frac{1}{2}(\sqrt{33} - 1)y_3^2$$

as given in Theorem 1. This means that the index is $k = 2$, the rank is $r = 3$, and the signature is $2k - r = 1$, so new variables z_1, z_2, and z_3 can be found such that $q = z_1^2 + z_2^2 - z_3^2$.

EXERCISES 7.4.1

1. In each case, find a symmetric matrix A such that $q = \mathbf{x}^T B\mathbf{x}$ takes the form $q = \mathbf{x}^T A\mathbf{x}$.

(a) $B = \begin{bmatrix} 1 & 1 \\ 0 & 1 \end{bmatrix}$ **(b)** $B = \begin{bmatrix} 1 & 1 \\ -1 & 2 \end{bmatrix}$

(c) $B = \begin{bmatrix} 1 & 0 & 1 \\ 1 & 1 & 0 \\ 0 & 1 & 1 \end{bmatrix}$ **(d)** $B = \begin{bmatrix} 1 & 2 & -1 \\ 4 & 1 & 0 \\ 5 & -2 & 3 \end{bmatrix}$

2. In each case, find a change of variables that will diagonalize the quadratic form q. Determine the index, rank, and signature of q.

(a) $q = x_1^2 + 2x_1x_2 + x_2^2$

(b) $q = x_1^2 + 4x_1x_2 + x_2^2$

(c) $q = x_1^2 + x_2^2 + x_3^2 - 4(x_1x_2 + x_1x_3 + x_2x_3)$

(d) $q = 7x_1^2 + x_2^2 + x_3^2 + 8x_1x_2 + 8x_1x_3 - 16x_2x_3$

(e) $q = 2(x_1^2 + x_2^2 + x_3^2 - x_1x_2 + x_1x_3 - x_2x_3)$

(f) $q = 5x_1^2 + 8x_2^2 + 5x_3^2 - 4(x_1x_2 + 2x_1x_3 + x_2x_3)$

(g) $q = x_1^2 - x_3^2 - 4x_1x_2 + 4x_2x_3$

(h) $q = x_1^2 + x_3^2 - 2x_1x_2 + 2x_2x_3$

3. For each of the following, write the equation in terms of new variables so that it is in standard position, and identify the curve.

(a) $xy = 1$

(b) $3x^2 - 4xy = 2$

(c) $6x^2 + 6xy - 2y^2 = 5$

(d) $2x^2 + 4xy + 5y^2 = 1$

4. Show that, if a quadratic form q is positive definite, there exists an inner product $\langle\ ,\ \rangle$ on $\mathbb{R}^n$ such that $q(\mathbf{x}) = \langle \mathbf{x},\mathbf{x} \rangle = \|\mathbf{x}\|^2$ for all $\mathbf{x}$.

5. A **bilinear form** β on a vector space V is a function that assigns to every pair $\mathbf{v},\mathbf{w}$ of vectors a number denoted $\beta(\mathbf{v},\mathbf{w})$ in such a way that

$$\beta(r\mathbf{v} + s\mathbf{w},\mathbf{u}) = r\beta(\mathbf{v},\mathbf{u}) + s\beta(\mathbf{w},\mathbf{u})$$
$$\beta(\mathbf{u},r\mathbf{v} + s\mathbf{w}) = r\beta(\mathbf{u},\mathbf{v}) + s\beta(\mathbf{u},\mathbf{w})$$

hold for all $\mathbf{v},\mathbf{w}$ and $\mathbf{u}$ in V and all r and s in $\mathbb{R}$. The form β is said to be **symmetric** if $\beta(\mathbf{v},\mathbf{w}) = \beta(\mathbf{w},\mathbf{v})$ holds for all $\mathbf{v},\mathbf{w}$ in V. It is called **positive** if $\beta(\mathbf{v},\mathbf{v}) \geq 0$ for all $\mathbf{v}$ in V and **positive definite** if $\beta(\mathbf{v},\mathbf{v}) > 0$ for all $\mathbf{v} \neq \mathbf{0}$ in V. Show that:

(a) Inner products are just positive definite bilinear forms.

(b) If β is any bilinear form on $\mathbb{R}^n$ (written as columns), there exists an $n \times n$ matrix A such that $\beta(\mathbf{x},\mathbf{y}) = \mathbf{x}^TA\mathbf{y}$ for all $\mathbf{x},\mathbf{y}$ in $\mathbb{R}^n$. Conversely, this defines a bilinear form for every matrix A. [*Hint:* Theorem 7 in Section 6.1.]

(c) The bilinear form in part (**b**) is symmetric (an inner product) if and only if the matrix A is symmetric (positive definite).

6. Two $n \times n$ matrices A and B are called **congruent** if $B = P^TAP$ holds for some invertible matrix P. Write $B \overset{c}{\sim} A$ in this case.

(a) Show that congruence is an equivalence relation:

(i) $A \overset{c}{\sim} A$ for all A.

(ii) If $B \overset{c}{\sim} A$, then $A \overset{c}{\sim} B$.

(iii) If $A \overset{c}{\sim} B$ and $B \overset{c}{\sim} C$, then $A \overset{c}{\sim} C$.

(b) Show that the following are congruence invariants.

(i) Being symmetric **(ii)** Being invertible **(iii)** Having rank r

(c) If $k \leq r \leq n$, let $D(k,r)$ denote the $n \times n$ diagonal matrix whose main diagonal consists of k ones, followed by $r - k$ minus ones, followed by $n - r$ zeros. If A is an $n \times n$ symmetric matrix, show that $A \overset{c}{\sim} D(k,r)$ if and only if it has index k and rank r. [*Hint:* If A has index k and rank r, a change of variables $\mathbf{x}^* = Q\mathbf{z}$ can be found such that $\mathbf{z}^T(Q^TAQ)\mathbf{z} = \mathbf{z}^TD(k,r)\mathbf{z}$ holds for all $\mathbf{z}$. Use Exercise 31 in Section 6.1. For the converse, use Theorem 2.]

(d) Show that two $n \times n$ symmetric matrices are congruent if and only if they have the same index and the same rank. [*Hint:* See part (**c**).]

Hence there is exactly one matrix $D(k,r)$ in each congruence class of symmetric matrices, so these $D(k,r)$ are canonical forms for symmetric matrices under the congruence relation. By virtue of part (**d**), index and rank are called a **complete set of congruence invariants** for symmetric matrices.

7. Given a symmetric matrix A, define $q_A(\mathbf{x}) = \mathbf{x}^TA\mathbf{x}$.

(a) If $A' = Q^TAQ$, show that A' is symmetric and $q_{A'}(\mathbf{x}) = q_A(Q\mathbf{x})$ for all $\mathbf{x}$.

(b) Show that $A' \overset{c}{\sim} A$ if and only if A' is symmetric and there is an invertible matrix Q such that $q_{A'}(\mathbf{x}) = q_A(Q\mathbf{x})$ for all $\mathbf{x}$. [*Hint:* See Exercise 31 in Section 6.1.]

7.4.2 Systems of Differential Equations[†]

Solving a wide variety of problems, particularly in science and engineering, comes down to solving a differential equation or a system of such equations. In this section, vector spaces and matrix multiplication will be used to describe systems of differential equations, and diagonalization will be used to solve such systems. Of course, our methods really are only a first step into the vast theory of differential equations, but, at least for linear systems of first-order differential equations, the techniques do solve the problem and provide a basis for further work.

If f is a function of a real variable x, and if f' and f'' denote the first and second derivatives of f, then equations of the form

$$f' + af = 0 \qquad f'' + af' + bf = 0 \qquad (a \text{ and } b \text{ numbers})$$

are called **differential equations of order 1 and 2,** respectively. One approach to such equations is to reduce those of order greater than one to *systems* of first-order equations. Then matrix diagonalization techniques can be applied. Consequently, the first task is to treat such systems.

The general problem is to find differentiable functions $f_1, f_2, \dots, f_n$ that satisfy a system of equations of the form

$$\begin{aligned} f_1' &= a_{11}f_1 + a_{12}f_2 + \cdots + a_{1n}f_n \\ f_2' &= a_{21}f_1 + a_{22}f_2 + \cdots + a_{2n}f_n \\ \vdots \quad & \quad \vdots \\ f_n' &= a_{n1}f_1 + a_{n2}f_2 + \cdots + a_{nn}f_n \end{aligned}$$

where the a_{ij} are constants. This is called a **linear system of differential equations.** The first step is to put it in matrix form. Write

$$\mathbf{f} = \begin{bmatrix} f_1 \\ f_2 \\ \vdots \\ f_n \end{bmatrix}, \mathbf{f}' = \begin{bmatrix} f_1' \\ f_2' \\ \vdots \\ f_n' \end{bmatrix} \qquad A = \begin{bmatrix} a_{11} & a_{12} & \dots & a_{1n} \\ a_{21} & a_{22} & \dots & a_{2n} \\ \vdots & \vdots & & \vdots \\ a_{n1} & a_{n2} & \dots & a_{nn} \end{bmatrix}$$

Then the system can be written compactly as

$$\mathbf{f}' = A\mathbf{f}$$

and, given the matrix A, the problem is to find a column $\mathbf{f}$ of differentiable functions that satisfies this condition.

[†]This section requires calculus.

Linear algebra enters into this as follows: These columns of functions become a vector space if "matrix" addition and scalar multiplication are used.

$$\begin{bmatrix} f_1 \\ f_2 \\ \vdots \\ f_n \end{bmatrix} + \begin{bmatrix} g_1 \\ g_2 \\ \vdots \\ g_n \end{bmatrix} = \begin{bmatrix} f_1 + g_1 \\ f_2 + g_2 \\ \vdots \\ f_n + g_n \end{bmatrix} \qquad a\begin{bmatrix} f_1 \\ f_2 \\ \vdots \\ f_n \end{bmatrix} = \begin{bmatrix} af_1 \\ af_2 \\ \vdots \\ af_n \end{bmatrix}$$

Of course, addition $f_i + g_i$ and scalar multiplication af_i of the individual functions are defined pointwise (as in Example 7 in Section 5.1). That is, the actions of $f_i + g_i$ and af_i are given by

$$(f_i + g_i)(x) = f_i(x) + g_i(x) \qquad \text{and} \qquad (af_i)(x) = af_i(x)$$

for all x. With these definitions, the set V of all n-columns of functions becomes a vector space, as the reader can verify. The zero vector and the negative of a vector are

$$\mathbf{0} = \begin{bmatrix} 0 \\ 0 \\ \vdots \\ 0 \end{bmatrix} \qquad -\begin{bmatrix} f_1 \\ f_2 \\ \vdots \\ f_n \end{bmatrix} = \begin{bmatrix} -f_1 \\ -f_2 \\ \vdots \\ -f_n \end{bmatrix}$$

just as for matrices. This vector space will be denoted F^n.

Our concern here is not for F^n but for those columns of functions in it that satisfy the linear system

$$U = \{\mathbf{f} \mid \mathbf{f} \text{ lies in } F^n \text{ and } \mathbf{f}' = A\mathbf{f}\}$$

Now recall the following three basic facts about differentiation:

$$\mathbf{0}' = \mathbf{0} \qquad (\mathbf{f} + \mathbf{g})' = \mathbf{f}' + \mathbf{g}' \qquad (c\mathbf{f})' = c\mathbf{f}'$$

Hence U is a subspace of F^n. The problem now is to find a convenient basis for U.

The case $n = 1$ has been discussed earlier. The problem is to find all functions f satisfying

$$f' = af \qquad a \text{ a constant}$$

and it follows from Theorem 4 in Section 5.6 that the space of solutions has dimension 1 and that $\{e^{ax}\}$ is a basis. If the matrix A is diagonalizable, this case can be used in the general situation. The following example provides an illustration.

EXAMPLE 6 Find a solution to the system

$$\begin{aligned} f_1' &= f_1 + 3f_2 \\ f_2' &= 2f_1 + 2f_2 \end{aligned}$$

that satisfies $f_1(0) = 0, f_2(0) = 5$.

Solution This is $\mathbf{f}' = A\mathbf{f}$, where $\mathbf{f} = \begin{bmatrix} f_1 \\ f_2 \end{bmatrix}$ and $A = \begin{bmatrix} 1 & 3 \\ 2 & 2 \end{bmatrix}$. The reader can verify that $c_A(x) = (x - 4)(x + 1)$, and $\mathbf{p}_1 = \begin{bmatrix} 1 \\ 1 \end{bmatrix}$ and $\mathbf{p}_2 = \begin{bmatrix} 3 \\ -2 \end{bmatrix}$ are eigenvectors corresponding to the eigenvalues 4 and -1, respectively. Hence the diagonalization algorithm gives $P^{-1}AP = \begin{bmatrix} 4 & 0 \\ 0 & -1 \end{bmatrix}$, where $P = [\mathbf{p}_1\ \mathbf{p}_2] = \begin{bmatrix} 1 & 3 \\ 1 & -2 \end{bmatrix}$. Now consider new functions g_1 and g_2 given by $\mathbf{f} = P\mathbf{g}$ (equivalently, $\mathbf{g} = P^{-1}\mathbf{f}$):

$$\begin{bmatrix} f_1 \\ f_2 \end{bmatrix} = \begin{bmatrix} 1 & 3 \\ 1 & -2 \end{bmatrix} \begin{bmatrix} g_1 \\ g_2 \end{bmatrix} \qquad \text{that is,} \qquad \begin{aligned} f_1 &= g_1 + 3g_2 \\ f_2 &= g_1 - 2g_2 \end{aligned}$$

Then $f_1' = g_1' + 3g_2'$ and $f_2' = g_1' - 2g_2'$ so that

$$\mathbf{f}' = \begin{bmatrix} f_1' \\ f_2' \end{bmatrix} = \begin{bmatrix} 1 & 3 \\ 1 & -2 \end{bmatrix} \begin{bmatrix} g_1' \\ g_2' \end{bmatrix} = P\mathbf{g}'$$

If this is substituted in $\mathbf{f}' = A\mathbf{f}$, the result is $P\mathbf{g}' = AP\mathbf{g}$, whence

$$\mathbf{g}' = P^{-1}AP\mathbf{g}$$

But this means that

$$\begin{bmatrix} g_1' \\ g_2' \end{bmatrix} = \begin{bmatrix} 4 & 0 \\ 0 & -1 \end{bmatrix} \begin{bmatrix} g_1 \\ g_2 \end{bmatrix} \qquad \text{so} \qquad \begin{aligned} g_1' &= 4g_1 \\ g_2' &= -g_2 \end{aligned}$$

Then the case $n = 1$ gives $g_1(x) = ce^{4x}$, $g_2(x) = de^{-x}$, where c and d are constants. Finally, then,

$$\begin{bmatrix} f_1(x) \\ f_2(x) \end{bmatrix} = P\begin{bmatrix} g_1(x) \\ g_2(x) \end{bmatrix} = \begin{bmatrix} 1 & 3 \\ 1 & -2 \end{bmatrix} \begin{bmatrix} ce^{4x} \\ de^{-x} \end{bmatrix} = \begin{bmatrix} ce^{4x} + 3de^{-x} \\ ce^{4x} - 2de^{-x} \end{bmatrix}$$

so the *general solution* is

$$\begin{aligned} f_1(x) &= ce^{4x} + 3de^{-x} \\ f_2(x) &= ce^{4x} - 2de^{-x} \end{aligned} \qquad c \text{ and } d \text{ constants}$$

It is worth observing that this can be written in matrix form as

$$\begin{bmatrix} f_1(x) \\ f_2(x) \end{bmatrix} = c\begin{bmatrix} 1 \\ 1 \end{bmatrix} e^{4x} + d\begin{bmatrix} 3 \\ -2 \end{bmatrix} e^{-x}$$

$$\mathbf{f}(x) = c\mathbf{p}_1 e^{4x} + d\mathbf{p}_2 e^{-x}$$

This form of the solution works more generally, as will be shown below.

Finally, the requirement in this example that $f_1(0) = 0$ and $f_2(0) = 5$ determines the constants c and d:

$$0 = f_1(0) = ce^0 + 3de^0 = c + 3d$$
$$5 = f_2(0) = ce^0 - 2de^0 = c - 2d$$

These equations give $c = 3$ and $d = -1$, so

$$f_1(x) = 3e^{4x} - 3e^{-x}$$
$$f_2(x) = 3e^{4x} + 2e^{-x}$$

satisfy all the requirements.

The technique of this example works in general.

THEOREM 3 Consider a linear system

$$\mathbf{f}' = A\mathbf{f}$$

of differential equations, where A is an $n \times n$ diagonalizable matrix. Let $P^{-1}AP$ be diagonal, where P is given in terms of its columns

$$P = [\mathbf{p}_1\ \mathbf{p}_2 \cdots \mathbf{p}_n]$$

and $\{\mathbf{p}_1, \mathbf{p}_2, \ldots, \mathbf{p}_n\}$ are independent eigenvectors of A. If $\mathbf{p}_i$ corresponds to the eigenvalue λ_i for each i, then

$$\{\mathbf{p}_1 e^{\lambda_1 x}, \mathbf{p}_2 e^{\lambda_2 x}, \ldots, \mathbf{p}_n e^{\lambda_n x}\}$$

is a basis for the space of solutions of $\mathbf{f}' = A\mathbf{f}$.

Proof Such $\mathbf{p}_i$ exist by virtue of the assumption that A is diagonalizable, and their independence guarantees that P is invertible. Let

$$P^{-1}AP = \begin{bmatrix} \lambda_1 & 0 & \dots & 0 \\ 0 & \lambda_2 & \dots & 0 \\ \vdots & \vdots & & \vdots \\ 0 & 0 & \dots & \lambda_n \end{bmatrix}$$

where $\lambda_1, \lambda_2, \ldots, \lambda_n$ are the (not necessarily distinct) eigenvalues of

A. As in the example, define a new column of functions $\mathbf{g} = \begin{bmatrix} g_1 \\ g_2 \\ \vdots \\ g_n \end{bmatrix}$ by $\mathbf{g} = P^{-1}\mathbf{f}$; equivalently, $\mathbf{f} = P\mathbf{g}$. If f_i is the ith component of $\mathbf{f}$ and $P = [p_{ij}]$, this gives

$$f_i = p_{i1}g_1 + p_{i2}g_2 + \cdots + p_{in}g_n$$

Differentiation preserves this relationship:

$$f_i' = p_{i1}g_1' + p_{i2}g_2' + \cdots + p_{in}g_n'$$

so $\mathbf{f}' = P\mathbf{g}'$. Substituting this into $\mathbf{f}' = A\mathbf{f}$ gives $P\mathbf{g}' = AP\mathbf{g}$. But then multiplication by P^{-1} gives $\mathbf{g}' = P^{-1}AP\mathbf{g}$, so the original system of equations for $\mathbf{f}$ becomes much simpler in terms of $\mathbf{g}$.

$$\begin{bmatrix} g_1' \\ g_2' \\ \vdots \\ g_n' \end{bmatrix} = \begin{bmatrix} \lambda_1 & 0 & \dots & 0 \\ 0 & \lambda_2 & \dots & 0 \\ \vdots & \vdots & & \vdots \\ 0 & 0 & \dots & \lambda_n \end{bmatrix} \begin{bmatrix} g_1 \\ g_2 \\ \vdots \\ g_n \end{bmatrix}$$

Hence $g_i' = \lambda_i g_i$ holds for each i, and Theorem 4 in Section 5.6 implies that the only solutions are

$$g_i(x) = c_i e^{\lambda_i x} \qquad c_i \text{ some constant}$$

Then the relationship $\mathbf{f} = P\mathbf{g}$ gives the functions $f_1, f_2, \dots, f_n$ as follows:

$$\mathbf{f}(x) = [\mathbf{p}_1\ \mathbf{p}_2 \dots \mathbf{p}_n] \begin{bmatrix} c_1 e^{\lambda_1 x} \\ c_2 e^{\lambda_2 x} \\ \vdots \\ c_n e^{\lambda_n x} \end{bmatrix} = c_1\mathbf{p}_1 e^{\lambda_1 x} + c_2\mathbf{p}_2 e^{\lambda_2 x} + \cdots + c_n\mathbf{p}_n e^{\lambda_n x}$$

Hence the columns $\{\mathbf{p}_1 e^{\lambda_1 x}, \mathbf{p}_2 e^{\lambda_2 x}, \dots, \mathbf{p}_n e^{\lambda_n x}\}$ span the space U of solutions. They are independent because, if

$$c_1 e^{\lambda_1 x}\mathbf{p}_1 + c_2 e^{\lambda_2 x}\mathbf{p}_2 + \cdots + c_n e^{\lambda_n x}\mathbf{p}_n = \mathbf{0}$$

in F^n, then the left side vanishes for all x. In particular, taking $x = 0$ gives $c_1\mathbf{p}_1 + \cdots + c_n\mathbf{p}_n = \mathbf{0}$, so the independence of the $\mathbf{p}_i$ gives $c_1 = \cdots = c_n = 0$. ■

The theorem shows that *every* solution to $\mathbf{f}' = A\mathbf{f}$ is a linear combination

$$\mathbf{f}(x) = c_1\mathbf{p}_1 e^{\lambda_1 x} + c_2\mathbf{p}_2 e^{\lambda_2 x} + \cdots + c_n\mathbf{p}_n e^{\lambda_n x}$$

where the coefficients c_i are arbitrary. Hence this is called the **general**

solution to the system of differential equations. In most cases the solution functions $f_i(x)$ are required to satisfy boundary conditions, often of the form $f_i(a) = b_i$, where $a, b_1, \ldots, b_n$ are prescribed numbers. These conditions determine the constants c_i. The following example illustrates this and displays a situation where one eigenvalue has multiplicity greater than 1.

EXAMPLE 7 Find the general solution to the system

$$\begin{aligned} f_1' &= 5f_1 + 8f_2 + 16f_3 \\ f_2' &= 4f_1 + f_2 + 8f_3 \\ f_3' &= -4f_1 - 4f_2 - 11f_3 \end{aligned}$$

Then find a solution satisfying the boundary conditions $f_1(0) = f_2(0) = f_3(0) = 1$.

Solution The system has the form $\mathbf{f}' = A\mathbf{f}$, where $A = \begin{bmatrix} 5 & 8 & 16 \\ 4 & 1 & 8 \\ -4 & -4 & -11 \end{bmatrix}$. This matrix was considered in Example 1 in Section 7.2, where it was found that $c_A(x) = (x + 3)^2(x - 1)$ and that independent eigenvectors corresponding to the eigenvalues -3, -3, and 1 are, respectively,

$$\mathbf{p}_1 = \begin{bmatrix} -1 \\ 1 \\ 0 \end{bmatrix} \qquad \mathbf{p}_2 = \begin{bmatrix} -2 \\ 0 \\ 1 \end{bmatrix} \qquad \mathbf{p}_3 = \begin{bmatrix} 2 \\ 1 \\ -1 \end{bmatrix}$$

Hence $\{\mathbf{p}_1e^{-3x}, \mathbf{p}_2e^{-3x}, \mathbf{p}_3e^{x}\}$ spans the space of solutions, so the general solution is

$$\mathbf{f}(x) = c_1\begin{bmatrix} -1 \\ 1 \\ 0 \end{bmatrix}e^{-3x} + c_2\begin{bmatrix} -2 \\ 0 \\ 1 \end{bmatrix}e^{-3x} + c_3\begin{bmatrix} 2 \\ 1 \\ -1 \end{bmatrix}e^{x} \qquad c_i \text{ constants}$$

The boundary conditions $f_1(0) = f_2(0) = f_3(0) = 1$ determine the constants c_i.

$$\begin{aligned} \begin{bmatrix} 1 \\ 1 \\ 1 \end{bmatrix} = \mathbf{f}(0) &= c_1\begin{bmatrix} -1 \\ 1 \\ 0 \end{bmatrix} + c_2\begin{bmatrix} -2 \\ 0 \\ 1 \end{bmatrix} + c_3\begin{bmatrix} 2 \\ 1 \\ -1 \end{bmatrix} \\ &= \begin{bmatrix} -1 & -2 & 2 \\ 1 & 0 & 1 \\ 0 & 1 & -1 \end{bmatrix}\begin{bmatrix} c_1 \\ c_2 \\ c_3 \end{bmatrix} \end{aligned}$$

The solution is $c_1 = -3$, $c_2 = 5$, $c_3 = 4$, so the required specific solution is

$$\begin{aligned} f_1(x) &= -7e^{-3x} + 8e^x \\ f_2(x) &= -3e^{-3x} + 4e^x \\ f_3(x) &= 5e^{-3x} - 4e^x \end{aligned}$$

The foregoing analysis fails if A is not diagonalizable, a situation that will not be treated in this book.

EXERCISES 7.4.2

1. Use Theorem 3 to find the general solution to each of the following systems. Then find a specific solution satisfying the given boundary condition.

(a) $f_1' = 2f_1 + 4f_2$, $f_1(0) = 0$
$f_2' = 3f_1 + 3f_2$, $f_2(0) = 1$

(b) $f_1' = -f_1 + 5f_2$, $f_1(0) = 1$
$f_2' = f_1 + 3f_2$, $f_2(0) = -1$

(c) $f_1' = 4f_2 + 4f_3$
$f_2' = f_1 + f_2 - 2f_3$
$f_3' = -f_1 + f_2 + 4f_3$
$f_1(0) = f_2(0) = f_3(0) = 1$

(d) $f_1' = 2f_1 + f_2 + 2f_3$
$f_2' = 2f_1 + 2f_2 - 2f_3$
$f_3' = 3f_1 + f_2 + f_3$
$f_1(0) = f_2(0) = f_3(0) = 1$

2. **(a)** Show that $e^{\lambda x}$, $e^{\mu x}$, and $e^{\delta x}$ are linearly independent functions if λ, μ, and δ are distinct. [*Hint:* If $re^{\lambda x} + se^{\mu x} + te^{\delta x} = 0$, differentiate twice and use Theorem 2 in Section 3.4.]

(b) Generalize part **(a)** to $\{e^{\lambda_1 x}, e^{\lambda_2 x}, \ldots, e^{\lambda_n x}\}$.

8 Linear Transformations

SECTION 8.1 Basic Properties

8.1.1 Examples and Elementary Properties

Much of mathematics is concerned with the study of functions. Polynomial functions such as $p(x) = 3x^2 - 5x + 1$ come up in a wide variety of situations. Functions such as the exponential function e^x, the logarithm $\ln x$, and the trigonometric functions $\sin x$ and $\cos x$ play a fundamental role in calculus as well as in other areas of mathematics. If X and Y are sets, a **function** f from X to Y (written $f : X \to Y$ or $X \xrightarrow{f} Y$) is a rule that associates with every element x of X an element $f(x)$ of Y. In all these examples $X = \mathbb{R}$ and $Y = \mathbb{R}$, but we shall be considering functions where X and Y are both vector spaces.

DEFINITION If V and W are two vector spaces, a function $T : V \to W$ is called a **linear transformation** if it satisfies the following axioms.

T1. $T(\mathbf{v} + \mathbf{v}_1) = T(\mathbf{v}) + T(\mathbf{v}_1)$ for all $\mathbf{v}$ and $\mathbf{v}_1$ in V

T2. $T(r\mathbf{v}) = rT(\mathbf{v})$ for all $\mathbf{v}$ in V and all r in $\mathbb{R}$

Axiom T1 is just the requirement that T *preserves* vector addition. It asserts that the result $T(\mathbf{v} + \mathbf{v}_1)$ of adding $\mathbf{v}$ and $\mathbf{v}_1$ first and then applying T is the same as applying T first to get $T(\mathbf{v})$ and $T(\mathbf{v}_1)$ and then adding. In other words, the operations of adding and applying T can be done in either order, and the result will be the same if T is a linear transformation. Similarly, axiom T2 means that T preserves scalar multiplication: The operations of applying T and multiplying by r can be done in either order.

Note that, even though the additions in axiom T1 are both denoted by the same symbol $+$, the addition on the left forming $\mathbf{v} + \mathbf{v}_1$ is carried out in V, whereas the addition $T(\mathbf{v}) + T(\mathbf{v}_1)$ is done in W. Similarly, the scalar multiplications $r\mathbf{v}$ and $rT(\mathbf{v})$ in axiom T2 refer to the spaces V and W, respectively. We shall continue to use this ambiguous notation. It is much simpler, and it causes no confusion because it is always clear from the context which space is involved.

Linear transformations occur in a wide variety of situations, as the following examples indicate.

EXAMPLE 1 Define a function $T : \mathbb{R}^2 \to \mathbb{R}^3$ by $T\begin{bmatrix} x \\ y \end{bmatrix} = \begin{bmatrix} x + y \\ x - 2y \\ 3x \end{bmatrix}$ for all $\begin{bmatrix} x \\ y \end{bmatrix}$ in $\mathbb{R}^2$. Show that T is a linear transformation.

Solution We verify the axioms. Given $\begin{bmatrix} x \\ y \end{bmatrix}$ and $\begin{bmatrix} x_1 \\ y_1 \end{bmatrix}$ in $\mathbb{R}^2$, compute

$$T\left(\begin{bmatrix} x \\ y \end{bmatrix} + \begin{bmatrix} x_1 \\ y_1 \end{bmatrix}\right) = T\begin{bmatrix} x + x_1 \\ y + y_1 \end{bmatrix} = \begin{bmatrix} (x + x_1) + (y + y_1) \\ (x + x_1) - 2(y + y_1) \\ 3(x + x_1) \end{bmatrix}$$

$$= \begin{bmatrix} x + y \\ x - 2y \\ 3x \end{bmatrix} + \begin{bmatrix} x_1 + y_1 \\ x_1 - 2y_1 \\ 3x_1 \end{bmatrix} = T\begin{bmatrix} x \\ y \end{bmatrix} + T\begin{bmatrix} x_1 \\ y_1 \end{bmatrix}$$

This proves axiom T1, and axiom T2 is proved as follows:

$$T\left(r\begin{bmatrix} x \\ y \end{bmatrix}\right) = T\begin{bmatrix} rx \\ ry \end{bmatrix} = \begin{bmatrix} rx + ry \\ rx - 2ry \\ 3rx \end{bmatrix} = r\begin{bmatrix} x + y \\ x - 2y \\ 3x \end{bmatrix} = rT\begin{bmatrix} x \\ y \end{bmatrix}$$

Hence T is a linear transformation.

The next example is the prototype of all linear transformations $\mathbb{R}^n \to \mathbb{R}^m$.

EXAMPLE 2 If A is any $m \times n$ matrix, define a function $T_A : \mathbb{R}^n \to \mathbb{R}^m$ by $T_A(\mathbf{v}) = A\mathbf{v}$ for all columns $\mathbf{v}$ in $\mathbb{R}^n$. Show that T_A is a linear transformation.

Solution Given $\mathbf{v}$ and $\mathbf{v}_1$ in $\mathbb{R}^n$ and r in $\mathbb{R}$, matrix arithmetic yields

$$T_A(\mathbf{v} + \mathbf{v}_1) = A(\mathbf{v} + \mathbf{v}_1) = A\mathbf{v} + A\mathbf{v}_1 = T_A(\mathbf{v}) + T_A(\mathbf{v}_1)$$

$$T_A(r\mathbf{v}) = A(r\mathbf{v}) = r(A\mathbf{v}) = rT_A(\mathbf{v})$$

Hence T_A is a linear transformation.

Given an $m \times n$ matrix A, the transformation $T_A : \mathbb{R}^n \to \mathbb{R}^m$ in Example 2 will be called the **matrix transformation** corresponding to

A. It is important for two reasons. First, *every* transformation from $\mathbb{R}^n$ to $\mathbb{R}^m$ arises in this way from an $m \times n$ matrix (this is Theorem 4 below). Second, we achieve a useful perspective on matrices when we view them as linear transformations in this way.

The next example lists three important linear transformations that will be referred to later. The verification of axioms T1 and T2 is left to the reader.

EXAMPLE 3 If V and W are vector spaces, the following are linear transformations:

Identity transformation on V	$1_V : V \to V$ where $1_V(\mathbf{v}) = \mathbf{v}$ for all $\mathbf{v}$ in V
Zero transformation V → W	$0 \ : V \to W$ where $0(\mathbf{v}) = \mathbf{0}$ for all $\mathbf{v}$ in V
Scalar transformation V → V	$a \ : V \to V$ where $a(\mathbf{v}) = a\mathbf{v}$ for all $\mathbf{v}$ in V

(Here a is any real number.)

The symbol 0 will be used to denote the zero transformation from V to W for *any* spaces V and W. It was also used earlier (Example 8 in Section 5.1) to denote the zero function $[a, b] \to \mathbb{R}$.

The next example gives two important transformations of matrices. Recall that the trace tr A of an $n \times n$ matrix A is the sum of the entries on the main diagonal.

EXAMPLE 4 Transposition and trace are linear transformations. More precisely,

$$T : \mathbf{M}_{mn} \to \mathbf{M}_{nm} \qquad \text{where } T(A) = A^T \text{ for all } A \text{ in } \mathbf{M}_{mn}$$

$$T : \mathbf{M}_{nn} \to \mathbb{R} \qquad \text{where } T(A) = \operatorname{tr} A \text{ for all } A \text{ in } \mathbf{M}_{nn}$$

are both linear transformations.

Solution Axioms T1 and T2 for transposition are $(A + B)^T = A^T + B^T$ and $(rA)^T = r(A^T)$, respectively. The verifications for trace are left to the reader.

Many important geometric transformations are linear. As before, identify the vector $\mathbf{v} = (x,y)$ in $\mathbb{R}^2$ with the geometric vector from the origin to the point (x,y) in the plane. Then the usual vector addition in $\mathbb{R}^2$ corresponds to the parallelogram law of addition for the geometric vectors, and a similar correspondence holds for scalar multiplication.

EXAMPLE 5 Let $R_\theta : \mathbb{R}^2 \to \mathbb{R}^2$ denote a counterclockwise rotation of the plane about the origin through an angle θ, and let $S : \mathbb{R}^2 \to \mathbb{R}^2$ denote a reflection in a line through the origin. Show that R_θ and S are linear transformations.

Solution Given $\mathbf{v} = (x,y)$ and $\mathbf{w} = (x',y')$, they are represented in the diagram as geometric vectors (arrows) emanating from the origin, and

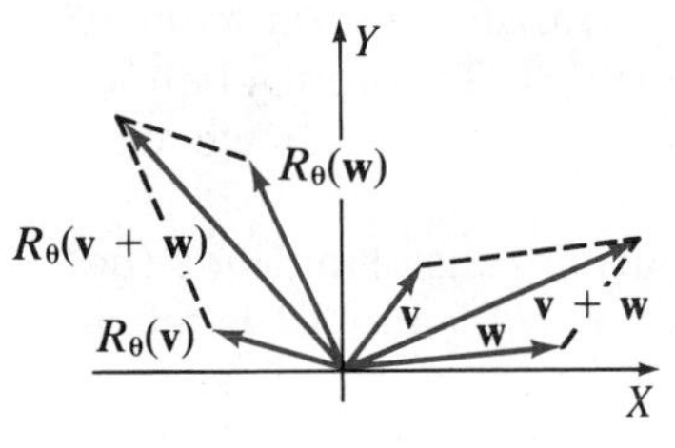

their sum $\mathbf{v} + \mathbf{w}$ is the diagonal of the parallelogram they determine. The effect of R_θ is as follows: $R_\theta(\mathbf{v})$ is the vector obtained by rotating $\mathbf{v}$ counterclockwise about the origin through an angle θ. Hence the vectors $R_\theta(\mathbf{v})$, $R_\theta(\mathbf{w})$, and $R_\theta(\mathbf{v} + \mathbf{w})$ can be obtained simultaneously by rotating the *entire parallelogram*. The result is the parallelogram determined by $R_\theta(\mathbf{v})$ and $R_\theta(\mathbf{w})$, and the diagonal is $R_\theta(\mathbf{v} + \mathbf{w})$. But the diagonal is $R_\theta(\mathbf{v}) + R_\theta(\mathbf{w})$ by the parallelogram law, so it follows that $R_\theta(\mathbf{v} + \mathbf{w}) = R_\theta(\mathbf{v}) + R_\theta(\mathbf{w})$. A similar argument shows that $R_\theta(r\mathbf{v}) = rR_\theta(\mathbf{v})$, so R_θ is linear.

The proof for S is analogous, because S flips the entire parallelogram over the line of reflection.

As in Example 5, rotations about lines through the origin and reflections in planes through the origin can be seen to be linear transformations $\mathbb{R}^3 \to \mathbb{R}^3$. Other geometric linear transformations are given in Exercise 12.

The projections we studied in Chapter 4 are also linear transformations, and the next example shows this for any inner product space.

EXAMPLE 6 Let U be a finite dimensional subspace of an inner product space V. Then projection on U is a linear transformation. More precisely, T is a linear transformation, where

$$T : V \to U \text{ is defined by } T(\mathbf{v}) = \text{proj}_U(\mathbf{v}) \text{ for all } \mathbf{v} \text{ in } V$$

Solution Let $\{\mathbf{e}_1, \mathbf{e}_2, \ldots, \mathbf{e}_m\}$ be an orthonormal basis of U. Then Theorem 9 in Section 6.2 gives

$$T(\mathbf{v}) = \langle \mathbf{v},\mathbf{e}_1\rangle \mathbf{e}_1 + \langle \mathbf{v},\mathbf{e}_2\rangle \mathbf{e}_2 + \cdots + \langle \mathbf{v},\mathbf{e}_m\rangle \mathbf{e}_m$$

The result now follows from $\langle \mathbf{u} + \mathbf{v},\mathbf{e}_i\rangle = \langle \mathbf{u},\mathbf{e}_i\rangle + \langle \mathbf{v},\mathbf{e}_i\rangle$ and $\langle r\mathbf{v},\mathbf{e}_i\rangle = r\langle \mathbf{v},\mathbf{e}_i\rangle$ for each i.

The next example involves some calculus.

EXAMPLE 7 The differentiation and integration operations are linear transformations. More precisely,

$$D : \mathbf{P}_n \to \mathbf{P}_{n-1} \quad \text{where } D[p(x)] = p'(x) \text{ for all } p(x) \text{ in } \mathbf{P}_n$$

$$I : \mathbf{P}_n \to \mathbf{P}_{n+1} \quad \text{where } I[p(x)] = \int_0^x p(t)\,dt \text{ for all } p(x) \text{ in } \mathbf{P}_n$$

are linear transformations.

Solution These restate the following fundamental properties of differentiation and integration.

$$[p(x) + q(x)]' = p'(x) + q'(x) \quad \text{and} \quad [rp(x)]' = rp'(x)$$

$$\int_0^x [p(t) + q(t)]\, dt = \int_0^x p(t)\, dt + \int_0^x q(t)\, dt \quad \text{and}$$

$$\int_0^x rp(t)\, dt = r\int_0^x p(t)\, dt$$

The list of examples could go on and on; linear transformations are very common. Because of this it may come as a surprise that *all* linear transformations have so many properties in common (in addition to the axioms). The next theorem collects three useful properties of this type. They may be described by saying that in addition to preserving addition and scalar multiplication (these are the axioms), linear transformations preserve the zero vector, negatives, and linear combinations.

THEOREM 1

Let $T : V \to W$ be a linear transformation.

(1) $T(\mathbf{0}) = \mathbf{0}$

(2) $T(-\mathbf{v}) = -T(\mathbf{v})$ for all $\mathbf{v}$ in V

(3) $T(r_1\mathbf{v}_1 + r_2\mathbf{v}_2 + \cdots + r_k\mathbf{v}_k) = r_1T(\mathbf{v}_1) + r_2T(\mathbf{v}_2) + \cdots + r_kT(\mathbf{v}_k)$ for all $\mathbf{v}_i$ in V and all r_i in $\mathbb{R}$

Proof Given any $\mathbf{v}$ in V,

(1) $T(\mathbf{0}) = T(0\mathbf{v}) = 0T(\mathbf{v}) = \mathbf{0}$

(2) $T(-\mathbf{v}) = T[(-1)\mathbf{v}] = (-1)T(\mathbf{v}) = -T(\mathbf{v})$

(3) If $k = 1$, this is $T(r_1\mathbf{v}_1) = r_1T(\mathbf{v}_1)$ by axiom T2. The general result is proved by induction on k: If it holds for a particular $k \geq 1$, then, using axiom T1 and the induction assumption,

$$\begin{aligned} T(r_1\mathbf{v}_1 + \cdots + r_k\mathbf{v}_k + r_{k+1}\mathbf{v}_{k+1}) &= T(r_1\mathbf{v}_1 + \cdots + r_k\mathbf{v}_k) + T(r_{k+1}\mathbf{v}_{k+1}) \\ &= r_1T(\mathbf{v}_1) + \cdots + r_kT(\mathbf{v}_k) + r_{k+1}T(\mathbf{v}_{k+1}) \end{aligned}$$

This completes the induction and so proves property (3). ■

The ability to use the last part of Theorem 1 effectively is vital to any facility with linear transformations. The next two examples provide illustrations.

EXAMPLE 8

Let $T : V \to W$ be a linear transformation. If $T(\mathbf{v} - 3\mathbf{v}_1) = \mathbf{w}$ and $T(2\mathbf{v} - \mathbf{v}_1) = \mathbf{w}_1$, find $T(\mathbf{v})$ and $T(\mathbf{v}_1)$ in terms of $\mathbf{w}$ and $\mathbf{w}_1$.

Solution The given relations imply that

$$T(\mathbf{v}) - 3T(\mathbf{v}_1) = \mathbf{w}$$

$$2T(\mathbf{v}) - T(\mathbf{v}_1) = \mathbf{w}_1$$

by Theorem 1. Subtracting twice the first from the second gives $T(\mathbf{v}_1) = \frac{1}{5}(\mathbf{w}_1 - 2\mathbf{w})$. Then substitution gives $T(\mathbf{v}) = \frac{1}{5}(3\mathbf{w}_1 - \mathbf{w})$.

EXAMPLE 9 If $T : \mathbb{R}^3 \to \mathbb{R}$ is a linear transformation and $T(3,-1,2) = 5$, $T(1,0,1) = 2$, compute $T(-1,1,0)$.

Solution This can be done by Theorem 1, provided that $(-1,1,0)$ can be expressed as a linear combination of $(3,-1,2)$ and $(1,0,1)$. This is indeed possible: $(-1,1,0) = -(3,-1,2) + 2(1,0,1)$, so

$$T(-1,1,0) = -T(3,-1,2) + 2T(1,0,1) = -5 + 4 = -1$$

The full effect of property (3) of Theorem 1 is this: If $T : V \to W$ is a linear transformation and $T(\mathbf{v}_1), T(\mathbf{v}_2), \ldots, T(\mathbf{v}_n)$ are known, then $T(\mathbf{v})$ can be computed for *every* $\mathbf{v}$ in $\text{span}\{\mathbf{v}_1, \mathbf{v}_2, \ldots, \mathbf{v}_n\}$. In particular, if $\{\mathbf{v}_1, \mathbf{v}_2, \ldots, \mathbf{v}_n\}$ spans V, then $T(\mathbf{v})$ is determined for all $\mathbf{v}$ in V by $T(\mathbf{v}_1), T(\mathbf{v}_2), \ldots, T(\mathbf{v}_n)$. The next theorem states this somewhat differently. As for functions in general, two linear transformations $T : V \to W$ and $S : V \to W$ are called **equal** (written $T = S$) if they have the same **action**; that is, $T(\mathbf{v}) = S(\mathbf{v})$ for all $\mathbf{v}$ in V.

THEOREM 2 Let $T : V \to W$ and $S : V \to W$ be two linear transformations. Suppose that $V = \text{span}\{\mathbf{v}_1, \mathbf{v}_2, \ldots, \mathbf{v}_n\}$. If $T(\mathbf{v}_i) = S(\mathbf{v}_i)$ for each i, then $S = T$.

Proof Given $\mathbf{v}$ in V, write $\mathbf{v} = t_1\mathbf{v}_1 + t_2\mathbf{v}_2 + \cdots + t_n\mathbf{v}_n$. Then Theorem 1 gives

$$\begin{aligned} T(\mathbf{v}) &= t_1T(\mathbf{v}_1) + t_2T(\mathbf{v}_2) + \cdots + t_nT(\mathbf{v}_n) \\ &= t_1S(\mathbf{v}_1) + t_2S(\mathbf{v}_2) + \cdots + t_nS(\mathbf{v}_n) \\ &= S(\mathbf{v}) \end{aligned}$$

Hence S and T have the same action, so $S = T$. ■

EXAMPLE 10 Let $V = \text{span}\{\mathbf{v}_1, \ldots, \mathbf{v}_n\}$. If $T : V \to W$ is a linear transformation and $T(\mathbf{v}_1) = \cdots = T(\mathbf{v}_n) = \mathbf{0}$, then $T = 0$, the zero transformation from V to W.

Solution The zero transformation $0 : V \to W$ is defined by $0(\mathbf{v}) = \mathbf{0}$ for all $\mathbf{v}$ in V (Example 3), so $T(\mathbf{v}_i) = 0(\mathbf{v}_i)$ holds for each i. Hence $T = 0$ by Theorem 2.

Theorem 2 can be expressed as follows: A linear transformation $T : V \to W$ is completely determined by its effect on a spanning set for V. If the spanning set is a basis, we can say more.

THEOREM 3

Let V and W be vector spaces and let $\{\mathbf{e}_1, \mathbf{e}_2, \ldots, \mathbf{e}_n\}$ be a basis of V. Given any vectors $\mathbf{w}_1, \mathbf{w}_2, \ldots, \mathbf{w}_n$ in W (they needn't be distinct), there exists a unique linear transformation $T : V \to W$ satisfying $T(\mathbf{e}_i) = \mathbf{w}_i$ for each $i = 1, 2, \ldots, n$. In fact, the action of T is as follows: Given $\mathbf{v} = v_1\mathbf{e}_1 + v_2\mathbf{e}_2 + \cdots + v_n\mathbf{e}_n$ in V, then

$$T(\mathbf{v}) = T(v_1\mathbf{e}_1 + v_2\mathbf{e}_2 + \cdots + v_n\mathbf{e}_n) = v_1\mathbf{w}_1 + v_2\mathbf{w}_2 + \cdots + v_n\mathbf{w}_n$$

Proof The uniqueness part is easily dealt with. If such a transformation T *does* exist, and if S is any other such transformation, then $T(\mathbf{e}_i) = \mathbf{w}_i = S(\mathbf{e}_i)$ holds for each i, so $S = T$ by Theorem 2. Hence T is unique if it exists, and it remains to show that there really is such a linear transformation. Given $\mathbf{v}$ in V, we must specify $T(\mathbf{v})$ in W. Because $\{\mathbf{e}_1, \ldots, \mathbf{e}_n\}$ is a basis of V, we have $\mathbf{v} = v_1\mathbf{e}_1 + \cdots + v_n\mathbf{e}_n$, where $v_1, \ldots, v_n$ are *uniquely* determined by $\mathbf{v}$ (this is Theorem 1 in Section 5.3). Hence we may define $T : V \to W$ by

$$T(\mathbf{v}) = T(v_1\mathbf{e}_1 + \cdots + v_n\mathbf{e}_n) = v_1\mathbf{w}_1 + v_2\mathbf{w}_2 + \cdots + v_n\mathbf{w}_n$$

for all $\mathbf{v} = v_1\mathbf{e}_1 + \cdots + v_n\mathbf{e}_n$ in V. This clearly satisfies $T(\mathbf{e}_i) = \mathbf{w}_i$; the verification that T is linear is left to the reader. ■

This theorem shows that linear transformations are defined almost at will: Simply specify where the basis vectors are to be taken, and the rest of the action is dictated by the linearity. Moreover, Theorem 2 shows that deciding whether two linear transformations are equal comes down to determining whether they have the same effect on the basis vectors. So, given a basis $\{\mathbf{e}_1, \ldots, \mathbf{e}_n\}$ of a vector space V, there are exactly as many linear transformations $V \to W$ as there are selections $\mathbf{w}_1, \mathbf{w}_2, \ldots, \mathbf{w}_n$ of vectors in W (not necessarily distinct).

EXAMPLE 11

Find a linear transformation $T : \mathbb{R}^3 \to \mathbb{R}^2$ such that

$$T\begin{bmatrix}1\\1\\0\end{bmatrix} = \begin{bmatrix}2\\1\end{bmatrix} \quad T\begin{bmatrix}1\\0\\1\end{bmatrix} = \begin{bmatrix}1\\-1\end{bmatrix} \quad T\begin{bmatrix}0\\1\\1\end{bmatrix} = \begin{bmatrix}0\\0\end{bmatrix}$$

Solution The set $\left\{\begin{bmatrix}1\\1\\0\end{bmatrix}, \begin{bmatrix}1\\0\\1\end{bmatrix}, \begin{bmatrix}0\\1\\1\end{bmatrix}\right\}$ is a basis of $\mathbb{R}^3$, so Theorem 3 applies. The expansion of an arbitrary vector in $\mathbb{R}^3$ as a linear combination of these vectors is

$$\begin{bmatrix}x\\y\\z\end{bmatrix} = \frac{1}{2}(x + y - z)\begin{bmatrix}1\\1\\0\end{bmatrix} + \frac{1}{2}(x - y + z)\begin{bmatrix}1\\0\\1\end{bmatrix} + \frac{1}{2}(-x + y + z)\begin{bmatrix}0\\1\\1\end{bmatrix}$$

Hence the transformation T must be given by

$$\begin{aligned} T\begin{bmatrix}x\\y\\z\end{bmatrix} &= \frac{1}{2}(x + y - z)T\begin{bmatrix}1\\1\\0\end{bmatrix} + \frac{1}{2}(x - y + z)T\begin{bmatrix}1\\0\\1\end{bmatrix} + \frac{1}{2}(-x + y + z)T\begin{bmatrix}0\\1\\1\end{bmatrix} \\ &= \frac{1}{2}(x + y - z)\begin{bmatrix}2\\1\end{bmatrix} + \frac{1}{2}(x - y + z)\begin{bmatrix}1\\-1\end{bmatrix} + \frac{1}{2}(-x + y + z)\begin{bmatrix}0\\0\end{bmatrix} \\ &= \frac{1}{2}\begin{bmatrix}3x + y - z\\2(y - z)\end{bmatrix}\end{aligned}$$

Recall (Example 2) that every $m \times n$ matrix gives rise to the matrix transformation $T_A : \mathbb{R}^n \to \mathbb{R}^m$ given by $T_A(\mathbf{v}) = A\mathbf{v}$ for all columns $\mathbf{v}$ in $\mathbb{R}^n$. In fact *all* linear transformations $\mathbb{R}^n \to \mathbb{R}^m$ arise in this way.

THEOREM 4

Let $T : \mathbb{R}^n \to \mathbb{R}^m$ be a linear transformation. Write vectors in $\mathbb{R}^n$ as columns.

(1) There exists an $m \times n$ matrix A such that $T(\mathbf{v}) = A\mathbf{v}$ for all columns $\mathbf{v}$ in $\mathbb{R}^n$.

(2) The columns of A are respectively $T(\mathbf{e}_1), T(\mathbf{e}_2), \ldots, T(\mathbf{e}_n)$, where $\{\mathbf{e}_1, \ldots, \mathbf{e}_n\}$ is the standard basis of $\mathbb{R}^n$. Hence A can be written in terms of its columns as

$$A = [T(\mathbf{e}_1)\ T(\mathbf{e}_2) \ldots T(\mathbf{e}_n)]$$

Proof Let $\{\mathbf{e}_1, \mathbf{e}_2, \ldots, \mathbf{e}_n\}$ be the standard basis of $\mathbb{R}^n$ and write

$$T(\mathbf{e}_1) = \begin{bmatrix}a_{11}\\a_{21}\\\vdots\\a_{m1}\end{bmatrix}, T(\mathbf{e}_2) = \begin{bmatrix}a_{12}\\a_{22}\\\vdots\\a_{m2}\end{bmatrix}, \ldots, T(\mathbf{e}_n) = \begin{bmatrix}a_{1n}\\a_{2n}\\\vdots\\a_{mn}\end{bmatrix}$$

Then $A = [a_{ij}]$ is an $m \times n$ matrix whose jth column is $T(\mathbf{e}_j)$. Given $\mathbf{v}$ in $\mathbb{R}^n$, write

$$\mathbf{v} = v_1\mathbf{e}_1 + v_2\mathbf{e}_2 + \cdots + v_n\mathbf{e}_n$$

Now compute $T(\mathbf{v})$, using Theorem 1.

$$\begin{aligned} T(\mathbf{v}) &= v_1 T(\mathbf{e}_1) + v_2 T(\mathbf{e}_2) + \cdots + v_n T(\mathbf{e}_n) \\ &= v_1\begin{bmatrix} a_{11} \\ a_{21} \\ \vdots \\ a_{m1} \end{bmatrix} + v_2\begin{bmatrix} a_{12} \\ a_{22} \\ \vdots \\ a_{m2} \end{bmatrix} + \cdots + v_n\begin{bmatrix} a_{1n} \\ a_{2n} \\ \vdots \\ a_{mn} \end{bmatrix} \\ &= \begin{bmatrix} a_{11}v_1 + a_{12}v_2 + \cdots + a_{1n}v_n \\ a_{21}v_1 + a_{22}v_2 + \cdots + a_{2n}v_n \\ \vdots \\ a_{m1}v_1 + a_{m2}v_2 + \cdots + a_{mn}v_n \end{bmatrix} \\ &= A\mathbf{v} \quad \blacksquare \end{aligned}$$

The matrix A in Theorem 4 is called the **standard matrix** of T.

EXAMPLE 12

Find the standard matrix of $T : \mathbb{R}^3 \to \mathbb{R}^2$ when $T\begin{bmatrix} x \\ y \\ z \end{bmatrix} = \begin{bmatrix} x - 2y + z \\ x - z \end{bmatrix}$.

Solution The desired matrix can be observed directly.

$$T\begin{bmatrix} x \\ y \\ z \end{bmatrix} = \begin{bmatrix} 1 & -2 & 1 \\ 1 & 0 & -1 \end{bmatrix}\begin{bmatrix} x \\ y \\ z \end{bmatrix}$$

However, the second part of Theorem 4 also gives the matrix. If $\{\mathbf{e}_1, \mathbf{e}_2, \mathbf{e}_3\}$ is the standard basis of $\mathbb{R}^3$, then the columns are indeed $T(\mathbf{e}_1)$, $T(\mathbf{e}_2)$, and $T(\mathbf{e}_3)$.

EXAMPLE 13

Find the standard matrix of the following linear transformations $\mathbb{R}^2 \to \mathbb{R}^2$.

(1) **Rotation** R_θ about the origin through the angle θ.

(2) **Projection** P_m on the line $y = mx$.

(3) **Reflection** S_m in the line $y = mx$.

Solution

(1) If $\{\mathbf{e}_1, \mathbf{e}_2\}$ is the standard basis, the diagram gives $R_\theta(\mathbf{e}_1) = \begin{bmatrix} \cos\theta \\ \sin\theta \end{bmatrix}$ and $R_\theta(\mathbf{e}_2) = \begin{bmatrix} -\sin\theta \\ \cos\theta \end{bmatrix}$, so the matrix is $\begin{bmatrix} \cos\theta & -\sin\theta \\ \sin\theta & \cos\theta \end{bmatrix}$

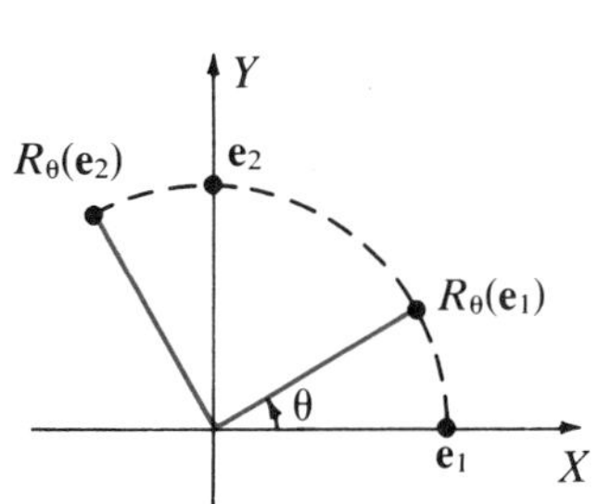

(2) Write $\mathbf{v} = \begin{bmatrix} x \\ y \end{bmatrix}$ and $\mathbf{d} = \begin{bmatrix} 1 \\ m \end{bmatrix}$. Then $\mathbf{d}$ is a direction vector for the line, so Theorem 6 of Section 4.2 gives

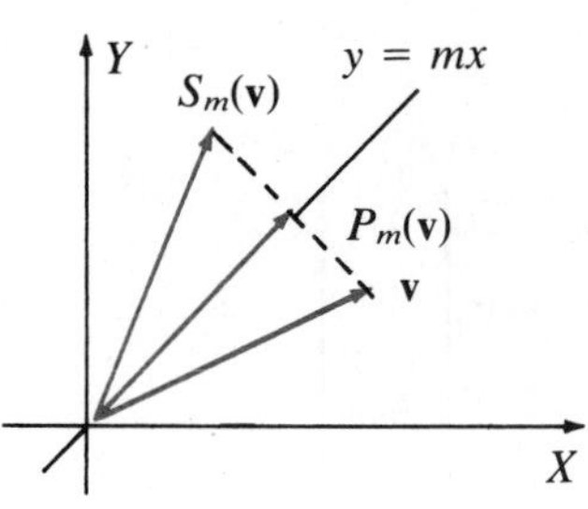

$$P_m(\mathbf{v}) = \frac{\mathbf{v}\cdot\mathbf{d}}{\|\mathbf{d}\|^2}\mathbf{d} = \frac{x+ym}{1+m^2}\begin{bmatrix}1\\m\end{bmatrix} = \frac{1}{1+m^2}\begin{bmatrix}x+ym\\xm+ym^2\end{bmatrix}$$
$$= \frac{1}{1+m^2}\begin{bmatrix}1 & m\\m & m^2\end{bmatrix}\begin{bmatrix}x\\y\end{bmatrix}$$

Hence the matrix is $\frac{1}{1+m^2}\begin{bmatrix}1 & m\\m & m^2\end{bmatrix}$.

(3) The second diagram gives $S_m(\mathbf{v}) = \mathbf{v} + 2[P_m(\mathbf{v}) - \mathbf{v}] = 2P_m(\mathbf{v}) - \mathbf{v}$. Hence (2) gives

$$S_m(\mathbf{v}) = \frac{2}{1+m^2}\begin{bmatrix}1 & m\\m & m^2\end{bmatrix}\begin{bmatrix}x\\y\end{bmatrix} - \begin{bmatrix}x\\y\end{bmatrix}$$
$$= \frac{1}{1+m^2}\begin{bmatrix}1-m^2 & 2m\\2m & m^2-1\end{bmatrix}\begin{bmatrix}x\\y\end{bmatrix}$$

Hence the matrix is $\frac{1}{1+m^2}\begin{bmatrix}1-m^2 & 2m\\2m & m^2-1\end{bmatrix}$.

EXERCISES 8.1.1

1. Show that each of the following functions is a linear transformation.

(a) $T:\mathbb{R}^2\to\mathbb{R}^3;\ T\begin{bmatrix}x\\y\end{bmatrix} = \begin{bmatrix}x-y\\x+2y\\3y\end{bmatrix}$

(b) $T:\mathbb{R}^3\to\mathbb{R}^2;\ T\begin{bmatrix}x\\y\\z\end{bmatrix} = \begin{bmatrix}2x-3y+5z\\0\end{bmatrix}$

(c) $T:\mathbb{R}^2\to\mathbb{R}^2;\ T(x,y) = (x,-y)$ (reflection in the X-axis)

(d) $T:\mathbb{R}^3\to\mathbb{R}^3;\ T(x,y,z) = (x,y,-z)$ (reflection in the X-Y plane)

(e) $T:\mathbb{C}\to\mathbb{C};\ T(z) = \bar{z}$ (conjugation)

(f) $T:\mathbf{M}_{mn}\to\mathbf{M}_{kl};\ T(A) = PAQ$, P a $k\times m$ matrix, Q an $n\times l$ matrix

(g) $T:\mathbf{M}_{nn}\to\mathbf{M}_{nn};\ T(A) = A^T + A$

(h) $T:\mathbf{P}_n\to\mathbb{R};\ T[p(x)] = p(0)$

(i) $T:\mathbf{P}_n\to\mathbb{R};\ T(r_0 + r_1x + \cdots + r_nx^n) = r_n$

(j) $T:V\to V;\ T(\mathbf{v}) = R(\mathbf{v}) + S(\mathbf{v})$, where $R: V\to V$ and $S: V\to V$ are linear transformations

(k) $T:V\to V;\ T(\mathbf{v}) = a[S(\mathbf{v})]$ where a is a scalar and $S: V\to V$ is a linear transformation

(l) $T:V\to\mathbb{R};\ T(r_1\mathbf{e}_1 + \cdots + r_n\mathbf{e}_n) = r_1$, where $\{\mathbf{e}_1, \ldots, \mathbf{e}_n\}$ is a fixed basis of V.

2. In each case, show that T is *not* a linear transformation.

(a) $T:\mathbf{M}_{nn}\to\mathbb{R};\ T(A) = \det A$

(b) $T : \mathbf{M}_{nm} \to \mathbb{R}$; $T(A) = \text{rank } A$
(c) $T : \mathbb{R} \to \mathbb{R}$; $T(x) = x^2$
(d) $T : V \to V$; $T(\mathbf{v}) = \mathbf{v} + \mathbf{u}$ where $\mathbf{u} \neq \mathbf{0}$ is a fixed vector in V (T is called the **translation** by $\mathbf{u}$)

3. In each case, assume that T is a linear transformation.
(a) If $T : V \to \mathbb{R}$ and $T(\mathbf{v}_1) = 1$, $T(\mathbf{v}_2) = -1$, find $T(3\mathbf{v}_1 - 5\mathbf{v}_2)$.
(b) If $T : V \to \mathbb{R}$ and $T(\mathbf{v}_1) = 2$, $T(\mathbf{v}_2) = -3$, find $T(3\mathbf{v}_1 + 2\mathbf{v}_2)$.
(c) If $T : \mathbb{R}^2 \to \mathbb{R}^2$ and $T\begin{bmatrix} 1 \\ 3 \end{bmatrix} = \begin{bmatrix} 1 \\ 1 \end{bmatrix}$, $T\begin{bmatrix} 1 \\ 1 \end{bmatrix} = \begin{bmatrix} 0 \\ 1 \end{bmatrix}$, find $T\begin{bmatrix} -1 \\ 3 \end{bmatrix}$.
(d) If $T : \mathbb{R}^2 \to \mathbb{R}^2$ and $T\begin{bmatrix} 1 \\ -1 \end{bmatrix} = \begin{bmatrix} 0 \\ 1 \end{bmatrix}$, $T\begin{bmatrix} 1 \\ 1 \end{bmatrix} = \begin{bmatrix} 1 \\ 0 \end{bmatrix}$, find $T\begin{bmatrix} 1 \\ -7 \end{bmatrix}$.
(e) If $T : \mathbf{P}_2 \to \mathbf{P}_2$ and $T(x + 1) = x$, $T(x - 1) = 1$, $T(x^2) = 0$, find $T(2 + 3x - x^2)$.
(f) If $T : \mathbf{P}_2 \to \mathbb{R}$ and $T(x + 2) = 1$, $T(1) = 5$, $T(x^2 + x) = 0$, find $T(2 - x + 3x^2)$.

4. In each case, find a linear transformation with the given properties and compute $T(\mathbf{v})$.
(a) $T : \mathbb{R}^2 \to \mathbb{R}^3$; $T(1,2) = (1,0,1)$, $T(-1,0) = (0,1,1)$; $\mathbf{v} = (2,1)$
(b) $T : \mathbb{R}^2 \to \mathbb{R}^3$; $T(2,-1) = (1,-1,1)$, $T(1,1) = (0,1,0)$; $\mathbf{v} = (-1,2)$
(c) $T : \mathbf{P}_2 \to \mathbf{P}_3$; $T(x^2) = x^3$, $T(x + 1) = 0$, $T(x - 1) = x$; $\mathbf{v} = x^2 + x + 1$
(d) $T : \mathbf{M}_{22} \to \mathbb{R}$; $T\begin{bmatrix} 1 & 0 \\ 0 & 0 \end{bmatrix} = 3$, $T\begin{bmatrix} 0 & 1 \\ 1 & 0 \end{bmatrix} = -1$, $T\begin{bmatrix} 1 & 0 \\ 1 & 0 \end{bmatrix} = 0 = T\begin{bmatrix} 0 & 0 \\ 0 & 1 \end{bmatrix}$; $\mathbf{v} = \begin{bmatrix} a & b \\ c & d \end{bmatrix}$

5. If $T : V \to V$ is a linear transformation, find $T(\mathbf{v})$ and $T(\mathbf{w})$ if:
(a) $T(\mathbf{v} + \mathbf{w}) = \mathbf{v} - 2\mathbf{w}$ and $T(2\mathbf{v} - \mathbf{w}) = 2\mathbf{v}$
(b) $T(\mathbf{v} + 2\mathbf{w}) = 3\mathbf{v} - \mathbf{w}$ and $T(\mathbf{v} - \mathbf{w}) = 2\mathbf{v} - 4\mathbf{w}$

6. If $T : V \to W$ is a linear transformation, show that $T(\mathbf{v} - \mathbf{v}_1) = T(\mathbf{v}) - T(\mathbf{v}_1)$ for all $\mathbf{v}$ and $\mathbf{v}_1$ in V.

7. Let $\{\mathbf{e}_1, \mathbf{e}_2\}$ be the standard basis of $\mathbb{R}^2$. Is it possible to have a linear transformation T such that $T(\mathbf{e}_1)$ lies in $\mathbb{R}$ while $T(\mathbf{e}_2)$ lies in $\mathbb{R}^2$? Explain your answer.

8. If A is an $m \times n$ matrix, let T_A: $\mathbb{R}^n \to \mathbb{R}^m$ denote the matrix transformation given by $T_A(\mathbf{v}) = A\mathbf{v}$. If B is an $m \times n$ matrix and $T_A = T_B$, show that $A = B$.

9. (a) Show that every linear transformation $T : \mathbb{R} \to \mathbb{R}$ has the form $T(x) = ax$ for some fixed a in $\mathbb{R}$ (that is, T is a scalar transformation).
(b) Show that every linear transformation $T : \mathbb{R}^n \to \mathbb{R}$ has the form $T(x_1, x_2, \ldots, x_n) = a_1x_1 + a_2x_2 + \cdots + a_nx_n$ for fixed $a_1, a_2, \ldots, a_n$ in $\mathbb{R}$.

10. Let $T : V \to W$ be a linear transformation. Show that:
(a) If U is a subspace of V, then $T(U) = \{T(\mathbf{u}) \mid \mathbf{u} \text{ in } U\}$ is a subspace of W (called the **image** of U under T).
(b) If P is a subspace of W, then $T^{-1}(P) = \{\mathbf{v} \text{ in } V | T(\mathbf{v}) \text{ in } P\}$ is a subspace of V (called the **preimage** of P under T).

11. Let $T : \mathbb{R}^m \to \mathbb{R}^n$ be a linear transformation where the vectors are written as rows.

(a) Show that there is an $m \times n$ matrix A such that $T(\mathbf{v}) = \mathbf{v}A$ for all $\mathbf{v}$ in $\mathbb{R}^n$.

(b) Show that the rows of A are $T(\mathbf{e}_1), T(\mathbf{e}_2), \ldots, T(\mathbf{e}_m)$, respectively, where $\{\mathbf{e}_1, \ldots, \mathbf{e}_m\}$ is the standard basis of $\mathbb{R}^m$.

12. Find the linear transformation $T : \mathbb{R}^2 \to \mathbb{R}^2$ that has the given geometric action, and find the standard matrix of T.

(a) Given $a > 0$, each point at distance r from the origin is moved radially out if $a > 1$ (or in if $a < 1$) from the origin to the point at distance ar from the origin. (This is a **contraction** if $a < 1$ and a **dilation** if $a > 1$.)

(b) Given $a > 0$, the horizontal line through $\begin{bmatrix} 0 \\ b \end{bmatrix}$ becomes the horizontal line through $\begin{bmatrix} 0 \\ ab \end{bmatrix}$. (This is called **stretching** in the Y-direction.)

(c) Given k in $\mathbb{R}$, each point $\begin{bmatrix} x \\ y \end{bmatrix}$ is moved vertically to $\begin{bmatrix} x \\ y + kx \end{bmatrix}$. (This is a **shear** in the Y-direction.)

(d) Each point $\begin{bmatrix} x \\ y \end{bmatrix}$ is carried to $\begin{bmatrix} x \\ 0 \end{bmatrix}$. (This is **projection** on the X-axis.)

13. Find the linear transformation $T : \mathbb{R}^3 \to \mathbb{R}^3$ that has the given geometric action, and find the standard matrix of T.

(a) Rotation of θ about the Z-axis counterclockwise in the X-Y-plane.

(b) Reflection in the X-Y-plane.

14. Let $\mathbf{n} = \begin{bmatrix} a \\ b \\ c \end{bmatrix} \neq \mathbf{0}$ in $\mathbb{R}^3$, and let U denote the plane through the origin with normal $\mathbf{n}$ (as in Chapter 4). In each case, find the standard matrix of the linear transformation $\mathbb{R}^3 \to \mathbb{R}^3$.

(a) The projection on U.

(b) The reflection in U. [*Hint:* Example 13.]

15. Let $\mathbf{u}$, $\mathbf{v}$, and $\mathbf{w}$ denote vectors in $\mathbb{R}^3$ (as in Chapter 4).

(a) Show that $\mathbf{u} \times (\mathbf{v} \times \mathbf{w}) = (\mathbf{u} \cdot \mathbf{w})\mathbf{v} - (\mathbf{u} \cdot \mathbf{v})\,\mathbf{w}$ by applying Theorem 2 to S: $\mathbb{R}^3 \to \mathbb{R}^3$ and $T : \mathbb{R}^3 \to \mathbb{R}^3$, where $S(\mathbf{x}) = \mathbf{x} \times (\mathbf{v} \times \mathbf{w})$ and $T(\mathbf{x}) = (\mathbf{x} \cdot \mathbf{w})\,\mathbf{v} - (\mathbf{x} \cdot \mathbf{v})\mathbf{w}$.

(b) Consider $T_A : \mathbb{R}^3 \to \mathbb{R}^3$, where A is a 3×3 matrix. Show that the volume of the parallelepiped determined by $T_A(\mathbf{u})$, $T_A(\mathbf{v})$, and $T_A(\mathbf{w})$ equals $|\det A|$ times the volume of the parallelepiped determined by $\mathbf{u}$,$\mathbf{v}$, and $\mathbf{w}$. [*Hint:* Theorem 12 in Section 4.2.]

16. Show that differentiation is the only linear transformation $\mathbf{P}_n \to \mathbf{P}_n$ that satisfies $T(x^k) = kx^{k-1}$ for each $k = 0, 1, 2, \ldots, n$.

17. Let $T : V \to W$ be a linear transformation, and let $\mathbf{v}_1, \ldots, \mathbf{v}_n$ denote vectors in V.

(a) If $\{T(\mathbf{v}_1), \ldots, T(\mathbf{v}_n)\}$ is linearly independent, so also is $\{\mathbf{v}_1, \ldots, \mathbf{v}_n\}$.

(b) Show that the converse of part (**a**) is false for $T : \mathbb{R}^2 \to \mathbb{R}^2$, where $T(x,y) = (x,0)$.

18. Given a in $\mathbb{R}$, define the **evaluation** map $E_a : \mathbf{P}_n \to \mathbb{R}$ by $E_a[p(x)] = p(a)$ for all $p(x)$ in $\mathbf{P}_n$.

(a) Show that E_a is a linear transformation satisfying the additional condition that $E_a(x^k) = [E_a(x)]^k$ holds for all $k = 0, 1, 2, \ldots$. [*Note:* $x^0 = 1$.]

(b) If $T : \mathbf{P}_n \to \mathbb{R}$ is a linear transformation satisfying $T(x^k) = [T(x)]^k$ for all $k = 0, 1, 2, \ldots$, show that $T = E_a$ for some a in $\mathbb{R}$.

19. Define $T : \mathbf{M}_{nn} \to \mathbb{R}$ by $T(A) = \operatorname{tr} A$, the trace of A.

(a) Show that $T(AB) = T(BA)$ for all A and B in $\mathbf{M}_{nn}$. (See the proof of Theorem 3 in Section 7.1.)

(b) If $S : \mathbf{M}_{nn} \to \mathbb{R}$ is any linear transformation satisfying $S(AB) = S(BA)$ for all A and B in $\mathbf{M}_{nn}$, show that there exists a number k such that $S(A) = kT(A)$ for all A.

[*Hint* for part (**b**): Let E_{ij} denote the $n \times n$ matrix with 1 in the (i,j)-position and zeros elsewhere. Show that $E_{ik}E_{lj} = \begin{cases} 0 & \text{if } k \neq l \\ E_{ij} & \text{if } k = l \end{cases}$. Use this to show that $S(E_{ij}) = 0$ if $i \neq j$ and $S(E_{11}) = S(E_{22}) = \cdots = S(E_{nn})$. Put $k = S(E_{11})$ and use the fact the $\{E_{ij} | 1 \leq i, j \leq n\}$ is a basis of $\mathbf{M}_{nn}$.]

20. Let V and W be vector spaces, let V be finite dimensional, and let $\mathbf{v} \neq \mathbf{0}$ in V. Given any $\mathbf{w}$ in W, show that there exists a linear transformation $T : V \to W$ with $T(\mathbf{v}) = \mathbf{w}$. [*Hint:* Theorem 7 in Section 5.3 and Theorem 3.]

8.1.2 Kernel and Image of a Linear Transformation

This section is devoted to the study of two important subspaces associated with a linear transformation.

DEFINITION

Let $T : V \to W$ denote a linear transformation. The **kernel** of T (denoted $\ker T$) and the **image** of T (denoted $\operatorname{im} T$) are defined by

$$\ker T = \{\mathbf{v} \text{ in } V \mid T(\mathbf{v}) = \mathbf{0}\}$$
$$\operatorname{im} T = \{T(\mathbf{v}) \mid \mathbf{v} \text{ in } V\}$$

The kernel of T is often called the **null space** of T. It consists of all vectors $\mathbf{v}$ in V satisfying the *condition* that $T(\mathbf{v}) = \mathbf{0}$. The image of T is often called the **range** of T. It consists of all vectors $\mathbf{w}$ in W of the *form* $\mathbf{w} = T(\mathbf{v})$ for some $\mathbf{v}$ in V.

THEOREM 5 If $T : V \to W$ is a linear transformation, $\ker T$ is a subspace of V, and $\operatorname{im} T$ is a subspace of W.

Proof The fact that $T(\mathbf{0}) = \mathbf{0}$ shows that both $\ker T$ and $\operatorname{im} T$ contain the zero vector. If $\mathbf{v}$ and $\mathbf{v}_1$ lie in $\ker T$, then $T(\mathbf{v}) = \mathbf{0} = T(\mathbf{v}_1)$, so

$$T(\mathbf{v} + \mathbf{v}_1) = T(\mathbf{v}) + T(\mathbf{v}_1) = \mathbf{0} + \mathbf{0} = \mathbf{0}$$
$$T(r\mathbf{v}) = rT(\mathbf{v}) = \mathrm{r}\mathbf{0} = \mathbf{0} \qquad \text{for all } r \text{ in } \mathbb{R}$$

Hence $\mathbf{v} + \mathbf{v}_1$ and $r\mathbf{v}$ lie in $\ker T$ (they satisfy the required condition), so $\ker T$ is a subspace of V. If $\mathbf{w}$ and $\mathbf{w}_1$ lie in $\operatorname{im} T$, write $\mathbf{w} = T(\mathbf{v})$ and $\mathbf{w}_1 = T(\mathbf{v}_1)$, where $\mathbf{v}$ and $\mathbf{v}_1$ lie in V. Then

$$\mathbf{w} + \mathbf{w}_1 = T(\mathbf{v}) + T(\mathbf{v}_1) = T(\mathbf{v} + \mathbf{v}_1)$$
$$r\mathbf{w} = rT(\mathbf{v}) = T(r\mathbf{v}) \qquad \text{for all } r \text{ in } \mathbb{R}$$

Hence $\mathbf{w} + \mathbf{w}_1$ and $r\mathbf{w}$ both lie in $\operatorname{im} T$ (they have the required form), so $\operatorname{im} T$ is a subspace of W. ■

EXAMPLE 14 If $T : \mathbb{R}^3 \to \mathbb{R}^3$ is defined by $T(x,y,z) = (x - y, z, y - x)$, find $\ker T$ and $\operatorname{im} T$, and compute their dimensions.

Solution We use the definitions:

$$\ker T = \{(x,y,z) \mid (x - y, z, y - x) = (0, 0, 0)\} = \{(t,t,0) \mid t \text{ in } \mathbb{R}\}$$
$$\operatorname{im} T = \{(x - y, z, y - x) \mid x, y, z \text{ in } \mathbb{R}\} = \{(s, t, -s) \mid s, t \text{ in } \mathbb{R}\}$$

Hence $\dim(\ker T) = 1$ and $\dim(\operatorname{im} T) = 2$.

EXAMPLE 15 If A is an $m \times n$ matrix and $T_A : \mathbb{R}^n \to \mathbb{R}^m$ is defined by $T_A(\mathbf{v}) = A\mathbf{v}$ for every column $\mathbf{v}$ in $\mathbb{R}^n$, then:

$$\ker T_A = \{\mathbf{v} \mid A\mathbf{v} = \mathbf{0}\} \qquad \text{and} \qquad \operatorname{im} T_A = \{A\mathbf{v} \mid \mathbf{v} \text{ in } \mathbb{R}^n\}$$

are, respectively, the null space and range of A (see Examples 3 and 4 in Section 5.2).

DEFINITION Given a linear transformation $T : V \to W$,

$\dim(\ker T)$ is called the **nullity** of T and denoted as nullity (T).
$\dim(\operatorname{im} T)$ is called the **rank** of T and denoted as rank (T).

The rank of a matrix A was defined earlier to be the dimension of col A, the column space of A. The two usages of the word *rank* are consistent in the following sense:

EXAMPLE 16 Given an $m \times n$ matrix A, show that im T_A = col A, so rank T_A = rank A.

Solution Write $A = [\mathbf{c}_1 \ldots \mathbf{c}_n]$ in terms of its columns. Then

$$\text{im } T_A = \{A\mathbf{v} \mid \mathbf{v} \text{ in } \mathbb{R}^n\} = \left\{ [\mathbf{c}_1 \ldots \mathbf{c}_n] \begin{bmatrix} v_1 \\ \vdots \\ v_n \end{bmatrix} \;\middle|\; v_i \text{ in } \mathbb{R} \right\}$$

$$= \{v_1\mathbf{c}_1 + \cdots + v_n\mathbf{c}_n \mid v_i \text{ in } \mathbb{R}\}$$

Hence im $T_A = \text{span}\{\mathbf{c}_1, \ldots, \mathbf{c}_n\}$ is the column space of A.

EXAMPLE 17 Given the 4×3 matrix $A = \begin{bmatrix} 1 & -1 & 2 \\ 3 & 0 & 1 \\ 1 & 2 & -3 \\ -2 & -1 & 1 \end{bmatrix}$, compute the kernel and image of the corresponding matrix transformation $T_A : \mathbb{R}^3 \to \mathbb{R}^4$, and determine the rank and nullity of T_A.

Solution Bring A to reduced row-echelon form:

$$\begin{bmatrix} 1 & -1 & 2 \\ 3 & 0 & 1 \\ 1 & 2 & -3 \\ -2 & -1 & 1 \end{bmatrix} \to \begin{bmatrix} 1 & -1 & 2 \\ 0 & 3 & -5 \\ 0 & 3 & -5 \\ 0 & -3 & 5 \end{bmatrix} \to \begin{bmatrix} 1 & 0 & \frac{1}{3} \\ 0 & 1 & -\frac{5}{3} \\ 0 & 0 & 0 \\ 0 & 0 & 0 \end{bmatrix}$$

so rank T_A = rank A = 2. Moreover, the solutions to $A\mathbf{v} = \mathbf{0}$ are $[-t\ 5t\ 3t]^T$, where t is a parameter. Because ker $T_A = \{\mathbf{v} \text{ in } \mathbb{R}^3 \mid A\mathbf{v} = \mathbf{0}\}$, this means that nullity T_A = dim (ker T_A) = 1.

A useful way to study a subspace of a vector space is often to exhibit it as the kernel or image of a linear transformation. Here is an example.

EXAMPLE 18 Define a transformation $T : \mathbf{M}_{nn} \to \mathbf{M}_{nn}$ by $T(A) = A - A^T$ for all A in $\mathbf{M}_{nn}$. Show that T is linear and that:

(a) ker T consists of all symmetric matrices.

(b) im T consists of all skew-symmetric matrices.

Solution The verification that T is linear is omitted. To prove part (a), note that a matrix A lies in ker T just when $0 = T(A) = A - A^T$, and this occurs if and only if $A = A^T$—that is, A is symmetric. Turning to part (b), the space im T consists of all matrices $T(A)$, A in $\mathbf{M}_{nn}$. Every such matrix is skew-symmetric because

$$T(A)^T = (A - A^T)^T = A^T - A = -T(A)$$

On the other hand, if S is skew-symmetric (that is, $S^T = -S$), then S lies in im T. In fact,

$$T\left[\frac{1}{2}S\right] = \frac{1}{2}S - \left[\frac{1}{2}S\right]^T = \frac{1}{2}(S + S) = S$$

DEFINITION

Let $T : V \to W$ be a linear transformation.

(1) T is said to be **onto** if im $T = W$.

(2) T is said to be **one-to-one** if $T(\mathbf{v}) = T(\mathbf{v}_1)$ implies $\mathbf{v} = \mathbf{v}_1$.

Thus T is onto if *every* vector $\mathbf{w}$ in W has the form $\mathbf{w} = T(\mathbf{v})$ for some (not necessarily unique) vector $\mathbf{v}$ in V,whereas T is one-to-one if two distinct vectors $\mathbf{v} \neq \mathbf{v}_1$ in V *cannot* be carried to the same image $T(\mathbf{v}) = T(\mathbf{v}_1)$ in W. The onto transformations T are those for which im T is as large a subspace of W as possible. By contrast, the one-to-one transformations T are the ones with ker T as *small* as possible.

THEOREM 6

If $T : V \to W$ is a linear transformation, then T is one-to-one if and only if ker $T = 0$.

Proof If T is one-to-one, let $\mathbf{v}$ be any vector in ker T. Then $T(\mathbf{v}) = \mathbf{0}$, so $T(\mathbf{v}) = T(\mathbf{0})$. Hence $\mathbf{v} = \mathbf{0}$ because T is one-to-one. Conversely, assume that ker $T = 0$ and let $T(\mathbf{v}) = T(\mathbf{v}_1)$ with $\mathbf{v}$ and $\mathbf{v}_1$ in V. Then $T(\mathbf{v} - \mathbf{v}_1) = T(\mathbf{v}) - T(\mathbf{v}_1) = \mathbf{0}$, so $\mathbf{v} - \mathbf{v}_1$ lies in ker $T = 0$. This means that $\mathbf{v} - \mathbf{v}_1 = \mathbf{0}$, so $\mathbf{v} = \mathbf{v}_1$. This proves that T is one-to-one. ■

EXAMPLE 19 The identity transformation $1_V : V \to V$ is both one-to-one and onto for any vector space V.

EXAMPLE 20 Consider the linear transformations

$$S : \mathbb{R}^3 \to \mathbb{R}^2 \quad \text{given by } S(x,y,z) = (x + y, x - y)$$
$$T : \mathbb{R}^2 \to \mathbb{R}^3 \quad \text{given by } T(x,y) = (x + y, x - y, x)$$

Show that T is one-to-one but not onto, whereas S is onto but not one-to-one.

Solution The verification that they are linear is omitted. T is one-to-one because

$$\ker T = \{(x,y) \mid x + y = x - y = x = 0\} = \{(0,0)\} = 0$$

However, it is not onto. For example $(0,0,1)$ does not lie in $\operatorname{im} T$, because if $(0,0,1) = (x + y, x - y, x)$ for some x and y, then $x + y = 0 = x - y$ and $x = 1$, an impossibility. Turning to S, it is not one-to-one because $(0,0,1)$ lies in $\ker S$. But every element (s,t) in $\mathbb{R}^2$ lies in $\operatorname{im} S$, because $(s,t) = (x + y, x - y)$ for some x and y (in fact $x = \frac{1}{2}(s + t)$ and $y = \frac{1}{2}(s - t)$). Hence S is onto.

EXAMPLE 21 Let U be an invertible $m \times m$ matrix and define

$$T : \mathbf{M}_{mn} \to \mathbf{M}_{mn} \text{ by } T(X) = UX \text{ for all } X \text{ in } \mathbf{M}_{mn}$$

Show that T is a linear transformation that is both one-to-one and onto.

Solution The verification that T is linear is left to the reader. To see that T is one-to-one, let $T(X) = 0$. Then $UX = 0$, so left-multiplication by U^{-1} gives $X = 0$. Hence $\ker T = 0$, so T is one-to-one. Finally, if Y is any member of $\mathbf{M}_{mn}$, then $U^{-1}Y$ lies in $\mathbf{M}_{mn}$ too, and $T(U^{-1}Y) = U(U^{-1}Y) = Y$. This shows that T is onto.

The linear transformations $\mathbb{R}^n \to \mathbb{R}^m$ all have the form T_A for some $m \times n$ matrix A (Theorem 4). The next theorem gives conditions under which they are onto or one-to-one.

THEOREM 7

Let A be an $m \times n$ matrix, and let $T_A : \mathbb{R}^n \to \mathbb{R}^m$ be the matrix transformation defined by $T_A(\mathbf{v}) = A\mathbf{v}$.

(1) T_A is onto if and only if $\operatorname{rank} A = m$ (linearly independent rows).

(2) T_A is one-to-one if and only if $\operatorname{rank} A = n$ (linearly independent columns)

Proof (1) We have that $\operatorname{im} T_A$ is the column space of A (see Example 16), so T_A is onto if and only if the column space of A is $\mathbb{R}^m$. Because the rank of A is the dimension of the column space, this holds if and only if $\operatorname{rank} A = m$; this is equivalent to A having independent rows by Theorem 5 in Section 5.4.

(2) $\ker T_A = \{\mathbf{v} \text{ in } \mathbb{R}^n \mid A\mathbf{v} = \mathbf{0}\}$, so (using Theorem 6) T_A is one-to-one if and only if $A\mathbf{v} = \mathbf{0}$ implies $\mathbf{v} = \mathbf{0}$. But $A\mathbf{v}$ is a linear combina-

tion of the columns of A, the coefficients being the components of $\mathbf{v}$, so this proves (2). ■

The following theorem is the main result of this section.

THEOREM 8
Dimension Theorem

Let $T : V \to W$ be any linear transformation, and assume that $\ker T$ and $\operatorname{im} T$ are both finite dimensional. Then V is also finite dimensional and

$$\dim V = \dim(\ker T) + \dim(\operatorname{im} T)$$

In other words, $\dim V = \text{nullity}(T) + \text{rank}(T)$.

Proof Every vector in $\operatorname{im} T$ has the form $T(\mathbf{v})$ for some $\mathbf{v}$ in V. Hence let $\{T(\mathbf{e}_1), T(\mathbf{e}_2), \ldots, T(\mathbf{e}_r)\}$ be a basis of $\operatorname{im} T$, where the $\mathbf{e}_i$ lie in V. Let $\{\mathbf{e}_{r+1}, \mathbf{e}_{r+2}, \ldots, \mathbf{e}_n\}$ be any basis of $\ker T$. Then $\dim(\operatorname{im} T) = r$ and $\dim(\ker T) = n - r$, so it suffices to show that $B = \{\mathbf{e}_1, \mathbf{e}_2, \ldots, \mathbf{e}_n\}$ is a basis of V.

(1) *B spans V.* If $\mathbf{v}$ lies in V, then $T(\mathbf{v})$ lies in $\operatorname{im} V$, so

$$T(\mathbf{v}) = t_1T(\mathbf{e}_1) + t_2T(\mathbf{e}_2) + \cdots + t_rT(\mathbf{e}_r)$$

This implies that $\mathbf{v} - t_1\mathbf{e}_1 - t_2\mathbf{e}_2 - \cdots - t_r\mathbf{e}_r$ lies in $\ker T$ and so is a linear combination of $\mathbf{e}_{r+1}, \ldots, \mathbf{e}_n$. Hence $\mathbf{v}$ is a linear combination of the vectors in B.

(2) *B is linearly independent.* Suppose that

$$t_1\mathbf{e}_1 + \cdots + t_r\mathbf{e}_r + t_{r+1}\mathbf{e}_{r+1} + \cdots + t_n\mathbf{e}_n = \mathbf{0} \qquad (*)$$

Applying T gives $t_1T(\mathbf{e}_1) + \cdots + t_rT(\mathbf{e}_r) = \mathbf{0}$ (because $T(\mathbf{e}_i) = \mathbf{0}$ for $i = r + 1, \ldots, n$), so the independence of $\{T(\mathbf{e}_1), \ldots, T(\mathbf{e}_r)\}$ yields $t_1 = \cdots = t_r = 0$. Hence (*) becomes

$$t_{r+1}\mathbf{e}_{r+1} + \cdots + t_n\mathbf{e}_n = \mathbf{0}$$

so $t_{r+1} = \cdots = t_n = 0$ by the independence of $\{\mathbf{e}_{r+1}, \ldots, \mathbf{e}_n\}$. This proves that B is linearly independent. ■

Note that we end up with a basis $B = \{\mathbf{e}_1, \mathbf{e}_2, \ldots, \mathbf{e}_r, \mathbf{e}_{r+1}, \ldots, \mathbf{e}_n\}$ of V with the property that $\{\mathbf{e}_{r+1}, \ldots, \mathbf{e}_n\}$ is a basis of $\ker T$ and $\{T(\mathbf{e}_1), \ldots, T(\mathbf{e}_r)\}$ is a basis of $\operatorname{im} T$. In fact, if V is known in advance to be finite dimensional, then *any* basis $\{\mathbf{e}_{r+1}, \ldots, \mathbf{e}_n\}$ of $\ker T$ can be extended to a basis $\{\mathbf{e}_1, \mathbf{e}_2, \ldots, \mathbf{e}_r, \mathbf{e}_{r+1}, \ldots, \mathbf{e}_n\}$ of V by Theorem 7 in Section 5.3 and, no matter how this is done, the vectors $\{T(\mathbf{e}_1), \ldots, T(\mathbf{e}_r)\}$ will turn out to be a basis of $\operatorname{im} T$. This result is useful, and we record it for reference. The proof is much like that of Theorem 8 and is left as Exercise 21.

COROLLARY Let $T : V \to W$ be a linear transformation, and let $\{\mathbf{e}_1, \ldots, \mathbf{e}_r, \mathbf{e}_{r+1}, \ldots, \mathbf{e}_n\}$ be a basis of V such that $\{\mathbf{e}_{r+1}, \ldots, \mathbf{e}_n\}$ is a basis of ker T. Then $\{T(\mathbf{e}_1), \ldots, T(\mathbf{e}_r)\}$ is a basis of im T, and hence $r = \text{rank } T$.

The dimension theorem is one of the most useful results in all of linear algebra. It shows that if one of dim(ker T) and dim(im T) can be found, then the other is automatically known. In many cases it is easier to compute one than the other, so the theorem is a real asset. The rest of this section is devoted to illustrations of this. The next example uses the dimension theorem to give a different proof of Theorem 4 in Section 5.4.

EXAMPLE 22 Let A be an $m \times n$ matrix of rank r. Show that the space of all solutions of the system $A\mathbf{x} = \mathbf{0}$ of m homogeneous equations in n variables has dimension $n - r$.

Solution The space in question is just ker T_A, where $T_A : \mathbb{R}^n \to \mathbb{R}^m$ is defined by $T_A(\mathbf{v}) = A\mathbf{v}$ for all columns $\mathbf{v}$ in $\mathbb{R}^n$. But dim(im T_A) $=$ rank T_A $=$ rank $A = r$ by Example 16, so dim(ker T_A) $= n - r$ by the dimension theorem.

EXAMPLE 23[†] Let $D : \mathbf{P}_n \to \mathbf{P}_{n-1}$ be the differentiation map defined by $D[p(x)] = p'(x)$. Compute ker D and hence conclude that D is onto.

Solution Because $p'(x) = 0$ means $p(x)$ is constant, we have dim(ker D) $= 1$. Because dim $\mathbf{P}_n = n + 1$, the dimension theorem gives

$$\dim(\text{im } D) = (n+1) - \dim(\ker D) = n = \dim(\mathbf{P}_{n-1})$$

This implies that im $D = \mathbf{P}_{n-1}$, so D is onto.

Of course it is not difficult to verify directly that each polynomial $q(x)$ in $\mathbf{P}_{n-1}$ is the derivative of some polynomial in $\mathbf{P}_n$ (simply integrate $q(x)$!), so the dimension theorem is not needed in this case. However, in many situations it is difficult to see directly that a linear transformation is onto, and the method given in this example may be the easiest way by far to proceed. Here is another illustration.

EXAMPLE 24 Given a in $\mathbb{R}$, define the **evaluation map** $E_a : \mathbf{P}_n \to \mathbb{R}$ by $E_a[p(x)] = p(a)$. Show that E_a is linear and onto, and hence conclude that $\{(x - a), (x - a)^2, \ldots, (x - a)^n\}$ is a basis of ker E_a, the subspace of all polynomials $p(x)$ for which $p(a) = 0$.

[†]This example uses calculus and can be omitted with no loss of continuity.

Solution The verification that E_a is linear and onto is left to the reader. Hence $\dim(\text{im } E_a) = \dim(\mathbb{R}) = 1$, so $\dim(\ker E_a) = (n + 1) - 1 = n$ by the dimension theorem. Now each of the n polynomials $(x - a)$, $(x - a)^2, \ldots, (x - a)^n$ clearly lies in $\ker E_a$, so they are a basis because they are linearly independent (they have distinct degrees).

The next example deduces an important result on inner product spaces (Theorem 10 in Section 6.2) from the dimension theorem.

EXAMPLE 25 Let U be a subspace of a finite dimensional inner product space V. Show that $\dim V = \dim U + \dim U^\perp$ by showing that $U = \text{im}(\text{proj}_U)$ and $U^\perp = \ker(\text{proj}_U)$.

Solution Write $T = \text{proj}_U$ for convenience. Then $T: V \to V$ is linear by Example 6, and the projection theorem (Theorem 9 in Section 6.2) shows that, given $\mathbf{v}$ in V,

$$T(\mathbf{v}) \in U \qquad \text{and} \qquad \mathbf{v} - T(\mathbf{v}) \in U^\perp$$

These show immediately that $\text{im } T \subseteq U$ and that $\ker T \subseteq U^\perp$. But if $\mathbf{v}$ is in U, then $\mathbf{v} - T(\mathbf{v})$ is in $U \cap U^\perp = 0$, so $\mathbf{v} = T(\mathbf{v})$ is in $\text{im } T$. Hence $\text{im } T = U$. Finally, if $\mathbf{v}$ is in $U^\perp$, then $T(\mathbf{v}) = \mathbf{v} - (\mathbf{v} - T(\mathbf{v}))$ is in $U \cap U^\perp = 0$, so $\mathbf{v}$ is in $\ker T$. Hence $\ker T = U^\perp$.

We now apply the dimension theorem to the rank of a matrix.

EXAMPLE 26 If A is any $m \times n$ matrix, show that $\text{rank } A = \text{rank } A^TA = \text{rank } AA^T$.

Solution It suffices to show that $\text{rank } A = \text{rank } A^TA$ (the rest follows by replacing A with A^T). Write $B = A^TA$, and consider the associated matrix transformations

$$T_A : \mathbb{R}^n \to \mathbb{R}^m \quad \text{and} \quad T_B : \mathbb{R}^n \to \mathbb{R}^n$$

The dimension theorem and Example 16 give

$$\text{rank } A = \text{rank } T_A = \dim(\text{im } T_A) = n - \dim(\ker T_A)$$
$$\text{rank } B = \text{rank } T_B = \dim(\text{im } T_B) = n - \dim(\ker T_B)$$

so it suffices to show that $\ker T_A = \ker T_B$. Now $A\mathbf{v} = \mathbf{0}$ implies that $B\mathbf{v} = A^TA\mathbf{v} = \mathbf{0}$, so $\ker T_A$ is contained in $\ker T_B$. On the other hand, if $B\mathbf{v} = \mathbf{0}$, then $A^TA\mathbf{v} = \mathbf{0}$, so

$$(A\mathbf{v}) \cdot (A\mathbf{v}) = \mathbf{v}^TA^TA\mathbf{v} = \mathbf{v}^T\mathbf{0} = 0$$

This implies that $A\mathbf{v} = \mathbf{0}$.

We conclude with a very useful consequence of the dimension theorem.

THEOREM 9 Let $T : V \to W$ be a linear transformation and assume that $\dim V = \dim W = n$. Then T is onto if and only if T is one-to-one.

Proof T is one-to-one if and only if $\dim(\ker T) = 0$ (this is Theorem 6), and T is onto if and only if $\dim(\operatorname{im} T) = n$. The dimension theorem asserts that $\dim(\ker T) + \dim(\operatorname{im} T) = n$, so the result follows. ■

EXERCISES 8.1.2

1. For each matrix A, find a basis for the kernel and image of T_A, and find the rank and nullity of T_A.

(a) $\begin{bmatrix} 1 & 2 & -1 & 1 \\ 3 & 1 & 0 & 2 \\ 1 & -3 & 2 & 0 \end{bmatrix}$ **(b)** $\begin{bmatrix} 2 & 1 & -1 & 3 \\ 1 & 0 & 3 & 1 \\ 1 & 1 & -4 & 2 \end{bmatrix}$

(c) $\begin{bmatrix} 1 & 2 & -1 \\ 3 & 1 & 2 \\ 4 & -1 & 5 \\ 0 & 2 & -2 \end{bmatrix}$ **(d)** $\begin{bmatrix} 2 & 1 & 0 \\ 1 & -1 & 3 \\ 1 & 2 & -3 \\ 0 & 3 & -6 \end{bmatrix}$

2. In each case, **(i)** find a basis of ker T, and **(ii)** find a basis of im T.

(a) $T : \mathbf{P}_2 \to \mathbb{R}^2$; $T(a + bx + cx^2) = (a,b)$ **(b)** $T : \mathbf{P}_2 \to \mathbb{R}^2$; $T[p(x)] = [p(0),p(1)]$

(c) $T : \mathbb{R}^3 \to \mathbb{R}^3$; $T(x,y,z) = (x + y, x + y, 0)$ **(d)** $T : \mathbb{R}^3 \to \mathbb{R}^4$; $T(x,y,z) = (x,x,y,y)$

(e) $T : \mathbf{M}_{22} \to \mathbf{M}_{22}$; $T\begin{bmatrix} a & b \\ c & d \end{bmatrix} = \begin{bmatrix} a + b & b + c \\ c + d & d + a \end{bmatrix}$

(f) $T : \mathbf{M}_{22} \to \mathbb{R}$; $T\begin{bmatrix} a & b \\ c & d \end{bmatrix} = a + d$

(g) $T : \mathbf{P}_n \to \mathbb{R}$; $T(r_0 + r_1x + \cdots + r_nx^n) = r_n$

(h) $T : \mathbb{R}^n \to \mathbb{R}$; $T(r_1, r_2, \ldots, r_n) = r_1 + r_2 + \cdots + r_n$

(i) T: $\mathbf{M}_{22} \to \mathbf{M}_{22}$; $T(X) = XA - AX$, where $A = \begin{bmatrix} 0 & 1 \\ 1 & 0 \end{bmatrix}$

(j) T: $\mathbf{M}_{22} \to \mathbf{M}_{22}$; $T(X) = XA$, where $A = \begin{bmatrix} 1 & 1 \\ 0 & 0 \end{bmatrix}$

3. Let $P : V \to \mathbb{R}$ and $Q : V \to \mathbb{R}$ be linear transformations, where V is a vector space. Define $T : V \to \mathbb{R}^2$ by $T(\mathbf{v}) = (P(\mathbf{v}),Q(\mathbf{v}))$.

(a) Show that T is a linear transformation.

(b) Show that $\ker T = \ker P \cap \ker Q$, the set of vectors in both ker P and ker Q.

4. In each case, find a basis $B = \{\mathbf{e}_1, \ldots, \mathbf{e}_r, \mathbf{e}_{r+1}, \ldots, \mathbf{e}_n\}$ of V such that $\{\mathbf{e}_{r+1}, \ldots, \mathbf{e}_n\}$ is a basis of ker T, and verify the corollary to Theorem 8.

(a) $T : \mathbb{R}^3 \to \mathbb{R}^4$; $T(x,y,z) = (x - y + 2z, x + y - z, 2x + z, 2y - 3z)$

(b) $T : \mathbb{R}^3 \to \mathbb{R}^4$; $T(x,y,z) = (x + y + z, 2x - y + 3z, z - 3y, 3x + 4z)$

5. Show that the following are equivalent for a linear transformation $T : V \to W$.
(a) $\ker T = V$ (b) $\operatorname{im} T = 0$ (c) $T = 0$
6. Let A and B be $m \times n$ and $k \times n$ matrices, respectively. Assume that $A\mathbf{x} = \mathbf{0}$ implies $B\mathbf{x} = \mathbf{0}$ for every n-column $\mathbf{x}$. Show that rank $A \geq$ rank B.
7. Can a linear transformation $T : \mathbb{R}^n \to \mathbb{R}^m$ be one-to-one if $m < n$? Justify your answer.
8. Let $T : V \to W$ be a linear transformation where V and W are finite dimensional.
(a) If $\dim V < \dim W$, show that T is not onto.
(b) If $\dim V > \dim W$, show that T is not one-to-one.
(c) If T is onto, show that $\dim V \geq \dim W$.
(d) If T is one-to-one, show that $\dim V \leq \dim W$.
9. Let A be an $m \times n$ matrix of rank r. Define $V = \{\mathbf{v} \text{ in } \mathbb{R}^m \mid \mathbf{v}A = 0\}$. Show that $\dim V = m - r$.
10. Given $\{\mathbf{v}_1, \ldots, \mathbf{v}_n\}$ in a vector space V, define $T : \mathbb{R}^n \to V$ by $T(r_1, \ldots, r_n) = r_1\mathbf{v}_1 + \cdots + r_n\mathbf{v}_n$. Show that T is linear, and that:
(a) T is one-to-one if and only if $\{\mathbf{v}_1, \ldots, \mathbf{v}_n\}$ is linearly independent.
(b) T is onto if and only if $V = \operatorname{span}\{\mathbf{v}_1, \ldots, \mathbf{v}_n\}$.
11. Consider $V = \left\{\begin{bmatrix} a & b \\ c & d \end{bmatrix} \middle| \; a + c = b + d\right\}$.
(a) Consider $S : \mathbf{M}_{22} \to \mathbb{R}$ with $S\begin{bmatrix} a & b \\ c & d \end{bmatrix} = a + c - b - d$. Show that S is linear and onto and that $V = \ker S$, and so conclude that V is a subspace of $\mathbf{M}_{22}$ and that $\dim V = 3$.
(b) Consider $T : V \to \mathbb{R}$ with $T\begin{bmatrix} a & b \\ c & d \end{bmatrix} = a + c$. Show that T is linear and onto, and use this information to compute $\dim(\ker T)$.
12. Define $T : \mathbf{P}_n \to \mathbb{R}$ by $T[p(x)] =$ the sum of all the coefficients of $p(x)$.
(a) Show that T is linear and onto.
(b) Deduce that $\dim(\ker T) = n$.
(c) Use part **(b)** to conclude that $\{x - 1, x^2 - 1, \ldots, x^n - 1\}$ is a basis of $\ker T$.
13. Use the dimension theorem to prove Theorem 1 in Section 1.3: If A is an $m \times n$ matrix with $m < n$, the system $A\mathbf{x} = \mathbf{0}$ of m homogeneous equations in n variables always has a nontrivial solution.
14. Let B be an $n \times n$ matrix, and consider the subspaces $U = \{A \mid A \text{ in } \mathbf{M}_{mn}, AB = 0\}$ and $V = \{AB \mid A \text{ in } \mathbf{M}_{mn}\}$. Show that $\dim U + \dim V = mn$.
15. Call a polynomial p **even** if $p(x) = p(-x)$ and **odd** if $p(x) = -p(-x)$. Let U and V denote, respectively, the sets of even and odd polynomials in $\mathbf{P}_n$. Define $T : \mathbf{P}_n \to \mathbf{P}_n$ by $T(p) = p(x) - p(-x)$.
(a) Show that T is linear, $\ker T = U$, and $\operatorname{im} T = V$.
(b) Show that $\dim U + \dim V = n + 1$.

(c) If $n = 2k$ or $n = 2k + 1$, show that $\dim U = k + 1$, and hence find $\dim V$ from part (b). [*Hint:* Show that a polynomial is even just when it has the form $p(x^2)$ for some polynomial $p(x)$.]

16. Define $T : \mathbf{P}_n \to \mathbf{P}_{n-1}$ by $T[p(x)] = p(x + 1) - p(x)$.
    (a) Verify that $T[p(x)]$ really does lie in $\mathbf{P}_{n-1}$ for all $p(x)$ in $\mathbf{P}_n$.
    (b) Show that T is a linear transformation.
    (c) Show that ker T is the set of all constant polynomials. [*Hint:* If $p(x)$ is in ker T, define $q(x) = p(x) - p(0)$. Show that $q(1) = q(2) = q(3) = \cdots = 0$ and deduce that $q(x)$ is the zero polynomial.]
    (d) Show that every polynomial $f(x)$ in $\mathbf{P}_{n-1}$ can be written as $f(x) = p(x + 1) - p(x)$ for some polynomial $p(x)$ in $\mathbf{P}_n$.
17. Let U and V denote the spaces of symmetric and skew-symmetric $n \times n$ matrices. Show that $\dim U + \dim V = n^2$.
18. Let K be a subspace of a vector space V, and let $\dim V = n$ and $\dim K = k$.
    (a) If $K = \ker T$ for some linear transformation $T : V \to W$, show that $\dim W \geq n - k$.
    (b) If W is any vector space of dimension $n - k$, use Theorem 3 to construct an onto linear transformation $T : V \to W$ such that $\ker T = K$. [*Hint:* Use a basis $\{\mathbf{u}_1, \mathbf{u}_2, \ldots, \mathbf{u}_k, \mathbf{w}_1, \mathbf{w}_2, \ldots, \mathbf{w}_{n-k}\}$ of V, where $\{\mathbf{u}_1, \ldots, \mathbf{u}_k\}$ is a basis of K (Theorem 9 in Section 5.3).]
19. Show that linear independence is "preserved" by one-to-one transformations and that spanning sets are "preserved" by onto transformations. More precisely, if $T : V \to W$ is a linear transformation, show that:
    (a) If T is one-to-one and $\{\mathbf{v}_1, \ldots, \mathbf{v}_n\}$ is linearly independent in V, then $\{T(\mathbf{v}_1), \ldots, T(\mathbf{v}_n)\}$ is linearly independent in W.
    (b) If T is onto and $V = \operatorname{span}\{\mathbf{v}_1, \ldots, \mathbf{v}_n\}$ then $W = \operatorname{span}\{T(\mathbf{v}_1), \ldots, T(\mathbf{v}_n)\}$.
20. If $T : \mathbb{R}^n \to \mathbb{R}^n$ is a linear transformation of rank 1, show that there exist numbers $a_1, a_2, \ldots, a_n$ and $b_1, b_2, \ldots, b_n$ such that $T(\mathbf{v}) = \mathbf{v}A$ for all rows $\mathbf{v}$ in $\mathbb{R}^n$, where

$$A = \begin{bmatrix} a_1b_1 & a_1b_2 & \cdots & a_1b_n \\ a_2b_1 & a_2b_2 & \cdots & a_2b_n \\ \vdots & \vdots & & \vdots \\ a_nb_1 & a_nb_2 & \cdots & a_nb_n \end{bmatrix}$$

    [*Hint:* Exercise 9 in Section 8.1.1.]
21. Prove the corollary to Theorem 8.
22. Let $T : V \to \mathbb{R}$ be a nonzero linear transformation, where $\dim V = n$. Show that there is a basis $\{\mathbf{e}_1, \ldots, \mathbf{e}_n\}$ of V such that $T(r_1\mathbf{e}_1 + r_2\mathbf{e}_2 + \cdots + r_n\mathbf{e}_n) = r_1$.

8.1.3 Isomorphisms and Composition

Often vector spaces can "look" quite different but, at bottom, be the same vector space displayed in different symbols. The notion of isomorphism clarifies this.

DEFINITION A linear transformation $T : V \to W$ is called an **ismorphism** if it is both onto and one-to-one. The vector spaces V and W are called **isomorphic** if there exists an isomorphism $T : V \to W$; when this is the case, we write $V \cong W$.

The word *isomorphism* comes from two Greek roots: *iso*, meaning "same," and *morphos*, meaning "form." The isomorphism T induces a pairing

$$\mathbf{v} \to T(\mathbf{v})$$

between the vector $\mathbf{v}$ in V and the vector $T(\mathbf{v})$ in W that preserves vector addition and scalar multiplication. Hence, *as far as their vector space properties are concerned,* the spaces V and W are identical except for notation. Because addition and scalar multiplication in either space are completely determined by the same operations in the other space—all *vector space* properties of either space are completely determined by those of the other. This means a lot. For example, it implies that the dimensions of the spaces must be the same, because the whole notion of dimension was defined in terms of the vector addition and scalar multiplication. We shall return to this later.

EXAMPLE 27 The identity transformation $1_V : V \to V$ is an isomorphism for any vector space V.

EXAMPLE 28 Let $T : \mathbf{M}_{mn} \to \mathbf{M}_{nm}$ be defined by $T(A) = A^T$ for all A in $\mathbf{M}_{mn}$. Show that T is an isomorphism, so $\mathbf{M}_{mn} \cong \mathbf{M}_{nm}$.

Solution Part **(h)** of Exercise 1.

EXAMPLE 29 If U is any invertible $m \times m$ matrix, show that the map $T : \mathbf{M}_{mn} \to \mathbf{M}_{mn}$ given by $T(X) = UX$ is an isomorphism.

Solution See Example 21.

We considered one of the most important examples of isomorphic spaces in Chapter 4. There Euclidean space $\mathbb{R}^3$ was *identified* with the space of geometric vectors by pairing each 3-tuple (x,y,z) with the "arrow" $\mathbf{v}$ from the origin to the point $P(x,y,z)$—called the *position vector* of the point P. These arrows form a vector space using the parallelogram law of vector addition and the scalar multiplication described in Section 4.1. The remarkable thing is that the function T from $\mathbb{R}^3$ to this space given by $T(x,y,z) = \mathbf{v}$ is an isomorphism (this is Theorem 4 in Section 4.1), so $\mathbb{R}^3$ and the space of arrows are isomorphic. This fact justifies the *identification*

$$\mathbf{v} = (x,y,z)$$

of the geometric arrows with the algebraic 3-tuples that was made in Chapter 4. This identification is very useful. The arrows give a "picture" of the vectors and so bring geometric intuition into $\mathbb{R}^3$; the 3-tuples are useful for doing detailed calculations and so bring analytic power into geometry. This is one of the best examples of the power of an isomorphism to shed light on *both* spaces being considered.

The following theorem gives a very useful characterization of isomorphisms: they are the linear transformations that preserve bases.

THEOREM 10 If V and W are finite dimensional spaces, the following conditions are equivalent for a linear transformation $T : V \to W$.

(1) T is an isomorphism.

(2) If $\{\mathbf{e}_1, \mathbf{e}_2, \ldots, \mathbf{e}_n\}$ is any basis of V, then $\{T(\mathbf{e}_1), T(\mathbf{e}_2), \ldots, T(\mathbf{e}_n)\}$ is a basis of W.

(3) There exists a basis $\{\mathbf{e}_1, \mathbf{e}_2, \ldots, \mathbf{e}_n\}$ of V such that $\{T(\mathbf{e}_1), T(\mathbf{e}_2), \ldots, T(\mathbf{e}_n)\}$ is a basis of W.

Proof (1) implies (2). Let $\{\mathbf{e}_1, \ldots, \mathbf{e}_n\}$ be a basis of V. If $t_1T(\mathbf{e}_1) + \cdots + t_nT(\mathbf{e}_n) = \mathbf{0}$ with t_i in $\mathbb{R}$, then $T[t_1\mathbf{e}_1 + \cdots + t_n\mathbf{e}_n] = \mathbf{0}$, so $t_1\mathbf{e}_1 + \cdots + t_n\mathbf{e}_n = \mathbf{0}$ (because $\ker T = 0$). But then each $t_i = 0$ by the independence of the $\mathbf{v}_i$, so $\{T(\mathbf{e}_1), \ldots, T(\mathbf{e}_n)\}$ is linearly independent. To show that it spans W, choose $\mathbf{w}$ in W. Because T is onto, $\mathbf{w} = T(\mathbf{v})$ for some $\mathbf{v}$ in V, so write $\mathbf{v} = t_1\mathbf{v}_1 + \cdots + t_n\mathbf{v}_n$. Then $\mathbf{w} = T(\mathbf{v}) = t_1T(\mathbf{v}_1) + \cdots + t_nT(\mathbf{v}_n)$, so $\{T(\mathbf{v}_1), \ldots, T(\mathbf{v}_n)\}$ spans W.

(2) implies (3). This is clear. V is finite dimensional and so *has* a basis.

(3) implies (1). If $T(\mathbf{v}) = \mathbf{0}$, write $\mathbf{v} = v_1\mathbf{e}_1 + \cdots + v_n\mathbf{e}_n$. Then $\mathbf{0} = T(\mathbf{v}) = v_1T(\mathbf{e}_1) + \cdots + v_nT(\mathbf{e}_n)$, so $v_1 = \cdots = v_n = 0$ by (3). Hence $\mathbf{v} = \mathbf{0}$, so $\ker T = 0$ and T is one-to-one. To show that T is onto, let $\mathbf{w}$ be any vector in W. By (3) there exist $w_1, \ldots, w_n$ in $\mathbb{R}$ such that $\mathbf{w} = w_1T(\mathbf{e}_1) + \cdots + w_nT(\mathbf{e}_n) = T(w_1\mathbf{e}_1 + \cdots + w_n\mathbf{e}_n)$. This proves that T is onto. ■

This theorem dovetails nicely with Theorem 3 as follows. Let V and W be vector spaces of dimension n, and suppose that $\{\mathbf{e}_1, \mathbf{e}_2, \ldots, \mathbf{e}_n\}$ and $\{\mathbf{f}_1, \mathbf{f}_2, \ldots, \mathbf{f}_n\}$ are bases of V and W, respectively. Theorem 3 asserts that there exists a linear transformation $T : V \to W$ such that

$$T(\mathbf{e}_i) = \mathbf{f}_i \qquad \text{for each } i = 1, 2, \ldots, n$$

Then $\{T(\mathbf{e}_1), \ldots, T(\mathbf{e}_n)\}$ is evidently a basis of W, so T is an isomorphism. Furthermore, the action of T is prescribed by

$$T(r_1\mathbf{e}_1 + \cdots + r_n\mathbf{e}_n) = r_1\mathbf{f}_1 + \cdots + r_n\mathbf{f}_n$$

so isomorphisms between spaces of equal dimension can be written down at will as soon as bases are known. In particular, we have proved half of the following theorem.

THEOREM 11 The following are equivalent for two finite dimensional vector spaces V and W.

(1) $V \cong W$; that is, V and W are isomorphic

(2) $\dim V = \dim W$

Proof The foregoing discussion shows that (2) implies (1). Assume that $V \cong W$ and let $T : V \to W$ be an isomorphism. If $\{\mathbf{e}_1, \ldots, \mathbf{e}_n\}$ is a basis of V, then $\{T(\mathbf{e}_1), \ldots, T(\mathbf{e}_n)\}$ is a basis of W by Theorem 10. This shows that $\dim W = n = \dim V$ and so proves that (1) implies (2). ■

Hence the dimension of a vector space completely determines the spaces to which it is isomorphic.

EXAMPLE 30 Let V denote the space of all 2×2 symmetric matrices. Find an isomorphism $T : \mathbf{P}_2 \to V$ such that $T(1) = I$.

Solution $\{1, x, x^2\}$ is a basis of $\mathbf{P}_2$, and we want a basis of V containing I. The set $\left\{I = \begin{bmatrix} 1 & 0 \\ 0 & 1 \end{bmatrix}, \begin{bmatrix} 0 & 1 \\ 1 & 0 \end{bmatrix}, \begin{bmatrix} 0 & 0 \\ 0 & 1 \end{bmatrix}\right\}$ is independent in V, so it is a basis because $\dim V = 3$ (by Example 13 in Section 5.3). Hence define $T : \mathbf{P}_2 \to V$ by taking $T(1) = \begin{bmatrix} 1 & 0 \\ 0 & 1 \end{bmatrix}$, $T(x) = \begin{bmatrix} 0 & 1 \\ 1 & 0 \end{bmatrix}$, and $T(x^2) = \begin{bmatrix} 0 & 0 \\ 0 & 1 \end{bmatrix}$ and extending linearly as in Theorem 3. Then T is an isomorphism by Theorem 10, and its action is given by $T(a + bx + cx^2) = aT(1) + bT(x) + cT(x^2) = \begin{bmatrix} a & b \\ b & a + c \end{bmatrix}$.

Theorem 11 shows that every vector space of dimension n is isomorphic to the space $\mathbb{R}^n$ of columns (because $\dim \mathbb{R}^n = n$). Here is a specific isomorphism.

EXAMPLE 31 If V has dimension n and B is any ordered basis of V, the coordinate transformation $C_B : V \to \mathbb{R}^n$ is an isomorphism.

Solution If $B = \{\mathbf{e}_1, \mathbf{e}_2, \ldots, \mathbf{e}_n\}$, then C_B is given by $C_B(\mathbf{v}) = \begin{bmatrix} v_1 \\ v_2 \\ \vdots \\ v_n \end{bmatrix}$, where $\mathbf{v} = v_1\mathbf{e}_1 + \cdots + v_n\mathbf{e}_n$. This is evidently onto and one-to-one, and it was shown to be linear in Theorem 4 in Section 5.5.

DEFINITION

Given linear transformations $V \xrightarrow{T} W \xrightarrow{S} U$, the **composite** $ST : V \to U$ of T and S is defined by

$$ST(\mathbf{v}) = S[T(\mathbf{v})] \qquad \text{for all } \mathbf{v} \text{ in } V$$

The operation of forming the new function ST is called **composition.**

The action of ST can be described compactly as follows: ST means first T then S. (Incidentally, some authors write $S \circ T$ in place of ST, but we shall stick to the simpler notation.)

EXAMPLE 32 Let $T : \mathbb{R}^3 \to \mathbb{R}^2$ and $S : \mathbb{R}^2 \to \mathbb{R}^4$ be defined by $T(x,y,z) = (x + y, y + z)$ and $S(x,y) = (x - y, x + y, y, x)$. Describe the action of ST.

Solution Given (x,y,z) in $\mathbb{R}^3$, the definition yields

$$ST(x,y,z) = S[T(x,y,z)] = S(x + y, y + z) = (x - z, x + 2y + z, y + z, x + y)$$

This describes $ST(x,y,z)$ for all (x,y,z) in $\mathbb{R}^3$.

Not all pairs of linear transformations can be composed. In fact, if S and T are the transformations in Example 32, then ST is defined as was shown but TS *cannot* be formed, because

$$\mathbb{R}^2 \xrightarrow{S} \mathbb{R}^4 \qquad \text{and} \qquad \mathbb{R}^3 \xrightarrow{T} \mathbb{R}^2$$

do not "link" in this order. And, even if ST and TS *can* both be formed, the new functions ST and TS need not be equal.

EXAMPLE 33 Define $S : \mathbf{M}_{22} \to \mathbf{M}_{22}$ and $T : \mathbf{M}_{22} \to \mathbf{M}_{22}$ by $S\begin{bmatrix} a & b \\ c & d \end{bmatrix} = \begin{bmatrix} c & d \\ a & b \end{bmatrix}$ and $T(A) = A^T$. Describe the action of ST and TS, and show that $ST \neq TS$.

Solution $ST\begin{bmatrix} a & b \\ c & d \end{bmatrix} = S\begin{bmatrix} a & c \\ b & d \end{bmatrix} = \begin{bmatrix} b & d \\ a & c \end{bmatrix}$, whereas $TS\begin{bmatrix} a & b \\ c & d \end{bmatrix} = T\begin{bmatrix} c & d \\ a & b \end{bmatrix} = \begin{bmatrix} c & a \\ d & b \end{bmatrix}$. It is clear that $TS\begin{bmatrix} a & b \\ c & d \end{bmatrix}$ need not equal $ST\begin{bmatrix} a & b \\ c & d \end{bmatrix}$, so $TS \neq ST$.

THEOREM 12 Let V and W be finite dimensional vector spaces. The following conditions are equivalent for a linear transformation $T : V \to W$.

(1) T is an isomorphism.

(2) There exists a linear transformation $S : W \to V$ such that $ST = 1_V$ and $TS = 1_W$.

Moreover, S is an isomorphism and is uniquely determined by T: If $\mathbf{w}$ in W is written as $\mathbf{w} = T(\mathbf{v})$, then $S(\mathbf{w}) = \mathbf{v}$.

Proof (1) implies (2). If $B = \{\mathbf{e}_1, \ldots, \mathbf{e}_n\}$ is a basis of V, then $D = \{T(\mathbf{e}_1), \ldots, T(\mathbf{e}_n)\}$ is a basis of W by Theorem 10. Hence (using Theorem 3), define $S : W \to V$ by

$$S[T(\mathbf{e}_i)] = \mathbf{e}_i \qquad \text{for each } i \qquad (*)$$

This gives $ST = 1_V$ by Theorem 2. But applying T gives $T[S[T(\mathbf{e}_i)]] = T(\mathbf{e}_i)$ for each i, so $TS = 1_W$ (again by Theorem 2, using the basis D of W).

(2) implies (1). If $T(\mathbf{v}) = T(\mathbf{v}_1)$, then $S[T(\mathbf{v})] = S[T(\mathbf{v}_1)]$. Because $ST = 1_V$, this reads $\mathbf{v} = \mathbf{v}_1$; that is, T is one-to-one. Given $\mathbf{w}$ in W, the fact that $TS = 1_W$ means that $\mathbf{w} = T[S(\mathbf{w})]$, and T is onto.

S is uniquely determined by the condition $ST = 1_V$, because this condition implies (*). It is an isomorphism because it carries the basis D to B. Finally, given $\mathbf{w}$ in W, write $\mathbf{w} = r_1T(\mathbf{e}_1) + \cdots + r_nT(\mathbf{e}_n) = T(\mathbf{v})$, where $\mathbf{v} = r_1\mathbf{e}_1 + \cdots + r_n\mathbf{e}_n$. Then $S(\mathbf{w}) = \mathbf{v}$ by (*). ■

DEFINITION Given an isomorphism $T : V \to W$, the isomorphism $S : W \to V$ satisfying condition (2) of Theorem 12 is called the **inverse** of T and is denoted by T^{-1}.

Equation (*) in the proof of Theorem 12 shows how to define T^{-1} using the image of a basis under the isomorphism T. Here is an example.

EXAMPLE 34 Define $T : P_1 \to P_1$ by $T(a + bx) = (a - b) + ax$. Show that T has an inverse, and find the action of T^{-1}.

Solution Because $T(1) = 1 + x$ and $T(x) = -1$, T carries the basis $B = \{1, x\}$ to the basis $D = \{1 + x, -1\}$. Hence T is an isomorphism, and T^{-1} is defined by $T^{-1}(1 + x) = 1$ and $T^{-1}(-1) = x$. Because $a + bx = b(1 + x) + (b - a)(-1)$, we obtain $T^{-1}(a + bx) = b + (b - a)x$.

EXAMPLE 35 If $B = \{\mathbf{e}_1, \ldots, \mathbf{e}_n\}$ is an ordered basis of a vector space V, the coordinate map $C_B : V \to \mathbb{R}^n$ is an isomorphism (Example 31). Find the action of C_B^{-1}.

Solution The action of C_B is given by $C_B(v_1\mathbf{e}_1 + \cdots + v_n\mathbf{e}_n) = \begin{bmatrix} v_1 \\ \vdots \\ v_n \end{bmatrix}$ for all $\mathbf{v} = v_1\mathbf{e}_1 + \cdots + v_n\mathbf{e}_n$ in V. Then the action of C_B^{-1} "reverses" this.

$$C_B^{-1}\begin{bmatrix} v_1 \\ \vdots \\ v_n \end{bmatrix} = v_1\mathbf{e}_1 + \cdots + v_n\mathbf{e}_n \qquad \text{for all } \begin{bmatrix} v_1 \\ \vdots \\ v_n \end{bmatrix} \text{ in } \mathbb{R}^n$$

Condition (2) in Theorem 12 characterizes the inverse of a linear transformation $T : V \to W$ as the (unique) transformation $S : W \to V$ that satisfies $ST = 1_V$ and $TS = 1_W$. This often determines the inverse.

EXAMPLE 36 Define $T : \mathbb{R}^3 \to \mathbb{R}^3$ by $T(x,y,z) = (z,x,y)$. Show that $T^3 = 1_{\mathbb{R}^3}$, and hence find T^{-1}.

Solution $T^2(x,y,z) = T[T(x,y,z)] = T(z,x,y) = (y,z,x)$. Hence

$$T^3(x,y,z) = T[T^2(x,y,z)] = T(y,z,x) = (x,y,z)$$

This shows that $T^3 = 1_{\mathbb{R}^3}$, so $T(T^2) = 1_{\mathbb{R}^3} = (T^2)T$. Thus $T^{-1} = T^2$ by (2) of Theorem 12.

EXAMPLE 37 Define $T : \mathbf{P}_n \to \mathbb{R}^{n+1}$ by $T(p) = (p(0), p(1), \ldots, p(n))$ for all p in $\mathbf{P}_n$. Show that T^{-1} exists.

Solution The verification that T is linear is left to the reader. If $T(p) = 0$ then $p(k) = 0$ for $k = 0, 1, \ldots, n$, so p has $n + 1$ distinct roots. Because p has degree at most n, this implies that $p = 0$ is the zero polynomial (corollary to Theorem 3 in Section 5.6) and hence that T is one-to-one. But $\dim \mathbf{P}_n = n + 1 = \dim \mathbb{R}^{n+1}$, so this means that T is also onto and hence is an isomorphism. Thus T^{-1} exists by Theorem 12. Note that we have not given an explicit description of the action of T^{-1}. We have merely shown that such a description exists. To give it requires some ingenuity; one method involves the Lagrange interpolation formula (Theorem 3 in Section 5.6).

The following theorem collects several properties of the matrix transformations T_A for reference.

THEOREM 13 Let A and B denote matrices.

(1) If $T_A = T_B$, then $A = B$.

(2) $T_I = 1_{\mathbb{R}^n}$

(3) $T_A T_B = T_{AB}$

(4) T_A has an inverse if and only if A is invertible, and then $(T_A)^{-1} = T_{A^{-1}}$.

(5) $\operatorname{im} T_A = \operatorname{col} A$, so $\operatorname{rank} T_A = \operatorname{rank} A$.

Proof If $\mathbf{e}_j$ is column j of I_n, then $T_A(\mathbf{e}_j) = A\mathbf{e}_j$ is column j of A. This implies (1), (2) is obvious, and (5) is Example 16. Property (3) follows from the fact that

$$T_A[T_B(\mathbf{v})] = A[B\mathbf{v}] = (AB)\mathbf{v} = T_{AB}(\mathbf{v})$$

holds for all $\mathbf{v}$. To prove (4), assume first that A^{-1} exists. Then $T_A T_{A^{-1}} = T_{AA^{-1}} = T_I$ is the identity map, using (2) and (3). Similarly $T_{A^{-1}} T_A = T_I$, so $T_{A^{-1}}$ is the inverse of T_A by (2) of Theorem 12. Conversely, assume that A is $n \times n$ and that $(T_A)^{-1} : \mathbb{R}^n \to \mathbb{R}^n$ exists. Then $(T_A)^{-1} = T_B$ for some $n \times n$ matrix B (Theorem 4), and so (2) and (3) give

$$T_{AB} = T_A T_B = T_A (T_A)^{-1} = 1_{\mathbb{R}^n} = T_I$$

Hence $AB = I$ by (1). Similarly $BA = I$, so $B = A^{-1}$. ■

We conclude with some further properties of inverses and isomorphisms. The proofs are left as Exercise 26.

THEOREM 14 Let $V \xrightarrow{T} W \xrightarrow{S} U \xrightarrow{R} Z$ be linear transformations.

(1) The composite ST is again a linear transformation.

(2) $T1_V = T$ and $1_W T = T$

(3) $(RS)T = R(ST)$

Theorem 14 is valid even if the vector spaces are not finite dimensional. Moreover, condition (2) in Theorem 12 makes sense in general: If $T : V \to W$ is any linear transformation, suppose there exists a linear transformation $S : W \to V$ satisfying $ST = 1_V$ and $TS = 1_W$. This uniquely determines S. Indeed, if $S' : W \to V$ satisfies $S'T = 1_V$ and $TS' = 1_W$, then Theorem 14 gives

$$S' = S'1_W = S'(TS) = (S'T)S = 1_VS = S$$

As in the finite dimensional case, this unique transformation S is called the **inverse** of T, and it is denoted by $S = T^{-1}$. It is thus uniquely determined by the conditions

$$T^{-1}T = 1_V \qquad \text{and} \qquad TT^{-1} = 1_W$$

These conditions, together with Theorem 14, can be used to prove the following result. The proof parallels the corresponding proof for matrices (Theorem 2 in Section 2.3) and is left as Exercise 26.

THEOREM 15

Let V, W, and U denote vector spaces.

(1) $1_V : V \to V$ has an inverse and $1_V^{-1} = 1_V$.

(2) If $T : V \to W$ has an inverse, then $T^{-1} : W \to V$ also has an inverse and $(T^{-1})^{-1} = T$.

(3) If $V \xrightarrow{T} W \xrightarrow{S} U$ and T and S both have inverses, then $ST : V \to U$ also has an inverse and $(ST)^{-1} = T^{-1}S^{-1}$.

Theorem 12 has the following extension to arbitrary spaces (the proof is omitted).

THEOREM 16

A linear transformation is an isomorphism if and only if it has an inverse.

Recall that two vector spaces V and W are called *isomorphic* (written $V \cong W$) if there exists an isomorphism $T : V \to W$. The following properties follow from Theorem 15; the proofs are left to the reader (Exercise 26).

THEOREM 17

(1) $V \cong V$ for every vector space V.

(2) If $V \cong W$, then $W \cong V$.

(3) If $V \cong W$ and $W \cong U$, then $V \cong U$.

Thus $\cong$ is an equivalence relation on the class of all vector spaces.

EXERCISES 8.1.3

1. Verify that each of the following is an isomorphism (Theorem 9 is useful).
   (a) $T : \mathbb{R}^3 \to \mathbb{R}^3$; $T(x,y,z) = (x + y, y + z, z + x)$
   (b) $T : \mathbb{R}^3 \to \mathbb{R}^3$; $T(x,y,z) = (x, x + y, x + y + z)$
   (c) $T : \mathbb{C} \to \mathbb{C}$; $T(z) = \bar{z}$
   (d) $T : \mathbf{M}_{mn} \to \mathbf{M}_{mn}$; $T(X) = UXV$, U and V invertible
   (e) $T : \mathbf{P}_1 \to \mathbb{R}^2$; $T[p(x)] = [p(0),p(1)]$
   (f) $T : V \to V$; $T(\mathbf{v}) = k\mathbf{v}$, $k \neq 0$ a fixed number, V any vector space
   (g) $T : \mathbf{M}_{22} \to \mathbb{R}^4$; $T\begin{bmatrix} a & b \\ c & d \end{bmatrix} = (a + b, d, c, a - b)$
   (h) $T : \mathbf{M}_{mn} \to \mathbf{M}_{nm}$; $T(A) = A^T$
2. Show that $\{a + bx + cx^2, a_1 + b_1x + c_1x^2, a_2 + b_2x + c_2x^2\}$ is a basis of $\mathbf{P}_2$ if and only if $\{(a,b,c), (a_1,b_1,c_1), (a_2,b_2,c_2)\}$ is a basis of $\mathbb{R}^3$.
3. If V is any vector space, let V^n denote the space of all n-tuples $(\mathbf{v}_1, \mathbf{v}_2, \ldots, \mathbf{v}_n)$, where each $\mathbf{v}_i$ lies in V. (This is a vector space with componentwise operations, see Exercise 21 in Section 5.1.) If $C_j(A)$ denotes the jth column of the $m \times n$ matrix A, show that $T : \mathbf{M}_{mn} \to (\mathbb{R}^m)^n$ is an isomorphism if $T(A) = [C_1(A), C_2(A), \ldots, C_n(A)]$. (Here $\mathbb{R}^m$ consists of columns.)
4. In each case, compute the action of ST and TS, and show that $ST \neq TS$.
   (a) $S : \mathbb{R}^2 \to \mathbb{R}^2$ with $S(x,y) = (y,x)$; $T : \mathbb{R}^2 \to \mathbb{R}^2$ with $T(x,y) = (x,0)$
   (b) $S : \mathbb{R}^3 \to \mathbb{R}^3$ with $S(x,y,z) = (x,0,z)$; $T : \mathbb{R}^3 \to \mathbb{R}^3$ with $T(x,y,z) = (x + y, 0, y + z)$
5. In each case, show that the linear transformation T satisfies $T^2 = T$.
   (a) $T : \mathbb{R}^4 \to \mathbb{R}^4$; $T(x,y,z,w) = (x,0,z,0)$
   (b) $T : \mathbb{R}^2 \to \mathbb{R}^2$; $T(x,y) = (x + y, 0)$
6. Determine whether each of the following transformations T has an inverse and, if so, determine the action of T^{-1}.
   (a) $T : \mathbb{R}^3 \to \mathbb{R}^3$; $T(x,y,z) = (x + y, y + z, z + x)$
   (b) $T : \mathbb{R}^4 \to \mathbb{R}^4$; $T(x,y,z,t) = (x + y, y + z, z + t, t + x)$
   (c) $T : \mathbf{M}_{22} \to \mathbf{M}_{22}$; $T\begin{bmatrix} a & b \\ c & d \end{bmatrix} = \begin{bmatrix} a - c & b - d \\ 2a - c & 2b - d \end{bmatrix}$
   (d) $T : \mathbf{M}_{22} \to \mathbf{M}_{22}$; $T\begin{bmatrix} a & b \\ c & d \end{bmatrix} = \begin{bmatrix} a + 2c & b + 2d \\ 3c - a & 3d - b \end{bmatrix}$
   (e) $T : \mathbf{P}_2 \to \mathbb{R}^3$; $T(a + bx + cx^2) = (a - c, 2b, a + c)$
   (f) $T : \mathbf{P}_2 \to \mathbb{R}^3$; $T(p) = [p(0),p(1),p(-1)]$
7. In each case, show that T is self-inverse: $T^{-1} = T$.
   (a) $T : \mathbb{R}^4 \to \mathbb{R}^4$; $T(x,y,z,w) = (x,-y,-z,w)$
   (b) $T : \mathbb{R}^2 \to \mathbb{R}^2$; $T(x,y) = (ky - x, y)$, k any fixed number
8. Using the notation of Example 13, show that $R_\theta R_\phi = R_{\theta+\phi}$ for all angles θ and ϕ.
9. In each case, show that $T^6 = 1_{\mathbb{R}^4}$ and so determine T^{-1}.

(a) $T : \mathbb{R}^4 \to \mathbb{R}^4$; $T(x,y,z,w) = (-x,z,w,y)$

(b) $T : \mathbb{R}^4 \to \mathbb{R}^4$; $T(x,y,z,w) = (-y, x - y, z, -w)$

10. Given linear transformations $V \xrightarrow{T} W \xrightarrow{S} U$:

(a) If S and T are both one-to-one, show that ST is one-to-one.

(b) If S and T are both onto, show that ST is onto.

11. Let $T : V \to W$ be a linear transformation.

(a) If T is one-to-one and $TR = TR_1$ for transformations R and $R_1 : U \to V$, show that $R = R_1$.

(b) If T is onto and $ST = S_1T$ for transformations S and $S_1 : W \to U$, show that $S = S_1$.

12. Consider the linear transformations $V \xrightarrow{T} W \xrightarrow{R} U$.

(a) Show that $\ker T \subseteq \ker RT$.

(b) Show that $\operatorname{im} RT \subseteq \operatorname{im} R$.

13. Let $V \xrightarrow{T} U \xrightarrow{S} W$ be linear transformations.

(a) If ST is one-to-one, show that T is one-to-one and that $\dim V \le \dim U$.

(b) If ST is onto, show that S is onto and that $\dim W \le \dim U$.

14. Let $T : V \to V$ be a linear transformation. Show that $T^2 = 1_V$ if and only if T is invertible and $T = T^{-1}$.

15. Let N be a nilpotent $n \times n$ matrix (that is, $N^k = 0$ for some k). Show that $T : \mathbf{M}_{nm} \to \mathbf{M}_{nm}$ is an isomorphism if $T(X) = X - NX$. [*Hint:* If X is in $\ker T$, show that $X = NX = N^2X = \cdots$. Then use Theorem 9.]

16. Let $T : V \to W$ be a linear transformation, and let $\{\mathbf{e}_1, \ldots, \mathbf{e}_r, \mathbf{e}_{r+1}, \ldots, \mathbf{e}_n\}$ be any basis of V such that $\{\mathbf{e}_{r+1}, \ldots, \mathbf{e}_n\}$ is a basis of $\ker T$. Show that $\operatorname{im} T \cong \operatorname{span}\{\mathbf{e}_1, \ldots, \mathbf{e}_r\}$. [*Hint:* See the corollary to Theorem 8.]

17. Is every isomorphism $T : \mathbf{M}_{22} \to \mathbf{M}_{22}$ given (as in Example 29) by an invertible matrix U such that $T(X) = UX$ for all X in $\mathbf{M}_{22}$? Prove your answer.

18. Let $\mathbf{D}_n$ denote the space of all functions f from $\{1, 2, \ldots, n\}$ to $\mathbb{R}$ (see Exercise 13 in Section 5.3.2). If $T : \mathbf{D}_n \to \mathbb{R}^n$ is defined by $T(f) = (f(1), f(2), \ldots, f(n))$, show that T is an isomorphism.

19. (a) Let V be the vector space of Exercise 7 in Section 5.1. Find an isomorphism $T : V \to \mathbb{R}^1$.

(b) Let V be the vector space of Exercise 8 in Section 5.1. Find an isomorphism $T : V \to \mathbb{R}^2$.

20. Let $V \xrightarrow{T} W \xrightarrow{S} V$ be linear transformations such that $ST = 1_V$. If $\dim V = \dim W = n$, show that $S = T^{-1}$ and $T = S^{-1}$. [*Hint:* Exercise 13 and Theorems 9, 12, and 14.]

21. Let $V \xrightarrow{T} W \xrightarrow{S} V$ be functions such that $TS = 1_W$ and $ST = 1_V$. If T is linear, show that S is also linear.

22. Use Theorems 13 and 14 to prove that $(AB)C = A(BC)$ for matrices of sizes $m \times n$, $n \times k$, and $k \times p$, respectively.

23. Let A and B be matrices of size $p \times m$ and $n \times q$. Assume that $mn = pq$. Define $R : \mathbf{M}_{mn} \to \mathbf{M}_{pq}$ by $R(X) = AXB$.

(a) Show that $\mathbf{M}_{mn} \cong \mathbf{M}_{pq}$ by comparing dimensions.

(b) Show that R is a linear transformation.

(c) Show that if R is an isomorphism, then $m = p$ and $n = q$. [*Hint:* Show that $T : \mathbf{M}_{mn} \to \mathbf{M}_{pn}$ given by $T(X) = AX$ and $S : \mathbf{M}_{mn} \to \mathbf{M}_{mq}$ given by $S(X) = XB$ are both one-to-one, and use Theorem 8.]

24. Let $T : V \to V$ be a linear transformation such that $T^2 = 0$ is the zero transformation.

(a) If $V \neq 0$, show that T cannot be invertible.

(b) If $R : V \to V$ is defined by $R(\mathbf{v}) = \mathbf{v} + T(\mathbf{v})$ for all $\mathbf{v}$ in V, show that R is linear and invertible.

25. Let V consist of all sequences $[x_0, x_1, x_2, \ldots)$ of numbers, and define vector operations

$$[x_0, x_1, \ldots) + [y_0, y_1, \ldots) = [x_0 + y_0, x_1 + y_1, \ldots)$$
$$r[x_0, x_1, \ldots) = [rx_0, rx_1, \ldots)$$

(a) Show that V is a vector space of infinite dimension.

(b) Define $T : V \to V$ and $S : V \to V$ by $T[x_0, x_1, \ldots) = [x_1, x_2, \ldots)$ and $S[x_0, x_1, \ldots) = [0, x_0, x_1, \ldots)$. Show that $TS = 1_V$, so TS is one-to-one and onto, but T is not one-to-one and S is not onto.

26. **(a)** Prove Theorem 14.

(b) Prove Theorem 15.

(c) Prove Theorem 17.

(d) Use part **(b)** and Example 29 to show that, if $\dim V = \dim W$, then $V \cong W$ (half of Theorem 11)

27.[†] Define $T : \mathbf{P}_n \to \mathbf{P}_n$ by $T(p) = p(x) + xp'(x)$ for all p in $\mathbf{P}_n$.

(a) Show that T is linear.

(b) Show that $\ker T = 0$ and conclude that T is an isomorphism. [*Hint:* Write $p(x) = a_0 + a_1x + \cdots + a_nx^n$, and compare coefficients if $p(x) = -xp'(x)$.]

(c) Conclude that each $q(x)$ in $\mathbf{P}_n$ has the form $q(x) = p(x) + xp'(x)$ for some unique polynomial $p(x)$.

(d) Does this remain valid if T is defined by $T[p(x)] = p(x) - xp'(x)$?

28. Let $T : V \to W$ be a linear transformation, where V and W are finite dimensional.

(a) Show that T is one-to-one if and only if there exists a linear transformation $S : W \to V$ with $ST = 1_V$. [*Hint:* If $\{\mathbf{e}_1, \ldots, \mathbf{e}_n\}$ is a basis of V and T is one-to-one, show that W has a basis $\{T(\mathbf{e}_1), \ldots, T(\mathbf{e}_n), \mathbf{f}_{n+1}, \ldots, \mathbf{f}_{n+k}\}$ and use Theorem 3.]

[†]This exercise requires calculus.

(b) Show that T is onto if and only if there exists a linear transformation $S : W \to V$ with $TS = 1_W$. [*Hint:* Let $\{\mathbf{e}_1, \ldots, \mathbf{e}_r, \ldots, \mathbf{e}_n\}$ be a basis of V such that $\{\mathbf{e}_{r+1}, \ldots, \mathbf{e}_n\}$ is a basis of ker T. Use the corollary to Theorem 8 and Theorem 3.]

29. If $T : V \to V$ is a linear transformation where dim $V = n$, show that $TST = T$ for some isomorphism $S : V \to V$. [*Hint:* Let $\{\mathbf{e}_1, \ldots, \mathbf{e}_r, \mathbf{e}_{r+1}, \ldots, \mathbf{e}_n\}$ be as in the corollary to Theorem 8. Extend $\{T(\mathbf{e}_1), \ldots, T(\mathbf{e}_r)\}$ to a basis of V, and use Theorems 2, 3, and 10.]

30. Let S and T be linear transformations $V \to W$, where dim $V = n$ and dim $W = m$.

(a) Show that ker S = ker T if and only if $T = RS$ for some isomorphism $R : W \to W$.

(b) Show that im S = im T if and only if $T = SR$ for some isomorphism $R : V \to V$. [*Hint:* This is similar to the preceding two exercises.]

31. Let A and B denote $m \times n$ matrices. In each case show that (1) and (2) are equivalent.

(a) (1) A and B have the same null space.
(2) $B = PA$ for some invertible $m \times m$ matrix P.

(b) (1) A and B have the same range.
(2) $B = AQ$ for some invertible $n \times n$ matrix Q.
[*Hint:* Use the preceding exercise and Theorem 13.]

SECTION 8.2 Matrices and Linear Transformations

8.2.1 The Matrix of a Linear Transformation

It was shown in Theorem 4 in Section 8.1 that every linear transformation from $\mathbb{R}^n$ to $\mathbb{R}^m$ is a matrix transformation

$$T_A : \mathbb{R}^n \to \mathbb{R}^m \text{ given by } T_A(\mathbf{v}) = A\mathbf{v} \text{ for all columns } \mathbf{v} \text{ in } \mathbb{R}^n$$

Furthermore, the $m \times n$ matrix A is uniquely determined by T; in fact, its columns (in order) are $T(\mathbf{e}_1), T(\mathbf{e}_2), \ldots, T(\mathbf{e}_n)$, where $\{\mathbf{e}_1, \ldots, \mathbf{e}_n\}$ is the standard ordered basis of $\mathbb{R}^n$. In general, let $T : V \to W$ be *any* linear transformation where dim $V = n$ and dim $W = m$. The aim now is to describe the action of T as multiplication by a matrix, and the idea is to convert the vectors in V and W into columns in $\mathbb{R}^n$ and $\mathbb{R}^m$, respectively, and then represent the corresponding transformation $\mathbb{R}^n \to \mathbb{R}^m$ by a matrix. Moreover, a method of converting vectors to columns is already available: Simply replace a vector with its coordinates relative to some ordered basis.

More explicitly, let $B = \{\mathbf{e}_1, \mathbf{e}_2, \ldots, \mathbf{e}_n\}$ and D be ordered bases of V and W, respectively. Recall (Section 5.5) that, given a vector $\mathbf{v} = v_1\mathbf{e}_1 + \cdots + v_n\mathbf{e}_n$ in V, its coordinate vector relative to B is defined as

$$C_B(\mathbf{v}) = \begin{bmatrix} v_1 \\ \vdots \\ v_n \end{bmatrix}$$

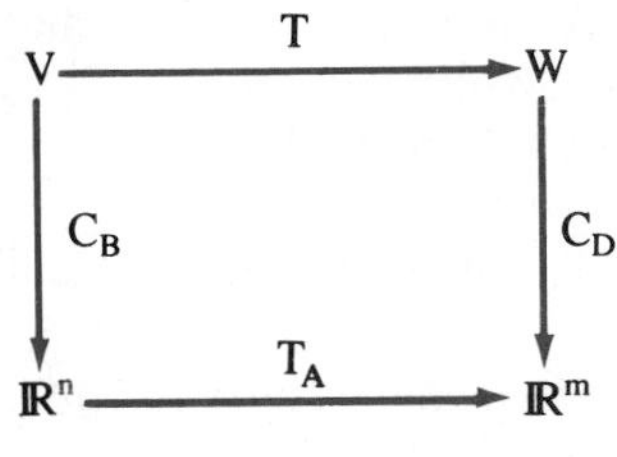

FIGURE 8.1

Furthermore, this defines an isomorphism $C_B : V \to \mathbb{R}^n$. Similarly, $C_D : W \to \mathbb{R}^m$ is an isomorphism and we have the situation shown in Figure 8.1, where A is an $m \times n$ matrix (to be determined). Our hope is that the matrix A can be chosen in such a way that the action of T on a vector $\mathbf{v}$ in V can be performed by first taking coordinates (that is, applying C_B to $\mathbf{v}$), then multiplying by A (applying T_A), and finally converting the resulting m-tuple back to a vector in W (applying $C_D{}^{-1}$). In symbols, we wish to choose A such that

$$T = C_D{}^{-1}T_AC_B \qquad \text{equivalently,} \qquad C_DT = T_AC_B$$

T_A acts by left multiplication by A, so the condition is

$$C_D[T(\mathbf{v})] = AC_B(\mathbf{v}) \qquad \text{for all } \mathbf{v} \text{ in } V$$

This requirement completely determines A. Write $A = [\mathbf{c}_1\ \mathbf{c}_2\ \dots\ \mathbf{c}_n]$ in terms of its columns. The fact that $C_B(\mathbf{e}_j)$ is column j of the identity matrix gives

$$C_D[T(\mathbf{e}_j)] = AC_B(\mathbf{e}_j) = \mathbf{c}_j \qquad \text{for all } j$$

Hence

$$A = [C_D[T(\mathbf{e}_1)]\ C_D[T(\mathbf{e}_2)] \dots C_D[T(\mathbf{e}_n)]]$$

This prompts the following definition:

DEFINITION

Given $T : V \to W$, $B = \{\mathbf{e}_1, \dots, \mathbf{e}_n\}$ and D as above, write

$$M_{DB}(T) = [C_D[T(\mathbf{e}_1)]\ C_D[T(\mathbf{e}_2)] \dots C_D[T(\mathbf{e}_n)]]$$

This is called the **matrix of T corresponding to the ordered bases B and D.**

This gives $M_{DB}(T)$ in terms of its columns. The discussion is summarized in the following important theorem.

THEOREM 1

Let $T : V \to W$ be a linear transformation where $\dim V = n$ and $\dim W = m$, and let $B = \{\mathbf{e}_1, \dots, \mathbf{e}_n\}$ and D be ordered bases of V and

W, respectively. Then the matrix $M_{DB}(T)$ given above is the unique $m \times n$ matrix A that satisfies

$$C_D T = T_A C_B$$

Hence the defining property of $M_{DB}(T)$ is

$$C_D[T(\mathbf{v})] = M_{DB}(T)C_B(\mathbf{v}) \quad \text{for all } \mathbf{v} \text{ in } V$$

EXAMPLE 1 Define $T : \mathbf{P}_2 \to \mathbb{R}^2$ by $T(a + bx + cx^2) = (a + c, b - a - c)$ for all polynomials $a + bx + cx^2$. If

$$B = \{\mathbf{e}_1 = 1, \mathbf{e}_2 = x, \mathbf{e}_3 = x^2\} \qquad \text{and} \qquad D = \{\mathbf{f}_1 = (1,0), \mathbf{f}_2 = (0,1)\}$$

compute $M_{DB}(T)$ and verify Theorem 1.

Solution We have $T(\mathbf{e}_1) = \mathbf{f}_1 - \mathbf{f}_2$, $T(\mathbf{e}_2) = \mathbf{f}_2$, and $T(\mathbf{e}_3) = \mathbf{f}_1 - \mathbf{f}_2$. Hence

$$M_{DB}(T) = [C_D[T(\mathbf{e}_1)]\ C_D[T(\mathbf{e}_2)]\ C_D[T(\mathbf{e}_3)]] = \begin{bmatrix} 1 & 0 & 1 \\ -1 & 1 & -1 \end{bmatrix}$$

If $\mathbf{v} = a + bx + cx^2 = a\mathbf{e}_1 + b\mathbf{e}_2 + c\mathbf{e}_3$, then $T(\mathbf{v}) = (a + c)\mathbf{f}_1 + (b - a - c)\mathbf{f}_2$, so

$$C_D[T(\mathbf{v})] = \begin{bmatrix} a + c \\ b - a - c \end{bmatrix} = \begin{bmatrix} 1 & 0 & 1 \\ -1 & 1 & -1 \end{bmatrix} \begin{bmatrix} a \\ b \\ c \end{bmatrix} = M_{DB}(T)C_B(\mathbf{v})$$

as Theorem 1 asserts.

The next example will be referred to below.

EXAMPLE 2 Let A be an $m \times n$ matrix, and let $T_A : \mathbb{R}^n \to \mathbb{R}^m$ be the matrix transformation induced by A: $T_A(\mathbf{v}) = A\mathbf{v}$ for all $\mathbf{v}$ in $\mathbb{R}^n$. If B and D are the standard bases of $\mathbb{R}^n$ and $\mathbb{R}^m$, respectively (in the usual order), then

$$M_{DB}(T_A) = A$$

In other words, the matrix of T_A corresponding to the standard bases is A itself.

Solution Write $B = \{\mathbf{e}_1, \ldots, \mathbf{e}_n\}$. Because D is the standard basis of $\mathbb{R}^m$, it is easy to verify that $C_D(\mathbf{w}) = \mathbf{w}$ for all columns $\mathbf{w}$ in $\mathbb{R}^m$. Hence

$$M_{DB}(T_A) = [T_A(\mathbf{e}_1)\ T_A(\mathbf{e}_2) \ldots T_A(\mathbf{e}_n)] = [A\mathbf{e}_1\ A\mathbf{e}_2 \ldots A\mathbf{e}_n] = A$$

because $A\mathbf{e}_j$ is the jth column of A.

EXAMPLE 3 Let V and W have ordered bases B and D, respectively.

1. The identity transformation $1_V : V \to V$ has matrix $M_{BB}(1_V) = I_n$—the identity matrix.
2. The zero transformation $0 : V \to W$ has matrix $M_{DB}(0) = 0$—the zero matrix.

Solution Exercise 12.

The first result in Example 3 is false if the two bases of V are not equal. In fact, if B is the standard basis of $\mathbb{R}^n$, then the basis D of $\mathbb{R}^n$ can be chosen so that $M_{BD}(1_{\mathbb{R}^n})$ turns out to be any invertible matrix at all (Exercise 13).

The matrix $M_{DB}(T)$ of a linear transformation $T : V \to W$ can be used to compute the action of T. Given $\mathbf{v}$ in V, the idea is to compute $C_D[T(\mathbf{v})] = M_{DB}(T)C_B(\mathbf{v})$ first (a known quantity if the matrix $M_{DB}(T)$ is known) and then get $T(\mathbf{v})$.

EXAMPLE 4 Define a linear transformation $T : \mathbf{P}_2 \to \mathbb{R}^2$ by $T[p(x)] = (p(1), p(-1))$. Compute the matrix of T corresponding to $B = \{x + 1, x - 1, x^2\}$ and $D = \{(1,0), (1,-1)\}$, and use it to compute $T(x)$.

Solution

$$\begin{aligned} M_{DB}(T) &= [C_D[T(x+1)]\ C_D[T(x-1)]\ C_D[T(x^2)]] \\ &= [C_D(2,0)\ C_D(0,-2)\ C_D(1,1)] \\ &= \begin{bmatrix} 2 & -2 & 2 \\ 0 & 2 & -1 \end{bmatrix} \end{aligned}$$

Now $x = \frac{1}{2}(x + 1) + \frac{1}{2}(x - 1) + 0x^2$ in terms of the basis B, so

$$C_D[T(x)] = M_{DB}(T)C_B(x) = \begin{bmatrix} 2 & -2 & 2 \\ 0 & 2 & -1 \end{bmatrix}\begin{bmatrix} \frac{1}{2} \\ \frac{1}{2} \\ 0 \end{bmatrix} = \begin{bmatrix} 0 \\ 1 \end{bmatrix}$$

Finally, then, $T(x) = 0(1,0) + 1(1,-1) = (1,-1)$, in accordance with the original definition of T.

The next theorem shows that composition of linear transformations is compatible with multiplication of the corresponding matrices.

THEOREM 2 Let $V \xrightarrow{T} W \xrightarrow{S} U$ be linear transformations, and let B, D, and E be finite ordered bases of V, W, and U, respectively. Then

$$M_{EB}(ST) = M_{ED}(S) \cdot M_{DB}(T)$$

Proof It is possible to prove this by simply computing the matrices in question and multiplying by "brute force." However it is easier (and more instructive) to use the *uniqueness* part of Theorem 1. For simplicity of notation, write

$$X = M_{DB}(T),\ Y = M_{ED}(S), \text{ and } Z = M_{EB}(ST)$$

The aim is to show that $Z = YX$.

Assume that the dimensions of V, W, and U are n, m, and k, respectively, and consider the accompanying diagrams:

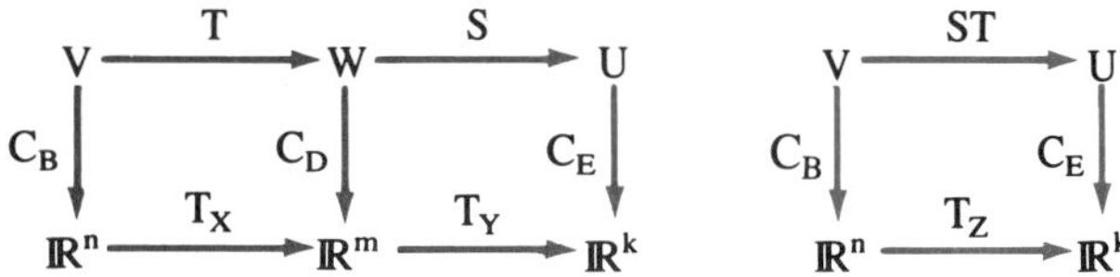

where, as usual, the actions of T_X, T_Y, and T_Z are left multiplication by X, Y, and Z, respectively. By Theorem 1, the matrices X, Y, and Z are uniquely determined by the following properties:

$$C_DT = T_XC_B, \quad C_ES = T_YC_D, \quad \text{and} \quad C_EST = T_ZC_B$$

Now ST appears as the composite map along the top of the first diagram. Compute the composite C_EST as follows (using the first two properties above and Theorem 14 in Section 8.1).

$$C_EST = (C_ES)T = (T_YC_D)T = T_Y(C_DT) = T_Y(T_XC_B) = (T_YT_X)C_B$$

Now $T_YT_X = T_{YX}$ by Theorem 13 in Section 8.1, so this gives $C_EST = T_{YX}C_B$. Comparing this with $C_EST = T_ZC_B$, and noting the *uniqueness* of Z with this property, the conclusion is $Z = YX$ as required. ■

THEOREM 3

Let $T : V \to W$ be a linear transformation, where $\dim V = \dim W = n$. The following are equivalent.

(1) T is an isomorphism.

(2) $M_{DB}(T)$ is invertible for all ordered bases B and D of V and W.

(3) $M_{DB}(T)$ is invertible for some pair of ordered bases B and D of V and W.

When this is the case, $M_{DB}(T)^{-1} = M_{BD}(T^{-1})$.

Proof (1) implies (2). We have $V \xrightarrow{T} W \xrightarrow{T^{-1}} V$, so Theorem 2 and Example 3 give

$$M_{BD}(T^{-1})M_{DB}(T) = M_{BB}(T^{-1}T) = M_{BB}(1_V) = I_n$$

Similarly, $M_{DB}(T)M_{BD}(T^{-1}) = I_n$, proving (2) (and the last statement in the theorem).

(2) implies (3). This is clear.

(3) implies (1). Let $M_{DB}(T)$ be invertible for some B and D and, for convenience, write $A = M_{DB}(T)$. Then $C_DT = T_AC_B$ by Theorem 1. Now C_D and C_B are isomorphisms (Example 31 in Section 8.1), so this gives $T = C_D{}^{-1}T_AC_B$. But the composite of isomorphisms is again an isomorphism, so it suffices to show that T_A is an isomorphism. However, this follows from Theorem 13 in Section 8.1. ■

EXAMPLE 5 If $B = \{\mathbf{e}_1, \mathbf{e}_2, \mathbf{e}_3, \mathbf{e}_4\}$ is an ordered basis of a vector space V, let $T : V \to V$ be the linear transformation satisfying

$$T(\mathbf{e}_1) = \mathbf{e}_1 + \mathbf{e}_2$$
$$T(\mathbf{e}_2) = \mathbf{e}_2 + \mathbf{e}_3$$
$$T(\mathbf{e}_3) = \mathbf{e}_3 + \mathbf{e}_4$$
$$T(\mathbf{e}_4) = \mathbf{e}_4 + \mathbf{e}_1$$

Show that T is not an isomorphism.

Solution $M_{BB}(T) = [C_B(\mathbf{e}_1 + \mathbf{e}_2)\, C_B(\mathbf{e}_2 + \mathbf{e}_3)\, C_B(\mathbf{e}_3 + \mathbf{e}_4)\, C_B(\mathbf{e}_4 + \mathbf{e}_1)] = \begin{bmatrix} 1 & 0 & 0 & 1 \\ 1 & 1 & 0 & 0 \\ 0 & 1 & 1 & 0 \\ 0 & 0 & 1 & 1 \end{bmatrix}$. This matrix has zero determinant, so T cannot be an isomorphism by Theorem 3. Of course this can be seen in other ways (for example, $T(\mathbf{e}_1 - \mathbf{e}_2 + \mathbf{e}_3 - \mathbf{e}_4) = \mathbf{0}$), but Theorem 3 provides a systematic method to check it, suitable for use on a computer.

EXAMPLE 6 Let $p_0, p_1, \ldots, p_n$ be polynomials in $\mathbf{P}_n$ and assume that p_k has degree k for each $k = 0, 1, 2, \ldots, n$. Let $T : \mathbf{P}_n \to \mathbf{P}_n$ be the linear transformation such that $T(x^k) = p_k$ for each $k = 0, 1, 2, \ldots, n$. Show that T is an isomorphism and hence that $\{p_0, p_1, \ldots, p_n\}$ is a basis of $\mathbf{P}_n$.

Solution Let $B = \{1, x, \ldots, x^n\}$, and let a_k denote the coefficient of x^k in p_k. The fact that deg $(p_k) = k$ shows that $a_k \neq 0$ and that the coefficients of higher powers of x in p_k are zero. Hence

$$M_{BB}(T) = [C_B[p_0]\, C_B[p_1] \ldots C_B[p_n]] = \begin{bmatrix} a_0 & * & * & \ldots & * \\ 0 & a_1 & * & \ldots & * \\ 0 & 0 & a_2 & \ldots & * \\ \vdots & \vdots & \vdots & & \vdots \\ 0 & 0 & 0 & \ldots & a_n \end{bmatrix}$$

where the * entries may or may not be zero. This matrix has determinant $a_0 a_1 \ldots a_n \neq 0$ so T is an isomorphism by Theorem 3. The rest follows by Theorem 10 in Section 8.1.

Of course, this can be accomplished by verifying directly that $\{p_0, p_1, \ldots, p_n\}$ is a basis of $\mathbf{P}_n$ (this was done in Theorem 1 in Section 5.6.1). However, the foregoing argument is easy and provides a *proof* of Theorem 1 in Section 5.6.1.

In view of Theorem 3, it is not surprising that there is a connection between the rank of a linear transformation and the rank of the corresponding matrices.

THEOREM 4 Let $T : V \to W$ be a linear transformation where $\dim V = n$ and $\dim W = m$. If B and D are any ordered bases of V and W, then rank $T = \text{rank}\,[M_{DB}(T)]$.

Proof Write $A = M_{DB}(T)$ for convenience. The column space of A is $U = \{A\mathbf{x} \mid \mathbf{x} \text{ in } \mathbb{R}^n\}$. Hence rank $A = \dim U$ and so, because rank $T = \dim(\text{im } T)$, it suffices to find an isomorphism $S : \text{im } T \to U$. Now every vector in im T has the form $T(\mathbf{v})$, $\mathbf{v}$ in V, and by Theorem 1, $C_D[T(\mathbf{v})] = AC_B(\mathbf{v})$ lies in U. So define $S : \text{im } T \to U$ by

$$S[T(\mathbf{v})] = C_D[T(\mathbf{v})] \qquad \text{for all vectors } T(\mathbf{v}) \text{ in im } T$$

The fact that C_D is linear and one-to-one implies immediately that S is linear and one-to-one. To see that S is onto, let $A\mathbf{x}$ be any member of U, $\mathbf{x}$ in $\mathbb{R}^n$. Then $\mathbf{x} = C_B(\mathbf{v})$ for some $\mathbf{v}$ in V, because C_B is onto. Hence $A\mathbf{x} = AC_B(\mathbf{v}) = C_D[T(\mathbf{v})] = S[T(\mathbf{v})]$, so S is onto. This means S is an isomorphism. ■

We conclude with an example showing that the matrix of a linear transformation can be made very simple by a careful choice of the two bases.

EXAMPLE 7 Let $T : V \to W$ be a linear transformation where $\dim V = n$ and $\dim W = m$. Choose an ordered basis $B = \{\mathbf{e}_1, \ldots, \mathbf{e}_r, \mathbf{e}_{r+1}, \ldots, \mathbf{e}_n\}$ of V, where $\{\mathbf{e}_{r+1}, \ldots, \mathbf{e}_n\}$ is a basis of ker T. Then $\{T(\mathbf{e}_1), \ldots, T(\mathbf{e}_r)\}$ is a basis of im T (corollary to Theorem 8 in Section 8.1), so extend it to an ordered basis $D = \{T(\mathbf{e}_1), \ldots, T(\mathbf{e}_r), \mathbf{f}_{r+1}, \ldots, \mathbf{f}_m\}$ of W. Show that

$$M_{DB}(T) = \begin{bmatrix} I_r & 0 \\ 0 & 0 \end{bmatrix}$$

in block form. Moreover $r = \text{rank } T$.

Solution $M_{DB}(T) = [C_D[T(\mathbf{e}_1)] \ldots C_D[T(\mathbf{e}_r)]\; C_D[T(\mathbf{e}_{r+1})] \ldots C_D[T(\mathbf{e}_n)]] = \begin{bmatrix} I_r & 0 \\ 0 & 0 \end{bmatrix}$ because $T(\mathbf{e}_{r+1}) = \cdots = T(\mathbf{e}_n) = \mathbf{0}$. Then rank $T = r$ by Theorem 4.

EXERCISES 8.2.1

1. Suppose $T : \mathbf{P}_2 \to \mathbb{R}^2$ is a linear transformation. If $B = \{1, x, x^2\}$ and $D = \{(1,1), (0,1)\}$, find the action of T given:
   (a) $M_{DB}(T) = \begin{bmatrix} 1 & 2 & -1 \\ -1 & 0 & 1 \end{bmatrix}$ **(b)** $M_{DB}(T) = \begin{bmatrix} 2 & 1 & 3 \\ -1 & 0 & -2 \end{bmatrix}$
2. In each case, find the matrix of $T : V \to W$ corresponding to the bases B and D of V and W, respectively.
   (a) $T : \mathbf{M}_{22} \to \mathbb{R}$, $T(A) = \text{tr}A$; $B = \left\{\begin{bmatrix} 1 & 0 \\ 0 & 0 \end{bmatrix}, \begin{bmatrix} 0 & 1 \\ 0 & 0 \end{bmatrix}, \begin{bmatrix} 0 & 0 \\ 1 & 0 \end{bmatrix}, \begin{bmatrix} 0 & 0 \\ 0 & 1 \end{bmatrix}\right\}$, $D = \{1\}$
   (b) $T : \mathbf{M}_{22} \to \mathbf{M}_{22}$, $T(A) = A^T$; $B = D = \left\{\begin{bmatrix} 1 & 0 \\ 0 & 0 \end{bmatrix}, \begin{bmatrix} 0 & 1 \\ 0 & 0 \end{bmatrix}, \begin{bmatrix} 0 & 0 \\ 1 & 0 \end{bmatrix}, \begin{bmatrix} 0 & 0 \\ 0 & 1 \end{bmatrix}\right\}$
   (c) $T : \mathbf{P}_2 \to \mathbf{P}_3$, $T[p(x)] = xp(x)$; $B = \{1, x, x^2\}$, $D = \{1, x, x^2, x^3\}$
   (d) $T : \mathbf{P}_2 \to \mathbf{P}_2$, $T[p(x)] = p(x + 1)$; $B = D = \{1, x, x^2\}$
3. In each case, find the matrix of $T : V \to W$ corresponding to the bases B and D, respectively, and use it to compute $C_D[T(\mathbf{v})]$, and hence $T(\mathbf{v})$.
   (a) $T : \mathbb{R}^3 \to \mathbb{R}^4$, $T(x,y,z) = (x + z, 2z, y - z, x + 2y)$; B and D standard; $\mathbf{v} = (1, -1, 3)$
   (b) $T : \mathbb{R}^2 \to \mathbb{R}^4$, $T(x,y) = (2x - y, 3x + 2y, 4y, x)$; $B = \{(1,1), (1,0)\}$, D standard; $\mathbf{v} = (a, b)$
   (c) $T : \mathbf{P}_2 \to \mathbb{R}^2$, $T(a + bx + cx^2) = (a + c, 2b)$; $B = \{1, x, x^2\}$, $D = \{(1,0), (1,-1)\}$; $\mathbf{v} = a + bx + cx^2$
   (d) $T : \mathbf{P}_2 \to \mathbb{R}^2$, $T(a + bx + cx^2) = (a + b, c)$; $B = \{1, x, x^2\}$, $D = \{(1,-1), (1,1)\}$; $\mathbf{v} = a + bx + cx^2$
   (e) $T : \mathbf{M}_{22} \to \mathbb{R}$, $T\begin{bmatrix} a & b \\ c & d \end{bmatrix} = a + b + c + d$; $B = \left\{\begin{bmatrix} 1 & 0 \\ 0 & 0 \end{bmatrix}, \begin{bmatrix} 0 & 1 \\ 0 & 0 \end{bmatrix}, \begin{bmatrix} 0 & 0 \\ 1 & 0 \end{bmatrix}, \begin{bmatrix} 0 & 0 \\ 0 & 1 \end{bmatrix}\right\}$, $D = \{1\}$; $\mathbf{v} = \begin{bmatrix} a & b \\ c & d \end{bmatrix}$
   (f) $T : \mathbf{M}_{22} \to \mathbf{M}_{22}$, $T\begin{bmatrix} a & b \\ c & d \end{bmatrix} = \begin{bmatrix} a & b + c \\ b + c & d \end{bmatrix}$; $B = D = \left\{\begin{bmatrix} 1 & 0 \\ 0 & 0 \end{bmatrix}, \begin{bmatrix} 0 & 1 \\ 0 & 0 \end{bmatrix}, \begin{bmatrix} 0 & 0 \\ 1 & 0 \end{bmatrix}, \begin{bmatrix} 0 & 0 \\ 0 & 1 \end{bmatrix}\right\}$; $\mathbf{v} = \begin{bmatrix} a & b \\ c & d \end{bmatrix}$
4. In each case, verify Theorem 2. Use the standard basis in $\mathbb{R}^n$ and $\{1, x, x^2\}$ in $\mathbf{P}_2$.
   (a) $\mathbb{R}^3 \xrightarrow{T} \mathbb{R}^2 \xrightarrow{S} \mathbb{R}^4$; $T(a,b,c) = (a + b, b - c)$, $S(a,b) = (a, b - 2a, 3b, a + b)$

(b) $\mathbb{R}^3 \xrightarrow{T} \mathbb{R}^4 \xrightarrow{S} \mathbb{R}^2$; $T(a,b,c) = (a + b, c + b, a + c, b - a)$, $S(a,b,c,d) = (a + b, c - d)$

(c) $\mathbf{P}_2 \xrightarrow{T} \mathbb{R}^3 \xrightarrow{S} \mathbf{P}_2$; $T(a + bx + cx^2) = (a, b - c, c - a)$, $S(a,b,c) = b + cx + (a - c)x^2$

(d) $\mathbb{R}^3 \xrightarrow{T} \mathbf{P}_2 \xrightarrow{S} \mathbb{R}^2$; $T(a,b,c) = (a - b) + (c - a)x + bx^2$, $S(a + bx + cx^2) = (a - b, c)$

5. Verify Theorem 2 for $\mathbf{M}_{22} \xrightarrow{T} \mathbf{M}_{22} \xrightarrow{S} \mathbf{P}_2$, where $T(A) = A^T$ and $S\begin{bmatrix} a & b \\ c & d \end{bmatrix} = b + (a + d)x + cx^2$. Use the bases $B = D = \left\{\begin{bmatrix} 1 & 0 \\ 0 & 0 \end{bmatrix}, \begin{bmatrix} 0 & 1 \\ 0 & 0 \end{bmatrix}, \begin{bmatrix} 0 & 0 \\ 1 & 0 \end{bmatrix}, \begin{bmatrix} 0 & 0 \\ 0 & 1 \end{bmatrix}\right\}$ and $E = \{1, x, x^2\}$.

6. In each case, find T^{-1} and verify $M_{DB}(T)^{-1} = M_{BD}(T^{-1})$.

(a) $T : \mathbb{R}^2 \to \mathbb{R}^2$, $T(a,b) = (a + 2b, 2a + 5b)$; $B = D =$ standard

(b) $T : \mathbb{R}^3 \to \mathbb{R}^3$, $T(a,b,c) = (b + c, a + c, a + b)$; $B = D =$ standard

(c) $T : \mathbf{P}_2 \to \mathbb{R}^3$, $T(a + bx + cx^2) = (a - c, b, 2a - c)$; $B = \{1, x, x^2\}$, $D =$ standard

(d) $T : \mathbf{P}_2 \to \mathbb{R}^3$, $T(a + bx + cx^2) = (a + b + c, b + c, c)$; $B = \{1, x, x^2\}$, $D =$ standard

7. In each case, show that $M_{DB}(T)$ is invertible and use the fact that $M_{BD}(T^{-1}) = [M_{DB}(T)]^{-1}$ to determine the action of T^{-1}.

(a) $T : \mathbf{P}_2 \to \mathbb{R}^3$, $T(a + bx + cx^2) = (a + c, c, b - c)$; $B = \{1, x, x^2\}$, $D =$ standard

(b) $T : \mathbf{M}_{22} \to \mathbb{R}^4$; $T\begin{bmatrix} a & b \\ c & d \end{bmatrix} = (a + b + c, b + c, c, d)$;

$$B = \left\{\begin{bmatrix} 1 & 0 \\ 0 & 0 \end{bmatrix}, \begin{bmatrix} 0 & 1 \\ 0 & 0 \end{bmatrix}, \begin{bmatrix} 0 & 0 \\ 1 & 0 \end{bmatrix}, \begin{bmatrix} 0 & 0 \\ 0 & 1 \end{bmatrix}\right\}, D = \text{standard}$$

8.[†] Let $D : \mathbf{P}_3 \to \mathbf{P}_2$ be the differentiation map given by $D[p(x)] = p'(x)$. Find the matrix of D corresponding to the bases $B = \{1, x, x^2, x^3\}$ and $E = \{1, x, x^2\}$, and use it to compute $D(a + bx + cx^2 + dx^3)$.

9. Use Theorem 3 to show that $T : V \to V$ is not an isomorphism if $\ker T \neq 0$ (assume $\dim V = n$). [*Hint:* Choose any ordered basis B containing a vector in $\ker T$.]

10. Let $T : V \to \mathbb{R}$ be a linear transformation, and let $D = \{1\}$ be the basis of $\mathbb{R}$. Given any ordered basis $B = \{\mathbf{e}_1 \ldots \mathbf{e}_n\}$ of V, show that $M_{DB}(T) = [T(\mathbf{e}_1) \ldots T(\mathbf{e}_n)]$.

11. Let $T : V \to W$ be an isomorphism, let $B = \{\mathbf{e}_1, \ldots, \mathbf{e}_n\}$ be an ordered basis of V, and let $D = \{T(\mathbf{e}_1), \ldots, T(\mathbf{e}_n)\}$. Show that $M_{DB}(T) = I_n$—the $n \times n$ identity matrix.

12. Complete the solution to Example 3.

[†]This exercise requires calculus.

13. Let U be any invertible $n \times n$ matrix, and let $D = \{\mathbf{f}_1, \mathbf{f}_2, \ldots, \mathbf{f}_n\}$ where $\mathbf{f}_j$ is column j of U. Show that $M_{BD}(1_{\mathbb{R}^n}) = U$ when B is the standard basis of $\mathbb{R}^n$.
14. Let B be an ordered basis of the n-dimensional space V, and let $C_B : V \to \mathbb{R}^n$ be the coordinate transformation. If D is the standard basis of $\mathbb{R}^n$, show that $M_{DB}(C_B) = I_n$.
15. Let $T : \mathbf{P}_2 \to \mathbb{R}^3$ be defined by $T(p) = (p(0), p(1), p(2))$ for all p in $\mathbf{P}_2$. Let $B = \{1, x, x^2\}$ and $D = \{(1,0,0), (0,1,0), (0,0,1)\}$.
    (a) Show that $M_{DB}(T) = \begin{bmatrix} 1 & 0 & 0 \\ 1 & 1 & 1 \\ 1 & 2 & 4 \end{bmatrix}$ and conclude that T is an isomorphism.
    (b) Generalize to $T : \mathbf{P}_n \to \mathbb{R}^{n+1}$ where $T(p) = (p(a_0), p(a_1), \ldots, p(a_n))$ and $a_0, a_1, \ldots, a_n$ are distinct real numbers. [*Hint:* Theorem 2 in Section 3.4.]
16.† Let $T : \mathbf{P}_n \to \mathbf{P}_n$ be defined by $T[p(x)] = p(x) + xp'(x)$, where $p'(x)$ denotes the derivative. Show that T is an isomorphism by finding $M_{BB}(T)$ when $B = \{1, x, x^2, \ldots, x^n\}$.
17. If k is any number, define $T_k : \mathbf{M}_{22} \to \mathbf{M}_{22}$ by $T_k(A) = A + kA^T$.
    (a) If $B = \left\{\begin{bmatrix} 1 & 0 \\ 0 & 0 \end{bmatrix}, \begin{bmatrix} 0 & 0 \\ 0 & 1 \end{bmatrix}, \begin{bmatrix} 0 & 1 \\ 1 & 0 \end{bmatrix}, \begin{bmatrix} 0 & 1 \\ -1 & 0 \end{bmatrix}\right\}$ find $M_{BB}(T_k)$, and conclude that T_k is invertible if $k \neq 1$ and $k \neq -1$.
    (b) Repeat for $T_k : \mathbf{M}_{33} \to \mathbf{M}_{33}$. Can you generalize?

8.2.2 The Vector Space of Linear Transformations

If V and W are vector spaces of dimension n and m, respectively, we have shown above that every linear transformation $T : V \to W$ determines a unique $m \times n$ matrix and that composition of transformations corresponds to multiplication of matrices. In fact, much more is true: The correspondence preserves addition and scalar multiplication as well, and the set of all linear transformations from V to W can be made into a vector space in such a way that the correspondence is an isomorphism. This section is devoted to a discussion of this.

The set of all linear transformations from V to W will be denoted by

$$\mathbf{L}(V,W) = \{T \mid T : V \to W \text{ is a linear transformation}\}$$

Define addition and scalar multiplication of linear transformations as follows: Given S and T in $\mathbf{L}(V,W)$ and a in $\mathbb{R}$, define $S + T : V \to W$ and $aT : V \to W$ by

†This exercise requires calculus.

$$(S + T)(\mathbf{v}) = S(\mathbf{v}) + T(\mathbf{v}) \qquad \text{for all } \mathbf{v} \text{ in } V$$
$$(aT)(\mathbf{v}) = aT(\mathbf{v}) \qquad \text{for all } \mathbf{v} \text{ in } V$$

These are called **pointwise** addition and scalar multiplication, respectively, and the new functions are linear transformations. For example,

$$\begin{aligned}(S + T)(\mathbf{v} + \mathbf{v}_1) &= S(\mathbf{v} + \mathbf{v}_1) + T(\mathbf{v} + \mathbf{v}_1)\\ &= S(\mathbf{v}) + S(\mathbf{v}_1) + T(\mathbf{v}) + T(\mathbf{v}_1)\\ &= S(\mathbf{v}) + T(\mathbf{v}) + S(\mathbf{v}_1) + T(\mathbf{v}_1)\\ &= (S + T)(\mathbf{v}) + (S + T)(\mathbf{v}_1)\end{aligned}$$

holds for all $\mathbf{v}$ and $\mathbf{v}_1$ in V. Similarly, $(S + T)(r\mathbf{v}) = r(S + T)(\mathbf{v})$, so the map $S + T$ is linear. The verification that aT is linear is left to the reader.

EXAMPLE 8 Let T and S in $\mathbf{L}(\mathbb{R}^2,\mathbb{R}^3)$ be defined by

$$T(x,y) = (x - y, x + 2y, x)$$
$$S(x,y) = (y, x - y, x)$$

for all (x,y) in $\mathbb{R}^2$. Determine the actions of $S + T$ and $3T$.

Solution Given (x,y) in $\mathbb{R}^2$, compute

$$\begin{aligned}(S + T)(x,y) = S(x,y) + T(x,y) &= (y, x - y, x) + (x - y, x + 2y, x)\\ &= (x, 2x + y, 2x)\\ (3T)(x,y) = 3T(x,y) &= 3(x - y, x + 2y, x)\\ &= (3x - 3y, 3x + 6y, 3x)\end{aligned}$$

The set $\mathbf{L}(V,W)$ is closed under pointwise addition and scalar multiplication, and, as might be expected, the other vector-space axioms hold. For example, $a(S + T) = aS + aT$ is verified as follows. Given any $\mathbf{v}$ in V, compute

$$\begin{aligned}[a(S + T)](\mathbf{v}) &= a[(S + T)(\mathbf{v})]\\ &= a[S(\mathbf{v}) + T(\mathbf{v})]\\ &= aS(\mathbf{v}) + aT(\mathbf{v})\\ &= (aS)(\mathbf{v}) + (aT)(\mathbf{v})\\ &= (aS + aT)(\mathbf{v})\end{aligned}$$

This shows that $a(S + T)$ and $aS + aT$ have the same action, so $a(S + T) = aS + aT$. The other vector-space axioms are verified in the same way. In particular, the **zero** transformation $0 : V \to W$ is the zero vector in $\mathbf{L}(V,W)$:

$$0 + T = T \qquad \text{for all } T \text{ in } \mathbf{L}(V,W)$$

and the **negative** of a transformation T is the transformation $-T = (-1)T$. The verifications are analogous to those in Example 7 in Section 5.1.

These vector spaces $\mathbf{L}(V,W)$ have additional structure. For example, if T lies in $\mathbf{L}(V,W)$ and S lies in $\mathbf{L}(W,U)$, then their composite ST lies in $\mathbf{L}(V,U)$. The relationship between composition (when defined) and addition and scalar multiplication is analogous to the situation for matrices. Theorem 5 collects the basic facts.

THEOREM 5 Let V and W be vector spaces, and let $\mathbf{L}(V,W)$ denote the set of all linear transformations $V \to W$.

(1) $\mathbf{L}(V,W)$ is a vector space under pointwise addition and scalar multiplication.

(2) The following properties hold, provided that the transformations link together in such a way that all the operations are defined.

(a) $R(ST) = (RS)T$
(b) $1_W T = T = T1_V$
(c) $R(S + T) = RS + RT$
(d) $(S + T)R = SR + TR$
(e) $(aS)T = a(ST) = S(aT)$

Proof (1) was discussed above, and properties (a) and (b) in (2) restate Theorem 14 in Section 8.1. Of the last three properties we prove only (c). Let $V \xrightarrow{S,T} W \xrightarrow{R} U$. If $\mathbf{v}$ lies in V,

$$\begin{aligned}[R(S + T)](\mathbf{v}) &= R[(S + T)(\mathbf{v})] \\ &= R[S(\mathbf{v}) + T(\mathbf{v})] \\ &= R[S(\mathbf{v})] + R[T(\mathbf{v})] \\ &= (RS)(\mathbf{v}) + (RT)(\mathbf{v}) \\ &= (RS + RT)(\mathbf{v})\end{aligned}$$

Hence $R(S + T) = RS + RT$. ■

The close parallel between the properties in Theorem 5 and the analogous properties for matrices suggests that an isomorphism is at work. In fact, a natural isomorphism has already been discussed. Recall that if $T : V \to W$ is a linear transformation where $\dim V = n$ and $\dim W = m$, and if B and D are ordered bases of V and W, respectively, the $m \times n$ matrix corresponding to T is denoted $M_{DB}(T)$. Hence $M_{DB} : \mathbf{L}(V,W) \to \mathbf{M}_{mn}$ is a function. It turns out to be an isomorphism.

THEOREM 6 Let V and W denote vector spaces of dimensions n and m, respectively. Given ordered bases B of V and D of W,

$$M_{DB} : \mathbf{L}(V,W) \to \mathbf{M}_{mn}$$

is an isomorphism of vector spaces.

Proof Let $B = \{\mathbf{e}_1, \mathbf{e}_2, \ldots, \mathbf{e}_n\}$. Given S and T in $\mathbf{L}(V,W)$, column j of $M_{DB}(S + T)$ is

$$C_D[(S + T)(\mathbf{e}_j)] = C_D[S(\mathbf{e}_j) + T(\mathbf{e}_j)] = C_D[S(\mathbf{e}_j)] + C_D[T(\mathbf{e}_j)]$$

because C_D is linear. The right side is the sum of the jth columns of $M_{DB}(S)$ and $M_{DB}(T)$, so we have $M_{DB}(S + T) = M_{DB}(S) + M_{DB}(T)$. The proof that $M_{DB}(aT) = aM_{DB}(T)$ is similar, so M_{DB} is a linear transformation.

To show that M_{DB} is one-to-one, let $M_{DB}(T) = 0$. Then each column is zero, so $C_D[T(\mathbf{e}_j)] = \mathbf{0}$ for $j = 1, 2, \ldots, n$. But C_D is one-to-one, so $T(\mathbf{e}_j) = \mathbf{0}$ for each j. Because the $\mathbf{e}_j$ are a basis of V, this shows that $T = 0$ by Example 10 in Section 8.1. This shows that $\ker M_{DB} = 0$, so M_{DB} is one-to-one.

Finally, to show that M_{DB} is onto, let $A = [a_{ij}]$ be any matrix in $\mathbf{M}_{mn}$. We want $T : V \to W$ such that $M_{DB}(T) = A$; that is, $C_D[T(\mathbf{e}_j)]$ must equal column j of A for each j. If $D = \{\mathbf{f}_1, \ldots, \mathbf{f}_m\}$, this means

$$T(\mathbf{e}_j) = a_{1j}\mathbf{f}_1 + a_{2j}\mathbf{f}_2 + \cdots + a_{mj}\mathbf{f}_m \qquad \text{for each } j$$

Such a transformation exists by Theorem 3 in Section 8.1. ■

The fact that isomorphic spaces have the same dimension immediately gives the following corollary.

COROLLARY If $\dim V = n$ and $\dim W = m$, then $\dim[\mathbf{L}(V,W)] = nm$.

EXERCISES 8.2.2

1. If $T : V \to W$ is a linear transformation, verify that $aT : V \to W$ is also a linear transformation.
2. Complete the verification that $\mathbf{L}(V,W)$ is a vector space.
3. Prove properties (d) and (e) of Theorem 5.
4. Given S and T in $\mathbf{L}(V,W)$, show that:

(a) $\ker S \cap \ker T \subseteq \ker(S + T)$

(b) $\operatorname{im}(S + T) \subseteq \operatorname{im} S + \operatorname{im} T$ [See Exercise 21 in Section 5.3.3.]

5. Let V and W be vector spaces. If X is a subset of V, define

$$X^0 = \{T \text{ in } \mathbf{L}(V,W) \mid T(\mathbf{v}) = \mathbf{0} \text{ for all } \mathbf{v} \text{ in } X\}$$

(a) Show that X^0 is a subspace of $\mathbf{L}(V,W)$.

(b) If $X \subseteq X_1$, show that $X_1{}^0 \subseteq X^0$.

(c) If U and U_1 are subspaces of V, show that $(U + U_1)^0 = U^0 \cap U_1{}^0$. [See Exercise 21 in Section 5.3.3.]

6. Define $R : \mathbf{M}_{mn} \to \mathbf{L}(\mathbb{R}^n,\mathbb{R}^m)$ by $R(A) = T_A$ for each $m \times n$ matrix A, where $T_A : \mathbb{R}^n \to \mathbb{R}^m$ is the matrix transformation given by $T_A(\mathbf{v}) = A\mathbf{v}$ for all columns $\mathbf{v}$ in $\mathbb{R}^n$. Show that R is an isomorphism two ways:

(a) By showing directly that R is linear, onto, and one-to-one.

(b) By showing that $R = M_{DB}{}^{-1}$, where B and D are the standard bases in $\mathbb{R}^n$ and $\mathbb{R}^m$.[*Hint:* Example 2 and Exercise 20 in Section 8.1.3.]

7. Let V and W be vector spaces, and let $B = \{\mathbf{e}_1, \ldots, \mathbf{e}_n\}$ and $D = \{\mathbf{f}_1, \ldots, \mathbf{f}_m\}$ be bases of V and W, respectively. Given an $m \times n$ matrix $A = [a_{ij}]$, let $T^A : V \to W$ be the linear transformation satisfying

$$T^A(\mathbf{e}_j) = a_{1j}\mathbf{f}_1 + a_{2j}\mathbf{f}_2 + \cdots + a_{mj}\mathbf{f}_m$$

for each $j = 1, 2, \ldots, n$. If $R : \mathbf{M}_{mn} \to \mathbf{L}(V,W)$ is defined by $R(A) = T^A$, show that $R = M_{DB}{}^{-1}$. [*Hint:* See the proof of Theorem 6, and use Exercise 20 in Section 8.1.3.]

8. Let V be any vector space (we do not assume it is finite dimensional). Given $\mathbf{v}$ in V, define $S_\mathbf{v} : \mathbb{R} \to V$ by $S_\mathbf{v}(r) = r\mathbf{v}$.

(a) Show that $S_\mathbf{v}$ lies in $\mathbf{L}(\mathbb{R},V)$ for each $\mathbf{v}$ in V.

(b) Show that the map $R : V \to \mathbf{L}(\mathbb{R},V)$ given by $R(\mathbf{v}) = S_\mathbf{v}$ is an isomorphism. [*Hint:* To show that R is onto, if T lies in $\mathbf{L}(\mathbb{R},V)$, show that $T = S_\mathbf{v}$ where $\mathbf{v} = T(1)$.]

9. Let V be a vector space with ordered basis $B = \{\mathbf{e}_1, \mathbf{e}_2, \ldots, \mathbf{e}_n\}$. For each $i = 1, 2, \ldots, m$, define $S_i : \mathbb{R} \to V$ by $S_i(r) = r\mathbf{e}_i$ for all r in $\mathbb{R}$.

(a) Show that each S_i lies in $\mathbf{L}(\mathbb{R},V)$ and $S_i(1) = \mathbf{e}_i$.

(b) Given T in $\mathbf{L}(\mathbb{R},V)$, let $T(1) = a_1\mathbf{e}_1 + a_2\mathbf{e}_2 + \cdots + a_n\mathbf{e}_n$, a_i in $\mathbb{R}$. Show that $T = a_1S_1 + a_2S_2 + \cdots + a_nS_n$.

(c) Show that $\{S_1, S_2, \ldots, S_n\}$ is a basis of $\mathbf{L}(\mathbb{R},V)$ and so confirm the corollary to Theorem 6 in this case.

10. If V is a vector space, the space $V^* = \mathbf{L}(V,\mathbb{R})$ is called the **dual** of V. Given a basis $B = \{\mathbf{e}_1, \mathbf{e}_2, \ldots, \mathbf{e}_n\}$ of V, let $E_i : V \to \mathbb{R}$ for each $i = 1, 2, \ldots, m$ be the linear transformation satisfying

$$E_i(\mathbf{e}_j) = \begin{cases} 0 & \text{if } i \neq j \\ 1 & \text{if } i = j \end{cases}$$

(Each E_i exists by Theorem 3 in Section 8.1.) Prove the following:

(a) $E_i(r_1\mathbf{e}_1 + \cdots + r_n\mathbf{e}_n) = r_i$ for each $i = 1, 2, \ldots, n$

(b) $\mathbf{v} = E_1(\mathbf{v})\mathbf{e}_1 + E_2(\mathbf{v})\mathbf{e}_2 + \cdots + E_n(\mathbf{v})\mathbf{e}_n$ for all $\mathbf{v}$ in V

(c) $T = T(\mathbf{e}_1)E_1 + T(\mathbf{e}_2)E_2 + \cdots + T(\mathbf{e}_n)E_n$ for all T in V^*

(d) $\{E_1, E_2, \ldots, E_n\}$ is a basis of V^* (called the **dual basis** of B).
Given $\mathbf{v}$ in V, define $\mathbf{v}^* : V \to \mathbb{R}$ by

$$\mathbf{v}^*(\mathbf{w}) = E_1(\mathbf{v})E_1(\mathbf{w}) + E_2(\mathbf{v})E_2(\mathbf{w}) + \cdots + E_n(\mathbf{v})E_n(\mathbf{w}) \quad \text{for all } \mathbf{w} \text{ in } V$$

Show that:

(e) $\mathbf{v}^* : V \to \mathbb{R}$ is linear so $\mathbf{v}^*$ lies in V^*

(f) $\mathbf{e}_i^* = E_i$ for each $i = 1, 2, \ldots, n$

(g) The map $R : V \to V^*$ with $R(\mathbf{v}) = \mathbf{v}^*$ is an isomorphism. [*Hint:* Show that R is linear and one-to-one and use Theorem 9 in Section 8.1. Alternatively, show that $R^{-1}(T) = T(\mathbf{e}_1)\mathbf{e}_1 + \cdots + T(\mathbf{e}_n)\mathbf{e}_n$.]

SECTION 8.3 An Application to Linear Recurrence Relations (Optional)

In many applications it is required to compute numbers $x_0, x_1, x_2, \ldots, x_n, \ldots$ having the property that each is determined by those that come before. The simplest example of such a situation occurs when each x_{n+1} is a fixed multiple of x_n—say, $x_{n+1} = 3x_n$ for all $n = 0, 1, 2, \ldots$. If x_0 is given, the remaining x_n can be computed successively.

$$\begin{aligned} x_1 &= 3x_0 \\ x_2 &= 3x_1 = 3^2x_0 \\ x_3 &= 3x_2 = 3^3x_0 \end{aligned}$$

Clearly $x_n = 3^nx_0$ holds for all $n \geq 0$ and gives an explicit formula for x_n as a function of n, provided that x_0 is stipulated.

Other situations are possible. For example, each x_n could be determined by the two preceding numbers in the sequence:

$$x_{n+2} = 6x_n - x_{n+1} \qquad \text{for all } n \geq 0$$

Then the whole sequence $x_0, x_1, x_2, x_3, \ldots$ is determined once x_0 and x_1 are given. For example, if $x_0 = 1$ and $x_1 = 2$, then

$$\begin{aligned} x_2 &= 6x_0 - x_1 = 4 \\ x_3 &= 6x_1 - x_2 = 8 \\ x_4 &= 6x_2 - x_3 = 16 \end{aligned}$$

In this case it appears that $x_n = 2^n$ holds for all $n \geq 0$. This certainly works for $0 \leq n \leq 4$, and it satisfies the recurrence formula:

$$6x_n - x_{n+1} = 6 \cdot 2^n - 2^{n+1} = 2^{n+1}(3 - 1) = 2^{n+2} = x_{n+2}$$

However, the reader should not get the idea that it is always easy to guess the formula for x_n. For example, if $x_0 = 1$ and $x_1 = 1$, then $x_{n+2} = 6x_n - x_{n+1}$ generates the sequence

$$1, 1, 5, 1, 29, -23, 197, \ldots$$

No formula for the nth term of *this* sequence is apparent!

Nonetheless, it is possible to use vector-space techniques to analyze such sequences. To begin, consider the very simple linear recurrence

$$x_{n+2} = 2x_n + x_{n+1}$$

We are interested in finding all sequences $x_0, x_1, x_2, \ldots$, such that the relation holds for all $n \geq 0$. Clearly each such sequence is completely determined by its first two terms. In fact, it is useful to write down the most general such sequence where $x_0 = a_0$ and $x_1 = a_1$. The first few terms of this "general" sequence are

$$\begin{aligned}
x_0 &= a_0 &&= 1a_0 + 0a_1 \\
x_1 &= a_1 &&= 0a_0 + 1a_1 \\
x_2 &= 2x_0 + x_1 &&= 2a_0 + 1a_1 \\
x_3 &= 2x_1 + x_2 &&= 2a_0 + 3a_1 \\
x_4 &= 2x_2 + x_3 &&= 6a_0 + 5a_1 \\
x_5 &= 2x_3 + x_4 &&= 10a_0 + 11a_1
\end{aligned}$$

Hence each term of the "general" sequence is a linear combination of a_0 and a_1. The important observation is that the coefficients of a_0 and a_1 are *sequences satisfying* the recurrence

Coefficients of a_0: $1, 0, 2, 2, 6, 10, \ldots$

Coefficients of a_1: $0, 1, 1, 3, 5, 11, \ldots$

(take $(a_0,a_1) = (1,0)$ and $(0,1)$, respectively). If we denote these sequences by $y_0, y_1, y_2, \ldots$ and $z_0, z_1, z_2, \ldots$, respectively, then *every* sequence $x_0, x_1, \ldots$ satisfying the recurrence is given by

$$x_n = a_0y_n + a_1z_n \qquad \text{for all } n$$

This suggests that the "general" sequence $x_0, x_1, x_2, \ldots$ is a "linear combination" of the specific sequences $y_0, y_1, y_2, \ldots$ and $z_0, z_1, z_2, \ldots$ where the coefficients a_0 and a_1 are independent of n. This further suggests that the set of all sequences that satisfy the recurrence is a vector space and that the sequences y_0, y_1, y_2 and z_0, z_1, z_2 span this vector space. In order to make all this precise, we must frame a few definitions.

Sequences will be considered entities in their own right, so it is useful to have a special notation for them. Let

$$[x_n) \quad \text{denote the sequence } x_0, x_1, x_2, \ldots, x_n, \ldots$$

EXAMPLE 1

$[n)$	is the sequence $0, 1, 2, 3, \ldots$
$[n + 1)$	is the sequence $1, 2, 3, 4, \ldots$
$[2^n)$	is the sequence $1, 2, 2^2, 2^3, \ldots$
$[(-1)^n)$	is the sequence $1, -1, 1, -1, \ldots$
$[5)$	is the sequence $5, 5, 5, 5, \ldots$

Sequences of the form $[c)$ for a fixed number c will be referred to as **constant sequences,** and those of the form $[\lambda^n)$, λ some number, are **power sequences.**

Two sequences are regarded as **equal** when they are identical:

$$[x_n) = [y_n) \qquad \text{means} \qquad x_n = y_n \quad \text{for all } n = 0, 1, 2, \ldots$$

As suggested above, addition and scalar multiplication of sequences are defined by

$$[x_n) + [y_n) = [x_n + y_n)$$
$$r[x_n) = [rx_n)$$

These operations are analogous to the addition and scalar multiplication in $\mathbb{R}^n$, and it is easy to check that the vector-space axioms are satisfied. The zero vector is the constant sequence $[0)$, and the negative of a sequence $[x_n)$ is given by $-[x_n) = [-x_n)$.

Now suppose k real numbers $r_0, r_1, \ldots, r_{k-1}$ are given, and consider the **linear recurrence relation** determined by these numbers.

$$x_{n+k} = r_0 x_n + r_1 x_{n+1} + \cdots + r_{k-1} x_{n+k-1} \qquad (*)$$

When $r_0 \neq 0$, we say this recurrence has **length** k. (We shall usually assume that $r_0 \neq 0$; otherwise, we are essentially dealing with a recurrence of shorter length than k.) For example, the relation $x_{n+2} = 2x_n + x_{n+1}$ considered above is of length 2.

A sequence $[x_n)$ is said to **satisfy** the relation (*) if (*) holds for all $n \geq 0$. Let V denote the set of all sequences that satisfy the relation. In symbols,

$$V = \{[x_n) \mid x_{n+k} = r_0 x_n + r_1 x_{n+1} + \cdots + r_{k-1} x_{n+k-1} \quad \text{holds for all } n \geq 0\}$$

It is easy to see that the constant sequence $[0)$ lies in V and that V is closed under addition and scalar multiplication of sequences. Hence V

is vector space (being a subspace of the space of all sequences). The following important observation about V is needed (it was used implicitly above): If the first k terms of two sequences agree, then the sequences are identical. More formally,

LEMMA Let $[x_n)$ and $[y_n)$ denote two sequences in V. Then

$$[x_n) = [y_n) \text{ if and only if } x_0 = y_0, x_1 = y_1, \ldots, x_{k-1} = y_{k-1}$$

Proof If $[x_n) = [y_n)$, then $x_n = y_n$ for *all* $n = 0, 1, 2, \ldots$. Conversely, if $x_i = y_i$ for $i = 0, 1, \ldots, k - 1$, use the recurrence (*) for $n = 0$.

$$x_k = r_0x_0 + r_1x_1 + \cdots + r_{k-1}x_{k-1} = r_0y_0 + r_1y_1 + \cdots + r_{k-1}y_{k-1} = y_k$$

Next the recurrence for $n = 1$ establishes $x_{k+1} = y_{k+1}$. The process continues to show that $x_{n+k} = y_{n+k}$ holds for *all* $n \geq 0$ by induction on n. Hence $[x_n) = [y_n)$. ■

This shows that a sequence in V is completely determined by its first k terms. In particular, given a k-tuple

$$\mathbf{v} = (v_0, v_1, \ldots, v_{k-1})$$

in $\mathbb{R}^k$, define $T(\mathbf{v})$ to be the sequence in V whose first k terms are $v_0, v_1, \ldots, v_{k-1}$. The rest of the sequence $T(\mathbf{v})$ is determined by the recurrence, so $T : \mathbb{R}^k \to V$ is a function. In fact, it is an isomorphism.

THEOREM 1 Given real numbers $r_0, r_1, \ldots, r_{k-1}$ let

$$V = \{[x_n) \mid x_{n+k} = r_0x_n + r_1r_{n+1} + \cdots + r_{k-1}x_{n+k-1}, n \geq 0\}$$

denote the vector space of all sequences satisfying the linear recurrence relation determined by $r_0, r_1, \ldots, r_{k-1}$. Then the function

$$T : \mathbb{R}^k \to V$$

defined above is an isomorphism. In particular:

(1) $\dim V = k$.

(2) If $\{\mathbf{v}_1, \ldots, \mathbf{v}_k\}$ is any basis of $\mathbb{R}^k$, then $\{T(\mathbf{v}_1), \ldots, T(\mathbf{v}_k)\}$ is a basis of V.

Proof (1) and (2) will follow from Theorems 10 and 11 in Section 8.1 as soon as we show that T is an isomorphism. Given $\mathbf{v}$ and $\mathbf{w}$ in $\mathbb{R}^k$, write $\mathbf{v} = (v_0, v_1, \ldots, v_{k-1})$ and $\mathbf{w} = (w_0, w_1, \ldots, w_{k-1})$. The first k terms of

$T(\mathbf{v})$ and $T(\mathbf{w})$ are $v_0, v_1, \ldots, v_{k-1}$ and $w_0, w_1, \ldots, w_{k-1}$, respectively, so the first k terms of $T(\mathbf{v}) + T(\mathbf{w})$ are $v_0 + w_0, v_1 + w_1, \ldots, v_{k-1} + w_{k-1}$. Because these terms agree with the first k terms of $T(\mathbf{v} + \mathbf{w})$, the lemma implies that $T(\mathbf{v} + \mathbf{w}) = T(\mathbf{v}) + T(\mathbf{w})$. The proof that $T(r\mathbf{v}) = rT(\mathbf{v})$ is similar, so T is linear.

Now let $[x_n)$ be any sequence in V, and let $\mathbf{v} = (x_0, x_1, \ldots, x_{k-1})$. Then the first k terms of $[x_n)$ and $T(\mathbf{v})$ agree, so $T(\mathbf{v}) = [x_n)$. Hence T is onto. Finally, if $T(\mathbf{v}) = [0)$ is the zero sequence, then the first k terms of $T(\mathbf{v})$ are all zero (*all* terms of $T(\mathbf{v})$ are zero!) so $\mathbf{v} = \mathbf{0}$. This means that $\ker T = 0$, so T is one-to-one. ■

EXAMPLE 2 Show that the sequences $[1)$, $[n)$, and $[(-1)^n)$ are a basis of the space V of all solutions of the recurrence

$$x_{n+3} = -x_n + x_{n+1} + x_{n+2}$$

Then find the solution satisfying $x_0 = 1, x_1 = 2, x_2 = 5$.

Solution The verifications that these sequences satisfy the recurrence (and hence lie in V) are left to the reader. They are a basis because $[1) = T(1,1,1)$, $[n) = T(0,1,2)$, and $[(-1)^n) = T(1,-1,1)$; and $\{(1,1,1), (0,1,2), (1,-1,1)\}$ is a basis of $\mathbb{R}^3$. Finally, the sequence $[x_n)$ in V satisfying $x_0 = 1, x_1 = 2, x_2 = 5$ is a linear combination of this basis:

$$[x_n) = t_1[1) + t_2[n) + t_3[(-1)^n)$$

The nth term is $x_n = t_1 + nt_2 + (-1)^n t_3$, so taking $n = 0, 1, 2$ gives

$$\begin{aligned} 1 &= x_0 = t_1 + 0 + t_3 \\ 2 &= x_1 = t_1 + t_2 - t_3 \\ 5 &= x_2 = t_1 + 2t_2 + t_3 \end{aligned}$$

This has the solution $t_1 = t_3 = \frac{1}{2}, t_2 = 2$, so $x_n = \frac{1}{2} + 2n + \frac{1}{2}(-1)^n$.

This technique clearly works for any linear recurrence of length k: Simply take your favorite basis $\{\mathbf{v}_1, \ldots, \mathbf{v}_k\}$ of $\mathbb{R}^k$—perhaps the standard basis—and compute $T(\mathbf{v}_1), \ldots, T(\mathbf{v}_k)$. This is a basis of V all right, but the nth term of $T(\mathbf{v}_i)$ is not usually given as an explicit function of n. (The basis in Example 2 was carefully chosen so that the nth terms of the three sequences were 1, n, and $(-1)^n$, respectively, each a simple function of n.)

It turns out that an explicit basis of V can be written down in the general situation. Given the recurrence

$$x_{n+k} = r_0 x_n + r_1 x_{n+1} + \cdots + r_{k-1} x_{n+k-1} \qquad (*)$$

the idea is to look for numbers λ such that the power sequence $[\lambda^n)$ satisfies (*). This happens if and only if

$$\lambda^{n+k} = r_0\lambda^n + r_1\lambda^{n+1} + \cdots + r_{k-1}\lambda^{n+k-1}$$

holds for all $n \geq 0$. This is true just when the case $n = 0$ holds; that is,

$$\lambda^k = r_0 + r_1\lambda + \cdots + r_{k-1}\lambda^{k-1}$$

The polynomial

$$p(x) = x^k - r_{k-1}x^{k-1} - \cdots - r_1x - r_0$$

is called the polynomial **associated** with the linear recurrence (*). Thus every root λ of $p(x)$ provides a sequence $[\lambda^n)$ satisfying (*). If there are k distinct roots, the power sequences provide a basis. Incidentally, if $\lambda = 0$, the sequence $[\lambda^n)$ is $1, 0, 0, \ldots$; that is, we accept the convention that $0^0 = 1$.

THEOREM 2 Let $r_0, r_1, \ldots, r_{k-1}$ be real numbers; let

$$V = \{[x_n) \mid x_{n+k} = r_0x_n + \cdots + r_{k-1}x_{n+k-1} \text{ for all } n \geq 0\}$$

denote the vector space of all sequences satisfying the linear recurrence relation determined by $r_0, r_1, \ldots, r_{k-1}$; and let

$$p(x) = x^k - r_{k-1}x^{k-1} - \cdots - r_1x - r_0$$

denote the polynomial associated with the recurrence relation. Then

(1) $[\lambda^n)$ lies in V if and only if λ is a root of $p(x)$.

(2) If $\lambda_1, \lambda_2, \ldots, \lambda_k$ are distinct real roots of $p(x)$, then $\{[\lambda_1{}^n), [\lambda_2{}^n), \ldots, [\lambda_k{}^n)\}$ is a basis of V.

Proof It remains to prove (2). But $[\lambda_i{}^n) = T(\mathbf{v}_i)$ where $\mathbf{v}_i = (1, \lambda_i, \lambda_i{}^2, \ldots, \lambda_i{}^{k-1})$, so (2) follows by Theorem 1, provided that $\{\mathbf{v}_1, \mathbf{v}_2, \ldots, \mathbf{v}_n\}$ is a basis of $\mathbb{R}^k$. This is true (by Theorem 10 in Section 5.3), provided that the matrix

$$\begin{bmatrix} 1 & 1 & \cdots & 1 \\ \lambda_1 & \lambda_2 & \cdots & \lambda_k \\ \lambda_1{}^2 & \lambda_2{}^2 & \cdots & \lambda_k{}^2 \\ \vdots & \vdots & & \vdots \\ \lambda_1{}^{k-1} & \lambda_2{}^{k-1} & \cdots & \lambda_k{}^{k-1} \end{bmatrix}$$

is invertible. But this is a Vandermonde matrix and so is invertible provided that the λ_i are distinct (Theorem 2 in Section 3.4). This proves (2). ■

EXAMPLE 3 Find the solution of $x_{n+2} = 2x_n + x_{n+1}$ that satisfies $x_0 = a, x_1 = b$.

Solution The associated polynomial is $p(x) = x^2 - x - 2 = (x - 2) \cdot (x + 1)$. The roots are $\lambda_1 = 2$ and $\lambda_2 = -1$, so the sequences $[2^n)$ and $[(-1)^n)$ are a basis for the space of solutions. Hence every solution $[x_n)$ is a linear combination

$$[x_n) = t_1[2^n) + t_2[(-1)^n)$$

This means that $x_n = t_1 2^n + t_2(-1)^n$ holds for $n = 0, 1, 2, \ldots$, so (taking $n = 0, 1$) $x_0 = a$ and $x_1 = b$ give

$$\begin{aligned} t_1 + t_2 &= a \\ 2t_1 - t_2 &= b \end{aligned}$$

These are easily solved: $t_1 = \frac{1}{3}(a + b)$ and $t_2 = \frac{1}{3}(2a - b)$, so

$$x_n = \frac{1}{3}[(a + b)2^n + (2a - b)(-1)^n]$$

The next example has historical interest.

EXAMPLE 4 How many pairs of rabbits will be produced in a year, beginning with a single pair, if in every month each pair brings forth a new pair that becomes productive from the second month on? Assume no pairs die.

Solution Let x_n be the number of pairs after n months. Then $x_0 = 1$ and $x_1 = 1$ (the first pair does not reproduce in the first month). Moreover,

$$x_{n+2} = x_{n+1} + x_n$$

holds for all $n \geq 0$, because the x_{n+2} pairs at the end of month $n + 2$ are made up of x_{n+1} pairs alive at the end of the preceding month, together with x_n babies. The associated polynomial is $p(x) = x^2 - x - 1$, and $\lambda_1 = \frac{1}{2}(1 + \sqrt{5})$ and $\lambda_2 = \frac{1}{2}(1 - \sqrt{5})$ are the roots. These are distinct, so $\{[\lambda_1{}^n), [\lambda_2{}^n)\}$ is a basis of the space of solutions. Hence $x_n = t_1\lambda_1{}^n + t_2\lambda_2{}^n$ holds for some constants t_1 and t_2, and taking $n = 0, 1$ gives

$$\begin{aligned} t_1 \quad + t_2 \quad &= x_0 = 1 \\ t_1\lambda_1 + t_2\lambda_2 &= x_1 = 1 \end{aligned}$$

The solution is $t_1 = \lambda_1/\sqrt{5}$, $t_2 = -\lambda_2/\sqrt{5}$, so

$$x_n = \frac{1}{\sqrt{5}}\left[\left(\frac{1 + \sqrt{5}}{2}\right)^{n+1} - \left(\frac{1 - \sqrt{5}}{2}\right)^{n+1}\right] \qquad (*)$$

It is worth noting that $\frac{1}{2}(1 - \sqrt{5}) = -0.618$; large powers of this number will be very small. So for values of n larger than 10, x_n is approximated by $x_n = \frac{1}{\sqrt{5}}\left[\frac{1 + \sqrt{5}}{2}\right]^{n+1}$. This gives $x_{12} = 233$ pairs, and after 60 months the population is $x_{60} = 2.5$ million million! Clearly the stipulation that none of these pairs die would be invalid by then!

The numbers obtained in Example 4 are called the **Fibonacci sequence** and are usually denoted $f_0, f_1, f_2, \ldots$. They are defined by the conditions

$$\begin{aligned} f_0 &= f_1 = 1 \\ f_{n+2} &= f_n + f_{n+1} \qquad n = 0, 1, 2, \ldots \end{aligned}$$

The first few numbers in the sequence are $1, 1, 2, 3, 5, 8, 13, 21, 34, \ldots$. They are named after Leonardo Fibonacci (c. 1170–1250), one of the few outstanding European mathematicians of the middle ages. Example 4 was stated in his main work *Liber Abaci* (1202). The formula (*) in Example 4 is called the **Binet formula.**

If $p(x)$ is the polynomial associated with a linear recurrence relation, and if $p(x)$ has k distinct roots $\lambda_1, \lambda_2, \ldots, \lambda_k$, then $p(x)$ factors completely:

$$p(x) = (x - \lambda_1)(x - \lambda_2)\ldots(x - \lambda_k)$$

Each root λ_i provides a sequence $[\lambda_n{}^n)$ satisfying the recurrence, and they are a basis of V by Theorem 2. In this case, each λ_i has multiplicity 1 as a root of $p(x)$. In general, a root λ has **multiplicity** m if $p(x) = (x - \lambda)^m q(x)$, where $q(\lambda) \neq 0$. In this case, there are fewer than k distinct roots and so fewer than k sequences $[\lambda^n)$ satisfying the recurrence. However, we can still obtain a basis, because if λ has multiplicity m (and $\lambda \neq 0$), it provides m linearly independent sequences that satisfy the recurrence. In order to prove this, it is convenient to give another way to describe the space V of all sequences satisfying a given linear recurrence relation.

Let $\mathbf{S}$ denote the vector space of *all* sequences, and define a function

$$S : \mathbf{S} \to \mathbf{S} \text{ by } S[x_n) = [x_{n+1})$$

S is clearly a linear transformation and is called the **shift operator** on $\mathbf{S}$. Note that powers of S shift the sequence further: $S^2[x_n) = S[x_{n+1}) = [x_{n+2})$. In general,

$$S^k[x_n) = [x_{n+k}) \qquad \text{for each } k = 0, 1, 2, \ldots$$

But then a linear recurrence relation

$$x_{n+k} = r_0 x_n + r_1 x_{n+1} + \cdots + r_{k-1} x_{n+k-1} \qquad \text{for } n = 0, 1, \ldots$$

can be written

$$S^k[x_n) = r_0[x_n) + r_1S[x_n) + \cdots + r_{k-1}S^{k-1}[x_n) \qquad (*)$$

Now let $p(x) = x^k - r_{k-1}x^{k-1} - \cdots - r_1x - r_0$ denote the polynomial associated with the recurrence relation. The set $\mathbf{L}[\mathbf{S}, \mathbf{S}]$ of all linear transformations from $\mathbf{S}$ to itself is a vector space (Theorem 5 in Section 8.2) that is closed under composition. In particular,

$$p(S) = S^k - r_{k-1}S^{k-1} - \cdots - r_1S - r_0 1_{\mathbf{S}}$$

is a linear transformation called the **evaluation** of p at S. The point is that condition (*) can be written as

$$p(S)[x_n) = 0$$

In other words, the space V of all sequences satisfying the recurrence relation is just ker $[p(S)]$. This is the first assertion in the following theorem.

THEOREM 3 Let $r_0, r_1, \ldots, r_{k-1}$ be real numbers, and let

$$V = \{[x_n) \mid x_{n+k} = r_0x_n + r_1x_{n+1} + \cdots + r_{k-1}x_{n+k-1} \quad \text{for all } n \geq 0\}$$

denote the space of all sequences satisfying the linear recurrence relation determined by $r_0, r_1, \ldots, r_{k-1}$. Let

$$p(x) = x^k - r_{k-1}x^{k-1} - \cdots - r_1x - r_0$$

denote the corresponding polynomial. Then:

(1) $V = \ker[p(S)]$, where S is the shift operator.

(2) If $p(x) = (x - \lambda)^m q(x)$, where $\lambda \neq 0$ and $m > 1$, then the sequences

$$\{[\lambda^n), [n\lambda^n), [n^2\lambda^n), \ldots, [n^{m-1}\lambda^n)\}$$

all lie in V and are linearly independent.

Proof (Sketch) It remains to prove (2). If $\binom{n}{k} = \dfrac{n(n-1)\cdots(n-k+1)}{k!}$ denotes the binomial coefficient, the idea is to use (1) to show that the sequence $s_k = \left[\binom{n}{k}\lambda^n\right)$ is a solution for each $k = 0, 1, \ldots, m - 1$. Then (2) of Theorem 1 can be applied to show that $\{s_0, s_1, \ldots, s_{m-1}\}$ is linearly independent. Finally, the sequences $t_k = [n^k\lambda^n)$, $k = 0, 1, \ldots, m-1$, in the present theorem can be given by $t_k = \sum_{j=0}^{m-1} a_{kj}s_j$, where $A = [a_{ij}]$ is an invertible matrix. Then (2) follows. ■

This theorem combines with Theorem 2 to give a basis for V when $p(x)$ has k real roots (not necessarily distinct) none of which is zero. This last requirement means $r_0 \neq 0$, a condition that is unimportant in practice (see Remark 1 below).

THEOREM 4 Let $r_0, r_1, \ldots, r_{k-1}$ be real numbers with $r_0 \neq 0$; let

$$V = \{[x_n) \mid x_{n+k} = r_0x_n + r_1x_{n+1} + \cdots + r_{k-1}x_{n+k-1} \quad \text{for all } n \geq 0\}$$

denote the space of all sequences satisfying the linear recurrence relation of length k determined by $r_0, \ldots, r_{n-1}$; and assume that the polynomial

$$p(x) = x^k - r_{k-1}x^{k-1} - \cdots - r_1x - r_0$$

factors completely as

$$p(x) = (x - \lambda_1)^{m_1}(x - \lambda_2)^{m_2} \ldots (x - \lambda_p)^{m_p}$$

where $\lambda_1, \lambda_2, \ldots, \lambda_p$ are distinct real numbers and each $m_i \geq 1$. Then $\lambda_i \neq 0$ for each i, and

$$\{[\lambda_1{}^n), [n\lambda_1{}^n), \ldots, [n^{m_1-1}\lambda_1{}^n); [\lambda_2{}^n), [n\lambda_2{}^n), \ldots, [n^{m_2-1}\lambda_2{}^n); \ldots; [\lambda_p{}^n), [n\lambda_p{}^n), \ldots, [n^{m_p-1}\lambda_p{}^n)\}$$

is a basis of V.

Proof There are $m_1 + m_2 + \cdots + m_p = k$ sequences in all, so because $\dim V = k$, it suffices to show that they are linearly independent. The assumption that $r_0 \neq 0$ implies that 0 is not a root of $p(x)$. Hence each $\lambda_i \neq 0$, so $\{[\lambda_i{}^n), [n\lambda_i{}^n), \ldots, [n^{m_i-1}\lambda_i{}^n)\}$ is linearly independent by Theorem 3. The proof that the whole set of sequences is linearly independent is omitted. ■

EXAMPLE 5 Find a basis for the space V of all sequences $[x_n)$ satisfying

$$x_{n+3} = -9x_n - 3x_{n+1} + 5x_{n+2}$$

Solution The associated polynomial is $p(x) = x^3 - 5x^2 + 3x + 9 = (x - 3)^2 (x + 1)$. Hence $\lambda = 3$ is a double root, so $[3^n)$ and $[n3^n)$ both lie in V by Theorem 3 (the reader should verify this). Similarly, $\lambda = -1$ is a root of multiplicity 1, so $[(-1)^n)$ lies in V. Hence $\{[3^n), [n3^n), [(-1)^n)\}$ is a basis by Theorem 4.

Remark 1: If $r_0 = 0$ [so $p(x)$ has 0 as a root], the recurrence reduces to one of shorter length. For example, consider

$$x_{n+4} = 0x_n + 0x_{n+1} + 3x_{n+2} + 2x_{n+3} \qquad (*)$$

If we set $y_n = x_{n+2}$, this recurrence becomes $y_{n+2} = 3y_n + 2y_{n+1}$, which has solutions $[3^n)$ and $[(-1)^n)$. These give the following solutions to (*):

$$[0, 0, 1, 3, 3^2, \ldots)$$
$$[0, 0, 1, -1, (-1)^2, \ldots)$$

In addition, it is easy to verify that

$$[1, 0, 0, 0, 0, \ldots)$$
$$[0, 1, 0, 0, 0, \ldots)$$

are also solutions to (*). The space of all solutions of (*) has dimension 4 (Theorem 1), so these sequences are a basis. This technique works whenever $r_0 = 0$.

Remark 2: Theorem 4 completely describes the space V of sequences that satisfy a linear recurrence relation for which the associated polynomial $p(x)$ has all real roots. However, in many cases of interest, $p(x)$ has complex roots that are not real. If $p(\mu) = 0$, μ complex, then $p(\overline{\mu}) = 0$ too ($\overline{\mu}$ the conjugate), and the main observation is that $[\mu^n + \overline{\mu}^n)$ and $[i(\mu^n - \overline{\mu}^n))$ are *real* solutions. Analogs of the above theorems can then be proved.

EXERCISES 8.3

1. Find a basis for the space V of sequences $[x_n)$ satisfying the following recurrences, and use it to find the sequence satisfying $x_0 = 1$, $x_1 = 2$, $x_2 = 1$.
   (a) $x_{n+3} = -2x_n + x_{n+1} + 2x_{n+2}$ (b) $x_{n+3} = -6x_n + 7x_{n+1}$
   (c) $x_{n+3} = -36x_n + 7x_{n+2}$
2. In each case, find a basis for the space V of all sequences $[x_n)$ satisfying the recurrence, and use it to find x_n if $x_0 = 1$, $x_1 = -1$, and $x_2 = 1$.
   (a) $x_{n+3} = x_n + x_{n+1} - x_{n+2}$ (b) $x_{n+3} = -2x_n + 3x_{n+1}$
   (c) $x_{n+3} = -4x_n + 3x_{n+2}$ (d) $x_{n+3} = x_n - 3x_{n+1} + 3x_{n+2}$
   (e) $x_{n+3} = 8x_n - 12x_{n+1} + 6x_{n+2}$

3. Find a basis for the space V of sequences $[x_n)$ satisfying each of the following recurrences.
   (a) $x_{n+2} = -a^2x_n + 2ax_{n+1}$ (b) $x_{n+2} = -abx_n + (a+b)x_{n+1}, (a \neq b)$
4. Define the **Lucas sequence** $x_0, x_1, x_2, \ldots$ by $x_0 = 1, x_1 = 3, x_{n+2} = x_n + x_{n+1}$, $n \geq 0$. Show that $x_n = \left[\frac{1+\sqrt{5}}{2}\right]^{n+1} + \left[\frac{1-\sqrt{5}}{2}\right]^{n+1}$ for $n = 0, 1, 2, \ldots$.
5. The **generalized Fibonacci sequence** $g_0, g_1, \ldots$ is defined by $g_0 = a, g_1 = b$, $g_{n+2} = g_n + g_{n+1}$. Show that $g_n = \frac{1}{\sqrt{5}}[(b - a\lambda_2)\lambda_1{}^n - (b - a\lambda_1)\lambda_2{}^n]$, where $\lambda_1 = \frac{1}{2}(1+\sqrt{5})$ and $\lambda_2 = \frac{1}{2}(1-\sqrt{5})$.
6. In each case, find a basis of V.
   (a) $V = \{[x_n) \mid x_{n+4} = 2x_{n+2} - x_{n+3}$ for $n \geq 0\}$
   (b) $V = \{[x_n) \mid x_{n+4} = -x_{n+2} + 2x_{n+3}$ for $n \geq 0\}$
7. Suppose that $[x_n)$ satisfies a linear recurrence relation of length k. If

$$\{\mathbf{e}_0 = (1, 0, \ldots, 0), \mathbf{e}_1 = (0, 1, \ldots, 0), \ldots, \mathbf{e}_{k-1} = (0, 0, \ldots, 1)\}$$

   is the standard basis of $\mathbb{R}^k$, show that $x_n = x_0T(\mathbf{e}_0) + x_1T(\mathbf{e}_1) + \cdots + x_{k-1}T(\mathbf{e}_{k-1})$ holds for all $n \geq k$. (Here T is as in Theorem 1.)
8. Show that the shift operator S is onto but not one-to-one. Find ker S.
9. Find a basis for the space V of all sequences $[x_n)$ satisfying $x_{n+2} = -x_n$.

9 Linear Operators

SECTION 9.1 Change of Basis and Similarity

A linear transformation from a vector space V to itself is called a **linear operator** on V. Many facts about the set $\mathbf{L}(V,V)$ of all linear operators on V are special cases of previous results about $\mathbf{L}(V,W)$. In particular, Theorem 5 in Section 8.2 has a simpler form for operators, and we repeat it here for reference.

THEOREM 1

If V is a vector space, the set $\mathbf{L}(V,V)$ of all linear operators on V has the following properties.

(1) $\mathbf{L}(V,V)$ is a vector space under pointwise addition and scalar multiplication.

(2) $\mathbf{L}(V,V)$ is closed under composition, and the following hold for all R, S, and T in $\mathbf{L}(V,V)$ and r in $\mathbb{R}$.

- (a) $R(ST) = (RS)T$
- (b) $1_V T = T = T 1_V$
- (c) $R(S + T) = RS + RT$
- (d) $(S + T)R = SR + TR$
- (e) $(rS)T = r(ST) = S(rT)$

In Section 8.2 it was shown how to associate a unique matrix $M_{DB}(T)$ with each linear transformation $T : V \to W$, where B and D are ordered bases of V and W, respectively. In the case of an operator $T : V \to V$, we are taking $W = V$ but are using *two* bases B and D of V, one at each end as it were. It certainly appears to be simpler to take $D = B$ (in fact, this is essential if we want to compute $M_{DB}(T^2)$ using Theorem 2 in Section 8.2), and the isomorphism $M_{BB} : \mathbf{L}(V,V) \to \mathbf{M}_{nn}$

preserves composition in the sense that

$$M_{BB}(ST) = M_{BB}(S)M_{BB}(T)$$

holds for any linear operators S and T in $\mathbf{L}(V,V)$ and for any basis B of V (this is Theorem 2 in Section 8.2). With this in mind, we introduce some terminology.

DEFINITION Given a linear operator $T : V \to V$ and an ordered basis B of V, the matrix $M_{BB}(T)$ will be called **the matrix of T with respect to B** and will be denoted by

$$M_B(T)$$

The following theorem embodies information from Theorems 1, 2, 3, and 6 and Example 3 in Section 8.2, specialized for the case of linear operators.

THEOREM 2 Let V be an n-dimensional vector space with ordered basis $B = \{\mathbf{e}_1, \mathbf{e}_2, \ldots, \mathbf{e}_n\}$.

(1) Given any linear operator T in $\mathbf{L}(V,V)$, there is a unique $n \times n$ matrix $M_B(T)$ such that

$$C_B[T(\mathbf{v})] = M_B(T)C_B(\mathbf{v})$$

holds for all $\mathbf{v}$ in V, where $C_B(\mathbf{v})$ is the coordinate vector of $\mathbf{v}$.

(2) The matrix $M_B(T)$ is given in terms of its columns by

$$M_B(T) = [C_B[T(\mathbf{e}_1)]\ C_B[T(\mathbf{e}_2)] \ldots C_B[T(\mathbf{e}_n)]]$$

(3) The function $M_B : \mathbf{L}(V,V) \to \mathbf{M}_{nn}$ is an isomorphism of vector spaces and preserves multiplication in the sense that
   (a) $M_B(ST) = M_B(S)M_B(T)$ for all S and T in $\mathbf{L}(V,V)$.
   (b) $M_B(1_V) = I_n$.

(4) The following conditions are equivalent for a linear operator $T : V \to V$.
   (a) T has an inverse.
   (b) $M_B(T)$ is invertible for every basis B of V.
   (c) $M_B(T)$ is invertible for some basis B of V.

In this case, $M_B(T^{-1}) = M_B(T)^{-1}$.

Now suppose $T : V \to V$ is any linear operator. If *different* bases B and D are admitted in the two copies of V, then they can be chosen such that $M_{DB}(T) = \begin{bmatrix} I_r & 0 \\ 0 & 0 \end{bmatrix}$ in block form (Example 7 in Section 8.2). However, we are insisting that $B = D$, so the question arises of just how simple $M_B(T)$ can be made by an appropriate choice of basis B.

The first task is to describe how the matrix $M_B(T)$ of an operator changes when the basis B is changed. In order to do this, it is necessary to recall some basic facts from Section 5.5 about transition matrices. If $B = \{\mathbf{e}_1, \mathbf{e}_2, \ldots, \mathbf{e}_n\}$ and B_0 are two ordered bases of a vector space V, the **transition matrix** $P_{B_0 \leftarrow B}$ from B to B_0 is the $n \times n$ matrix defined in terms of its columns by

$$P_{B_0 \leftarrow B} = [C_{B_0}(\mathbf{e}_1)\, C_{B_0}(\mathbf{e}_2) \ldots C_{B_0}(\mathbf{e}_n)]$$

This relates the coordinates of vectors in V as follows (Theorem 2 in Section 5.5):

$$C_{B_0}(\mathbf{v}) = P_{B_0 \leftarrow B}\, C_B(\mathbf{v}) \qquad \text{for all } \mathbf{v} \text{ in } V$$

On the other hand, Theorem 2 (1) asserts that

$$C_B[T(\mathbf{v})] = M_B(T)\, C_B(\mathbf{v}) \qquad \text{for all } \mathbf{v} \text{ in } V$$

Combining these (and writing $P = P_{B_0 \leftarrow B}$ for convenience) gives

$$\begin{aligned} PM_B(T)C_B(\mathbf{v}) &= PC_B[T(\mathbf{v})] \\ &= C_{B_0}[T(\mathbf{v})] \\ &= M_{B_0}(T)C_{B_0}(\mathbf{v}) \\ &= M_{B_0}(T)PC_B(\mathbf{v}) \end{aligned}$$

This holds for all $\mathbf{v}$ in V, and [because $C_B(\mathbf{e}_j)$ is the jth column of the identity matrix] it follows that

$$PM_B(T) = M_{B_0}(T)P$$

Now P is invertible (in fact, $P^{-1} = P_{B \leftarrow B_0}$ by Theorem 3 in Section 5.5), so this gives

$$M_B(T) = P^{-1}M_{B_0}(T)P$$

This asserts that $M_{B_0}(T)$ and $M_B(T)$ are **similar** matrices.

THEOREM 3 Let B_0 and B be two ordered bases of a finite dimensional vector space V. If $T : V \to V$ is any linear operator, the matrices $M_B(T)$ and

$M_{B_0}(T)$ of T with respect to these bases are similar. More precisely,

$$M_B(T) = P^{-1} M_{B_0}(T)P$$

where $P = P_{B_0 \leftarrow B}$ is the transition matrix from B to B_0. Here

$$P_{B_0 \leftarrow B} = [C_{B_0}(\mathbf{e}_1) \ldots C_{B_0}(\mathbf{e}_n)]$$

where $B = \{\mathbf{e}_1, \ldots, \mathbf{e}_n\}$.

EXAMPLE 1 Let $T : \mathbb{R}^3 \to \mathbb{R}^3$ be defined by $T(a,b,c) = (2a - b, b + c, c - 3a)$. If B_0 denotes the standard basis of $\mathbb{R}^3$ and $B = \{(1,1,0), (1,0,1), (0,1,0)\}$, find an invertible matrix P such that $P^{-1}M_{B_0}(T)P = M_B(T)$.

Solution Use Theorem 2 (2) to compute

$$M_{B_0}(T) = [C_{B_0}(2,0,-3)\ C_{B_0}(-1,1,0)\ C_{B_0}(0,1,1)] = \begin{bmatrix} 2 & -1 & 0 \\ 0 & 1 & 1 \\ -3 & 0 & 1 \end{bmatrix}$$

$$M_B(T) = [C_B(1,1,-3)\ C_B(2,1,-2)\ C_B(-1,1,0)] = \begin{bmatrix} 4 & 4 & -1 \\ -3 & -2 & 0 \\ -3 & -3 & 2 \end{bmatrix}$$

Using Theorem 3, we find that

$$P = P_{B_0 \leftarrow B} = [C_{B_0}(1,1,0)\ C_{B_0}(1,0,1)\ C_{B_0}(0,1,0)] = \begin{bmatrix} 1 & 1 & 0 \\ 1 & 0 & 1 \\ 0 & 1 & 0 \end{bmatrix}$$

Similar matrices were studied in Section 7.1.2 in preparation for the discussion of diagonalization in Section 7.2. Theorem 3 comes into this as follows: Suppose an $n \times n$ matrix $A = M_{B_0}(T)$ is the matrix of some operator $T : V \to V$ with respect to an ordered basis B_0. If another ordered basis B of V can be found such that $M_B(T) = D$ is diagonal, then Theorem 3 shows how to find an invertible P such that $P^{-1}AP = D$. In other words, the "algebraic" problem of choosing P such that $P^{-1}AP$ is diagonal comes down to the "geometric" problem of finding a basis B such that $M_B(T)$ is diagonal. This shift of emphasis is one of the most important techniques in linear algebra and is the main theme of this chapter.

Each $n \times n$ matrix A can easily be realized as the matrix of an operator. In fact (Example 2 in Section 8.2),

$$M_{B_0}(T_A) = A$$

where $T_A : \mathbb{R}^n \to \mathbb{R}^n$ is the matrix operator given by $T_A(\mathbf{v}) = A\mathbf{v}$, and B_0 is the standard basis of $\mathbb{R}^n$. The first part of the next theorem gives the converse of Theorem 3: Any pair of similar matrices can be realized as the matrices of the same linear transformation with respect to different bases.

THEOREM 4 Let A be an $n \times n$ matrix, and let B_0 be the standard basis of $\mathbb{R}^n$.

(1) If $A' = P^{-1}AP$, let B be the ordered basis of $\mathbb{R}^n$ consisting of the columns of P in order. Then

$$M_{B_0}(T_A) = A \qquad \text{and} \qquad M_B(T_A) = A'$$

(2) If B is any ordered basis of $\mathbb{R}^n$, let P be the (invertible) matrix whose columns are the vectors in B in order. Then

$$M_B(T_A) = P^{-1}AP$$

Proof Let B and P be as in (1), and let $\mathbf{p}_1, \ldots, \mathbf{p}_n$ be the columns of P. Then

$$P_{B_0 \leftarrow B} = [C_{B_0}(\mathbf{p}_1) \ldots C_{B_0}(\mathbf{p}_n)] = [\mathbf{p}_1 \ldots \mathbf{p}_n] = P$$

The theorem now follows from Theorem 3. ■

EXAMPLE 2 Given $A = \begin{bmatrix} 10 & 6 \\ -18 & -11 \end{bmatrix}$, $P = \begin{bmatrix} 2 & -1 \\ -3 & 2 \end{bmatrix}$, and $D = \begin{bmatrix} 1 & 0 \\ 0 & -2 \end{bmatrix}$, verify that $P^{-1}AP = D$, and use this fact to find a basis B of $\mathbb{R}^2$ such that $M_B(T_A) = D$.

Solution $P^{-1}AP = D$ holds if $AP = PD$; this verification is left to the reader. Let B consist of the columns of P: $B = \left\{ \begin{bmatrix} 2 \\ -3 \end{bmatrix}, \begin{bmatrix} -1 \\ 2 \end{bmatrix} \right\}$. Then Theorem 4 gives

$$\begin{aligned} M_B(T_A) &= \left[C_B\left[T_A\left[\begin{matrix} 2 \\ -3 \end{matrix} \right] \right] \; C_B\left[T_A\left[\begin{matrix} -1 \\ 2 \end{matrix} \right] \right] \right] \\ &= \left[C_B\begin{bmatrix} 2 \\ -3 \end{bmatrix} \; C_B\begin{bmatrix} 2 \\ -4 \end{bmatrix} \right] = \begin{bmatrix} 1 & 0 \\ 0 & -2 \end{bmatrix} \end{aligned}$$

Hence $M_B(T_A) = D$ as asserted.

Recall that a property of $n \times n$ matrices is called a **similarity invariant** if, whenever a given $n \times n$ matrix A has the property, every matrix similar to A also has the property. Theorem 3 in Section 7.1 shows that rank, determinant, and trace are all similarity invariants.

To illustrate how such similarity invariants are related to linear operators, consider the case of rank. If $T : V \to V$ is a linear operator, the matrices of T with respect to various bases of V all have the same rank (being similar), so it is natural to regard the common rank of all these matrices as a property of T itself and not of the particular matrix used to describe T. Hence the rank of T could be *defined* to be the rank of A, where A is *any* matrix of T. This would be unambiguous by the above discussion. Of course, this is all unnecessary in the case of rank, because rank T was defined earlier to be the dimension of im T, and this turned out to equal the rank of every matrix representing T (Theorem 4 in Section 8.2). This definition of rank T is said to be *intrinsic* because it makes no reference to the matrices representing T. However, the technique serves to identify a property of T with *every* similarity invariant, and some of these properties are not so easily defined intrinsically.

In particular, if $T : V \to V$ is a linear operator on a finite dimensional space V, define the **determinant** of T (denoted det T) by

$$\det T = \det M_B(T), \quad B \text{ any basis of } V$$

This is independent of the choice of basis B, because if B_0 is any other basis of V, the matrices $M_B(T)$ and $M_{B_0}(T)$ are similar and so have the same determinant. In the same way, the **trace** of T (denoted tr T) may be defined by

$$\operatorname{tr} T = \operatorname{tr} M_B(T), \quad B \text{ any basis of } V$$

This is unambiguous for the same reasons.

The fact (Theorem 2) that the function $M_B : \mathbf{L}(V,V) \to \mathbf{M}_{nn}$ is an isomorphism that preserves products enables us to translate properties of matrices to properties of linear operators. The next example translates a property of determinants.

EXAMPLE 3 Let S and T denote linear operators on the finite dimensional space V. Show that

$$\det(ST) = \det S \det T$$

Solution Choose a basis B of V and use Theorem 2 and Theorem 1 in Section 3.2:

$$\det(ST) = \det[M_B(ST)] = \det[M_B(S)] \det[M_B(T)] = \det S \det T$$

Recall next that the characteristic polynomial of a matrix is another similarity invariant: If A and A' are similar matrices, then $c_A(x) = c_{A'}(x)$ (Theorem 3 in Section 7.1). As discussed above, the discovery of a similarity invariant means the discovery of a property of linear operators. In this case, if $T : V \to V$ is a linear operator on the finite dimensional space V, define the **characteristic polynomial** of T by

$$c_T(x) = c_A(x) \qquad \text{where } A = M_B(T), \quad B \text{ any basis of } V$$

In other words, the characteristic polynomial of an operator T is the characteristic polynomial of *any* matrix representing T. This is unambiguous because any two such matrices are similar by Theorem 3.

EXAMPLE 4 Compute the characteristic polynomial $c_T(x)$ of the operator $T : \mathbf{P}_2 \to \mathbf{P}_2$ given by $T(a + bx + cx^2) = (b + c) + (a + c)x + (a + b)x^2$.

Solution If $B = \{1, x, x^2\}$, the corresponding matrix of T is

$$M_B(T) = [C_B[T(1)]\ C_B[T(x)]\ C_B[T(x^2)]] = \begin{bmatrix} 0 & 1 & 1 \\ 1 & 0 & 1 \\ 1 & 1 & 0 \end{bmatrix}$$

Hence the characteristic polynomial of T is

$$c_T(x) = \det[xI - M_B(T)] = \det\begin{bmatrix} x & -1 & -1 \\ -1 & x & -1 \\ -1 & -1 & x \end{bmatrix}$$

$$= (x + 1)^2(x - 2) = x^3 - 3x - 2$$

EXERCISES 9.1

1. In each case, find $P = P_{B_0 \leftarrow B}$ and verify that $P^{-1}M_{B_0}(T)P = M_B(T)$ for the given operator T.
   (a) $T : \mathbb{R}^3 \to \mathbb{R}^3$, $T(a,b,c) = (2a - b, b + c, c - 3a)$; $B_0 = \{(1,1,0), (1,0,1), (0,1,0)\}$ and B is the standard basis.
   (b) $T : \mathbf{P}_2 \to \mathbf{P}_2$, $T(a + bx + cx^2) = (a + b) + (b + c)x + (c + a)x^2$; $B_0 = \{1, x, x^2\}$ and $B = \{1 - x^2, 1 + x, 2x + x^2\}$.
   (c) $T : \mathbf{M}_{22} \to \mathbf{M}_{22}$, $T\begin{bmatrix} a & b \\ c & d \end{bmatrix} = \begin{bmatrix} a + d & b + c \\ a + c & b + d \end{bmatrix}$; $B_0 = \left\{\begin{bmatrix} 1 & 0 \\ 0 & 0 \end{bmatrix}, \begin{bmatrix} 0 & 1 \\ 0 & 0 \end{bmatrix}, \begin{bmatrix} 0 & 0 \\ 1 & 0 \end{bmatrix}, \begin{bmatrix} 0 & 0 \\ 0 & 1 \end{bmatrix}\right\}$ and $B = \left\{\begin{bmatrix} 1 & 1 \\ 0 & 0 \end{bmatrix}, \begin{bmatrix} 0 & 0 \\ 1 & 1 \end{bmatrix}, \begin{bmatrix} 1 & 0 \\ 0 & 1 \end{bmatrix}, \begin{bmatrix} 0 & 1 \\ 1 & 1 \end{bmatrix}\right\}$
2. In each case, verify that $P^{-1}AP = D$ and find a basis B of $\mathbb{R}^2$ such that $M_B(T_A) = D$.
   (a) $A = \begin{bmatrix} 11 & -6 \\ 12 & -6 \end{bmatrix}$ $\quad P = \begin{bmatrix} 2 & 3 \\ 3 & 4 \end{bmatrix}$ $\quad D = \begin{bmatrix} 2 & 0 \\ 0 & 3 \end{bmatrix}$
   (b) $A = \begin{bmatrix} 29 & -12 \\ 70 & -29 \end{bmatrix}$ $\quad P = \begin{bmatrix} 3 & 2 \\ 7 & 5 \end{bmatrix}$ $\quad D = \begin{bmatrix} 1 & 0 \\ 0 & -1 \end{bmatrix}$
3. In each case, compute the characteristic polynomial $c_T(x)$.
   (a) $T : \mathbb{R}^2 \to \mathbb{R}^2$, $T(a,b) = (a - b, 2b - a)$

(b) $T : \mathbb{R}^2 \to \mathbb{R}^2$, $T(a,b) = (3a + 5b, 2a + 3b)$

(c) $T : \mathbf{P}_2 \to \mathbf{P}_2$, $T(a + bx + cx^2) = (a - 2c) + (2a + b + c)x + (c - a)x^2$

(d) $T : \mathbf{P}_2 \to \mathbf{P}_2$, $T(a + bx + cx^2) = (a + b - 2c) + (a - 2b + c)x + (b - 2a)x^2$

(e) $T : \mathbb{R}^3 \to \mathbb{R}^3$, $T(a,b,c) = (b,c,a)$

(f) $T : \mathbf{M}_{22} \to \mathbf{M}_{22}$, $T\begin{bmatrix} a & b \\ c & d \end{bmatrix} = \begin{bmatrix} a - c & b - d \\ a - c & b - d \end{bmatrix}$

4.[†] Let $D : \mathbf{P}_n \to \mathbf{P}_n$ be the differentiation operator: $D[p(x)] = p'(x)$. Show that $D^{n+1} = 0$.

5. If V is finite dimensional, show that a linear operator T on V has an inverse if and only if $\det T \neq 0$.

6. Show that the trace function $\mathbf{L}(V,V) \to \mathbb{R}$ is a linear transformation that satisfies $\operatorname{tr}(ST) = \operatorname{tr}(TS)$ for all S and T in $\mathbf{L}(V,V)$. [*Hint:* Example 4 in Section 8.1 and the proof of Theorem 3 in Section 7.1.]

SECTION 9.2 Reducible Operators

9.2.1 Invariant Subspaces and Direct Sums

A fundamental question in linear algebra is the following: If $T : V \to V$ is a linear transformation, how can a basis B of V be chosen such as to make the matrix $M_B(T)$ as simple as possible? In particular, when can B be found such that $M_B(T)$ is diagonal? The basic technique for answering such questions will be explained in this section.

DEFINITION If $T : V \to V$ is a linear operator, a subspace U of V is said to be ***T*-invariant** if $T(\mathbf{u})$ lies in U for every vector $\mathbf{u}$ in U.

If $T : V \to V$ is a linear operator and U is a subspace of V, define the **image** of U under T by

$$T(U) = \{T(\mathbf{u}) \mid \mathbf{u} \text{ in } U\}$$

This is a subspace of V, and "U is T-invariant" means simply that $T(U) \subseteq U$. It is clear that $T(V) \subseteq V$ and $T(0) \subseteq 0$. Hence:

EXAMPLE 1 If $T : V \to V$ is any linear operator; then V and 0 are T-invariant subspaces.

[†]This exercise requires calculus.

EXAMPLE 2 Define $T : \mathbb{R}^3 \to \mathbb{R}^3$ by $T(a,b,c) = (3a + 2b, b - c, 4a + 2b - c)$. Show that $U = \{(a,b,a) \mid a,b \text{ in } \mathbb{R}\}$ is T-invariant.

Solution Given $\mathbf{u} = (a,b,a)$ in U, we must show that $T(\mathbf{u})$ lies in U. Compute

$$\begin{aligned} T(\mathbf{u}) = T(a,b,a) &= (3a + 2b, b - a, 4a + 2b - a) \\ &= (3a + 2b, b - a, 3a + 2b) \end{aligned}$$

This lies in U (because the first and last entries are equal), so U is T-invariant.

EXAMPLE 3 Define $T : \mathbb{R}^2 \to \mathbb{R}^2$ by $T(a,b) = (b, -a)$. Show that $\mathbb{R}^2$ contains no T-invariant subspace except 0 and $\mathbb{R}^2$.

Solution Suppose if possible that U is T-invariant but $U \neq 0$, $U \neq \mathbb{R}^2$. Then U has dimension 1—say, $U = \mathbb{R}\mathbf{u}$ where $\mathbf{u} \neq \mathbf{0}$. Then $T(\mathbf{u})$ lies in U—say, $T(\mathbf{u}) = r\mathbf{u}$, r in $\mathbb{R}$. If we write $\mathbf{u} = (a,b)$, this is $(b, -a) = r(a,b)$, which gives $b = ra$ and $-a = rb$. Eliminating b gives $r^2a = rb = -a$, whence $a = 0$. Then $b = ra = 0$ too, contrary to the assumption that $\mathbf{u} \neq \mathbf{0}$. Hence no one-dimensional T-invariant subspace exists.

If a spanning set for a subspace U is known, it is easy to check whether U is invariant.

EXAMPLE 4 Let $T : V \to V$ be a linear operator, and suppose that $U = \text{span}\{\mathbf{u}_1, \mathbf{u}_2, \ldots, \mathbf{u}_k\}$ is a subspace of V. Show that U is T-invariant if and only if $T(\mathbf{u}_i)$ lies in U for each $i = 1, 2, \ldots, k$.

Solution Given $\mathbf{u}$ in U, write it as $\mathbf{u} = r_1\mathbf{u}_1 + \cdots + r_k\mathbf{u}_k$, r_i in $\mathbb{R}$. Then

$$T(\mathbf{u}) = r_1T(\mathbf{u}_1) + \cdots + r_kT(\mathbf{u}_k)$$

and this lies in U if each $T(\mathbf{u}_i)$ lies in U. This shows that U is T-invariant if each $T(\mathbf{u}_i)$ lies in U; the converse is clear.

Let $T : V \to V$ be a linear operator. If U is any subspace, then the image $T(U)$ of U under T is another subspace of V, and

$$T : U \to T(U)$$

is clearly a linear transformation. Hence T is a linear *operator* on U precisely when $T(U) \subseteq U$—that is, when U is T-invariant. This is the reason for the importance of T-invariant subspaces, and in this case, the linear operator $T : U \to U$ is called the **restriction** of T to U.

It turns out that the existence of a T-invariant subspace provides the first step toward finding a basis that simplifies the matrix of T.

THEOREM 1 Let $T : V \to V$ be a linear operator where V has dimension n, and suppose that U is any T-invariant subspace of V. Let $B_1 = \{\mathbf{e}_1, \ldots, \mathbf{e}_k\}$ be any basis of U, and extend it to a basis $B = \{\mathbf{e}_1, \ldots, \mathbf{e}_k, \mathbf{e}_{k+1}, \ldots, \mathbf{e}_n\}$ of V in any way. Then $M_B(T)$ has the block triangular form

$$M_B(T) = \begin{bmatrix} M_{B_1}(T) & Y \\ 0 & Z \end{bmatrix}$$

where Z is $(n - k) \times (n - k)$ and $M_{B_1}(T)$ is the matrix of the restriction of T to U.

Proof The matrix of (the restriction of) $T : U \to U$ with respect to the basis B_1 is the $k \times k$ matrix

$$M_{B_1}(T) = [C_{B_1}[T(\mathbf{e}_1)]\ C_{B_1}[T(\mathbf{e}_2)] \ldots C_{B_1}[T(\mathbf{e}_k)]]$$

Now compare the first column $C_{B_1}[T(\mathbf{e}_1)]$ here with the first column $C_B[T(\mathbf{e}_1)]$ of $M_B(T)$. The fact that $T(\mathbf{e}_1)$ lies in U (because U is T-invariant) means that $T(\mathbf{e}_1)$ has the form

$$T(\mathbf{e}_1) = t_1\mathbf{e}_1 + t_2\mathbf{e}_2 + \cdots + t_k\mathbf{e}_k + 0\mathbf{e}_{k+1} + \cdots + 0\mathbf{e}_n$$

Consequently,

$$C_{B_1}[T(\mathbf{e}_1)] = \begin{bmatrix} t_1 \\ t_2 \\ \vdots \\ t_k \end{bmatrix} \text{ in } \mathbb{R}^k \quad \text{while} \quad C_B[T(\mathbf{e}_1)] = \begin{bmatrix} t_1 \\ t_2 \\ \vdots \\ t_k \\ 0 \\ \vdots \\ 0 \end{bmatrix} \text{ in } \mathbb{R}^n$$

This shows that $M_B(T)$ and $\begin{bmatrix} M_{B_1}(T) & Y \\ 0 & Z \end{bmatrix}$ have identical first columns. Similar statements apply to columns 2, 3, . . . , k, and this proves the theorem. ■

The block upper triangular form for the matrix $M_B(T)$ in Theorem 1 is very useful, because the determinant of such a matrix equals the product of the determinants of each of the diagonal blocks. This is recorded below for reference, together with an important application to characteristic polynomials.

THEOREM 2 Let A be a block upper triangular matrix, say

$$A = \begin{bmatrix} A_{11} & A_{12} & A_{13} & \cdots & A_{1n} \\ 0 & A_{22} & A_{23} & \cdots & A_{2n} \\ 0 & 0 & A_{33} & \cdots & A_{3n} \\ \vdots & \vdots & \vdots & & \vdots \\ 0 & 0 & 0 & \cdots & A_{nn} \end{bmatrix}$$

where the diagonal blocks are square. Then:

(1) $\det A = (\det A_{11})(\det A_{22})(\det A_{33}) \ldots \det(A_{nn})$

(2) $c_A(x) = c_{A_{11}}(x)\, c_{A_{22}}(x)\, c_{A_{33}}(x) \ldots c_{A_{nn}}(x)$

Proof (1) is a restatement of Theorem 5 in Section 3.1, and (2) follows from (1) because

$$xI - A = \begin{bmatrix} xI - A_{11} & -A_{12} & -A_{13} & \cdots & -A_{1n} \\ 0 & xI - A_{22} & -A_{23} & \cdots & -A_{2n} \\ 0 & 0 & xI - A_{33} & \cdots & -A_{3n} \\ \vdots & \vdots & \vdots & & \vdots \\ 0 & 0 & 0 & \cdots & xI - A_{nn} \end{bmatrix}$$

where, in each diagonal block, the symbol I stands for the identity matrix of the appropriate size. ■

EXAMPLE 5 Consider the linear operator $T : \mathbf{P}_2 \to \mathbf{P}_2$ given by

$$T(a + bx + cx^2) = (-2a - b + 2c) + (a + b)x + (-6a - 2b + 5c)x^2$$

Show that $U = \text{span}\{x, 1 + 2x^2\}$ is T-invariant, use it to find a block upper triangular matrix for T, and use that to compute $c_T(x)$.

Solution To show that U is T-invariant, it suffices (Example 4) to show that $T(x)$ and $T(1 + 2x^2)$ both lie in U.

$$T(x) = -1 + x - 2x^2 = x - (1 + 2x^2)$$
$$T(1 + 2x^2) = 2 + x + 4x^2 = x + 2(1 + 2x^2)$$

Hence U is T-invariant. Because $B_1 = \{x, 1 + 2x^2\}$ is a basis of U, we complete it to a basis B of $\mathbf{P}_2$ in any way at all—say, $B = \{x, 1 + 2x^2, x^2\}$. Then

$$M_B(T) = [C_B[T(x)]\ C_B[T(1 + 2x^2)]\ C_B[T(x^2)]]$$
$$= [C_B(-1 + x - 2x^2)\ C_B(2 + x + 4x^2)\ C_B(2 + 5x^2)]$$
$$= \left[\begin{array}{cc|c} 1 & 1 & 0 \\ -1 & 2 & 2 \\ \hline 0 & 0 & 1 \end{array}\right]$$

is in block upper triangular form. Finally,

$$c_T(x) = \det\left[\begin{array}{cc|c} x-1 & -1 & 0 \\ 1 & x-2 & -2 \\ \hline 0 & 0 & x-1 \end{array}\right]$$
$$= \det\begin{bmatrix} x-1 & -1 \\ 1 & x-2 \end{bmatrix}\det(x-1) = (x^2 - 3x + 3)(x - 1)$$

This completes the solution.

Before proceeding, it is necessary to introduce the notion of a direct-sum decomposition of a vector space V. If U and W are subspaces of V, their **sum** $U + W$ and their **intersection** $U \cap W$ are defined by

$$U + W = \{\mathbf{u} + \mathbf{w} \mid \mathbf{u} \text{ in } U \text{ and } \mathbf{w} \text{ in } W\}$$
$$U \cap W = \{\mathbf{v} \mid \mathbf{v} \text{ lies in both } U \text{ and } W\}$$

These are subspaces of V, the sum containing both U and W and the intersection contained in both U and W. It turns out that the most interesting pairs U and W of subspaces are those for which $U \cap W$ is as small as possible and $U + W$ is as large as possible.

DEFINITION

A vector space V is said to be the **direct sum** of subspaces U and W if

$$U \cap W = 0 \qquad \text{and} \qquad U + W = V$$

In this case we write $V = U \oplus W$. Given a subspace U, any subspace W such that $V = U \oplus W$ is called a **complement** of U in V.

EXAMPLE 6 In the space $\mathbb{R}^5$ let $U = \{(a,b,c,0,0) \mid a, b, \text{ and } c \text{ in } \mathbb{R}\}$ and $W = \{(0,0,0,d,e) \mid d \text{ and } e \text{ in } \mathbb{R}\}$. Show that U and W are subspaces of $\mathbb{R}^5$, and $\mathbb{R}^5 = U \oplus W$.

Solution If $\mathbf{v} = (a,b,c,d,e)$ is any vector in $\mathbb{R}^5$, then $\mathbf{v} = (a,b,c,0,0) + (0,0,0,d,e)$, so $\mathbf{v}$ lies in $U + W$. Hence $\mathbb{R}^5 = U + W$. To show that

$U \cap W = 0$, let $\mathbf{v} = (a,b,c,d,e)$ lie in $U \cap W$. Then $d = e = 0$ because $\mathbf{v}$ lies in U, and $a = b = c = 0$ because $\mathbf{v}$ lies in W. Thus $\mathbf{v} = (0,0,0,0,0) = \mathbf{0}$, so $\mathbf{0}$ is the only vector in $U \cap W$. Hence $U \cap W = 0$, and the verification that $\mathbb{R}^5 = U \oplus W$ is complete.

EXAMPLE 7 Let $\{\mathbf{e}_1, \mathbf{e}_2, \ldots, \mathbf{e}_n\}$ be a basis of a vector space V, and partition it into two parts: $\{\mathbf{e}_1, \ldots, \mathbf{e}_k\}$ and $\{\mathbf{e}_{k+1}, \ldots, \mathbf{e}_n\}$. If $U = \text{span}\{\mathbf{e}_1, \ldots, \mathbf{e}_k\}$ and $W = \{\mathbf{e}_{k+1}, \ldots, \mathbf{e}_n\}$, show that $V = U \oplus W$.

Solution If $\mathbf{v}$ lies in $U \cap W$, then $\mathbf{v} = a_1\mathbf{e}_1 + \cdots + a_k\mathbf{e}_k$, and $\mathbf{v} = b_{k+1}\mathbf{e}_{k+1} + \cdots + b_n\mathbf{e}_n$ holds for some a_i and b_j in $\mathbb{R}$. The fact that the $\mathbf{e}_i$ are linearly independent forces all $a_i = b_j = 0$, so $\mathbf{v} = \mathbf{0}$. Hence $U \cap W = 0$. Now, given $\mathbf{v}$ in V, write $\mathbf{v} = v_1\mathbf{e}_1 + \cdots + v_n\mathbf{e}_n$, where the v_i are in $\mathbb{R}$. Then $\mathbf{v} = \mathbf{u} + \mathbf{w}$, where $\mathbf{u} = v_1\mathbf{e}_1 + \cdots + v_k\mathbf{e}_k$ lies in U and $\mathbf{w} = v_{k+1}\mathbf{e}_{k+1} + \cdots + v_n\mathbf{e}_n$ lies in W. This proves that $V = U + W$.

THEOREM 3 Let U and W be subspaces of a finite dimensional vector space V. The following conditions are equivalent:

(1) $V = U \oplus V$.

(2) Each vector $\mathbf{v}$ in V can be written uniquely in the form

$$\mathbf{v} = \mathbf{u} + \mathbf{w}, \quad \mathbf{u} \text{ in } U, \mathbf{w} \text{ in } W$$

The uniqueness means that if $\mathbf{v} = \mathbf{u}_1 + \mathbf{w}_1$ is another such representation, then $\mathbf{u}_1 = \mathbf{u}$ and $\mathbf{w}_1 = \mathbf{w}$.

(3) If $\{\mathbf{u}_1, \ldots, \mathbf{u}_k\}$ and $\{\mathbf{w}_1, \ldots, \mathbf{w}_m\}$ are bases of U and W, respectively, then $B = \{\mathbf{u}_1, \ldots, \mathbf{u}_k, \mathbf{w}_1, \ldots, \mathbf{w}_m\}$ is a basis of V.

Proof (1) implies (2). Given $\mathbf{v}$ in V, we have $\mathbf{v} = \mathbf{u} + \mathbf{w}$, $\mathbf{u}$ in U, $\mathbf{w}$ in W, because $V = U + W$. If also $\mathbf{v} = \mathbf{u}_1 + \mathbf{w}_1$, then $\mathbf{u} - \mathbf{u}_1 = \mathbf{w}_1 - \mathbf{w}$ lies in $U \cap W = 0$, so $\mathbf{u} = \mathbf{u}_1$ and $\mathbf{w} = \mathbf{w}_1$.

(2) implies (3). Given $\mathbf{v}$ in V and $\mathbf{v} = \mathbf{u} + \mathbf{w}$, $\mathbf{u}$ in U, $\mathbf{w}$ in W, we know that $\mathbf{v}$ lies in span B. Hence B spans V. To see that B is independent, let $a_1\mathbf{u}_1 + \cdots + a_k\mathbf{u}_k + b_1\mathbf{w}_1 + \cdots + b_m\mathbf{w}_m = \mathbf{0}$. Write $\mathbf{u} = a_1\mathbf{u}_1 + \cdots + a_k\mathbf{u}_k$ and $\mathbf{w} = b_1\mathbf{w}_1 + \cdots + b_m\mathbf{w}_m$. Then $\mathbf{u} + \mathbf{w} = \mathbf{0} + \mathbf{0}$, and so $\mathbf{u} = \mathbf{0}$ and $\mathbf{w} = \mathbf{0}$ by the uniqueness in (2). Hence $a_i = 0$ and $b_j = 0$ for all i and j.

(3) implies (1). This is by Example 7. ■

Condition (3) gives the following useful result.

THEOREM 4 If a finite dimensional vector space V is the direct sum $V = U \oplus W$ of subspaces U and W, then

$$\dim V = \dim U + \dim W$$

The next example shows that the projection theorem (Theorem 9 in Section 6.2) is really a direct-sum decomposition.

EXAMPLE 8 Let V be an inner product space, and let U be a finite dimensional subspace. Then

$$V = U \oplus U^{\perp}$$

Solution The equation $V = U + U^{\perp}$ holds because, given $\mathbf{v}$ in V, the vector $\text{proj}_U(\mathbf{v})$ lies in U and $\mathbf{v} - \text{proj}_U(\mathbf{v})$ lies in $U^{\perp}$. To see that $U \cap U^{\perp} = 0$, observe that any vector in $U \cap U^{\perp}$ would be orthogonal to itself and hence must be zero.

These direct-sum decompositions of V play an important role in any discussion of invariant subspaces. If $T : V \to V$ is a linear operator and if U_1 is a T-invariant subspace, the block upper triangular matrix

$$M_B(T) = \begin{bmatrix} M_{B_1}(T) & Y \\ 0 & Z \end{bmatrix}$$

in Theorem 1 is achieved by choosing any basis $B_1 = \{\mathbf{e}_1, \ldots, \mathbf{e}_k\}$ of U_1 and completing it to a basis $B = \{\mathbf{e}_1, \ldots, \mathbf{e}_k, \mathbf{e}_{k+1}, \ldots, \mathbf{e}_n\}$ of V in any way at all. The fact that U_1 is T-invariant ensures that the first k columns of $M_B(T)$ have the given form, and the question arises whether the additional basis vectors $\mathbf{e}_{k+1}, \ldots, \mathbf{e}_n$ can be chosen such that

$$U_2 = \text{span}\{\mathbf{e}_{k+1}, \ldots, \mathbf{e}_n\}$$

is *also* T-invariant. In other words, does each T-invariant subspace of V have a T-invariant complement? Unfortunately the answer is "no" in general (see Example 10) but, when it is possible, the matrix $M_B(T)$ above simplifies further. The fact that the complement $U_2 = \text{span}\{\mathbf{e}_{k+1}, \ldots, \mathbf{e}_n\}$ is T-invariant too means that $Y = 0$ in the above notation and that $Z = M_{B_2}(T)$ is the matrix of the restriction of T to U_2 (where $B_2 = \{\mathbf{e}_{k+1}, \ldots, \mathbf{e}_n\}$). The verification is the same as in the proof of Theorem 1.

THEOREM 5 Let $T : V \to V$ be a linear operator where V has dimension n. Suppose $V = U_1 \oplus U_2$ where both U_1 and U_2 are T-invariant. If $B_1 = \{\mathbf{e}_1, \ldots, \mathbf{e}_k\}$

and $B_2 = \{\mathbf{e}_{k+1}, \ldots, \mathbf{e}_n\}$ are bases of U_1 and U_2, respectively, then

$$B = \{\mathbf{e}_1, \ldots, \mathbf{e}_k, \mathbf{e}_{k+1}, \ldots, \mathbf{e}_n\}$$

is a basis of V, and $M_B(T)$ has the block diagonal form

$$M_B(T) = \begin{bmatrix} M_{B_1}(T) & 0 \\ 0 & M_{B_2}(T) \end{bmatrix}$$

where $M_{B_1}(T)$ and $M_{B_2}(T)$ are the matrices of the restrictions of T to U_1 and to U_2, respectively.

The linear operator $T : V \to V$ is said to be **reducible** if nonzero T-invariant subspaces U_1 and U_2 can be found such that $V = U_1 \oplus U_2$. Then T has a matrix in block diagonal form as in Theorem 5, and the study of T is reduced to studying its restrictions to the lower-dimensional spaces U_1 and U_2. If these can be determined, so can T. Here is an example where the action of T on the invariant subspaces U_1 and U_2 is very simple indeed. The result for operators is used to derive the corresponding similarity theorem for matrices.

EXAMPLE 9 Let $T : V \to V$ be a linear operator satisfying $T^2 = 1$ (such operators are called **involutions**). Define

$$U_1 = \{\mathbf{v} \mid T(\mathbf{v}) = \mathbf{v}\} \qquad \text{and} \qquad U_2 = \{\mathbf{v} \mid T(\mathbf{v}) = -\mathbf{v}\}$$

(a) Show that $V = U_1 \oplus U_2$.

(b) If $\dim V = n$, find a basis B of V such that $M_B(T) = \begin{bmatrix} I_k & 0 \\ 0 & -I_{n-k} \end{bmatrix}$.

(c) Conclude that, if A is an $n \times n$ matrix such that $A^2 = I$, then A is similar to $\begin{bmatrix} I_k & 0 \\ 0 & -I_{n-k} \end{bmatrix}$ for some k.

Solution

(a) The verification that U_1 and U_2 are subspaces of V is left to the reader. If $\mathbf{v}$ lies in $U_1 \cap U_2$, then $\mathbf{v} = T(\mathbf{v}) = -\mathbf{v}$, and it follows that $\mathbf{v} = \mathbf{0}$. Hence $U_1 \cap U_2 = 0$. Given $\mathbf{v}$ in V, write

$$\mathbf{v} = \frac{1}{2}\{[\mathbf{v} + T(\mathbf{v})] + [\mathbf{v} - T(\mathbf{v})]\}$$

Then $\mathbf{v} + T(\mathbf{v})$ lies in U_1, because $T[\mathbf{v} + T(\mathbf{v})] = T(\mathbf{v}) + T^2(\mathbf{v}) = T(\mathbf{v}) + \mathbf{v} = \mathbf{v} + T(\mathbf{v})$. Similarly $\mathbf{v} - T(\mathbf{v})$ lies in U_2, and it follows that $V = U_1 + U_2$. This proves part (a).

(b) U_1 and U_2 are clearly T-invariant, so the result follows from Theorem 5 if bases $B_1 = \{\mathbf{e}_1, \ldots, \mathbf{e}_k\}$ and $B_2 = \{\mathbf{e}_{k+1}, \ldots, \mathbf{e}_n\}$ of U_1 and U_2 can be found such that $M_{B_1}(T) = I_k$ and $M_{B_2}(T) = -I_{n-k}$. But this is true for *any* choice of B_1 and B_2:

$$\begin{aligned} M_{B_1}(T) &= [C_{B_1}[T(\mathbf{e}_1)]\; C_{B_1}[T(\mathbf{e}_2)] \ldots C_{B_1}[T(\mathbf{e}_k)]] \\ &= [C_{B_1}(\mathbf{e}_1)\; C_{B_1}(\mathbf{e}_2) \ldots C_{B_1}(\mathbf{e}_k)] \\ &= I_k \end{aligned}$$

A similar argument shows that $M_{B_2}(T) = -I_{n-k}$, so part (b) follows with $B = \{\mathbf{e}_1, \mathbf{e}_2, \ldots, \mathbf{e}_n\}$.

(c) Given A such that $A^2 = I$, consider $T_A : \mathbb{R}^n \to \mathbb{R}^n$. Then $(T_A)^2 = T_{A^2} = T_I = 1$ (by Theorem 13 in Section 8.1) so, by part (b), there exists a basis B of $\mathbb{R}^n$ such that

$$M_B(T_A) = \begin{bmatrix} I_r & 0 \\ 0 & -I_{n-r} \end{bmatrix}$$

But Theorem 4 in Section 9.1 shows that $M_B(T_A) = P^{-1}AP$ for some invertible matrix P, and this proves part (c).

Note that the passage from the result for operators to the analogous result for matrices is routine and can be carried out in any situation, as in the verification of part (c). The key is the analysis of the operators. In this case, the involutions are just the operators satisfying $T^2 = 1$, and the simplicity of this condition means that the invariant subspaces U_1 and U_2 are easy to find.

Unfortunately, not every linear operator $T : V \to V$ is reducible. In fact, the linear operator in Example 3 has *no* invariant subspaces except 0 and V. On the other hand, one might expect that this is the only type of nonreducible operator; that is, if the operator *has* an invariant subspace that is not 0 or V, then *some* invariant complement must exist. The next example shows that even this is not valid.

EXAMPLE 10

Let $A = \begin{bmatrix} 1 & 1 \\ 0 & 1 \end{bmatrix}$ and consider the matrix operator $T_A : \mathbb{R}^2 \to \mathbb{R}^2$ given by $T_A\begin{bmatrix} a \\ b \end{bmatrix} = A\begin{bmatrix} a \\ b \end{bmatrix} = \begin{bmatrix} a+b \\ b \end{bmatrix}$. Show that $U_1 = \mathbb{R}\begin{bmatrix} 1 \\ 0 \end{bmatrix}$ is T_A-invariant but that U has no T_A-invariant complement in $\mathbb{R}^2$.

Solution Because $U_1 = \operatorname{span}\left\{\begin{bmatrix} 1 \\ 0 \end{bmatrix}\right\}$, and $T_A\begin{bmatrix} 1 \\ 0 \end{bmatrix} = \begin{bmatrix} 1 \\ 0 \end{bmatrix}$, it follows (by Example 4) that U_1 is T_A-invariant. Now assume, if possible, that U_1 has a T_A-invariant complement U_2 in $\mathbb{R}^2$. Then $U_1 \oplus U_2 = \mathbb{R}^2$ and $T_A(U_2) \subseteq U_2$. Now

$$2 = \dim \mathbb{R}^2 = \dim U_1 + \dim U_2 = 1 + \dim U_2$$

so $\dim U_2 = 1$. Let $U_2 = \mathbb{R}\begin{bmatrix} p \\ q \end{bmatrix}$. Then $\begin{bmatrix} p \\ q \end{bmatrix}$ is not in U_1 (because $U_1 \cap U_2 = 0$) and hence $q \neq 0$. On the other hand, $T_A\begin{bmatrix} p \\ q \end{bmatrix}$ lies in $U_2 = \mathbb{R}\begin{bmatrix} p \\ q \end{bmatrix}$, say

$$\begin{bmatrix} p + q \\ q \end{bmatrix} = T_A\begin{bmatrix} p \\ q \end{bmatrix} = \lambda\begin{bmatrix} p \\ q \end{bmatrix}, \quad \lambda \text{ in } \mathbb{R}$$

Hence $p + q = \lambda p$ and $q = \lambda q$. Because $q \neq 0$, the second of these implies that $\lambda = 1$, whence the first implies that $q = 0$, a contradiction. So a T_A-invariant complement of U_1 does not exist.

EXERCISES 9.2.1

1. Let T be a linear operator on V. If U and U_1 are T-invariant, show that $U \cap U_1$ and $U + U_1$ are also T-invariant.

2. If $T : V \to V$ is any linear operator, show that $\ker T$ and $\operatorname{im} T$ are T-invariant subspaces.

3. Let S and T be linear operators on V and assume that $ST = TS$.

(a) Show that $\operatorname{im} S$ and $\ker S$ are T-invariant.

(b) If U is T-invariant, show that $S(U)$ is T-invariant.

4. Let $T : V \to V$ be a linear operator. Given $\mathbf{e}$ in V, let U denote the set of vectors in V that lie in every T-invariant subspace that contains $\mathbf{e}$.

(a) Show that U is a T-invariant subspace of V containing $\mathbf{e}$.

(b) Show that U is contained in every T-invariant subspace of V that contains $\mathbf{e}$.

5. (a) Show that every subspace is T-invariant if T is a scalar operator.

(b) Conversely, if every subspace is T-invariant, show that T is scalar.

6. Show that the only subspaces of V that are T-invariant for every operator $T : V \to V$ are 0 and V. Assume that V is finite dimensional. [*Hint:* Theorem 3 in Section 8.1.1.]

7. Suppose that $T : V \to V$ is a linear operator and that U is a T-invariant subspace of V. If S is an invertible operator, put $T' = STS^{-1}$. Show that $S(U)$ is a T'-invariant subspace.

8. In each case, show that U is T-invariant, use it to find a block upper triangular matrix for T, and use that to compute $c_T(x)$.

(a) $T : \mathbf{P}_2 \to \mathbf{P}_2$, $T(a + bx + cx^2) = (-a + 2b + c) + (a + 3b + c)x + (a + 4b)x^2$, $U = \operatorname{span}\{1, x + x^2\}$

(b) $T : \mathbf{P}_2 \to \mathbf{P}_2$, $T(a + bx + cx^2) = (5a - 2b + c) + (5a - b + c)x + (a + 2c)x^2$, $U = \text{span}\{1 - 2x^2, x + x^2\}$

9. In each case, show that $T_A : \mathbb{R}^2 \to \mathbb{R}^2$ has no invariant subspaces except 0 and $\mathbb{R}^2$.

(a) $A = \begin{bmatrix} 1 & 2 \\ -1 & -1 \end{bmatrix}$ **(b)** $A = \begin{bmatrix} \cos\theta & \sin\theta \\ -\sin\theta & \cos\theta \end{bmatrix}, 0 < \theta < \pi.$

10. In each case, show that $V = U \oplus W$.

(a) $V = \mathbb{R}^4$, $U = \text{span}\{(1,1,0,0), (0,1,1,0)\}$, $W = \text{span}\{(0,1,0,1), (0,0,1,1)\}$

(b) $V = \mathbb{R}^4$, $U = \{(a,a,b,b) \mid a,b \text{ in } \mathbb{R}\}$, $W = \{(c, d, c, -d) \mid c,d \text{ in } \mathbb{R}\}$

(c) $V = \mathbf{P}_3$, $U = \{a + bx \mid a,b \text{ in } \mathbb{R}\}$, $W = \{ax^2 + bx^3 \mid a,b \text{ in } \mathbb{R}\}$

(d) $V = \mathbf{M}_{22}$, $U = \left\{\begin{bmatrix} a & a \\ b & b \end{bmatrix} \middle| a,b \text{ in } \mathbb{R}\right\}$, $W = \left\{\begin{bmatrix} a & b \\ -a & b \end{bmatrix} \middle| a,b \text{ in } \mathbb{R}\right\}$

11. Let $U = \text{span}\{(1,0,0,0), (0,1,0,0)\}$ in $\mathbb{R}^4$. Show that $\mathbb{R}^4 = U \oplus W_1$ and $\mathbb{R}^4 = U \oplus W_2$, where $W_1 = \text{span}\{(0,0,1,0), (0,0,0,1)\}$ and $W_2 = \text{span}\{(1,1,1,1), (1,1,1,-1)\}$.

12. Let U be a subspace of a finite dimensional space V, and suppose that $V = U \oplus W_1$ and $V = U \oplus W_2$ hold for subspaces W_1 and W_2. Show that $\dim W_1 = \dim W_2$.

13. Call a polynomial $f(x)$ **even** if $f(-x) = f(x)$ and **odd** if $f(-x) = -f(x)$.

(a) Show that $f(x) + f(-x)$ is even and $f(x) - f(-x)$ is odd for every polynomial $f(x)$.

(b) If U and W denote the sets of even and odd polynomials in $\mathbf{P}_n$, respectively, show that U and W are subspaces and that $\mathbf{P}_n = U \oplus W$.

14. Consider the following sets of symmetric and skew-symmetric $n \times n$ matrices.

$$U = \{A \mid A \text{ in } \mathbf{M}_{nn}, A^T = A\}$$
$$W = \{A \mid A \text{ in } \mathbf{M}_{nn}, A^T = -A\}$$

Show that these are subspaces of $\mathbf{M}_{nn}$ and that $\mathbf{M}_{nn} = U \oplus W$. [*Hint:* Given X in $\mathbf{M}_{nn}$, $X + X^T$ is symmetric and $X - X^T$ is skew-symmetric.]

15. Let E be a 2×2 matrix such that $E^2 = E$. Show that $\mathbf{M}_{22} = U \oplus W$, where $U = \{A \mid AE = A\}$ and $W = \{B \mid BE = 0\}$. [*Hint: XE* lies in U for every matrix X.]

16. Let $V \xrightarrow{T} W \xrightarrow{S} V$ be linear transformations, and assume that $\dim V$ and $\dim W$ are finite.

(a) If $ST = 1_V$, show that $W = \text{im } T \oplus \ker S$ (see Section 8.1.2). [*Hint:* Given $\mathbf{w}$ in W, show that $\mathbf{w} - TS(\mathbf{w})$ lies in $\ker S$.]

(b) Illustrate with $\mathbb{R}^2 \xrightarrow{T} \mathbb{R}^3 \xrightarrow{S} \mathbb{R}^2$ where $T(x,y) = (x,y,0)$ and $S(x,y,z) = (x,y)$.

17. Let V be a finite dimensional vector space, and let U and W be subspaces such that $U \cap W = 0$. If $\dim U + \dim W = \dim V$, show that $V = U \oplus W$.

18. In each case, show that $T^2 = 1$ and find (as in Example 9) an ordered basis B such that $M_B(T)$ has the form given.

(a) $T : \mathbf{M}_{22} \to \mathbf{M}_{22}$ where $T(A) = A^T$, $M_B(T) = \begin{bmatrix} I_3 & 0 \\ 0 & -1 \end{bmatrix}$

(b) $T : \mathbf{P}_3 \to \mathbf{P}_3$ where $T[p(x)] = p(-x)$, $M_B(T) = \begin{bmatrix} I_2 & 0 \\ 0 & -I_2 \end{bmatrix}$

(c) $T : \mathbb{C} \to \mathbb{C}$ where $T(a + bi) = a - bi$, $M_B(T) = \begin{bmatrix} 1 & 0 \\ 0 & -1 \end{bmatrix}$

(d) $T : \mathbb{R}^3 \to \mathbb{R}^3$ where $T(a,b,c) = (-a + 2b + c, b + c, -c)$, $M_B(T) = \begin{bmatrix} 1 & 0 \\ 0 & -I_2 \end{bmatrix}$

(e) $T : V \to V$ where $T(\mathbf{v}) = -\mathbf{v}$, $\dim V = n$, $M_B(T) = -I_n$

19. Let U and W denote subspaces of a vector space V.

(a) If $V = U \oplus W$, define $T : V \to V$ by $T(\mathbf{v}) = \mathbf{w}$ where $\mathbf{v}$ is written (uniquely) as $\mathbf{v} = \mathbf{u} + \mathbf{w}$ with $\mathbf{u}$ in U and $\mathbf{w}$ in W. Show that T is a linear transformation, $U = \ker T$, $W = \operatorname{im} T$, and $T^2 = T$.

(b) Conversely, if $T : V \to V$ is a linear transformation such that $T^2 = T$, show that $V = \ker T \oplus \operatorname{im} T$. [*Hint:* $\mathbf{v} - T(\mathbf{v})$ lies in $\ker T$ and $T(\mathbf{v})$ lies in $\operatorname{im} T$ for all $\mathbf{v}$ in V.]

20. Let $T : V \to V$ be a linear operator satisfying $T^2 = T$ (such operators are called **idempotents**). Define $U_1 = \{\mathbf{v} \mid T(\mathbf{v}) = \mathbf{v}\}$ and $U_2 = \ker T = \{\mathbf{v} \mid T(\mathbf{v}) = \mathbf{0}\}$.

(a) Show that $V = U_1 \oplus U_2$. [*Hint:* Exercise 19.]

(b) If $\dim V = n$, find a basis B of V such that $M_B(T) = \begin{bmatrix} I_r & 0 \\ 0 & 0 \end{bmatrix}$, where $r = \operatorname{rank} T$.

(c) If A is an $n \times n$ matrix such that $A^2 = A$, show that A is similar to $\begin{bmatrix} I_r & 0 \\ 0 & 0 \end{bmatrix}$, where $r = \operatorname{rank} A$. [*Hint:* Example 9.]

21. In each case, show that $T^2 = T$ and find (as in the preceding exercise) an ordered basis B such that $M_B(T)$ has the form given (0_k is the $k \times k$ zero matrix).

(a) $T : \mathbf{P}_2 \to \mathbf{P}_2$ where $T(a + bx + cx^2) = (a - b + c)(1 + x + x^2)$, $M_B(T) = \begin{bmatrix} 1 & 0 \\ 0 & 0_2 \end{bmatrix}$

(b) $T : \mathbb{R}^3 \to \mathbb{R}^3$ where $T(a,b,c) = (a + 2b, 0, 4b + c)$, $M_B(T) = \begin{bmatrix} I_2 & 0 \\ 0 & 0 \end{bmatrix}$

(c) $T : \mathbf{M}_{22} \to \mathbf{M}_{22}$ where $T\begin{bmatrix} a & b \\ c & d \end{bmatrix} = \begin{bmatrix} -5 & -15 \\ 2 & 6 \end{bmatrix}\begin{bmatrix} a & b \\ c & d \end{bmatrix}$, $M_B(T) = \begin{bmatrix} I_2 & 0 \\ 0 & 0_2 \end{bmatrix}$

22. Let $T : V \to V$ be an operator satisfying $T^2 = cT$, $c \neq 0$.

(a) Show that $V = U \oplus \ker T$, where $U = \{\mathbf{u} \mid T(\mathbf{u}) = c\mathbf{u}\}$. [*Hint:* Compute $T\left[\mathbf{v} - \frac{1}{c}T(\mathbf{v})\right]$.]

(b) If $\dim V = n$, show that V has a basis B such that $M_B(T) = \begin{bmatrix} cI_r & 0 \\ 0 & 0 \end{bmatrix}$, where $r = \operatorname{rank} T$.

(c) If A is any $n \times n$ matrix of rank r such that $A^2 = cA$, $c \neq 0$, show that A is similar to $\begin{bmatrix} cI_r & 0 \\ 0 & 0 \end{bmatrix}$.

23. Let $T : V \to V$ be an operator such that $T^2 = c^2$, $c \neq 0$.

(a) Show that $V = U_1 \oplus U_2$, where $U_1 = \{\mathbf{v} \mid T(\mathbf{v}) = c\mathbf{v}\}$ and $U_2 = \{\mathbf{v} \mid T(\mathbf{v}) = -c\mathbf{v}\}$. [*Hint:* $\mathbf{v} = \frac{1}{2c}\{[T(\mathbf{v}) + c\mathbf{v}] - [T(\mathbf{v}) - c\mathbf{v}]\}$.]

(b) If $\dim V = n$, show that V has a basis B such that $M_B(T) = \begin{bmatrix} cI_k & 0 \\ 0 & -cI_{n-k} \end{bmatrix}$ for some k.

(c) If A is an $n \times n$ matrix such that $A^2 = c^2 I$, $c \neq 0$, show that A is similar to $\begin{bmatrix} cI_k & 0 \\ 0 & -cI_{n-k} \end{bmatrix}$ for some k.

24. If P is a fixed $n \times n$ matrix, define $T : \mathbf{M}_{nn} \to \mathbf{M}_{nn}$ by $T(A) = PA$. Let U_j denote the subspace of $\mathbf{M}_{nn}$ consisting of all matrices with all columns zero except possibly column j.

(a) Show that each U_j is T-invariant.

(b) Show that $\mathbf{M}_{nn}$ has a basis B such that $M_B(T)$ is block diagonal with each block on the diagonal equal to P.

25. Let V be a vector space. If $f : V \to \mathbb{R}$ is a linear transformation and $\mathbf{z}$ is a vector in V, define $T_{f,z} : V \to V$ by $T_{f,z}(\mathbf{v}) = f(\mathbf{v})\mathbf{z}$ for all $\mathbf{v}$ in V. Assume that $f \neq 0$ and $\mathbf{z} \neq \mathbf{0}$.

(a) Show that $T_{f,z}$ is a linear operator of rank 1.

(b) Show that $T_{f,z}$ is an idempotent if and only if $f(\mathbf{z}) = 1$. (Recall that $T : V \to V$ is an idempotent if $T^2 = T$.)

(c) Show that every idempotent $T : V \to V$ of rank 1 has the form $T = T_{f,z}$ for some $f : V \to \mathbb{R}$ and some $\mathbf{z}$ in V with $f(\mathbf{z}) = 1$. [*Hint:* Write $\operatorname{im} T = \mathbb{R}\mathbf{z}$ and show that $T(\mathbf{z}) = \mathbf{z}$. Then use Exercise 19.]

9.2.2 Eigenvalues and Diagonalization

In the discussion of invariant subspaces in Section 9.2.1, the reader may have thought it an oversight that the simplest examples of such spaces (aside from 0 and V) were not discussed—namely, those of dimension 1. Of course, such a space has the form $\mathbb{R}\mathbf{v}$ for a nonzero vector $\mathbf{v}$, and the condition for it to be invariant is as follows:

THEOREM 6

Let $T : V \to V$ be a linear operator. A one-dimensional subspace $\mathbb{R}\mathbf{v}$, $\mathbf{v} \neq \mathbf{0}$, is T-invariant if and only if $T(\mathbf{v}) = \lambda\mathbf{v}$ for some number λ.

Proof A typical vector in $\mathbb{R}\mathbf{v}$ has the form $r\mathbf{v}$, r in $\mathbb{R}$, and $T(r\mathbf{v}) = rT(\mathbf{v})$ lies in $\mathbb{R}\mathbf{v}$ for all r if and only if $T(\mathbf{v})$ lies in $\mathbb{R}\mathbf{v}$. This means that $T(\mathbf{v}) = \lambda\mathbf{v}$ for some λ in $\mathbb{R}$. ■

DEFINITION

A real number λ is called an **eigenvalue** of an operator $T : V \to V$ if

$$T(\mathbf{v}) = \lambda\mathbf{v}$$

holds for some nonzero vector $\mathbf{v}$ in V. In this case, $\mathbf{v}$ is called an **eigenvector** of T corresponding to λ. The subspace

$$E_\lambda(T) = \{\mathbf{v} \text{ in } V \mid T(\mathbf{v}) = \lambda\mathbf{v}\}$$

is called the **eigenspace** of T corresponding to λ.

Eigenvalues and eigenvectors arose in Section 7.1 in connection with matrices. An eigenvalue of an $n \times n$ matrix A was defined to be a real number λ such that $A\mathbf{p} = \lambda\mathbf{p}$ holds for some nonzero column $\mathbf{p}$ in $\mathbb{R}^n$, in which case $\mathbf{p}$ was called an eigenvector of A. Clearly λ and $\mathbf{p}$ are an eigenvalue and an eigenvector, respectively, of the operator $T_A : \mathbb{R}^n \to \mathbb{R}^n$, so the matrix terminology of Section 7.1 is consistent with the present terminology for operators. Similarly, the eigenspace

$$E_\lambda(A) = \{\mathbf{p} \text{ in } \mathbb{R}^n \mid A\mathbf{p} = \lambda\mathbf{p}\}$$

of A corresponding to λ (as defined in Section 7.1) is the eigenspace of the matrix operator T_A. These spaces can be determined for a given matrix A by solving systems of linear equations (see Section 7.1).

The following theorem reveals a fundamental connection between the eigenspaces of an operator T and those of the matrices representing T. Recall that the characteristic polynomial $c_T(x)$ of an operator $T : V \to V$ on a finite dimensional space V is defined by $c_T(x) = \det[xI - M_B(T)]$, where B is any ordered basis of V.

THEOREM 7

Let $T : V \to V$ be a linear operator where $\dim V = n$; let B denote any ordered basis of V; and let $C_B : V \to \mathbb{R}^n$ denote the coordinate isomorphism. Then:

(1) The eigenvalues λ of T are precisely the eigenvalues of the matrix $M_B(T)$ and thus are the roots of the characteristic polynomial $c_T(x)$.

(2) In this case the eigenspaces $E_\lambda(T)$ and $E_\lambda[M_B(T)]$ are iso-

morphic. In fact, the restriction of C_B,

$$C_B : E_\lambda(T) \to E_\lambda[M_B(T)]$$

is an isomorphism.

Proof Write $A = M_B(T)$. If $T(\mathbf{v}) = \lambda\mathbf{v}$, then applying C_B gives $\lambda C_B(\mathbf{v}) = C_B[T(\mathbf{v})] = AC_B(\mathbf{v})$, using Theorem 2 in Section 9.1. Hence $C_B(\mathbf{v})$ lies in $E_\lambda(A)$, so we do indeed have a function $C_B : E_\lambda(T) \to E_\lambda(A)$. It is clearly linear and one-to-one; we claim it is onto. If $\mathbf{p}$ is in $E_\lambda(A)$, write $\mathbf{p} = C_B(\mathbf{v})$ for some $\mathbf{v}$ in V (C_B is onto). This $\mathbf{v}$ actually lies in $E_\lambda(T)$. In fact,

$$C_B[T(\mathbf{v})] = AC_B(\mathbf{v}) = A\mathbf{p} = \lambda\mathbf{p} = \lambda C_B(\mathbf{v}) = C_B(\lambda\mathbf{v})$$

so $T(\mathbf{v}) = \lambda\mathbf{v}$ (C_B is one-to-one). This proves (2). As to (1), we have already shown that eigenvalues of T are eigenvalues of A. The converse follows, as in the foregoing proof that C_B is onto [part (a) of Exercise 4]. ■

Theorem 7 shows how to pass back and forth between the eigenvectors of an operator T and the eigenvectors of any matrix $M_B(T)$ of T. The eigenvectors of T are just the vectors $\mathbf{v}$ in V whose coordinate vector $C_B(\mathbf{v})$ is an eigenvector of the matrix $M_B(T)$:

$$E_\lambda(T) = \{\mathbf{v} \mid C_B(\mathbf{v}) \text{ lies in } E_\lambda[M_B(T)]\}$$

Similarly, the eigenvectors of $M_B(T)$ are precisely the coordinate vectors $C_B(\mathbf{v})$ of eigenvectors $\mathbf{v}$ of T:

$$E_\lambda[M_B(T)] = \{C_B(\mathbf{v}) \mid \mathbf{v} \text{ lies in } E_\lambda(T)\}$$

EXAMPLE 11 Find the eigenvalues and eigenspaces for $T : \mathbf{P}_3 \to \mathbf{P}_3$ given by

$$T(a + bx + cx^2) = (2a + b + c) + (2a + b - 2c)x - (a + 2c)x^2$$

Solution If $B = \{1, x, x^2\}$, then

$$M_B(T) = [C_B[T(1)]\ C_B[T(x)]\ C_B[T(x^2)]] = \begin{bmatrix} 2 & 1 & 1 \\ 2 & 1 & -2 \\ -1 & 0 & -2 \end{bmatrix}$$

This is the matrix considered in Example 3 of Section 7.1, and

$$c_T(x) = \det[xI - M_B(T)] = (x + 1)^2(x - 3)$$

Hence the eigenvalues are $\lambda = -1$ and $\lambda = 3$. Theorem 1 in Section 7.1 shows that the eigenspace for any square matrix A corresponding to λ is given by solving homogeneous equations.

$$E_\lambda(A) = \{\mathbf{p} \text{ in } \mathbb{R}^n \mid (\lambda I - A)\mathbf{p} = \mathbf{0}\}$$

Hence $E_{-1}[M_B(T)] = \mathbb{R}\begin{bmatrix} -1 \\ 2 \\ 1 \end{bmatrix}$ and $E_3[M_B(T)] = \mathbb{R}\begin{bmatrix} 5 \\ 6 \\ -1 \end{bmatrix}$. Then Theorem 7 gives the eigenspaces of T:

$$E_{-1}(T) = \mathbb{R}(-1 + 2x + x^2) \qquad \text{and} \qquad E_3(T) = \mathbb{R}(5 + 6x - x^2)$$

Another important feature of eigenspaces is that they are invariant subspaces, a fact that may not be too surprising in view of Theorem 6.

THEOREM 8 Each eigenspace of a linear operator $T : V \to V$ is a T-invariant subspace of V.

Proof Suppose the eigenspace is $E_\lambda(T)$. If $\mathbf{v}$ lies in $E_\lambda(T)$, then $T(\mathbf{v}) = \lambda\mathbf{v}$, so $T[T(\mathbf{v})] = T(\lambda\mathbf{v}) = \lambda T(\mathbf{v})$. This shows that $T(\mathbf{v})$ lies in $E_\lambda(T)$ too. ■

If T is an operator on an n-dimensional space V, and if λ is an eigenvalue of T, then λ is a root of the characteristic polynomial $c_T(x)$ by Theorem 7. As for matrices, λ is said to be an eigenvalue of **multiplicity** m if

$$c_T(x) = (x - \lambda)^m q(x), \quad q(\lambda) \neq 0$$

The multiplicity m is the highest power of $x - \lambda$ that is a factor of $c_T(x)$. The next theorem shows that m is an upper bound on the dimension of the eigenspace $E_\lambda(T)$ determined by T. The matrix version of this result was stated as Theorem 4 in Section 7.2, but the matrix techniques then available could not easily provide a proof.

THEOREM 9 Let $T : V \to V$ be a linear operator where $\dim V = n$. Let λ be an eigenvalue of T of multiplicity m. Then $m \geq \dim\,[E_\lambda(T)]$.

Proof Write $d = \dim\,[E_\lambda(T)]$, and let $B = \{\mathbf{e}_1, \ldots, \mathbf{e}_d, \ldots, \mathbf{e}_n\}$ be an ordered basis of V such that $B_1 = \{\mathbf{e}_1, \ldots, \mathbf{e}_d\}$ is a basis of $E_\lambda(T)$. Theorem 8 shows that $E_\lambda(T)$ is T-invariant, so $T : E_\lambda(T) \to E_\lambda(T)$ is a linear operator and $M_{B_1}(T) = \lambda I_d$ because each vector in B_1 is an eigenvector corresponding to λ. Hence Theorem 1 shows that $M_B(T)$ has the block form

$$M_B(T) = \begin{bmatrix} \lambda I_d & Y \\ 0 & Z \end{bmatrix}$$

But then $c_T(x)$ has the form

$$c_T(x) = \det [xI - M_B(T)] = (x - \lambda)^d p(x)$$

The multiplicity m is the *highest* power of $x - \lambda$ that is a factor of $c_T(x)$, so it follows that $m \geq d$, as required. ■

It can happen that the multiplicity of an eigenvalue is strictly greater than the dimension of the corresponding eigenspace. In fact, $\lambda = -1$ has multiplicity 2 as an eigenvalue of the operator T in Example 11, but the corresponding eigenspace $E_{-1}(T)$ has dimension 1. The next theorem identifies one class of operators that have the property that the multiplicity of each eigenvalue equals the dimension of the corresponding eigenspace.

THEOREM 10 Let $T : V \to V$ be a linear operator on a finite dimensional space V. Then the following conditions are equivalent.

(1) V has a basis consisting of eigenvectors of T.

(2) There exists a basis B of V such that $M_B(T)$ is diagonal.

Moreover, if this is the case, the characteristic polynomial $c_T(x)$ factors completely over $\mathbb{R}$ as follows:

$$c_T(x) = (x - \lambda_1)^{m_1} (x - \lambda_2)^{m_2} \dots (x - \lambda_k)^{m_k}$$

where $\lambda_1, \lambda_2, \dots, \lambda_k$ are the distinct real eigenvalues of T. If B is a basis of eigenvectors, then $M_B(T)$ is diagonal with λ_1 repeated m_1 times on the main diagonal, λ_2 repeated m_2 times, and so on.

Proof Let $B = \{\mathbf{e}_1, \mathbf{e}_2, \dots, \mathbf{e}_n\}$ be a basis of V such that $M_B(T)$ is diagonal, say

$$\begin{bmatrix} \lambda_1 & 0 & \dots & 0 \\ 0 & \lambda_2 & \dots & 0 \\ \vdots & \vdots & & \vdots \\ 0 & 0 & \dots & \lambda_n \end{bmatrix} = M_B(T) = [C_B[T(\mathbf{e}_1)]\ C_B[T(\mathbf{e}_2)] \dots C_B[T(\mathbf{e}_n)]]$$

where the λ_i are real numbers. Comparing columns shows that $T(\mathbf{e}_i) = \lambda_i\mathbf{e}_i$ for $i = 1, 2, \dots, n$. Hence each λ_i is an eigenvalue with $\mathbf{e}_i$ as an eigenvector, so B consists of eigenvectors. This proves that (2) implies (1). Conversely, if $B = \{\mathbf{e}_1, \mathbf{e}_2, \dots, \mathbf{e}_n\}$ is a basis of eigenvectors and $\lambda_1, \lambda_2, \dots, \lambda_n$ are the corresponding eigenvalues (not necessarily distinct), then $T(\mathbf{e}_i) = \lambda_i\mathbf{e}_i$ for each i, so $M_B(T)$ has the above diagonal form. This proves that (1) implies (2); the rest is clear, because $c_T(x) = \det [xI - M_B(T)]$. ■

DEFINITION

A linear operator T on a finite dimensional space V is called **diagonalizable** if V has a basis of eigenvectors of T.

EXAMPLE 12

Let $T : \mathbf{P}_2 \to \mathbf{P}_2$ be given by

$$T(a + bx + cx^2) = (a + 4c) - 2bx + (3a + 2c)x^2$$

Find the eigenspaces of T, and hence find a basis of eigenvectors.

Solution If $B_0 = \{1, x, x^2\}$, then

$$M_{B_0}(T) = \begin{bmatrix} 1 & 0 & 4 \\ 0 & -2 & 0 \\ 3 & 0 & 2 \end{bmatrix} \quad \text{so}$$

$$c_T(x) = \det\begin{bmatrix} x-1 & 0 & -4 \\ 0 & x+2 & 0 \\ -3 & 0 & x-2 \end{bmatrix} = (x+2)^2(x-5)$$

The eigenvalues of T are thus $\lambda = -2$ and $\lambda = 5$. The eigenspaces of $M_{B_0}(T)$ are

$$E_{-2}[M_{B_0}(T)] = \operatorname{span}\left\{\begin{bmatrix} 0 \\ 1 \\ 0 \end{bmatrix}, \begin{bmatrix} 4 \\ 0 \\ -3 \end{bmatrix}\right\} \quad \text{and} \quad E_5[M_{B_0}M(T)] = \operatorname{span}\left\{\begin{bmatrix} 1 \\ 0 \\ 1 \end{bmatrix}\right\}$$

Hence $E_{-2}(T) = \operatorname{span}\{x, 4 - 3x^2\}$ and $E_5(T) = \operatorname{span}\{1 + x^2\}$. Now $\{x, 4 - 3x^2\}$ and $\{1 + x^2\}$ are bases of these eigenspaces, and the reader can verify that the set $B = \{x, 4 - 3x^2, 1 + x^2\}$ is a basis of $\mathbf{P}_2$ (consisting of eigenvectors).

The actual construction of a basis of eigenvectors (when it exists) is always accomplished, as in Example 12, by putting together bases of all the eigenspaces. The next theorem and its corollary guarantee that this larger set of vectors will be linearly independent. The proof of the theorem is omitted; it is the analog of the proof of Theorem 2 in Section 7.2. The proof of the corollary is left as part (b) of Exercise 4.

THEOREM 11

Let $T : V \to V$ be a linear operator. If $\lambda_1, \ldots, \lambda_k$ are distinct eigenvalues of T, and $\mathbf{v}_1, \ldots, \mathbf{v}_k$ are corresponding eigenvectors, then $\{\mathbf{v}_1, \ldots, \mathbf{v}_k\}$ is linearly independent.

COROLLARY Let $\lambda_1, \ldots, \lambda_k$ be distinct eigenvalues of a linear operator $T : V \to V$, where $\dim V = n$. Write $d_i = \dim[E_{\lambda_i}(T)]$ for each i, and let B_i be a basis of $E_{\lambda_i}(T)$. If B consists of all vectors lying in at least one of the B_i, then B is linearly independent and contains $d_1 + \cdots + d_k$ vectors.

EXAMPLE 13 Let $T : \mathbf{M}_{22} \to \mathbf{M}_{22}$ be the transposition operator, $T(A) = A^T$ for all A in $\mathbf{M}_{22}$. Find a basis of $\mathbf{M}_{22}$ that diagonalizes T.

Solution Let $B_0 = \left\{ \begin{bmatrix} 1 & 0 \\ 0 & 0 \end{bmatrix}, \begin{bmatrix} 0 & 1 \\ 0 & 0 \end{bmatrix}, \begin{bmatrix} 0 & 0 \\ 1 & 0 \end{bmatrix}, \begin{bmatrix} 0 & 0 \\ 0 & 1 \end{bmatrix} \right\}$. Then

$$M_{B_0}(T) = \begin{bmatrix} 1 & 0 & 0 & 0 \\ 0 & 0 & 1 & 0 \\ 0 & 1 & 0 & 0 \\ 0 & 0 & 0 & 1 \end{bmatrix} \quad \text{so}$$

$$c_T(x) = \det \begin{bmatrix} x-1 & 0 & 0 & 0 \\ 0 & x & -1 & 0 \\ 0 & -1 & x & 0 \\ 0 & 0 & 0 & x-1 \end{bmatrix} = (x-1)^3(x+1)$$

The eigenspaces then turn out to be

$$E_1(T) = \operatorname{span}\left\{ \begin{bmatrix} 1 & 0 \\ 0 & 0 \end{bmatrix}, \begin{bmatrix} 0 & 1 \\ 1 & 0 \end{bmatrix}, \begin{bmatrix} 0 & 0 \\ 0 & 1 \end{bmatrix} \right\} \quad \text{and}$$

$$E_{-1}(T) = \operatorname{span}\left\{ \begin{bmatrix} 0 & 1 \\ -1 & 0 \end{bmatrix} \right\}$$

so the required basis is

$$B = \left\{ \begin{bmatrix} 1 & 0 \\ 0 & 0 \end{bmatrix}, \begin{bmatrix} 0 & 1 \\ 1 & 0 \end{bmatrix}, \begin{bmatrix} 0 & 0 \\ 0 & 1 \end{bmatrix}, \begin{bmatrix} 0 & 1 \\ -1 & 0 \end{bmatrix} \right\}$$

Note that $E_1(T) = \{A \mid T(A) = A\}$ is the set of symmetric matrices, whereas $E_{-1}(T) = \{A \mid T(A) = -A\}$ is the set of skew-symmetric matrices.

Theorem 10 shows that the characteristic polynomial of any diagonalizable operator factors completely over $\mathbb{R}$. However, the converse is not true, as Example 11 shows. The next theorem characterizes the diagonalizable operators T among those for which $c_T(x)$ does factor completely. It is the analog of Theorem 5 in Section 7.2, and the same proof applies.

THEOREM 12 Let $T : V \to V$ be a linear operator where $\dim V = n$, and assume that the characteristic polynomial $c_T(x)$ factors completely over $\mathbb{R}$:

$$c_T(x) = (x - \lambda_1)^{m_1}(x - \lambda_2)^{m_2} \dots (x - \lambda_k)^{m_k}$$

where $\lambda_1, \lambda_2, \dots, \lambda_k$ are the distinct eigenvalues of T. Write $d_i = \dim[E_{\lambda_i}(T)]$ for each i. Then the following conditions are equivalent.

(1) T is diagonalizable.

(2) $d_1 + d_2 + \cdots + d_k = n$.

(3) $d_i = m_i$ for each i.

EXAMPLE 14 Show that $T : \mathbf{M}_{22} \to \mathbf{M}_{22}$ is not diagonalizable if

$$T\begin{bmatrix} a & b \\ c & d \end{bmatrix} = \begin{bmatrix} a + 3b & 2a + 2b + d \\ 4c & c - d \end{bmatrix}$$

Solution If $B = \left\{\begin{bmatrix} 1 & 0 \\ 0 & 0 \end{bmatrix}, \begin{bmatrix} 0 & 1 \\ 0 & 0 \end{bmatrix}, \begin{bmatrix} 0 & 0 \\ 1 & 0 \end{bmatrix}, \begin{bmatrix} 0 & 0 \\ 0 & 1 \end{bmatrix}\right\}$, then

$$M_B(T) = \begin{bmatrix} 1 & 3 & 0 & 0 \\ 2 & 2 & 0 & 1 \\ 0 & 0 & 4 & 0 \\ 0 & 0 & 1 & -1 \end{bmatrix}, \quad \text{whence}$$

$$c_T(x) = \det\begin{bmatrix} x - 1 & -3 & 0 & 0 \\ -2 & x - 2 & 0 & -1 \\ 0 & 0 & x - 4 & 0 \\ 0 & 0 & -1 & x + 1 \end{bmatrix}$$

$$= (x - 4)^2(x + 1)^2$$

The reader can verify that $E_4(T) = \operatorname{span}\left\{\begin{bmatrix} 1 & 1 \\ 0 & 0 \end{bmatrix}\right\}$ and hence has dimension 1. Because 4 has multiplicity 2, T is not diagonalizable by Theorem 12.

EXERCISES 9.2.2

1. In each case, find a basis of eigenvectors for T and the corresponding diagonal matrix.

(a) $T : \mathbf{P}_2 \to \mathbf{P}_2$, $T(a + bx + cx^2) = (a + 3b + 7c) - (3a + 5b + 7c)x + (3a + 3b + 5c)x^2$

(b) $T : \mathbf{M}_{22} \to \mathbf{M}_{22}$, $T\begin{bmatrix} a & b \\ c & d \end{bmatrix} = \begin{bmatrix} 2a + 2c & 4b + 6d \\ -2a - 3c & -3b - 5d \end{bmatrix}$

2. Let $A = \begin{bmatrix} 0 & 1 \\ 0 & 0 \end{bmatrix}$ and consider $T_A : \mathbb{R}^2 \to \mathbb{R}^2$.

(a) Show that the only eigenvalue of T_A is $\lambda = 0$.

(b) Show that $\ker(T_A) = \mathbb{R}\begin{bmatrix} 1 \\ 0 \end{bmatrix}$ is the unique T_A-invariant subspace of $\mathbb{R}^2$ (except for 0 and $\mathbb{R}^2$).

3. In each case, find the eigenvalues of T_A and use them to diagonalize A.

(a) $A = \begin{bmatrix} 1 & 0 & 2 \\ 0 & 1 & 0 \\ 2 & 0 & 1 \end{bmatrix}$ (b) $A = \begin{bmatrix} -1 & 2 & 0 \\ 2 & 0 & -2 \\ 0 & -2 & 1 \end{bmatrix}$

4. (a) Complete the proof of Theorem 7.

(b) Prove the corollary to Theorem 11.

5. If an operator T on an n-dimensional vector space V has n distinct eigenvalues, show that T is diagonalizable.

6. If $A = \begin{bmatrix} 2 & -5 & 0 & 0 \\ 1 & -2 & 0 & 0 \\ 0 & 0 & -1 & -2 \\ 0 & 0 & 1 & 1 \end{bmatrix}$, show that $T_A : \mathbb{R}^4 \to \mathbb{R}^4$ has two-dimensional T-invariant subspaces U and W such that $\mathbb{R}^4 = U \oplus W$, but A has no real eigenvalue.

7. In each case, use Theorem 12 to show that T is not diagonalizable.

(a) $T : \mathbf{M}_{22} \to \mathbf{M}_{22}$, $T\begin{bmatrix} a & b \\ c & d \end{bmatrix} = \begin{bmatrix} 2a + b + 4c & -b \\ 3(a + c) & 5d \end{bmatrix}$

(b) $T : \mathbf{P}_2 \to \mathbf{P}_2$, $T(a + bx + cx^2) = (3c - 4a) + 2bx + (2a + b + c)x^2$

8. Let $T : V \to V$ be an idempotent; that is, $T^2 = T$. Show that the only eigenvalues of T are 1 and 0 and that T is diagonalizable. [*Hint:* Exercise 20 in Section 9.2.1.]

9. Let $T : V \to V$ be an involution; that is, $T^2 = 1$. Show that the only eigenvalues of T are 1 and -1 and that T is diagonalizable. [*Hint:* Example 9.]

10. Define $T : \mathbf{P}_2 \to \mathbf{P}_2$ by $T(a + bx + cx^2) = (a + 2c) + (ka + 6b)x + (5a + 4c)x^2$. Show that T is diagonalizable if and only if $k = 0$.

11. Let U be a fixed $n \times n$ matrix, and consider the operator $T : \mathbf{M}_{nn} \to \mathbf{M}_{nn}$ given by $T(A) = UA$.

(a) Show that λ is an eigenvalue of T if and only if it is an eigenvalue of U.

(b) If λ is an eigenvalue of T, show that $E_\lambda(T)$ consists of all matrices whose columns lie in $E_\lambda(U)$:

$$E_\lambda(T) = \{[\mathbf{p}_1\ \mathbf{p}_2 \cdots \mathbf{p}_n] \mid \mathbf{p}_i \text{ in } E_\lambda(U) \quad \text{for each } i\}$$

(c) Show that if $\dim[E_\lambda(U)] = d$, then $\dim[E_\lambda(T)] = nd$. [*Hint:* If $B = \{\mathbf{e}_1, \ldots, \mathbf{e}_d\}$ is a basis of $E_\lambda(U)$, consider the set of all matrices with one column from B and the other columns zero.]

(d) Use part (c) to show that if U is diagonalizable, then T is diagonalizable.

SECTION 9.3 Orthogonal Diagonalization

The principal axes theorem (Theorem 6 in Section 7.2) asserts that if A is a symmetric $n \times n$ matrix, then $\mathbb{R}^n$ has an orthonormal basis consisting of eigenvectors of A. This fact is very important, not only as a theoretical tool in the study of matrices, but also in the applications of matrices to other things (for example, quadratic forms and differential equations—see Section 7.4). The reason for the importance of the theorem is that it takes a large and easily identifiable class of matrices (the symmetric ones) and reveals a significant and useful fact about them.

It turns out that there is a natural way to define a symmetric linear operator T on an inner product space V. These operators are examined in this section, the main goal being to prove that if V is finite dimensional, not only is T diagonalizable, but V also has an *orthogonal* basis of eigenvectors of T. This yields another proof of the principal axes theorem.

An important reason for using an *orthogonal* basis B of V is the fact that the representation of a vector $\mathbf{v}$ in V as a linear combination of the vectors in B is easy to obtain. If $B = \{\mathbf{e}_1, \ldots, \mathbf{e}_n\}$, the expansion theorem (Theorem 4 in Section 6.2) gives

$$\mathbf{v} = \frac{\langle \mathbf{e}_1, \mathbf{v} \rangle}{\|\mathbf{e}_1\|^2}\mathbf{e}_1 + \frac{\langle \mathbf{e}_2, \mathbf{v} \rangle}{\|\mathbf{e}_2\|^2}\mathbf{e}_2 + \cdots + \frac{\langle \mathbf{e}_n, \mathbf{v} \rangle}{\|\mathbf{e}_n\|^2}\mathbf{e}_n$$

In particular, if B is orthonormal, then $\|\mathbf{e}_i\| = 1$ for each i, so the coefficients are simply $\langle \mathbf{e}_i, \mathbf{v} \rangle$. The next theorem uses this information to generate formulas for the (i,j)-entry of the matrix of a linear operator with respect to an orthogonal basis.

THEOREM 1

Let $T : V \to V$ be a linear operator on an inner product space V. If $B = \{\mathbf{e}_1, \mathbf{e}_2, \ldots, \mathbf{e}_n\}$ is an orthogonal basis of V, then

$$M_B(T) = \left[\frac{\langle \mathbf{e}_i, T(\mathbf{e}_j) \rangle}{\|\mathbf{e}_i\|^2}\right]$$

In particular, if B is orthonormal, then

$$M_B(T) = [\langle \mathbf{e}_i, T(\mathbf{e}_j) \rangle]$$

Proof Write $M_B(T) = [a_{ij}]$. The jth column of $M_B(T)$ is $C_B[T(\mathbf{e}_j)]$, so

$$T(\mathbf{e}_j) = a_{1j}\mathbf{e}_1 + \cdots + a_{ij}\mathbf{e}_i + \cdots + a_{nj}\mathbf{e}_n$$

by the definition of C_B. The expansion theorem gives the ith coefficient as

$$a_{ij} = \frac{\langle \mathbf{e}_i, T(\mathbf{e}_j)\rangle}{\|\mathbf{e}_i\|^2} \quad \blacksquare$$

EXAMPLE 1 Let $T : \mathbb{R}^3 \to \mathbb{R}^3$ be given by

$$T(a,b,c) = (a + 2b - c, 2a + 3c, -a + 3b + 2c)$$

If the dot product in $\mathbb{R}^3$ is used, find the matrix of T with respect to the standard basis $B = \{\mathbf{e}_1 = (1, 0, 0), \mathbf{e}_2 = (0, 1, 0), \mathbf{e}_3 = (0, 0, 1)\}$

Solution The basis B is orthonormal, so

$$M_B(T) = \begin{bmatrix} \mathbf{e}_1 \cdot T(\mathbf{e}_1) & \mathbf{e}_1 \cdot T(\mathbf{e}_2) & \mathbf{e}_1 \cdot T(\mathbf{e}_3) \\ \mathbf{e}_2 \cdot T(\mathbf{e}_1) & \mathbf{e}_2 \cdot T(\mathbf{e}_2) & \mathbf{e}_2 \cdot T(\mathbf{e}_3) \\ \mathbf{e}_3 \cdot T(\mathbf{e}_1) & \mathbf{e}_3 \cdot T(\mathbf{e}_2) & \mathbf{e}_3 \cdot T(\mathbf{e}_3) \end{bmatrix} = \begin{bmatrix} 1 & 2 & -1 \\ 2 & 0 & 3 \\ -1 & 3 & 2 \end{bmatrix}$$

as the reader can verify. Of course, this can also be found in the usual way.

It is not difficult to verify that an $n \times n$ matrix A is symmetric if and only if $\mathbf{v} \cdot (A\mathbf{w}) = (A\mathbf{v}) \cdot \mathbf{w}$ holds for all columns $\mathbf{v}$ and $\mathbf{w}$ in $\mathbb{R}^n$. The analog for operators is as follows:

THEOREM 2 Let V be a finite dimensional inner product space. The following conditions are equivalent for a linear operator $T : V \to V$.

(1) $\langle \mathbf{v}, T(\mathbf{w})\rangle = \langle T(\mathbf{v}), \mathbf{w}\rangle$ for all $\mathbf{v}$ and $\mathbf{w}$ in V.

(2) The matrix of T is symmetric with respect to every orthonormal basis of V.

(3) The matrix of T is symmetric with respect to some orthonormal basis of V.

(4) There is an orthonormal basis $B = \{\mathbf{e}_1, \mathbf{e}_2, \ldots, \mathbf{e}_n\}$ of V such that $\langle \mathbf{e}_i, T(\mathbf{e}_j)\rangle = \langle T(\mathbf{e}_i), \mathbf{e}_j\rangle$ holds for all i and j.

Proof

(1) implies (2). Given (1), let $B = \{\mathbf{e}_1, \ldots, \mathbf{e}_n\}$ be an orthonormal basis of V and write $M_B(T) = [a_{ij}]$. The aim is to prove that $a_{ij} = a_{ji}$ for all i and j. The fact that $\langle \mathbf{v}, \mathbf{w}\rangle = \langle \mathbf{w}, \mathbf{v}\rangle$ for all $\mathbf{v}$ and $\mathbf{w}$, together with Theorem 1, gives

$$a_{ij} = \langle \mathbf{e}_i, T(\mathbf{e}_j)\rangle = \langle T(\mathbf{e}_i), \mathbf{e}_j\rangle = \langle \mathbf{e}_j, T(\mathbf{e}_i)\rangle = a_{ji}$$

(2) implies (3). This is clear.

(3) implies (4). Assume that $B = \{\mathbf{e}_1, \ldots, \mathbf{e}_n\}$ is an orthonormal basis of V such that $M_B(T)$ is symmetric. We have $M_B(T) = [\langle \mathbf{e}_i, T(\mathbf{e}_j)\rangle]$ by Theorem 1, so

$$\langle \mathbf{e}_i, T(\mathbf{e}_j)\rangle = \langle \mathbf{e}_j, T(\mathbf{e}_i)\rangle$$

holds for all i and j. Then (4) follows, because $\langle\ ,\ \rangle$ is symmetric.

(4) implies (1). Let $\mathbf{v}$ and $\mathbf{w}$ be vectors in V, and write them as $\mathbf{v} = \sum_{i=1}^{n} v_i\mathbf{e}_i$, $\mathbf{w} = \sum_{j=1}^{n} w_j\mathbf{e}_j$. Theorem 2 in Section 6.1 then gives

$$\begin{aligned}\langle \mathbf{v}, T(\mathbf{w})\rangle &= \left\langle \sum_i v_i\mathbf{e}_i, \sum_j w_jT(\mathbf{e}_j)\right\rangle \\ &= \sum_i \sum_j v_iw_j\,\langle \mathbf{e}_i, T(\mathbf{e}_j)\rangle \\ &= \sum_i \sum_j v_iw_j\,\langle T(\mathbf{e}_i), \mathbf{e}_j\rangle \\ &= \left\langle \sum_i v_iT(\mathbf{e}_i), \sum_j w_j\mathbf{e}_j\right\rangle \\ &= \langle T(\mathbf{v}), \mathbf{w}\rangle\end{aligned}$$

This proves (1). ■

DEFINITION

> A linear operator T on an inner product space V is called **symmetric** if $\langle \mathbf{v}, T(\mathbf{w})\rangle = \langle T(\mathbf{v}), \mathbf{w}\rangle$ holds for all $\mathbf{v}$ and $\mathbf{w}$ in V.

EXAMPLE 2 If A is an $n \times n$ matrix, let $T_A : \mathbb{R}^n \to \mathbb{R}^n$ be the matrix operator given by $T_A(\mathbf{v}) = A\mathbf{v}$ for all columns $\mathbf{v}$. If the dot product is used in $\mathbb{R}^n$, then T_A is a symmetric operator if and only if A is a symmetric matrix.

Solution If B is the standard basis of $\mathbb{R}^n$, then B is orthonormal if the dot product is used. We have $M_B(T_A) = A$ (by Example 2 in Section 8.2), so the result follows immediately from Theorem 2.

It is important to note that whether an operator is symmetric depends on which inner product is being used. Here is an example.

EXAMPLE 3 Let $T : \mathbb{R}^2 \to \mathbb{R}^2$ be given by $T(a,b) = (b - a, a + 2b)$. Show that T is symmetric if the dot product is used in $\mathbb{R}^2$ but that it is not symmetric if the following inner product is used:

$$\langle (a,b), (a',b') \rangle = (a,b)\begin{bmatrix} 1 & -1 \\ -1 & 2 \end{bmatrix}\begin{bmatrix} a' \\ b' \end{bmatrix}$$
$$= aa' - ba' - ab' + 2bb'$$

Solution $B_0 = \{(1,0), (0,1)\}$ is an orthonormal basis with respect to the dot product, and $M_{B_0}(T) = \begin{bmatrix} -1 & 1 \\ 1 & 2 \end{bmatrix}$ is symmetric. Now B_0 is *not* orthogonal with respect to $\langle\ ,\ \rangle$; in fact, $\langle (1,0), (0,1) \rangle = -1$. However, $B = \{(1,0), (1,1)\}$ is orthonormal with respect to $\langle\ ,\ \rangle$ as the reader can verify, and

$$M_B(T) = [C_B[T(1,0)]\ C_B[T(1,1)]]$$
$$= [C_B(-1,1)\ C_B(0,3)] = \begin{bmatrix} -2 & -3 \\ 1 & 3 \end{bmatrix}$$

This is *not* symmetric, so T is *not* a symmetric operator with respect to $\langle\ ,\ \rangle$.

Observe that this can also be tested without having to find a new orthonormal basis. In fact,

$$\langle (a,b), T(a',b') \rangle = 3bb' + 3ba' - ab' - 2aa'$$
$$\langle T(a,b), (a',b') \rangle = 3bb' - ba' + 3ab' - 2aa'$$

and the fact that these are not equal for all (a,b) and (a',b') shows again that T is not symmetric with respect to $\langle\ ,\ \rangle$.

EXAMPLE 4 Show that the set of symmetric linear operators on an inner product space V is a subspace of $\mathbf{L}(V,V)$.

Solution First, the zero operator 0 is symmetric because

$$\langle \mathbf{v}, 0(\mathbf{w}) \rangle = \langle \mathbf{v}, \mathbf{0} \rangle = 0 = \langle \mathbf{0}, \mathbf{w} \rangle = \langle 0(\mathbf{v}), \mathbf{w} \rangle$$

holds for all $\mathbf{v}$ and $\mathbf{w}$ in V. If S and T are both symmetric, then

$$\langle \mathbf{v}, (S + T)(\mathbf{w}) \rangle = \langle \mathbf{v}, S(\mathbf{w}) \rangle + \langle \mathbf{v}, T(\mathbf{w}) \rangle$$
$$= \langle S(\mathbf{v}), \mathbf{w} \rangle + \langle T(\mathbf{v}), \mathbf{w} \rangle$$
$$= \langle (S + T)(\mathbf{v}), \mathbf{w} \rangle$$

so $S + T$ is symmetric. Similarly, rT is symmetric for all r in $\mathbb{R}$ (Exercise 4).

If V is a finite dimensional inner product space, the eigenvalues of an operator $T : V \to V$ are the same as those of $M_B(T)$ for any orthonormal basis B. If T is symmetric, this is a symmetric matrix and so has real eigenvalues by Theorem 2 in Section 7.1. Hence:

THEOREM 3

> A symmetric linear operator on a finite dimensional inner product space has a real eigenvalue.

If U is a subspace of an inner product space V, recall that its orthogonal complement is the subspace $U^{\perp}$ of V defined by

$$U^{\perp} = \{\mathbf{v} \text{ in } V \mid \langle \mathbf{v},\mathbf{u}\rangle = 0 \text{ for all } \mathbf{u} \text{ in } U\}$$

THEOREM 4

> Let $T : V \to V$ be a symmetric linear operator on an inner product space V, and let U be a T-invariant subspace of V. Then:
>
> (1) The restriction of T to U is a symmetric linear operator on U.
>
> (2) $U^{\perp}$ is also T-invariant.

Proof

(1) U is itself an inner product space using the same inner product, and the condition that T is symmetric is clearly preserved.

(2) If $\mathbf{v}$ is in $U^{\perp}$, our task is to show that $T(\mathbf{v})$ is also in $U^{\perp}$; that is, $\langle T(\mathbf{v}),\mathbf{u}\rangle = 0$ for all $\mathbf{u}$ in U. But if $\mathbf{u}$ is in U, then $T(\mathbf{u})$ also lies in U because U is T-invariant, so

$$\langle T(\mathbf{v}),\mathbf{u}\rangle = \langle \mathbf{v},T(\mathbf{u})\rangle = 0$$

using the symmetry of T and the definition of $U^{\perp}$. ■

These preliminary results enable us to prove the following theorem, which is probably the most important result in this book.

THEOREM 5
Principal Axes Theorem

> The following conditions are equivalent for a linear operator T on a finite dimensional inner product space V.
>
> (1) T is symmetric.
>
> (2) V has an orthogonal basis consisting of eigenvectors of T.
>
> If B is a basis as in (2), then
>
> $$M_B(T) = \begin{bmatrix} \lambda_1 & 0 & \cdots & 0 \\ 0 & \lambda_2 & \cdots & 0 \\ \vdots & \vdots & & \vdots \\ 0 & 0 & \cdots & \lambda_n \end{bmatrix}$$
>
> where $\lambda_1, \lambda_2, \ldots, \lambda_n$ are the (not necessarily distinct) eigenvalues of T.

Proof

(1) implies (2). Assume that T is symmetric, and proceed by induction on $n = \dim V$. If $n = 1$, *every* nonzero vector in V is an eigenvector of T, so there is nothing to prove. If $n \geq 2$, assume inductively that the theorem holds for spaces of dimension less than n. Let λ_1 be a real eigenvalue of T (by Theorem 3) and choose an eigenvector $\mathbf{e}_1$ corresponding to λ_1. Then $U = \mathbb{R}\mathbf{e}_1$ is T-invariant, so $U^\perp$ is also T-invariant by Theorem 4 (T is symmetric). Because $\dim U^\perp = n - 1$ (Theorem 10 in Section 6.2), and because the restriction of T to $U^\perp$ is a symmetric operator (Theorem 4), it follows by induction that $U^\perp$ has an orthogonal basis $\{\mathbf{e}_2, \ldots, \mathbf{e}_n\}$ of eigenvectors of T. Hence $B = \{\mathbf{e}_1, \mathbf{e}_2, \ldots, \mathbf{e}_n\}$ is an orthogonal basis of V, which proves (2).

(2) implies (1). If $B = \{\mathbf{e}_1, \ldots, \mathbf{e}_n\}$ is a basis as in (2), then we may assume B is orthonormal (normalize if necessary). If $\lambda_1, \lambda_2, \ldots, \lambda_n$ are the corresponding eigenvalues, then $M_B(T)$ is diagonal as in the theorem. In particular, $M_B(T)$ is symmetric, so T is symmetric by Theorem 2. ■

Using the notation of the theorem, it is clear that $c_T(x) = (x - \lambda_1) \cdot (x - \lambda_2) \ldots (x - \lambda_n)$, so we obtain the following corollary.

COROLLARY

If T is a symmetric operator on a finite dimensional inner product space, then $c_T(x)$ factors completely over $\mathbb{R}$.

This property does not characterize symmetric operators (but see Exercise 11).

If A is a symmetric $n \times n$ matrix, Section 7.2.2 explains how to find an orthonormal basis B of eigenvectors: Simply find an orthonormal basis of each eigenspace of A, and take B to be the set of all vectors so obtained. If $B = \{\mathbf{p}_1, \mathbf{p}_2, \ldots, \mathbf{p}_n\}$, the $n \times n$ matrix $P = [\mathbf{p}_1\ \mathbf{p}_2 \ldots \mathbf{p}_n]$ with the $\mathbf{p}_j$ as columns is invertible and $P^{-1}AP$ is diagonal. Furthermore, the fact that the columns of P are orthonormal means that $P^{-1} = P^T$ (see Theorem 5 in Section 6.2), and such matrices are called *orthogonal*. The reason for recalling all this in the present section is that the technique can be used to compute an orthonormal basis of eigenvectors of any symmetric linear operator on a finite dimensional inner product space.

THEOREM 6

Let $T : V \to V$ be a symmetric linear operator on the n-dimensional inner product space V, and let B_0 be any convenient orthonormal basis of V. Let $\{\mathbf{p}_1, \ldots, \mathbf{p}_n\}$ be an orthonormal basis of $\mathbb{R}^n$ consisting of eigenvectors of $M_{B_0}(T)$ (computed as outlined above), and let $B = \{\mathbf{e}_1, \ldots, \mathbf{e}_n\}$ be the vectors in V such that $C_{B_0}(\mathbf{e}_j) = \mathbf{p}_j$ holds for each j. Then:

(1) B is an orthonormal basis of V consisting of eigenvectors of T.

(2) The transition matrix $P = P_{B_0 \leftarrow B} = [\mathbf{p}_1 \ldots \mathbf{p}_n]$ from B to B_0 is an orthogonal matrix, and

$$M_B(T) = P^{-1}M_{B_0}(T)P$$

is diagonal, the diagonal elements being the eigenvalues of T.

Proof B is orthonormal by Theorem 6 in Section 6.2 and consists of eigenvectors of T by (the remark following) Theorem 7 in Section 9.2. The transition matrix $P_{B_0 \leftarrow B}$ is given in terms of its columns by

$$P_{B_0 \leftarrow B} = [C_{B_0}(\mathbf{e}_1)\, C_{B_0}(\mathbf{e}_2) \ldots C_{B_0}(\mathbf{e}_n)] = [\mathbf{p}_1\, \mathbf{p}_2 \ldots \mathbf{p}_n]$$

This is an orthogonal matrix because the $\mathbf{p}_j$ are orthonormal. The rest is Theorem 3 in Section 9.1. ■

EXAMPLE 5 Let $T : \mathbf{P}_2 \to \mathbf{P}_2$ be given by

$$T(a + bx + cx^2) = (8a - 2b + 2c) + (-2a + 5b + 4c)x + (2a + 4b + 5c)x^2$$

Using the inner product $\langle a + bx + cx^2, a' + b'x + c'x^2 \rangle = aa' + bb' + cc'$, show that T is symmetric and find an orthonormal basis of $\mathbf{P}_2$ consisting of eigenvectors.

Solution If $B_0 = \{1, x, x^2\}$, then $M_{B_0}(T) = \begin{bmatrix} 8 & -2 & 2 \\ -2 & 5 & 4 \\ 2 & 4 & 5 \end{bmatrix}$ is symmetric, so T is symmetric. This matrix was analyzed in Example 6 in Section 7.2, where it was found that an *orthonormal* basis of eigenvectors is

$$\left\{ \frac{1}{3}\begin{bmatrix} 1 \\ 2 \\ -2 \end{bmatrix}, \frac{1}{3}\begin{bmatrix} 2 \\ 1 \\ 2 \end{bmatrix}, \frac{1}{3}\begin{bmatrix} -2 \\ 2 \\ 1 \end{bmatrix} \right\}$$

The corresponding orthonormal basis of $\mathbf{P}_2$ is thus

$$B = \left\{ \frac{1}{3}(1 + 2x - 2x^2),\ \frac{1}{3}(2 + x + 2x^2),\ \frac{1}{3}(-2 + 2x + x^2) \right\}$$

EXERCISES 9.3

1. In each case, show that T is symmetric by calculating $M_B(T)$ for some orthonormal basis B.
   (a) $T : \mathbb{R}^3 \to \mathbb{R}^3$; $T(a,b,c) = (a - 2b, -2a + 2b + 2c, 2b - c)$; dot product
   (b) $T : \mathbf{M}_{22} \to \mathbf{M}_{22}$; $T\begin{bmatrix} a & b \\ c & d \end{bmatrix} = \begin{bmatrix} c - a & d - b \\ a + 2c & b + 2d \end{bmatrix}$;
   inner product: $\left\langle \begin{bmatrix} x & y \\ z & w \end{bmatrix}, \begin{bmatrix} x' & y' \\ z' & w' \end{bmatrix} \right\rangle = xx' + yy' + zz' + ww'$
   (c) $T : \mathbf{P}_2 \to \mathbf{P}_2$; $T(a + bx + cx^2) = (b + c) + (a + c)x + (a + b)x^2$; inner product $\langle a + bx + cx^2, a' + b'x + c'x^2 \rangle = aa' + bb' + cc'$
2. Let $T : \mathbb{R}^2 \to \mathbb{R}^2$ be given by $T(a,b) = (2a + b, a - b)$.
   (a) Show that T is symmetric if the dot product is used.
   (b) Show that T is *not* symmetric if $\langle \mathbf{v},\mathbf{w} \rangle = \mathbf{v}A\mathbf{w}^T$, where $A = \begin{bmatrix} 1 & 1 \\ 1 & 2 \end{bmatrix}$.
   [*Hint:* Check that $B = \{(1,0), (1,-1)\}$ is an orthonormal basis.]
3. Let $T : \mathbb{R}^2 \to \mathbb{R}^2$ be given by $T(a,b) = (a - b, b - a)$. Use the dot product in $\mathbb{R}^2$.
   (a) Show that T is symmetric.
   (b) Show that $M_B(T)$ is *not* symmetric if the orthogonal basis $B = \{(1,0), (0,2)\}$ is used. Why does this not contradict Theorem 2?
4. Let V be an n-dimensional inner product space, and let T and S denote symmetric linear operators on V. Show that:
   (a) The identity operator is symmetric.
   (b) rT is symmetric for all r in $\mathbb{R}$.
   (c) If $ST = TS$, then ST is symmetric.
5. In each case, show that T is symmetric and find an orthonormal basis of eigenvectors of T.
   (a) $T : \mathbb{R}^3 \to \mathbb{R}^3$; $T(a,b,c) = (2a + 2c, 3b, 2a + 5c)$; use the dot product
   (b) $T : \mathbb{R}^3 \to \mathbb{R}^3$; $T(a,b,c) = (7a - b, -a + 7b, 2c)$; use the dot product
   (c) $T : \mathbf{P}_2 \to \mathbf{P}_2$; $T(a + bx + cx^2) = 3b + (3a + 4c)x + 4bx^2$; inner product $\langle a + bx + cx^2, a' + b'x + c'x^2 \rangle = aa' + bb' + cc'$
   (d) $T : \mathbf{P}_2 \to \mathbf{P}_2$; $T(a + bx + cx^2) = (c - a) + 3bx + (a - c)x^2$; inner product as in part **(c)**
6. If A is any $n \times n$ matrix, let $T_A : \mathbb{R}^n \to \mathbb{R}^n$ be given by $T_A(\mathbf{v}) = A\mathbf{v}$. Suppose an inner product on $\mathbb{R}^n$ is given by $\langle \mathbf{v},\mathbf{w} \rangle = \mathbf{v}^T P \mathbf{w}$, where P is a positive definite matrix (see Section 6.1).
   (a) Show that T_A is symmetric if and only if $PA = A^T P$.
   (b) Use part **(a)** to deduce Example 2.
7. Let $T : \mathbf{M}_{22} \to \mathbf{M}_{22}$ be given by $T(X) = AX$, where A is a fixed 2×2 matrix.
   (a) Compute $M_B(T)$, where $B = \left\{ \begin{bmatrix} 1 & 0 \\ 0 & 0 \end{bmatrix}, \begin{bmatrix} 0 & 0 \\ 1 & 0 \end{bmatrix}, \begin{bmatrix} 0 & 1 \\ 0 & 0 \end{bmatrix}, \begin{bmatrix} 0 & 0 \\ 0 & 1 \end{bmatrix} \right\}$. Note the order!

(b) Show that $c_T(x) = [c_A(x)]^2$.

(c) If the inner product on $\mathbf{M}_{22}$ is $\left\langle \begin{bmatrix} a & b \\ c & d \end{bmatrix}, \begin{bmatrix} a' & b' \\ c' & d' \end{bmatrix} \right\rangle = aa' + bb' + cc' + dd'$, show that T is symmetric if and only if A is a symmetric matrix.

8. Let $T : \mathbf{M}_{22} \to \mathbf{M}_{22}$ be defined by $T(X) = PXQ$, where P and Q are nonzero 2×2 matrices. Use the inner product in part **(c)** of Exercise 7. Show that T is symmetric if and only if either P and Q are both symmetric or both are scalar multiples of $\begin{bmatrix} 0 & 1 \\ -1 & 0 \end{bmatrix}$. [*Hint:* If B is as in part **(a)** of Exercise 7, then $M_B(T) = \begin{bmatrix} aP & cP \\ bP & dP \end{bmatrix}$ in block form, where $Q = \begin{bmatrix} a & b \\ c & d \end{bmatrix}$. If $B_0 = \left\{ \begin{bmatrix} 1 & 0 \\ 0 & 0 \end{bmatrix}, \begin{bmatrix} 0 & 1 \\ 0 & 0 \end{bmatrix}, \begin{bmatrix} 0 & 0 \\ 1 & 0 \end{bmatrix}, \begin{bmatrix} 0 & 0 \\ 0 & 1 \end{bmatrix} \right\}$, then $M_{B_0}(T) = \begin{bmatrix} pQ^T & qQ^T \\ rQ^T & sQ^T \end{bmatrix}$, where $P = \begin{bmatrix} p & q \\ r & s \end{bmatrix}$. Use the fact that $cP = bP^T$ implies that $(c^2 - b^2)P = 0$.]

9. Let $T : V \to W$ be any linear transformation, and let $B = \{\mathbf{b}_1, \ldots, \mathbf{b}_n\}$ and $D = \{\mathbf{d}_1, \ldots, \mathbf{d}_m\}$ be bases of V and W, respectively. If W is an inner product space and D is orthogonal, show that

$$M_{DB}(T) = \left[\frac{\langle \mathbf{d}_i, T(\mathbf{b}_j) \rangle}{\|\mathbf{d}_i\|^2} \right]$$

This is a generalization of Theorem 1.

10. Let $T : V \to V$ be a linear operator on an inner product space V of finite dimension. Show that the following are equivalent.

(a) $\langle \mathbf{v}, T(\mathbf{w}) \rangle = -\langle T(\mathbf{v}), \mathbf{w} \rangle$ for all $\mathbf{v}$ and $\mathbf{w}$ in V.

(b) $M_B(T)$ is skew-symmetric for every orthonormal basis B.

(c) $M_B(T)$ is skew-symmetric for some orthonormal basis B.

Such operators T are called **skew-symmetric** operators.

11. Let $T : V \to V$ be a linear operator on an n-dimensional inner product space V.

(a) Show that T is symmetric if and only if it satisfies the following two conditions.

(i) $c_T(x)$ factors completely over $\mathbb{R}$.

(ii) If U is a T-invariant subspace of V, then $U^\perp$ is also T-invariant.

(b) Using the standard inner product in $\mathbb{R}^2$, show that $T : \mathbb{R}^2 \to \mathbb{R}^2$ with $T(a,b) = (a, a + b)$ satisfies condition **(i)** and that $S : \mathbb{R}^2 \to \mathbb{R}^2$ with $S(a,b) = (b, -a)$ satisfies condition **(ii)** but that neither is symmetric. (Example 3 in Section 9.2 is useful for S.)

[*Hint* for part **(a)**: If conditions **(i)** and **(ii)** hold, proceed by induction on n. By condition **(i)**, let $\mathbf{e}_1$ be an eigenvector of T. If $U = \mathbb{R}\mathbf{e}_1$, then $U^\perp$ is T-invariant by condition **(ii)**, so show that the restriction of T to $U^\perp$ satisfies conditions **(i)** and **(ii)**. (Theorem 5 in Section 9.2 is helpful for part (i)). Then apply induction to show that V has an orthogonal basis of eigenvectors (as in Theorem 5).]

12. Let $B = \{\mathbf{e}_1, \mathbf{e}_2, \ldots, \mathbf{e}_n\}$ be an orthonormal basis of an inner product space V. Given a linear operator $T : V \to V$, define the linear operator $T' : V \to V$ by

$$T'(\mathbf{e}_j) = \langle \mathbf{e}_j, T(\mathbf{e}_1)\rangle \mathbf{e}_1 + \langle \mathbf{e}_j, T(\mathbf{e}_2)\rangle \mathbf{e}_2 + \cdots + \langle \mathbf{e}_j, T(\mathbf{e}_n)\rangle \mathbf{e}_n = \sum_{i=1}^{n} \langle \mathbf{e}_j, T(\mathbf{e}_i)\rangle \mathbf{e}_i$$

(a) Show that $(aT)' = aT'$.

(b) Show that $(S + T)' = S' + T'$.

(c) Show that $M_B(T')$ is the transpose of $M_B(T)$.

(d) Show that $(T')' = T$, using part **(c)**. [*Hint:* $M_B(S) = M_B(T)$ implies that $S = T$.]

(e) Show that $(ST)' = T'S'$, using part **(c)**.

(f) Show that T is symmetric if and only if $T = T'$. [*Hint:* See Theorem 4 in Section 6.2 and Theorem 2.]

(g) Show that $T + T'$ and TT' are symmetric, using parts **(b)** through **(e)**.

13. Let V be a finite dimensional inner product space. Show that the following conditions are equivalent for a linear operator $T : V \to V$.

(a) T is symmetric and $T^2 = T$.

(b) $M_B(T) = \begin{bmatrix} I_r & 0 \\ 0 & 0 \end{bmatrix}$ for some orthonormal basis B of V.

An operator is called a **projection** if it satisfies these conditions. [*Hint:* If $T^2 = T$ and $T(\mathbf{v}) = \lambda\mathbf{v}$, apply T to get $\lambda\mathbf{v} = \lambda^2\mathbf{v}$. Hence show that 0, 1 are the only eigenvalues of T.]

14. Let V denote a finite dimensional inner product space. Given a subspace U, define $\text{proj}_U : V \to V$ as in Theorem 9 in Section 6.2.

(a) Show that proj_U is a projection in the sense of Exercise 13.

(b) If T is any projection, show that $T = \text{proj}_U$, where $U = \text{im } T$. [*Hint:* Use $T^2 = T$ to show that $V = \text{im } T \oplus \ker T$ and $T(\mathbf{u}) = \mathbf{u}$ for all $\mathbf{u}$ in $\text{im } T$. Use the fact that T is symmetric to show that $\ker T \subseteq (\text{im } T)^{\perp}$ and hence that these are equal because they have the same dimension.]

APPENDIX A Complex Numbers

The fact that the square of every real number is positive shows that the equation $x^2 + 1 = 0$ has no real root; in other words, there is no real number u such that $u^2 = -1$. So the set of real numbers is inadequate for finding all roots of all polynomials. This kind of problem arises with other number systems as well. The set of integers contains no solution of the equation $3x + 2 = 0$, and the rational numbers had to be invented to solve such equations. But the set of rational numbers is also incomplete because, for example, it contains no root of the polynomial $x^2 - 2$. Hence the real numbers were invented. In the same way, the set of complex numbers was invented, which contains all real numbers together with a root of the equation $x^2 + 1 = 0$. The process ends here, however: the complex numbers have the property that every polynomial with complex coefficients has a (complex) root. This fact is known as the fundamental theorem of algebra.

One pleasant aspect of the complex numbers is that, whereas describing the real numbers in terms of the rationals is a rather complicated business, the complex numbers are quite easy to describe in terms of real numbers. Every **complex number** has the form

$$a + bi$$

where a and b are real numbers, and i is a root of the polynomial $x^2 + 1$. Here a and b are called the **real part** and the **imaginary part** of the complex number, respectively. The real numbers are now regarded as special complex numbers of the form $a + 0i = a$, with zero imaginary part. The complex numbers of the form $0 + bi = bi$ with zero real part are called **pure imaginary** numbers. The complex number i itself is called the **imaginary unit** and is distinguished by the fact that

$$i^2 = -1$$

As the terms *complex* and *imaginary* suggest, these numbers met with some resistance when they were first used. This has changed; now they are essential in science and engineering as well as mathematics, and they are used extensively. The names persist, however, and continue to be a bit misleading: These numbers are no more "complex" than the real numbers, and the number i is no more "imaginary" than -1.

Much as for polynomials, two complex numbers are declared to be **equal** if and only if they have the same real parts and the same imaginary parts. In symbols,

$$a + bi = a' + b'i \qquad \text{if and only if} \qquad a = a' \text{ and } b = b'$$

The **addition** and **subtraction** of complex numbers is accomplished by adding and subtracting real and imaginary parts:

$$(a + bi) + (a' + b'i) = (a + a') + (b + b')i$$
$$(a + bi) - (a' + b'i) = (a - a') + (b - b')i$$

This is analogous to these operations for linear polynomials $a + bx$ and $a' + b'x$, and the **multiplication** of complex numbers is also analogous with one difference: $i^2 = -1$. The definition is

$$(a + bi)(a' + b'i) = (aa' - bb') + (ab' + ba')i$$

With these definitions of equality, addition, and multiplication, the complex numbers satisfy all the basic arithmetical axioms adhered to by the real numbers (the verifications are omitted). One consequence of this is that they may be manipulated in the obvious fashion, except that i^2 is replaced by -1 wherever it occurs and the rule for equality must be observed.

EXAMPLE 1 If $z = 2 - 3i$ and $w = -1 + i$, write each of the following in the form $a + bi$: $z + w$, $z - w$, zw, $\frac{1}{3}z$, and z^2.

Solution

$$z + w = (2 - 3i) + (-1 + i) = (2 - 1) + (-3 + 1)i = 1 - 2i$$
$$z - w = (2 - 3i) - (-1 + i) = (2 + 1) + (-3 - 1)i = 3 - 4i$$
$$zw = (2 - 3i)(-1 + i) = (-2 - 3i^2) + (2 + 3)i = 1 + 5i$$
$$\frac{1}{3}z = \frac{1}{3}(2 - 3i) = \frac{2}{3} - i$$
$$z^2 = (2 - 3i)(2 - 3i) = (4 + 9i^2) + (-6 - 6)i = -5 - 12i$$

EXAMPLE 2 Find all complex numbers z such that $z^2 = i$.

Solution Write $z = a + bi$; we must determine a and b. Now $z^2 = (a^2 - b^2) + (2ab)i$, so the condition $z^2 = i$ becomes

$$(a^2 - b^2) + (2ab)i = 0 + i$$

Equating real and imaginary parts, we find that $a^2 = b^2$ and $2ab = 1$. The solution is $a = b = \pm\frac{1}{\sqrt{2}}$, so the complex numbers required are $z = \frac{1}{\sqrt{2}} + \frac{1}{\sqrt{2}}i$ and $z = -\frac{1}{\sqrt{2}} - \frac{1}{\sqrt{2}}i$.

As for real numbers, it is possible to divide by every nonzero complex number z. That is, there exists a complex number w such that $wz = 1$. As in the real case, this number w is called the **inverse** of z and is denoted by z^{-1} or $\frac{1}{z}$. Moreover, if $z = a + bi$, the fact that $z \neq 0$ means that $a \neq 0$ or $b \neq 0$. Hence $a^2 + b^2 \neq 0$, and an explicit formula for the inverse is

$$\frac{1}{z} = \frac{a}{a^2 + b^2} - \frac{b}{a^2 + b^2} \cdot i$$

In actual calculation, the work is facilitated by two useful notions: the conjugate and the absolute value of a complex number. The next example illustrates the technique.

EXAMPLE 3 Write $\frac{3 + 2i}{2 + 5i}$ in the form $a + bi$.

Solution Multiply top and bottom by the complex number $2 - 5i$ (obtained from the denominator by negating the imaginary part). The result is

$$\frac{3 + 2i}{2 + 5i} = \frac{(2 - 5i)(3 + 2i)}{(2 - 5i)(2 + 5i)} = \frac{(6 + 10) + (4 - 15)i}{(4 + 25) + 0i} = \frac{16}{29} - \frac{11}{29}i$$

Hence the simplified form is $\frac{16}{29} - \frac{11}{29}i$, as required.

The key to this technique is that the product $(2 - 5i)(2 + 5i) = 29$ in the denominator turned out to be a *real* number. The situation in general leads to the following notation: If $z = a + bi$ is a complex number, the **conjugate** of z is the complex number, denoted $\overline{z}$, given by

$$\overline{z} = a - bi \qquad \text{where } z = a + bi$$

Hence $\overline{z}$ is obtained from z by negating the imaginary part. For example, $\overline{(2 + 3i)} = 2 - 3i$ and $\overline{(1 - i)} = 1 + i$. If we multiply z by $\overline{z}$, we obtain

$$z\overline{z} = a^2 + b^2 \qquad \text{where } z = a + bi$$

The real number $a^2 + b^2$ is always nonnegative, so we can state the following definition: The **absolute value** or **modulus** of a complex number $z = a + bi$, denoted by $|z|$, is the positive square root $\sqrt{a^2 + b^2}$; that is,

$$|z| = \sqrt{a^2 + b^2} \qquad \text{where } z = a + bi$$

For example,

$$|2 - 3i| = \sqrt{2^2 + (-3)^2} = \sqrt{13} \text{ and } |1 + i| = \sqrt{1^2 + 1^2} = \sqrt{2}.$$

Note that if a real number a is viewed as the complex number $a + 0i$, its absolute value (as a complex number) is $|a| = \sqrt{a^2}$, which agrees with its absolute value as a *real* number.

With these notions in hand, we can describe the technique applied in Example 3 as follows: When converting a quotient z/w of complex numbers to the form $a + bi$, multiply top and bottom by the conjugate $\overline{w}$ of the denominator.

The following list contains the most important properties of conjugates and absolute values. Throughout, z and w denote complex numbers.

C1. $\overline{z \pm w} = \overline{z} \pm \overline{w}$

C2. $\overline{zw} = \overline{z}\,\overline{w}$

C3. $\overline{(z/w)} = \overline{z}/\overline{w}$

C4. $\overline{(\overline{z})} = z$

C5. z is real if and only if $\overline{z} = z$

C6. $z\overline{z} = |z|^2$

C7. $\dfrac{1}{z} = \dfrac{\overline{z}}{|z|^2}$

C8. $|z| \geq 0$ for all complex numbers z

C9. $|z| = 0$ if and only if $z = 0$

C10. $|zw| = |z|\,|w|$

C11. $\left|\dfrac{z}{w}\right| = \dfrac{|z|}{|w|}$

C12. $|z + w| \leq |z| + |w|$ **(triangle inequality)**

All of these properties (except property C12) can (and should) be verified by the reader for arbitrary complex numbers $z = a + bi$ and $w = c + di$. They are not independent; for example, property C10 follows from properties C2 and C6.

The triangle inequality comes, as its name suggests, from a geometrical representation of the complex numbers analogous to identifi-

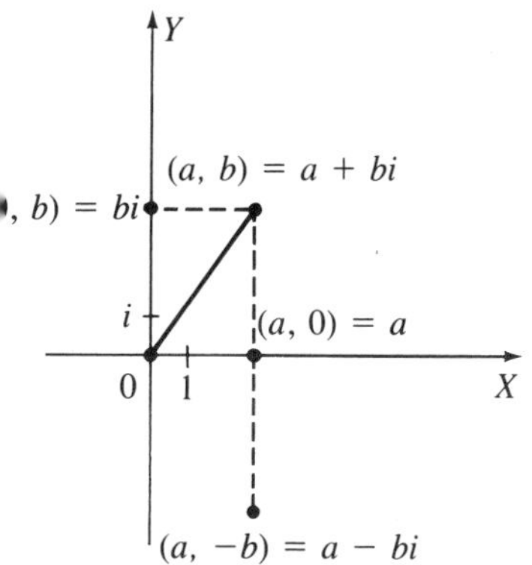

FIGURE A.1

cation of the real numbers with the points of a line. The representation is achieved as follows: Introduce a rectangular coordinate system in the plane (Figure A.1), and identify the complex number $a + bi$ with the point (a,b). When this is done, the plane is called the **complex plane.** Note that the point $(a,0)$ on the X-axis now represents the *real* number $a = a + 0i$, and for this reason, the X-axis is called the **real axis.** Similarly, the Y-axis is called the **imaginary axis.** The identification $(a,b) = a + bi$ of the geometrical point (a,b) and the complex number $a + bi$ will be used without comment below. For example, the origin will be referred to as 0.

This representation of the complex numbers in the complex plane gives a useful way of describing the absolute value and conjugate of a complex number $z = a + bi$. The absolute value $|z| = \sqrt{a^2 + b^2}$ is just the distance from z to the origin. This makes properties C8 and C9 quite obvious. The conjugate $\bar{z} = a - bi$ of z is just the reflection of z in the real axis (X-axis), a fact that makes properties C4 and C5 clear.

Given two complex numbers $z_1 = a_1 + b_1 i = (a_1,b_1)$ and $z_2 = a_2 + b_2 i = (a_2,b_2)$, the absolute value of their difference

$$|z_1 - z_2| = \sqrt{(a_1 - a_2)^2 + (b_1 - b_2)^2}$$

is just the distance between them. This gives the **complex distance formula:**

$$|z_1 - z_2| \text{ is the distance between } z_1 \text{ and } z_2$$

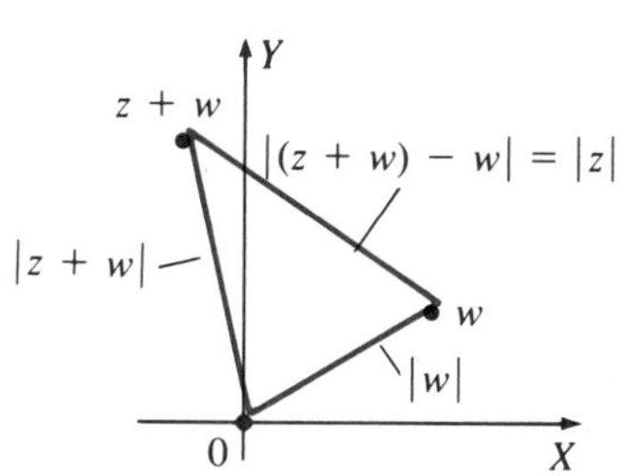

FIGURE A.2

This useful fact yields a simple verification of the triangle inequality, property C12. Suppose z and w are given complex numbers. Consider the triangle in Figure A.2 whose vertices are 0, w, and $z + w$. The three sides have lengths $|z|$, $|w|$, and $|z + w|$ by the complex distance formula, so the inequality

$$|z + w| \leq |z| + |w|$$

expresses the obvious geometrical fact that the sum of the lengths of two sides of a triangle is at least as great as the length of the third side.

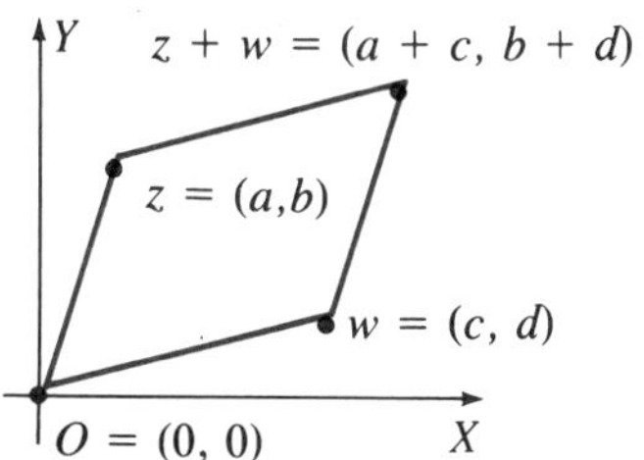

FIGURE A.3

The representation of complex numbers as points in the complex plane has another very useful property: It allows us to give a geometrical description of the sum and product of two complex numbers. To obtain the description for the sum, let

$$z = a + bi = (a,b)$$
$$w = c + di = (c,d)$$

denote two complex numbers. We claim that the four points 0, z, w, and $z + w$ form the vertices of a parallelogram. In fact, in Figure A.3 the lines from 0 to z and from w to $z + w$ have slopes

$$\frac{b-0}{a-0}=\frac{b}{a} \quad \text{and} \quad \frac{(b+d)-d}{(a+c)-c}=\frac{b}{a}$$

respectively, so these lines are parallel. (If it happens that $a = 0$, then both these lines are vertical.) Similarly, the lines from z to $z + w$ and from 0 to w are also parallel, so the figure with vertices 0, z, w, and $z + w$ is indeed a parallelogram. Hence the complex number $z + w$ can be obtained geometrically from z and w by *completing* the parallelogram. This is sometimes called the **parallelogram law** of complex addition. Readers who have studied mechanics will recall that velocities and accelerations add in the same way; in fact, these are all special cases of *vector* addition.

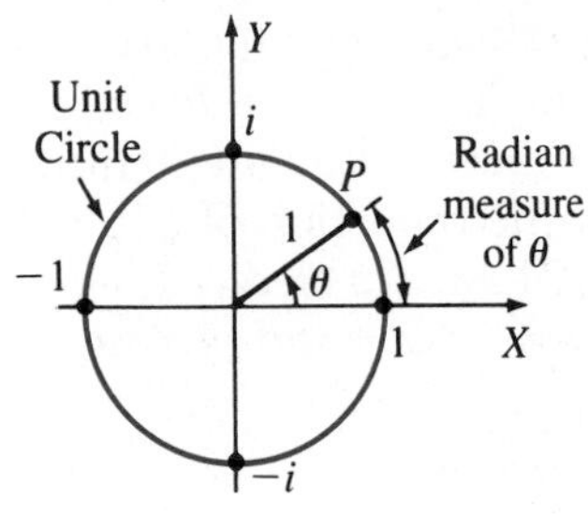

FIGURE A.4

The geometric description of what happens when two complex numbers are multiplied is at least as elegant as the parallelogram law of addition, but it requires that the complex numbers be represented in polar form. Before giving this, we pause to recall the general definition of the trigonometric functions sine and cosine. An angle θ in the complex plane is in **standard position** if it is measured counterclockwise from the real axis as indicated in Figure A.4. Rather than using degrees to measure angles, it is more natural to use radian measure. This is defined as follows: The circle with its center at the origin and radius 1 (called the **unit circle**) is drawn in Figure A.4. It has circumference 2π, and the **radian measure** of θ is the length of the arc on the unit circle from 1 to the point P on the unit circle determined by θ. Hence $90° = \frac{\pi}{2}$, $45° = \frac{\pi}{4}$, $180° = \pi$, and a full circle has the angle $360° = 2\pi$.

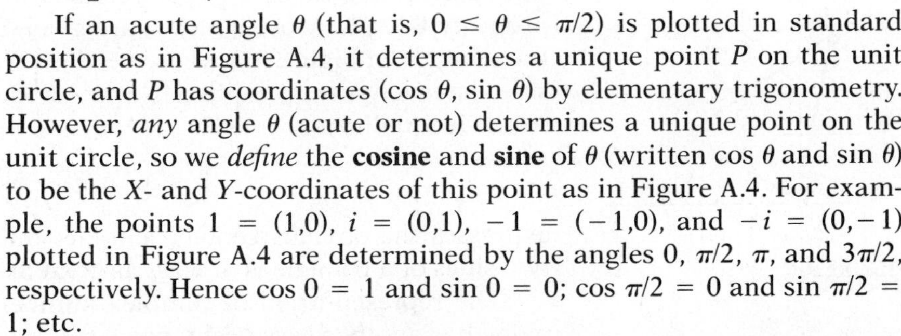

If an acute angle θ (that is, $0 \le \theta \le \pi/2$) is plotted in standard position as in Figure A.4, it determines a unique point P on the unit circle, and P has coordinates $(\cos\theta, \sin\theta)$ by elementary trigonometry. However, *any* angle θ (acute or not) determines a unique point on the unit circle, so we *define* the **cosine** and **sine** of θ (written $\cos\theta$ and $\sin\theta$) to be the X- and Y-coordinates of this point as in Figure A.4. For example, the points $1 = (1,0)$, $i = (0,1)$, $-1 = (-1,0)$, and $-i = (0,-1)$ plotted in Figure A.4 are determined by the angles 0, $\pi/2$, π, and $3\pi/2$, respectively. Hence $\cos 0 = 1$ and $\sin 0 = 0$; $\cos \pi/2 = 0$ and $\sin \pi/2 = 1$; etc.

Now we can describe the polar form of a complex number. Let $z = a + bi$ be a complex number, and write the absolute value of z as

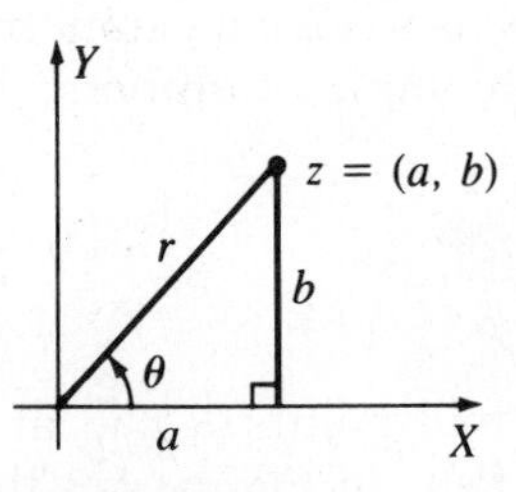

FIGURE A.5

$$r = |z| = \sqrt{a^2 + b^2}$$

If $z \neq 0$, the angle θ shown in Figure A.5 is called an **argument** of z and is denoted

$$\theta = \arg z$$

This angle is not unique ($\theta + 2\pi k$ would do as well for any $k = 0, \pm 1,$

$\pm 2, \ldots$). However, there is only one argument θ in the range $-\pi < \theta \leq \pi$, and this is sometimes called the **principal argument** of z.

Returning to Figure A.5, we find that the real and imaginary parts a and b of z are related to r and θ by

$$a = r\cos\theta$$
$$b = r\sin\theta$$

Hence the complex number $z = a + bi$ has the form

$$z = r(\cos\theta + i\sin\theta) \qquad r = |z|, \theta = \arg(z)$$

The combination $\cos\theta + i\sin\theta$ is so important that a special notation is used. This is

$$e^{i\theta} = \cos\theta + i\sin\theta$$

With this notation, z is written

$$z = re^{i\theta} \qquad r = |z|, \theta = \arg(z)$$

This is a **polar form** of the complex number z. Of course it is not unique, because the argument can be changed by adding a multiple of 2π.

EXAMPLE 4 Write $z_1 = -2 + 2i$ and $z_2 = -i$ in polar form.

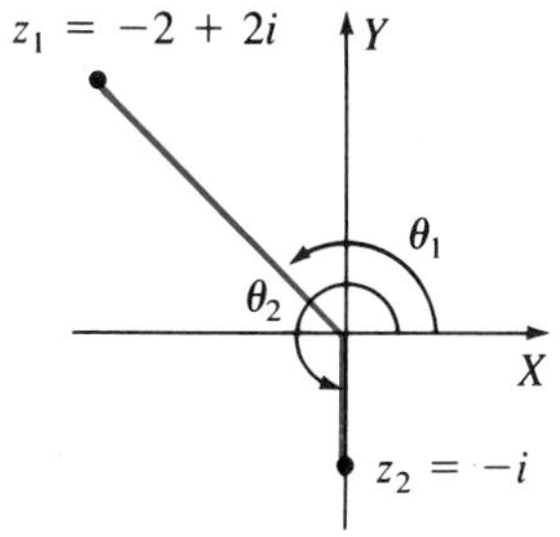

$z_1 = -2 + 2i$ Y θ_1 θ_2 X $z_2 = -i$

FIGURE A.6

Solution The two numbers are plotted in the complex plane in Figure A.6. The absolute values are

$$r_1 = |-2 + 2i| = \sqrt{(-2)^2 + 2^2} = 2\sqrt{2}$$
$$r_2 = |-i| = \sqrt{0^2 + (-1)^2} = 1$$

By inspection of Figure A.6, arguments of z_1 and z_2 are

$$\theta_1 = \arg(-2 + 2i) = 3\pi/4$$
$$\theta_2 = \arg(-i) = 3\pi/2$$

The corresponding polar forms are $z_1 = -2 + 2i = 2\sqrt{2}\, e^{3\pi i/4}$ and $z_2 = -i = e^{3\pi i/2}$. Of course, we could have taken the argument $-\pi/2$ for z_2 and obtained the polar form $z_2 = e^{-\pi i/2}$.

In the notation $e^{i\theta} = \cos\theta + i\sin\theta$, the number e is the familiar constant $e = 2.71828\ldots$ from calculus. The reason for using e will not be given here; the reason why $\cos\theta + i\sin\theta$ is written as an *exponential* function of θ is that the **law of exponents** holds:

$$e^{i\theta} \cdot e^{i\phi} = e^{i(\theta + \phi)}$$

where θ and ϕ are any two angles. In fact, this is an immediate consequence of the addition identities for $\sin(\theta + \phi)$ and $\cos(\theta + \phi)$:

$$\begin{aligned} e^{i\theta}e^{i\phi} &= (\cos\theta + i\sin\theta)(\cos\phi + i\sin\phi) \\ &= (\cos\theta\cos\phi - \sin\theta\sin\phi) + i(\cos\theta\sin\phi + \sin\theta\cos\phi) \\ &= \cos(\theta + \phi) + i\sin(\theta + \phi) \\ &= e^{i(\theta+\phi)} \end{aligned}$$

This is analogous to the rule $e^a e^b = e^{a+b}$, which holds for real numbers a and b, so it is not unnatural to use the exponential notation $e^{i\theta}$ for the expression $\cos\theta + i\sin\theta$. In fact, a whole theory exists wherein functions such as e^z, $\sin z$, and $\cos z$ are studied, where z is a *complex* variable. Many deep and beautiful theorems can be proved in this theory, one of which is the so-called fundamental theorem of algebra mentioned below. We shall not pursue this here.

The geometric description of the multiplication of two complex numbers follows from the law of exponents.

THEOREM 1
Multiplication Rule

If $z_1 = r_1 e^{i\theta_1}$ and $z_2 = r_2 e^{i\theta_2}$ are complex numbers in polar form, then

$$z_1 z_2 = r_1 r_2 e^{i(\theta_1+\theta_2)}$$

In other words, to multiply two complex numbers, simply multiply the absolute values and add the arguments. This simplifies calculations considerably, particularly when we observe that it is valid for *any* arguments θ_1 and θ_2.

EXAMPLE 5

Multiply $(1 - i)(1 + \sqrt{3}i)$.

Solution We have $|1 - i| = \sqrt{2}$ and $|1 + \sqrt{3}i| = 2$ and, from Figure A.7,

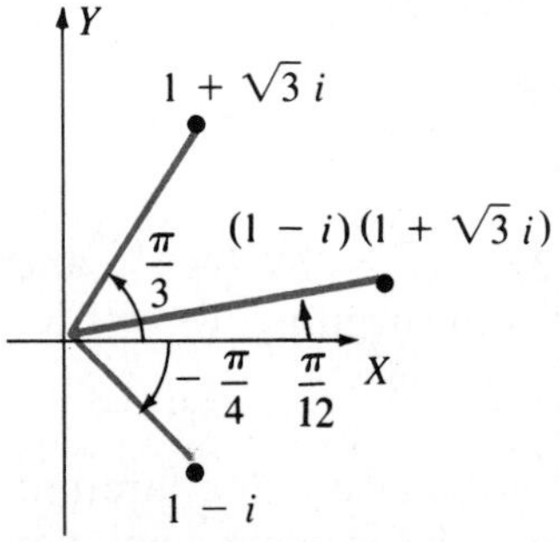

$$\begin{aligned} 1 - i &= \sqrt{2}e^{-i\pi/4} \\ 1 + \sqrt{3}i &= 2e^{i\pi/3} \end{aligned}$$

FIGURE A.7

Hence, by the multiplication rule,

$$\begin{aligned} (1 - i)(1 + \sqrt{3}i) &= (\sqrt{2}\, e^{-i\pi/4})(2e^{i\pi/3}) \\ &= 2\sqrt{2}\, e^{i(-\pi/4 + \pi/3)} \\ &= 2\sqrt{2}\, e^{i\pi/12} \end{aligned}$$

This gives the required product in polar form. Of course, direct multiplication gives $(1 - i)(1 + \sqrt{3}i) = (\sqrt{3} + 1) + (\sqrt{3} - 1)i$. Hence, equating real and imaginary parts gives $\cos(\pi/12) = (\sqrt{3} + 1)/2\sqrt{2}$ and $\sin(\pi/12) = (\sqrt{3} - 1)/2\sqrt{2}$.

If a complex number $z = re^{i\theta}$ is given in polar form, the powers assume a particularly simple form. In fact, $z^2 = (re^{i\theta})(re^{i\theta}) = r^2e^{2i\theta}$, $z^3 = z^2 \cdot z = (r^2e^{2i\theta})(re^{i\theta}) = r^3e^{3i\theta}$, and so on. Continuing in this way, it follows by induction that the following theorem holds for any positive integer n. The name honors Abraham De Moivre (1667–1754).

THEOREM 2
De Moivre's Theorem

If θ is any angle, then $(e^{i\theta})^n = e^{in\theta}$ holds for all integers n.

Proof It clearly holds for $n = 0$. To derive it for $n < 0$, first observe that

$$\text{if } z = re^{i\theta} \neq 0 \qquad \text{then} \qquad z^{-1} = \frac{1}{r}e^{-i\theta}$$

In fact, $(re^{i\theta})\left(\frac{1}{r}e^{-i\theta}\right) = 1e^{i0} = 1$ by the multiplication rule. If n is negative, write it as $n = -m$, $m > 0$. Then

$$(re^{i\theta})^n = [(re^{i\theta})^{-1}]^m = \left(\frac{1}{r}e^{-i\theta}\right)^m = r^{-m}e^{i(-m\theta)} = r^ne^{in\theta}$$

If $r = 1$, this is De Moivre's theorem for negative n. ■

EXAMPLE 6 Verify that $(-1 + \sqrt{3}i)^3 = 8$.

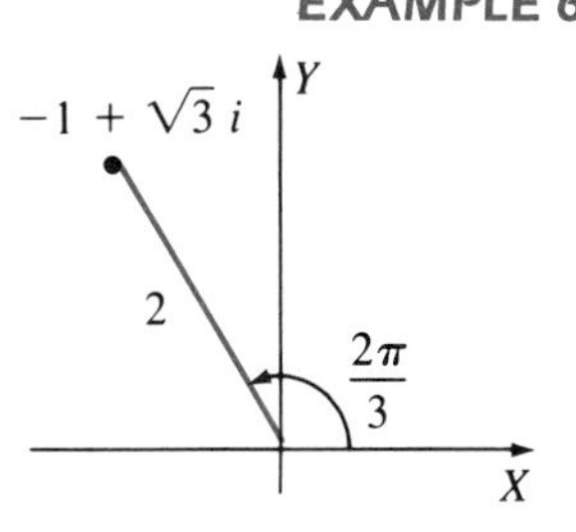

FIGURE A.8

Solution See Figure A.8. Because $|-1 + \sqrt{3}i| = 2$ and $\arg(-1 + \sqrt{3}i) = \frac{2\pi}{3}$, we have $(-1 + \sqrt{3}i)^3 = (2e^{2\pi i/3})^3 = 8e^{3(2\pi/3)i} = 8$, because $e^{2\pi i} = 1$.

De Moivre's theorem can be used to find nth roots of complex numbers where n is positive. The next example illustrates this technique.

EXAMPLE 7 Find the cube roots of unity; that is, find all complex numbers z such that $z^3 = 1$.

Solution First write $z = re^{i\theta}$ and $1 = 1e^{i \cdot 0}$ in polar form. We must use the condition $z^3 = 1$ to determine r and θ. Because $z^3 = r^3e^{3i\theta}$ by De Moivre's theorem, this is

$$r^3e^{3\theta i} = 1e^{0 \cdot i}$$

These two complex numbers are equal, so their absolute values must be equal and the arguments must either be equal or differ by an integral multiple of 2π:

$$r^3 = 1$$
$$3\theta = 0 + 2k\pi, \qquad k \text{ some integer}$$

Because r is real and positive, the condition $r^3 = 1$ implies that $r = 1$. However,

$$\theta = \frac{2k\pi}{3}, \qquad k \text{ some integer}$$

seems at first glance to yield infinitely many different angles for z. However, choosing $k = 0, 1, 2$ gives three possible arguments θ (where $0 \le \theta < 2\pi$), and the corresponding roots are

$$1e^{0i} = 1$$
$$1e^{2\pi i/3} = -\frac{1}{2} + \frac{\sqrt{3}}{2}i$$
$$1e^{4\pi i/3} = -\frac{1}{2} - \frac{\sqrt{3}}{2}i$$

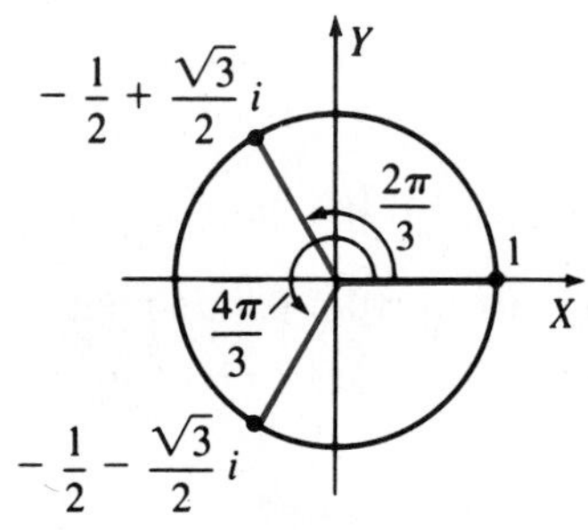

FIGURE A.9

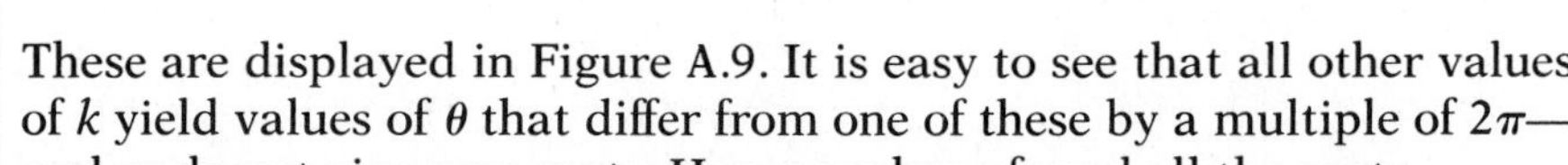

These are displayed in Figure A.9. It is easy to see that all other values of k yield values of θ that differ from one of these by a multiple of 2π—and so do not give new roots. Hence we have found all the roots.

The same type of calculation gives all complex nth roots of unity; that is, all complex numbers z such that $z^n = 1$. As before, write $1 = 1e^{0 \cdot i}$ and

$$z = re^{i\theta}$$

in polar form. Then $z^n = 1$ takes the form

$$r^n e^{n\theta i} = 1e^{0i}$$

using De Moivre's theorem. Comparing absolute values and arguments yields

$$r^n = 1$$
$$n\theta = 0 + 2k\pi, \qquad k \text{ some integer}$$

Hence $r = 1$, and the n values

$$\theta = \frac{2k\pi}{n}, \qquad k = 0, 1, 2, \ldots, n - 1$$

of θ all lie in the range $0 \le \theta < 2\pi$. As before, *every* choice of k yields a value of θ that differs from one of these by a multiple of 2π, so these

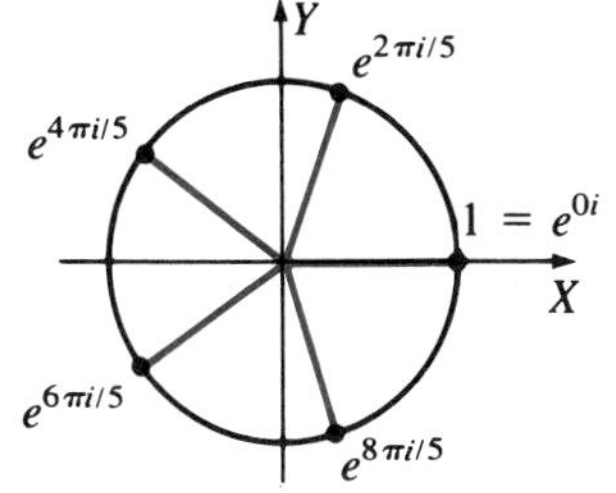

FIGURE A.10

give the arguments of *all* the possible roots. Hence the **nth roots of unity** are given by

$$z = e^{2\pi ki/n}, k = 0, 1, 2, \ldots, n - 1.$$

There are n of them, and they can be found geometrically as the points on the unit circle that cut it into n equal sectors, starting at 1. The case $n = 5$ is shown in Figure A.10, where the five fifth roots of unity are plotted.

The method used above to find the nth roots of unity works equally well to find the nth roots of any complex number in polar form. We give one example.

EXAMPLE 8 Find the fourth roots of $\sqrt{2} + \sqrt{2}i$.

Solution First write $\sqrt{2} + \sqrt{2}i = 2e^{\pi i/4}$ in polar form. If $z = re^{i\theta}$ satisfies $z^4 = \sqrt{2} + \sqrt{2}i$, then

$$r^4e^{i(4\theta)} = 2e^{\pi i/4}$$

Hence $r^4 = 2$ and $4\theta = \dfrac{\pi}{4} + 2k\pi$, k an integer. We obtain four distinct roots (and hence all) by

$$r = \sqrt[4]{2}, \theta = \frac{\pi}{16} + \frac{8k\pi}{16}, \qquad k = 0, 1, 2, 3$$

Thus the four roots are

$$\sqrt[4]{2}e^{\pi i/16} \qquad \sqrt[4]{2}e^{9\pi i/16} \qquad \sqrt[4]{2}e^{17\pi i/16} \qquad \sqrt[4]{2}e^{25\pi i/16}$$

Of course, reducing these roots to the form $a + bi$ would require the computation of $\sqrt[4]{2}$ and the sine and cosine of the various angles.

An expression of the form $ax^2 + bx + c$, where the coefficients $a \neq 0$, b, and c are real numbers, is called a **real quadratic**. A complex number u is called a **root** of the quadratic if $au^2 + bu + c = 0$. The roots are given by the famous **quadratic formula**:

$$u = \frac{-b \pm \sqrt{b^2 - 4ac}}{2a}$$

The quantity $d = b^2 - 4ac$ is called the **discriminant** of the quadratic $ax^2 + bx + c$, and there is no real root if and only if $d < 0$. In this case the quadratic is said to be **irreducible**. Moreover, the fact that $d < 0$ means that $\sqrt{d} = i\sqrt{|d|}$, so the two (complex) roots are conjugates of each other:

$$u = \frac{1}{2a}(-b + i\sqrt{|d|}) \quad \text{and} \quad \overline{u} = \frac{1}{2a}(-b - i\sqrt{|d|}).$$

The converse of this is true too: Given any nonreal complex number u, then u and $\overline{u}$ are the roots of a real irreducible quadratic. Indeed, the quadratic

$$x^2 - (u + \overline{u})x + u\overline{u} = (x - u)(x - \overline{u})$$

has real coefficients ($u\overline{u} = |u|^2$ and $u + \overline{u}$ is twice the real part of u) and so is irreducible because its roots u and $\overline{u}$ are not real.

EXAMPLE 9 Find a real irreducible quadratic with $u = 3 - 4i$ as a root.

Solution We have $u + \overline{u} = 6$ and $|u|^2 = 25$, so $x^2 - 6x + 25$ is irreducible with u and $\overline{u} = 3 + 4i$ as roots.

The quadratic formula works for quadratics with complex coefficients. If $p \neq 0$ and v and w are complex numbers, the equation

$$px^2 + vx + w = 0$$

can be solved by an old technique called **completing the square** (it was known to the Moslem mathematician Al-Khowarizmi in the ninth century A.D.). The idea is to write the equation as $x^2 + \frac{v}{p}x = -\frac{w}{p}$ and then complete the square on the left by adding $(\frac{v}{2p})^2$ to each side:

$$\left(x + \frac{v}{2p}\right)^2 = x^2 + \frac{v}{p}x + \left(\frac{v}{2p}\right)^2 = -\frac{w}{p} + \left(\frac{v}{2p}\right)^2 = \frac{v^2 - 4pw}{4p^2}$$

Taking square roots gives the complex version of the quadratic formula

$$x = \frac{-v \pm \sqrt{v^2 - 4pw}}{2p}$$

Of course, the discriminant $v^2 - 4pw$ is now a complex number, and we need the foregoing methods to find its square roots. Here is an example.

EXAMPLE 10 Find all complex numbers z such that $z^2 - iz + (1 + 3i) = 0$.

Solution The quadratic formula gives

$$z = \frac{1}{2}[i \pm \sqrt{i^2 - 4(1 + 3i)}] = \frac{1}{2}[i \pm \sqrt{-5 - 12i}] = \frac{1}{2}[i \pm w]$$

where $w = \sqrt{-5 - 12i}$. Hence $w^2 = -5 - 12i$, so if $w = a + bi$, equating real and imaginary parts gives $a^2 - b^2 = -5$ and $2ab = -12$. Hence $b = -6/a$, so $a^2 - 36/a^2 = -5$. This gives a quadratic in a^2: $a^4 + 5a^2 - 36 = 0$, which factors as $(a^2 - 4)(a^2 + 9) = 0$. Thus $a = \pm 2$ and

$b = -6/a = \mp 3$, so $w = \pm(2 - 3i)$. Finally,

$$z = \frac{1}{2}[i \pm w] = \frac{1}{2}[i \pm (2 - 3i)]$$

Hence the roots are $z = 1 - i$ and $-1 + 2i$.

If one root of a quadratic equation $px^2 + vx + w = 0$ is known, it is easy to find the other root. Because we can divide both sides of the equation by p, we state the result for quadratics with 1 as the coefficient of x^2.

THEOREM 3

If u_1 and u_2 are the roots of the quadratic equation

$$x^2 + vx + w = 0$$

then $u_1 + u_2 = -v$ and $u_1u_2 = w$.

Proof Because u_1 and u_2 are roots of $x^2 + vx + w = 0$, the factor theorem asserts that the quadratic factors as

$$x^2 + vx + w = (x - u_1)(x - u_2)$$

The right side is $x^2 - (u_1 + u_2)x + u_1u_2$, so the result follows because corresponding coefficients must be equal. ■

EXAMPLE 11 Show that $u_1 = 1 + i$ is a root of $x^2 + (1 - 2i)x - (3 + i) = 0$, and then find the other root.

Solution $u_1{}^2 + (1 - 2i)u_1 - (3 + i) = (1 - 1 + 2i) + (1 + 2 - i) - (3 + i) = 0$, so u_1 is a root. If u_2 is the other root, then $u_1 + u_2 = -(1 - 2i)$ by Theorem 3, so $u_2 = -(1 - 2i) - u_1 = -2 + i$. Of course, this also follows from $u_1u_2 = -(3 + i)$.

As we mentioned earlier, the complex numbers are the culmination of a long search by mathematicians to find a set of numbers large enough to contain a root of every polynomial. The fact that the complex numbers have this property was first proved by Gauss in 1797 when he was 20 years old. The proof is omitted.

THEOREM 4 Fundamental Theorem of Algebra

Every polynomial of positive degree with complex coefficients has a complex root.

If $f(x)$ is a polynomial with complex coefficients, and if u_1 is a root, then the factor theorem asserts that

$$f(x) = (x - u_1)g(x)$$

where $g(x)$ is a polynomial with complex coefficients and with degree one less than the degree of $f(x)$. Suppose that u_2 is a root of $g(x)$, again by the fundamental theorem. Then $g(x) = (x - u_2)h(x)$, so

$$f(x) = (x - u_1)(x - u_2)h(x)$$

This process continues until the last polynomial to appear is linear. Thus $f(x)$ has been expressed as a product of linear factors. The last of these factors can be written in the form $u(x - u_n)$, where u and u_n are complex (verify this), so the fundamental theorem takes the following form.

THEOREM 5 Every polynomial $f(x)$ of degree $n \geq 1$ has the form

$$f(x) = u(x - u_1)(x - u_2) \cdots (x - u_n)$$

where $u, u_1, \ldots, u_n$ are complex numbers and $u \neq 0$. The numbers $u_1, u_2, \ldots, u_n$ are the roots of $f(x)$ (and need not all be distinct), and u is the coefficient of x^n.

This form of the fundamental theorem, when applied to a polynomial $f(x)$ with *real* coefficients, can be used to deduce the following result.

THEOREM 6 Every polynomial $f(x)$ of positive degree with real coefficients can be factored as a product of linear and irreducible quadratic factors.

In fact, suppose $f(x)$ has the form

$$f(x) = a_n x^n + a_{n-1}x^{n-1} + \cdots + a_1 x + a_0$$

where the coefficients a_i are real. If u is a complex root of $f(x)$, then we claim first that $\overline{u}$ is also a root. In fact, we have $f(u) = 0$, so

$$\begin{aligned} 0 = \overline{0} = \overline{f(u)} &= \overline{a_n u^n + a_{n-1}u^{n-1} + \cdots + a_1 u + a_0} \\ &= \overline{a_n u^n} + \overline{a_{n-1}u^{n-1}} + \cdots + \overline{a_1 u} + \overline{a_0} \\ &= \overline{a}_n \overline{u}^n + \overline{a}_{n-1}\overline{u}^{n-1} + \cdots + \overline{a}_1 \overline{u} + \overline{a}_0 \\ &= a_n \overline{u}^n + a_{n-1}\overline{u}^{n-1} + \cdots + a_1 \overline{u} + a_0 \\ &= f(\overline{u}) \end{aligned}$$

where $\overline{a}_i = a_i$ for each i because the coefficients a_i are real. Thus, if u is a root of $f(x)$, so is its conjugate $\overline{u}$. Now some of the roots of $f(x)$ are real (and so equal their conjugates), but the nonreal roots come in pairs, u and $\overline{u}$. We can thus write $f(x)$ as a product:

$$f(x) = a_n(x - r_1) \cdots (x - r_k)(x - u_1)(x - \overline{u}_1) \cdots (x - u_m)(x - \overline{u}_m) \quad (*)$$

where a_n is the coefficient of x_n in $f(x)$; $r_1, r_2, \ldots, r_k$ are the real roots; and $u_1,\overline{u}_1, u_2,\overline{u}_2, \ldots, u_m,\overline{u}_m$ are the nonreal roots. But the product

$$(x - u_j)(x - \overline{u}_j) = x^2 - (u_j + \overline{u}_j)x + u_j\overline{u}_j$$

is a real irreducible quadratic for each j (see the discussion preceding Example 9). Hence (*) shows that $f(x)$ is a product of linear and irreducible quadratic factors, each with real coefficients. This is the desired result.

EXERCISES

1. Solve each of the following for the real number x.

(a) $x - 4i = (2 - i)^2$ **(b)** $(2 + xi)(3 - 2i) = 12 + 5i$

(c) $(2 + xi)^2 = 4$ **(d)** $(2 + xi)(2 - xi) = 5$

2. Convert each of the following to the form $a + bi$.

(a) $(2 - 3i) - 2(2 - 3i) + 9$ **(b)** $(3 - 2i)(1 + i) + |3 + 4i|$

(c) $\dfrac{1 + i}{2 - 3i} + \dfrac{1 - i}{-2 + 3i}$ **(d)** $\dfrac{3 - 2i}{1 - i} - \dfrac{3 - 7i}{2 - 3i}$

(e) i^{131} **(f)** $(2 - i)^3$

(g) $(1 + i)^4$ **(h)** $(1 - i)^2(2 + i)^2$

3. In each case, find the complex number z.

(a) $iz - (1 + i)^2 = 3 - i$ **(b)** $(i + z) - 3i(2 - z) = iz + 1$

(c) $z^2 = -i$ **(d)** $z^2 = 3 - 4i$

4. In each case, find the roots of the real quadratic equation.

(a) $x^2 - 2x + 3 = 0$ **(b)** $x^2 - x + 1 = 0$

(c) $3x^2 - 4x + 2 = 0$ **(d)** $2x^2 - 5x + 2 = 0$

5. Find all numbers x in each case.

(a) $x^3 = 8$ **(b)** $x^3 = -8$ **(c)** $x^4 = 16$ **(d)** $x^4 = 64$

6. In each case, find a real quadratic with u as a root, and find the other root.

(a) $u = 1 + i$ **(b)** $u = 2 - 3i$ **(c)** $u = -i$ **(d)** $u = 3 - 4i$

7. Find the roots of $x^2 - 2\cos\theta\, x + 1 = 0$, θ any angle.

8. Find a real polynomial of degree 4 with $2 - i$ and $3 - 2i$ as roots.

9. Let re z and im z denote, respectively, the real and imaginary parts of z. Show that:

(a) $\text{im}(iz) = \text{re } z$ **(b)** $\text{re}(iz) = -\text{im } z$

(c) $z + \bar{z} = 2 \text{ re } z$
(d) $z - \bar{z} = 2 \text{ im } z$
(e) $\text{re}(z + w) = \text{re } z + \text{re } w$, and $\text{re}(tz) = t \cdot \text{re } z$ if t is real
(f) $\text{im}(z + w) = \text{im } z + \text{im } w$, and $\text{im}(tz) = t \cdot \text{im } z$ if t is real

10. In each case, show that u is a root of the quadratic equation, and find the other root.
(a) $x^2 - 3ix + (-3 + i) = 0$; $u = 1 + i$
(b) $x^2 + ix - (4 - 2i) = 0$; $u = -2$
(c) $x^2 - (3 - 2i)x + (5 - i) = 0$; $u = 2 - 3i$
(d) $x^2 + 3(1 - i)x - 5i = 0$; $u = -2 + i$

11. Find the roots of each of the following complex quadratic equations.
(a) $x^2 + 2x + (1 + i) = 0$
(b) $x^2 - x + (1 - i) = 0$
(c) $x^2 - (2 - i)x + (3 - i) = 0$
(d) $x^2 - 3(1 - i)x - 5i = 0$

12. In each case, describe the graph of the equation (where z denotes a complex number).
(a) $|z| = 1$
(b) $|z - 1| = 2$
(c) $z = i\bar{z}$
(d) $z = -\bar{z}$
(e) $z = |z|$
(f) $\text{im } z = m \cdot \text{re } z$, m a real number

13. (a) Verify $|zw| = |z| \cdot |w|$ directly for $z = a + bi$ and $w = c + di$.
(b) Deduce (a) from properties C2 and C6.

14. Prove that $|w + z|^2 = |w|^2 + |z|^2 + w\bar{z} + \bar{w}z$ for all complex numbers w and z.

15. Use property C5 to show that $(1 + i)^n + (1 - i)^n$ is real for all n.

16. Express each of the following in polar form (use the principal argument).
(a) $3 - 3i$
(b) $-4i$
(c) $-\sqrt{3} + i$
(d) $-4 + 4\sqrt{3}i$
(e) $-7i$
(f) $-6 + 6i$

17. Express each of the following in the form $a + bi$.
(a) $3e^{\pi i}$
(b) $e^{7\pi i/3}$
(c) $2e^{3\pi i/4}$
(d) $\sqrt{2}e^{-\pi i/4}$
(e) $e^{5\pi i/4}$
(f) $2\sqrt{3}e^{-2\pi i/6}$

18. Express each of the following in the form $a + bi$.
(a) $(-1 + \sqrt{3}i)^2$
(b) $(1 + \sqrt{3}i)^{-4}$
(c) $(1 + i)^8$
(d) $(1 - i)^{10}$
(e) $(1 - i)^6(\sqrt{3} + i)^3$
(f) $(\sqrt{3} - i)^9(2 - 2i)^5$

19. Use De Moivre's theorem to show that:
(a) $\cos 2\theta = \cos^2\theta - \sin^2\theta$; $\sin 2\theta = 2 \cos \theta \sin \theta$
(b) $\cos 3\theta = \cos^3\theta - 3 \cos \theta \sin^2\theta$; $\sin 3\theta = 3 \cos^2\theta \sin \theta - \sin^3\theta$

20. (a) Find the fourth roots of unity.
(b) Find the sixth roots of unity.

21. Find all complex numbers z such that:
(a) $z^4 = -1$
(b) $z^4 = 2(\sqrt{3}i - 1)$
(c) $z^3 = -27i$
(d) $z^6 = -64$

22. If $z = re^{i\theta}$ in polar form, show that:
(a) $\bar{z} = re^{-i\theta}$;
(b) $z^{-1} = \frac{1}{r}e^{-i\theta}$

23. (a) Suppose z_1, z_2, z_3, z_4, and z_5 are equally spaced around the unit circle. Show that $z_1 + z_2 + z_3 + z_4 + z_5 = 0$. [*Hint:* $(1 - z)(1 + z + z^2 + z^3 + z^4) = 1 - z^5$ for any complex number z.]

(b) Repeat **(a)** for any $n \geq 2$ points equally spaced around the unit circle.

24. If $z = a + bi$, show that $|a| + |b| \leq \sqrt{2} \cdot |z|$. [*Hint:* $(|a| - |b|)^2 \geq 0$]

25. Let $z \neq 0$ be a complex number. If t is real, describe tz geometrically in terms of z if: **(a)** $t > 0$; **(b)** $t < 0$.

26. If z and w are nonzero complex numbers, show that $|z + w| = |z| + |w|$ if and only if one is a positive real multiple of the other. [*Hint:* Consider the parallelogram with vertices 0, w, z, and $z + w$. Use the preceding exercise and the fact that if t is real, $|1 + t| = 1 + |t|$ is impossible if $t < 0$.]

27. If a and b are *rational* numbers, let p and q denote numbers of the form $a + b\sqrt{2}$. If $p = a + b\sqrt{2}$, define $\tilde{p} = a - b\sqrt{2}$ and $[p] = a^2 - 2b^2$. Show that each of the following holds.

(a) $a + b\sqrt{2} = a_1 + b_1\sqrt{2}$ only if $a = a_1$ and $b = b_1$

(b) $\widetilde{p \pm q} = \tilde{p} \pm \tilde{q}$

(c) $\widetilde{pq} = \tilde{p}\tilde{q}$

(d) $[p] = p\tilde{p}$

(e) $[pq] = [p][q]$

(f) If $f(x)$ is a polynomial with rational coefficients and $p = a + b\sqrt{2}$ is a root of $f(x)$, then $\tilde{p}$ is also a root of $f(x)$.

APPENDIX B Introduction to Linear Programming

Many important problems involve *linear inequalities* rather than *linear equations.* In other words, a condition on the variables x and y might take the form of an inequality $2x - 3y \le 4$ rather than that of an equality $2x - 3y = 4$. Linear programming is a method of finding a solution to a system of such inequalities that maximizes a function of the form $p = ax + by$, where a and b are constants. The general method of solving such problems (called the simplex method) involves Gaussian elimination techniques, so it is natural to include a discussion of it here. However, the proofs of the main theorems are omitted. The interested reader should consult a text on linear programming. [For example, S. I. Gass, *Linear Programming,* 4th ed. (New York: McGraw-Hill, 1975) gives a thorough treatment. J. G. Kemeny, J. L. Snell, and G. L. Thompson, *Introduction to Finite Mathematics* (Englewood Cliffs, N.J.: Prentice Hall, 1974) gives a more elementary treatment and relates linear programming to the theory of games.]

SECTION B.1 Graphical Methods

When only two variables are present, there is a geometrical method of solution that, although it is not useful as a practical tool when more variables are involved, *is* useful in illustrating how solutions to these problems arise. Before giving an example, we must clarify what an inequality of the form

$$2x_1 + 3x_2 \le 5$$

means in geometrical form. (We use x_1 and x_2 in place of x and y to conform to later notations.) Of course, the graph of the corresponding equation $2x_1 + 3x_2 = 5$ is well known; it is a line and consists of all points $P(x_1,x_2)$ in the plane whose coordinates x_1 and x_2 satisfy the equation. The lines parallel to this one all have equations $2x_1 + 3x_2 = c$ for

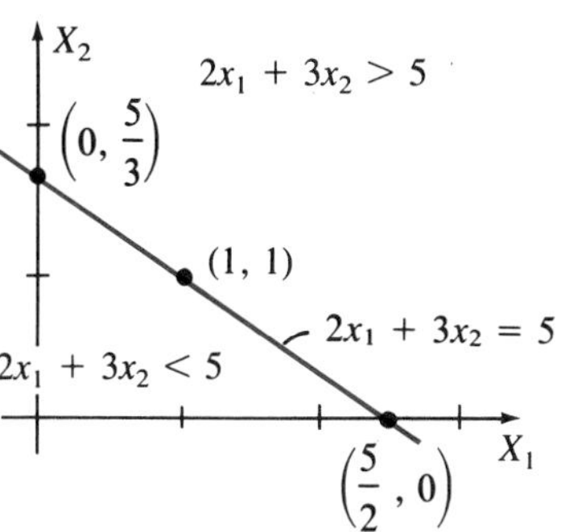

FIGURE B.1

some value of c, so the points $P(x_1,x_2)$ whose coordinates satisfy the *inequality* $2x_1 + 3x_2 < 5$ are just those points lying on one side of the line $2x_1 + 3x_2 = 5$ (points on the other side are those satisfying $2x_1 + 3x_2 > 5$). The situation is illustrated in Figure B.1. The points $P(x_1,x_2)$ satisfying $2x_1 + 3x_2 \le 5$ are just the points on or below the line. In general, the region on or to one side of a straight line is called a **half-plane.** We shall loosely speak of the half-plane $ax_1 + bx_2 \le c$ (where a, b, and c are constants). The line $ax_1 + bx_2 = c$ will be called the **boundary** of the half-plane.

When two or more inequalities are given, the set of points that satisfies all of them is the region common to all the half-planes involved, and such regions are important in linear programming. Here is an example.

EXAMPLE 1 Determine the region in the plane given by

$$x_1 \ge 0$$
$$x_1 - x_2 \le 0$$
$$x_1 + x_2 \le 4$$

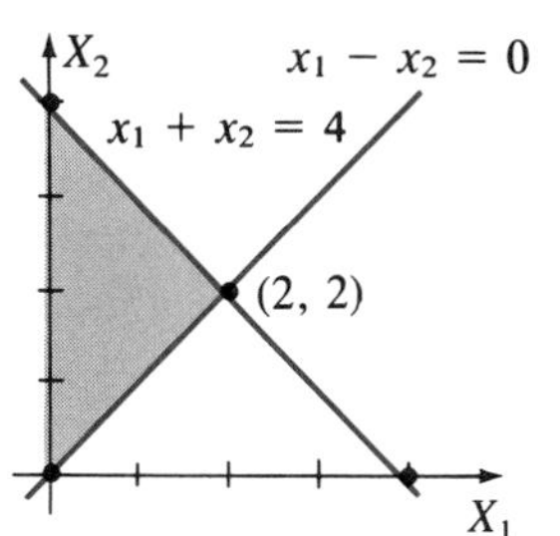

Solution The half-plane $x_1 \ge 0$ consists of all points on or to the right of the X_2-axis, the half-plane $x_1 - x_2 \le 0$ consists of all points on or above the line $x_2 = x_1$, and the half-plane $x_1 + x_2 \le 4$ consists of all points on or below the line $x_2 = -x_1 + 4$. The lines in question are plotted in the diagram, and the region common to all these half-planes is just the shaded portion.

The general linear programming problem (with two variables) can now be stated: Suppose a region in the plane (called the **feasible** region) is given as in Example 1 by a set of linear inequalities (called **constraints**) in two variables x_1 and x_2 (so the feasible region consists of all points common to all the corresponding half-planes). The problem is to find all points in this region at which a linear function of the form $p = ax_1 + bx_2$ is as large as possible. This function p is called the **objective function,** and these points are said to **maximize** p over the feasible region. In applications, p might be the profit in some commercial venture, or some other quantity that is desired to be large. The precise nature of p plays no part in the solution, except that it should be a linear function of the variables x_1 and x_2 (hence the name *linear* programming). The following example illustrates how the method works.

EXAMPLE 2 Find the point (or points) $P(x_1,x_2)$ in the region

$$4x_1 + x_2 \le 16$$

$$\begin{aligned} x_1 + x_2 &\leq 6 \\ x_1 + 3x_2 &\leq 15 \\ x_1 &\geq 0 \\ x_2 &\geq 0 \end{aligned}$$

for which the quantity

$$p = 2x_1 + 3x_2$$

is as large as possible.

Solution The lines $4x_1 + x_2 = 16$, $x_1 + x_2 = 6$, and $x_1 + 3x_2 = 15$ are plotted in the first diagram, and the corresponding half-planes lie below these lines. Hence the feasible region in question is the shaded part. The second diagram exhibits this region again (but with a larger scale) and also shows the line $2x_1 + 3x_2 = p$, plotted for various values of p. Because p has the same value at any point on one of these lines, they are sometimes called **level lines** for p. The aim is to find a point in the shaded region at which p has as large a value as possible. These values increase as the level lines rise, so the vertex (1.5,4.5)—the intersection of the lines $x_1 + 3x_2 = 15$ and $x_1 + x_2 = 6$—clearly gives the largest value of p. This value is

$$p = 2(1.5) + 3(4.5) = 16.3$$

and it is the desired maximal value.

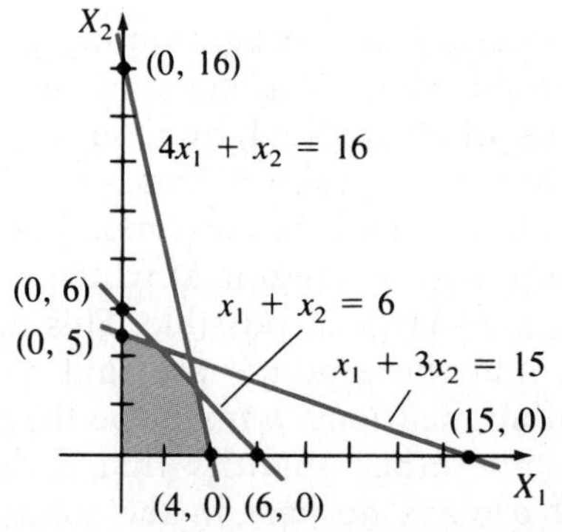

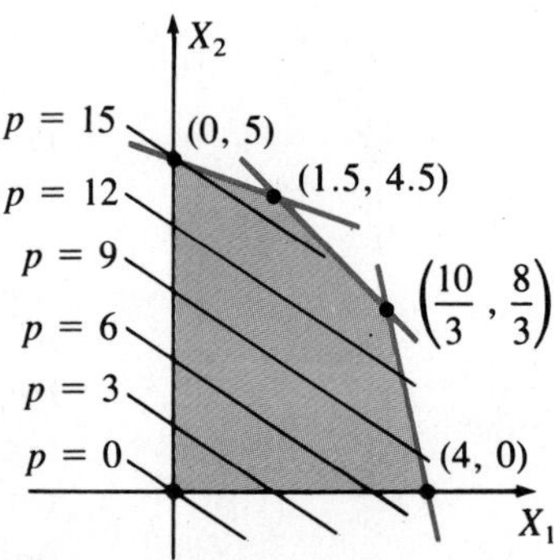

It is quite clear that the method used in Example 2 will work in a variety of similar situations. However, before we attempt to say anything in general, consider the following example.

EXAMPLE 3 Maximize $p = 2x_1 + x_2$ over the region

$$\begin{aligned} x_1 + x_2 &\geq 3 \\ -x_1 + x_2 &\leq 3 \\ -x_1 + 2x_2 &\geq 0 \end{aligned}$$

Solution The region in question is sketched in the diagram, where the main difference from the previous example emerges: The feasible region in this case is unbounded. Three of the level lines (corresponding to $p =$ 6, 9, and 12) are also plotted, and it is clear that there are points in the feasible region at which the objective function p is as large as we please. Consequently p has *no maximum* over the feasible region in question.

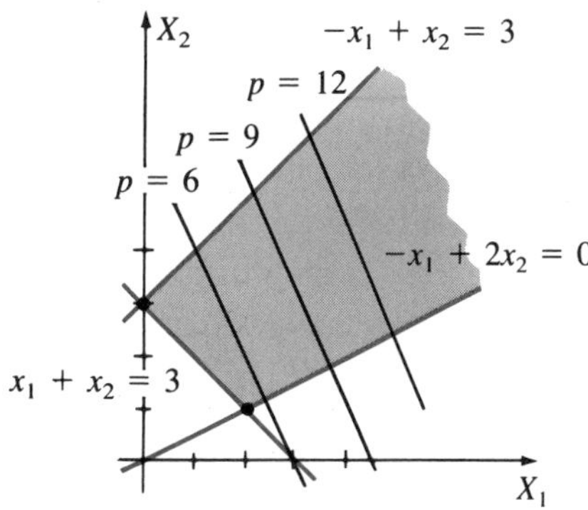

With this example in mind, the reader might guess that, in general, the objective function has no maximum just when the feasible region is unbounded. But this is not the case. Exercise 2 gives a situation wherein the feasible region is unbounded but the objective function does indeed have a maximum (the reader should try to construct such an example before working this exercise).

What is true is that, if the feasible region is bounded (that is, can be contained in some circle), then the objective function p has a maximum value. In fact, the level lines for p form a set of parallel lines, each corresponding to a fixed value of p. If p were to increase continuously, the corresponding level line would move continuously in a direction perpendicular to itself. As this moving level line crosses the feasible region, it is clear that a largest value of p can be found so that the corresponding line intersects the feasible region. Because the edges of the feasible region are line segments, even more can be said: Either this level line corresponding to the largest value of p will intersect the feasible region at a vertex, or, possibly, the intersection will consist of an entire edge of the feasible region. Either way, the maximum value of p will be achieved at a vertex of the feasible region. This is good news. There are at most a finite number of vertices (there are only finitely many constraints), so only a finite number of feasible points (the vertices) need be looked at. If p is evaluated at each of them, the vertex yielding the largest value of p will be the desired feasible point.

Finally, the same argument shows that p achieves a minimum value over a bounded feasible region, and that minimum is found at a vertex. The following theorem summarizes this discussion.

THEOREM 1 Let p be a linear function of two variables x_1 and x_2:

$$p = ax_1 + bx_2$$

If a finite set of linear inequalities in x_1 and x_2 determine a bounded feasible region, then there is a point in this feasible region (in fact, a vertex point) that maximizes p, and there is a vertex point that minimizes p.

EXAMPLE 4 A manufacturer wants to make two types of toys. The large toy requires 4 square feet of plywood and 50 milliliters (ml) of paint, whereas the smaller toy requires 3 square feet of plywood and only 20 ml of paint. There are 1800 square feet of plywood and 16 liters of paint available. If the large toys sell for a profit of \$21 each and each of the small toys yields an \$18 profit, determine the number of toys of each size that the manufacturer must make to maximize its total profit.

Solution Let x_1 and x_2 denote the number of large and small toys, respectively, to be made. These variables satisfy the following constraints:

$$\begin{aligned} 4x_1 + 3x_2 &\le 1800 \quad \text{(plywood)} \\ 5x_1 + 2x_2 &\le 1600 \quad \text{(paint)} \\ x_1 &\ge 0 \\ x_2 &\ge 0 \end{aligned}$$

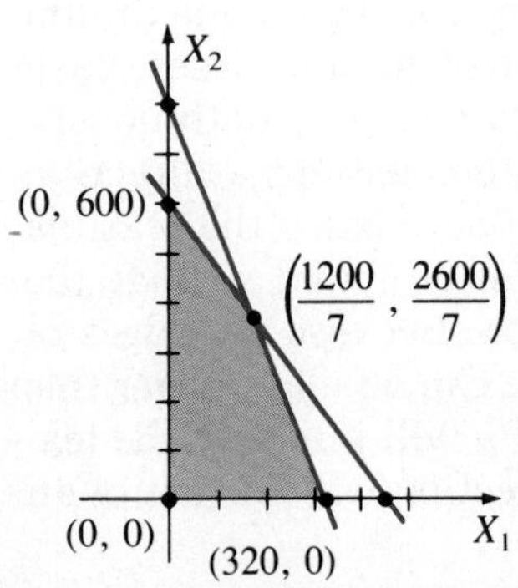

The feasible region corresponding to these inequalities is plotted in the diagram. The four vertices have coordinates (0,0), (0,600), $\left(\frac{1200}{7}, \frac{2600}{7}\right)$, and (320,0). The total profit p of the enterprise is

$$p = 21x_1 + 18x_2$$

and the values of p at each of the vertices are

$$\begin{aligned} p &= 0 && \text{at } (0,0) \\ p &= 10{,}800 && \text{at } (0,600) \\ p &= 10{,}285.71 && \text{at } \left(\frac{1200}{7}, \frac{2600}{7}\right) \\ p &= 6{,}720 && \text{at } (320,0). \end{aligned}$$

Hence the manufacturer will maximize its profits by producing 600 small toys and no large toys at all.

Theorem 1 can be extended. First of all, the objective function p may very well have a maximum over the feasible region even if that region is unbounded. Moreover, the argument leading to Theorem 1 can be modified to show that, *if the objective function p has a maximum over the feasible region, that maximum will be attained at a vertex.*

Second, Theorem 1 can be extended to more than two variables x_1, $x_2, \ldots, x_n$. A function of the form

$$p = c_1x_1 + c_2x_2 + \cdots + c_nx_n$$

is called a **linear function** of these variables, and a condition of the form

$$a_1x_1 + a_2x_2 + \cdots + a_nx_n \le b$$

is called a **linear constraint** on the variables. The set of n-tuples $(x_1, x_2, \ldots, x_n)$ satisfying a finite number of such linear constraints is called the **feasible region** determined by these constraints, and the n-tuples themselves are called **feasible points.** Consider the equation

$$a_1x_1 + a_2x_2 + \cdots + a_nx_n = b$$

which was obtained from the foregoing constraint by replacing the inequality by equality. The set of all n-tuples $(x_1, x_2, \ldots, x_n)$ satisfying this equation is called a **bounding hyperplane** for the feasible region. In the case of two variables these are lines; they are actual planes when $n = 3$. By analogy with the two-variable situation, an n-variable feasible point is called an **extreme point** (or **corner point**) of the feasible region if it lies on n (or more) bounding hyperplanes.

The general linear programming problem is to find a feasible point such that the **objective function** p is as large as possible, in which case the point is said to **maximize** p. (Similarly, we could seek a feasible point that **minimizes** p.) The extended theorem is stated below. The proof (though similar in spirit to that of Theorem 1) is omitted. The feasible region is said to be **bounded** if there exists a number M such that $|x_i| \le M$ holds for every feasible point $(x_1, x_2, \ldots, x_n)$ and each $i = 1, 2, \ldots, n$.

THEOREM 2 Let p be a linear function of the variables $x_1, x_2, \ldots, x_n$:

$$p = c_1x_1 + c_2x_2 + \cdots + c_nx_n$$

and consider the feasible region determined by a finite number of linear constraints on these variables.

(1) If p has a maximum value in the feasible region, then that maximum occurs at an extreme point.

(2) If p has a minimum value in the feasible region, then that minimum occurs at an extreme point.

(3) If the feasible region is bounded, then p has both a maximum and a minimum.

EXAMPLE 5 Find the maximum and minimum value of

$$p = 4x_1 - 3x_2 + 7x_3$$

subject to the following constraints:

$$\begin{aligned} 5x_1 + 2x_2 + 4x_3 &\le 20 \\ x_1 &\ge 0 \\ x_2 &\ge 0 \\ x_3 &\ge 0 \end{aligned}$$

Solution These constraints are sufficiently simple that a picture of the feasible region can easily be drawn. The bounding hyperplanes in this case are ordinary planes, and on the diagram, $x_1 \ge 0$ represents the region in front of the X_2X_3-plane, $x_2 \ge 0$ gives the region to the right of the X_1X_3-plane, and $x_3 \ge 0$ yields the region above the X_1X_2-plane. The fourth bounding hyperplane is

$$5x_1 + 2x_2 + 4x_3 = 20$$

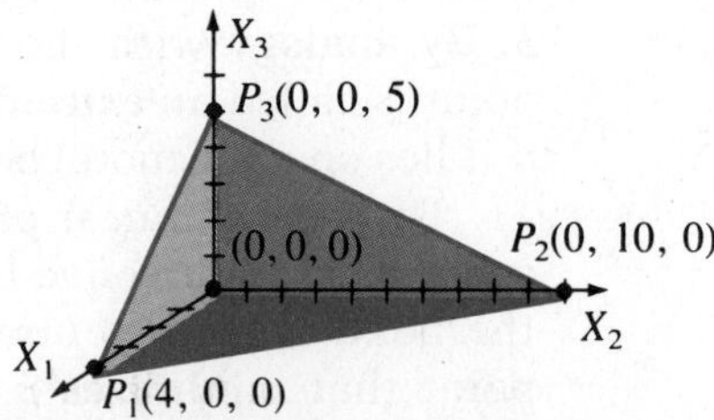

and the intersections of this plane with the X_1-, X_2-, and X_3-axes are plotted as P_1, P_2, and P_3, respectively. The constraint

$$5x_1 + 2x_2 + 4x_3 \le 20$$

determines the region below this plane, so the feasible region is the tetrahedron with vertices P_1, P_2, P_3, and the origin. Hence these are the extreme points. If the objective function P is evaluated at the extreme points, the results are

$$\begin{aligned} p &= 0 && \text{at the origin} \\ p &= 16 && \text{at } P_1(4,0,0) \\ p &= -30 && \text{at } P_2(0,10,0) \\ p &= 35 && \text{at } P_3(0,0,5) \end{aligned}$$

Because the feasible region is bounded, p has both a maximum and a minimum; they are $p = 35$ [at $P_3(0,0,5)$] and $p = -30$ [at $P_2(0,10,0)$].

The procedure we just used has two drawbacks. First, it is not always easy to determine whether a maximum exists. But even when this is known (if the feasible set is bounded, for example), the number of extreme points can be very large and the amount of computation required to find them can be excessive, even for a computer. This is why the method will not be pursued here.

A much more efficient procedure exists that reduces the number of extreme points that must be examined. The idea is quite simple. To get some insight into how it works, consider the general linear programming problem with three variables. Then the bounding hyperplanes are real planes, and the edges of the feasible region are the lines of intersection of pairs of bounding planes. Now suppose the objective function p is evaluated at some extreme point. Choose an edge emanating from this point along which the function p increases (or decreases if a minimum is desired). If no such edge exists, it can be shown that the extreme point gives the maximum. Otherwise there are two possibilities: (1) We encounter another extreme point on that edge at which p is larger. Then repeat the process. (2) There is no other vertex along this edge so p increases without bound, and no maximum exists. At each stage we discover there is no maximum, or we find the maximum, or we are led to another extreme point at which p is larger. Clearly the same extreme point cannot be encountered twice in this fashion (p increases), and so, because there are only finitely many extreme points, the process is effective: It either shows that no maximum exists, or, if there *is* a maximum, it leads to an extreme point yielding that maximum. Furthermore, in the types of problems usually found in practice, the method converges quickly.

This description of the algorithm is geometrical in nature. However, the whole thing can be cast in algebraic form. This will be described in the next section.

EXERCISES B.1

1. In each case, find the maximum and minimum values of p by finding the feasible region and examining p at the vertex points.

(a) $p = 3x_1 + 2x_2$
$$\begin{aligned} 3x_1 + x_2 &\le 9 \\ 2x_1 + 5x_2 &\le 10 \\ x_1 &\ge 0 \\ x_2 &\ge 0 \end{aligned}$$

(b) $p = 4x_1 + 3x_2$
$$\begin{aligned} x_2 &\le 5 \\ x_1 + x_2 &\le 8 \\ x_1 - x_2 &\le 2 \\ x_1 &\ge 0 \\ x_2 &\ge 0 \end{aligned}$$

(c) $p = 2x_1 - x_2$
$$\begin{aligned} x_1 + x_2 &\ge 2 \\ x_1 + 2x_2 &\le 14 \\ -x_1 + x_2 &\le 4 \\ x_1 &\le 6 \\ x_1 &\ge 0 \\ x_2 &\ge 0 \end{aligned}$$

(d) $p = x_2 - x_1$
$$\begin{aligned} 3x_1 + 2x_2 &\ge 12 \\ x_2 - x_1 &\le 6 \\ 2x_1 + 3x_2 &\le 38 \\ x_1 &\le 10 \\ x_1 &\ge 0 \\ x_2 &\ge 0 \end{aligned}$$

2. Consider the problem of maximizing $p = x_2 - 2x_1$ subject to:

$$\begin{aligned} -x_1 + x_2 &\le 6 \\ x_1 - 2x_2 &\le 0 \\ x_1 &\ge 0 \\ x_2 &\ge 0 \end{aligned}$$

Show that the feasible region is unbounded but that p still has a maximum.

3. Show that there are no points in the following feasible region.

$$\begin{aligned} x_1 + x_2 &\le 4 \\ x_2 - x_1 &\ge 4 \\ 3x_1 + x_2 &\ge 12 \end{aligned}$$

4. In Example 4, assume that small toys continue to earn a profit of \$18 per toy but that profits for large toys increase. Find the number of toys that should be produced to maximize profits in each of the following cases.

(a) Large-toy profit is \$23 per toy.

(b) Large-toy profit is \$24 per toy.

(c) Large-toy profit is \$25 per toy.

5. A man wishes to invest a portion of \$100,000 in two stocks A and B. He feels that at most \$70,000 should go into A and at most \$60,000 should go into B. If A and B pay 10% and 8% dividends, respectively, how much should he invest in each stock to maximize his dividends?

6. Repeat Exercise 5 where A pays 8% and B pays 10%.

7. A vitamin pill manufacturer uses two ingredients P and Q. The amounts of vitamins A, C, and D per gram of ingredient are given in the table. The ingredients are mixed with at least 85 grams of filler to make batches of 100 grams that are then pressed into pills. The law requires that each batch contain at least 12 units of vitamin A, at least 12 units of vitamin C, and at least 10 units of vitamin D. If P costs \$5 per gram and Q costs \$2 per gram, how many grams of each should be used per batch to minimize the cost?

	Vitamin		
	A	*C*	*D*
P	2 units	1 unit	5 units
Q	2 units	4 units	1 unit

8. Repeat Exercise 7 where P costs \$2 per gram and Q costs \$3 per gram.
9. An oil company produces two grades of heating oil, grade 1 and grade 2, and makes a profit of \$8 per barrel on grade 1 oil and \$5 per barrel on grade 2 oil. The refinery operates 100 hours per week and grade 1 oil takes $\frac{1}{4}$ hour per barrel to produce, whereas grade 2 oil takes only $\frac{1}{8}$ hour per barrel. The pipeline into the refinery can supply only enough crude to make 500 barrels of oil (either grade). Finally, warehouse constraints dictate that no more than 400 barrels of either type of oil can be produced per week. What production levels should be maintained to maximize profit?
10. Repeat Exercise 9 where the profits are \$7 on grade 1 oil and \$6 on grade 2 oil.
11. A small bakery makes white bread and brown bread in batches, and it has the capacity to make 8 batches per day. Each batch of white bread requires 1 unit of yeast, but the brown bread takes 2 units of yeast per batch. On the other hand, white bread costs \$30 per day for marketing, whereas brown bread only costs \$10 per day. If \$180 per day are available for marketing, and if 13 units of yeast are available per day, find the number of batches of each type that the bakery should make per day to maximize profits if it makes \$300 profit per batch of white bread and \$200 profit per batch of brown bread.

SECTION B.2 The Simplex Algorithm

This is a simple, straightforward method for solving linear programming problems that was discovered in the 1940s by George Dantzig. The idea is to identify certain "basic" feasible points and to prove that the maximum value (if it exists) of the objective function p occurs at one of these points. Then the algorithm proceeds roughly as follows: If a basic feasible point is at hand, a procedure is given for deciding whether it yields the maximum value of the objective function and, if not, how to find a basic feasible point that produces a larger value of the objective function. The process continues until a maximum is reached (or until it is established that there is no maximum).

We will develop the algorithm only for the **standard** linear programming problem:

Maximize the linear **objective function**

$$p = c_1x_1 + c_2x_2 + \cdots + c_nx_n$$

of the variables $x_1, x_2, \ldots, x_n$ subject to a finite collection of **constraints**

$$\begin{aligned} a_{11}x_1 + a_{12}x_2 + \cdots + a_{1n}x_n &\leq b_1 \\ a_{21}x_1 + a_{22}x_2 + \cdots + a_{2n}x_n &\leq b_2 \\ \vdots \qquad \vdots \qquad\qquad \vdots \quad &\quad \vdots \\ a_{m1}x_1 + a_{m2}x_2 + \cdots + a_{mn}x_n &\leq b_m \end{aligned}$$

Furthermore, the variables x_i and the constants b_j are all required to be nonnegative:

$$\begin{aligned} x_i &\geq 0 \qquad \text{for } i = 1, 2, \ldots, n \\ b_j &\geq 0 \qquad \text{for } j = 1, 2, \ldots, m \end{aligned}$$

The requirement that p be a linear function of the variables is vital (nonlinear programming is much more difficult), but the condition that the x_i be nonnegative and the fact that we are maximizing p (not minimizing) are not severe restrictions. On the other hand, the requirement that the constants b_j be nonnegative is a real restriction (although it is satisfied in many practical applications). We refer the reader to texts on linear programming for ways in which the algorithm can be used in the nonstandard case.

The various steps in the algorithm are best explained by working a specific example in detail.

PROTOTYPE EXAMPLE Maximize $p = 2x_1 + 3x_2 - x_3$ subject to:

$$\begin{aligned} x_1 + 2x_2 + 2x_3 &\leq 6 \\ 3x_1 - x_2 + x_3 &\leq 9 \\ 2x_1 + 3x_2 + 5x_3 &\leq 20 \\ x_i &\geq 0 \qquad \text{for } i = 1, 2, 3 \end{aligned}$$

The first step in the procedure is to convert the constraints from inequalities to equalities. This is achieved by introducing new variables x_4, x_5, and x_6 (called **slack variables**), one for each constraint. The new problem is to maximize $p = 2x_1 + 3x_2 - x_3 + 0x_4 + 0x_5 + 0x_6$ subject to:

$$\begin{aligned} x_1 + 2x_2 + 2x_3 + x_4 \qquad\qquad &= 6 \\ 3x_1 - x_2 + x_3 \qquad + x_5 \qquad &= 9 \\ 2x_1 + 3x_2 + 5x_3 \qquad\qquad + x_6 &= 20 \\ x_i \geq 0 \qquad \text{for } i = 1, 2, 3, 4, 5, 6 \end{aligned}$$

The claim is that if $(x_1,x_2,x_3,x_4,x_5,x_6)$ is a solution to this problem, then (x_1,x_2,x_3) is a solution to the original problem. In fact, the constraints are satisfied (because $x_4 \geq 0$, $x_5 \geq 0$, and $x_6 \geq 0$), so (x_1,x_2,x_3) is a feasible solution for the original problem. Moreover, if (x_1',x_2',x_3') were another feasible point for the original problem yielding a larger value of p, then taking

$$\begin{aligned} x_4' &= 6 - x_1' - 2x_2' - 2x_3' \\ x_5' &= 9 - 3x_1' + x_2' - x_3' \\ x_6' &= 20 - 2x_1' - 3x_2' - 5x_3' \end{aligned}$$

would give a feasible point $(x_1',x_2',x_3',x_4',x_5',x_6')$ for the new problem yielding a higher value of p, a contradiction.

So it suffices to solve the new problem. To do so, write the relationship $p = 2x_1 + 3x_2 - x_3$ as a fourth equation to get

$$\begin{aligned} x_1 + 2x_2 + 2x_3 + x_4 \qquad\qquad\qquad &= 6 \\ 3x_1 - x_2 + x_3 \qquad + x_5 \qquad\qquad &= 9 \\ 2x_1 + 3x_2 + 5x_3 \qquad\qquad + x_6 \qquad &= 20 \\ -2x_1 - 3x_2 + x_3 \qquad\qquad\qquad + p &= 0 \end{aligned}$$

This amounts to considering p as yet another variable. The augmented matrix (Section 1.2.1) for this system of equations is

$$\begin{array}{cccccccc} x_1 & x_2 & x_3 & \boxed{x_4} & \boxed{x_5} & \boxed{x_6} & p & \\ \hline \end{array}$$
$$\begin{bmatrix} 1 & 2 & 2 & 1 & 0 & 0 & 0 & 6 \\ 3 & -1 & 1 & 0 & 1 & 0 & 0 & 9 \\ 2 & 3 & 5 & 0 & 0 & 1 & 0 & 20 \\ -2 & -3 & 1 & 0 & 0 & 0 & 1 & 0 \end{bmatrix}$$

This is called the initial **simplex tableau** for the problem. The idea is to use elementary row operations to create a sequence of such tableaux (keeping p in the bottom row) that will lead to a solution. This is analogous to the modification of the augmented matrix in Gaussian elimination, except that here we allow only feasible solutions.

Note that the columns corresponding to the slack variables x_4, x_5, and x_6 all consist of zeros and a single 1 and that the 1's are in different rows (the way these variables were introduced guarantees this). These will be called **basic columns,** and the slack variables are called the **basic variables** in the initial tableau. (They are indicated by a box.) Because of this, there is one obvious solution to the equations: Set all the nonbasic variables equal to zero and solve for the basic variables: $x_4 = 6$, $x_5 = 9$, $x_6 = 20$ (and $p = 0$). In other words,

$$(0,0,0,6,9,20) \text{ is a feasible solution yielding } p = 0$$

Such a solution (with all nonbasic variables zero) is called a **basic feasible solution.** Note the role of the last column in all this: The numbers 6, 9, and 20 are the constants in the original constraints. It is the fact that these are *positive* that makes (0,0,0,6,9,20) a *feasible* solution. Also, the bottom entry in the last column is the value of p at the basic feasible solution (0 in this case).

The key to the whole algorithm is the following theorem. The proof is not difficult but would require some preliminary discussion of convex sets. Hence we omit it and refer the reader to texts on linear programming, such as S. I. Gass, *Linear Programming,* 4th ed. (New York: McGraw-Hill, 1975).

THEOREM 3 If a standard linear programming problem has a solution, then there is a basic feasible solution that yields the maximum value of the objective function. (Such basic feasible solutions are called **optimal**.)

Hence our goal is to find an optimal basic feasible point. Our construction using slack variables guarantees an initial basic feasible solution; the next step is to see whether it is optimal.

The bottom row of the initial tableau gives p in terms of the nonbasic variables (the original expression for p in this case):

$$p = 2x_1 + 3x_2 - x_3$$

The fact that some of the coefficients here are positive suggests that this value of p is *not* optimal, because increasing x_1 or x_2 at all will increase p. In fact, it would seem better to try to increase x_2; it has the larger of the two positive coefficients (equivalently the most *negative* entry in the last row of the tableau). This in turn suggests that we try to modify the tableau so that x_2 becomes a new basic variable. For this reason, x_2 is called the **entering** variable. Its column is called the **pivot column.**

This is accomplished by doing elementary row operations to convert the pivot column into a basic column. The question is where to locate the 1. We do not put the 1 in the last row because we do not want to disturb p, but it can be placed at any other location in the pivot column where the present entry is nonzero (they all qualify in this example). The entry chosen is called the **pivot**, and it is chosen as follows:

1. The pivot entry must be positive.
2. Among the positive entries available, the pivot is the one that produces the smallest ratio when divided into the right-most entry in its row.

These are chosen so that the basic feasible solution in the tableau we are creating will indeed be feasible. (The situation where no unique pivot entry is determined by conditions 1 and 2 will be discussed below.)

Returning to the prototype example, we have rewritten the initial tableau below and circled the pivotal entry. The ratios corresponding to the two positive entries in the pivot column (column 2 here) are shown at the right. No ratio is computed for row 2, because the corresponding entry in the pivot column is negative.

$$\begin{array}{cccccccc} x_1 & x_2 & x_3 & \boxed{x_4} & \boxed{x_5} & \boxed{x_6} & p & \\ \hline \end{array}$$
$$\begin{bmatrix} 1 & \textcircled{2} & 2 & 1 & 0 & 0 & 0 & 6 \\ 3 & -1 & 1 & 0 & 1 & 0 & 0 & 9 \\ 2 & 3 & 5 & 0 & 0 & 1 & 0 & 20 \\ -2 & -3 & 1 & 0 & 0 & 0 & 1 & 0 \end{bmatrix} \quad \begin{array}{l} \text{ratio: } 6/2 = 3 \\ \\ \text{ratio: } 20/3 = 6.7 \\ \\ \end{array}$$

Now do elementary row operations to convert the pivot to 1 and all other entries in its column to 0. The result is

$$\begin{array}{cccccccc} x_1 & \boxed{x_2} & x_3 & x_4 & \boxed{x_5} & \boxed{x_6} & p & \\ \hline \end{array}$$
$$\begin{bmatrix} \frac{1}{2} & 1 & 1 & \frac{1}{2} & 0 & 0 & 0 & 3 \\ \frac{7}{2} & 0 & 2 & \frac{1}{2} & 1 & 0 & 0 & 12 \\ \frac{1}{2} & 0 & 2 & \frac{-3}{2} & 0 & 1 & 0 & 11 \\ \frac{-1}{2} & 0 & 4 & \frac{3}{2} & 0 & 0 & 1 & 9 \end{bmatrix}$$

Note that the former basic variable x_4 is no longer basic (this is because it had a 1 in the same row as the pivot), and it is sometimes called the **departing** variable. The new basic variables are x_2, x_5, and x_6, and the new basic feasible solution (taking the new nonbasic variables equal to zero) is $x_2 = 3, x_5 = 12, x_6 = 11$, and $p = 9$. In other words,

$$(0,3,0,0,12,11) \text{ is a feasible solution yielding } p = 9$$

This is better than before; p has increased from 0 to 9.

Now repeat the process. The last row here yields

$$p = 9 + \frac{1}{2}x_1 - 4x_3 - \frac{3}{2}x_4$$

so there is still hope of increasing p by making x_1 basic (it has a positive coefficient). Hence the first column is the pivot column and all three entries (above the bottom row) are positive. The tableau is displayed once more below with the ratios given and the next pivot (with the smallest ratio) circled.

$$\begin{array}{ccccccc} x_1 & \boxed{x_2} & x_3 & x_4 & \boxed{x_5} & \boxed{x_6} & p \end{array}$$

$$\left[\begin{array}{cccccccc} \frac{1}{2} & 1 & 1 & \frac{1}{2} & 0 & 0 & 0 & 3 \\ \textcircled{\frac{7}{2}} & 0 & 2 & \frac{1}{2} & 1 & 0 & 0 & 12 \\ \frac{1}{2} & 0 & 2 & \frac{-3}{2} & 0 & 1 & 0 & 11 \\ \frac{-1}{2} & 0 & 4 & \frac{3}{2} & 0 & 0 & 1 & 9 \end{array}\right] \quad \begin{array}{l} \text{ratio: } \frac{3}{1/2} = 6 \\ \text{ratio: } \frac{12}{7/2} = \frac{24}{7} \\ \text{ratio: } \frac{11}{1/2} = 22 \\ \\ \end{array}$$

Row operations give the third tableau with x_1, x_2, and x_6 as basic variables.

$$\begin{array}{ccccccc} \boxed{x_1} & \boxed{x_2} & x_3 & x_4 & x_5 & \boxed{x_6} & p \end{array}$$

$$\left[\begin{array}{cccccccc} 0 & 1 & \frac{5}{7} & \frac{3}{7} & \frac{-1}{7} & 0 & 0 & \frac{9}{7} \\ 1 & 0 & \frac{4}{7} & \frac{1}{7} & \frac{2}{7} & 0 & 0 & \frac{24}{7} \\ 0 & 0 & \frac{12}{7} & \frac{-11}{7} & \frac{-1}{7} & 1 & 0 & \frac{65}{7} \\ 0 & 0 & \frac{30}{7} & \frac{11}{7} & \frac{1}{7} & 0 & 1 & \frac{75}{7} \end{array}\right]$$

The corresponding basic feasible solution (setting $x_3 = x_4 = x_5 = 0$) is $x_1 = 24/7, x_2 = 9/7, x_6 = 65/7$, and $p = 75/7$. In other words,

$$\left(\frac{24}{7},\frac{9}{7},0,0,0,\frac{65}{7}\right) \text{ is a feasible solution yielding } p = \frac{75}{7}$$

However, we claim that this is optimal. The last row of the third tableau gives

$$p = \frac{75}{7} - \frac{30}{7}x_3 - \frac{11}{7}x_4 - \frac{1}{7}x_5$$

so because x_3, x_4, and x_5 are nonnegative, p can be no greater than 75/7. The above solution achieves $p = 75/7$, so it must be optimal. This completes the solution of the prototype example.

Note that the test for optimality is clear: If the last row of a tableau has only *positive* entries, then the corresponding basic feasible solution is optimal. If not, the column corresponding to the *most negative* entry in the last row is the pivot column, the corresponding variable is the entering variable, and a new tableau is constructed with that variable as a new basic variable.

Of course not every standard linear programming problem has a solution (see Exercise 1). It can happen that feasible points can be found that make the objective function p as large as we wish. In this case p has no maximum and we say p is **unbounded.** The algorithm provides a way to determine whether this is the case (step 3 below).

The algorithm works in exactly the same way for any standard linear programming problem. Suppose that

$$p = c_1x_1 + c_2x_2 + \cdots + c_nx_n$$

is to be maximized subject to the following m constraints:

$$\begin{array}{l} a_{11}x_1 + a_{12}x_2 + \cdots + a_{1n}x_n \le b_1 \\ a_{21}x_1 + a_{22}x_2 + \cdots + a_{2n}x_n \le b_2 \\ \quad\vdots \qquad\quad \vdots \qquad\qquad\quad \vdots \qquad\quad \vdots \\ a_{m1}x_1 + a_{m2}x_2 + \cdots + a_{mn}x_n \le b_m \end{array} \qquad \begin{array}{l} x_i \ge 0 \text{ for } i = 1, 2, \ldots, n \\ b_j \ge 0 \text{ for } j = 1, 2, \ldots, m \end{array}$$

Introduce m slack variables $x_{n+1}, \ldots, x_{n+m}$ to make the inequalities into equalities. The new problem is to maximize

$$p = c_1x_1 + c_2x_2 + \cdots + c_nx_n + 0x_{n+1} + \cdots + 0x_{n+m}$$

subject to:

$$\begin{array}{lr} a_{11}x_1 + a_{12}x_2 + \cdots + a_{1n}x_n + x_{n+1} & = b_1 \\ a_{21}x_1 + a_{22}x_2 + \cdots + a_{2n}x_n + \cdots + x_{n+2} & = b_2 \\ \quad\vdots \qquad\quad \vdots \qquad\qquad\quad \vdots & \vdots \\ a_{m1}x_1 + a_{m2}x_2 + \cdots + a_{mn}x_n + \cdots \qquad + x_{n+m} & = b_m \end{array}$$

$$x_i \geq 0 \quad \text{for } i = 1, 2, \ldots, n+m$$
$$b_j \geq 0 \quad \text{for } j = 1, 2, \ldots, m$$

Any optimal solution $(x_1, \ldots, x_n, x_{n+1}, \ldots, x_{n+m})$ to this new problem yields an optimal solution $(x_1, \ldots, x_n)$ to the original problem (the argument in the prototype example works), so we solve the new problem. Because the constraints are now equations, some of the methods of Gaussian elimination apply. For convenience, write the expression for p as another equation:

$$-c_1x_1 - c_2x_2 - \cdots - c_nx_n + \cdots \qquad + p = 0$$

The augmented matrix for this larger system of equations is

$$\begin{array}{cccccccccc} x_1 & x_2 & \cdots & x_n & x_{n+1} & x_{n+2} & & x_{n+m} & p & \\ \hline \end{array}$$
$$\left[\begin{array}{cccccccccc} a_{11} & a_{12} & \cdots & a_{1n} & 1 & 0 & \cdots & 0 & 0 & b_1 \\ a_{21} & a_{22} & \cdots & a_{2n} & 0 & 1 & \cdots & 0 & 0 & b_2 \\ \vdots & \vdots & & \vdots & \vdots & \vdots & & \vdots & \vdots & \vdots \\ a_{m1} & a_{m2} & & a_{mn} & 0 & 0 & \cdots & 1 & 0 & b_m \\ -c_1 & -c_2 & & -c_n & 0 & 0 & \cdots & 0 & 1 & 0 \end{array}\right]$$

and this is called the initial **simplex tableau** for the problem. The slack variables are called **basic variables** because their columns are **basic columns** (all entries are zero except a single one in each column). The fact that $b_j \geq 0$ holds for each j means that we can obtain a feasible solution by setting all the nonbasic variables equal to zero. The result:

$(0, 0, \ldots, 0, b_1, b_2, \ldots, b_m)$ is a feasible solution yielding $p = 0$

This is the initial basic feasible solution. Of course, it may not be optimal.

Now the algorithm starts. Suppose a tableau has been constructed with m basic variables (that is, m variables corresponding to basic columns in the tableau) such that the last column contains no negative entries (for example, the initial tableau). Then, if the nonbasic variables are set equal to zero, the values of the basic variables are determined (they are just the entries of the last column, the value of p being the last entry). Hence this gives the **basic feasible solution** corresponding to the tableau.

The actual execution of the algorithm is best described in the following steps. For convenience, we display them first as a flow chart.

THE SIMPLEX ALGORITHM

STEP 0
Prepare the initial tableau.

↓

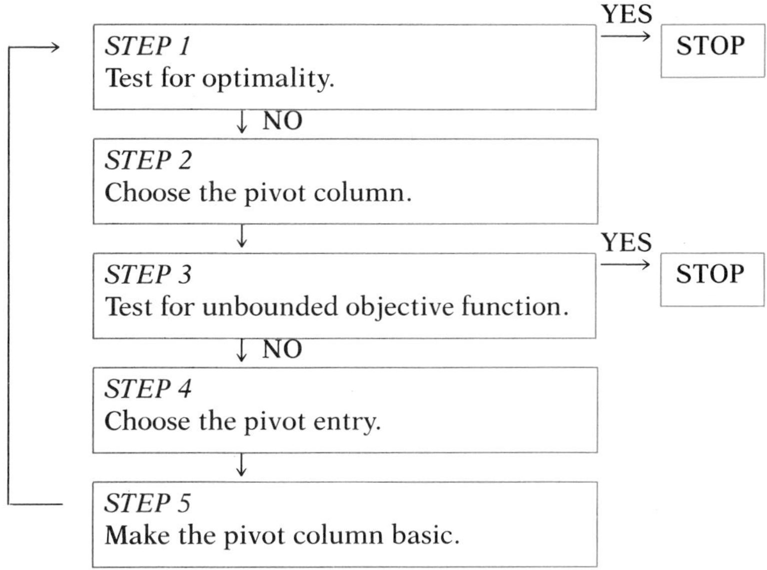

The details of the steps are as follows. As above, we assume that there are n variables and m constraints.

STEP 0. Prepare the initial tableau.
Introduce slack variables, one for each constraint, and convert each inequality into an equation. Then write the expression for p as another equation (as above). The augmented matrix for the resulting system of $m + 1$ equations (the equation for p last) is the initial tableau.

STEP 1. Test for optimality.
Given a tableau, the corresponding basic feasible solution is optimal if no entry in the last row (except the last) is negative. (The argument in the prototype example works.) In this case, stop—the maximum value of p is the lower right entry. Otherwise go on to step 2.

STEP 2. Choose the pivot column.
This is the column (not the last) whose bottom entry is the most negative (the worst offender, as it were). If there is a tie, choose either possibility.

STEP 3. Test for unbounded objective function.
This occurs if no entry in the pivot column is positive. (We omit the proof.) In this case, stop—the objective function has no maximum. Otherwise go on to step 4.

STEP 4. Choose the pivot entry.
Among the positive entries in the pivot column, choose the one that has the smallest ratio when divided into the last entry in its row. If

two ratios are equal, choose either. (This may lead to cycling; see Remark 2 below.)

STEP 5. Make the pivot column basic.
Use elementary row operations to make the pivot entry 1 and every other entry in the pivot column (including the last) zero.

EXAMPLE 6 Maximize $p = 3x_1 + x_2 + 2x_3$ subject to:

$$2x_1 - x_2 + 3x_3 \leq 2$$
$$3x_1 + x_2 + x_3 \leq 5 \qquad x_i \geq 0 \text{ for } i = 1, 2, 3$$

Solution Introduce slack variables x_4 and x_5, and rewrite the equation for p.

$$\begin{aligned} 2x_1 - x_2 + 3x_3 + x_4 \qquad &= 2 \\ 3x_1 + x_2 + x_3 \qquad + x_5 &= 5 \qquad x_i \geq 0 \text{ for } i = 1, 2, 3, 4, 5 \\ -3x_1 - x_2 - 2x_3 \qquad + p &= 0 \end{aligned}$$

Hence the initial tableau (with the basic variables boxed) is

$$\begin{array}{ccccccc} x_1 & x_2 & x_3 & \boxed{x_4} & \boxed{x_5} & p & \\ \hline \end{array}$$
$$\left[\begin{array}{ccccccc} \textcircled{2} & -1 & 3 & 1 & 0 & 0 & 2 \\ 3 & 1 & 1 & 0 & 1 & 0 & 5 \\ -3 & -1 & -2 & 0 & 0 & 1 & 0 \end{array}\right] \begin{array}{l} \text{ratio: } 2/2 = 1 \\ \text{ratio: } 5/3 \\ \\ \end{array}$$

The basic feasible solution here is not optimal (the last row has negative entries), and the pivot column is the first (-3 is the most negative). The ratios are computed as above and the pivot is circled. Hence row operations give the next tableau (the new basic variables are boxed).

$$\begin{array}{ccccccc} \boxed{x_1} & x_2 & x_3 & x_4 & \boxed{x_5} & p & \\ \hline \end{array}$$
$$\left[\begin{array}{ccccccc} 1 & \frac{-1}{2} & \frac{3}{2} & \frac{1}{2} & 0 & 0 & 1 \\ 0 & \textcircled{\frac{5}{2}} & \frac{-7}{2} & \frac{-3}{2} & 1 & 0 & 2 \\ 0 & \frac{-5}{2} & \frac{5}{2} & \frac{3}{2} & 0 & 1 & 3 \end{array}\right]$$

This is still not optimal, and the pivot column is the second. Here the pivot is the only positive entry, so no ratios need be computed. Row operations give

$$\begin{array}{ccccccc} \boxed{x_1} & \boxed{x_2} & x_3 & x_4 & x_5 & p & \\ \hline \end{array}$$

$$\begin{bmatrix} 1 & 0 & \textcircled{\frac{4}{5}} & \frac{1}{5} & \frac{1}{5} & 0 & \frac{7}{5} \\ 0 & 1 & \frac{-7}{5} & \frac{-3}{5} & \frac{2}{5} & 0 & \frac{4}{5} \\ 0 & 0 & -1 & 0 & 1 & 1 & 5 \end{bmatrix}$$

Again no optimal solution exists, but p has increased to 5 (lower right entry). The next tableau is

$$\begin{array}{ccccccc} x_1 & \boxed{x_2} & \boxed{x_3} & x_4 & x_5 & p & \\ \hline \end{array}$$

$$\begin{bmatrix} \frac{5}{4} & 0 & 1 & \frac{1}{4} & \frac{1}{4} & 0 & \frac{7}{4} \\ \frac{7}{4} & 1 & 0 & \frac{-1}{4} & \frac{3}{4} & 0 & \frac{13}{4} \\ \frac{5}{4} & 0 & 0 & \frac{1}{4} & \frac{5}{4} & 1 & \frac{27}{4} \end{bmatrix}.$$

Hence p has a maximum of 27/4 when $x_1 = 0$, $x_2 = 13/4$, and $x_3 = 7/4$.

We conclude with two remarks on the algorithm.

Remark 1: Rationale for selecting the pivot entry (step 4).

Suppose the rth column is the pivot column. Write it as

$$\begin{bmatrix} t_{1r} \\ \vdots \\ t_{qr} \\ \vdots \\ t_{mr} \\ s_r \end{bmatrix}$$

where $s_r < 0$ (from step 2) and at least one other entry is positive. Suppose we decide to take t_{qr} as the pivot. Write its row as follows

$$(t_{q1}, t_{q2}, \ldots, t_{qr}, \ldots, 0, d_q)$$

where the last entry $d_q \geq 0$ (because the tableau produced a feasible solution when all nonbasic variables were set equal to zero, so d_q is the value of one of the basic variables). We want to divide this row by t_{qr} (so $t_{qr} \neq 0$), and the last entry of the new row, d_q/t_{qr}, is required to be nonnegative too (we want a new *tableau*). Hence the pivot t_{qr} must be *positive.*

Then we convert each other entry t_{ir} in the pivot column to zero by subtracting t_{ir} times the pivot row. If the right entry of row i is d_i, the new right entry is

$$d_i - t_{ir}\left[\frac{d_q}{t_{qr}}\right] = t_{ir}\left[\frac{d_i}{t_{ir}} - \frac{d_q}{t_{qr}}\right]$$

This is clearly positive if t_{ir} is negative or zero (use the left side). If $t_{ir} > 0$, it is positive if the ratio d_q/t_{qr} for the pivot is less than the ratio for t_{ir}. This shows why we choose the pivot with minimal ratio (in step 4).

Remark 2: Degeneracy and cycling.

If two ratios in step 4 are equal, the argument in Remark 1 shows that in the next tableau, some entry in the last column will be zero (so some basic variable will take the value 0). In this case the algorithm is said to **degenerate,** and it may lead to cycling (that is, the sequence of basic feasible solutions we are creating may contain the same solution twice and so continue to loop indefinitely). This is rare in practical problems (computer round-off error tends to eliminate it), and algorithms exist to deal with it.

EXERCISES B.2

1. Consider the following standard linear programming problem: Maximize $p = x_1 + x_2$ subject to:

$$\begin{aligned} -x_1 + x_2 &\leq 1 \\ x_1 - 2x_2 &\leq 2 \\ x_1 \geq 0, x_2 &\geq 0 \end{aligned}$$

(a) Using the methods of Section B.1, show that p is unbounded over the feasible region.

(b) Use the simplex algorithm to arrive at the same conclusion.

2. In each case, maximize p subject to the given constraints, and find values of the x_i that yield the maximum. Assume $x_i \geq 0$ for all i.

(a) $p = x_1 + 2x_2 + 3x_3$

$$\begin{aligned} 3x_1 - x_2 - x_3 &\leq 3 \\ x_1 + x_2 + 2x_3 &\leq 2 \end{aligned}$$

(b) $p = 2x_1 + x_2 + x_3$

$$\begin{aligned} 3x_1 + x_2 + 2x_3 &\leq 2 \\ x_1 + x_2 + 3x_3 &\leq 5 \end{aligned}$$

(c) $p = x_1 + 2x_2$

$$3x_1 + x_2 \leq 4$$

(d) $p = 3x_1 + 2x_2$

$$4x_1 + 3x_2 \leq 5$$

$$x_1 + 2x_2 \le 3$$
$$2x_1 + 3x_2 \le 5$$

$$x_1 + x_2 \le 2$$
$$3x_1 + 4x_2 \le 4$$

(e) $p = 3x_1 + x_2 + 2x_3$
$$2x_1 + x_2 - x_3 \le 3$$
$$x_1 + x_2 + x_3 \le 4$$
$$x_1 + 2x_2 + x_3 \le 5$$

(f) $p = 2x_1 + 3x_2 + 2x_3$
$$2x_1 + 3x_2 - x_3 \le 4$$
$$x_1 - x_2 + x_3 \le 2$$
$$3x_1 + 4x_2 + x_3 \le 5$$

(g) $p = x_1 + x_2 + 2x_3 + x_4$
$$3x_1 + x_2 - x_3 + 2x_4 \le 5$$
$$x_1 + 2x_2 + x_3 - x_4 \le 4$$

(h) $p = 2x_1 + x_2 + 3x_3 + 2x_4$
$$3x_1 + 4x_2 + x_3 - 2x_4 \le 6$$
$$2x_1 + 3x_2 - x_3 + 3x_4 \le 5$$

3. Can the maximum of the objective function ever be negative in a standard linear programming problem? Explain.

4. Suppose a standard linear programming problem has more variables than constraints. If the objective function has a maximum, show that this must have at least one optimal solution with one original variable zero.

5. An automobile company makes three types of cars: compact, sports, and full-size; the profits per unit are \$500, \$700, and \$600, respectively. Transportation costs per vehicle are \$300, \$400, and \$500, respectively. And labor costs are \$500, \$500, and \$400, respectively. If the total transportation cost is not to exceed \$40,000 and the total labor cost is not to exceed \$30,000, find the maximum profit.

6. A short-order restaurant sells three dinners—regular, diet, and super—on which it makes profits of \$1.00, \$2.00, and \$1.50, respectively. The restaurant cannot serve more than 300 dinners daily. The three dinners require 2, 4, and 2 minutes to prepare, and at most 1000 minutes of preparation time are available per day. The dinners require 2, 0, and 3 minutes to cook, and 1000 minutes of cooking time are available daily. Finally, the dinners require 50, 300, and 100 grams of fresh produce, and this commodity is limited to 45 kilograms daily. Find the numbers of dinners of each type that will maximize profits.

7. A trucking company has 100 trucks, which are dispatched from three locations: A, B, and C. Each truck at A, B, and C uses 40, 30, and 30 units of fuel daily, and 2500 units per day are available. The costs of labor to operate and maintain each truck are \$70, \$80, and \$70 per truck per day at the three locations, and \$8000 per day is the maximum that the company can pay for labor. How many trucks should be allocated to each location if the daily profits per truck are \$300, \$250, and \$200 at locations A, B, and C?

8. A lawn mower company makes three models: standard, deluxe, and super. The construction of each mower involves three stages: motor construction, frame construction, and final assembly. The following table gives the number of hours of labor required per mower for each stage and the total number of hours of labor available per week for each stage. It also gives the profit per week. Find the weekly production schedule that maximizes profit.

	Standard	*Deluxe*	*Super*	*Hours Available*
motor	1	1	2	2500
frame	1	2	2	2000
assembly	1	1	1	1800
PROFIT	$30	$40	$55	

APPENDIX C Mathematical Induction

Suppose one is presented with the following sequence of equations:

$$
\begin{aligned}
1 &= 1 \\
1 + 3 &= 4 \\
1 + 3 + 5 &= 9 \\
1 + 3 + 5 + 7 &= 16 \\
1 + 3 + 5 + 7 + 9 &= 25
\end{aligned}
$$

It is clear that there is a pattern. The numbers on the right side of the equations are the squares $1^2, 2^2, 3^2, 4^2$, and 5^2, and, in the equation with n^2 on the right side, the left side is the sum of the first n odd numbers. The odd numbers are

$$
\begin{aligned}
1 &= 2 \cdot 1 - 1 \\
3 &= 2 \cdot 2 - 1 \\
5 &= 2 \cdot 3 - 1 \\
7 &= 2 \cdot 4 - 1 \\
9 &= 2 \cdot 5 - 1
\end{aligned}
$$

and from this it is clear that the nth odd number is $2n - 1$. Hence, at least for $n = 1, 2, 3, 4$, or 5, the following is true:

$$1 + 3 + \cdots + (2n - 1) = n^2 \qquad (S_n)$$

The question arises whether the statement S_n is true for *every* n. There is no hope of separately verifying all these statements because there are infinitely many of them. A more subtle approach is required.

The idea is as follows: Suppose it is verified that the statement S_{n+1} will be true whenever S_n is true. That is, suppose we prove that, *if* S_n is true, then it necessarily follows that S_{n+1} is also true. Then, if we can show that S_1 is true, it follows that S_2 is true, and from this that S_3 is true, hence that S_4 is true, and so on and on. This is the principle of induction. In order to express it more compactly, it is useful to have a short way to express the assertion "If S_n is true, then S_{n+1} is true." We write this assertion as

$$S_n \Rightarrow S_{n+1}$$

and read it as "S_n implies S_{n+1}." We can now state the principle of mathematical induction.

THE PRINCIPLE OF MATHEMATICAL INDUCTION

Suppose S_n is a statement about the natural number n for each $n = 1, 2, 3, \ldots$. Suppose further that

(1) S_1 is true.

(2) $S_n \Rightarrow S_{n+1}$ for every $n \geq 1$.

Then S_n is true for every $n \geq 1$.

This is one of the most useful techniques in all of mathematics. It applies in a very wide number of situations. The following examples will illustrate this.

EXAMPLE 1 Show that $1 + 2 + \cdots + n = \frac{1}{2}n(n + 1)$ for $n \geq 1$.

Solution Let S_n be the statement: $1 + 2 + \cdots + n = \frac{1}{2}n(n + 1)$. We apply induction.

1. S_1 is true. The statement S_1 is $1 = \frac{1}{2}1(1 + 1)$, which is true.
2. $S_n \Rightarrow S_{n+1}$. We *assume* that S_n is true for some $n \geq 1$; that is, that $1 + 2 + \cdots + n = \frac{1}{2}n(n + 1)$.

We must prove that

$$S_{n+1}\text{:} \quad 1 + 2 + \cdots + (n + 1) = \frac{1}{2}(n + 1)(n + 2)$$

is also true, and we are entitled to use S_n to do so. Now the left side of S_{n+1} is the sum of the first $n + 1$ positive integers. Hence the second-to-last term is n, so we can write

$$\begin{aligned} 1 + 2 + \cdots + (n + 1) &= (1 + 2 + \cdots + n) + (n + 1) \\ &= \frac{1}{2}n(n + 1) + (n + 1) \qquad \text{using } S_n \\ &= \frac{1}{2}(n + 1)[n + 2] \end{aligned}$$

This shows that S_{n+1} is true and so completes the induction.

In the verification that $S_n \Rightarrow S_{n+1}$, we *assume* that S_n is true and use it to deduce that S_{n+1} is true. The assumption that S_n is true is sometimes called the **induction hypothesis**.

EXAMPLE 2 If x is any number such that $x \neq 1$, show that $1 + x + x^2 + \cdots + x^n = \dfrac{x^{n+1} - 1}{x - 1}$ for $n \geq 1$.

Solution Let S_n be the statement: $1 + x + x^2 + \cdots + x^n = \dfrac{x^{n+1} - 1}{x - 1}$.

1. S_1 is true. S_1 reads $1 + x = \dfrac{x^2 - 1}{x - 1}$, which is true because $x^2 - 1 = (x - 1)(x + 1)$.
2. $S_n \Rightarrow S_{n+1}$. Assume the truth of S_n: $1 + x + x^2 + \cdots + x^n = \dfrac{x^{n+1} - 1}{x - 1}$.

We must *deduce* from this the truth of S_{n+1}: $1 + x + x^2 + \cdots + x^{n+1} = \dfrac{x^{n+2} - 1}{x - 1}$. Starting with the left side of S_{n+1}, we find

$$\begin{aligned} 1 + x + x^2 + \cdots + x^{n+1} &= (1 + x + x^2 + \cdots + x^n) + x^{n+1} \\ &= \frac{x^{n+1} - 1}{x - 1} + x^{n+1} \\ &= \frac{x^{n+1} - 1 + x^{n+1}(x - 1)}{x - 1} \\ &= \frac{x^{n+2} - 1}{x + 1} \end{aligned}$$

This shows that S_{n+1} is true and so completes the induction.

Both of these examples involve formulas for a certain sum, and it is often convenient to use summation notation. For example, $\sum_{k=1}^{n} (2k - 1)$ means that in the expression $(2k - 1)$, k is to be given the values $k = 1$, $k = 2$, $k = 3$, . . . , $k = n$, and then the resulting n numbers are to be added. The same thing applies to other expressions involving k. For example,

$$\sum_{k=1}^{n} k^3 = 1^3 + 2^3 + \cdots + n^3$$

$$\sum_{k=1}^{5} (3k - 1) = (3 \cdot 1 - 1) + (3 \cdot 2 - 1) + (3 \cdot 3 - 1) + (3 \cdot 4 - 1) + (3 \cdot 5 - 1)$$

The next example involves this notation.

EXAMPLE 3 Show that $\sum_{k=1}^{n}(3k^2 - k) = n^2(n + 1)$ for each $n \geq 1$.

Solution Let S_n be the statement: $\sum_{k=1}^{n}(3k^2 - k) = n^2(n + 1)$.

1. S_1 is true. S_1 reads $(3 \cdot 1^2 - 1) = 1^2(1 + 1)$, which is true.
2. $S_n \Rightarrow S_{n+1}$. Assume that S_n is true. We must prove S_{n+1}:

$$\begin{aligned}\sum_{k=1}^{n+1}(3k^2 - k) &= \sum_{k=1}^{n}(3k^2 - k) + [3(n + 1)^2 - (n + 1)] \\ &= n^2(n + 1) + (n + 1)[3(n + 1) - 1] \\ &= (n + 1)[n^2 + 3n + 2] \\ &= (n + 1)[(n + 1)(n + 2)] \\ &= (n + 1)^2(n + 2)\end{aligned}$$

This proves that S_{n+1} is true.

We now turn to examples wherein induction is used to prove propositions that do not involve sums.

EXAMPLE 4 Show that $7^n + 2$ is a multiple of 3 for all $n \geq 1$.

Solution

1. S_1 is true. $7^1 + 2 = 9$ is a multiple of 3.
2. $S_n \Rightarrow S_{n+1}$. Assume that $7^n + 2$ is a multiple of 3 for some $n \geq 1$; say, $7^n + 2 = 3m$ for some integer m. Then

$$7^{n+1} + 2 = 7(7^n) + 2 = 7(3m - 2) + 2 = 21m - 12 = 3(7m - 4)$$

so $7^{n+1} + 2$ is also a multiple of 3. This proves that S_{n+1} is true.

In all the foregoing examples, we have used the principle of induction starting at 1; that is, we have verified that S_1 is true and that $S_n \Rightarrow S_{n+1}$ for each $n \geq 1$, and then we have concluded that S_n is true for every $n \geq 1$. But there is nothing special about 1 here: If m is some fixed integer and we verify that

1. S_m is true.
2. $S_n \Rightarrow S_{n+1}$ for every $n \geq m$.

then it follows that S_n is true for every $n \geq m$. This "extended" induction principle is just as plausible as the induction principle and can, in fact, be proved by induction. The next example will illustrate it. Recall that if n is a positive integer, the number $n!$ (which is read "n-factorial") is the product

$$n! = n(n-1)(n-2)\ldots 3\cdot 2\cdot 1$$

of all the numbers from n to 1. Thus $2! = 2$, $3! = 6$, and so on.

EXAMPLE 5 Show that $2^n < n!$ for all $n \geq 4$.

Solution Observe that $2^n < n!$ is actually false if $n = 1, 2, 3$.

1. S_4 is true. $2^4 = 16 < 24 = 4!$.
2. $S_n \Rightarrow S_{n+1}$ if $n \geq 4$. Assume that S_n is true; that is $2^n < n!$. Then

$$\begin{aligned} 2^{n+1} &= 2\cdot 2^n \\ &< 2\cdot n! && \text{because } 2^n < n! \\ &< (n+1)n! && \text{because } 2 < n+1 \\ &= (n+1)! \end{aligned}$$

Hence S_{n+1} is true.

EXERCISES

In Exercises 1–19, prove the given statement by induction for all $n \geq 1$.

1. $1 + 3 + 5 + 7 + \cdots + (2n-1) = n^2$

2. $1^2 + 2^2 + \cdots + n^2 = \frac{1}{6}n(n+1)(2n+1)$

3. $1^3 + 2^3 + \cdots + n^3 = (1 + 2 + \cdots + n)^2$

4. $1\cdot 2 + 2\cdot 3 + \cdots + n(n+1) = \frac{1}{3}n(n+1)(n+2)$

5. $1\cdot 2^2 + 2\cdot 3^2 + \cdots + n(n+1)^2 = \frac{1}{12}n(n+1)(n+2)(3n+5)$

6. $\frac{1}{1\cdot 2} + \frac{1}{2\cdot 3} + \cdots + \frac{1}{n(n+1)} = \frac{n}{n+1}$

7. $1^2 + 3^2 + \cdots + (2n-1)^2 = \frac{n}{3}(4n^2 - 1)$

8. $\frac{1}{1\cdot 2\cdot 3} + \frac{1}{2\cdot 3\cdot 4} + \cdots + \frac{1}{n(n+1)(n+2)} = \frac{n(n+3)}{4(n+1)(n+2)}$

9. $1 + 2 + 2^2 + \cdots + 2^{n-1} = 2^n - 1$

10. $3 + 3^3 + 3^5 + \cdots + 3^{2n-1} = \frac{3}{8}(9^n - 1)$

11. $\frac{1}{1^2} + \frac{1}{2^2} + \cdots + \frac{1}{n^2} \leq 2 - \frac{1}{n}$

12. $n < 2^n$

13. For any integer $m > 0$, $m!\, n! < (m+n)!$

14. $\frac{1}{1^2} + \frac{1}{2^2} + \cdots + \frac{1}{n^2} \le 2 - \frac{1}{n}$

15. $\frac{1}{\sqrt{1}} + \frac{1}{\sqrt{2}} + \cdots + \frac{1}{\sqrt{n}} \ge \sqrt{n}$

16. $n^3 + (n + 1)^3 + (n + 2)^3$ is a multiple of 9.

17. $5^n + 3$ is a multiple of 4.

18. $n^3 - n$ is a multiple of 3.

19. $3^{2n+1} + 2^{n+2}$ is a multiple of 7.

20. Let $A_n = \left(1 - \frac{1}{2}\right)\left(1 - \frac{1}{3}\right)\left(1 - \frac{1}{4}\right)\cdots\left(1 - \frac{1}{n}\right)$. Find a formula for A_n and prove it.

21. Let $B_n = 1 \cdot 1! + 2 \cdot 2! + 3 \cdot 3! + \cdots + n \cdot n!$ Find a formula for B_n and prove it.

22. Suppose S_n is a statement about n for each $n \ge 1$. Explain what must be done to prove that S_n is true for all $n \ge 1$ if:
    (a) It is known that $S_n \Rightarrow S_{n+2}$ for each $n \ge 1$.
    (b) It is known that $S_n \Rightarrow S_{n+8}$ for each $n \ge 1$.
    (c) It is known that $S_n \Rightarrow S_{n+1}$ for each $n \ge 10$.
    (d) It is known that both S_n and $S_{n+1} \Rightarrow S_{n+2}$ for each $n \ge 1$.

23. If S_n is a statement for each $n \ge 1$, argue that S_n is true for all $n \ge 1$ if it is known that the following two conditions hold:
    (a) $S_n \Rightarrow S_{n-1}$ for each $n \ge 2$.
    (b) S_n is true for infinitely many values of n.

24. Suppose a sequence $a_1, a_2, \ldots$ of numbers is given that satisfies:
    (a) $a_1 = 2$ (b) $a_{n+1} = 2a_n$ for each $n \ge 1$
    Formulate a theorem giving a_n in terms of n, and prove your result by induction.

25. Suppose a sequence $a_1, a_2, \ldots$ of numbers is given that satisfies:
    (a) $a_1 = b$ (b) $a_{n+1} = ca_n + b$ for $n = 1, 2, 3, \ldots$
    Formulate a theorem giving a_n in terms of n, and prove your result by induction.

26. (a) Show that $n^2 \le 2^n$ for all $n \ge 4$.
    (b) Show that $n^3 \le 2^n$ for all $n \ge 10$.

Selected Answers

Exercises 1.1 (Page 7)

2. (b) $\begin{bmatrix} t \\ \frac{1}{3}(1-2t) \end{bmatrix}$ or $\begin{bmatrix} \frac{1}{2}(1-3s) \\ s \end{bmatrix}$ **(d)** $\begin{bmatrix} 1+2s-5t \\ s \\ t \end{bmatrix}$ or $\begin{bmatrix} q \\ p \\ \frac{1}{5}(1-p+2q) \end{bmatrix}$

3. $\begin{bmatrix} \frac{5}{2} \\ t \end{bmatrix}$ **4.** $\begin{bmatrix} \frac{1}{4}(3+2s) \\ s \\ t \end{bmatrix}$ **5. (a)** No solution if $b \neq 0$. If $b = 0$, *any* x is a solution. **(b)** $x = \dfrac{b}{a}$

9. No. If the corresponding planes are parallel and distinct, there is no solution. Otherwise they either coincide or have a whole common line of solutions. **10.** $x' = 5, y' = 1$, so $x = 23, y = -32$

Exercises 1.2.1 (Page 16)

1. (b) $\begin{bmatrix} 1 & 2 & 0 \\ 0 & 1 & 1 \end{bmatrix}$ **(d)** $\begin{bmatrix} 1 & 1 & 0 & 1 \\ 0 & 1 & 1 & 0 \\ -1 & 0 & 1 & 2 \end{bmatrix}$ **2. (b)** $\begin{aligned} 2x - y \quad\quad &= -1 \\ -3x + 2y + z &= 0 \\ y + z &= 3 \end{aligned}$ or $\begin{aligned} 2x_1 - x_2 \quad\quad &= -1 \\ -3x_1 + 2x_2 + x_3 &= 0 \\ x_2 + x_3 &= 3 \end{aligned}$

3. (b) $\begin{bmatrix} -\frac{1}{7} \\ -\frac{3}{7} \end{bmatrix}$ **(d)** $\begin{bmatrix} \frac{1}{3}(t+2) \\ t \end{bmatrix}$ **(f)** no solution **4. (b)** $\begin{bmatrix} -15t-21 \\ -11t-17 \\ t \end{bmatrix}$ **(d)** no solution

(f) $\begin{bmatrix} -7 \\ -9 \\ 1 \end{bmatrix}$ **(h)** $\begin{bmatrix} 4 \\ 3+2t \\ t \end{bmatrix}$ **5. (b)** $\begin{bmatrix} 2t+8 \\ -t-19 \\ t \end{bmatrix}$ $R_3 = 5R_1 - 4R_2$

6. (b) If $ab \neq 2$, unique solution $\begin{bmatrix} \dfrac{-2-5b}{2-ab} \\ \dfrac{a+5}{2-ab} \end{bmatrix}$. If $ab = 2$: no solution if $a \neq -5$;

if $a = -5$, the solutions are $\begin{bmatrix} -1 + \frac{2}{5}t \\ t \end{bmatrix}$.

(d) If $a \neq 2$, unique solution $\frac{1}{a-2}\begin{bmatrix} 1-b \\ ab-2 \end{bmatrix}$. If $a = 2$: no solution if $b \neq 1$;

if $b = 1$, the solutions are $\frac{1}{2}\begin{bmatrix} 1-t \\ 2t \end{bmatrix}$.

7. (b) Unique solution $\begin{bmatrix} -2a+b+5c \\ 3a-b-6c \\ -2a+b+4c \end{bmatrix}$ for any a, b, c

(d) If $abc \neq -1$, unique solution $\begin{bmatrix} 0 \\ 0 \\ 0 \end{bmatrix}$; if $abc = -1$ the solutions are $\begin{bmatrix} abt \\ -bt \\ t \end{bmatrix}$.

(f) If $a = 1$, solutions $\begin{bmatrix} -t \\ t \\ -1 \end{bmatrix}$. If $a = 0$, there is no solution. If $a \neq 1$ and $a \neq 0$, unique solution $\frac{1}{a}\begin{bmatrix} a-1 \\ 0 \\ -1 \end{bmatrix}$.

(h) If $a = 1$, no solution. If $a = 2$, the solutions are $\begin{bmatrix} 2-2t \\ -t \\ t \end{bmatrix}$. If $a \neq 1$ and $a \neq 2$,

unique solution $\frac{1}{3(a-1)}\begin{bmatrix} 8-5a \\ -2-a \\ a^2+a-2 \end{bmatrix}$

8. (b) $x^2 + y^2 - 2x + 6y - 6 = 0$ **9.** (b) $y = x^2 - 6$

10. (b) 5 of brand 1, 0 of brand 2, 3 of brand 3

12. Nissans, \$10 per day; Fords, \$12 per day; Chevrolets, \$13 per day

14. 2 adults, 8 youths, 140 children **15.** $\begin{bmatrix} 2 \\ -1 \end{bmatrix}$ and $\begin{bmatrix} -2 \\ 1 \end{bmatrix}$

Exercises 1.2.2 (Page 24)

1. (b) no, no (d) no, yes (f) no, no **2.** (b) $\begin{bmatrix} 0 & 1 & -3 & 0 & 0 & 0 & 0 \\ 0 & 0 & 0 & 1 & 0 & 0 & -1 \\ 0 & 0 & 0 & 0 & 1 & 0 & 0 \\ 0 & 0 & 0 & 0 & 0 & 1 & 1 \end{bmatrix}$

3. (b) $\begin{bmatrix} 2r-2s-t+1 \\ r \\ -5s+3t-1 \\ s \\ -6t+1 \\ t \end{bmatrix}$ (d) $\begin{bmatrix} -4s-5t-4 \\ -2s+t-2 \\ s \\ 1 \\ t \end{bmatrix}$

4. (b) $\begin{bmatrix} 0 \\ -t \\ 0 \\ t \end{bmatrix}$ (d) $\begin{bmatrix} 1 \\ 1-t \\ 1+t \\ t \end{bmatrix}$ (f) $\frac{1}{10}\begin{bmatrix} -6s-6t+16 \\ 4s-t+1 \\ 10s \\ 10t \end{bmatrix}$ **7.** (b) 1 (d) 3 (f) 1

8. (b) 2 (d) 3 **9.** $\begin{bmatrix} R_1 \\ R_2 \end{bmatrix} \to \begin{bmatrix} R_1+R_2 \\ R_2 \end{bmatrix} \to \begin{bmatrix} R_1+R_2 \\ -R_1 \end{bmatrix} \to \begin{bmatrix} R_2 \\ -R_1 \end{bmatrix} \to \begin{bmatrix} R_2 \\ R_1 \end{bmatrix}$ **12.** (b) no

Exercises 1.3 (Page 30)

1. (b) $a = -3, \begin{bmatrix} 9t \\ -5t \\ t \end{bmatrix}$ **(d)** $a = 1, \begin{bmatrix} -t \\ t \\ 0 \end{bmatrix}$ or $a = -1, \begin{bmatrix} t \\ 0 \\ t \end{bmatrix}$ **5. (b)** $t\begin{bmatrix} -3 \\ 3 \\ 1 \end{bmatrix}$

(d) $s\begin{bmatrix} -7 \\ 15 \\ 7 \\ 2 \\ 0 \end{bmatrix} + t\begin{bmatrix} -7 \\ 17 \\ 9 \\ 0 \\ 4 \end{bmatrix}$ **(f)** $r\begin{bmatrix} 7 \\ 9 \\ 1 \\ 0 \\ 0 \end{bmatrix} + s\begin{bmatrix} -4 \\ -5 \\ 0 \\ 1 \\ 0 \end{bmatrix} + t\begin{bmatrix} 5 \\ 6 \\ 0 \\ 0 \\ 1 \end{bmatrix}$

Exercises 1.4.1 (Page 33)

1. (b) $\begin{aligned} f_1 &= 85 - f_4 - f_7 \\ f_2 &= 60 - f_4 - f_7 \\ f_3 &= -75 + f_4 + f_6 \\ f_5 &= 40 - f_6 + f_7 \end{aligned}$ **2. (a)** $\begin{aligned} f_1 &= 55 - f_4 \\ f_2 &= 20 - f_4 + f_5 \\ f_3 &= 15 - f_5 \end{aligned}$ **(b)** $f_5 = 15$, $25 \le f_4 \le 30$ **3. (a)** $\begin{aligned} f_1 &= 50 + f_5 \\ f_2 &= 20 + f_5 \\ f_3 &= 60 + f_5 \\ f_4 &= 35 + f_5 \end{aligned}$

Exercises 1.4.2 (Page 35)

2. $\begin{bmatrix} I_1 \\ I_2 \\ I_3 \end{bmatrix} = \begin{bmatrix} -\frac{1}{5} \\ \frac{3}{5} \\ \frac{4}{5} \end{bmatrix}$ **4.** $\begin{bmatrix} I_1 \\ I_2 \\ I_3 \\ I_4 \\ I_5 \\ I_6 \end{bmatrix} = \begin{bmatrix} 2 \\ 1 \\ \frac{1}{2} \\ \frac{3}{2} \\ \frac{3}{2} \\ \frac{1}{2} \end{bmatrix}$

Exercises 2.1 (Page 46)

1. (b) $(a, b, c, d) = (-2, -4, -6, 0) + t(1, 1, 1, 1)$, t arbitrary **(d)** $a = b = c = d = t$, t arbitrary

2. (b) $\begin{bmatrix} -14 \\ -20 \end{bmatrix}$ **(d)** $(-12, 4, -12)$ **(f)** $\begin{bmatrix} 0 & 1 & -2 \\ -1 & 0 & 4 \\ 2 & -4 & 0 \end{bmatrix}$ **(h)** $\begin{bmatrix} 4 & -1 \\ -1 & -6 \end{bmatrix}$

3. (b) $\begin{bmatrix} 15 & -5 \\ 10 & 0 \end{bmatrix}$ **(d)** impossible **(f)** $\begin{bmatrix} 5 & 2 \\ 0 & -1 \end{bmatrix}$ **(h)** impossible

4. (b) $\begin{bmatrix} 4 \\ \frac{1}{2} \end{bmatrix}$ **5. (b)** $A = -\frac{11}{3}B$ **6. (b)** $X = 4A - 3B, Y = 4B - 5A$

7. (b) $Y = (s, t), X = \frac{1}{2}(1 + 5s, 2 + 5t)$; s and t arbitrary **8. (b)** $20A - 7B + 2C$

9. (b) If $A = \begin{bmatrix} a & b \\ c & d \end{bmatrix}$, then $(p, q, r, s) = \frac{1}{2}(2d, a + b - c - d, a - b + c - d, -a + b + c + d)$.

11. (b) If $A + A' = 0$, then $-A = -A + 0 = -A + (A + A') = (-A + A) + A' = 0 + A' = A'$

14. (b) $A = A^T$, so using Theorem 2 in Section 2.1, $(kA)^T = kA^T = kA$.

16. (c) Suppose $A = S + W$, where $S = S^T$ and $W = -W^T$. Then $A^T = S^T + W^T = S - W$, so $A + A^T = 2S$ and $A - A^T = 2W$. Hence $S = \frac{1}{2}(A + A^T)$ and $W = \frac{1}{2}(A - A^T)$ are uniquely determined by A.

Exercises 2.2 (Page 59)

1. (b) $\begin{bmatrix} -1 & -6 & -2 \\ 0 & 6 & 10 \end{bmatrix}$ **(d)** $(-3, -15)$ **(f)** -23 **(h)** $\begin{bmatrix} 1 & 0 \\ 0 & 1 \end{bmatrix}$

2. (b) $BA = \begin{bmatrix} -1 & 4 & -10 \\ 1 & 2 & 4 \end{bmatrix}$ $B^2 = \begin{bmatrix} 7 & -6 \\ -1 & 6 \end{bmatrix}$ $CB = \begin{bmatrix} -2 & 12 \\ 2 & -6 \\ 1 & 6 \end{bmatrix}$ $AC = \begin{bmatrix} 4 & 10 \\ -2 & -1 \end{bmatrix}$ $CA = \begin{bmatrix} 2 & 4 & 8 \\ -1 & -1 & -5 \\ 1 & 4 & 2 \end{bmatrix}$

3. (b) $(a, b, a_1, b_1) = (3, 0, 1, 2)$ **4. (b)** $A^2 - A - 6I = \begin{bmatrix} 8 & 2 \\ 2 & 5 \end{bmatrix} - \begin{bmatrix} 2 & 2 \\ 2 & -1 \end{bmatrix} - \begin{bmatrix} 6 & 0 \\ 0 & 6 \end{bmatrix} = \begin{bmatrix} 0 & 0 \\ 0 & 0 \end{bmatrix}$

5. (b) $A(BC) = \begin{bmatrix} 1 & -1 \\ 0 & 1 \end{bmatrix}\begin{bmatrix} -9 & -16 \\ 5 & 1 \end{bmatrix} = \begin{bmatrix} -14 & -17 \\ 5 & 1 \end{bmatrix} = \begin{bmatrix} -2 & -1 & -2 \\ 3 & 1 & 0 \end{bmatrix}\begin{bmatrix} 1 & 0 \\ 2 & 1 \\ 5 & 8 \end{bmatrix} = (AB)C$

6. (b) $A\mathbf{x} = \mathbf{b}$, where $A = \begin{bmatrix} -1 & 2 & -1 & 1 \\ 2 & 1 & -1 & 2 \\ 3 & -2 & 0 & 1 \end{bmatrix}$, $\mathbf{x} = \begin{bmatrix} x_1 \\ x_2 \\ x_3 \\ x_4 \end{bmatrix}$, and $\mathbf{b} = \begin{bmatrix} 6 \\ 1 \\ 0 \end{bmatrix}$

7. (b) $\begin{bmatrix} -2 \\ 2 \\ 0 \end{bmatrix} + t\begin{bmatrix} 1 \\ -3 \\ 1 \end{bmatrix}$ **(d)** $\begin{bmatrix} 3 \\ -9 \\ -2 \\ 0 \end{bmatrix} + t\begin{bmatrix} -1 \\ 4 \\ 1 \\ 1 \end{bmatrix}$

10. (b) If $\mathbf{s}_k = \mathbf{s}_m$, then $\mathbf{s} + k(\mathbf{s} - \mathbf{t}) = \mathbf{s} + m(\mathbf{s} - \mathbf{t})$. So $(k - m)(\mathbf{s} - \mathbf{t}) = \mathbf{0}$. But $\mathbf{s} - \mathbf{t}$ is not zero (because $\mathbf{s}$ and $\mathbf{t}$ are distinct), so $k - m = 0$ by Example 7 in Section 2.1.

11. (b) $m \times n$ and $n \times m$ for some m and n

12. (b) (i) $\begin{bmatrix} 1 & 0 \\ 0 & 1 \end{bmatrix}, \begin{bmatrix} 1 & 0 \\ 0 & -1 \end{bmatrix}, \begin{bmatrix} 1 & 1 \\ 0 & -1 \end{bmatrix}$ **(ii)** $\begin{bmatrix} 1 & 0 \\ 0 & 0 \end{bmatrix}, \begin{bmatrix} 1 & 0 \\ 0 & 1 \end{bmatrix}, \begin{bmatrix} 1 & 1 \\ 0 & 0 \end{bmatrix}$

13. (b)
$$AB = \left[\begin{array}{ccc|c} 2 & -1 & 3 & 1 \\ 1 & 0 & 1 & 2 \\ 0 & 0 & 1 & 0 \\ \hline 0 & 0 & 0 & 1 \end{array}\right]\left[\begin{array}{cc|c} 1 & 2 & 0 \\ -1 & 0 & 0 \\ 0 & 5 & 1 \\ \hline 1 & -1 & 0 \end{array}\right]$$
$$= \left[\begin{array}{c|c} \begin{bmatrix} 3 & 19 \\ 1 & 7 \\ 0 & 5 \end{bmatrix} + \begin{bmatrix} 1 & -1 \\ 2 & -2 \\ 0 & 0 \end{bmatrix} & \begin{bmatrix} 3 \\ 3 \\ 1 \end{bmatrix} + \begin{bmatrix} 0 \\ 0 \\ 0 \end{bmatrix} \\ \hline (1 \ \ -1) & 0 \end{array}\right]$$
$$= \left[\begin{array}{cc|c} 4 & 18 & 3 \\ 3 & 5 & 1 \\ 0 & 5 & 1 \\ \hline 1 & -1 & 0 \end{array}\right]$$

14. (b) $A^{2k} = \left[\begin{array}{cc|cc} 1 & -2k & 0 & 0 \\ 0 & 1 & 0 & 0 \\ \hline 0 & 0 & 1 & 0 \\ 0 & 0 & 0 & 1 \end{array}\right]$ for $k = 0, 1, 2, \ldots$ $A^{2k+1} = A^kA = \left[\begin{array}{cc|cc} 1 & -(2k+1) & 2 & -1 \\ 0 & 1 & 0 & 0 \\ \hline 0 & 0 & -1 & 1 \\ 0 & 0 & 0 & 1 \end{array}\right]$ for $k = 0, 1, 2, \ldots$

15. (b) The entries of column j of QA are formed by taking the dot product of each row of Q with C_j, and these are precisely the entries of QC_j.

17. (b) If Y is row i of the identity matrix I, then YA is row i of $IA = A$. **18. (b)** $CA^2C - ABCB$

19. (b) $(kA)C = k(AC) = k(CA) = C(kA)$

20. (b) We have $A^T = A$ and $B^T = B$, so $(AB)^T = B^TA^T = BA$. Hence AB is symmetric if and only if $AB = BA$.

22. If $BC = I$, then $AB = 0$ gives $0 = 0C = (AB)C = A(BC) = AI = A$, contrary to assumption.

25. (b) If $A = [a_{ij}]$, then $\text{tr}(kA) = \text{tr}[ka_{ij}] = \sum_{i=1}^{n} ka_{ii} = k\sum_{i=1}^{n} a_{ii} = k\,\text{tr}(A)$.

(e) Write $A^T = [a'_{ij}]$, where $a'_{ij} = a_{ji}$. Then $AA^T = \left(\sum_{k=1}^{n} a_{ik}a'_{kj}\right)$, so $\text{tr}(AA^T) = \sum_{i=1}^{n}\left[\sum_{k=1}^{n} a_{ik}a'_{ki}\right] = \sum_{i=1}^{n}\sum_{k=1}^{n} a_{ik}^2$.

26. (e) Observe that $PQ = P^2 + PAP - P^2AP = P$, so $Q^2 = PQ + APQ - PAPQ = P + AP - PAP = Q$.

28. (b) $(A + B)(A - B) = A^2 - AB + BA - B^2$, and $(A - B)(A + B) = A^2 + AB - BA - B^2$. These are equal if and only if $-AB + BA = AB - BA$ (that is, $2BA = 2AB$—that is, $BA = AB$).

30. (d) Using parts **(c)** and **(b)**: $I_{pq}AI_{rs} = \sum_{i=1}^{n}\sum_{j=1}^{n} a_{ij}I_{pq}I_{ij}I_{rs}$. The only nonzero term occurs when $i = q$ and $j = r$, so $I_{pq}AI_{rs} = a_{qr}I_{ps}$.

Exercises 2.3.1 (Page 71)

2. (b) $\frac{1}{5}\begin{bmatrix} 2 & -1 \\ -3 & 4 \end{bmatrix}$ **(d)** $\begin{bmatrix} 2 & -1 & 3 \\ 3 & 1 & -1 \\ 1 & 1 & -2 \end{bmatrix}$ **(f)** $\frac{1}{10}\begin{bmatrix} 1 & 4 & -1 \\ -2 & 2 & 2 \\ -9 & 14 & -1 \end{bmatrix}$ **(h)** $\frac{1}{4}\begin{bmatrix} 2 & 0 & -2 \\ -5 & 2 & 5 \\ -3 & 2 & -1 \end{bmatrix}$

(j) $\begin{bmatrix} 0 & 0 & 1 & -2 \\ -1 & -2 & -1 & -3 \\ 1 & 2 & 1 & 2 \\ 0 & -1 & 0 & 0 \end{bmatrix}$ **(l)** $\begin{bmatrix} 1 & -2 & 6 & -30 & 210 \\ 0 & 1 & -3 & 15 & -105 \\ 0 & 0 & 1 & -5 & 35 \\ 0 & 0 & 0 & 1 & -7 \\ 0 & 0 & 0 & 0 & 1 \end{bmatrix}$

3. (b) $\begin{bmatrix} x \\ y \end{bmatrix} = \frac{1}{5}\begin{bmatrix} 4 & -3 \\ 1 & -2 \end{bmatrix}\begin{bmatrix} 0 \\ 1 \end{bmatrix} = \frac{1}{5}\begin{bmatrix} -3 \\ -2 \end{bmatrix}$ **(d)** $\begin{bmatrix} x \\ y \\ z \end{bmatrix} = \frac{1}{5}\begin{bmatrix} 9 & -14 & 6 \\ 4 & -4 & 1 \\ -10 & 15 & -5 \end{bmatrix}\begin{bmatrix} 1 \\ -1 \\ 0 \end{bmatrix} = \frac{1}{5}\begin{bmatrix} 23 \\ 8 \\ -25 \end{bmatrix}$

(f) $\begin{bmatrix} x \\ y \\ z \\ w \end{bmatrix} = \frac{1}{2}\begin{bmatrix} 0 & 1 & -1 & 1 \\ 0 & 1 & 1 & -1 \\ 2 & -1 & -1 & -1 \\ 0 & -1 & 1 & 1 \end{bmatrix}\begin{bmatrix} 1 \\ 0 \\ -1 \\ 2 \end{bmatrix} = \frac{1}{2}\begin{bmatrix} 3 \\ -3 \\ 1 \\ 1 \end{bmatrix}$ **4. (b)** $B = A^{-1}(AB) = \begin{bmatrix} 4 & -2 & 1 \\ 7 & -2 & 4 \\ -1 & 2 & -1 \end{bmatrix}$

5. (b) $\frac{1}{10}\begin{bmatrix} 3 & -2 \\ 1 & 1 \end{bmatrix}$ **(d)** $\frac{1}{2}\begin{bmatrix} 0 & 1 \\ 1 & -1 \end{bmatrix}$ **(f)** $\frac{1}{2}\begin{bmatrix} 2 & 0 \\ -6 & 1 \end{bmatrix}$ **6. (b)** $A = \frac{1}{2}\begin{bmatrix} 2 & -1 & 3 \\ 0 & 1 & -1 \\ -2 & 1 & -1 \end{bmatrix}$

8. (b) $AB = I$ **10.** $B = IB = (CA)B = C(AB) = C$ **11. (b) (ii)** $\begin{bmatrix} x_1 \\ x_2 \end{bmatrix} = \begin{bmatrix} 2 \\ -1 \end{bmatrix}$

13. $Q^{-1} = \dfrac{1}{a^2 + b^2 + c^2 + d^2}Q^T$, provided that at least one of a, b, c, and d is nonzero.

15. (b) $B^4 = I$, so $B^{-1} = B^3 = \begin{bmatrix} 0 & 1 \\ -1 & 0 \end{bmatrix}$

16. (b) $\dfrac{1}{6 + c^2}\begin{bmatrix} 3 & c \\ -c & 2 \end{bmatrix}$ **(d)** $\begin{bmatrix} c^2 - 2 & -c & 1 \\ -c & 1 & 0 \\ 3 - c^2 & c & -1 \end{bmatrix}$ **17.** $\frac{1}{c}\begin{bmatrix} -c - 4 & c + 2 & -1 \\ -2c & c & 0 \\ 4 & -2 & 1 \end{bmatrix}$

19. (b) If column j of A is zero, $AY = 0$ where Y is column j of the identity matrix. Proceed as in Example 9.

(d) If each column of A sums to 0, $XA = 0$ where X is the row of 1's. Proceed as in Example 9.

20. (b) (ii) $(-1, 1, 1)A = 0$ **23. (d) (ii)** $\left[\begin{array}{cc|c} 2 & -1 & 0 \\ -5 & 3 & 0 \\ \hline 0 & 0 & -1 \end{array}\right]$ **(iv)** $\left[\begin{array}{cc|cc} 3 & -4 & 0 & 0 \\ -2 & 3 & 0 & 0 \\ \hline 0 & 0 & 1 & 3 \\ 0 & 0 & 0 & -1 \end{array}\right]$

24. (a) If it is invertible, there are no zero rows or columns, so $a \neq 0$, $b \neq 0$. $A^{-1} = \frac{1}{ab}\begin{bmatrix} b & -x \\ 0 & a \end{bmatrix}$

(c) (ii) $\left[\begin{array}{cc|cc} 1 & -1 & 7 & -13 \\ -2 & 3 & -18 & 33 \\ \hline 0 & 0 & -1 & 2 \\ 0 & 0 & 3 & -5 \end{array}\right]$ **26. (b)** $P = \begin{bmatrix} 1 & 0 \\ 0 & 0 \end{bmatrix}, Q = \begin{bmatrix} 1 & 0 \\ 0 & 0 \end{bmatrix}, R = \begin{bmatrix} 1 & 0 \\ 1 & 0 \end{bmatrix}$

27. (d) If $A^n = 0$, $(I - A)^{-1} = I + A + \cdots + A^{n-1}$.

29. (b) $A[B(AB)^{-1}] = I = [(BA)^{-1}B]A$, so A is invertible by Exercise 10.

31. (a) Have $AC = CA$. Left-multiply by A^{-1} to get $C = A^{-1}CA$. Then right-multiply by A^{-1} to get $CA^{-1} = A^{-1}C$.

33. (b) $(I - 2P)^2 = I - 4P + 4P^2$, and this equals I if and only if $P^2 = P$.

35. (b) $(A^{-1} + B^{-1})^{-1} = B(A + B)^{-1}A$

Exercises 2.3.2 (Page 84)

1. (b) Interchange rows 1 and 3 of I. $E^{-1} = E$ **(d)** Add (-2) times row 1 of I to row 2. $E^{-1} = \begin{bmatrix} 1 & 0 & 0 \\ 2 & 1 & 0 \\ 0 & 0 & 1 \end{bmatrix}$

(f) Multiply row 3 of I by 5. $E^{-1} = \begin{bmatrix} 1 & 0 & 0 \\ 0 & 1 & 0 \\ 0 & 0 & \frac{1}{5} \end{bmatrix}$ **2. (b)** $\begin{bmatrix} -1 & 0 \\ 0 & 1 \end{bmatrix}$ **(d)** $\begin{bmatrix} 1 & -1 \\ 0 & 1 \end{bmatrix}$ **(f)** $\begin{bmatrix} 0 & 1 \\ 1 & 0 \end{bmatrix}$

3. (b) The only possibilities for E are $\begin{bmatrix} 0 & 1 \\ 1 & 0 \end{bmatrix}, \begin{bmatrix} k & 0 \\ 0 & 1 \end{bmatrix}, \begin{bmatrix} 1 & 0 \\ 0 & k \end{bmatrix}, \begin{bmatrix} 1 & k \\ 0 & 1 \end{bmatrix}$, and $\begin{bmatrix} 1 & 0 \\ k & 1 \end{bmatrix}$. In each case, EA has a row different from C. **5. (b)** 0 is not invertible.

6. (b) $\begin{bmatrix} 1 & -2 \\ 0 & 1 \end{bmatrix}\begin{bmatrix} 1 & 0 \\ 0 & \frac{1}{2} \end{bmatrix}\begin{bmatrix} 1 & 0 \\ -5 & 1 \end{bmatrix}A = \begin{bmatrix} 1 & 0 & 7 \\ 0 & 1 & -3 \end{bmatrix}$. Alternatively, $\begin{bmatrix} 1 & 0 \\ 0 & \frac{1}{2} \end{bmatrix}\begin{bmatrix} 1 & -1 \\ 0 & 1 \end{bmatrix}\begin{bmatrix} 1 & 0 \\ -5 & 1 \end{bmatrix}A = \begin{bmatrix} 1 & 0 & 7 \\ 0 & 1 & -3 \end{bmatrix}$.

(d) $\begin{bmatrix} 1 & 2 & 0 \\ 0 & 1 & 0 \\ 0 & 0 & 1 \end{bmatrix}\begin{bmatrix} 1 & 0 & 0 \\ 0 & \frac{1}{5} & 0 \\ 0 & 0 & 1 \end{bmatrix}\begin{bmatrix} 1 & 0 & 0 \\ 0 & 1 & 0 \\ 0 & -1 & 1 \end{bmatrix}\begin{bmatrix} 1 & 0 & 0 \\ 0 & 1 & 0 \\ -2 & 0 & 1 \end{bmatrix}\begin{bmatrix} 1 & 0 & 0 \\ -3 & 1 & 0 \\ 0 & 0 & 1 \end{bmatrix}\begin{bmatrix} 0 & 0 & 1 \\ 0 & 1 & 0 \\ 1 & 0 & 0 \end{bmatrix}A = \begin{bmatrix} 1 & 0 & \frac{1}{5} & \frac{1}{5} \\ 0 & 1 & -\frac{7}{5} & -\frac{2}{5} \\ 0 & 0 & 0 & 0 \end{bmatrix}$

7. (b) $U = \begin{bmatrix} 1 & 1 \\ 1 & 0 \end{bmatrix} = \begin{bmatrix} 1 & 1 \\ 0 & 1 \end{bmatrix}\begin{bmatrix} 0 & 1 \\ 1 & 0 \end{bmatrix}$

8. (b) $A = \begin{bmatrix} 0 & 1 \\ 1 & 0 \end{bmatrix}\begin{bmatrix} 1 & 0 \\ 2 & 1 \end{bmatrix}\begin{bmatrix} 1 & 0 \\ 0 & -1 \end{bmatrix}\begin{bmatrix} 1 & 2 \\ 0 & 1 \end{bmatrix}$ **(d)** $A = \begin{bmatrix} 1 & 0 & 0 \\ 0 & 1 & 0 \\ -2 & 0 & 1 \end{bmatrix}\begin{bmatrix} 1 & 0 & 0 \\ 0 & 1 & 0 \\ 0 & 2 & 1 \end{bmatrix}\begin{bmatrix} 1 & 0 & -3 \\ 0 & 1 & 0 \\ 0 & 0 & 1 \end{bmatrix}\begin{bmatrix} 1 & 0 & 0 \\ 0 & 1 & 4 \\ 0 & 0 & 1 \end{bmatrix}$

14. (b) implies part **(a)**. If $YA = 0$ and Y is $1 \times n$, then $YAB = 0$ so $Y = 0$ (AB is invertible). Hence A is invertible (Theorem 8), so $B = A^{-1}(AB)$ is invertible (Theorem 2).

15. (b) $B\begin{bmatrix} -1 \\ 3 \\ -1 \end{bmatrix} = \mathbf{0}$, so B is not invertible by Theorem 10. **18. (b)** Multiply column i by $1/k$.

20. (a) If $A \overset{r}{\sim} B$, then $B = UA$ by Theorem 7, so $A = U^{-1}B$. Conversely, if $A = UB$, U invertible, then $B = U^{-1}A$. By Theorem 8, write $U^{-1} = E_kE_{k-1} \cdots E_2E_1$ where each E_i is elementary. Then the sequence of row operations

$$A \to E_1A \to E_2E_1A \to \cdots \to E_k \cdots E_1A = B$$

carries A to B. Thus $A \overset{r}{\sim} B$.

(d) Both are row-equivalent to $\begin{bmatrix} 1 & 0 & 0 & -18 \\ 0 & 1 & 0 & -11 \\ 0 & 0 & 1 & 3 \end{bmatrix}$, their reduced row-echelon form.

(e) (ii) $\begin{bmatrix} 0 & 0 & a \\ 0 & 0 & b \end{bmatrix}$ a and b not both zero **(g) (ii)** equivalent **(iv)** not equivalent

Exercises 2.4 (Page 98)

1. (b) $\begin{bmatrix} 2 & 0 & 0 \\ 1 & -3 & 0 \\ -1 & 9 & 1 \end{bmatrix} \begin{bmatrix} 1 & 2 & 1 \\ 0 & 1 & -\frac{2}{3} \\ 0 & 0 & 0 \end{bmatrix}$ **(d)** $\begin{bmatrix} -1 & 0 & 0 & 0 \\ 1 & 1 & 0 & 0 \\ 1 & -1 & 1 & 0 \\ 0 & -2 & 0 & 1 \end{bmatrix} \begin{bmatrix} 1 & 3 & -1 & 0 & 1 \\ 0 & 1 & 2 & 1 & 0 \\ 0 & 0 & 0 & 0 & 0 \\ 0 & 0 & 0 & 0 & 0 \end{bmatrix}$

(f) $\begin{bmatrix} 2 & 0 & 0 & 0 \\ 1 & -2 & 0 & 0 \\ 3 & -2 & 1 & 0 \\ 1 & 2 & 0 & 1 \end{bmatrix} \begin{bmatrix} 1 & 1 & -1 & 2 & 1 \\ 0 & 1 & -\frac{1}{2} & 0 & 0 \\ 0 & 0 & 0 & 0 & 0 \\ 0 & 0 & 0 & 0 & 0 \end{bmatrix}$

2. (b) $P = \begin{bmatrix} 0 & 0 & 1 \\ 1 & 0 & 0 \\ 0 & 1 & 0 \end{bmatrix}$ $PA = \begin{bmatrix} -1 & 2 & 1 \\ 0 & -1 & 2 \\ 0 & 0 & 4 \end{bmatrix} = \begin{bmatrix} -1 & 0 & 0 \\ 0 & -1 & 0 \\ 0 & 0 & 4 \end{bmatrix} \begin{bmatrix} 1 & -2 & -1 \\ 0 & 1 & -2 \\ 0 & 0 & 1 \end{bmatrix}$

(d) $P = \begin{bmatrix} 1 & 0 & 0 & 0 \\ 0 & 0 & 1 & 0 \\ 0 & 0 & 0 & 1 \\ 0 & 1 & 0 & 0 \end{bmatrix}$ $PA = \begin{bmatrix} -1 & -2 & 3 & 0 \\ 1 & 1 & -1 & 3 \\ 2 & 5 & -10 & 1 \\ 2 & 4 & -6 & 5 \end{bmatrix} = \begin{bmatrix} -1 & 0 & 0 & 0 \\ 1 & -1 & 0 & 0 \\ 2 & 1 & -2 & 0 \\ 2 & 0 & 0 & 5 \end{bmatrix} \begin{bmatrix} 1 & 2 & -3 & 0 \\ 0 & 1 & -2 & -3 \\ 0 & 0 & 1 & -2 \\ 0 & 0 & 0 & 1 \end{bmatrix}$

3. (b) $\mathbf{y} = \begin{bmatrix} -1 \\ 0 \\ 0 \end{bmatrix}$ $\mathbf{x} = \begin{bmatrix} -1+2t \\ -t \\ s \\ t \end{bmatrix}$ s and t arbitrary **(d)** $\mathbf{y} = \begin{bmatrix} 2 \\ 8 \\ -1 \\ 0 \end{bmatrix}$ $\mathbf{x} = \begin{bmatrix} 8-2t \\ 6-t \\ -1-t \\ t \end{bmatrix}$ t arbitrary

5. $\begin{bmatrix} \mathbf{r}_1 \\ \mathbf{r}_2 \end{bmatrix} \to \begin{bmatrix} \mathbf{r}_1+\mathbf{r}_2 \\ \mathbf{r}_2 \end{bmatrix} \to \begin{bmatrix} \mathbf{r}_1+\mathbf{r}_2 \\ -\mathbf{r}_1 \end{bmatrix} \to \begin{bmatrix} \mathbf{r}_2 \\ \mathbf{r}_1 \end{bmatrix} \to \begin{bmatrix} \mathbf{r}_2 \\ \mathbf{r}_1 \end{bmatrix}$

7. (b) Let $A = LU = L_1U_1$ be LU-factorizations of the invertible matrix A. Then U and U_1 have no row of zeros and so (being row-echelon) are upper triangular with 1's on the main diagonal. Thus, using **(a)**, the diagonal matrix $D = UU_1^{-1}$ has 1's on the main diagonal. Thus $D = I$, $U = U_1$, and $L = L_1$.

Exercises 2.5.1 (Page 103)

1. (b) $\begin{bmatrix} t \\ 3t \\ t \end{bmatrix}$ **(d)** $\begin{bmatrix} 14t \\ 17t \\ 47t \\ 23t \end{bmatrix}$ **2.** $\begin{bmatrix} t \\ t \\ t \end{bmatrix}$

4. $P = \begin{bmatrix} bt \\ (1-a)t \end{bmatrix}$ is nonzero (for some t) unless $b = 0$ and $a = 1$. In that case, $\begin{bmatrix} 1 \\ 1 \end{bmatrix}$ is a solution. If the entries of A are positive, then $P = \begin{bmatrix} b \\ 1-a \end{bmatrix}$ has positive entries.

Exercises 2.5.2 (Page 111)

1. (b) not regular **2. (b)** $\frac{1}{3}\begin{bmatrix} 2 \\ 1 \end{bmatrix}$, $\frac{3}{8}$ **(d)** $\frac{1}{3}\begin{bmatrix} 1 \\ 1 \\ 1 \end{bmatrix}$, .312 **(f)** $\frac{1}{20}\begin{bmatrix} 5 \\ 7 \\ 8 \end{bmatrix}$, .306

3. (b) $0, \frac{7}{27}$ **4. (b)** 50% middle, 25% upper, 25% lower **6.** $\frac{7}{16}, \frac{9}{16}$

8. (a) $\frac{94}{450}$ **(b)** He spends equal time in compartments 3 and 4. The steady-state vector is $\frac{1}{18}\begin{bmatrix} 3 \\ 2 \\ 5 \\ 5 \\ 3 \end{bmatrix}$.

Exercises 3.1 (Page 126)

1. (b) 0 **(d)** -1 **(f)** -39 **(h)** 0 **(j)** $2\,abc$ **(l)** 0 **(n)** -56 **(p)** $abcd$
5. (b) -17 **(d)** 106 **6. (b)** 0 **7. (b)** 12 **9. (b)** 35 **(d)** -16

10. (b) $-(x-2)(x^2+2x-12)$ **11. (b)** -7 **12. (b)** $\pm\frac{\sqrt{6}}{2}$ **(d)** $x = \pm y$

19. If A is $n \times n$ then $\det B = (-1)^{n(n-1)/2} \det A$.

Exercises 3.2 (Page 138)

1. (b) $\begin{bmatrix} 1 & -1 & -2 \\ -3 & 1 & 6 \\ -3 & 1 & 4 \end{bmatrix}$ **(d)** $\frac{1}{3}\begin{bmatrix} -1 & 2 & 2 \\ 2 & -1 & 2 \\ 2 & 2 & -1 \end{bmatrix}$ **2. (b)** $c \neq 0$ **(d)** any c **(f)** $c \neq -1$

3. (b) -2 **4. (b)** 1 **5. (b)** $\frac{4}{9}$ **6. (b)** $\frac{1}{11}\begin{bmatrix} 5 \\ 21 \end{bmatrix}$ **(d)** $\frac{1}{79}\begin{bmatrix} 12 \\ -37 \\ -2 \end{bmatrix}$ **7. (b)** $\frac{4}{51}$

8. (b) $\det A = 1, -1$ **(d)** $\det A = 1$ **(f)** $\det A = 0$ if n is odd; nothing can be said if n is even

17. (b) $\frac{1}{c^2}\begin{bmatrix} c & 0 & c \\ 0 & c^2 & c \\ -c & c^2 & c \end{bmatrix}, c \neq 0$ **(d)** $\frac{1}{2}\begin{bmatrix} 8-c^2 & -c & c^2-6 \\ c & 1 & -c \\ c^2-10 & c & 8-c^2 \end{bmatrix}$

(f) $\frac{1}{c^3+1}\begin{bmatrix} 1-c & c^2+1 & -c-1 \\ c^2 & -c & c+1 \\ -c & 1 & c^2-1 \end{bmatrix}, c \neq -1$ **21. (b)** $-21\begin{bmatrix} 3 & 0 & 1 \\ 0 & 2 & 3 \\ 3 & 1 & -1 \end{bmatrix}$

23. (b) If $\det A = 0$, it follows from part **(a)**. If $\det A \neq 0$, take determinants of both sides of $A \cdot \text{adj}\, A = (\det A)I$. A is $n \times n$, so this yields $\det A \det(\text{adj}\, A) = (\det A)^n$ and part **(b)** follows.

24. (b) Have $(\text{adj}\, A)A = (\det A)I$ so, taking inverses, $A^{-1} \cdot (\text{adj}\, A)^{-1} = \frac{1}{\det A}I$. On the other hand, $A^{-1}\,\text{adj}(A^{-1}) = \det(A^{-1})I = \frac{1}{\det A}I$. Comparison yields $A^{-1}(\text{adj}\, A)^{-1} = A^{-1}\,\text{adj}(A^{-1})$, and part **(b)** follows.

Exercises 3.4 (Page 148)

1. (b) $5 - 4x + 2x^2$ **2. (b)** $1 - \frac{5}{3}x + \frac{1}{2}x^2 + \frac{7}{6}x^3$ **3. (b)** $p(x) = 1 - 0.51x + 2.1x^2 - 1.1x^3$; 1.25
4. (b) $p(.7) = .6432$ [$\sin(.7) = .6442$] **5.** $p(x) = 1 + 1.974x + 9.98x^2$; $p(\frac{3}{8}) = 3.14$ [$e^{3(3/8)} = 3.08$]

Exercises 4.1.1 (Page 157)

1. $\|\mathbf{v}\| = \sqrt{2 + \sqrt{2}} = 1.85$ km; $\tan\theta = \sqrt{2} - 1$, $\theta = .3927 = \pi/8$, or 22.5°
2. (b) $150\sqrt{5 + 2\sqrt{3}} = 436.4$ km/hr **3.** θ north of east, where $\cos\theta = \frac{5}{13}$ ($\theta = 1.176$ or 67.4°); 13 knots
5. (b) $\vec{FE} = \vec{FC} + \vec{CE} = \frac{1}{2}\vec{AC} + \frac{1}{2}\vec{CB} = \frac{1}{2}(\vec{AC} + \vec{CB}) = \frac{1}{2}\vec{AB}$ **6. (b)** $\mathbf{v}$
9. (b) $\vec{CP}_k = -\vec{CP}_{n+k}$ if $1 \le k \le n$, where there are $2n$ points.
11. $\vec{AB} = \vec{AE} + \vec{EB} = \frac{1}{2}\vec{AC} + \frac{1}{2}\vec{DB} = \frac{1}{2}(\vec{AB} + \vec{BC} + \vec{DC} + \vec{CB}) = \frac{1}{2}\vec{AB} + \frac{1}{2}\vec{DC}$; hence $\vec{AB} = \vec{DC}$.
12. If E is the intersection of the diagonals, let $\vec{AE} = a\vec{AC}$ and $\vec{BE} = b\vec{BD}$. Then $\vec{AB} = \vec{AE} - \vec{BE} = a\vec{AC} - b\vec{BD} = a(\vec{AD} + \vec{AB}) - b(\vec{AD} - \vec{AB})$. Hence $(1 - a - b)\vec{AB} = (a - b)\vec{AD}$, so $a = b$, $1 = a + b$. Hence $a = b = \frac{1}{2}$.

Exercises 4.1.2 (Page 167)

1. (b) $(-3,5,13)$ **(d)** $(10,-12,-27)$ **(f)** $(0,0,0)$ **2. (b)** Yes **(d)** Yes
3. (b) $\mathbf{u}$ **(d)** $-(\mathbf{u} + \mathbf{v})$ **4. (b)** $(-1,-1,5)$ **(d)** $(0,0,0)$ **(f)** $(-2,2,-2)$
5. (b) (i) $Q(5,-1,2)$ **(ii)** $Q(1,1,-4)$ **6. (b)** $\mathbf{x} = \mathbf{u} - 6\mathbf{v} + 5\mathbf{w} = (-16,4,9)$
7. (b) $(a,b,c) = (-5,8,6)$ **9. (b)** $\frac{1}{4}(5,-5,-2)$
12. (b) $(3,-1,4) + t(2,-1,5)$; $x = 3 + 2t$, $y = -1 - t$, $z = 4 + 5t$
(d) $(1,1,1) + t(1,1,1)$; $x = y = z = 1 + t$ **(f)** $(2,-1,1) + t(-1,0,1)$; $x = 2 - t$, $y = -1$, $z = 1 + t$
13. (b) P corresponds to $t = 2$; Q corresponds to $t = 5$.
14. (b) no intersection **(d)** $P(2,-1,3)$; $t = -2$, $s = -3$ **19.** $\mathbf{w} = (15,8)$; speed $= \|\mathbf{w}\| = 17$ km/hr

Exercises 4.2.1 (Page 176)

1. (b) $\sqrt{6}$ **(d)** $\sqrt{5}$ **(f)** $3\sqrt{6}$ **2. (b)** $\frac{1}{3}(-2,-1,2)$ **4. (b)** $\sqrt{2}$ **(d)** 3
5. (b) 6 **(d)** 0 **(f)** 0 **6. (b)** π or 180° **(d)** $\frac{\pi}{3}$ or 60° **(f)** $\frac{2\pi}{3}$ or 120°
7. (b) 1 or -17 **8. (b)** $t(-1,1,2)$ **(d)** $s(1,2,0) + t(0,3,1)$
10. (b) $29 + 57 = 86$ **12. (b)** $A = B = C = \frac{\pi}{3}$ or 60° **14. (b)** $\frac{11}{18}\mathbf{v}$ **(d)** $-\frac{1}{2}\mathbf{v}$
15. (b) $\frac{5}{21}(2,-1,-4) + \frac{1}{21}(53,26,20)$ **(d)** $\frac{27}{53}(6,-4,1) + \frac{1}{53}(-3,2,26)$ **16. (b)** $\frac{1}{26}\sqrt{5642}$, $Q(\frac{71}{26}, \frac{15}{26}, \frac{34}{26})$
18. (b) The four diagonals are (a,b,c), $(-a,b,c)$, $(a,-b,c)$, and $(a,b,-c)$ or their negatives. The dot products are $\pm(-a^2 + b^2 + c^2)$, $\pm(a^2 - b^2 + c^2)$, and $\pm(a^2 + b^2 - c^2)$.
22. (b) The sum of the squares of the lengths of the diagonals equals the sum of the squares of the lengths of the four sides.
26. (b) The angle θ between $\mathbf{u}$ and $(\mathbf{u} + \mathbf{v} + \mathbf{w})$ is given by $\cos\theta = \frac{\mathbf{u}\cdot(\mathbf{u} + \mathbf{v} + \mathbf{w})}{\|\mathbf{u}\|\,\|\mathbf{u} + \mathbf{v} + \mathbf{w}\|} = \frac{\|\mathbf{u}\|}{\sqrt{\|\mathbf{u}\|^2 + \|\mathbf{v}\|^2 + \|\mathbf{w}\|^2}} = \frac{1}{\sqrt{3}}$, because $\|\mathbf{u}\| = \|\mathbf{v}\| = \|\mathbf{w}\|$. Similar remarks apply to the other angles.
28. (b) This follows from **(a)** because $\|\mathbf{v}\|^2 = a^2 + b^2 + c^2$.
36. Using the hint, $a^2 + b^2 = hp + hq = h(p + q) = h^2$.

Exercises 4.2.2 (Page 187)

1. (b) (0,0,0) **(d)** (4, −15,8) **4. (b)** $\pm\frac{\sqrt{3}}{3}(1,-1,-1)$

5. (b) $-23x + 32y + 11z = 11$ **(d)** $2x - y + z = 5$ **(f)** $2x + 3y + 2z = 7$
(h) $2x - 7y - 3z = -1$ **(j)** $x - y - z = 3$

6. (b) $(x,y,z) = (2,-1,3) + t(2,1,0)$ **(d)** $(x,y,z) = (1,1,-1) + t(1,1,1)$ **(f)** $(x,y,z) = (1,1,2) + t(4,1,-5)$

7. (b) $\frac{\sqrt{6}}{3}, Q(\frac{7}{3},\frac{2}{3},-\frac{2}{3})$ **8. (b)** Yes. The equation is $5x - 3y - 4z = 0$. **9. (b)** $(-2,7,0) + t(3,-5,2)$

10. (b) $3x + 2z = d$, d arbitrary **(d)** $a(x - 3) + b(y - 2) + c(z + 4) = 0$ a,b, and c not all zero
(f) $ax + by + (b - a)z = a$ a and b not both zero
(h) $ax + by + (a + 2b)z = 5a + 4b$ a and b not both zero

11. (b) $\sqrt{10}$ **12. (b)** $\frac{\sqrt{14}}{2}, A(3,1,2), B(\frac{7}{2},\frac{-1}{2},3)$ **(d)** $\frac{\sqrt{6}}{6}, A(\frac{19}{3},2,\frac{1}{3}), B(\frac{37}{6},\frac{13}{6},0)$

13. (b) 0 **(d)** $\sqrt{5}$

14. (b) 7 **15. (b)** distance is $\|\mathbf{v} - \mathbf{v}_0\|$

22. (b) If $\mathbf{u} = (u_1,u_2,u_3)$, $\mathbf{v} = (v_1,v_2,v_3)$, and $\mathbf{w} = (w_1,w_2,w_3)$, then

$$\mathbf{u} \times (\mathbf{v} + \mathbf{w}) = \det\begin{bmatrix} \mathbf{i} & \mathbf{j} & \mathbf{k} \\ u_1 & u_2 & u_3 \\ v_1 + w_1 & v_2 + w_2 & v_3 + w_3 \end{bmatrix} = \det\begin{bmatrix} \mathbf{i} & \mathbf{j} & \mathbf{k} \\ u_1 & u_2 & u_3 \\ v_1 & v_2 & v_3 \end{bmatrix}$$

$$+ \det\begin{bmatrix} \mathbf{i} & \mathbf{j} & \mathbf{k} \\ u_1 & u_2 & u_3 \\ w_1 & w_2 & w_3 \end{bmatrix} = (\mathbf{u} \times \mathbf{v}) + (\mathbf{u} \times \mathbf{w}),$$ where we used Exercise 17 in Section 3.1

29. By block multiplication, $\begin{bmatrix} \mathbf{u}A \\ \mathbf{v}A \\ \mathbf{w}A \end{bmatrix} = \begin{bmatrix} \mathbf{u} \\ \mathbf{v} \\ \mathbf{w} \end{bmatrix} A$. Take determinants and use Theorem 1 in Section 3.2 and Theorem 12 in Section 4.2.

Exercises 4.3 (Page 198)

1. (b) $y = \frac{64}{13} - \frac{6}{13}x$ **(d)** $y = -\frac{4}{10} - \frac{17}{10}x$

2. (a) $(M^TM)^{-1} = \frac{1}{440}\begin{bmatrix} 436 & -306 & 50 \\ -306 & 571 & -135 \\ 50 & -135 & 35 \end{bmatrix}, y = 1.02 - 0.11x + 0.27x^2$

(b) $(M^TM)^{-1} = \frac{1}{4248}\begin{bmatrix} 3348 & 642 & -426 \\ 642 & 571 & -187 \\ -426 & -187 & 91 \end{bmatrix}, y = .127 - .024x + .194x^2$

4. $s = 101.76 - 4.87x$, $g = 9.74$ (the true value of g is 9.81). If a quadratic in s is fit, $(M^TM)^{-1} =$
$\frac{1}{4}\begin{bmatrix} 76 & -84 & 20 \\ -84 & 98 & -24 \\ 20 & -24 & 6 \end{bmatrix}, s = 101 - \frac{3}{2}t - \frac{9}{2}t^2, g = 9$

5. $y = (.75)x - 1.87$, 5.6 meters

Exercises 5.1 (Page 210)

1. (b) (−1, −8, −13, −6, 16) **(d)** (12, 51, 51, 27, −42)

2. (b) Not linearly independent (for example $2\mathbf{u} - 3\mathbf{v} - \mathbf{w} = \mathbf{0}$). **(d)** Linearly independent.

3. (b) There are no such numbers, a, b and c.

4. (b) A1. $(x,-x) + (y,-y) = [x + y, -(x + y)]$ and so lies in V.

A2. $(x,-x) + (y,-y) = [x + y, -(x + y)] = [y + x, -(y + x)] = (y,-y) + (x,-x)$

A3. $[(x,-x) + (y,-y)] + (z,-z) = (x + y + z, -x - y - z) = (x,-x) + [(y,-y) + (z,-z)]$

A4. $(x,-x) + (0,-0) = (x,-x)$ for all $(x,-x)$, so $(0,0)$ is the zero vector.

A5. $(x,-x) + (-x,x) = (0,0)$ so $(-x,x) = (-x,-(-x))$ in V is the negative of $(x,-x)$.

S1. $a(x,-x) = [ax, -(ax)]$ and so lies in V.

S2. $a[(x,-x) + (y,-y)] = (ax + ay, -ax - ay) = (ax, -ax) + (ay, -ay) = a(x,-x) + a(y,-y)$

S3. $(a + b)(x,-x) = [(a + b)x, -(a + b)x] = (ax, -ax) + (bx, -bx) = a(x,-x) + b(x,-x)$

S4. $(ab)(x,-x) = (abx, -abx) = a(bx, -bx) = a[b(x,-x)]$

S5. $1(x,-x) = (1x, -1x) = (x,-x)$

5. (b) No; S5 fails. **(d)** No; S4 and S5 fail.

6. (b) No; only A1 fails. **(d)** Yes. **(f)** Yes. **(h)** No; only S3 fails. **(j)** No; only S4 fails.

8. The zero vector is $(0,-1)$; the negative of (x,y) is $(-x,-2-y)$. **10. (b)** $\mathbf{x} = \frac{1}{7}(5\mathbf{u} - 2\mathbf{v})$, $\mathbf{y} = \frac{1}{7}(4\mathbf{u} - 3\mathbf{v})$

11. (b) $\mathbf{x} = 11\mathbf{u} - 3\mathbf{v} - 7\mathbf{t}$, $\mathbf{y} = 8\mathbf{u} - 2\mathbf{v} - 5\mathbf{t}$, $\mathbf{z} = \mathbf{t}$; $\mathbf{t}$ an arbitrary vector

(d) $\mathbf{x} = \mathbf{t}$, $\mathbf{y} = 19\mathbf{t}$, $\mathbf{z} = 4\mathbf{t}$; $\mathbf{t}$ an arbitrary vector **12. (b)** Equating entries gives

$$\begin{aligned} a \quad\; + c &= 0 \\ b + c &= 0 \\ b + c &= 0 \\ a \quad\; - c &= 0 \end{aligned}$$

The solution is $a = b = c = 0$.

(d) If $a \sin x + b \cos y + c = 0$ in $\mathbf{F}[0, \pi]$, then this must hold for *every* x in $[0, \pi]$. Taking $x = 0, \pi/2$, and π, respectively, gives

$$\begin{aligned} b + c &= 0 \\ a \quad\; + c &= 0 \\ -b + c &= 0 \end{aligned}$$

whence $a = b = c = 0$.

13. (b) $4\mathbf{w}$

16. (d) $a(-\mathbf{v}) + a\mathbf{v} = a(-\mathbf{v} + \mathbf{v}) = a\mathbf{0} = \mathbf{0}$ by Theorem 4. Because also $-(a\mathbf{v}) + a\mathbf{v} = \mathbf{0}$ (by the definition of $-(a\mathbf{v})$ in axiom A.5), this means that $a(-\mathbf{v}) = -(a\mathbf{v})$ by cancellation. Alternatively, use Theorem 4 to give $a(-\mathbf{v}) = a[(-1)\mathbf{v}] = [a(-1)]\mathbf{v} = [(-1)a]\mathbf{v} = (-1)(a\mathbf{v}) = -(a\mathbf{v})$.

Exercises 5.2 (Page 222)

1. (b) Yes. **(d)** No; not closed under addition. **(f)** No; not closed under addition.

2. (b) Yes. **(d)** Yes. **(f)** No; not closed under addition or scalar multiplication, and 0 is not in the set.

3. (b) Yes. **(d)** Yes. **(f)** No; not closed under addition.

4. (b) No; not closed under addition. **(d)** No; not closed under scalar multiplication. **(f)** Yes.

6. (b) No. Take $\mathbf{v} = \mathbf{0}$.

7. (b) $-3(x + 1) + 0(x^2 + x) + 2(x^2 + 2)$ **(d)** $\frac{2}{3}(x + 1) + \frac{1}{3}(x^2 + x) - \frac{1}{3}(x^2 + 2)$

8. (b) $3(1,-1,1) + 2(1,0,1) - 4(1,1,0)$ **(d)** $0(1,-1,1) + 0(1,0,1) + 0(1,1,0)$

9. (b) Yes; $\mathbf{v} = 3\mathbf{u} - 2\mathbf{w}$ **(d)** No **(f)** Yes; $\mathbf{v} = 3\mathbf{u} - \mathbf{w}$

10. (b) Yes; $1 = \cos^2 x + \sin^2 x$

(d) No. If $1 + x^2 = a \cos^2 x + b \sin^2 x$, then, taking $x = 0$ and $x = \pi$ gives $a = 1$ and $a = 1 + \pi^2$.

11. **(b)** Because $\mathbf{P}_2 = \text{span}\{1, x, x^2\}$, it suffices to show that $\{1, x, x^2\} \subseteq \text{span}\{1 + 2x^2, 3x, 1 + x\}$. But $x = \frac{1}{3}(3x)$; $1 = (1 + x) - x$ and $x^2 = \frac{1}{2}[(1 + 2x^2) - 1]$.

13. **(b)** $\mathbf{u} = (\mathbf{u} + \mathbf{w}) - \mathbf{w}$, $\mathbf{v} = -(\mathbf{u} - \mathbf{v}) + (\mathbf{u} + \mathbf{w}) - \mathbf{w}$, and $\mathbf{w} = \mathbf{w}$

17. span $\{\mathbf{0}\} = \{\mathbf{0}\}$

Exercises 5.3.1 (Page 228)

2. **(b)** Independent **(d)** Dependent; $(1,1,0,0) - (1,0,1,0) + (0,0,1,1) - (0,1,0,1) = (0,0,0,0)$

(f) $3(x^2 - x + 3) - 2(2x^2 + x + 5) + (x^2 + 5x + 1) = 0$

(h) $2\begin{bmatrix} -1 & 0 \\ 0 & -1 \end{bmatrix} + \begin{bmatrix} 1 & -1 \\ -1 & 1 \end{bmatrix} + \begin{bmatrix} 1 & 1 \\ 1 & 1 \end{bmatrix} = \begin{bmatrix} 0 & 0 \\ 0 & 0 \end{bmatrix}$

(j) $\dfrac{5}{x^2 + x - 6} + \dfrac{1}{x^2 - 5x + 6} - \dfrac{6}{x^2 - 9} = 0$

3. **(b)** Dependent; $1 - \sin^2 x - \cos^2 x = 0$ **4.** **(b)** $x \neq -\frac{1}{3}$

6. If a linear combination of the subset vanishes, it is a linear combination of the vectors in the larger set (coefficients outside the subset are zero) so it is trivial.

14. Because $\{\mathbf{u}, \mathbf{v}\}$ is linearly independent, $s\mathbf{u}' + t\mathbf{v}' = \mathbf{0}$ is equivalent to $(s,t)\begin{bmatrix} a & b \\ c & d \end{bmatrix} = (0,0)$. Now apply Theorem 8 in Section 2.3.

17. **(b)** Independent. **(d)** Dependent. For example, $(\mathbf{u} + \mathbf{v}) - (\mathbf{v} + \mathbf{w}) + (\mathbf{w} + \mathbf{z}) - (\mathbf{z} + \mathbf{u}) = \mathbf{0}$.

Exercises 5.3.2 (Page 235)

1. **(b)** If $r(-1,1,1) + s(1,-1,1) + t(1,1,-1) = (0,0,0)$, then

$$\begin{aligned} -r + s + t &= 0 \\ r - s + t &= 0 \\ r + s - t &= 0 \end{aligned}$$

and this implies that $r = s = t = 0$. This proves independence. To prove that they span $\mathbb{R}^3$, observe that

$$(0,0,1) = \frac{1}{2}[(-1,1,1) + (1,-1,1)]$$

so $(0,0,1)$ lies in span$\{(-1,1,1), (1,-1,1), (1,1,-1)\}$. The proof is similar for $(0,1,0)$ and $(1,0,0)$.

(d) If $r(1 + x) + s(x + x^2) + t(x^2 + x^3) + ux^3 = 0$, then $r = 0$, $r + s = 0$, $s + t = 0$, and $t + u = 0$, so $r = s = t = u = 0$. This proves independence. To show that they span $\mathbf{P}_3$, observe that $x^2 = (x^2 + x^3) - x^3$, $x = (x + x^2) - x^2$, and $1 = (1 + x) - x$, so $\{1, x, x^2, x^3\} \subseteq \text{span}\{1 + x, x + x^2, x^2 + x^3, x^3\}$.

2. **(b)** $\{(1,1,1,0), (1,-1,0,1)\}$; dimension $= 2$ **(d)** $\{(1,0,1,0), (-1,1,0,1), (0,1,0,1)\}$; dimension $= 3$

(f) $\{(1,0,1,0), (0,1,0,1), (1,-1,0,0)\}$; dimension $= 3$

3. **(b)** $\{1, x + x^2\}$; dimension $= 2$ **(d)** $\{1, x^2\}$; dimension $= 2$

4. **(b)** $\left\{\begin{bmatrix} 1 & 1 \\ -1 & 0 \end{bmatrix}, \begin{bmatrix} 1 & 0 \\ 0 & 1 \end{bmatrix}\right\}$; dimension $= 2$ **(d)** $\left\{\begin{bmatrix} 1 & 0 \\ 1 & 1 \end{bmatrix}, \begin{bmatrix} 0 & 1 \\ -1 & 0 \end{bmatrix}\right\}$; dimension $= 2$

6. **(b)** $\left\{\begin{bmatrix} 1 & 0 \\ 0 & 0 \end{bmatrix}, \begin{bmatrix} 0 & 1 \\ 0 & 0 \end{bmatrix}\right\}$

7. (b) $\left\{\begin{bmatrix}-1\\7\\10\end{bmatrix}\right\}$ **(d)** $\left\{\begin{bmatrix}-1\\5\\4\\0\end{bmatrix},\begin{bmatrix}1\\7\\0\\-4\end{bmatrix}\right\}$

8. (b) $\dim V = 7$ **(c)** $\dim V = n^2 - n + 1$ **9. (b)** $\{x^2 - x, x(x^2 - x), x^2(x^2 - x), x^3(x^2 - x)\}$; $\dim V = 4$

14. (b) $\dim 0_n = \frac{n}{2}$ if n is even and $\dim 0_n = \frac{n+1}{2}$ if n is odd

Exercises 5.3.3 (Page 242)

1. (b) $\{(0,1,1), (1,0,0), (0,1,0)\}$ **(d)** $\{x^2 - x + 1, 1, x\}$

2. (b) Any three except $\{x^2 + 3, x + 2, x^2 - 2x - 1\}$

3. (b) Add (0,1,0,0) and (0,0,1,0). **(d)** Add 1 and x^3.

4. (b) If $z = a + bi$, then $a \neq 0$ and $b \neq 0$. If $rz + s\bar{z} = 0$ then $(r + s)a = 0$ and $(r - s)b = 0$. This means that $r + s = 0 = r - s$, so $r = s = 0$. Thus $\{z, \bar{z}\}$ is independent; it is a basis because $\dim \mathbb{C} = 2$.

6. (b) Impossible. These polynomials span a subspace with basis $\{x - 1, x^2 - 1, x^3 - 1\}$.

7. (b) Not a basis **(d)** Not a basis **8. (b)** Yes; No **11. (b)** Invertible

12. (b) Not independent

19. (b) Let $V = [\mathbf{v}_1 \dots \mathbf{v}_n]$ and $W = [A\mathbf{v}_1 \dots A\mathbf{v}_n]$ in block form. Then V and W are invertible (Theorem 10) and $AV = W$. Hence $A = WV^{-1}$ is invertible.

21. (b) **(ii)** $U + W = U$ and $U \cap W = W$. **(iv)** $U + W = \mathbb{R}^3$ and $U \cap W = \mathbb{R}(2,3,1)$

Exercises 5.4 (Page 253)

1. (b) $\{(1,0,0), (0,1,0), (0,0,1)\}$; $\left\{\begin{bmatrix}1\\0\\-1\\1\end{bmatrix},\begin{bmatrix}0\\1\\2\\0\end{bmatrix},\begin{bmatrix}0\\0\\2\\-1\end{bmatrix}\right\}$; 3 **(d)** $\{(1,2,-1,3), (0,3,1,1)\}$; $\left\{\begin{bmatrix}1\\0\end{bmatrix},\begin{bmatrix}0\\1\end{bmatrix}\right\}$; 2

2. (b) $\{(1,1,0,0,0), (0,-2,2,5,1), (0,0,2,-3,6)\}$ **(d)** $\left\{\begin{bmatrix}1\\5\\-6\end{bmatrix},\begin{bmatrix}0\\1\\-1\end{bmatrix},\begin{bmatrix}0\\0\\1\end{bmatrix}\right\}$ **3. (b)** No; No **(d)** No

6. $\text{col}(AV) = \text{row}[(AV)^T] = \text{row}[V^TA^T] \subseteq \text{row}(A^T) = \text{col } A$. If V is invertible, the reverse inclusion follows by applying this to $A = (AV)V^{-1}$.

9. (b) The basis is $\left\{\begin{bmatrix}6\\0\\-4\\1\\0\end{bmatrix},\begin{bmatrix}5\\0\\-3\\0\\1\end{bmatrix}\right\}$, so the dimension is 2. Have rank $A = 3$ and $n - 3 = 2$.

13. (b) Let $\{\mathbf{u}_1, \dots, \mathbf{u}_r\}$ be a basis of col(A). Then $\mathbf{b}$ is *not* in col(A), so $\{\mathbf{u}_1, \dots, \mathbf{u}_r, \mathbf{b}\}$ is linearly independent. Show that $\text{col}[A\ \mathbf{b}] = \text{span}\{\mathbf{u}_1, \dots, \mathbf{u}_r, \mathbf{b}\}$.

15. (b) The r columns containing leading 1's are independent because the leading 1's are in different rows. If R is $m \times n$, col R is contained in the subspace of all columns in $\mathbb{R}^m$ with the last $m - r$ entries zero. This space has dimension r, so the (independent) columns containing leading 1's are a basis.

16. (b) $U = A^{-1}, V = I_2$; rank $A = 2$ **(d)** $U = \begin{bmatrix}-2 & 1 & 0\\3 & -1 & 0\\2 & -1 & 1\end{bmatrix}, V = \begin{bmatrix}1 & 0 & -1 & -3\\0 & 1 & 1 & 4\\0 & 0 & 1 & 0\\0 & 0 & 0 & 1\end{bmatrix}$; rank $A = 2$

20. (b) Clearly $B = \{V^{-1}\mathbf{x}_1, \dots, V^{-1}\mathbf{x}_k\} \subseteq N(AV)$; we show it is a basis. If $\Sigma r_i(V^{-1}\mathbf{x}_i) = \mathbf{0}$, then $V^{-1}[\Sigma r_i\mathbf{x}_i] = \mathbf{0}$, so $\Sigma r_i\mathbf{x}_i = \mathbf{0}$, whence $r_i = 0$. Thus B is independent. Given $\mathbf{y}$ in $N(AV)$, we have $AV\mathbf{y} = 0$ so $V\mathbf{y}$ is in $N(A)$. Thus $V\mathbf{y} = \Sigma r_i\mathbf{x}_i$, so $\mathbf{y} = \Sigma r_i(V^{-1}\mathbf{x}_i)$.

22. (b) $P = A, Q = I_2$ **(d)** $P = \begin{bmatrix} 1 & 1 \\ 3 & 2 \\ 1 & 0 \end{bmatrix}, Q = \begin{bmatrix} 1 & 0 & 1 & 3 \\ 0 & 1 & -1 & -4 \end{bmatrix}$

Exercises 5.5 (Page 263)

1. (b) $\frac{3\pi}{8}$ **(d)** $\frac{\pi}{6}$ **2. (b)** $\begin{bmatrix} a \\ 2b - c \\ c - b \end{bmatrix}$ **(d)** $\frac{1}{2}\begin{bmatrix} a - b \\ a + b \\ -a + 3b + 2c \end{bmatrix}$

4. (b) $\frac{1}{2}\begin{bmatrix} -3 & -2 & 1 \\ 2 & 2 & 0 \\ 0 & 0 & 2 \end{bmatrix}$

6. $\begin{bmatrix} 1 & 1 & 1 & 1 \\ 0 & -1 & -2 & -3 \\ 0 & 0 & 1 & 3 \\ 0 & 0 & 0 & -1 \end{bmatrix}$; $p = (a + b + c + d)1 + (-b - 2c - 3d)(1 - x) + (c + 3d)(1 - x)^2 + (-d)(1 - x)^3$

8. (b) $P_{B \leftarrow D} = \begin{bmatrix} 1 & 1 & -1 \\ 1 & -1 & 0 \\ 1 & 0 & 1 \end{bmatrix}, P_{D \leftarrow B} = \frac{1}{3}\begin{bmatrix} 1 & 1 & 1 \\ 1 & -2 & 1 \\ -1 & -1 & 2 \end{bmatrix}$

9. (b) $A = P_{D \leftarrow B}$, where $B = \{(1,2,-1), (2,3,0), (1,0,2)\}$. Hence

$$A^{-1} = P_{B \leftarrow D} = \begin{bmatrix} 6 & -4 & -3 \\ -4 & 3 & 2 \\ 3 & -2 & -1 \end{bmatrix}$$

12. If $\mathbf{b}_j = \sum_{i=1}^{n} p_{ij}\mathbf{d}_i$, then $C_D(\mathbf{b}_j) = \begin{bmatrix} p_{1j} \\ p_{2j} \\ \vdots \\ p_{nj} \end{bmatrix}$ = column j of P. Hence $P_{D \leftarrow B} = [C_D(\mathbf{b}_1) \ldots C_D(\mathbf{b}_n)] = P$. Given $\mathbf{v}$,

write $\mathbf{v} = \sum_{j=1}^{n} v_j\mathbf{b}_j$, so $C_B(\mathbf{v}) = \begin{bmatrix} v_1 \\ \vdots \\ v_n \end{bmatrix}$. Then

$$\mathbf{v} = \sum_{j=1}^{n} v_j\left[\sum_{i=1}^{n} p_{ij}\mathbf{d}_i\right] = \sum_{i=1}^{n}\left[\sum_{j=1}^{n} p_{ij}v_j\right]\mathbf{d}_i$$

Hence $\sum_{j=1}^{n} p_{ij}v_j$ is the ith entry of $C_D(\mathbf{v})$. But it is also the ith entry of $PC_B(\mathbf{v})$.

Exercises 5.6.1 (Page 270)

2. (b) $3 + 4(x - 1) + 3(x - 1)^2 + (x - 1)^3$ **(d)** $1 + (x - 1)^3$

6. (b) The polynomials are $(x - 1)(x - 2)$, $(x - 1)(x - 3)$, $(x - 2)(x - 3)$. Use $a_0 = 3$, $a_1 = 2$, and $a_2 = 1$.

7. (b) $\frac{3}{2}(x - 2)(x - 3) - 7(x - 1)(x - 3) + \frac{13}{2}(x - 1)(x - 2)$

10. (b) If $r(x - a)^2 + s(x - a)(x - b) + t(x - b)^2 = 0$, then evaluation at $x = a$ $(x = b)$ gives $t = 0$ $(r = 0)$. Thus $s(x - a)(x - b) = 0$, so $s = 0$. Use Theorem 8 in Section 5.3.

11. (b) Suppose $\{p_0(x), p_1(x), \ldots, p_{n-2}(x)\}$ is a basis of $\mathbf{P}_{n-2}$. We show that

$$\{(x - a)(x - b)p_0(x), (x - a)(x - b)p_1(x), \ldots, (x - a)(x - b)p_{n-2}(x)\}$$

is a basis of U_n. It is a spanning set by part **(a)**, so assume that a linear combination vanishes with coefficients $r_0, r_1, \ldots, r_{n-2}$. Then $(x-a)(x-b)[r_0p_0(x) + \cdots + r_{n-2}p_{n-2}(x)] = 0$, so $r_0p_0(x) + \cdots + r_{n-2}p_{n-2}(x) = 0$. This implies that $r_0 = \cdots = r_{n-2} = 0$.

Exercises 5.6.2 (Page 277)

1. (b) e^{1-x} **(d)** $\dfrac{e^{-3x}-e^{2x}}{e^{-3}-e^{2}}$ **(f)** $2e^{2x}(1+x)$ **(h)** $\dfrac{e^{ax}-e^{a(2-x)}}{1-e^{2a}}$ **(j)** $e^{\pi-2x}\sin x$

5. (b) $ce^{-x}+2$, c a constant **6. (b)** $ce^{-3x}+de^{2x}-\dfrac{x^3}{3}$

7. (a) $10\left(\dfrac{4}{5}\right)^{t/3}$ **(b)** $t = \dfrac{3\ln(1/2)}{\ln(4/5)} = 9.32$ hours **9.** $k = \left(\dfrac{\pi}{15}\right)^2 = 0.044$

Exercises 6.1 (Page 288)

1. (b) P5 fails **(d)** P5 fails **(f)** P5 fails

3. (b) $\dfrac{1}{\sqrt{15}}(2,-3,1,1)$ **(d)** $\dfrac{1}{\sqrt{17}}\begin{bmatrix} 3 \\ -1 \end{bmatrix}$ **4. (b)** $\sqrt{2}$ **(d)** $\sqrt{3}$

10. (b) $15\|\mathbf{u}\|^2 - 17\langle\mathbf{u},\mathbf{v}\rangle - 4\|\mathbf{v}\|^2$ **14. (b)** $\{(1,1,0), (0,2,1)\}$

16. $\langle\mathbf{v}-\mathbf{w},\mathbf{v}_i\rangle = \langle\mathbf{v},\mathbf{v}_i\rangle - \langle\mathbf{w},\mathbf{v}_i\rangle = 0$ for each i, so $\mathbf{v} = \mathbf{w}$ by Exercise 15

21. (b) If $\|(x,y)\|^2 = \langle(x,y)(x,y)\rangle$ for an inner product $\langle\,,\rangle$, then Theorem 7 gives a matrix $\begin{bmatrix} a & b \\ b & c \end{bmatrix}$ with

$$(|x|+|y|)^2 = \|(x,y)\|^2 = (x,y)\begin{bmatrix} a & b \\ b & c \end{bmatrix}\begin{bmatrix} x \\ y \end{bmatrix} = ax^2 + 2bxy + cy^2.$$ Hence $a = \|(1,0)\|^2 = 1$, $c = \|(0,1)\|^2 = 1$,

and $a + 2b + c = \|(1,1)\|^2 = 4$. Thus $b = 1$ and $(|x|+|y|)^2 = (x+y)^2$. But this is false if $x = 1$ and $y = -1$.

24. (b) $\langle\mathbf{v},\mathbf{v}\rangle = 5v_1{}^2 - 6v_1v_2 + 2v_2{}^2 = \frac{1}{5}[(5v_1 - 3v_2)^2 + v_2{}^2]$

(d) $\langle\mathbf{v},\mathbf{v}\rangle = 3v_1{}^2 + 8v_1v_2 + 6v_2{}^2 = \frac{1}{3}[(3v_1 + 4v_2)^2 + 2v_2{}^2]$ **25. (b)** $\begin{bmatrix} 1 & -1 \\ -1 & 2 \end{bmatrix}$ **(d)** $\begin{bmatrix} 1 & 0 & -2 \\ 0 & 2 & 0 \\ -2 & 0 & 5 \end{bmatrix}$

26. If $\mathbf{v}A = \mathbf{0}$, where $\mathbf{v}$ is in $\mathbb{R}^n$, then $(\mathbf{v}^T)^TA\mathbf{v}^T = \mathbf{v}A\mathbf{v}^T = 0$. This means that $\mathbf{v}^T = \mathbf{0}$, so $\mathbf{v} = \mathbf{0}$. Hence A is invertible by Theorem 8 in Section 2.3.

27. (b) kA is symmetric (A is), and $\mathbf{v}^T(kA)\mathbf{v} = \frac{1}{k}(k\mathbf{v})^TA(k\mathbf{v}) > 0$ if $\mathbf{v} \neq \mathbf{0}$, because $A\mathbf{v}$ is not 0.

(d) A^3 is symmetric and $\mathbf{v}^TA^3\mathbf{v} = (A\mathbf{v})^TA(A\mathbf{v}) > 0$ if $\mathbf{v} \neq \mathbf{0}$, because $A\mathbf{v} \neq \mathbf{0}$ (A is invertible by Exercise 26).

29. (b) A^2 is symmetric, and $\mathbf{v}^TA^2\mathbf{v} = (A\mathbf{v})^T(A\mathbf{v}) = (A\mathbf{v})\cdot(A\mathbf{v}) > 0$ because $A\mathbf{v} \neq \mathbf{0}$ (A is invertible).

Exercises 6.2.1 (Page 298)

1. (b) $\left\{\dfrac{1}{\sqrt{3}}(1,1,1), \dfrac{1}{\sqrt{42}}(4,1,-5), \dfrac{1}{\sqrt{14}}(2,-3,1)\right\}$

3. (b) $(a,b,c) = \frac{1}{2}(a-c)(1,0,-1) + \frac{1}{18}(a+4b+c)(1,4,1) + \frac{1}{9}(2a-b+2c)(2,-1,2)$

(d) $(a,b,c) = \frac{1}{3}(a+b+c)(1,1,1) + \frac{1}{2}(a-b)(1,-1,0) + \frac{1}{6}(a+b-2c)(1,1,-2)$

4. (b) $(14,1,-8,5) = 3(2,-1,0,3) + 4(2,1,-2,-1)$ **5. (b)** $\frac{1}{2}$ **(d)** $11/5\sqrt{6}$

7. (b) This follows from $\langle\mathbf{v}+\mathbf{w},\mathbf{v}-\mathbf{w}\rangle = \|\mathbf{v}\|^2 - \|\mathbf{w}\|^2$.

8. (b) $\frac{1}{5}\begin{bmatrix} 3 & -4 \\ 4 & 3 \end{bmatrix}$ **(d)** $\frac{1}{\sqrt{a^2+b^2}}\begin{bmatrix} a & b \\ -b & a \end{bmatrix}$ **(f)** $\begin{bmatrix} 2/\sqrt{6} & 1/\sqrt{6} & -1/\sqrt{6} \\ 1/\sqrt{3} & -1/\sqrt{3} & 1/\sqrt{3} \\ 0 & 1/\sqrt{2} & 1/\sqrt{2} \end{bmatrix}$ **(h)** $\frac{1}{7}\begin{bmatrix} 2 & 6 & -3 \\ 3 & 2 & 6 \\ -6 & 3 & 2 \end{bmatrix}$

12. (b) $\det\begin{bmatrix} \cos\theta & \sin\theta \\ -\sin\theta & \cos\theta \end{bmatrix} = 1$ and $\det\begin{bmatrix} \cos\theta & \sin\theta \\ \sin\theta & -\cos\theta \end{bmatrix} = -1$ (Remark: Exercise 24 shows that these give *all* possibilities.)

13. $\pm(\frac{2}{3}, \frac{-2}{3}, \frac{1}{3})$ **16.** $\left\{1, \frac{1}{\sqrt{12}}\left(x - \frac{1}{2}\right), \frac{1}{\sqrt{180}}\left(x^2 - x + \frac{1}{6}\right)\right\}$

17. Let P be orthogonal. If Q is obtained from P by writing its rows in a different order, then the rows of Q are still an orthonormal set, so Q is an orthogonal matrix by Theorem 5.

18. (b) Part (a) shows that $AA^T = D$, where D is diagonal with main diagonal entries $\|\mathbf{r}_1\|^2, \ldots, \|\mathbf{r}_n\|^2$. Hence $A^{-1} = A^TD^{-1}$, and the result follows because D^{-1} has diagonal entries $1/\|\mathbf{r}_1\|^2, \ldots, 1/\|\mathbf{r}_n\|^2$.

20. (b) Because $I - A$ and $I + A$ commute, $PP^T = (I - A)(I + A)^{-1}[(I + A)^{-1}]^T(I - A)^T = (I - A)(I + A)^{-1}(I - A)^{-1}(I + A) = I$.

Exercises 6.2.2 (Page 310)

1. (b) $\{(2,1), \frac{3}{5}(-1,2)\}$ **(d)** $\{(0,1,1), (1,0,0), (0,-2,2)\}$

2. (b) $\mathbf{v} = \frac{1}{182}(271,-221,1030) + \frac{1}{182}(93,403,62)$ **(d)** $\mathbf{v} = \frac{1}{4}(1,7,11,17) + \frac{1}{4}(7,-7,-7,7)$

(f) $\mathbf{v} = \frac{1}{12}(5a - 5b + c - 3d, -5a + 5b - c + 3d, a - b + 11c + 3d, -3a + 3b + 3c + 3d) + \frac{1}{12}(7a + 5b - c + 3d, 5a + 7b + c - 3d, -a + b + c - 3d, 3a - 3b - 3c + 9d)$

3. (a) $\frac{1}{10}(-9,3,-21,33) = \frac{3}{10}(-3,1,-7,11)$ **(c)** $\frac{1}{70}(-63,21,-147,231) = \frac{3}{10}(-3,1,-7,11)$

4. (b) $\{(1,-1,0), \frac{1}{2}(-1,-1,2)\}$; $\text{proj}_U(\mathbf{v}) = (1,0,-1)$

(d) $\{(1,-1,0,1), (1,1,0,0), \frac{1}{3}(-1,1,0,2)\}$; $\text{proj}_U(\mathbf{v}) = (2,0,0,1)$ **5. (b)** $\{1, x - 1, x^2 - 2x + \frac{2}{3}\}$

6. (b) Observe that $U = \mathbf{P}_1$. Starting with the basis $\{1, x\}$ of U, the Gram–Schmidt process yields the orthogonal basis $\{1, x - \frac{1}{2}\}$. Hence $\text{proj}_U(x^2 + 1) = x + \frac{5}{6}$, so $x^2 + 1 = (x + \frac{5}{6}) + (x^2 - x + \frac{1}{6})$.

7. (a) $A\mathbf{x}^T = \begin{bmatrix} \mathbf{v}_1 \cdot \mathbf{x} \\ \mathbf{v}_2 \cdot \mathbf{x} \\ \vdots \\ \mathbf{v}_k \cdot \mathbf{x} \end{bmatrix}$ so $A\mathbf{x}^T = \mathbf{0}$ if and only if $\mathbf{x}$ is in $U^\perp$

(b) $A = \begin{bmatrix} 1 & -1 & 2 & 1 \\ 1 & 0 & -1 & 1 \end{bmatrix} \to \begin{bmatrix} 1 & 0 & -1 & 1 \\ 0 & 1 & -3 & 0 \end{bmatrix}$, so $A\mathbf{x}^T = \mathbf{0}$ has solution $\mathbf{x} = (s - t, 3s, s, t) = s(1,3,1,0) + t(-1,0,0,1)$. Hence $U^\perp = \text{span}\{(1,3,1,0), (-1,0,0,1)\}$.

8. (b) $\{\mathbf{e}_{m+1}, \ldots, \mathbf{e}_n\} \subseteq U^\perp$, because each vector in U is orthogonal to every $\mathbf{e}_i$, $1 \le i \le m$. Because $\dim U^\perp = \dim V - \dim U = n - m$, the set is a basis by Theorem 8 in Section 5.3.

14. (b) $\text{proj}_U(-5,4,-3) = (-5,4,-3)$; $\text{proj}_U(-1,0,2) = \frac{1}{38}(-17,24,73)$

19. (b) $U^\perp \subseteq \{\mathbf{u}_1, \ldots, \mathbf{u}_m\}^\perp$ because each $\mathbf{u}_i$ is in U. Conversely, if $\langle \mathbf{v}, \mathbf{u}_i \rangle = 0$ for each i, and $\mathbf{u} = r_1\mathbf{u}_1 + \cdots + r_m\mathbf{u}_m$ is any vector in U, then $\langle \mathbf{v}, \mathbf{u} \rangle = r_1 \langle \mathbf{v}, \mathbf{u}_1 \rangle + \cdots + r_m \langle \mathbf{v}, \mathbf{u}_m \rangle = 0$.

20. (d) $E^T = A^T[(AA^T)^{-1}]^T(A^T)^T = A^T[(AA^T)^T]^{-1}A = A^T[AA^T]^{-1}A = E$

$E^2 = A^T(AA^T)^{-1}AA^T(AA^T)^{-1}A = A^T(AA^T)^{-1}TA = E$

Exercises 6.2.3 (Page 317)

1. (b) $L = \frac{1}{\sqrt{5}}\begin{bmatrix} 5 & 0 \\ 3 & 1 \end{bmatrix}, P = \frac{1}{\sqrt{5}}\begin{bmatrix} 2 & 1 \\ -1 & 2 \end{bmatrix}, (AA^T)^{-1} = \begin{bmatrix} 2 & -3 \\ -3 & 5 \end{bmatrix} = (L^{-1})^T L^{-1}$

   (d) $L = \frac{1}{\sqrt{3}}\begin{bmatrix} 3 & 0 & 0 \\ 0 & 3 & 0 \\ -1 & 1 & 2 \end{bmatrix}, P = \frac{1}{\sqrt{3}}\begin{bmatrix} 1 & -1 & 0 & 1 \\ 1 & 0 & 1 & -1 \\ 0 & 1 & 1 & 1 \end{bmatrix},$

   $(AA^T)^{-1} = \frac{1}{12}\begin{bmatrix} 5 & -1 & 3 \\ -1 & 5 & -3 \\ 3 & -3 & 9 \end{bmatrix} = (L^{-1})^T L^{-1}$

2. If $D = L_1^{-1}L = P_1P^{-1}$, then D is both lower and upper triangular, hence diagonal. It is also orthogonal ($D = P_1P^{-1}$), so $D^{-1} = D^T = D$. The rest follows.

Exercises 6.3.1 (Page 324)

1. (b) $\frac{1}{444}\begin{bmatrix} -240 \\ 552 \\ 1140 \end{bmatrix}, (A^TA)^{-1} = \frac{1}{144}\begin{bmatrix} 96 & -120 & -216 \\ -120 & 168 & 288 \\ -216 & 288 & 516 \end{bmatrix}$

2. (b) $\frac{1}{68}[11x + 75x^2 + 33(-1)^x], (M^TM)^{-1} = \frac{1}{68}\begin{bmatrix} 89 & -33 & -5 \\ -33 & 13 & 3 \\ -5 & 3 & 19 \end{bmatrix}$

3. (b) $\frac{1}{4}\left(9x^2 + 7\sin\left[\frac{\pi x}{2}\right]\right), (M^TM)^{-1} = \frac{1}{2}\begin{bmatrix} 1 & 0 \\ 0 & 1 \end{bmatrix}$

5. $y = -5.19 + 0.34x_1 + 0.51x_2 + 0.71x_3$ $\qquad (A^TA)^{-1} = \frac{1}{50160}\begin{bmatrix} 1035720 & -16032 & 10080 & -45300 \\ -16032 & 416 & -632 & 800 \\ 10080 & -632 & 2600 & -2180 \\ -45300 & 800 & -2180 & 3950 \end{bmatrix}$

6. (b) It suffices to show that the columns of $M = \begin{bmatrix} 1 & e^{x_1} \\ \vdots & \vdots \\ 1 & e^{x_n} \end{bmatrix}$ are independent. If $r_0\begin{bmatrix} 1 \\ \vdots \\ 1 \end{bmatrix} + r_1\begin{bmatrix} e^{x_1} \\ \vdots \\ e^{x_n} \end{bmatrix} = \begin{bmatrix} 0 \\ \vdots \\ 0 \end{bmatrix}$,

   then $r_0 + r_1e^{x_i} = 0$ for each i. Thus $r_1(e^{x_i} - e^{x_j}) = 0$ for all i and j. Hence $r_1 = 0$ because two x_i are distinct. Then $r_0 = r_0 + r_1e^{x_i} = 0$ too.

Exercises 6.3.2 (Page 329)

1. (b) $\frac{\pi}{2} - \frac{4}{\pi}\left[\cos x + \frac{\cos 3x}{3^2} + \frac{\cos 5x}{5^2}\right]$

   (d) $\frac{\pi}{4} + \left[\sin x - \frac{\sin 2x}{2} + \frac{\sin 3x}{3} - \frac{\sin 4x}{4} + \frac{\sin 5x}{5}\right] - \frac{2}{\pi}\left[\cos x + \frac{\cos 3x}{3^2} + \frac{\cos 5x}{5^2}\right]$

2. (b) $\frac{2}{\pi} - \frac{8}{\pi}\left[\frac{\cos 2x}{2^2 - 1} + \frac{\cos 4x}{4^2 - 1} + \frac{\cos 6x}{6^2 - 1}\right]$

4. $\int \cos kx \cos lx\, dx = \frac{1}{2}\left[\frac{\sin[(k + l)x]}{k + l} - \frac{\sin[(k - l)x]}{k - l}\right]_0^{\pi} = 0$ $\qquad$ provided that $k \neq l$

Exercises 7.1.1 (Page 335)

1. **(b)** $(x-3)(x+2)$; $E_3 = \text{span}\{(4,-1)^T\}$, $E_{-2} = \text{span}\{(1,1)^T\}$ **(d)** $(x-2)^3$; $E_2 = \text{span}\{(1,1,0)^T, (0,3,1)^T\}$
(f) $(x+1)^2(x-2)$; $E_{-1} = \text{span}\{(1,-1,0)^T, (1,0,-1)^T\}$, $E_2 = \text{span}\{(1,1,1)^T\}$
(h) $(x-1)^2(x-3)$; $E_1 = \text{span}\{(1,0,-1)^T\}$, $E_3 = \text{span}\{(1,0,1)^T\}$

4. $A\mathbf{p} = \lambda\mathbf{p}$ if and only if $(A - \alpha I)\mathbf{p} = (\lambda - \alpha)\mathbf{p}$. Same eigenvectors.

10. **(b)** $c_{rA}(x) = \det[xI - rA] = r^n \det\left[\frac{x}{r}I - A\right] = r^n c_A\left[\frac{x}{r}\right]$

12. **(b)** If $\lambda \neq 0$, $A\mathbf{p} = \lambda\mathbf{p}$ if and only if $A^{-1}\mathbf{p} = \frac{1}{\lambda}\mathbf{p}$. The result follows.

15. **(b)** If $A^n = 0$ and $A\mathbf{p} = \lambda\mathbf{p}$, $\mathbf{p} \neq \mathbf{0}$, then $A^2\mathbf{p} = A(\lambda\mathbf{p}) = \lambda A\mathbf{p} = \lambda^2\mathbf{p}$. In general, $A^k\mathbf{p} = \lambda^k\mathbf{p}$ for all $k \geq 1$. Hence, $\lambda^n\mathbf{p} = A^n\mathbf{p} = 0\mathbf{p} = \mathbf{0}$, so $\lambda = 0$ (because $\mathbf{p} \neq \mathbf{0}$.)

Exercises 7.1.2 (Page 340)

1. **(b)** traces = 2, ranks = 2, but $\det A = -5$, $\det B = -1$
(d) ranks = 2, determinants = 7, but $\text{tr } A = 5$, $\text{tr } B = 4$
(f) traces = -5, determinants = 0, but rank $A = 2$, rank $B = 1$

4. **(b)** If $B = P^{-1}AP$, then $B^{-1} = P^{-1}A^{-1}(P^{-1})^{-1} = P^{-1}A^{-1}P$.

5. **(b)** $P^{-1}AP = \begin{bmatrix} 1 & 0 \\ 0 & 2 \end{bmatrix}$, so $A^n = P\begin{bmatrix} 1 & 0 \\ 0 & 2^n \end{bmatrix}P^{-1} = \begin{bmatrix} 9 - 8 \cdot 2^n & 12(1 - 2^n) \\ 6(2^n - 1) & 9 \cdot 2^n - 8 \end{bmatrix}$.

6. **(b)** $p(A) = A^3 + 3A^2 + A - I = \begin{bmatrix} 0 & 0 \\ 0 & 21 \end{bmatrix}$. Hence $p(B) = P^{-1}p(A)P = \begin{bmatrix} 84 & 42 \\ -126 & -63 \end{bmatrix}$.

Exercises 7.1.3 (Page 346)

1. **(b)** Eigenvalues 4, -1; eigenvectors $\begin{bmatrix} 2 \\ -1 \end{bmatrix}, \begin{bmatrix} 1 \\ -3 \end{bmatrix}$; $\mathbf{v}_4 = \begin{bmatrix} 409 \\ -203 \end{bmatrix}$; $r_2 = 3.94$

(d) Eigenvalues $\lambda_1 = \frac{1}{2}(3 + \sqrt{13})$, $\lambda_2 = \frac{1}{2}(3 - \sqrt{13})$; eigenvectors $\begin{bmatrix} \lambda_1 \\ 1 \end{bmatrix}, \begin{bmatrix} \lambda_2 \\ 1 \end{bmatrix}$; $\mathbf{v}_4 = \begin{bmatrix} 142 \\ 43 \end{bmatrix}$; $r_3 =$ 3.3027750 (The true value is $\lambda_1 = 3.3027756$ to seven decimal places.)

2. **(b)** Eigenvalues $\lambda_1 = \frac{1}{2}(3 + \sqrt{13}) = 3.302776$, $\lambda_2 = \frac{1}{2}(3 - \sqrt{13}) = -0.302776$

$$A_1 = \begin{bmatrix} 3 & 1 \\ 1 & 0 \end{bmatrix}, Q_1 = \frac{1}{\sqrt{10}}\begin{bmatrix} 3 & -1 \\ 1 & 3 \end{bmatrix}, R_1 = \frac{1}{\sqrt{10}}\begin{bmatrix} 10 & 3 \\ 0 & -1 \end{bmatrix}$$
$$A_2 = \frac{1}{10}\begin{bmatrix} 33 & -1 \\ -1 & -3 \end{bmatrix}, Q_2 = \frac{1}{\sqrt{1090}}\begin{bmatrix} 33 & 1 \\ -1 & 33 \end{bmatrix}, R_2 = \frac{1}{\sqrt{1090}}\begin{bmatrix} 109 & -3 \\ 0 & -10 \end{bmatrix}$$
$$A_3 = \frac{1}{109}\begin{bmatrix} 360 & 1 \\ 1 & -33 \end{bmatrix} = \begin{bmatrix} 3.302775 & 0.009174 \\ 0.009174 & -0.302775 \end{bmatrix}$$

4. Use induction on k. If $k = 1$, $A_1 = A$. In general $A_{k+1} = Q_k^{-1}A_kQ_k = Q_k^TA_kQ_k$, so the fact that $A_k^T = A_k$ implies $A^T_{k+1} = A_{k+1}$. The eigenvalues of A are all real (Theorem 2), so the A_k converge to an upper triangular matrix T. But T must also be symmetric (it is the limit of symmetric matrices), so it is diagonal.

Exercises 7.2.1 (Page 354)

1. (b) Yes. $P = \begin{bmatrix} 2 & -2 \\ 1 & 1 \end{bmatrix}$, $P^{-1}AP = \begin{bmatrix} 0 & 0 \\ 0 & 4 \end{bmatrix}$ **(d)** Yes. $P = \begin{bmatrix} 1 & 0 & 6 \\ 0 & 1 & 0 \\ -1 & 0 & 5 \end{bmatrix}$, $P^{-1}AP = \begin{bmatrix} -3 & 0 & 0 \\ 0 & -3 & 0 \\ 0 & 0 & 8 \end{bmatrix}$
(f) No. $c_A(x) = (x - 1)(x^2 + 1)$ has roots $1, i$, and $-i$.
2. AB has only one eigenvalue, 14, and $\dim(E_{14}) = 1$, so Theorem 5 applies.
5. (b) If $P^{-1}AP = D$ is diagonal, then $P^{-1}(kA)P = k(P^{-1}AP) = kD$ is also diagonal.
6. $\begin{bmatrix} 1 & 1 \\ 0 & 1 \end{bmatrix} = \begin{bmatrix} 2 & 1 \\ 0 & -1 \end{bmatrix} + \begin{bmatrix} -1 & 0 \\ 0 & 2 \end{bmatrix}$. Both $\begin{bmatrix} 2 & 1 \\ 0 & -1 \end{bmatrix}$ and $\begin{bmatrix} -1 & 0 \\ 0 & 2 \end{bmatrix}$ are diagonalizable by Theorem 3, but for $\begin{bmatrix} 1 & 1 \\ 0 & 1 \end{bmatrix}$, $\dim(E_1) = 1$. So Theorem 5 applies.
8. We have $P^{-1}AP = \lambda I$ by the diagonalization algorithm, so $A = P(\lambda I)P^{-1} = \lambda I$.
13. (b) $x_n = \frac{1}{3}[4 - (-2)^n]$

Exercises 7.2.2 (Page 361)

1. (b) $\frac{1}{\sqrt{2}}\begin{bmatrix} 0 & 1 & 1 \\ \sqrt{2} & 0 & 0 \\ 0 & 1 & -1 \end{bmatrix}$ **(d)** $\frac{1}{3}\begin{bmatrix} 2 & -2 & 1 \\ 1 & 2 & 2 \\ 2 & 1 & -2 \end{bmatrix}$ **(f)** $\frac{1}{2}\begin{bmatrix} 1 & -1 & \sqrt{2} & 0 \\ -1 & 1 & \sqrt{2} & 0 \\ -1 & -1 & 0 & \sqrt{2} \\ 1 & 1 & 0 & \sqrt{2} \end{bmatrix}$
2. $P = \frac{1}{\sqrt{2}k}\begin{bmatrix} c\sqrt{2} & a & a \\ 0 & k & -k \\ -a\sqrt{2} & c & c \end{bmatrix}$ **4.** $P = \frac{1}{\sqrt{2}}\begin{bmatrix} 1 & 1 \\ 1 & -1 \end{bmatrix}$ **5.** $P = \frac{1}{\sqrt{2}}\begin{bmatrix} 0 & 1 & 1 \\ \sqrt{2} & 0 & 0 \\ 0 & 1 & -1 \end{bmatrix}$
9. (b) If A is positive definite, then part **(a)** and Theorem 9 show that $A = B^2$ where B is symmetric, and B is invertible because A is. Conversely, if $A = B^2$, $B = B^T$, then $A^T = (B^T)^2 = B^2 = A$. Moreover, A is positive definite by Exercise 29(b) in Section 6.1.

Exercises 7.3 (Page 372)

1. (b) $\sqrt{6}$ **(d)** $\sqrt{13}$ **2. (b)** Not orthogonal **(d)** Orthogonal
3. (b) Not a subspace. For example, $i(0,0,1) = (0,0,i)$ is not in U. **(d)** This is a subspace.
4. (b) Basis $\{(i,0,2), (1,0,-1)\}$; dimension 2 **(d)** Basis $\{(1,0,-2i), (0,1,1-i)\}$; dimension 2
5. (b) Normal only **(d)** Hermitian (and normal) **(f)** Normal only
(h) Unitary (and normal); Hermitian if z is real
6. (b) $U = \frac{1}{\sqrt{14}}\begin{bmatrix} -2 & 3-i \\ 3+i & 2 \end{bmatrix}$, $U^*ZU = \begin{bmatrix} -1 & 0 \\ 0 & 6 \end{bmatrix}$ **(d)** $U = \frac{1}{\sqrt{3}}\begin{bmatrix} 1 & 1+i \\ 1-i & -1 \end{bmatrix}$, $U^*ZU = \begin{bmatrix} 4 & 0 \\ 0 & 1 \end{bmatrix}$
(f) $U = \frac{1}{\sqrt{3}}\begin{bmatrix} \sqrt{3} & 0 & 0 \\ 0 & 1+i & 1 \\ 0 & -1 & 1-i \end{bmatrix}$, $U^*ZU = \begin{bmatrix} 1 & 0 & 0 \\ 0 & 0 & 0 \\ 0 & 0 & 3 \end{bmatrix}$
8. (b) $\|\lambda\mathbf{z}\|^2 = \langle\lambda\mathbf{z}, \lambda\mathbf{z}\rangle = \lambda\bar{\lambda}\langle\mathbf{z},\mathbf{z}\rangle = |\lambda|^2\,\|\mathbf{z}\|^2$
9. (b) If the (k,k)-entry of Z is z_{kk}, then the (k,k)-entry of $\bar{Z}$ is $\bar{z}_{kk}$, so the (k,k)-entry of $\mathbf{Z}^* = (\bar{Z})^T$ is $\bar{z}_{kk}$. This equals Z, so z_{kk} is real.
12. (b) $(S^2)^* = S^*S^* = (-S)(-S) = S^2$; $(iS)^* = \bar{\imath}S^* = (-i)(-S) = iS$
(d) If $Z = H + S$, as given, then $Z^* = H^* + S^* = H - S$, so $H = \frac{1}{2}(Z + Z^*)$ and $S = \frac{1}{2}(Z - Z^*)$. Hence the

representation is unique if it exists. But, always, $Z = \frac{1}{2}(Z + Z^*) + \frac{1}{2}(Z - Z^*)$, and these are Hermitian and skew-Hermitian, respectively.

14. (b) If $U^{-1} = U^*$, then using **(a)**, $(U^*)^{-1} = (U^{-1})^* = U^*$.

19. (b) Let $U = \begin{bmatrix} a & b \\ c & d \end{bmatrix}$ be real and invertible, and assume that $U^{-1}AU = \begin{bmatrix} \lambda & \mu \\ 0 & \nu \end{bmatrix}$. Then $AU = U\begin{bmatrix} \lambda & \mu \\ 0 & \nu \end{bmatrix}$, and first column entries are $c = a\lambda$ and $-a = c\lambda$. Hence λ is real (c and a are both real and are not both 0), and $(1 + \lambda^2)a = 0$. Thus $a = 0$, $c = a\lambda = 0$, a contradiction.

Exercises 7.4.1 (Page 384)

1. (b) $A = \begin{bmatrix} 1 & 0 \\ 0 & 2 \end{bmatrix}$ **(d)** $A = \begin{bmatrix} 1 & 3 & 2 \\ 3 & 1 & -1 \\ 2 & -1 & 3 \end{bmatrix}$

2. (b) $P = \frac{1}{\sqrt{2}}\begin{bmatrix} 1 & 1 \\ 1 & -1 \end{bmatrix}$ $\mathbf{y} = \frac{1}{\sqrt{2}}\begin{bmatrix} x_1 + x_2 \\ x_1 - x_2 \end{bmatrix}$; $q = 3y_1^2 - y_2^2$; 1,2,0

(d) $P = \frac{1}{3}\begin{bmatrix} 2 & 2 & -1 \\ 2 & -1 & 2 \\ -1 & 2 & 2 \end{bmatrix}$; $\mathbf{y} = \frac{1}{3}\begin{bmatrix} 2x_1 + 2x_2 - x_3 \\ 2x_1 - x_2 + 2x_3 \\ -x_1 + 2x_2 + 2x_3 \end{bmatrix}$; $q = 9y_1^2 + 9y_2^2 - 9y_3^2$; 2,3,1

(f) $P = \frac{1}{3}\begin{bmatrix} -2 & 1 & 2 \\ 2 & 2 & 1 \\ 1 & -2 & 2 \end{bmatrix}$; $\mathbf{y} = \frac{1}{3}\begin{bmatrix} -2x_1 + 2x_2 + x_3 \\ x_1 + 2x_2 - 2x_3 \\ 2x_1 + x_2 + 2x_3 \end{bmatrix}$; $q = 9y_1^2 + 9y_2^2$; 2,2,2

(h) $P = \frac{1}{\sqrt{6}}\begin{bmatrix} -\sqrt{2} & \sqrt{3} & 1 \\ \sqrt{2} & 0 & 2 \\ \sqrt{2} & \sqrt{3} & -1 \end{bmatrix}$; $\mathbf{y} = \frac{1}{\sqrt{6}}\begin{bmatrix} -\sqrt{2}x_1 + \sqrt{2}x_2 + \sqrt{2}x_3 \\ \sqrt{3}x_1 + \sqrt{3}x_3 \\ x_1 + 2x_2 - x_3 \end{bmatrix}$; $q = 2y_1^2 + y_2^2 - y_3^2$; 2, 3, 1

3. (b) $x_1 = \frac{1}{\sqrt{5}}(2x - y)$, $y_1 = \frac{1}{\sqrt{5}}(x + 2y)$; $4x_1^2 - y_1^2 = 2$; hyperbola

(d) $x_1 = \frac{1}{\sqrt{5}}(x + 2y)$, $y_1 = \frac{1}{\sqrt{5}}(2x - y)$; $6x_1^2 + y_1^2 = 1$; ellipse

Exercises 7.4.2 (Page 392)

1. (b) $c_1\begin{bmatrix} 1 \\ 1 \end{bmatrix}e^{4x} + c_2\begin{bmatrix} 5 \\ -1 \end{bmatrix}e^{-2x}$; $c_1 = -\frac{2}{3}$, $c_2 = \frac{1}{3}$

(d) $c_1\begin{bmatrix} -8 \\ 10 \\ 7 \end{bmatrix}e^{-x} + c_2\begin{bmatrix} 1 \\ -2 \\ 1 \end{bmatrix}e^{2x} + c_3\begin{bmatrix} 1 \\ 0 \\ 1 \end{bmatrix}e^{4x}$; $c_1 = 0$, $c_2 = -\frac{1}{2}$, $c_3 = \frac{3}{2}$

Exercises 8.1.1 (Page 402)

1. (b) $T(\mathbf{v}) = A\mathbf{v}$ where $A = \begin{bmatrix} 2 & -3 & 5 \\ 0 & 0 & 0 \end{bmatrix}$ **(d)** $T(\mathbf{v}) = \mathbf{v}A$ where $A = \begin{bmatrix} 1 & 0 & 0 \\ 0 & 1 & 0 \\ 0 & 0 & -1 \end{bmatrix}$

(f) $T(A + B) = P(A + B)Q = PAQ + PBQ = T(A) + T(B)$; $T(rA) = P(rA)Q = rPAQ = rT(A)$

(h) $T[(p + q)(x)] = (p + q)(0) = p(0) + q(0) = T[p(x)] + T[q(x)]$
$T[(rp)(x)] = (rp)(0) = rp(0) = rT[p(x)]$

(j) $T(\mathbf{v} + \mathbf{w}) = R(\mathbf{v} + \mathbf{w}) + S(\mathbf{v} + \mathbf{w}) = R(\mathbf{v}) + R(\mathbf{w}) + S(\mathbf{v}) + S(\mathbf{w}) = T(\mathbf{v}) + T(\mathbf{w})$
$T(r\mathbf{v}) = R(r\mathbf{v}) + S(r\mathbf{v}) = rR(\mathbf{v}) + rS(\mathbf{v}) = rT(\mathbf{v})$

(l) If $\mathbf{v} = r_1\mathbf{e}_1 + \cdots + r_n\mathbf{e}_n$ and $\mathbf{w} = s_1\mathbf{e}_1 + \cdots + s_n\mathbf{e}_n$, then $T(\mathbf{v} + \mathbf{w}) = r_1 + s_1 = T(\mathbf{v}) + T(\mathbf{w})$, and $T(r\mathbf{v}) = rr_1 = rT(\mathbf{v})$

2. (b) $\text{rank}(A + B) \neq \text{rank } A + \text{rank } B$ in general. For example, $A = \begin{bmatrix} 1 & 0 \\ 0 & 1 \end{bmatrix}$ and $B = \begin{bmatrix} 1 & 0 \\ 0 & -1 \end{bmatrix}$.

(d) $T(\mathbf{0}) = \mathbf{0} + \mathbf{u} = \mathbf{u} \neq \mathbf{0}$, so T is not linear by Theorem 1.

3. (b) $T(3\mathbf{v}_1 + 2\mathbf{v}_2) = 0$ **(d)** $T\begin{bmatrix} 1 \\ -7 \end{bmatrix} = \begin{bmatrix} -3 \\ 4 \end{bmatrix}$ **(f)** $T(2 - x + 3x^2) = 46$

4. (b) $T(x,y) = \frac{1}{3}(x - y, 3y, x - y)$; $T(-1,2) = (-1,2,-1)$ **(d)** $T\begin{bmatrix} a & b \\ c & d \end{bmatrix} = 3a - 3c + 2b$

5. (b) $T(\mathbf{v}) = \frac{1}{3}(7\mathbf{v} - 9\mathbf{w})$, $T(\mathbf{w}) = \frac{1}{3}(\mathbf{v} + 3\mathbf{w})$ **12. (b)** $A = \begin{bmatrix} 1 & 0 \\ 0 & a \end{bmatrix}$ **(d)** $A = \begin{bmatrix} 1 & 0 \\ 0 & 0 \end{bmatrix}$

13. (a) $T(x,y,z) = \begin{bmatrix} x\cos\theta - y\sin\theta \\ x\sin\theta + y\cos\theta \\ z \end{bmatrix}$, $A = \begin{bmatrix} \cos\theta & -\sin\theta & 0 \\ \sin\theta & \cos\theta & 0 \\ 0 & 0 & 1 \end{bmatrix}$

(b) $T(x,y,z) = (x,y,-z)$, $A = \begin{bmatrix} 1 & 0 & 0 \\ 0 & 1 & 0 \\ 0 & 0 & -1 \end{bmatrix}$

14. (a) $\frac{1}{a^2+b^2+c^2}\begin{bmatrix} b^2+c^2 & -ab & -ac \\ -ab & a^2+c^2 & -bc \\ -ac & -bc & a^2+b^2 \end{bmatrix}$ **(b)** $\frac{1}{a^2+b^2+c^2}\begin{bmatrix} -a^2+b^2+c^2 & -2ab & -2ac \\ -2ab & a^2-b^2+c^2 & -2bc \\ -2ac & -2bc & a^2+b^2-c^2 \end{bmatrix}$

15. (b) The matrix with columns $A\mathbf{u}$, $A\mathbf{v}$, and $A\mathbf{w}$ is $[A\mathbf{u}\ A\mathbf{v}\ A\mathbf{w}] = A[\mathbf{u}\ \mathbf{v}\ \mathbf{w}]$. Taking determinants gives $\det[A\mathbf{u}\ A\mathbf{v}\ A\mathbf{w}] = \det A \cdot \det[\mathbf{u}\ \mathbf{v}\ \mathbf{w}]$. Apply Theorem 12 in Section 4.2

Exercises 8.1.2 (Page 413)

1. (b) $\left\{\begin{bmatrix} -3 \\ 7 \\ 1 \\ 0 \end{bmatrix}, \begin{bmatrix} 1 \\ 1 \\ 0 \\ -1 \end{bmatrix}\right\}; \left\{\begin{bmatrix} 1 \\ 0 \\ 1 \end{bmatrix}, \begin{bmatrix} 0 \\ 1 \\ -1 \end{bmatrix}\right\}$; 2, 2 **(d)** $\left\{\begin{bmatrix} -1 \\ 2 \\ 1 \end{bmatrix}\right\}; \left\{\begin{bmatrix} 1 \\ 0 \\ 1 \\ 1 \end{bmatrix}, \begin{bmatrix} 0 \\ 1 \\ -1 \\ -2 \end{bmatrix}\right\}$; 2, 1

2. (b) $\{x^2 - x\}$; $\{(1,0), (0,1)\}$ **(d)** $\{(0,0,1)\}$; $\{(1,1,0,0), (0,0,1,1)\}$

(f) $\left\{\begin{bmatrix} 1 & 0 \\ 0 & -1 \end{bmatrix}, \begin{bmatrix} 0 & 1 \\ 0 & 0 \end{bmatrix}, \begin{bmatrix} 0 & 0 \\ 1 & 0 \end{bmatrix}\right\}$; $\{1\}$

(h) $\{(1,0,0,\ldots,0,-1), (0,1,0,\ldots,0,-1), \ldots, (0,0,0,\ldots,1,-1)\}$; $\{1\}$

(j) $\left\{\begin{bmatrix} 0 & 1 \\ 0 & 0 \end{bmatrix}, \begin{bmatrix} 0 & 0 \\ 0 & 1 \end{bmatrix}\right\}; \left\{\begin{bmatrix} 1 & 1 \\ 0 & 0 \end{bmatrix}, \begin{bmatrix} 0 & 0 \\ 1 & 1 \end{bmatrix}\right\}$

3. (b) $T(\mathbf{v}) = \mathbf{0} = (\mathbf{0},\mathbf{0})$ if and only if $P(\mathbf{v}) = \mathbf{0}$ and $Q(\mathbf{v}) = \mathbf{0}$; that is, if and only if $\mathbf{v}$ is in $\ker P \cap \ker Q$.

4. (b) $\ker T = \text{span}\{(-4,1,3)\}$; $B = \{(1,0,0), (0,1,0), (-4,1,3)\}$; $\text{im } T = \text{span}\{(1,2,0,3), (1,-1,-3,0)\}$

8. (b) and **(d)** If T is one-to-one, then $\dim V = \dim(\ker T) + \dim(\text{im } T) = \dim(\text{im } T) \leq \dim W$.

12. (b) $\dim(\ker T) = \dim \mathbf{P}_n - \dim(\text{im } T) = \dim \mathbf{P}_n - \dim \mathbb{R} = (n + 1) - 1 = n$

16. (b) $T[p + q] = (p + q)(x + 1) - (p + q)(x) = p(x + 1) + q(x + 1) - p(x) - q(x) = T(p) + T(q)$.
$T(rp) = (rp)(x + 1) - (rp)(x) = rp(x + 1) - rp(x) = rT(p)$.

(d) $\dim(\ker T) = 1$ by (c), so $\dim(\operatorname{im} T) = \dim \mathbf{P}_n - \dim(\ker T) = (n+1) - n = n$. But $\operatorname{im} T \subseteq \mathbf{P}_{n-1}$ and $\dim \mathbf{P}_{n-1} = n$, so $\operatorname{im} T = \mathbf{P}_{n-1}$. Thus T is onto, so every polynomial in $\mathbf{P}_{n-1}$ has the form $T(p) = p(x+1) - p(x)$ for some p in $\mathbf{P}_n$.

Exercises 8.1.3 (Page 424)

1. **(b)** T is onto because $T(1,-1,0) = (1,0,0)$, $T(0,1,-1) = (0,1,0)$, and $T(0,0,1) = (0,0,1)$. Use Theorem 9.
(d) T is one-to-one because $0 = T(X) = UXV$ implies that $X = 0$ (U and V are invertible). Use Theorem 9.
(f) T is one-to-one because $\mathbf{0} = T(\mathbf{v}) = k\mathbf{v}$ implies that $\mathbf{v} = \mathbf{0}$ because ($k \neq 0$). T is onto because $T(\frac{1}{k}\mathbf{v}) = \mathbf{v}$ for all $\mathbf{v}$. [Here Theorem 9 does not apply if dim V is not finite.]
(h) T is one-to-one because $T(A) = 0$ implies $A^T = 0$, whence $A = 0$. Use Theorem 9.

2. Define $T: \mathbb{R}^3 \to \mathbf{P}_2$ by $T(a_0,a_1,a_2) = a_0 + a_1x + a_2x^2$. This is an isomorphism, so every basis of $\mathbb{R}^3$ gives rise to a basis of $\mathbf{P}_2$ of the required type.

4. **(b)** $ST(x,y,z) = (x+y, 0, y+z)$, $TS(x,y,z) = (x,0,z)$

5. **(b)** $T^2(x,y) = T(x+y, 0) = (x+y, 0) = T(x,y)$. Hence $T^2 = T$.

6. **(b)** No inverse; $(1,-1,1,-1)$ is in $\ker T$.
(d) $T^{-1}\begin{bmatrix} a & b \\ c & d \end{bmatrix} = \frac{1}{5}\begin{bmatrix} 3a-2c & 3b-2d \\ a+c & b+d \end{bmatrix}$ **(f)** $T^{-1}(a,b,c) = \frac{1}{2}[2a + (b-c)x - (2a-b-c)x^2]$

7. **(b)** $T^2(x,y) = T(ky - x, y) = (ky - (ky - x), y) = (x, y)$ so $T^2 = 1_{\mathbb{R}^2}$

9. **(b)** $T^3(x,y,z,w) = (x,y,z,-w)$ so $T^6(x,y,z,w) = T^3[T^3(x,y,z,w)] = (x,y,z,w)$ Hence $T^{-1} = T^5$. So $T^{-1}(x,y,z,w) = (y-x, -x, z, -w)$.

10. **(b)** Given $\mathbf{u}$ in U, write $\mathbf{u} = S(\mathbf{w})$, $\mathbf{w}$ in W (because S is onto). Then write $\mathbf{w} = T(\mathbf{v})$, $\mathbf{v}$ in V (T is onto). Hence $\mathbf{u} = ST(\mathbf{v})$, so ST is onto.

13. **(b)** Given $\mathbf{w}$ in W, write $\mathbf{w} = ST(\mathbf{v})$, $\mathbf{v}$ in V (ST is onto). Then $\mathbf{w} = S[T(\mathbf{v})]$, $T(\mathbf{v})$ in U, so S is onto. But then $\operatorname{im} S = W$, so $\dim U = \dim(\ker S) + \dim(\operatorname{im} S) \geq \dim(\operatorname{im} S) = \dim W$.

19. **(b)** $T(x,y) = (x, y+1)$.

27. **(b)** If $T(p) = 0$, then $p(x) = -xp'(x)$. We write $p(x) = a_0 + a_1x + a_2x^2 + \cdots + a_nx^n$, and this becomes

$$a_0 + a_1x + a_2x^2 + \cdots + a_nx^n = -a_1x - 2a_2x^2 - \cdots - na_nx^n$$

Equating coefficients yields $a_0 = 0$, $2a_1 = 0$, $3a_2 = 0, \ldots, (n+1)a_n = 0$, whence $p(x) = 0$. This means that $\ker T = 0$, so T is one-to-one. But then T is an isomorphism by Theorem 9.

29. Let $B = \{\mathbf{e}_1, \ldots, \mathbf{e}_r, \mathbf{e}_{r+1}, \ldots, \mathbf{e}_n\}$ be a basis of V with $\{\mathbf{e}_{r+1}, \ldots, \mathbf{e}_n\}$ a basis of $\ker T$. If $\{T(\mathbf{e}_1), \ldots, T(\mathbf{e}_r), \mathbf{w}_{r+1}, \ldots, \mathbf{w}_n\}$ is a basis of V, define S by $S[T(\mathbf{e}_i)] = \mathbf{e}_i$ for $1 \leq i \leq r$, and $S(\mathbf{w}_j) = \mathbf{e}_j$ for $r+1 \leq j \leq n$. Then S is an isomorphism by Theorem 10, and $TST(\mathbf{e}_i) = T(\mathbf{e}_i)$ clearly holds for $1 \leq i \leq r$. But if $i \geq r+1$, then $T(\mathbf{e}_i) = 0 = TST(\mathbf{e}_i)$, so $T = TST$ by Theorem 2.

Exercises 8.2.1 (Page 434)

1. **(b)** Let $\mathbf{v} = a + bx + cx^2$. Then

$$C_D[T(\mathbf{v})] = M_{DB}(T)C_B(\mathbf{v}) = \begin{bmatrix} 2 & 1 & 3 \\ -1 & 0 & -2 \end{bmatrix}\begin{bmatrix} a \\ b \\ c \end{bmatrix} = \begin{bmatrix} 2a+b+3c \\ -a-2c \end{bmatrix}$$

Hence $T(\mathbf{v}) = (2a + b + 3c)(1,1) + (-a - 2c)(0,1) = (2a + b + 3c, a + b + c)$.

2. **(b)** $\begin{bmatrix} 1 & 0 & 0 & 0 \\ 0 & 0 & 1 & 0 \\ 0 & 1 & 0 & 0 \\ 0 & 0 & 0 & 1 \end{bmatrix}$ **(d)** $\begin{bmatrix} 1 & 1 & 1 \\ 0 & 1 & 2 \\ 0 & 0 & 1 \end{bmatrix}$

3. **(b)** $\begin{bmatrix} 1 & 2 \\ 5 & 3 \\ 4 & 0 \\ 1 & 1 \end{bmatrix}$; $C_D[T(a,b)] = \begin{bmatrix} 1 & 2 \\ 5 & 3 \\ 4 & 0 \\ 1 & 1 \end{bmatrix} \begin{bmatrix} b \\ a - b \end{bmatrix} = \begin{bmatrix} 2a - b \\ 3a + 2b \\ 4b \\ a \end{bmatrix}$

(d) $\frac{1}{2}\begin{bmatrix} 1 & 1 & -1 \\ 1 & 1 & 1 \end{bmatrix}$; $C_D[T(a + bx + cx^2)] = \frac{1}{2}\begin{bmatrix} 1 & 1 & -1 \\ 1 & 1 & 1 \end{bmatrix}\begin{bmatrix} a \\ b \\ c \end{bmatrix} = \frac{1}{2}\begin{bmatrix} a + b - c \\ a + b + c \end{bmatrix}$

(f) $\begin{bmatrix} 1 & 0 & 0 & 0 \\ 0 & 1 & 1 & 0 \\ 0 & 1 & 1 & 0 \\ 0 & 0 & 0 & 1 \end{bmatrix}$; $C_D\left(T\begin{bmatrix} a & b \\ c & d \end{bmatrix}\right) = \begin{bmatrix} 1 & 0 & 0 & 0 \\ 0 & 1 & 1 & 0 \\ 0 & 1 & 1 & 0 \\ 0 & 0 & 0 & 1 \end{bmatrix}\begin{bmatrix} a \\ b \\ c \\ d \end{bmatrix} = \begin{bmatrix} a \\ b + c \\ b + c \\ d \end{bmatrix}$

4. **(b)** $M_{ED}(S)M_{DB}(\mathrm{T}) = \begin{bmatrix} 1 & 1 & 0 & 0 \\ 0 & 0 & 1 & -1 \end{bmatrix}\begin{bmatrix} 1 & 1 & 0 \\ 0 & 1 & 1 \\ 1 & 0 & 1 \\ -1 & 1 & 0 \end{bmatrix} = \begin{bmatrix} 1 & 2 & 1 \\ 2 & -1 & 1 \end{bmatrix} = M_{EB}(ST)$

(d) $M_{ED}(S)M_{DB}(T) = \begin{bmatrix} 1 & -1 & 0 \\ 0 & 0 & 1 \end{bmatrix}\begin{bmatrix} 1 & -1 & 0 \\ -1 & 0 & 1 \\ 0 & 1 & 0 \end{bmatrix} = \begin{bmatrix} 2 & -1 & -1 \\ 0 & 1 & 0 \end{bmatrix} = M_{EB}(ST)$

5. $M_{ED}(S)M_{DB}(T) = \begin{bmatrix} 0 & 1 & 0 & 0 \\ 1 & 0 & 0 & 1 \\ 0 & 0 & 1 & 0 \end{bmatrix}\begin{bmatrix} 1 & 0 & 0 & 0 \\ 0 & 0 & 1 & 0 \\ 0 & 1 & 0 & 0 \\ 0 & 0 & 0 & 1 \end{bmatrix} = \begin{bmatrix} 0 & 0 & 1 & 0 \\ 1 & 0 & 0 & 1 \\ 0 & 1 & 0 & 0 \end{bmatrix} = M_{EB}(ST)$

6. **(b)** $T^{-1}(a,b,c) = \frac{1}{2}(b + c - a, a + c - b, a + b - c)$

$M_{DB}(T) = \begin{bmatrix} 0 & 1 & 1 \\ 1 & 0 & 1 \\ 1 & 1 & 0 \end{bmatrix}$; $M_{BD}(T^{-1}) = \frac{1}{2}\begin{bmatrix} -1 & 1 & 1 \\ 1 & -1 & 1 \\ 1 & 1 & -1 \end{bmatrix}$

(d) $T^{-1}(a,b,c) = (a - b) + (b - c)x + cx^2$

$M_{DB}(T) = \begin{bmatrix} 1 & 1 & 1 \\ 0 & 1 & 1 \\ 0 & 0 & 1 \end{bmatrix}$, $M_{BD}(T^{-1}) = \begin{bmatrix} 1 & -1 & 0 \\ 0 & 1 & -1 \\ 0 & 0 & 1 \end{bmatrix}$

7. **(b)** $M_{DB}(T^{-1}) = [M_{BD}(T)]^{-1} = \begin{bmatrix} 1 & 1 & 1 & 0 \\ 0 & 1 & 1 & 0 \\ 0 & 0 & 1 & 0 \\ 0 & 0 & 0 & 1 \end{bmatrix}^{-1} = \begin{bmatrix} 1 & -1 & 0 & 0 \\ 0 & 1 & -1 & 0 \\ 0 & 0 & 1 & 0 \\ 0 & 0 & 0 & 1 \end{bmatrix}$. Hence $C_B[T^{-1}(a,b,c,d)] =$

$M_{BD}(T^{-1})C_D(a,b,c,d) = \begin{bmatrix} 1 & -1 & 0 & 0 \\ 0 & 1 & -1 & 0 \\ 0 & 0 & 1 & 0 \\ 0 & 0 & 0 & 1 \end{bmatrix}\begin{bmatrix} a \\ b \\ c \\ d \end{bmatrix} = \begin{bmatrix} a - b \\ b - c \\ c \\ d \end{bmatrix}$ so $T^{-1}(a,b,c,d) = \begin{bmatrix} a - b & b - c \\ c & d \end{bmatrix}$.

8. $C_E[D(a + bx + cx^2 + dx^3)] = M_{EB}(D)C_B(a + bx + cx^2 + dx^3) = \begin{bmatrix} 0 & 1 & 0 & 0 \\ 0 & 0 & 2 & 0 \\ 0 & 0 & 0 & 3 \end{bmatrix}\begin{bmatrix} a \\ b \\ c \\ d \end{bmatrix} = \begin{bmatrix} b \\ 2c \\ 3d \end{bmatrix}$

Hence $D(a + bx + cx^2 + dx^3) = b + (2c)x + (3d)x^2$.

11. Have $C_D[T(\mathbf{e}_j)] =$ column j of I_n. Hence $M_{BD}(T) = [C_D[T(\mathbf{e}_1)]\ C_D[T(\mathbf{e}_2)] \ldots C_D[T(\mathbf{e}_n)]] = I_n$.

Exercises 8.2.2 (Page 439)

3. To show that $(aS)T = a(ST)$, let $\mathbf{v}$ lie in V. Then $[(aS)T](\mathbf{v}) = (aS)[T(\mathbf{v})] = a\{S[T(\mathbf{v})]\} = a\{(ST)(\mathbf{v})\} = [a(ST)](\mathbf{v})$. This holds for all $\mathbf{v}$ in V, so it shows that $(aS)T = a(ST)$.
4. **(b)** If $\mathbf{w}$ lies in im $(S + T)$, then $\mathbf{w} = (S + T)(\mathbf{v})$ for some $\mathbf{v}$ in V. But then $\mathbf{w} = S(\mathbf{v}) + T(\mathbf{v})$, so $\mathbf{w}$ lies in im S + im T.
5. **(b)** If $X \subseteq X_1$, let T lie in $X_1{}^0$. Then $T(\mathbf{v}) = \mathbf{0}$ for all $\mathbf{v}$ in X_1, whence $T(\mathbf{v}) = \mathbf{0}$ for all $\mathbf{v}$ in X. Thus T is in X^0 and we have shown that $X_1{}^0 \subseteq X^0$.
7. The proof that M_{DB} is onto in Theorem 6 shows that $M_{DB}[R(A)] = A$ for each A in $\mathbf{M}_{mn}$. Hence $M_{DB}R = 1_{\mathbf{M}_{mn}}$. Because dim $[\mathbf{L}(V,W)] = nm =$ dim $\mathbf{M}_{mn}$, Exercise 20 in Section 8.1.3 shows that $R = M_{DB}{}^{-1}$.

Exercises 8.3 (Page 451)

1. **(b)** $\{[1), [2^n), [(-3)^n)\}$; $x_n = \frac{1}{20}(15 + 2^{n+3} + (-3)^{n+1})$
2. **(b)** $\{[1), [n), [(-2)^n)\}$; $x_n = \frac{1}{9}(5 - 6n + (-2)^{n+2})$ **(d)** $\{[1), [n), [n^2)\}$; $x_n = 2(n - 1)^2 - 1$
3. **(b)** $\{[a^n), [b^n)\}$
6. **(b)** $[1,0,0,0,0, \ldots), [0,1,0,0,0, \ldots), [0,0,1,1,1, \ldots), [0,0,1,2,3, \ldots)$
9. By Remark 2,

$$[i^n + (-i)^n) = [2,0,-2,0,2,0,-2,0, \ldots)$$
$$[i(i^n - (-i)^n)) = [0,-2,0,2,0,-2,0,2, \ldots)$$

are solutions. They are linearly independent and so are a basis.

Exercises 9.1 (Page 459)

1. **(b)** $P = \begin{bmatrix} 1 & 1 & 0 \\ 0 & 1 & 2 \\ -1 & 0 & 1 \end{bmatrix}$
2. **(b)** $B = \left\{ \begin{bmatrix} 3 \\ 7 \end{bmatrix}, \begin{bmatrix} 2 \\ 5 \end{bmatrix} \right\}$
3. **(b)** $c_T(x) = x^2 - 6x - 1$ **(d)** $c_T(x) = x^3 + x^2 - 8x - 3$ **(f)** $c_T(x) = x^4$

Exercises 9.2.1 (Page 469)

3. **(b)** If $\mathbf{v}$ is in $S(U)$, write $\mathbf{v} = S(\mathbf{u})$, $\mathbf{u}$ in U. Then

$$T(\mathbf{v}) = T[S(\mathbf{u})] = (TS)(\mathbf{u}) = (ST)(\mathbf{u}) = S[T(\mathbf{u})]$$

and this lies in $S(U)$ because $T(\mathbf{u})$ lies in U (U is T-invariant).
6. Suppose U is T-invariant for every T. If $U \neq 0$, choose $\mathbf{u} \neq \mathbf{0}$ in U. Choose a basis $B = \{\mathbf{u}, \mathbf{u}_2, \ldots, \mathbf{u}_n\}$ of V containing $\mathbf{u}$. Given any $\mathbf{v}$ in V, there is (by Theorem 3 in Section 8.1) a linear transformation $T : V \to V$ such that $T(\mathbf{u}) = \mathbf{v}$, $T(\mathbf{u}_2) = \cdots = T(\mathbf{u}_n) = \mathbf{0}$. Then $\mathbf{v} = T(\mathbf{u})$ lies in U because U is T-invariant. This shows that $V = U$.
8. **(b)** $T(1 - 2x^2) = 3 + 3x - 3x^2 = 3(1 - 2x^2) + 3(x + x^2)$ and $T(x + x^2) = -(1 - 2x^2)$, so both are in U. Hence U is T-invariant by Example 4. If $B = \{1 - 2x^2, x + x^2, x^2\}$ then $M_B(T) = \begin{bmatrix} 3 & -1 & 1 \\ 3 & 0 & 1 \\ 0 & 0 & 3 \end{bmatrix}$, so

$$c_T(x) = \det \begin{bmatrix} x-3 & 1 & -1 \\ -3 & x & -1 \\ 0 & 0 & x-3 \end{bmatrix} = \det \begin{bmatrix} x-3 & 1 \\ -3 & x \end{bmatrix} \det(x-3) = (x^2 - 3x + 3)(x - 3)$$

9. (b) Suppose $\mathbb{R}\mathbf{u}$ is T_A-invariant where $\mathbf{u} \neq \mathbf{0}$. Then $T_A(\mathbf{u}) = r\mathbf{u}$ for some r in $\mathbb{R}$, so $(rI - A)\mathbf{u} = \mathbf{0}$. But $\det(rI - A) = (r - \cos\theta)^2 + \sin^2\theta \neq 0$ because $0 < \theta < \pi$. Hence $\mathbf{u} = \mathbf{0}$, a contradiction.

15. The fact that U and W are subspaces is easily verified using the subspace test. If A lies in $U \cap W$, then $A = AE = 0$; that is, $U \cap W = 0$. To show that $\mathbf{M}_{22} = U + V$, choose any A in $\mathbf{M}_{22}$. Then $A = AE + (A - AE)$ and AE lies in U [because $(AE)E = AE^2 = AE$], and $A - AE$ lies in W [because $(A - AE)E = AE - AE^2 = 0$].

18. (b) $T^2[p(x)] = p[-(-x)] = p(x)$, so $T^2 = 1$; $B = \{1, x^2; x, x^3\}$

(d) $T^2(a,b,c) = T(-a + 2b + c, b + c, -c) = (a,b,c)$, so $T^2 = 1$; $B = \{(1,1,0); (1,0,0), (0,-1,2)\}$

21. (b) $T^2(a,b,c) = T(a + 2b, 0, 4b + c) = (a + 2b, 0, 4b + c) = T(a,b,c)$ so $T^2 = T$; $B = \{(1,0,0), (0,0,1); (2,-1,4)\}$

Exercises 9.2.2 (Page 479)

1. (b) $B = \left\{\begin{bmatrix} 2 & 0 \\ -1 & 0 \end{bmatrix}, \begin{bmatrix} 0 & 2 \\ 0 & -1 \end{bmatrix}, \begin{bmatrix} 1 & 0 \\ -2 & 0 \end{bmatrix}, \begin{bmatrix} 0 & 1 \\ 0 & -1 \end{bmatrix}\right\}$; $M_B(T) = \begin{bmatrix} 1 & 0 & 0 & 0 \\ 0 & 1 & 0 & 0 \\ 0 & 0 & -2 & 0 \\ 0 & 0 & 0 & -2 \end{bmatrix}$

3. (b) $P = \begin{bmatrix} 2 & -2 & 1 \\ 1 & 2 & 2 \\ 2 & 1 & -2 \end{bmatrix}$; $P^{-1}AP = \begin{bmatrix} 0 & 0 & 0 \\ 0 & -3 & 0 \\ 0 & 0 & 3 \end{bmatrix}$

6. $U = \text{span}\left\{\begin{bmatrix} 1 \\ 0 \\ 0 \\ 0 \end{bmatrix}, \begin{bmatrix} 0 \\ 1 \\ 0 \\ 0 \end{bmatrix}\right\}$, $W = \text{span}\left\{\begin{bmatrix} 0 \\ 0 \\ 1 \\ 0 \end{bmatrix}, \begin{bmatrix} 0 \\ 0 \\ 0 \\ 1 \end{bmatrix}\right\}$; $c_A(x) = (x^2 + 1)^2$

7. (b) $c_T(x) = (x - 2)^2(x + 5)$ and $E_2(T) = \mathbb{R}(1 + 2x^2)$

Exercises 9.3 (Page 488)

1. (b) $B = \left\{\begin{bmatrix} 1 & 0 \\ 0 & 0 \end{bmatrix}, \begin{bmatrix} 0 & 1 \\ 0 & 0 \end{bmatrix}, \begin{bmatrix} 0 & 0 \\ 1 & 0 \end{bmatrix}, \begin{bmatrix} 0 & 0 \\ 0 & 1 \end{bmatrix}\right\}$; $M_B(T) \begin{bmatrix} -1 & 0 & 1 & 0 \\ 0 & -1 & 0 & 1 \\ 1 & 0 & 2 & 0 \\ 0 & 1 & 0 & 2 \end{bmatrix}$

4. (b) $\langle \mathbf{v}, (rT)(\mathbf{w})\rangle = \langle \mathbf{v}, rT(\mathbf{w})\rangle = r\langle \mathbf{v}, T(\mathbf{w})\rangle = r\langle T(\mathbf{v}), \mathbf{w}\rangle = \langle rT(\mathbf{v}), \mathbf{w}\rangle = \langle (rT)(\mathbf{v}), \mathbf{w}\rangle$

5. (b) If $B_0 = \{(1,0,0), (0,1,0), (0,0,1)\}$, then $M_{B_0}(T) = \begin{bmatrix} 7 & -1 & 0 \\ -1 & 7 & 0 \\ 0 & 0 & 2 \end{bmatrix}$ has an orthonormal basis of eigenvectors $\left\{\frac{1}{\sqrt{2}}\begin{bmatrix} 1 \\ 1 \\ 0 \end{bmatrix}, \frac{1}{\sqrt{2}}\begin{bmatrix} 1 \\ -1 \\ 0 \end{bmatrix}, \begin{bmatrix} 0 \\ 0 \\ 1 \end{bmatrix}\right\}$. Hence an orthonormal basis of eigenvectors of T is

$$\left\{\frac{1}{\sqrt{2}}(1,1,0), \frac{1}{\sqrt{2}}(1,-1,0), (0,0,1)\right\}$$

(d) If $B_0 = \{1, x, x^2\}$, then $M_{B_0}(T) = \begin{bmatrix} -1 & 0 & 1 \\ 0 & 3 & 0 \\ 1 & 0 & -1 \end{bmatrix}$ has an orthonormal basis of eigenvectors

$\left\{\begin{bmatrix}0\\1\\0\end{bmatrix}, \frac{1}{\sqrt{2}}\begin{bmatrix}1\\0\\1\end{bmatrix}, \frac{1}{\sqrt{2}}\begin{bmatrix}1\\0\\-1\end{bmatrix}\right\}$. Hence an orthonormal basis of eigenvectors of T is

$\left\{x, \frac{1}{\sqrt{2}}(1 + x^2), \frac{1}{\sqrt{2}}(1 - x^2)\right\}$.

7. (b) $M_B(T) = \begin{bmatrix}A & 0\\0 & A\end{bmatrix}$, so $c_T(x) = \det\begin{bmatrix}xI_2 - A & 0\\0 & xI_2 - A\end{bmatrix} = [c_A(x)]^2$.

12. (c) The coefficients in the definition of $T'(\mathbf{e}_j)$ are the entries in the jth column $C_B[T'(\mathbf{e}_j)]$ of $M_B(T')$. Hence $M_B(T') = [\langle \mathbf{e}_j, T(\mathbf{e}_i)\rangle]$, and this is the transpose of $M_B(T)$ by Theorem 1.

Appendix A (Page A15)

1. (b) $x = 3$ **(d)** $x = \pm 1$

2. (b) $10 + i$ **(d)** $\frac{11}{26} + \frac{23}{26}i$ **(f)** $2 - 11i$ **(h)** $8 - 6i$

3. (b) $\frac{11}{5} + \frac{3}{5}i$ **(d)** $\pm(2 - i)$

4. (b) $\frac{1}{2} \pm \frac{\sqrt{3}}{2}i$ **(d)** $2, \frac{1}{2}$

5. (b) $-2, 1 \pm \sqrt{3}i$ **(d)** $\pm 2\sqrt{2}, \pm 2\sqrt{2}i$

6. (b) $x^2 - 4x + 13; 2 + 3i$ **(d)** $x^2 - 6x + 25; 3 + 4i$

8. $x^4 - 10x^3 + 42x^2 - 82x + 65$

10. (b) $(-2)^2 + 2i - (4 - 2i) = 0; 2 - i$ **(d)** $(-2 + i)^2 + 3(1 - i)(-2 + i) - 5i = 0; -1 + 2i$

11. (b) $-i, 1 + i$ **(d)** $2 - i, 1 - 2i$

12. (b) Circle, center at 1, radius 2
(d) Imaginary axis
(f) Line $y = mx$

16. (b) $4e^{-\pi i/2}$ **(d)** $8e^{2\pi i/3}$ **(f)** $6\sqrt{2}e^{3\pi i/4}$

17. (b) $\frac{1}{2} + \frac{\sqrt{3}}{2}i$ **(d)** $1 - i$ **(f)** $\sqrt{3} - 3i$

18. (b) $-\frac{1}{32} + \frac{\sqrt{3}}{32}i$ **(d)** $-32i$ **(f)** $-2^{16}(1 + i)$

21. (b) $\pm\frac{\sqrt{2}}{2}(\sqrt{3} + i), \pm\frac{\sqrt{2}}{2}(-1 + \sqrt{3}i)$
(d) $\pm 2i, \pm(\sqrt{3} + i), \pm(\sqrt{3} - i)$

Appendix B.1 (Page A26)

1. (b) minimum 0, maximum 29 **(d)** minimum -10, maximum 6 (at two vertices)

4. (b) Either no large toys and 600 small toys, or 171 large toys and 372 small toys [*Note:* (172,372) fails to satisfy $5x_1 + 2x_2 \leq 1600$.] **6.** \$40,000 in A, 60,000 in B **8.** 4 grams of P, 2 grams of Q

10. 300 barrels of grade 1, 200 barrels of grade 2

Appendix B.2 (Page A38)

2. (b) $p = 2; x_2 = 2, x_1 = x_3 = 0$ **(d)** $p = \frac{15}{4}; x_1 = \frac{5}{4}, x_2 = 0$

(f) $p = 7; x_1 = 0, x_2 = \frac{3}{5}, x_3 = \frac{13}{5}$

(h) $p = 106; x_1 = x_2 = 0, x_3 = 28, x_4 = 11$

6. No regular dinners, 75 diet dinners, and 225 super dinners. Profit: \$487.50
8. 1600 standard mowers, no deluxe mowers, and 200 super mowers. Profit: \$59,000 per week

Appendix C (Page A45)

21. $B_n = (n + 1)! - 1$
22. **(b)** Verify each of $S_1, S_2, \ldots, S_8$.

Index

IMPORTANT SYMBOLS

Symbol	***Description***	***Page***
$\mathbb{C}$	complex numbers	**363**
row (A)	row space of matrix A	**245**
col(A)	column space of matrix A	**245**
rank (A)	rank of matrix A	**23, 246**
$C_B(\mathbf{v})$	coordinates of $\mathbf{v}$ with respect to B	**257**
$P_{D\leftarrow B}$	transition matrix from B to D	**260**
$\langle\mathbf{v},\mathbf{w}\rangle$	inner product of $\mathbf{v}$ and $\mathbf{w}$	**279**
$d(\mathbf{v},\mathbf{w})$	distance between $\mathbf{v}$ and $\mathbf{w}$	**285**
$U^{\perp}$	orthogonal complement of U	**305**
$\text{proj}_U(\mathbf{v})$	projection of $\mathbf{v}$ onto U	**307**
$E_\lambda(A)$	eigenspace of matrix A	**330**
$T{:}V\to W$	linear transformation	**393**
1_V	identity transformation on V	**395**
ker T	kernel of transformation T	**405**
im T	image of transformation T	**405**
nullity T	nullity of transformation T	**406**
rank T	rank of transformation T	**406**
$V\cong W$	isomorphic spaces	**416**
Z^*	conjugate transpose of matrix Z	**365**
$M_{DB}(T)$	matrix of transformation T	**428**
$M_B(T)$	matrix of operator T	**454**
$\mathbf{L}(V,W)$	space of transformations	**436**
det T	determinant of operator T	**458**
tr T	trace of operator T	**458**
$c_T(x)$	characteristic polynomial of T	**459**
$U\oplus W$	direct sum of U and W	**464**
$E_\lambda(T)$	eigenspace of transformation T	**473**
$\lvert z\rvert$	absolute value of complex number z	**A4**